Marine Ecology:
Selected Readings

Marine Ecology:
Selected Readings

Edited By

J. Stanley Cobb, Ph.D.

Department of Zoology
University of Rhode Island

And

Marilyn M. Harlin, Ph.D.

Department of Botany
University of Rhode Island

University Park Press

Baltimore • London • Tokyo

UNIVERSITY PARK PRESS
International Publishers in Science and Medicine
Chamber of Commerce Building
Baltimore, Maryland 21202

Portions typeset by The Composing Room of Michigan, Inc.
Manufactured in the United States of America by Universal Lithographers,
Inc., and The Maple Press Co.

Cover: Drawn from photograph by Mann in Science 182:975–981. (1973)

Library of Congress Cataloging in Publication Data
Main entry under title:

Marine ecology.

1. Marine ecology—Addresses, essays, lectures.
I. Cobb, J. Stanley.   II. Harlin, Marilyn M.
QH541.5.S3M27   574.5'2636   76-28371
ISBN 0-8391-0959-8

Contents

Preface

In the past several years interest in marine ecology has bloomed, along with the general increase in awareness of the environmental sciences. Recently, the study of marine ecosystems has contributed a great deal to the development of ecological theory, perhaps because of the relative newness of the field, perhaps also because of the challenge it presents.

This collection is part of a group of papers we have found useful in teaching a course in marine ecology to upper-level undergraduates at the University of Rhode Island. Instead of using a specific text, we assign readings from the primary literature, providing a unique opportunity for undergraduates to experience how research knowledge accumulates.

Our approach has been to investigate several topics in depth rather than to try to cover the entire range subsumed in the general term "marine ecology." The papers were chosen for their value in teaching, and several criteria were used in the selection process. Above all other considerations, they had to be readable by the undergraduate student unfamiliar with preceding research. We then tried to choose papers based on sound research that reflected current trends, but not ones that we suspected would become quickly dated. Some good papers were omitted because they were simply too long, others because they dealt with a side issue or too many concepts, and still others because our students found them unexciting. For many of the sections, an introductory overview paper has been selected to provide the necessary background for the papers that follow.

We have written an introduction for each unit in order to give the student a perspective on where the readings fit into the field being examined. We recognize that our comments and selections do not do justice to all the workers in that area, and we offer them as only a starting point that both the student and the instructor are encouraged to challenge.

Many people have offered valuable suggestions during the course of preparing this book. Ellsworth H. Wheeler, Jr., and Donna Watt reviewed early versions of the contents. H. Perry Jeffries and Candace A. Oviatt carefully read the text. Barbara Waters gave excellent secretarial assistance. Out heartfelt thanks to all the above. We would particularly like to express our appreciation to the students in the Marine Ecology course at the University of Rhode Island for their comments and criticisms of the papers selected.

Marine Ecology:
Selected Readings

Trophic Dynamics

Productivity

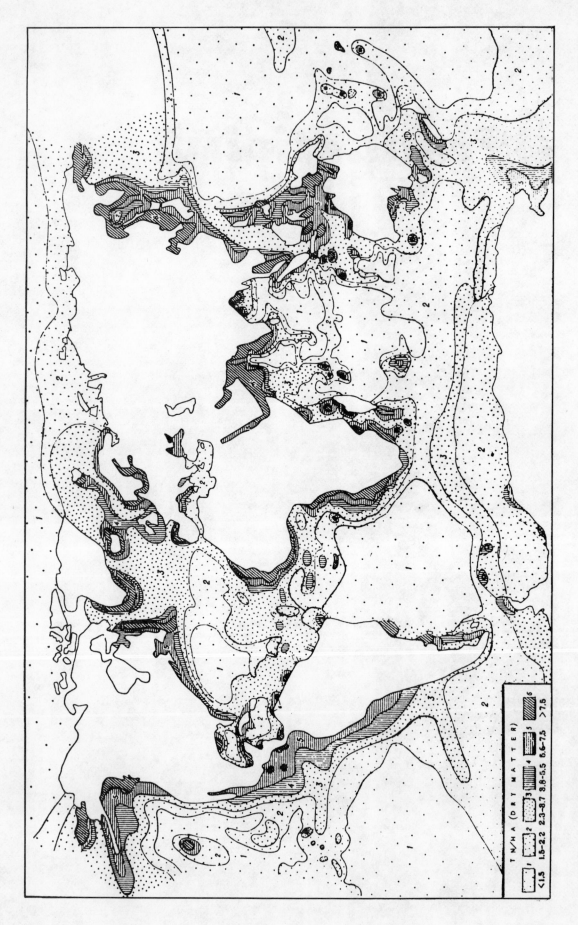

Figure 1. Pattern of distribution for primary production in the world's oceans. (From Rodin et al., 1975. Reproduced with permission of the National Academy of Sciences.)

Introduction

In ecological terms, productivity is the rate at which organic matter is formed. It is not to be confused with a measurement of the biomass of a standing crop, which does not take into consideration the time factor. Standing crop may be high and productivity low, and conversely, standing crop may be low with rapid turnover and consequent high productivity. It has become increasingly important to make these distinctions. In the fishing industry, for example, there may be a rich harvest, but the fisherman will not stay in operation unless there is a sustained yield.

To measure productivity, an organism or population may be weighed at the beginning and the end of a specific period of time, or the increase in fixed carbon may be assessed with a radioactive isotope. Sometimes rates may be determined from measured changes in the amount of net oxygen evolved or the content of chlorophyll a or ATP. Productivity is then expressed as weight per area per unit time (e.g., gC m^{-2} day^{-1}, kg dry wgt m^{-2} yr^{-1}, T ha^{-1} yr^{-1}). It can also be considered in terms of the energy changes involved and hence expressed in calories (Kcal m^{-2} yr^{-1}). Calories are determined directly by burning organic matter in a closed container in which heat output is assessed, or are calculated from the weights according to pre-established conversion factors (Crisp, 1975, p. 75): e.g., grams wet weight are multiplied by 0.5, grams dry weight by 4, grams ash-free dry weight by 5, and grams carbon by 10.

In the sea, the areas with the greatest productivity are estuaries, upwellings, and coral reefs. Figure 1 shows the pattern of distribution for primary production in the world's oceans. The pattern for secondary production corresponds closely to that for primary production because consumers depend upon photosynthetic organisms for their organic carbon. When we ask what determines whether an area will be highly productive, we must know what controls the growth of algae.

Although algae are limited in depth to the euphotic zone, there is sufficient light to sustain a population of some species almost anywhere on the surface of the earth. Species differ in their temperature tolerances and optima, but the geographic distribution of regions of high productivity is explained not so much by light and temperature differences as by the extent to which the water column is mixed. Mixing may be caused by upwellings, currents, and tides, or the disappearance of a thermocline in winter. This circulation of water brings to the euphotic zone the macronutrients (N, P, Si, Mg, K, Ca) and trace metals (Fe, Mn, Cu, Zn, B, Na, Mo, Cl, V, Co) needed by plants. Freshwater run-off from land brings a high nutrient load and provides mechanical mixing, with consequent high productivity, at the mouth of rivers. Most of the required nutrients are readily available in seawater but nitrogen or phosphorous may limit growth, especially during warmer seasons and at warmer latitudes. Silicon may limit spring diatom blooms. Nutrient limitation interacting with changes in light intensity and temperature can lead to seasonality in primary and secondary productivity.

Productivity varies throughout the year. Figure 2 shows three general patterns for planktonic algae and herbivores at different latitudes. When light is sufficient, algal growth tends to follow changes in nutrient concentration. Algae grow fast when using a large nutrient pool, but with the synthesis of protoplasm, nutrients are tied up. Herbivores grazing on these algae gain in number until their food supply falls. In the North Temperate Zone, it is typical to ob-

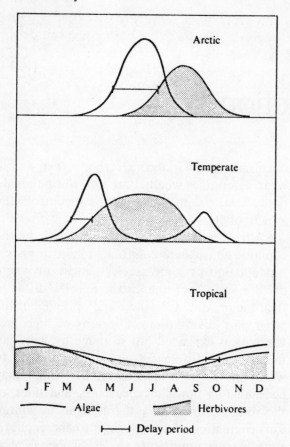

Figure 2. Seasonal patterns for algae and herbivore production cycles in different regions (Cushing, 1975).

serve a major peak of algal growth in spring and a second one in autumn, when waters mix and nutrients regenerated from organic matter that had accumulated in the sediment are returned to the water column. In the Artic, peak algal growth is later than in the temperature zones, and animals tend to have a longer life history. Consequently, only one peak appears before the days are too dark for optimal algal growth. In the tropics, the relatively uniform light and temperature cause algal and invertebrate populations to show little variation. Superimposed on this generalized curve are variations brought about by local conditions.

Many laboratories are presently defining parameters that would maximize the production of particular algal species. Toward this end the photosynthetic rate (Figure 3) is determined as a function of a specific parameter (light intensity, temperature, or nutrient concentration). The steeper the slope (A), the more markedly the factor being studied is assumed to affect the

growth of the organisms. Attention is given to maximum productivity (B), i.e., that level of intensity, temperature, or concentration beyond which no change in photosynthesis is seen. This dose-response curve is used to describe productivity in nature and predict events from changes in parameters. For detailed consideration of these relationships, the reader is referred to the text by Parsons and Takahashi (1973), and for methods of measuring productivity, to the handbook by Vollenweider (1969) and the recent review by Hall and Moll (1975).

Whereas algal growth is limited to the superficial euphotic zone, algal cells and debris rain downward through food chains. Some of the organic carbon rises with upward migrating plankton and fish, but the rest of it falls to the benthos, where it helps to support other life and where its building blocks can be regenerated.

In the first section of this collection of readings we introduce productivity at several levels. Yentsch reviews the primary producers and their requirements for growth, primarily light and nutrients. With increasing depth or dissolved organic matter, he points out that not only is light intensity reduced, but also the spectrum of light shifts toward shorter wavelengths, and only those algae survive that have accessory pigments capable of trapping the wavelengths available. Next, he describes how the movement of water regulates nutrient supply and thereby the size of the population that can develop. In most of the ocean, light energy is collected solely by single-

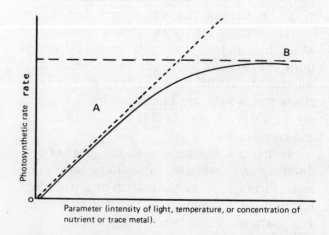

Figure 3. Variation of photosynthetic rate with changes in environmental variables. (Freely adapted from Parsons and Takahaski, 1973.)

celled algae, but in bays and coastal waters large seaweeds make an important contribution. In another paper, Mann shows that this component has been underestimated because turnover time in these algae had been incorrectly assessed.

Highly productive regions such as upwellings and estuaries are sometimes dotted with fishing boats. Additional watercraft have assembled oceanographers from several institutions to study these ecosystems. In a popularized publication from the National Science Foundation, the federal agency that has been supporting most of the research, the objectives of this investigation are explained and some of the cast of scientific characters introduced. Following this paper is one in which Walsh et al. explain in detail one aspect of this program, the changes in population structure over a 24 hr period. The next two papers describe part of a comprehensive investigation into another type of productive region, the Fraser River Estuary in British Columbia. This work is part of a set of three concurrent studies on different aspects of production and should be read for the experimental approach. The last paper in this set (LeBrasseur et al., 1969) was not included but is recommended to those wishing to follow up on the work.

One application of the study of productivity has been to get a handle on the total fish production in the sea and thus a basis for intelligent management. Many estimates have been made from assumptions that food webs and conversion efficiencies were uniform throughout the ocean, when in reality they vary greatly. The final paper in this unit is one in which Ryther calculates the world fish production as much less than had been predicted previously. He divided the ocean according to levels of primary production and considered the number of steps in the food chain and the surface area of each type of production. In the years since the publication

of his paper, Ryther's figures have been ardently challenged, and there has been evidence that the production of lake trout is independent of the trophic levels involved (Kerr and Martin, 1971). Nevertheless, it appears to us that Ryther's estimates were correct when one considers that most of the ocean is a biological desert. For a review of the discussion pursuant to Ryther's paper, as well as current figures on productivity in different marine ecosystems, see Bunt (1975).

LITERATURE CITED

Bunt, J. S. 1975. Primary productivity of marine ecosystems. *In* H. Lieth and R. H. Whittaker (eds.), Primary Productivity of the Biosphere. Springer-Verlag, New York.

Crisp, D. J. 1975. Secondary productivity in the sea. *In* Productivity of World Ecosystems; Proc. Symp. 31 Aug.—1 Sept. 1972, Seattle, Washington. National Academy of Sciences, Wash. D. C.

Cushing, D. H. 1975. Marine Ecology and Fisheries. Cambridge University Press, Cambridge. 278 pp.

Hall, C. A. S., and R. Moll. 1975. Methods of assessing aquatic primary productivity. *In* H. Lieth and R. H. Whittaker (eds.), Primary Productivity of the Biosphere, Springer-Verlag, New York.

Kerr, S. R., and N. V. Martin. 1970. Trophic-dynamics of lake trout production system. *In* J. H. Steele (ed.), Marine Food Chains. Univ. of California Press. pp. 365—376.

LeBrasseur, R. J., W. E. Barraclough, O. D. Kennedy, and T. R. Parsons. 1969. Production studies in the Strait of Georgia. Part III. Observations on the food of larval and juvenile fish in the Fraser River Plume, February to May, 1967. J. Exp. Mar. Biol. Ecol. 3: 51—61.

Parsons, T. R., and M. Takahashi. 1973. Biological Oceanographic Processes. Pergamon Press, New York 186 pp.

Rodin, L. E., N. E. Bazilevich, and N. N. Rozov. 1975. Productivity of the world's main ecosystems. *In* Productivity of World's Ecosystems; Proc. Symp. 31 Aug.—1 Sept. 1972, Seattle, Washington. National Academy of Sciences, Wash. D. C. p. 22.

Vollenweider, R. A. 1969. A manual on methods for measuring primary production in aquatic environments. IBP Handbook No. 12. Blackwell Scientific Publications, Oxford and Edinburgh. 213 pp.

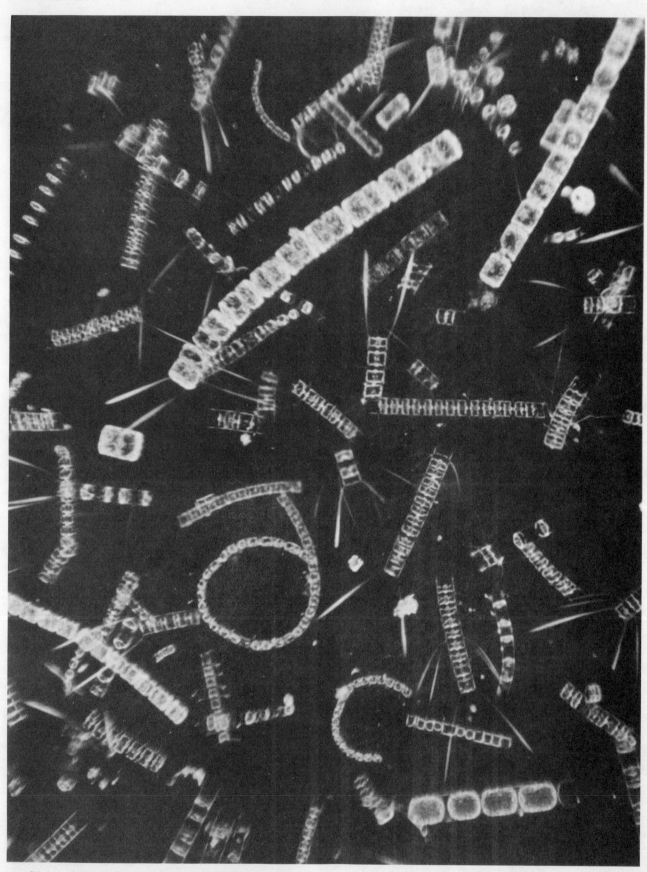

134. Living phytoplankton (× 175 approximately); chains of
cells of the diatoms *Chaetoceros* (with spines),
Thalassiosira condensata and *Lauderia borealis*

The Harvest – Primary Production *Charles S. Yentsch*

Primary production (defined as the rate at which organic matter is formed from inorganic substances) on this planet comes about through the process called photosynthesis. During this process, carbon dioxide and water are chemically converted to organic carbon compounds with the energy derived from sunlight. A by-product of photosynthesis is oxygen evolution. Two principal sites of primary production are terrestrial and aquatic; photosynthesis is carried on by the plants that inhabit these two environments. People who customarily walk along the seashore quickly notice that every tide washes in large seaweeds, while other similar algae grow on rocks or nearby seawalls. These are large plants, easily noticed, and can be important producers in estuaries and embayments. Hence, it is somewhat of a surprise to find that microscopic algae (phytoplankton) floating freely but invisibly in the sea-water are the principal primary producers and the beginning of the food chain in the world's oceans. Principal members of the phytoplankton are diatoms and dinoflagellates. Both groups of organisms (figures 134, 135) have chlorophyll and, of course, are capable of photosynthesis. The main distinction between the two groups is the composition of their cell walls. In diatoms, the material is housed with a silica cell wall, whereas in the dinoflagellates the wall is made from organic cellulose material.

Scientists began to study these microscopic unicellular organisms as an outgrowth of their continuing interest in plant ecology of the earth and its oceans. Plankton nets were pulled through the water collecting organisms which were then identified for the study of evolutionary relationships. Even to the most casual observer it became apparent that there were very large-scale variations in the distribution of numbers of phytoplankton. In many areas the number of organisms appeared to follow certain seasonal trends (i.e. high in spring; low in summer). Oceanographic scientists began to study the ocean in an attempt to understand the causes of these variations.

In the ocean – a medium that is in constant motion because of a dynamic current system – it soon became apparent, in any one area, that a variation in the abundance of phytoplankton does not result from biological changes alone; one must account for the transport of phytoplankton into and out of the area. For experimental work on phytoplankton growth, oceanographers try to overcome these difficulties by containing the environment inside bottles, large plastic spheres or circular tanks. This allows the study of biological change without the influence of ocean current transport.

Much of our present knowledge of factors influencing primary productivity comes from indirect sources such as laboratory experiments with cultured phytoplankton, or by semi-laboratory experiments carried out in the oceans. In the latter, there are great experimental difficulties. This is because the concentrations of plants in the ocean are low compared with laboratory cultures, and techniques must be specific and highly sensitive. If one were to choose the factor most relevant to primary production it would be *light – its intensity and quality*. Let us examine some of the characteristics of light in the sea.

THE NATURE AND DISTRIBUTION OF LIGHT IN THE EUPHOTIC ZONE

In the oceans only the upper 100 m or less is illuminated with sufficient light intensity for photosynthesis. This is called the euphotic zone. This radiation varies as a function of the time of day, the season and latitude. The depth of the euphotic zone is dependent upon the amount of solar radiation, light reflected from the sky, reflection from the sea surface, and reduction of light as it passes through the water. The fate of light rays entering the sea surface is to be scattered and absorbed.

The amount of radiation entering the sea surface depends upon the altitude of the sun. On a clear day when the sun is at a high altitude approximately 85 per cent of the radiation comes directly from the sun and 15 per cent is reflection from the sky. When the sun angle is low, such as in late afternoon, the proportion of sky light becomes greater. The altitude of the sun also influences the amount of radiation reflected from the sea surface. One notices this when standing on the deck of a boat at sunset; the glare from the water is much greater. As much as 40 per cent of the incoming light may be reflected from the surface whereas at a higher sun angle as little as three per cent may be reflected. Sea surface conditions are important in this respect, and much more of the light is reflected when waves are formed or when there is ice cover.

If one takes a light detecting device (photometer) and measures the amount of light penetrating the sea surface, and then turns the device over so that it measures the light coming back out of the sea, one finds that five to seven per cent of the incoming light is lost from the sea. In other words the great majority of the light that enters the sea stays there, and only a small amount escapes. This tells us that the ultimate fate of light rays entering the sea is absorption. Probably rays are scattered initially, then absorbed, either by the water itself or particles, or by coloured substances such as phytoplankton.

9

Reprinted from Deep Oceans ed. by P. J. Herring and M. R. Clarke. Praeger Publ., (1971)

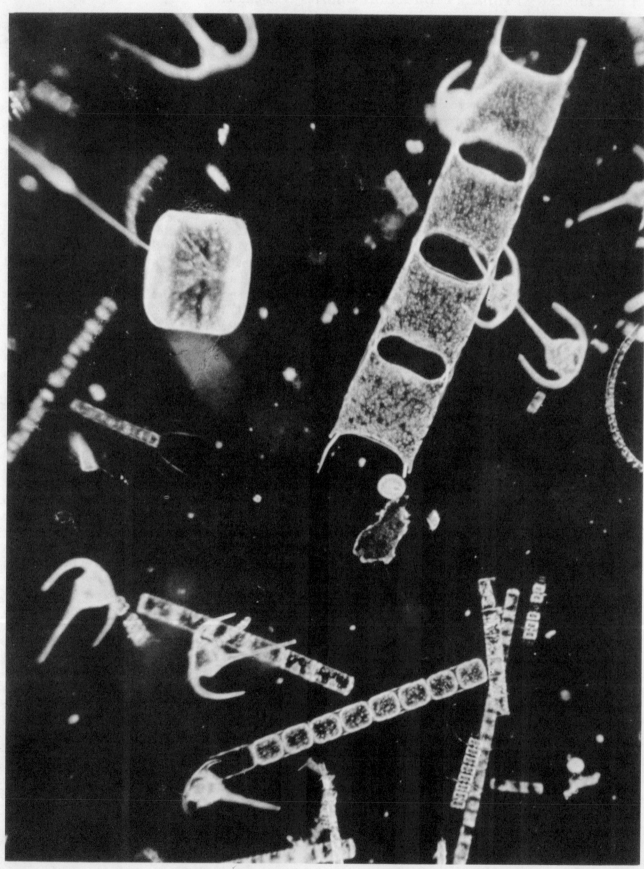

135. Living phytoplankton (× 220 approximately); chains of cells of the diatoms *Biddulphia sinensis, Rhizosolenia faeroense, Stephanopyxis borreri,* and *Chaetoceros* spp., the single pill-box like *Coscinodiscus conicinus,* and the anchor-like dinoflagellates *Ceratium* spp.

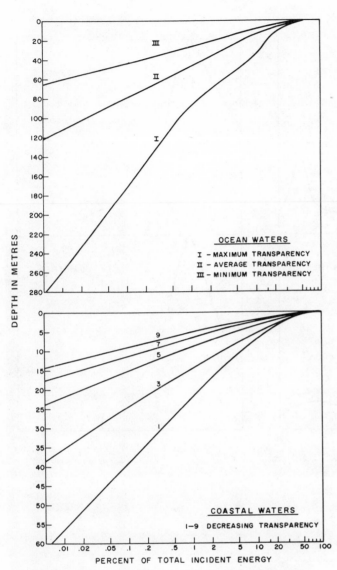

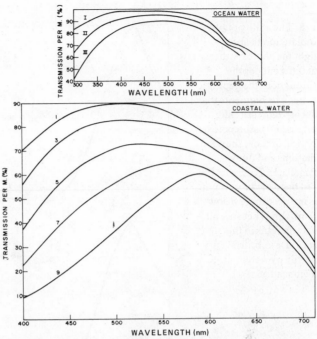

136. The penetration of light in ocean and coastal waters of different transparency (from Jerlov 1951). These measurements are made by lowering a waterproof photocell connected to a meter on deck by a wire. In full sunlight the depth of the euphotic zone occurs at about the one per cent light level. Note that in the clear oceans this is about 100 m, whereas in turbid coastal waters it would be not quite 10 m

137. The transmission of wavelengths of light over a path of 1 m in waters of different transparency (from Jerlov 1951). Note the broad equal transmission of wavelengths in clear ocean water and how the wavelength of maximum transmission is shifted with decreasing light transmission. Part of the green colour of coastal waters is due to their higher productivity and hence, greater numbers of phytoplankton (numbers as in figure 136)

Because of their proximity to land, coastal waters are generally less transparent than open ocean waters (see figure 136). Water itself greatly modifies the wavelengths which penetrate it. The red and dark blue ultraviolet wavelengths are quickly absorbed within the first few metres (figure 137). Clear water is extremely transparent to the blue-green region of the spectrum (see also Chapter 8).

The phytoplankton cell's ability to absorb submarine light depends upon the photosynthetic pigments located in special cell organelles, or chloroplasts of the algae. Some of the light absorption characteristics of phytoplankton and other photosynthetic organisms are shown in figure 102. Note that the reason green plants are visibly green is because chlorophyll absorbs red and blue light, and all photosynthetic organisms absorb the red and blue wavelengths. Photosynthetic bacteria, however, have the possibility of absorbing at the longer infrared wavelengths. In the blue-green region, absorption is principally by the pigments, chlorophylls and carotenoids. Absorption in the red and near infrared wavelengths is by chlorophyll alone. Red pigments of red and blue-green algae, called phycobilins, absorb the middle region of the visible spectrum.

Since diatoms and dinoflagellates are dominant organisms of the phytoplankton and their pigment composition is similar, it is important to consider their

138. The ability of phytoplankton to absorb light for photosynthesis at different depths throughout the euphotic zone. The pigments that absorb light are chlorophylls and carotenoids typical of those found in diatoms and dinoflagellates. Note that the red band of chlorophyll is not absorbing below 10 m, this is because all wavelengths in this region are absorbed primarily by water (after Yentsch 1962)

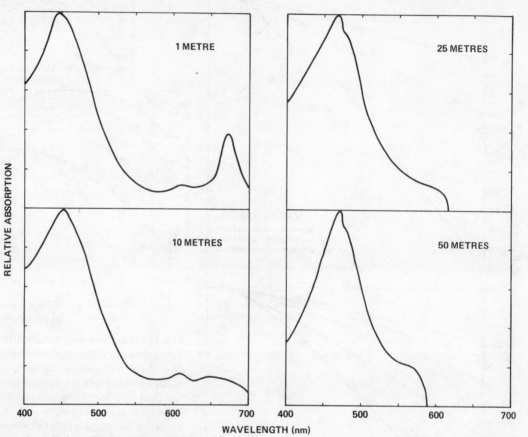

ability to absorb submarine light. Combining the absorption by algae with the wavelengths in the submarine light field shows that the red peak of chlorophyll cannot be active in photosynthesis much below 10 m (figure 138). This is because these wavelengths of red light have already been removed by water absorption. Blue absorption by chlorophyll is active to 50 m, and below 50 m it would appear that the carotenoid pigments are the principal absorbers of light. The utilization of light in photosynthesis depends, of course, upon its intensity; there must be sufficient intensity to supply the energy necessary for photosynthesis.

THE RELATIONSHIP OF LIGHT TO PHOTOSYNTHESIS
When light intensity is increased or decreased, the rate of photosynthesis (carbon dioxide fixation or oxygen evolution) responds according to the curve shown in figure 139. This is the so-called photosynthesis light curve and consists of two specific regions. In the first region, occurring at lower intensities, the photosynthetic rate increases in a linear measure with increasing light intensity. At a value of 10 – 15 per cent of full sunlight, the second region is entered, where further increase of light intensity increases photosynthesis only slightly; this is called 'saturation'. In some cases at very high light intensities, inhibition of the photosynthesis rate may occur.

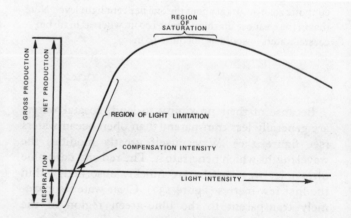

139. Relationship of the rate of photosynthesis to light intensity (heavy curve)

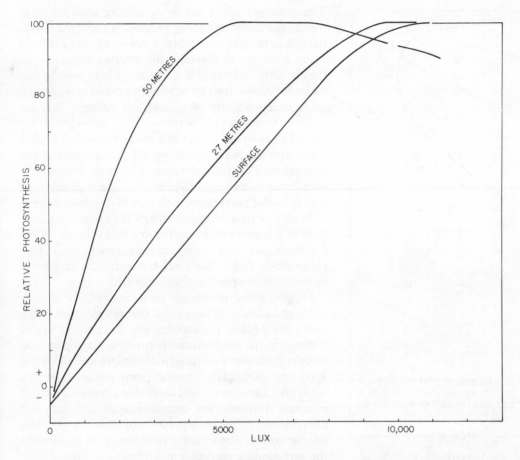

140. The relationship of photosynthetic rate to light intensity in phytoplankton populations at three points of the euphotic zone (after Steemann-Nielsen and Hansen 1959)

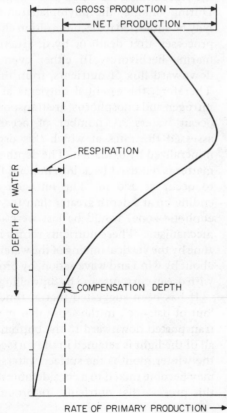

141. The rate of photosynthesis (heavy line) as a function of depth

The point where the photosynthesis curve intercepts the axis of the light intensity is called the compensation intensity. The amount of photosynthesis occurring below a certain light intensity (the compensation intensity) is the amount used for respiration. All photosynthesis above this intensity is called net photosynthesis. This means that this photosynthesis is in excess of that needed for respiration by the algae. The total of these two is termed gross photosynthesis. These parameters when considered with the exponential decrease of light in the water column are shown in figure 141.

At times, in the euphotic zone, environmental conditions arise which affect the maximum rate of photosynthesis. The effect of lowering the temperature is to decrease the maximum rate. The maximum photosynthetic rate is also depressed when there is a lack of carbon dioxide or nutrients. Moreover, if phytoplankton are at stationary levels in the euphotic zone, those at the surface are exposed to high light intensities, whereas the phytoplankton at depth are exposed to lower intensities. The result is that the plankton algae near the surface take on a photosynthesis light response similar to so-called 'sun plants' (figure 140). These plants have a high rate of photosynthesis at extremely high light intensities. On the other hand,

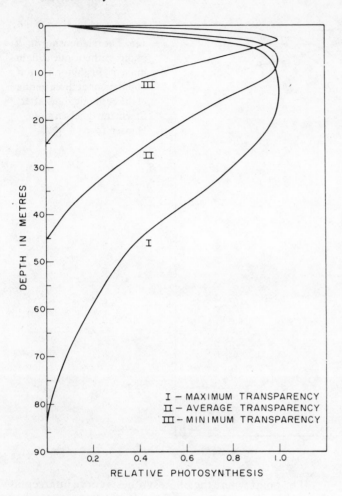

142. Integral or total photosynthesis in water columns of different clarity

The principal advantage in measuring chlorophyll is that as the major absorbing pigment it plays the essential role in the photosynthetic process. There have been many attempts to correlate the amount of photosynthesis with chlorophyll content. These values are variable because the rate of photosynthesis may change, quite independently of chlorophyll content. For example, temperature decreases the photosynthetic rate of algae in the Arctic and Antarctic, which photosynthesize less per unit chlorophyll than phytoplankton of temperate or tropical waters. Chlorophyll measurements may be hampered by the presence of chlorophyll debris in the particulate matter, which poses another difficult practical problem. Debris is a breakdown product that is generated with the death of the cell. These pigments are not easily distinguished from true chlorophyll, hence the total concentration of photosynthetic chlorophyll at times is in error.

Despite these problems, measurements of chlorophyll and photosynthesis show that not all areas of the ocean are equally productive, and the reasons are as follows: in the photosynthetic process, phytoplankton remove from water essential nutrients which, of course, make up particulate organic plant material. If the reader is a gardener he will know that amongst the most essential nutrients are compounds of nitrogen and phosphorus. In the incorporation of these compounds, the particulate plant matter becomes more dense than the surrounding sea-water and it begins to sink slowly. During sinking, the phytoplanktonic organic matter may be subjected to decomposition either by enzymatic processes after death or by destruction from grazing marine herbivores. In either event there is a net downward flux of nutrients, from the euphotic zone. Therefore, the essential nutrients at the surface, i.e. nitrogen and phosphorus, are transported to the deeper ocean waters. A number of oceanographers have assessed the rates at which this organic material is mineralized by oxidation. The depth at which all plant matter is oxidized back into mineral form is estimated to occur at 200 m. The nitrogen and phosphorus ending up at a depth greater than 100 m (i.e. below the euphotic zone) would be lost were it not for certain mechanisms. These nutrients return into the euphotic zone by the vertical mixing of the water column brought about by wind and wave action, by processes associated with ocean currents, and by 'upwelling' of deep waters.

It has been suggested that at times the oceans are 'out of balance', in the sense that many nutrients are transported downward to the bottom waters, whereas all of the light is retained in the surface waters. During the winter months the surface waters at mid-latitudes may become mixed to a considerable depth, and it is in this process that nutrients are brought back into the

phytoplankton near the base of the euphotic zone develop so-called 'shade characteristics' where maximum photosynthesis occurs at lower light intensities.

Despite these variations in the shape of the photosynthesis/light curve, accurate predictions of the vertical profile of photosynthesis in the euphotic zone can be made by using a general shape for the photosynthesis/light curve (similar in all phytoplankton) and taking into account the light intensity in the water column. These two factors are frequently used by aquatic ecologists in predicting integral or total photosynthesis in the euphotic zone. Examples of these predictions are shown in figure 142 and emphasize the importance of water clarity on the total amount of photosynthesis in a water column.

ENVIRONMENTAL FACTORS INFLUENCING PHOTOSYNTHESIS

Biomass estimates of phytoplankton are generally made by measuring chlorophyll in algae (biomass is the amount of living material expressed as weight).

143. Some of the principal factors controlling primary production on, and at the edge of, a continental shelf in temperate latitudes. Vertical enclosed arrows indicate intense vertical mixing. Horizontal enclosed arrows mean little vertical mixing. N means nutrients. The stipple represents phytoplankton growth. To the right is a schematic representation of temperature vs. depth (EZ = euphotic zone)

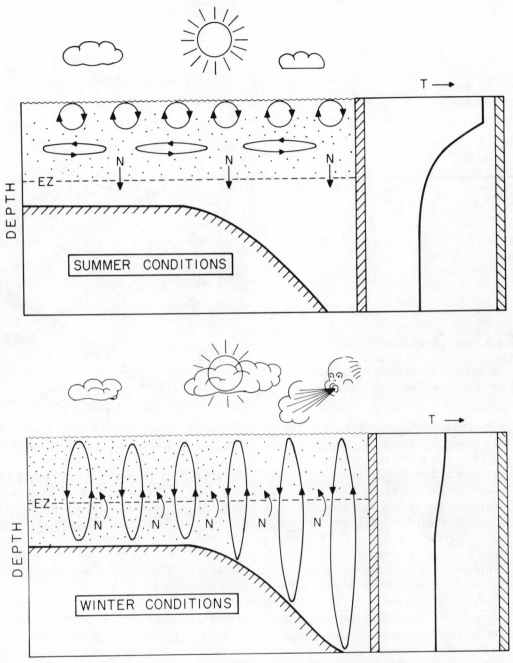

euphotic zone. Figure 143 will aid the reader in understanding this process. The mixing depth cannot be too great or the growth of the phytoplankton population will become limited by light. Generally speaking, vertical mixing is considered detrimental when it exceeds the depth of the euphotic zone (EZ). In some areas of the oceans, vertical mixing can easily go to depths as great as 200 – 300 m. 'Bloom' conditions in the mid-latitudes coincide with the formation of what is known as a 'seasonal thermocline', which is a region in the water column where the rate of temperature change with depth is greatest. It is this rate of tempera-

ture decrease that causes an increase in stability of the water column (with the denser water at the bottom), and the stability tends to slow down vertical mixing. During summer months, with increasing solar radiation, surface waters become heated and hence less dense; under these conditions the water-mass is considered stable. In most oceans, during this period, photosynthesis and standing crops of plants are quite low, largely because, as mentioned above, nutrients are lost to deeper waters and not returned to the euphotic zone. Under these conditions one must conclude that the phytoplankton have to maintain their nutrient

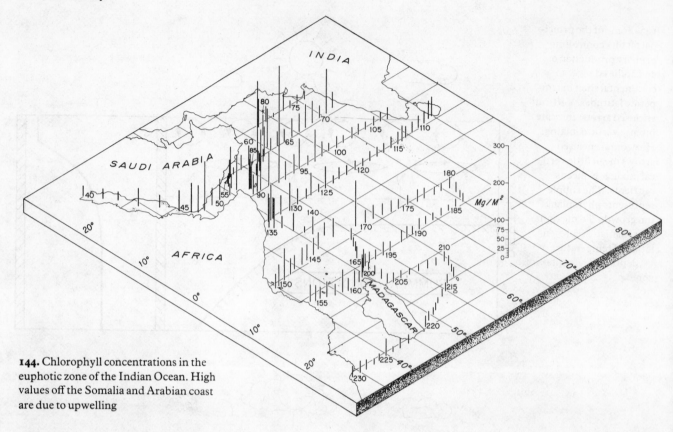

144. Chlorophyll concentrations in the euphotic zone of the Indian Ocean. High values off the Somalia and Arabian coast are due to upwelling

requirement by biochemical recycling as opposed to the physical processes of mixing water.

The cycle of vertical mixing starts when thermal stability is destroyed in the autumn. This is brought about by the decreasing solar radiation allowing the surface water to cool and then the wind mixing the water. As the thermoclinal layer deepens, the nutrient-rich waters at the base of the thermocline are mixed with the upper portion of the euphotic zone. In mid-latitudes, in winter months when solar radiation is low and mixing is intense, blooms still occur along the shallow coastal waters of some areas because, in these cases, depth of mixing is arrested by the bottom of the sea (figure 143). In deeper waters, however, mixing depths become much greater than the depth of the euphotic zone, and population growth is limited by the lack of light.

In the tropical oceans, variations in the intensity of solar energy are not as great as temperature variations at mid-latitudes, hence little or no seasonal variation in vertical mixing occurs. Production of phytoplankton in these areas is generally low for the same reasons as production is low during summer months at mid-latitudes.

The productive areas of tropical oceans are generally found in areas of large ocean currents and/or sites where water is 'upwelled'. In areas of large currents, certain conditions are set up near the high velocity edge of the current which aid in the upward transport of deep, nutrient-rich water. In certain areas of the oceans, upwelling of deeper waters appears to be a persistent feature, and these areas are very productive (i.e. Somalia, Arabian coast, West African coast and west coast of California) (figure 144). Upwelling is accomplished by the wind action transporting surface waters from near the coast back out to sea (figure 148). This displacement allows deep, nutrient-rich water to come to the surface.

So far in this discussion it has been emphasized that the factors augmenting primary production are light, nutrients and their interaction. Essential nutrients implicated are nitrogen and phosphorus; their occurrence and abundance is of importance.

Studies of the distribution of nitrogen and phosphorus compounds indicate that in the euphotic zone the removal by phytoplankton of the two elements runs parallel. Ratios of these compounds in sea-water are comparable to those in the elementary composition of live cultures of marine phytoplankton. Extremely low ratios of nitrogen to phosphorus indicate that phytoplankton have removed practically all of the inorganic nitrogen leaving only a residue of inorganic phosphorus.

Studies of cultures of marine phytoplankton show

145 The little squid *Pyroteuthis* (5 cm) has multi-coloured jewel-like light organs round its eyes, at the tip of its tentacles and in its body

146.   *above* The female paper nautilus *Argonauta*
(5 cm) carries its eggs in a shell secreted specially for this
purpose. Both sexes are sometimes to be found riding on the
backs of medusae such as *Pelagia* (figure 154)

147. *below* The gonostomatid fish *Bonapartia* (8 cm) has its
ventral surface almost completely covered with rows of light
organs

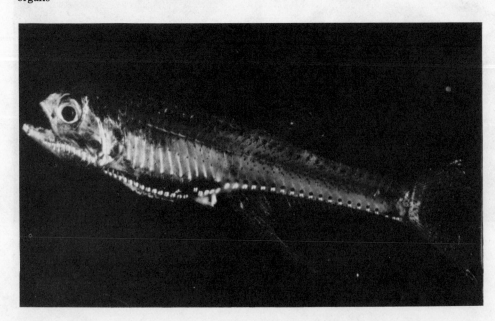

that when either nitrogen or phosphorus is exhausted from the culture media, the cells continue to grow and may divide for some time on their internal supplies, but 'deficient' cells develop in which the internal concentration of nitrogen and phosphorus becomes low enough to stop the cells dividing.

When all aspects of the phosphorus cycle in the sea are known, namely the concentration of particulate and dissolved organic and inorganic, it is possible to evaluate the phosphorus utilization of phytoplankton. Assuming a constant composition for phytoplankton, it becomes possible to estimate the phosphorus requirements for particular photosynthetic needs which occur throughout the year. When this computation is made it becomes apparent that phosphorus must be rapidly recycled in some fashion within the population to maintain the photosynthetic activity observed. The actual nature of this recycling or regeneration is poorly understood.

Nitrogen is available to the plants in the ocean in the form of inorganic ammonia, nitrate and nitrite. Whereas inorganic phosphorus is a sizeable fraction of the total phosphorus in the ocean, inorganic nitrogen is a minor factor in the distribution of the total nitrogen in sea-water. Of the inorganic fractions, nitrate is the most abundant and, as stated above, shows marked seasonal trends due to the photosynthetic activities of the phytoplankton. Recycling of the nitrogenous material also is poorly understood. It has been observed that marine animals (zooplankton) excrete nitrogen, largely ammonia, urea or uric acid (which are organic forms of nitrogenous material).

Extremely low concentration of nitrogen in the surface waters of tropical oceans has led to speculation as to the degree of biological nitrogen deficiency occurring there. In these cases, blue-green algae, such as *Trichodesmium*, are abundant largely because of their capacity to fix atmospheric nitrogen.

Other chemical compounds of importance to phytoplankton growth are silicate, iron, manganese and certain vitamins. The silicate requirement for diatoms, of course, is primarily to build diatom cell walls. A number of workers have shown that diatoms grown in water low in silicate develop a very thin cell wall. The availability of iron and manganese is sometimes very low in the open ocean, and their importance in the metabolism of certain organisms has led oceanographers to conclude that they may be responsible for limiting conditions for growth. Indirect evidence for growth limitation caused by the absence or small quantities of these metals can be shown by augmentation in growth after their addition to water already enriched with nitrogen and phosphorus. Other scientists have pointed out that most marine phytoplankton

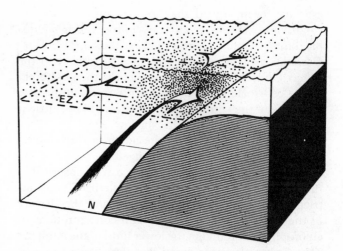

148. Schematic representation of upwelling along a coast. Wind blowing parallel to the coast creates a surface water transport 90° to the wind direction due to the rotation of the earth (Chapter 3). Deeper nutrient-rich water rises upward to replace that moved seaward. Stipple indicates high concentration of phytoplankton (EZ: euphotic zone; N: nutrients)

require vitamins, and principally vitamin B_{12}. The lack of direct measurements of the concentration of this vitamin has hampered our knowledge of its importance. Bio-assay techniques which can be used on a semi-quantitative basis have indicated that the concentration of this vitamin is indeed quite low.

SHORT-TERM FLUCTUATIONS IN THE NUMBERS OF PHYTOPLANKTON

The numbers of phytoplankton occurring at any given time are the result of the rate at which they are growing and the rate at which they are being removed. Phytoplankton are rapidly removed both by animals feeding on them and by sinking out of the euphotic zone. Because of the necessity of light for photosynthesis, the growth of phytoplankton is dependent upon adequate flotation within the euphotic zone. Buoyancy or the lack of it can be an active mechanism in changing the size of the population. Although some phytoplankton have mechanisms for swimming, a great majority of the species observed in marine populations are non-motile. Such cells, being slightly more dense than sea-water, tend to sink. Taxonomists and others have pointed out that the bizarre shape of diatoms – spirals and assemblies of long chains – are mechanisms to prevent or retard sinking. Since the amount of friction that a cell may exert on the surrounding medium is dependent upon the surface area, it is of utmost advantage for the cell to reduce its size if the surface area is to be large relative to the volume, and hence the density.

It has been argued that in very nutrient-starved conditions, sinking is an active process for phytoplankton, enabling them to enrich their metabolism. Consider a plant cell motionless with respect to the water. Under these conditions there is the possibility of the cell soaking up the nutrients in its immediate vicinity and thus building up a shadow of low nutrient water around the cell. The shadow effect would be overcome by sinking, since the medium surrounding the plant is being constantly changed.

Effects of phytoplankton sinking can be observed in the vertical distributions of chlorophyll. Where nutrients are low in the surface water and the density of the water is also low, maximum concentrations of chlorophyll are found in the lower limit of the euphotic zone. It has also been noted that these maxima occur where there is an increase in nutrients. Sinking experiments conducted by various workers have indicated that the buoyancy of the phytoplankton is somehow directly related to the physiological conditions of the cell. Hence a healthy cell, which is one enriched with nutrients, is more buoyant than a cell that is nutrient-starved. In contrast, when the stability of the water column is low and vertical mixing is active, maximum growth and concentration of chlorophyll is near the surface.

MODERN METHODS FOR ESTIMATING PRIMARY PRODUCTION

Estimates of the primary fixation of carbon are generally made by measuring the fixation of a radio-active tracer by the phytoplankton algae throughout the euphotic zone. Radio-active carbon 14 is added to water samples which are taken at specific depths throughout the euphotic zone. These samples are taken generally at the surface and at the depth to which 50, 25, 10 and 1 per cent of the surface light penetrates.

Water samples containing the phytoplankton population must be collected in water sampling bottles that are non-toxic, since small amounts of metals and trace impurities have been found to be toxic to the organisms. Samples of the population are placed in transparent glass bottles, and carbon 14 is then added to the bottles which are suspended within the euphotic zone on a wire or a rope for a period of some four to six hours in the middle of the day. The samples are recovered and the contents of the transparent bottles are filtered through a membrane filter, and the radio-activity in the particulate matter on the filter is counted and expressed as the quantity of carbon fixed during the period of exposure. Quite often the vertical profile of carbon fixation at each depth is integrated and the quantity of carbon fixed under a square metre of ocean surface is given. Carbon fixation is also

TABLE 9
Gross and net organic production of various natural and cultivated systems in grams dry weight produced per square metre per day. (From Ryther 1959; for data sources see original publication)

	Gross	Net
Theoretical potential		
Average radiation (200–400 g cal/cm²/day)	23–32	8–19
Maximum radiation (750 g cal/cm²/day)	38	27
Mass outdoor *Chlorella* culture		
Mean		12·4
Maximum		28·0
Land (maxima for entire growing seasons)		
Sugar		18·4
Rice		9·1
Wheat		4·6
Spartina marsh		9·0
Pine forest (best growing years)		6·0
Tall prairie		3·0
Short prairie		0·5
Desert		0·2
Marine (maxima for single days)		
Coral reef	24	(9·6)
Turtle grass flat	20·5	(11·3)
Polluted estuary	11·0	(8·0)
Grand Banks (April)	10·8	(6·5)
Walvis Bay	7·6	
Continental Shelf (May)	6·1	(3·7)
Sargasso Sea (April)	4·0	(2·8)
Marine (annual average)		
Long Island Sound	2·1	0·9
Continental Shelf	0·74	(0·40)
Sargasso Sea	0·74	0·35

frequently expressed as grams of dry algae.

Because of the high cost of ship operation, some oceanographers have found it profitable to utilize 'on deck' incubators. These are water boxes fitted with neutral density filters simulating the light intensities at different depths and flooded with sea-water. Other workers make estimates of carbon fixation using measurements of chlorophyll, light and water transparency.

COMPARATIVE VALUES FOR PRODUCTIVITY

In coastal waters, production values of carbon have been observed to exceed 3 grams per square metre per day (3 g/m²/day). Offshore, the carbon values range between 0·2 and 1 g/m²/day. In the open ocean, the values range near 0·3 g and probably do not exceed 0·8 g per day; however, not enough seasonal information has been obtained to establish this value clearly. The average annual production of carbon in the open ocean ranges between 25 and 100 g/m² of sea surface.

When primary production is considered in the

overall role of food chain efficiency in the sea, it becomes imperative to know what the efficiency of energy transfer is from one step in the food chain to another. This is one of the more poorly known aspects of the food chain dynamics in the ocean. It has been thought that in the open ocean a majority of the phytoplankton are consumed by zooplankton, including the very important small crustacean herbivores known as copepods. Most estimates of the food requirements for invertebrates living on the bottom of the ocean indicates that only a small fraction, one to ten per cent of the surface production, can ever reach the bottom and serve as a food source for these animals. It is a cardinal rule in discussing food chain dynamics to assign high efficiencies of food energy transfer whenever steps in the food chain are few. Characteristically, the largest vertebrate in the world, the blue or sulfur bottom whale, feeds mostly on a crustacean known as an euphausiid, which in turn feeds on diatoms or small copepods. This places the whale about three or four steps down the food chain sequence. Some of the larger fishes like tuna may be many more steps down. The decreasing efficiency with increasing number of steps in the food chain is largely due to the fact that there is only a small transfer of energy between each step, and the lion's share of the energy goes into maintenance of the organisms. Accurate estimates of the energy transfer from phytoplankton to other organisms have been badly hampered by a whole host of experimental problems. There is increasing evidence that one cannot generalize basically on this 'food chain length of step' thesis when studying energy transfer. The development of the organisms' feeding efficiency during its various growth stages must each be considered.

In Table 9 the productivity of some marine communities has been compared to that of land communities. Note that the maximum productivity of these communities is of the same order in practically all cases. However, it should be noted also that in the ocean planktonic communities, when the maximum is reached, it will only occur for an extremely short interval. Cultivated and cultured systems maintain a maximum for a much longer period. Also interesting to note is that coral reefs are highly productive. Yet they are found in ocean areas that are very poor in nutrients. Why? Maintenance of this high productivity must be due to the fact that new water is continually passing over the reef, and even though this water may be low in nutrients, the supply is virtually unlimited and continual.

On the whole one can say that the oceans, as compared to fertile regions of the earth, are virtually deserts. These data should serve as a warning to some who argue that production from the oceans will stem the famine situations arising from the increasing world population. *The oceans are only productive because of their size.* The total production of ocean is only two to three times that of land, whereas ocean covers 71 per cent of the planet. The real potential of food from the ocean is in exploitation of certain organisms that have a high potential yield of protein food, such as many bivalve molluscs. This manipulation of the ocean is now being called aquaculture and will eventually require the technical and scientific capability that has been applied to terrestrial agriculture.

Lobster with sea urchins in a forest of
seaweed. Sea urchins clear seaweeds
from large areas; lobsters help control
sea urchin populations by preying upon
them.

Seaweeds: Their Productivity and Strategy for Growth

The role of large marine algae in coastal productivity
is far more important than has been suspected.

K. H. Mann

The edge of the sea is one of the best habitats for plant growth in temperate latitudes. In favorable circumstances, net primary productivity may be as high as anywhere else on earth—comparable, for example, to a tropical rain forest. Seaweeds, which have successfully colonized this zone, are a unique form of life. They are attached to a hard substrate, not by a root system, but by a holdfast. Instead of relying on a rather localized supply of nutrients in the soil, they take their nutrients from the water that surrounds them. Because this water is kept in perpetual motion by tides and winds, the nutrient supply is virtually inexhaustible. Even if the seaweeds and plankton deplete the nutrients in surface waters, wind-induced or estuarine mixing renews the supply by causing upwelling of deeper water.

The growth of seaweeds below low-tide level, in the sublittoral, is far richer than in the intertidal areas. Before the advent of underwater vehicles and scuba gear, the algae of the intertidal zone were the focus of attention and provided elegant examples of species zonation in relation to gradients of environmental factors, such as degree of exposure to air or amount of wave action (1). While the intertidal zone is inhabited primarily by the fucoids, or rockweeds, the sublittoral is dominated by laminarians—that is, kelps. In clear water, kelps flourish from low-tide level to a depth of 20 to 30 meters. On gently sloping shores, they may extend 5 to 10 kilometers from the coastline. Several common species of kelp are strap-shaped and may be 2 to 3 m in length. Wave action keeps the blades in constant motion, providing maximum exposure to sunlight and contact with nutrients. Under these conditions, the vegetation may become extremely lush and justify the term "kelp forest," which is commonly applied. Kelp forests occur on all coasts in temperate latitudes and extend into the tropics where the cool waters of the Peru current extend northward (Fig. 1).

Seaweed Productivity in Nova Scotia

Although standing crops of seaweeds have been documented in various parts of the world (2) and short-term studies of photosynthesis have suggested high rates of production (3), the first year-round study of productivity in the sublittoral was carried out on the east coast of Canada (4). The pioneer work of MacFarlane had shown that in various parts of the Nova Scotian coast standing crops of Laminaria had a fresh weight of 20 to 29 kilograms per square meter (5). As part of a multidisciplinary study at St. Margaret's Bay, Nova Scotia, a systematic study of the seaweed zone along approximately 50 km of the shoreline was carried out with the aid of a research submarine and scuba gear. On 24 transects running at right angles to the shore, it was found that algal zones dominated by Laminaria and Agarum accounted for over 80 percent of the total biomass of seaweeds in the bay (Table 1) (6). The next step was to investigate the rate at which this biomass was turned over, in order to calculate the annual rate of tissue production.

The method used was basically very simple: 180 plants on five sites with different depths of water and exposure to wave action were identified by numbered tags. There were three species of plants—Laminaria longicruris, L. digitata, and Agarum cribrosum. Small holes were punched at intervals of 10 centimeters along the blades of these plants, and it was demonstrated by the movement of these holes that all growth in length occurred at the junction of the stipe and the blade (Fig. 2). After that, it was only necessary to punch one hole, 10 cm from the base of the blade, and record at intervals of a few weeks how far the hole had moved along the blade. It was found that the rate of movement of the holes was much greater than the net increase in length of the blade—growth at the base was almost balanced by erosion at the tips. The blades resembled moving belts of tissue, and the holes quickly moved from base to tip, "growing off the ends." Before a hole "grew off," a new hole was made 10 cm from the base. For 2 years a record was kept of the rate of growth at the bases of the blades of a large number of plants. Although this method was conceptually simple, the practical difficulties of finding and measuring numbered plants in dense kelp forests, in all kinds of sea conditions, and with sea ice as a hazard in winter, should not be underestimated.

Our finding was that all three species completely renewed the tissue in their blades between one and five times a year. Moreover, as the plants grew older they also grew wider and thicker. Plots of length against biomass showed that the increase in biomass was

The author is chairman, department of biology, Dalhousie University, Halifax. Nova Scotia, Canada.

Reprinted from Science 182: 975–981. (1973).

Table 1. Zonation and biomass (fresh weight) of seaweeds in St. Margaret's Bay, Nova Scotia, Canada, averaged from 24 transects.

Zone	Average width (m)	Biomass (kg m^{-2})	Biomass per meter of shore-line (kg)	Percent of total biomass
Fucus and *Ascophyllum*	15.5	10.67	124.9	8.7
Chorda and fine browns	87.9	1.08	95.3	6.5
Chondrus crispus	6.0	3.49	20.9	1.4
Zostera marina	4.9	1.02	5.0	0.3
Laminaria digitata and *L. longicruris*	22.7	16.01	363.5	25.0
L. longicruris	46.5	11.50	534.6	35.8
L. longicruris and *Agarum cribrosum*	36.7	4.88	179.2	11.6
A. cribrosum and *Ptilota serrata*	86.3	1.83	158.1	10.7

roughly proportional to the square of the increase in length. Hence, the biomass of new tissue produced annually was up to 20 times the initial biomass of the blade (Table 2). We were particularly surprised to find that growth in length was rapid throughout the winter, and that growth rate reached a peak in late winter or early spring, when the water temperature was close to 0°C (Fig. 3). The ratio of annual production to initial biomass was greatest in young plants and was generally higher for those nearer the surface of the sea. Weighted values of the ratio of production to biomass were calculated for each species, taking into account the relative proportions of young and old plants and the depth at which each species occurred. It was estimated that primary production in the seaweed zone averaged 1750 grams of carbon per square meter per year

(g C m^{-2} yr^{-1}) and that in St. Margaret's Bay, with a total area of 138 km^2, the total production of seaweed was about three times the total production of phytoplankton [191 g C m^{-2} yr^{-1} for phytoplankton (7) against 603 g C m^{-2} yr^{-1} for seaweed, averaged over the entire bay].

Other Marine Macrophytes

The productivity of *Laminaria* off Nova Scotia may be paralleled by that of other seaweeds. For example, giant kelp, *Macrocystis*, develops very large biomasses off the coast of California and in the Indian Ocean. The plants have very long stipes, reaching 10 m or more, and the blades form a dense mat near the surface of the sea. Although there may be up to 20 layers of blades, net assimilation cannot in-

crease very much above that of a single layer because of mutual shading. Biomasses up to 22 kg m^{-2} (fresh weight) have been reported off California, and 95 to 606 kg m^{-2}, with an average of 140 kg m^{-2}, in the Indian Ocean (8). The net annual productivity in California was 400 to 820 g C m^{-2} (9). If the ratio of production to biomass in the Indian Ocean were similar to that on the coast of California, production figures would be enormous, but severe self-shading probably prevents such high production. It seems likely that in giant kelp beds in the Indian Ocean there is a net annual production of about 2000 g C m^{-2}.

Intertidal seaweeds, such as *Fucus* and *Ascophyllum*, may occasionally have rates of production comparable with those of kelps. In Nova Scotia, the fresh weight may be as high as 32 kg m^{-2} (5), with an estimated productivity of 640 to 840 g C m^{-2} yr^{-1} (10). It has been shown that *Fucus* and *Ascophyllum* can double their weight in 5 to 10 days, and a natural population was able to fix more than 10 g C m^{-2} day^{-1} (3). Under conditions of rapid removal by wave action or browsing, it is likely that production rates in excess of 1000 g C m^{-2} yr^{-1} are attained. In sheltered areas of the coast, such as the mouths of estuaries and behind barrier beaches, seaweeds are often replaced by angiosperms rooted in the sediments. Common and well-studied species of angiosperms in temperate waters include *Spartina*, which forms salt marshes just below high-water level, and *Zostera*, which is found near and below low-tide level. *Spartina* production ranges from a maximum of 897 g C m^{-2} yr^{-1} in Georgia to less than 200 g C m^{-2} yr^{-1} in Delaware and New Jersey (11). *Zostera* was found to produce 340 g C m^{-2} yr^{-1} in Denmark, but up to 1500 g C m^{-2} yr^{-1} in various parts of the United States (12).

In tropical waters, a variety of sea grasses grow in sheltered, subtidal areas. *Thalassia* in the Caribbean Sea has been studied, and production figures of up to 5.8 g C m^{-2} day^{-1} have been recorded, suggesting an annual productivity on the order of 1000 g C m^{-2} (13). At the high-tide level in the tropics, the dominant vegetation type is the mangrove swamp. Species of *Rhizophora* and *Avicenna* are found on the tropical shores of several continents. Silt and organic matter accumulate among the roots and produce anaerobic conditions, but the man-

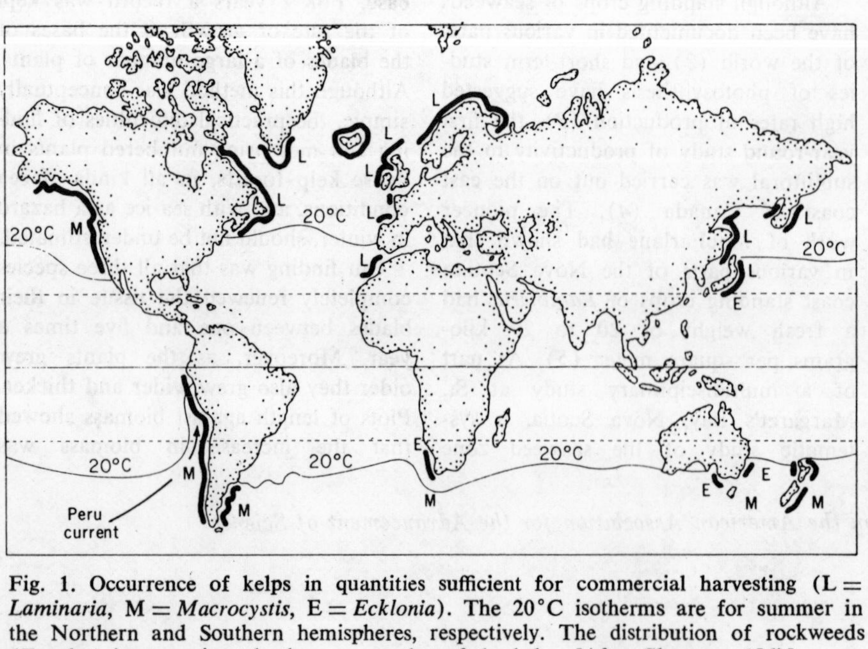

Fig. 1. Occurrence of kelps in quantities sufficient for commercial harvesting (L = *Laminaria*, M = *Macrocystis*, E = *Ecklonia*). The 20°C isotherms are for summer in the Northern and Southern hemispheres, respectively. The distribution of rockweeds (*Fucales*) is approximately the same as that of the kelps. [After Chapman (56)]

groves have aerial roots, which help overcome this difficulty. The conclusion to be drawn from a small number of productivity studies (*14, 15*) is that gross photosynthesis is high, but net productivity is modest, on the order of 300 to 400 g C m^{-2} yr^{-1}.

In Fig. 4, the productivity of various marine macrophytes is summarized. It is clear that seaweeds are among the most productive and that their productivity is as high as, or higher than, some of the most productive terrestrial systems. There exists, then, at the edge of the sea a source of intense primary production that helps create the conditions necessary for the abundant growth of organisms which form the food of young fish and that enables the coastal zone to perform its well-known role of nursery for many commercially important stocks of fish. What strategy do seaweeds adopt to enable them to grow so efficiently?

Growth Strategy of Seaweeds

It was shown earlier that *Laminaria* and *Agarum* in eastern Canada perform the surprising feat of growing rapidly throughout the winter, when temperatures are close to 0°C and light intensity is low (*4*). As Fig. 3 shows, growth rates are increasing at a time of year when temperature and light flux are decreasing. Other species of perennial, subtidal seaweeds have been shown to grow throughout the winter—for example, *Desmarestia aculeata* (*16*), *Cystoseira granulosa* (*17*), and *Hijikia fusiforme* (*18*).

Annual subtidal seaweeds behave

Table 2. Ratio of annual production to initial biomass (P/B) for three species of seaweed in Nova Scotia in two successive years.

Station	Species	Depth (m)	P/B
		1968 to 1969	
Laminaria longicruris	Strawberry Island	5	10.66
	Fox Point	5	6.14
		12	3.82
L. digitata	Fox Point	5	20.44
		12	5.50
Agarum cribrosum	Fox Point	12	4.09
		1969 to 1970	
L. longicruris	Luke Island	5	14.09
		10	13.24
	Fox Point	5	8.99
		8	7.76
		12	7.54
L. digitata	Fox Point	5	8.39
		8	12.48
		12	9.60
A. cribrosum	Fox Point	12	3.18

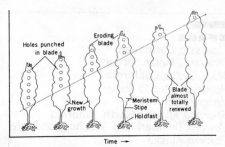

Fig. 2. Movement of punched holes on growing blades of *Laminaria longicruris* (not to scale).

differently. They usually produce the young sporophyte during the spring, grow through the summer, and reproduce during the fall. The spores form microscopic gametophyte stages that survive the winter and give rise to new macroscopic plants in the following spring. Examples are *Chorda filum* (*19*) and *Saccorhiza polyschides* (*20*). Hence, it appears that only large, perennial plants grow in the winter, which suggests that stored carbon may be necessary for the process of growth.

Storing food is a function of photosynthesis and respiration. Early work suggested that, with decreasing temperature, respiration fell more rapidly than did photosynthesis, so that, for a given light intensity, the difference between photosynthesis and respiration was greater at low temperatures than at higher temperatures (*21*). It is now known that, while photosynthesis is nearly temperature-independent at low light intensities, it is strongly influenced by temperature at the high light intensities required to build up a surplus (*22*). It has also been shown that *Laminaria* is capable of seasonal adaptation of respiration. *Laminaria hyperborea* in Scotland was found to have a respiration rate in August (at 16°C) that was only 40 percent of the rate in May (at 8°C) (*23*). The rate of respiration of *Laminaria* in northern Labrador was measured at the same temperature in winter and in summer, and the winter rate was lower (*3*). In this case, the lowering of respiration in winter may be an adaptation in order to survive the long, dark winter without depleting its energy reserves too much. Studies of respiration in a variety of seaweeds, in winter and in summer temperatures, showed summer depression of respiration in *Ascophyllum nodosum* and *Chondrus crispus*, but not in *Ulva*, *Enteromorpha*, or *Ceramium* (*3*).

Evidence is beginning to show that

the kelps are capable of storage, translocation, and mobilization of carbon reserves. For many years, the kelps were regarded as rather loose aggregations of cells with a limited ability to collaborate physiologically. They have no obvious bundles of vascular tissue comparable to those found in higher plants, but the kelps *Macrocystis* and *Nereocystis* have now been shown to have sieve cells (*24*), which almost certainly aid translocation. The so-called trumpet cells of *Laminaria* are very similar to sieve cells, except that they are nucleate (*25*). The dry matter of *Laminaria* varies markedly with season (*6, 26*). Mannitol reaches a peak in midsummer, and laminarin in

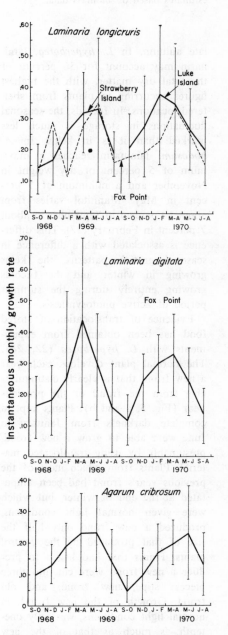

Fig. 3. Seasonal growth patterns of seaweeds in eastern Canada. Vertical lines are standard deviations.

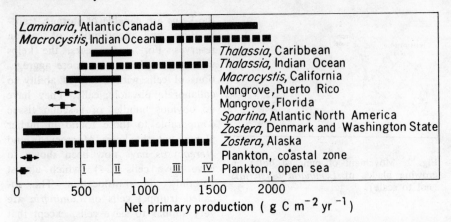

Fig. 4. The productivity of various marine macrophytes, compared with some terrestrial communities [I = medium-aged oak-pine forest, New York; II = young pine plantation, England; III = mature rain forest, Puerto Rico; IV = alfalfa field (an intensively managed system), United States]. Calculated as kilocalories × 0.1. [Source: E. P. Odum, *Fundamentals of Ecology* (Saunders, Philadelphia, ed. 3, 1971).] Broken lines are estimates based on biomass data.

late autumn. In *L. hyperborea*, laminarin may account for 36 percent of the total dry matter, with the highest figures occurring in plants from sheltered locations. In fucoids, the seasonal fluctuations in storage is much less marked than it is in kelps. In *A. nodosum*, laminarin reaches a maximum of 5 percent of dry weight in November and a minimum of 1 percent in May. Mannitol varies from about 10 percent in September to about 7 percent in February (*27*). This difference is associated with a difference in seasonal growth patterns, the kelps growing in winter and the fucoids growing entirely during the summer period of active photosynthesis.

Evidence of translocation of stored food has been obtained from experiments with *L. hyperborea* (*22, 28*). The normal plant develops each year a new frond that is clearly distinguishable from the frond of the previous year (Fig. 5, a and b). Plants kept in complete darkness from January to June were able to grow a new frond, presumably by using translocated material. Plants from which most of the previous year's frond had been amputated at the end of winter, but which were given normal light conditions, produced a new frond only half the size of that produced by the control plants. Plants that had begun to produce a new frond were cut into three pieces: stipe, new frond, and old frond. Growth of the new frond, in normal light conditions, was only one-tenth as much as that of the new fronds of intact plants (Fig. 5c) (*22, 28*). All this suggests that material

stored in the previous year's frond is translocated and used to produce a new frond in the winter and spring.

More direct evidence of translocation has been obtained by attaching a small, transparent container of radioactive sodium bicarbonate to the surface of a photosynthesizing kelp. Some hours later, an autoradiograph revealed a stream of radiocarbon products moving toward the base of the blade (*29*).

How Canadian Seaweeds Differ

The work on translocation in *L. hyperborea* does not explain the winter growth of the species observed in eastern Canada. In *L. hyperborea*, winter growth is confined to the production of a new frond, with the old frond still attached, and Luning's work has clearly shown that translocation from the old to the new takes place. In *L. longicruris* and *L. digitata*, however, the fronds are completely replaced at least once in the course of the winter, as evidenced by the growth of punched holes from base to tip and off the end. While it is probable that reserves of laminarin accumulated in the summer are mobilized to begin winter growth, it is inconceivable that they can be used to maintain it through several cycles of blade renewal. The alternative hypothesis, which has not been tested so far, is that these kelps are able to photosynthesize enough under winter conditions to provide the raw material for growth. It has been suggested that in high latitudes, where

winter conditions are particularly severe and prolonged, it is very unlikely that sufficient light energy reaches the plants to enable them to survive autotrophically. They must therefore practice heterotrophy, deriving their energy from the uptake of organic compounds from solution (*30*).

Whatever the basis for winter growth, one may ask what advantage accrues to the plants by reversing the seasonal pattern found in temperate latitudes. In part, it may compensate for the erosion that takes place at the tips of the blades as a result of wave action. On the open Atlantic Coast, wave action is felt at all depths at which kelp grows. While the severity of erosion varies according to weather conditions, it was found that winter growth generally makes up for the erosion and increases the size of the blades. In addition, the expansion of photosynthetic surface was most rapid when the concentrations of nutrients in surface waters were at their annual maximum; in this way, the plants were able to grow most rapidly before the phytoplankton bloomed and depleted the concentrations of nutrients.

The strategy adopted by these perennial sublittoral macrophytes is unique and appears to be highly successful. Their productivity is as high as that of many intensively managed crops. It is made possible by their constant immersion in a nutrient-containing medium and by their protection from freezing. In this respect, they have a great advantage over intertidal forms. *Ascophyllum*, in the intertidal zone, is intermittently subjected to freezing temperatures and accumulations of ice and snow. It has limited food reserves and, under these conditions, can do little more than minimize its respiratory losses and endure until conditions improve. Other species of seaweed adopt the strategy of passing the winter in the microscopic stage.

The marsh grasses, such as *Spartina alterniflora*, behave much like terrestrial grasses in this climate, translocating stored material to underground organs and allowing the aerial shoots to die and decompose. Such plants are unable to perform significant amounts of photosynthesis in Nova Scotia between November and May. The strategy of *Zostera* varies according to location. In subtidal situations it may remain green and apparently capable of photosynthesis throughout the winter, while plants exposed at low tide are damaged considerably by the frost.

Evidence now being obtained indicates that the anaerobic mud surrounding the roots of marsh grasses and sea grasses is the site of fixation of large amounts of atmospheric nitrogen (31). This renders the plants independent of supplies of dissolved nitrogen, which appear to limit primary production in many coastal areas. There is also evidence that salt marshes play an important role in overall coastal productivity. In a study of Petpeswick Inlet, Nova Scotia, which contains large areas of salt marsh, it was found that a great deal of dissolved nitrogen was exported on the ebb tide (32). These nutrients are made available for uptake by coastal algae, including the seaweeds.

The Fate of Seaweed Production

In St. Margaret's Bay, the main herbivores of the seaweed zone are the sea urchin, *Strongylocentrotus droebachiensis*, and the periwinkle, *Littorina littorea* (33). An energy budget was constructed for the sea urchin population (34), and rough calculations were made for the periwinkles (33). It was shown that the herbivores did not consume more than 10 percent of the net production of the seaweeds. The remaining 90 percent entered various detritus food chains, as particulate or dissolved organic matter.

Sieburth and Jensen, working in the United States and Norway, and Khailov and Burlakova, working in the Soviet Union, found that seaweeds in the laboratory release up to 40 percent of the products of gross photosynthesis in soluble form (35). Brylinsky measured in the field the photosynthesis, respiration, and production of dissolved organic matter by five species of marine macrophyte and found that none released more than 4 percent of the assimilated carbon (36). It is therefore possible that the high levels of production of organic matter observed by the other workers were induced by the experimental conditions.

Even if actively growing, healthy plants do not release soluble organic matter in large quantities, there is little doubt that senescent and dying plants do (37). In the case of the *Laminaria* discussed earlier, it is probable that release of dissolved organic matter takes place almost continuously, since

erosion at the tips is a continuous process and more than half the dry weight of the blades consists of soluble carbohydrates and ash (26).

There are two main routes by which dissolved organic matter may enter the particulate phase—uptake by microorganisms or physiochemical change. High concentrations of bacteria have been reported from the surfaces of seaweeds (38), and *Leucothrix*, a common algal epiphyte, has been shown to take up radioactive thymidine from the surrounding water (39). It seems probable that most dissolved organic material released by macrophytes is rapidly taken up by the bacteria on their surfaces or in the surrounding water. Dissolved organic matter that is free in the water is readily converted to particulate form at air-water interfaces (such as bubbles or the sea surface) or by adsorption on inorganic particles (40). Both microorganisms

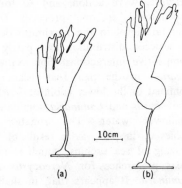

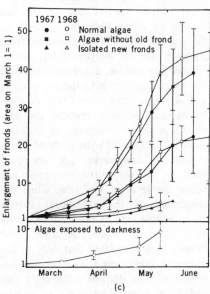

Fig. 5. *Laminaria hyperborea* (a) before and (b) after being kept in darkness from January to June; (c) relative growth of new season's frond under various experimental conditions. [From K. Luning (22, 28)]

and particulate organic matter are normal constituents of that diverse but ecologically important material known as detritus.

To summarize to this point, it is known that the kelps which occur commonly on the shores of the world's oceans in temperate climates are among the more productive plant systems known to man. It is estimated that less than 10 percent of this production normally enters grazing food chains and that the remainder enters detritus food chains, having been released as particulate or dissolved organic matter.

The subsequent fate of this material has yet to be investigated, but work on other forms of detritus give an idea of what to expect. Odum and de la Cruz have reported on the fate of *Spartina* leaves (41). The ash-free dry matter from fresh leaves was about 10 percent protein. That from dead leaves newly dropped was about 6 percent; as the leaves became broken up into successively finer particles, their protein content rose to about 24 percent. The change was attributed to a buildup of microorganisms on the leaf particles. Similar results were obtained in a study of the fate of mangrove leaves (15).

Detritus in this form is a highly nutritious food for planktonic and benthic invertebrates, and even for some kinds of fish (42). Snails and amphipods have been shown to strip the microorganisms from the detritus they consumed, releasing feces with a much reduced nitrogen content (43). Within a few days of liberation, the nitrogen in the fecal pellets had risen again, presumably because the pellets had been recolonized by microorganisms. In this way, even compounds that are refractory to the digestive systems of invertebrates may be progressively consumed by microorganisms and passed on to plankton or benthos.

Interactions with Other Organisms

Sea urchins graze on kelps and other seaweeds and have a major influence on their distribution. Within the *Laminaria* zone in St. Margaret's Bay are to be found various-sized patches of almost bare rock that are characterized by the presence of high densities of sea urchins (*S. droebachiensis*) (6). In the kelp forest, the average biomass density of sea urchins is 150 g m^{-2}, while in the bare patches it is 1200 g

m^{-2} (37). I have observed numerous instances of populations of sea urchins advancing the boundary of the cleared areas, at a rate of several meters per annum, by biting through the bases of the stipes and moving forward en masse. *Laminaria* is a preferred food of sea urchins (44), and sea urchins grow well in the laboratory when this is the sole source of food (34). Hence, there is little doubt that the sea urchins are responsible for the gaps in *Laminaria* cover. Bare areas from which all sea urchins are removed are rapidly recolonized by seaweeds (45).

Similar results have been obtained after removing other kinds of sea urchins in other areas: namely, *Strongylocentrotus purpuratus* and *S. franciscanus* (46), *Paracentrotus lividus* (47), and *Echinus esculentus* (48). The damage that *Strongylocentrotus* spp. inflict on beds of giant kelp (*Macrocystis*) along the coast of California is a matter of economic concern and has been extensively studied (49). The sea urchins have been controlled with quicklime, and *Macrocystis* has reestablished itself. There have also been observed natural cycles in which the sea urchin populations declined, presumably because they had overeaten their food supply. However, sewer outfalls appear to modify the interaction by providing nourishment for sea urchin populations in the absence of kelp, thus preventing the seaweed from returning to the area (50).

It is probable that outbreaks of high population density in sea urchins are triggered by reductions in the population density of their predators. The outbreaks in California have been attributed to a decline in the number of sea otters (*Enhydra lutris nereis*) as a result of hunting. A reversal of this trend in California was followed by a sharp decline in the population of sea urchins and expansion of kelp beds (49, 51). The starfish *Pycnopodia* was shown to be a key factor in the control of sea urchin populations near Seattle (46). In the case of the sea urchin populations of eastern Canada, the lobster *Homarus americanus* may well be the key predator. When lobsters are placed in cages on the sea floor and given their choice of several kinds of food, sea urchins are high on the list of preferred species (45). In the laboratory, two lobsters consumed 131 sea urchins, weighing a total of 342 g, in 7 months (44). Rock crabs are also active predators on sea urchins, but they eat smaller sizes and they eat less. Since lobsters are also predators of rock crabs, lobsters appear to be the controlling influence.

Human predation on lobsters is intense (52), and it is probable that the reduced population of lobsters on the east coast of Canada and the United States has permitted population explosions of sea urchins, with consequent overgrazing of the seaweed beds. An analogous situation would be the hunting of carnivores: on the Kaibab Plateau of Arizona (53), for example, such overhunting allowed deer populations to expand and overgraze their food supply. If my conclusions about the relationship of man, lobsters, sea urchins, and seaweed are correct, this is probably the first documented example of such a four-level interaction in the sea.

The zonation of seaweeds seems to be determined partly by the gradients of environmental factors (light, temperature, wave action, and so forth) encountered as one proceeds from high-tide level to the maximum depth of seaweed distribution and partly by competitive interactions. For example, *Agarum* in the San Juan islands is confined to the lower sublittoral, while *Nereocystis* and *Laminaria* dominate in shallower water. The zonation is achieved, in part, as a result of the grazing of sea urchins, which show a clear preference for *Nereocystis* and *Laminaria*. It appears that, in shallow water, these species can grow fast enough to offset both the grazing of the sea urchins and competition from *Agarum*, but in deeper water, where light is less intense, the *Laminaria* and *Nereocystis* succumb to the sea urchins and *Agarum* takes over (54). A similar explanation may account for part of the zonation observed on the east coast of Canada (6). In parts of Newfoundland, *Laminaria* spp. are confined to the very turbulent zone just below the level of low tide, while the remainder of the sublittoral carries dense populations of sea urchins, along with coralline algae, *Agarum*, and *Ptilota*. It is postulated that only in the turbulent sublittoral fringe are the activities of sea urchins sufficiently impeded to allow *Laminaria* to persist (55). It would be interesting to try to control the sea urchin populations by protecting and adding to the number of lobsters.

References and Notes

1. T. A. Stephenson and A. Stephenson, *J. Ecol.* **37**, 289 (1949).
2. G. Michanek, *F.A.O. Fish. Circ.* **128**, 37 (1971).
3. J. W. Kanwischer, in *Some Contemporary Studies in Marine Science*, H. Barnes, Ed. (Allen & Unwin, London, 1966), pp. 407–420.
4. K. H. Mann, *Mar. Biol. (Berl.)* **14**, 199 (1972).
5. C. MacFarlane, *Can. J. Bot.* **30**, 78 (1952).
6. K. H. Mann, *Mar. Biol. (Berl.)* **12**, 1 (1972).
7. T. Platt, *J. Cons. Cons. Int. Explor. Mer.* **33**, 324 (1971).
8. W. J. North, Ed. *Nova Hedwigia* **32** (Suppl.) 1 (1971); P. Grua, *Terre Vie* **2**, 215 (1964).
9. K. A. Clendenning, *Nova Hedwigia* .**32** (Suppl.) 259 (1971).
10. D. F. Westlake, *Biol. Rev.* **38**, 385 (1963).
11. J. M. Teal, *Ecology* **43**, 614 (1962); R. B. Williams, "Proceedings of the Atlantic Estuarine Research Society Meeting, Hampton, Virginia" (mimeographed, 1965); M. H. Morgan, unpublished data.
12. C. J. G. Petersen, *Rep. Dan. Biol. Sta.* **25**, 1 (1918); P. R. Burkholder and G. H. Bornside, *Bull. Torrey Bot. Club* **84**, 366 (1957); K. H. Mann, *Mem. Ist. Ital. Idrobiol. Dott Marco De Marchi Pallanza Italy* **29**, 353 (1972).
13. H. T. Odum, P. R. Burkholder, J. A. Rivero, *Univ. Texas Inst. Mar. Sci. Publ.* **9**, 404 (1959); J. A. Jones, thesis, University of Miami (1968); M. Brylinsky, unpublished data.
14. F. Golley, H. T. Odum, R. F. Wilson, *Ecology* **43**, 9 (1962).
15. E. J. Heald, thesis, University of Miami (1969).
16. A. R. O. Chapman, *Phycologia* **10**, 63 (1971).
17. ———, unpublished manuscript.
18. S. Suto, *Bull. Jap. Soc. Sci. Fish.* **17**, 13 (1951).
19. G. R. South and E. M. Burrows, *Br. Phycol. Bull.* **3**, 379 (1967).
20. T. A. Norton and E. M. Burrows, *Br. Phycol. J.* **4**, 19 (1969).
21. H. Kneip, *Int. Rev. Gesamten Hydrobiol. Hydrogr.* **7**, 1 (1914); R. Harder, *Jahrb. Wiss. Bot.* **56**, 254 (1915).
22. K. Luning, in *Proceedings of the Fourth European Marine Biology Symposium*, D. J. Crisp, Ed. (Cambridge Univ. Press, Cambridge, England, 1971), pp. 347–361.
23. W. A. P. Black, *J. Soc. Chem. Ind. (Lond.)* **69**, 161 (1950).
24. N. L. Nicholson, *J. Phycol.* **6**, 177 (1970).
25. B. C. Parker and J. Huber, *ibid.* **1**, 172 (1965); H. Ziegler and I. Ruck, *Planta* **73**, 72 (1967).
26. W. A. P. Black, *Nature (Lond.)* **161**, 174 (1948); *J. Soc. Chem. Ind. (Lond.)* **67**, 169 (1948); *ibid.*, p. 172; *ibid.* **69**, 161 (1950); *J. Mar. Biol. Ass. U.K.* **29**, 45 (1950); *ibid.* **33**, 49 (1954).
27. A. Jensen, *Rep. Norw. Inst. Seaweed Res.* **24**, 23 (1960).
28. K. Luning, *Mar. Biol. (Berl.)* **2**, 218 (1969).
29. K. Schmitz, K. Luning, J. Willenbrink, *Z. Pflanzenphysiol.* **67**, 418 (1972).
30. R. T. Wilce, *Bot. Mar.* **10**, 185 (1967).
31. D. G. Patriquin, *Mar. Biol. (Berl.)* **15**, 35 (1972); ——— and R. Knowles, *ibid.* **16**, 49 (1972).
32. W. H. Sutcliffe, Jr., personal communication.
33. R. J. Miller, K. H. Mann, D. J. Scarratt, *J. Fish. Res. Board Can.* **28**, 1733 (1971).
34. R. J. Miller and K. H. Mann, *Mar. Biol. (Berl.)* **18**, 99 (1973).
35. J. McN. Sieburth and A. Jensen, *J. Exp. Mar. Biol. Ecol.* **3**, 290 (1969); K. M. Khailov and Z. P. Burlakova, *Limnol. Oceanogr.* **14**, 521 (1969).
36. M. Brylinsky, thesis, University of Georgia (1971).
37. R. E. Johannes, *Advan. Microbiol. Sea* **1**, 203 (1968).
38. E. C. S. Chan and E. A. McManus, *Can. J. Microbiol.* **15**, 409 (1969).
39. T. D. Brock, *Science* **155**, 81 (1967).
40. E. R. Baylor and W. H. Sutcliffe, *Limnol. Oceanogr.* **8**, 369 (1963); G. A. Riley, *Advan. Mar. Biol.* **8**, 1 (1970).
41. E. P. Odum and A. A. de La Cruz, in *Estuaries*, G. H. Lauff, Ed. (American Association for the Advancement of Science, Washington, D.C., 1967), pp. 383–388.
42. W. E. Odum, in *Marine Food Chains*, J. H. Steele, Ed. (Univ. of California Press, Berkeley, 1970), pp. 222–240.

43. R. Newell, *Proc. Zool. Soc. Lond.* **144**, 25 (1965).
44. J. H. Himmelman and D. H. Steele, *Mar. Biol. (Berl.)* **9**, 315 (1971).
45. P. A. Breen, unpublished data.
46. R. T. Paine and R. L. Vadas, *Limnol. Oceanogr.* **14**, 710 (1969).
47. J. A. Kitching and F. J. Ebling, *J. Anim. Ecol.* **30**, 373 (1961).
48. J. M. Kain and N. S. Jones, in *Proceedings of the Fifth International Seaweed Symposium*, E. G. Young and J. L. McLachlan, Eds. (Pergamon, Oxford, 1966), pp. 139–140; N. S. Jones and J. M. Kain, *Helgol. Wiss. Meeresunters.* **15**, 460 (1967).
49. W. J. North, Ed., *Kelp Habitat Improvement Project, Annual Report 1964–65* (California Institute of Technology, Pasadena, 1965); *Kelp Habitat Improvement Project, Annual Report 1968–69* (California Institute of Technology, Pasadena, 1969).
50. J. S. Pearse, M. E. Clark, D. L. Leighton, C. T. Mitchell, W. J. North, *Kelp Habitat Improvement Project, Annual Report 1969–70*, W. J. North, Ed. (California Institute of Technology, Pasadena, 1970), appendix, pp. 1–93; M. E. Clark, in *Kelp Habitat Improvement Project, Annual Report 1968–69*, W. J. North, Ed. (California Institute of Technology, Pasadena, 1969), pp. 70–93.
51. J. H. McLean, *Biol. Bull. (Woods Hole)* **122**, 95 (1962).
52. D. G. Wilder, *Rapp. P.-V. Reun. Cons. Perm. Int. Explor. Mer.* **156**, 21 (1965).
53. A. Leopold, *Wisc. Conserv. Dep. Publ.* **321**, 3 (1943); G. Caughley, *Ecology* **51**, 53 (1970).
54. R. L. Vadas, thesis, University of Washington (1968).
55. J. H. Himmelman, unpublished data.
56. V. J. Chapman, *Salt Marshes and Salt Deserts of the World* (Interscience, New York, 1960).
57. I thank those who have assisted this investigation and review, by advice or loan of material, particularly A. R. O. Chapman, J. S. Craigie, A. C. Neish, R. L. Vadas, and J. H. Himmelman. Figures 1 and 4 are reproduced, with permission, from *Mem. Ist. Ital. Idrobiol. Dott Marco De Marchi Pallenza Italy*, **29**, 353 (1972); Fig. 3 from *Mar. Biol. (Berl.)* **14** (1972), p. 205; Fig. 5 from *ibid.* **2** (1969), pp. 221–222.

The Sea Turns Over

And where it does, nutrients upwelling from the depths create lush planktonic feeding zones for the world's fisheries.

On a summer day, when a stiff northerly wind blows along the Oregon coast, and there's a line in the sea where the color changes from a clear cobalt blue to a murky green—that's a day fishermen know they can make a good catch. And catch they do. The silver salmon, tuna, and other fish congregate in certain greenish areas where they feed on thriving pastures of marine plants and animals—and if the fishermen are there on the spot, the haul is good.

Fishermen worth their salt in all oceans have long known signals of wind, sea color, frontal lines, and temperatures in their special fishing areas—it's part of the lore of the sea. Yet, although they can recognize areas where the fish might be, they have been unable to predict these spots when winds and sea currents change.

Now, after several years of research, some of the factors that create good catches are becoming beautifully clear.

In many slow and ponderous ways—by action of the tides; the rotation of the Earth; and changes in seasons, winds, and temperatures—the waters of the world are constantly mixing, overturning, and welling up, bringing dissolved chemical nutrients from the sea

Fishing weather. Haze and fog shrouding the coastline south of Otter Crest, Oregon, indicate that chilled (and nutrient-rich) bottom waters are upwelling into warm surface waters.

depths to the surface, and churning oxygen-rich water from the surface down again to the deep sea.

These complex vertical and horizontal motions, vital to the living resources of the sea, may occur in the middle of the oceans, or at boundaries of different water masses, in eddies around the lees of island or land promontories projecting into a current, or over ridges and canyons beneath the open sea.

However, the most dynamic ocean turnover process is coastal upwelling, a phenomenon by which nutrients from the dark depths are periodically brought up in certain areas to the sunlit surface layers where photosynthesis can take place and where they can become available to microscopic plant life. These one-celled plants, phytoplankton, provide the base of the complex food chain of ocean life. Provided with nutrients and using energy from the Sun for photosynthesis, they multiply into large masses, offering feasting grounds for zooplankton and larger fish and creating the most productive fish-producing regions in the world.

Fishing yields in these upwelling areas and their immediate vicinity are at least a thousand times higher than in other oceanic areas. The coastal upwelling areas, comprising only one-tenth of one percent of the total area of the world's oceans, are estimated to contain more than half of the ocean's fish catch—a total fishery yield of more than 40 million metric tons a year.

Progress in understanding the theory of the upwelling phenomenon has reached a point of worldwide significance, states Richard Barber of Duke University's Marine Laboratory at Beaufort, North Carolina, and national coordinator of the Coastal Upwelling Ecosystem Analysis (CUEA) program. CUEA, part of the International Decade of Ocean Exploration, was started by the National Science Foundation to investigate the physical and biological aspects of upwelling. With more than 29 principal investigators and 13 U.S. research and educational institutions and organizations, CUEA has initiated a series of experiments and theoretical observations. Its purpose is to provide systems models for predicting on a daily basis the changing sites, courses, extent, temperatures, and various other factors of upwelling systems in particular locations. Eventually such system modeling

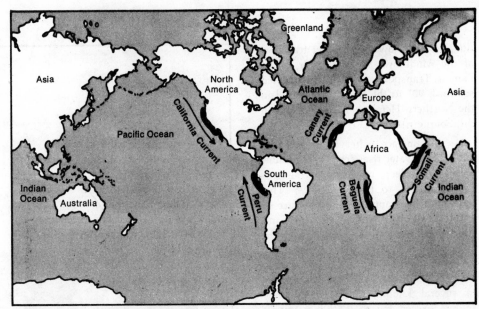

Upwelling areas. Strong persistent upwelling occurs along only a few coastlines, mostly on western continental edges. Some upwelling also occurs in mid-ocean areas around the Equator and in the Antarctic.

would be used to predict the production of the world fisheries on the basis of a few significant meteorological and oceanographic measurements.

Where it happens

Although upwelling may take place anywhere in the ocean, it occurs more regularly and most conspicuously along the western edges of continents in the low and mid latitudes, particularly along the western coasts of the Americas and of Africa. In these regions, the prevailing winds blow equatorward, and this, in combination with the Earth's rotation, causes the surface water to move away from the coast. The surface water is replaced by water from the depths.

There are only a few places in the world where conditions exist to form strong persistent upwelling over large regions. One is along the Peruvian coast, where the Peru Current flows northward west of Chile and Peru. Another region of marked upwelling occurs along the coasts of Baja California, California, Oregon, and Washington, where the California Current flows south. Here the upwelling peak moves up the coast with the warming weather in May, June, and July as the North Pacific atmospheric high pressure cell intensifies and north-

erly winds develop along the coast. By October it is finished. A third intensive upwelling area is along Southwestern Africa, with most intense activity in the southern spring months of September and October. A fourth dominant upwelling system occurs farther north—along Northwestern Africa.

Contrary to the "west coast rule," important coastal upwelling also develops in the region of Eastern Africa where, during the annual southwest monsoon, the Somali Current flows from the Southern Hemisphere up the east coast of Africa and along the Arabian coast into the Arabian Sea.

Upwelling also occurs in mid-ocean spots around the Equator and in the Antarctic. In the eastern equatorial region of the Pacific Ocean, for instance, the Cromwell Undercurrent creates an uprising in an area extending eastward along the Equator from the 180-degree meridian to the Galapagos Islands.

When the polar wind blows

The primary driving mechanism of a "typical" coastal upwelling system is the wind system blowing from an atmospheric high pressure toward the Equator and producing stress at the surface of the sea. Because of the effect of the Earth's rotation and fric-

Reprinted from Mosiac 5: 25–31, National Science Foundation (1974).

31

tional forces, however, the net transport of the windblown surface water, called Ekman Transport or Drift, is directed seaward, 90° to the right of the wind in the Northern Hemisphere (to the left in the Southern).

As the surface water is pushed offshore, cold water rises from several hundred meters deep, up and over the continental shelf, to take its place. This upwelling may appear at the surface in various patterns of tongues, plumes, and patches. Such tongues or plumes indicate intense upwelling locally and may assume many changing shapes, ranging in length and depth from a few meters to several kilometers. However, they seldom exceed ten to 30 kilometers in width, and ten to 20 meters in thickness. The tip of the tongue of intense upwelling usually can be found within ten to 20 kilometers off the coast. Frontal boundaries between the warm (old surface water) and cold (more newly upwelled) water masses are often sharp and distinct. Some occur over long distances, others over a space so small that a single ship may straddle the boundary.

An inherent feature of upwelling seems to be the presence of a narrow jetlike surface current that flows along the shore in the same direction as the prevailing winds on the seaward side of the upwelling front. At the same time as the surface current is flowing, a subsurface countercurrent is often observed flowing away from the Equator. This is considered a very important factor in the dynamics of upwelling and its associated ecosystem. Just how much these currents contribute to upwelling under various conditions and what effect they produce is not yet clear, points out physical oceanographer Robert Smith of Oregon State University, which was the center of extensive upwelling experiments during the summers of 1972 and 1973.

Each regional upwelling ecosystem is a separate, complex, and dynamic phenomenon that depends upon the physical conditions of its setting. Each is subject not only to variations within the system itself—such as the configuration of the coast and continental shelf, the strength and directions of ocean currents, and the local wind conditions—but also to influences external to it, such as the seasons and the world wind patterns.

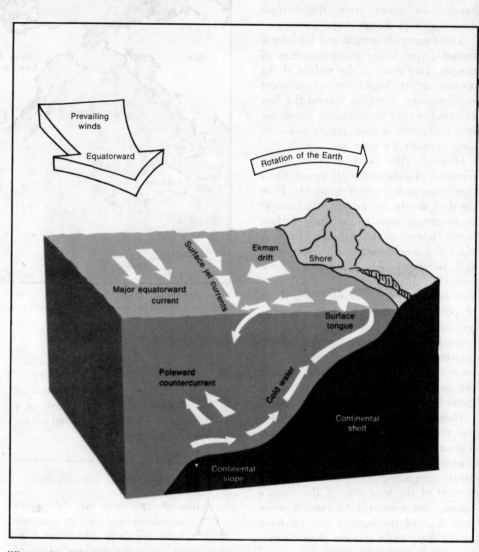

What makes it happen. Strong winds blowing along the shore toward the Equator, combined with the Earth's rotation and frictional forces, drive the warm surface water away from the shore. It is replaced by cold, nutrient-rich water from the lower depths; phytoplankton flourish, and fish congregate to feed on the plankton.

Unpredictable prodigy

Although the upwelling regions seem well defined, the process itself has, as yet, no constant or dependable schedule of when it might arrive or how long it will last. The length of time an upwelling season may prevail depends primarily upon how long the dominating winds blow along the coast. This in turn depends on the strength and positions of the atmospheric highs that occur as spring and summer arrive. During the season when conditions are generally right for upwelling, the process may vary from strong persistent upwelling to none at all to strong again over a period of weeks.

These modulations may occur several times during a season of upwelling conditions. They occur when the prevailing winds stop blowing in the direction favorable to upwelling—equatorward along west coasts—and come from other directions, hence stopping the offshore transportation of the warm surface water. This in turn shuts off the upwelling circulation of water from lower depths. The cold water remains below and the warm water above in horizontal layers. These wind changes of a few days' duration have large effects on water temperature over areas as wide as ten kilometers from the shore and as deep as 20 meters.

Once the source of nutrients is stopped or blocked, the growth of phytoplankton halts, as does the concentration of fish. The whole process collapses and the fish disperse to find their own food where they can, and fishing for the day, the week, perhaps longer, can become a haphazard affair for the fishermen.

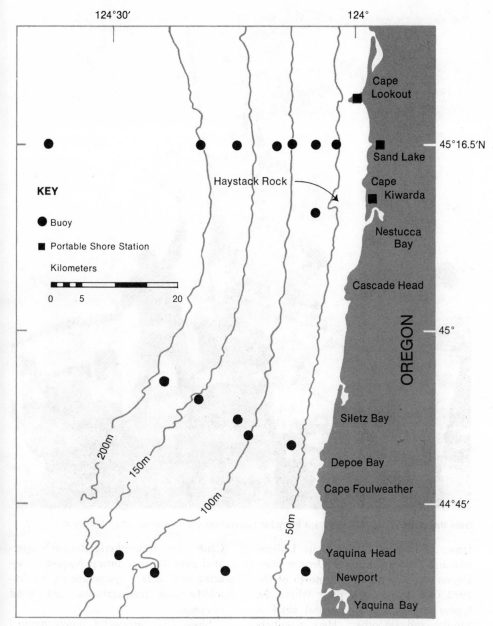

Buoy garden. Instruments for measuring temperatures and currents were positioned in rows extending from shore some 60 kilometers seaward during the CUE-I and CUE-II operations in the summers of 1973 and 1974.

the diminished numbers of anchovies are caused by overfishing during *el Niño*, or perhaps by a natural decrease in the biological cycle of the fish. Other tragic failures blamed on similar circumstances were the disappearance of the California sardine and Hokkaido herring.

The sunlit zone

Without doubt, coastal upwelling systems produce rich biological growth and activity. This high productivity of organic matter is limited to the sea surface layers where sunlight is sufficient for photosynthesis. Solar radiation penetrates through this upper sea layer, the euphotic zone, to depths of about 11 to 28 meters, depending on the concentration of the marine populations or on turbidity. Into this zone are brought up the deep sea's nitrogen, phosphorus, and silicon; in ion form or as compounds of nitrates, nitrites, phosphates, and silicates. Iron and traces of other minerals are also present.

At first, the cold, newly upwelled water is low in plant and animal population; it takes a while for primary biological activity to start, explains Richard Dugdale, a biological oceanographer from the University of Washington. As one-celled plankton multiply, they gradually consume and reduce the amount of nutrients. At the same time, oxygen content increases. Along the surface waters, oxygen is near saturation, and toward the seaward edge of the spreading plume it has been measured as high as 130 percent of saturation.

As an indication of the extent to which upwelling increases production of primary marine life, phytoplankton cell counts in the Peruvian upwelling ecosystem, in a good year, have been measured as high as 138,000 per liter. This compares to a normal ocean density of 1,000 to 10,000 per liter.

Feeding directly on the plant and animal plankton are the herbivorous zooplankton and fin fishes, such as tuna, salmon, anchovies, mullet, and herring, which are caught by fishermen or eaten by other carnivorous predators including birds, bonito, squid, and sea lions. In the Peruvian region, it is estimated that the annual numbers of fish captured by fishermen are probably matched by those consumed by marine predators.

As the upwelling tongue of cold water spreads over the sea surface and is

With the return of the prevailing northerly winds (or southerlies in the Southern Hemisphere), upwelling resumes, and plankton growth and concentration of marine life reoccurs.

Some upwelling systems take place year after year in specific areas. Others may continue for several years without interruption and then fail catastrophically—such as the failures of upwelling in 1965, 1971, and 1972 off the coast of Peru. In previous years, the Peruvian upwelling has created the world's most productive fishery area. In 1970, 22 percent of the total world fish catch—mostly anchovies—was harvested. With the absence of upwelling—*el Niño* or "The Child," as it is called—in 1971-72, the anchovy catch dropped from 12.3 million tons in 1970 to 4.5 million tons in 1972 —a tragedy for Peru which for decades had depended on the more than ten million tons of anchovy catch, from which nearly 70 percent of the world's fishmeal was produced. Scientists say the fish stocks are slowly recovering with the resumption of upwelling in 1973, but they have not yet multiplied to their former numbers. This disruption has raised several questions as to whether moved seaward by the wind, it carries the products of this biological activity—the excreta, dead and decaying plants, animals, and other organic compounds—which gradually sink toward the bottom of the sea. As this material drifts downward, it is acted upon by bacteria near the surface which reduce it again to inorganic nutrients that may be taken up by phytoplankton, dissolved into the sea, or come to brief rest on the sea floor before being upwelled again—a complete ecosystem recycling.

Moreover, though the processes are not yet well understood, there seems to be nutrient recycling within various layers and currents of the sea in an upwelling region—through local mixing and convection currents, or subsurface shoreward transport, or through injection at the source of upwelling. The proportion of regenerated nutrients in a fully developed upwelling ecosystem is quite high. One estimate is that a dense school of herbivorous fishes grazing on a phytoplankton crop, through its excreta alone, produces the daily nitrogen requirements of the phytoplankton in just two hours.

Expeditions for upwelling

Scientists have been aware of upwelling for a long time, particularly along the coast of Peru, where the occurrence of el Niño has been recorded by fishermen for more than 180 years. Scientific attempts to define and describe the upwelling system itself, however, did not occur until the early 1900's. In the late 1920's and 1930's, individual oceanographers began making major contributions to the studies. In 1968 an Upwelling Biome Program was included as part of the International Biological Program, and studies were made along the Peruvian coast and in the Mediterranean. With CUEA established in 1971, four expeditions took place in the succeeding two years to investigate the biological and physical effects of upwelling: MESCAL I and II off Baja California in the springs and summers of 1972 and 1973; and CUE I and II off the Oregon coast in the same years.

The latest major experiment, CUE II, operating from July through August 1973, took place on a site some 80 kilometers long along the Oregon coast from Newport to Cape Lookout, and extending some 60 kilometers into the Pacific Ocean. CUE II, co-directed by

Over the stern. Crewmen deploy a buoy for measuring ocean and weather conditions.

James O'Brien of Florida State University and Dale Pillsbury of Oregon State University, was set slightly north of the 1972 CUE I site, at a place where the topography of the continental shelf is simpler and smoother. Here scientists hoped to avoid the complicated sea circulations created by ridges and bumps of the sea bottom and to simplify the simulations of models.

Three research vessels—the "workhorse" Cayuse which can make a 180° turn "on a quarter" to place or pick up buoys and objects from the sea; and the more elegantly equipped Oceanographer and Yaquina—made repeated journeys back and forth over the area to measure temperatures, salinities, densities, and changes in currents. On another vessel, the Thomas G. Thompson, scientists made detailed studies on chemical nutrients, plankton, and fish, and compiled maps charting temperature zones, flow of currents, and areas of nutrients and biological activity. An instrumented aircraft from the National

Center for Atmospheric Research operated over the area during August, measuring sea surface temperatures, air humidity and temperatures, and wind systems.

Since 1965, arrays of buoys moored on Oregon's continental shelf have been used to record currents and temperatures. During CUE II more buoys were installed in arrays extending from land out to sea some 60 or more kilometers—across the continental shelf and over the edge of the continental slope. These buoys—spaced several kilometers apart, depending on the configuration of the sea floor or the currents—recorded temperatures and currents from the surface to the sea floor. They helped scientists "follow" various ebbs and flows of currents as the winds changed and the season progressed. Buoys from NOAA's Pacific Marine Environmental Laboratory were moored at single points along the array to measure the wind, solar radiation, and currents near the surface. Buoys from Oregon State University

Taking the sea's temperature. Equipment is placed at strategic points off the Oregon coast during CUE-II.

measured current velocities and directions at different depths.

Scientists found the 1972 data lacking in adequate information on wind directions and changes. Since wind is recognized as the major driving force in upwelling systems, emphasis was made in 1973 on increasing wind data from anemometers on land and sea stations.

In addition to new data-gathering techniques and equipment, a wholly new shipborne computerized data storage, processing, and display system was used in 1973, packed totally in a van about half a room large, and placed on board the *Thomas G. Thompson*. This system acquired data on the spot, processed it, and provided scientists with a nearly "real time" look at the area they were studying to help them update their plans during the cruise.

A step closer to prediction

Some parts of the upwelling process have been successfully simulated on two-dimensional, two-layer, time-dependent numerical models over coastal regions of simple topography under steady wind conditions. James O'Brien and John J. Walsh, University of Washington, are working on models that encompass basic features—the slope and configuration of the coastal shelf, the direction and speed of wind, biological variables, the flow of currents

and countercurrents, and the size and movement of the upwelling.

But since upwelling is a three-dimensional affair, actually four when one considers the time factor, O'Brien and his co-workers are developing a three-dimensional, time-variable circulation model to handle the many variables occurring during upwelling. "With this model," O'Brien says, "we hope to predict factors such as width of the upwelling tongue, the times and areas for upwelling to appear, and the flow of currents.

"Numerical models of time-dependent oceanographic phenomena have a particularly promising future," he says. "With models from experiments CUE I and II already feeding forecasts on a general scale, we hope numerical studies will be applied to other upwelling regions throughout the world."

The accumulation of data and testing of hardware and techniques during CUE II and other CUEA experiments have been building up to a large international project, JOINT I, scheduled for February through May 1974 off the northwestern coast of Africa.

"JOINT I is the first major experiment with all components of the CUEA program functioning together," says Barber. The international research project will be conducted in the area of Cape Blanc in cooperation with scientists from France, Germany (East and West), U.S.S.R.,

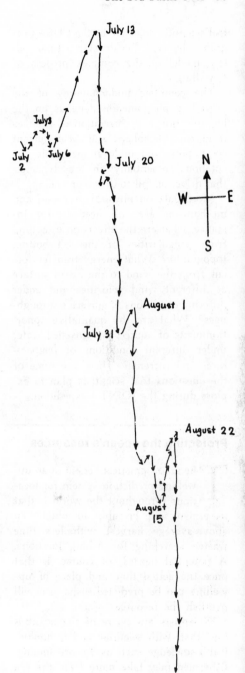

Tracks of winds. Arrows on this vector chart show how the wind blew daily along the Oregon coast during July and August of 1972. Prevailing winds blew from the north most of the time, but reversed around July 2, 6, 31, and August 15. During these reversals, upwelling slowed, shore water temperatures rose, and fishing was generally poor until north winds picked up again.

Spain, the United Kingdom, and Mauritania—all participants in an even larger international program called Cooperative Investigations of the Northern Part of the Eastern Central Atlantic.

"The primary research objective of JOINT I is exactly the same as the central scientific objective of the CUEA program," he says: "to develop a total system model of the complex process of upwelling."

The complete understanding of the upwelling phenomenon depends on the identification and correlation of changes in global atmospheric and ocean current circulations and in local mesoscale circulations—essentially on the dynamics that drive, or fail to drive, upwelling.

More data on physical processes and mechanisms are still needed. For instance, what are the effects on upwelling from irregularities of the sea bottom topography? What energy transfers occur from the wind to the ocean surface at different wind velocities· and under differing conditions of surface sea roughness? What are the qualitative apportionments of sources of upwelled water under different conditions of temperatures and currents? These are some of the questions that scientists plan to explore during the JOINT I expedition.

Protecting the ocean's resources

The major practical benefit of an upwelling prediction system for locations throughout the world is that fishermen will be able to obtain an above-average harvest with less time wasted searching for fishing locations. A potential hazard, of course, is that once the actual time and place of upwelling can be predicted, fishermen will overfish the resources of the sea.

"Scientists are aware of the possibilities that, with scientific and technological knowledge such as we are finding, fishermen may take more from the sea than can be naturally replaced," says Barber. Any kind of control over fishing rights and international legal codes is complex and hard to enforce, especially for the smaller coastal nations. Certainly, lack of regulation has contributed to the ruin of several fishing industries— the California sardines, for instance, or the menhaden on the East Coast.

Overfishing as well as other factors such as ocean pollution is a serious threat to be considered and controlled in order to maintain life and productivity of the sea. "Yet we have to be careful in interpreting what we think affects the marine population," points out Barber. "Other factors have to be considered—for instance, the natural biological cycles of

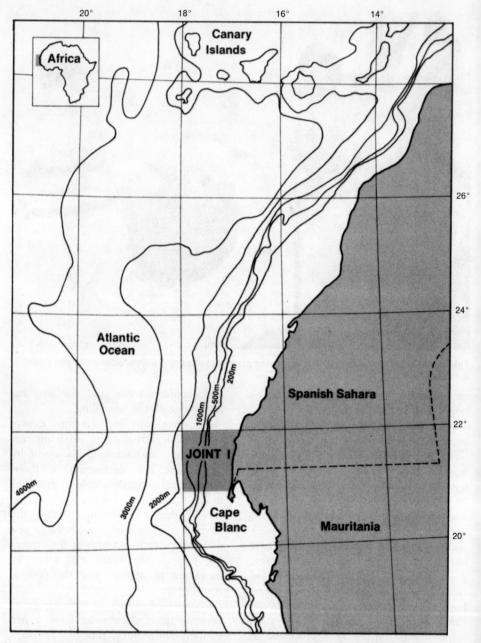

International rendezvous. In the summer of 1974 scientists from more than half a dozen nations will take part in JOINT I to study upwelling dynamics off the West African coast.

fish population. At times a species of fish may become abundant; at other times in their biological cycle there may be a natural decrease in their numbers."

By gaining a deeper understanding of the physical and biological dynamics of upwelling, scientists, fishermen, and national legislators will have more potential control over the supply of fish stock. "In other words," Barber says, "scientists can forecast the possibility of upwelling and hence productivity of fish in certain local areas. The fishing industry and national resources offices can then be alerted as to whether they should reduce their fishing quota or shorten the fishing season to let the fish

stock be replenished. Or in a good year they can increase their fishing harvest without damaging the stock."

Already, in some countries, fishing controls have been instituted. In Peru, for instance, fishing regulations are extremely strict. For the small, newly developing coastal countries it is important that their marine resources be under legal protection.

"For the world as a whole," says Barber, "there is a high payoff in the new scientific and technological knowledge of upwelling and fish harvest. Large and small nations need cooperation in order to preserve, protect, and manage these resources of the sea." ●

Gross features of the Peruvian upwelling system with special reference to possible diel variation*

by

JOHN J. WALSH,** JAMES C. KELLEY,**
RICHARD C. DUGDALE,** and BRUCE W. FROST **

INTRODUCTION

Interaction of simulation models, environmental data, experimental work, and data analysis is essential for investigations of total ecosystems. Flow of information is outlined in figure 1 where the initial inability of a simulation model to describe adequately experimental and field results is seen to serve as a necessary feedback to ongoing work and as a guide to future data acquisition. Rather than being an end in themselves, each activity complements the other, and with each pass through the cycle, a group of investigators is able to converge on an understanding of a complex system.

A marine upwelling system off the coast of Perú is the subject of such a cyclic analysis. In an initial field study, scientists from six U. S. laboratories and four other nations participated in Cruise 36 of the R/V *Thomas G. Thompson* to Peruvian waters during March, April, and May 1969. The project, Dynamics of Biological Production in Upwelling Ecosystems, was a part of the United States International Biological Program.

SMITH (1968) discussed the general pattern of water circulation in upwelling systems and reviewed the previous physical oceanographic studies of the Peru region. The Peru upwelling system supports the

* Contribution 568, Department of Oceanography, University of Washington, Seattle, Washington.
** Department of Oceanography, University of Washington, Seattle, Washington 98105.

37

Reprinted from Inv. Pesq. 35: 25–42. (1971).

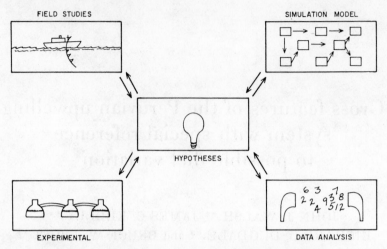

Fɪɢ. 1. — Systems analysis of aquatic ecosystems.

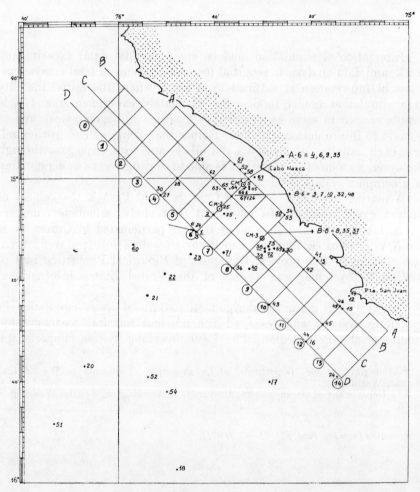

Fɪɢ. 2. — *Thompson* Cruise 36 grid area. CM denotes current meter locations; underlined numbers represent hydrographic stations; and numbers without underlines denote productivity stations.

world's largest single-species fishery (RYTHER, 1969) and has attracted numerous scientific investigations (CHIN, 1970). Definition of the upwelling system was the major goal of Cruise 36 of the *Thompson*, and, in this paper, distributions of temperature, nutrients, chlorophyll, and zooplankton are described and interpreted within the context of their use for a simulation model of daily productivity (WALSH and DUGDALE,

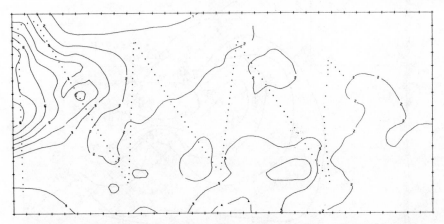

FIG. 3. — CalComp plot of chlorophyll distribution from an underway survey.

1971). Although data were not collected to compile a time series of diel variations in biological parameters, data acquired at different stations and on different days can be grouped into observations made during the hours of daylight and darkness. Diel differences inferred in this way from the distributions of some parameters must be considered highly tentative and will be further explored in another field study.

METHODS

The site chosen for the present study is located 300 km south of Lima, Perú, near Punta San Juan in an area of consistently cold surface temperatures. A five day drogue study was made in the same region by Dr. John Ryther and his associates in March and April 1966. A grid with squares 9.3 km (5 nautical miles) on a side (figure 2) provided orientation within the area.

Continuos underway surveys were made of temperature, salinity, and concentrations of chlorophyll and nutrients. These data were collected from a depth of 3 m while the *Thompson* steamed over an assigned path on the grid. The water was first pumped into a sea chest containing

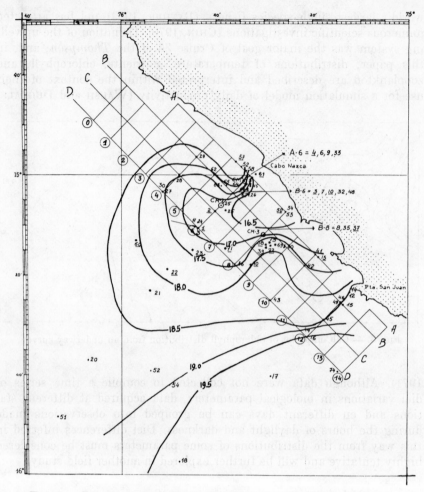

FIG. 4. — Day temperature distribution °C at 3 m depth (ship stations).

temperature and salinity sensors and then pumped through an Auto-Analyzer[R] array and a fluorometer.

Chlorophyll concentrations was determined with a Turner fluorometer (LORENZEN, 1966). Reactive phosphorus (HACER, GORDON, and PARK, 1968), reactive silicate (STRICKLAND and PARSONS, 1968), nitrate (adapted from WOOD, ARMSTRONG, and RICHARDS, 1967), and ammonia (MAC-ISAAC and OLUND, 1971) were measured by AutoAnalyzer[R].

Hydrographic and primary productivity stations and various experimental studies were conducted over the grid. All productivity stations were taken in daylight, while most underway surveys were made at night. Reduced hydrographic and productivity data have been reported (Anonymous, 1970).

Zooplankton samples were collected concurrently with the *Thompson* studies during the period, 5-10 April 1969, aboard the R/V *Gosnold* (Woods Hole Oceanographic Institution). A net with 70 cm mouth diameter and number 6 netting (0.24 mm between the threads) was towed vertically from 100 m (or from near the bottom in shallower depths) to the surface. A flowmeter was mounted in the mouth of the net. Zooplankton displacement volumes were determined by the method of FOXTON (1956).

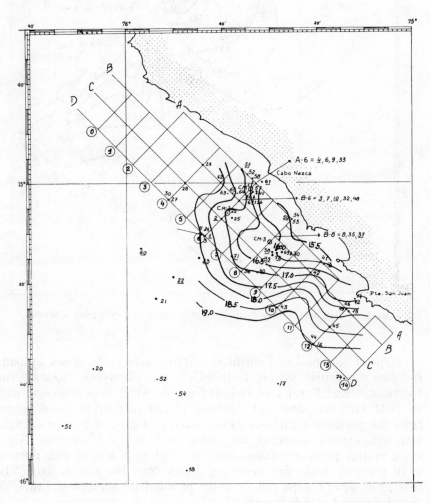

FIG. 5. — Night temperature distribution °C at 3 m depth (underway pass 2).

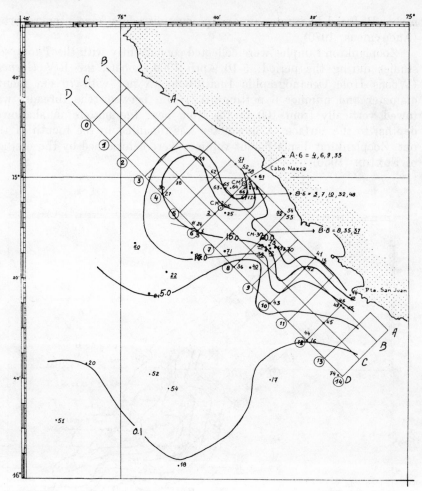

Fɪɢ. 6. — Day nitrate distribution mgAt/m³ at 3 m depth (ship stations).

Maps of biological and chemical features were made at sea utilizing the data acquisition system installed on the *Thompson* (Kᴇʟʟᴇʏ and Dᴜɢᴅᴀʟᴇ, 1969). Paper tape records from the DAS were processed with an IBM 1130 computer and CalComp plotter to provide visual output from the previous night's underway survey. Figure 3 is typical night map with crosses indicating the cruise track of the *Thompson*. Night maps in this paper are taken from the CalComp output and reduced to fit the grid, while day maps are drawn from the station data. The first 10 underway maps and all of the productivity stations of Cruise 36 have been used in this report.

RESULTS

In figures 4 and 5, the diel surface temperature distribution within the study area demonstrates a characteristic plume observed during the cruise. The plume of water colder than 19.5°C and present both in the contoured hydrographic station data and in the night underway passes, is the result of advection of the underlying cold waters to the surface and subsequent warming as it drifts to the northwest (SMITH *et al.*, 1969). Very little daily temperature difference or variation in position

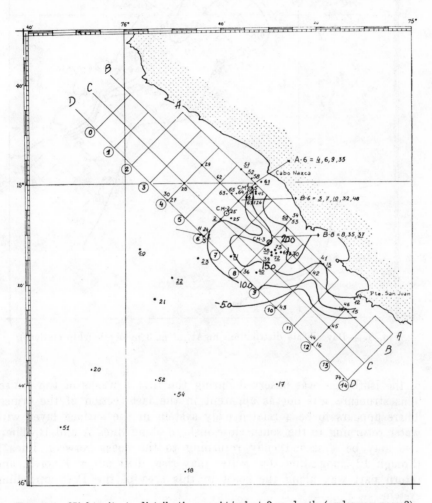

FIG. 7. — Night nitrate distribution mgAt/m³ at 3 m depth (underway pass 2).

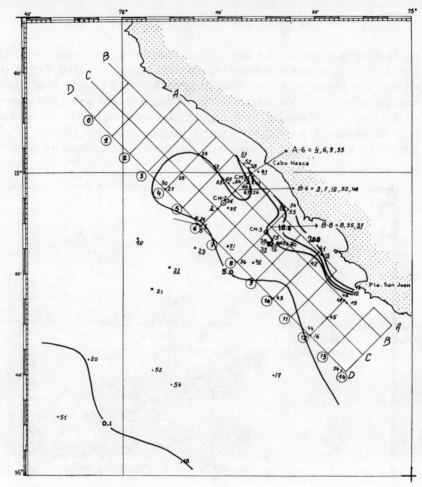

FIG. 8. — Day silicate distribution mgAt/m³ at 3 m depth (ship stations).

of the isotherms was observed during the first 3 weeks of the cruise. This structure was not as apparent in the last section of the cruise. There appears to be a related eddy system in the surface layer with water returning to the south close inshore along lines A and B. There also may be a second eddy returning to the coast between lines 9 through 12, suggesting that water may rise, drift out and north, and return part way along the coast in this area of the Perú upwelling system.

The offshore plume is repeated in the nutrient and chlorophyll distributions. High nitrate values (> 20 mg-at/m³) were found near shore in the recently upwelled water both night and day (figures 6 and 7). Low values (< 5 mg-at/m³) of nitrate were observed offshore. Little day and night difference in nitrate values is present, and both sets of contours reflect the possible eddy structure.

Silicate distribution also assumes the shape of a plume with high values inshore and low values offshore (figures 8 and 9). There appears to be a diel variation in silicate utilization in contrast to that of nitrate. The 10 mg-at/m³ isopleth of silicate extends farther downstream at night than in the day, suggesting that more silicate is used in the day and that this nutrient is replenished from upwelled water during the night. The apparent diurnal variation may not be real, however, and the

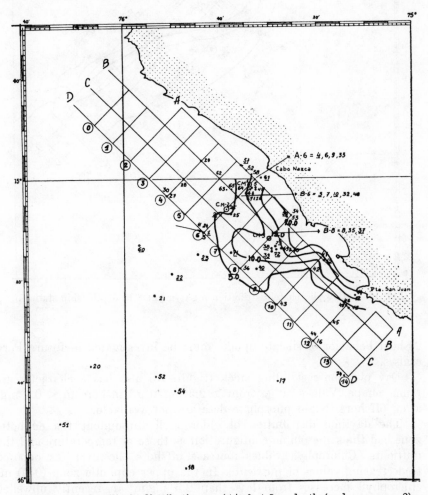

Fig. 9. — Night silicate distribution mgAt/m³ at 3 m depth (underway pass 2).

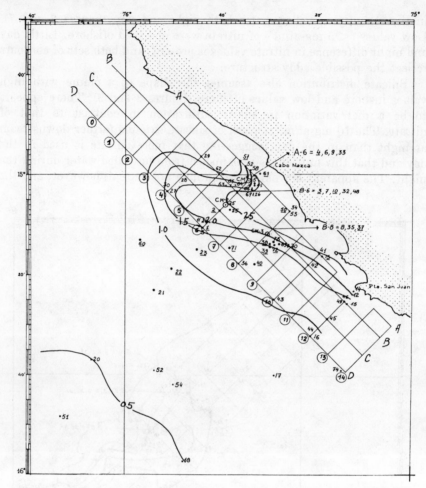

RIG. 10. — Day phosphate distribution mgAt/m³ at 3 m depth (ship stations).

24-hour behavior of silicate uptake must be investigated in future Peru cruises.

Day phosphate standing stock (figure 10) also has a characteristic plume shape. Values range from > 2.5 mg-at/m³ inshore to < 0.5 mg-at/m³ offshore. Night phosphate data are not available.

The daytime distribution of chlorophyll throughout the euphotic zone had the same plume configuration as those of temperature and the nutrients. Chlorophyll values increase offshore, however, as opposed to decreasing values of nutrients. In the upper euphotic zone (0-10 m) chlorophyll increases from less than 2.5 mg/m³ in recently upwelled water to greater than 10.0 mg/m³ at the downstream end of the plume (figure 11). The lower euphotic zone (10-20 m) does not exhibit as large a gradient (< 2.5 to 7.0 mg/m³) and has lower *in situ* values at each station (figure 12). The plume shape of the chlorophyll distribution in the lower euphotic zone suggests that the northward Ekman layer may extend down to at least 20 m.

Comparison of the day and night distribution of chlorophyll at 3 m depth (figures 13 and 14) suggests a diel variation in the size of the phytoplankton standing crop. No diel change is apparent in recently upwelled water where chlorophyll values are low. However, farther

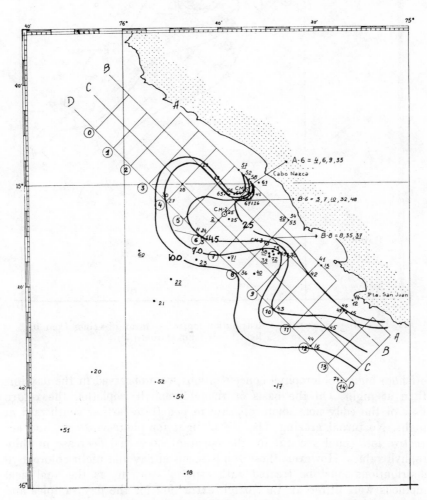

Fig. 11. — Day chlorophyll α distribution mg/m³ — mean of values from surface to 10 % light level (ship stations).

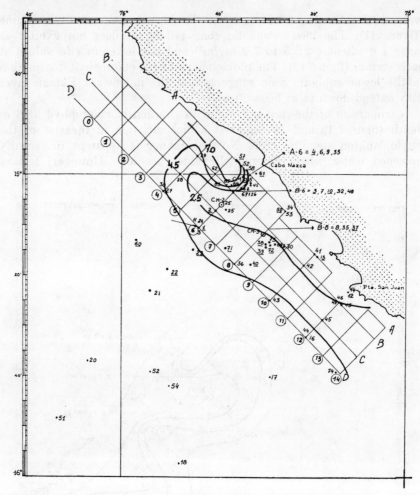

FIG. 12. — Day chlorophyll α distribution mg/m³ — mean of values from 10 % to 1 % light levels (ship stations).

offshore higher chlorophyll concentrations were observed in the daytime than at night. On the basis of the chlorophyll isopleths, the return flow of the eddy near shore appears to penetrate farther southward at night. Nocturnal grazing, absence of light for photosynthesis, and advective loss could account for the apparent nocturnal decrease in chlorophyll values. However, these comparisons of day and night chlorophyll distributions must be treated with especial caution, as the analytical methods were different, i.e., batch extraction for the day samples and continuous fluorometric for the night mapping.

The apparent low night chlorophyll values and the quasi-steady state nature of the plume over the first 3 weeks of the field study are supported in figure 15 which represents a composite of the 10 underway maps. Each map was overlaid over the preceding one and the 2.5 and 4.5 mg/m³ chlorophyll isopleths were superimposed to form the composite. The distribution of these isopleths from one map to the next was very similar. Figure 15 represents the grouped data, while figure 14 is one of the ten underway passes. Temperature and nutrient maps of the 10 nights also indicated that the system was essentially the same for 3 weeks.

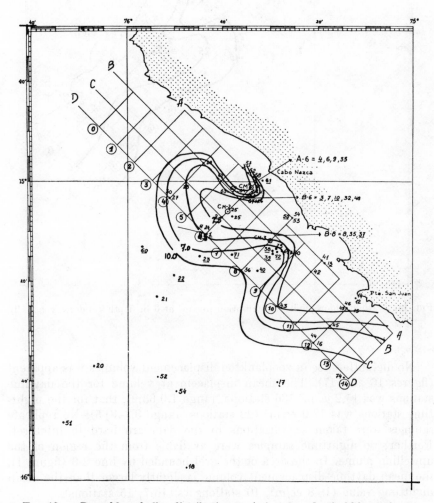

Fig. 13. — Day chlorophyll α distribution mg/m³ at 3 m depth (ship stations).

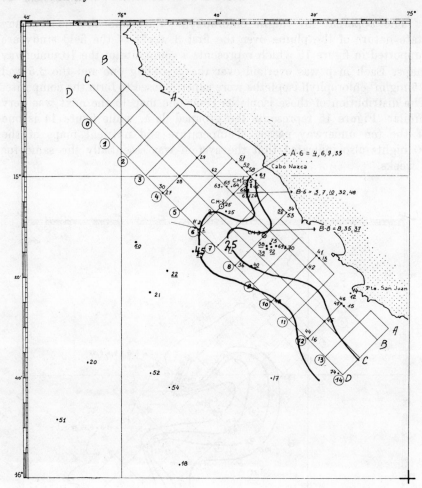

FIG. 14. — Night chlorophyll α distribution mg/m³ at 3 m depth (underway pass 2).

No diel difference in zooplankton displacement volumes was apparent (figures 16 and 17). The mean displacement volume for the daytime stations was 15.2 cc/m² (36 stations, range 1.0-63.2), that for the night-time stations was 17.9 cc/m² (22 stations, range 3.4-40.5). No replicate samples were taken so variations in the data are hard to interpret. Further, no nighttime samples were available from the region of the upwelling plume. In the area of the grid bounded by line 2-6 (figure 1), the mean daytime displacement volume is slightly larger than the mean nighttime value (18.8 cc/m², 10 stations vs. 16.7, 15 stations).

CONCLUSIONS

Over an interval of 3 weeks, the upwelling system at Pt. San Juan appeared to be in quasi-steady state. The position of upwelling water and resultant offshore plume of the Ekman drift were located in about the same place each day. A shoreward eddy appears to be associated with the plume off Pt. San Juan.

Phytoplankton carried in the upwelled water grow as the upwelled water moves downstream, depleting the phosphate, nitrate, and silicate

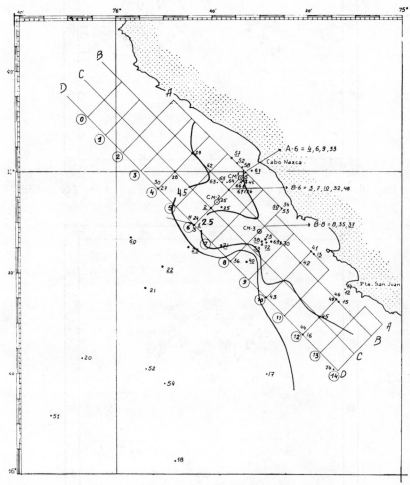

Fig. 15. — Night chorophyll α distribution mg/m³ at 3 m depth (composite of 10 underway passes).

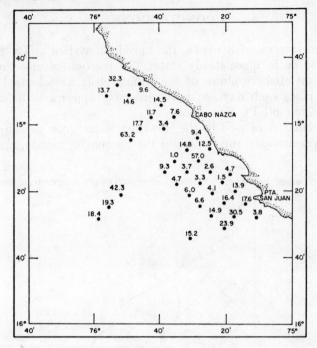

FIG. 16. — Zooplankton displacement volumes (cc/m²) of the day stations.

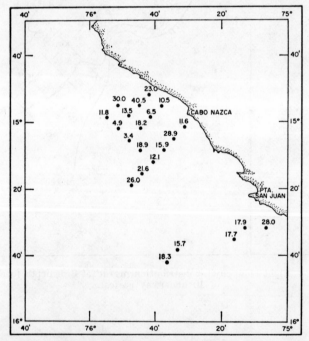

FIG. 17. — Zooplankton displacement values (cc/m²) of the night stations.

stocks and drifting offshore. There may be a diel cycle imposed on the system, since at least one of the nutrients, silicate, appears to be used more in the daytime than at night. Phytoplankton biomass, as indicated by chlorophyll, appears to decrease at night, perhaps as a result of grazing. The drogue study made by RYTHER *et al*. (1969) shows what may be a sinusoidal variation of the variables, suggesting a possible diel periodicity in the same area 3 years before. Anchoveta may be the major herbivores in the Peru upwelling system, and JORDÁN (1971) has shown definite diel variation in the distribution of anchoveta schools off Pt. San Juan. With the San Juan system defined, time and spatial continuity of sampling is now required to test the above interpretation of the Peru upwelling system. Such a sampling program could be accomplished in the next field study, i.e., the second pass through a cyclic systems analysis as outlined in figure 1.

REFERENCES

ANONYMOUS. — 1970. Biological production in upwelling ecosystems. Data report. Part 1: Hydrography and productivity. 97 pp. *Special Report 42, Department of Oceanography*, University of Washington, Seattle, Washington.

CHIN, E. (ed.). — 1970. Results of the *Anton Bruun* investigations along the west coast of South America. *Texas A & M University*. In press.

FOXTON, T. — 1956. The distribution of the standing crop of zooplankton in the Southern Ocean. *Discovery Report.*, 28 : 191-236.

HACER, S. W.; GORDON, L. I., and PARK, P. K. — 1968. A practical manual for use of Technicon AutoAnalyzer in seawater nutrient analysis. *A final report to BCF, Contract 14-17-0001-1759, October 1968*. Reference 68-33.

JORDÁN, R. — 1971. Distribution and behavior of the Peruvian anchoveta *Engraulis ringens* off the Peruvian coast in relation to hydrographyc conditions. *Inv. Pesq.*, 35 : 113-126.

KELLEY, J. C., and DUGDALE, R. C. — 1969. Adaptive sampling in upwelling studies. *Paper presented at ASLO meeting*, La Jolla, California, September 1969.

LORENZEN, C. J. — 1966. A method for the continuous measurement of *in vivo* chlorophyll concentration. *Deep-Sea Res.*, 13 : 223-228.

MACISAAC, J. J., and OLUND, R. K. — 1971. An automated extraction procedure for the determination of ammonia in sea water. *Inv. Pesq.*, 35 : 221-232.

RYTHER, J. H. — 1970. Photosynthesis and fish production in the sea. *Science*, 166 : 72-76.

RYTHER, J. H.; MENZEL, D. W.; HULBURT, E. M.; LORENZEN, C. J., and CORWIN, N. — 1970. The production and utilization of organic matter in the Peru Coastal Current. Results of the *Anton Bruun* investigations along the west coast of South America. *Texas A & M University*. In press.

SMITH, R. L.; MOOERS, C. N. K., and ENFIELD, D. B. — 1970. Mesoscale studies of the physical oceanography in two coastal upwelling regions: Oregon and Peru. *Paper presented at the International Conference on the Fertility of the Sea*, Sao Paulo, Brazil, December 1969.

STRICKLAND, J. D. H., and PARSONS, T. R. — 1968. A practical handbook of seawater analysis. *Fisheries Research Board of Canada Bull.*, 167, 311 pp.

WALSH, J. J., and DUGDALE, R. C. — 1971. A simulation model of the nitrogen flow in the Peruvian upwelling system. *Inv. Pesq.*, 35 : 309-330.

WOOD, E. O.; ARMSTRONG, F. A. J., and RICHARDS, F. A. — 1967. Determination of nitrate in seawater by cadmium-copper reduction to nitrate. *J. Mar. Biol. Assn. U. K.*, 47 :23-31.

PRODUCTION STUDIES IN THE STRAIT OF GEORGIA.
PART I. PRIMARY PRODUCTION UNDER THE FRASER RIVER PLUME, FEBRUARY TO MAY, 1967.

T. R. PARSONS, K. STEPHENS and R. J. LeBRASSEUR

Fisheries Research Board of Canada, Pacific Oceanographic Group, Nanaimo, B.C.

Abstract: Primary productivity measurements have been carried out during the period, February to May 1967, at two places in the general vicinity of the Fraser River estuary. At one station within the Fraser River plume the average radiation in the mixed layer was sufficient to allow for a net positive production throughout the period of study. An approximate estimate of this production gives a value of 50 g C/m² produced during the four month study. At the second station, which was generally outside of the immediate influence of the Fraser River plume, the average radiation in the mixed layer was such that on only two occasions could any appreciable production have taken place.

INTRODUCTION

The following presentation represents part of a continuing study of production processes in the Strait of Georgia, British Columbia. In this, the first of a series of three reports, special attention has been given to the level of primary production in an effort to explain the occurrence during the spring of very large numbers of zooplankton in the vicinity of the Fraser River estuary. Hutchinson, Lucas & McPhail (1929), and Tully & Dodimead (1957), have commented on the high production of this area, but no detailed explanation for this phenomenon has been given. Since the occurrence of high plankton production in the spring is followed in the same area by the arrival of large numbers of larval and juvenile fish, the area would seem to present an excellent one in which to study trophic relationships.

The general hydrography of the Strait of Georgia, including specific reference to the influence of Fraser River discharge in the area, has been discussed by Waldichuk (1957) and by Tully & Dodimead (1957). Major factors influencing the hydrographic conditions are wind, tide, insolation, and freshwater run-off. In the immediate vicinity of the Fraser estuary, low salinity surface waters form a plume in which a high degree of stability is maintained throughout the year. Outside this plume stability is reduced to zero during the winter months. The period chosen for this project was from February when a maximum difference in production was believed to occur between the waters of the plume and adjacent areas through to May when the annual flooding of the Fraser causes a marked increase in the extent of the plume and a decrease in the transparency of the water.[1]

[1] The following cruises were carried out in connection with this study:

Cruise 1 *CNAV Laymore* Feb. 13–19, 1967   Cruise 5 *CNAV Endeavour* Apr. 10–16, 1967
Cruise 2 *CNAV Laymore* Feb. 27 - Mar. 5, 1967   Cruise 6 *CNAV Laymore* Apr. 24–30, 1967
Cruise 3 *CNAV Laymore* Mar. 13–19, 1967   Cruise 7 *CNAV Laymore* May 15–21, 1967
Cruise 4 *CNAV Endeavour* Mar. 27 - Apr. 2, 1967

Reprinted from J. Exp. Mar. Biol. Ecol. 3: 27–38 (1969).

From the point of view of its subject matter, the work has been divided into three sections each dealing primarily with a single trophic level. It has not been possible, however, and indeed it would defeat part of the purpose, completely to isolate data on the three trophic levels in reporting the results. In some cases it has been necessary, therefore, to discuss data regarding one trophic level in several places.

METHODS

SAMPLING PROCEDURE

Samples were collected at two Stations identified as Station 1 (49° 18′ N: 123° 51.5′ W) and Station 7 (49° 10′ N: 123° 24′ W) in the Strait of Georgia. The latter was generally under the influence of the Fraser River plume during the period February to May while the former was outside of the plume until the middle of May when the Fraser River discharge reaches its maximum. The positions of the two stations are shown in Fig. 1. Samples for chemical analyses and Coulter counts were collected with 7-litre Van Dorn bottles at standard depths of 0, 5, 10, 20, 30, and 50 m. Temperature and salinity data were collected at the same depths and on a number of occasions at intermediate depths.

LIGHT

A continuous recording of solar radiation was obtained with an Eppley pyranometer located at Departure Bay (49° 12.6′ N: 123° 57.3′ W). Hourly integrals of radiation in langleys (1langley (ly) = 1 gcal/cm^2) were calculated using a disc integrator and from these values the average radiation for a 24 h interval was calculated for each cruise. These values were corrected for reflection at the sea surface using the tables given by von Arx (1962). The amount of energy available for photosynthesis was then determined by reducing the total radiation by a factor of 0.2 to allow for the absorption of non-photosynthetic energy in the first metre of sea water.

Light attenuation was measured as the percentage transmission at various depths using a Weston cell masked with a filter giving maximum response at 430 mμ. The extinction coefficient for 430 mμ light obtained from semi-log plots of light transmission with depth was then employed to characterize the water mass using Jerlov's coastal water-types (Jerlov, 1957); the extinction coefficients for *total* incident radiation below one metre were then found for each water-type from tables given by Jerlov.

CHEMICAL ANALYSES, TEMPERATURE, SALINITY

Chlorophyll *a* and nitrate were determined as described by Strickland & Parsons (1965), and particulate carbon by total combustion of GFC fibre glass filters on which particulate material was collected from a known volume of seawater. A Coleman carbon analyzer was used for the combustion. Temperature and salinity were determined using an *in situ* salinometer (Beckman Portable, RS5-2) and mixed layer depths were taken as the depth to the bottom of the halocline.

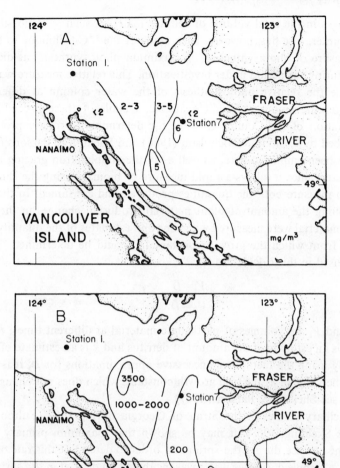

Fig. 1. Region of investigation. A, distribution of chlorophyll *a*, 13th Feb., 1967; B, distribution of zooplankton, April 24th, 1967.

COULTER VOLUMES AND PRIMARY PRODUCTION

Two techniques for determining primary production have been used. The actual net increase in cellular material was measured directly using a Coulter Counter. This technique has an advantage over the ^{14}C-technique in that it is possible to record both increases and decreases in the volume of cellular material, while with the ^{14}C-technique it is impossible to record a negative production; in addition there is no ambiguity as to what is actually being measured when an absolute measure is made of the changes in the total volume of cellular material, such as is possible with the Coulter Counter. Use has, however, been made of the ^{14}C-technique to determine a relative measure of photosynthesis (% maximum photosynthesis) at different light intensities during the period under investigation. This relative measure was then used to determine the relative photosynthesis of the water column at different depths depending on the amount of light available. Details of these procedures follow.

The operation of the Coulter Counter for determining primary production has been described by Cushing & Nicholson (1966) and Sheldon & Parsons (1967). This method has been employed here, as well as in the zooplankton grazing experiments described later, since it allows a rapid measure of biomass using the same property (volume) to measure both the quantities produced and consumed. In the procedure for determining the amount of plant material produced, changes in the volume of particulate material were measured with time and estimates of the quantity of detritus were made from which the growth constant (k) could be determined. The general expression used in this calculation is,

$$\frac{V_t - \hat{D}}{V_0 - \hat{D}} = e^{\hat{k}t}, \tag{1}$$

where V_0 and V_t are volumes of particulate material at different times, t is the time interval, $\hat{D}$ is an estimate of the amount of detritus and $\hat{k}$ is an estimate of the growth constant. By using the method of successive approximations for $\hat{D}$, it is possible to determine the value of k, the true growth constant, which does not change with time (for details see Sheldon & Parsons, 1967).

From ancillary studies in the Strait of Georgia during 1967 (Fulton, Kennedy, Stephens & Skelding, 1968), it may be shown that maximum primary production occurs at the surface during the spring. In order to determine this rate of maximum production, a sample of surface water was incubated on deck at *in situ* surface temperatures and under natural light conditions. Coulter volumes of particulate material were then determined over a period of at least 36 h.

From the rate of maximum production (determined from eq. 1), the rate of production at each depth in the euphotic zone down to the compensation depth was determined from relative photosynthesis (P) *vs* light intensity (I) curves, and the light intensity at each depth. The *P–I* curves employed in this study and the 24 h compensation light intensities were obtained at various times throughout the given period at a shore-based laboratory. These experiments were carried out using an incubator in which natural populations of phytoplankton from the area were incubated for time intervals of 4 h at different light intensities from 0.001 to 0.14 ly/min. The incubator has previously been described by McAllister, Shah & Strickland (1964). The light source was provided by 500 watt General Electric T 3Q tungsten-iodine vapour lamps and the light was filtered through 2.0 cm of a chemical filter described by Jitts, McAllister, Stephens & Strickland (1964). The uptake of $^{14}CO_2$ was used as a relative measure of photosynthesis and the results plotted as percentage maximum photo-

synthesis against light intensity. The curves obtained were then used to determine compensation light intensities following the procedure described by Steemann Nielsen & Hansen (1959).

The actual production at each depth was determined from the amount of plankton carbon multiplied by the production rate (generations/day) determined as described above. The quantity of phytoplankton carbon was assumed to be equal to the amount of chlorophyll a present multiplied by a factor. The factor employed for the conversion of chlorophyll a to carbon was obtained from a regression equation determined during the course of the investigation and given later (p. 34). From the rate of maximum production and the average quantity of non-detrital carbon in the mixed layer, the productivity (mg C/m^2/day) of the euphotic zone, was determined.

DISTRIBUTIONAL STUDIES

Changes in the concentration of chlorophyll a and suspended material, and differences in temperatures and salinity over the area were assessed during the first 24 h of each cruise using automated and semi-automated equipment. Chlorophyll a was determined using a fluorometer (Lorenzen, 1966); a relative measure of turbidity was obtained with a second fluorometer (Stephens, 1967), and temperature and salinity were monitored at the surface using an *in situ* salinometer. In addition, surface zooplankton samples were collected under way using a Miller high speed sampler. Data obtained during these rapid transects were used to map the area of high productivity, particularly in order to obtain samples for grazing experiments, which were carried out in the area, as described subsequently (Parsons, LeBrasseur, Fulton & Kennedy, 1969).

RESULTS

A general assessment of the differences in production over the area off the Fraser River is shown in Fig. 1; Fig. 1A shows the differences in standing stock of primary producers as indicated by chlorophyll a values obtained with a recording fluorometer during February; Fig. 1B shows the high standing stock of zooplankton which subsequently developed during April.

TABLE I

Compensation light intensities (I_c) of natural phytoplankton populations.

Month	I_c (4 h) (ly/min)	Average h darkness	I_c (24 h) (ly/min)
February	0.004	14	0.0063
March	0.005	12	0.0075
April	0.005	10	0.0071
May	0.007	9	0.0096

Fig. 2 shows the results of two typical experiments on the change of relative photosynthetic rate with light intensity (P against I curves on January 24th and April

26th, 1967). Compensation light intensities for each month are given in Table I together with values corrected for relative day-length. The latter results were obtained by adding a fraction of the light intensity to each 4 h value, depending on the hours of darkness at 50° N on the 15th day of each month. Values for the compensation light intensity for different phytoplankton crops are given in Table II for comparison with the calculated values for the mean amount of radiation in the mixed layer at Stations 1 and 7. Separate values for the compensation light intensity of the standing stock at Stations 1 and 7 were not considered necessary since the same phytoplankton species were present at both stations.

Differences in the amount of light available for photosynthesis in the water column within the Fraser River plume (Station 7) and outside (Station 1) have been determined using an approach similar to that described by Riley (1957). If I_0 is the incident

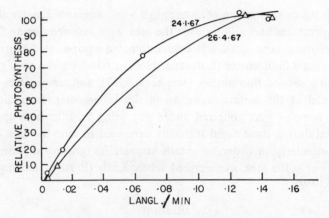

Fig. 2. Relative photosynthesis and light intensity.

radiation available for photosynthesis, K the extinction coefficient of photosynthetic radiation, and d_m the mixed layer depth, then an approximation of the amount of light in the water column can be given as

$$I_m = \frac{I_0(1 - e^{-Kd_m})}{d_m K}.$$ (2)

Using values of I_0, K and d_m determined as described in the previous section, the values for I_m during different cruises and at both Stations 1 and 7 were calculated and are given in Table II.

The growth characteristics of phytoplankton in surface incubated samples during successive cruises are shown in Fig. 3. From these curves, the growth rates of the phytoplankton in generations/day were determined as described previously and are given in Table III. The amount of phytoplankton carbon in the mixed layer was determined from the relation between total particulate carbon and chlorophyll a; the regression equation based on 81 estimations in this area, is;

TABLE II

Photosynthetic radiation in the mixed layer at Stations 1 and 7.

Period of investigation	Feb. 16–27th		Feb.28th–March 13th		March 14–28th		March 29th–April 9th		April 10–24th		April 25th–May 14th		May 15–25th	
I_0 (ly/min)	0.030		0.031		0.038		0.069		0.063		0.098		0.097	
I_c (ly/min)	0.0063		0.0075		0.0075		0.0071		0.0071		0.0096		0.0096	
Station No.	1	7	1	7	1	7	1	7	1	7	1	7	1	7
Mixed layer depth (d_m)	25	2.5	40	7.5	40	2.5	25	15	50	2.5	37	2.5	2.5	2.5
Extinction coefficient (K_e)	0.35	0.52	0.19	0.43	0.19	0.35	0.29	0.35	0.29	0.35	0.29	0.43	0.62	0.73
I_m (ly/min)	0.0029	0.017	0.0041	0.0092	0.0050	0.025	0.0095	0.013	0.0043	0.043	0.0092	0.059	0.050	0.044

$$\text{Carbon } (\mu g/l) = 59.5 \pm 7.3 \text{ Chl } a \ (\mu g/l) + 77.0 \pm 15.6,$$

where the upper and lower limits of the slope and intercept are the 95 % probable ranges of these values.

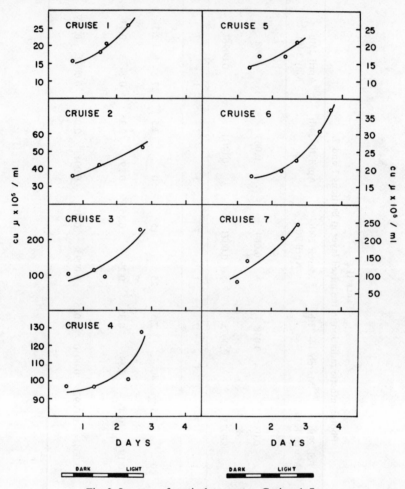

Fig. 3. Increase of particulate matter, Cruises 1–7.

The primary production per m² for the plume waters was then determined for each cruise and the results are shown in Table III together with the amount of nitrate in the surface layers during the same period.

DISCUSSION

The distribution of chlorophyll *a* during February and zooplankton during April (Fig. 1) have been included to show the approximate extent of differences in standing stock at Stations 1 and 7. In Fig. 1A it may be seen from chlorophyll *a* values that maximum concentrations of phytoplankton occurred at some distance from the river mouth in a crescent shaped distribution. The zooplankton which developed as a result of this area of high phytoplankton production occurred in a similar pattern (Fig. 1B) but at a greater distance from the delta. The extent to which Stations 1 and 7 were representative of waters inside and outside the plume depended in part on the effect of wind and tide on surface circulation and mixing. In general, however, the production initiated within the plume at Station 7 would tend to be transported in a northerly direction toward Station 1 (Waldichuk, 1957), where no appreciable increase in the production of the water column occurred until the end of March. It may also be seen from Fig. 1A that little production occurred in the immediate vicinity of the river mouth which is believed to be due to the very low transparency of the river water as it enters the sea.

TABLE III

Primary production within the Fraser River plume, February to May, 1967.

Period of investigation	Feb. 16–27th	Feb. 28th–March 13th	March 14–28th	March 29th–April 9th	April 10–24th	April 25th–May 14th	May 15–25th
Maximum algal productive rates (generations/day)	1.16	0.413	1.14	2.27	0.77	0.56	1.10
Plankton carbon (mg/m³)	71.5	131	405	255	95.0	136	160
Production (mg C/m²/day)	140	124	1135	2132	210	199	353
Nitrate (μg-at/l)	21.9	20.8	16.6	12.7	16.0	13.1	7.5

The maximum biomass of zooplankton developed in a general area between Stations 1 and 7. The principal zooplankton species in this area, *Calanus plumchrus*, is known to migrate vertically during the spring from depths of greater than 200 m. Thus, the distribution of zooplankton northwest and southeast of Station 7 (depth 240 m) may have been to some extent associated with the greater depth of the water column (maximum 400 m) in these areas.

From Table II it may be seen that water at Station 7 was characterized by mixed layer depths between 2.5 and 15 m, and transparencies which correspond to Jerlov's coastal water-types 4 to 6. Outside this area of high primary productivity, mixed layer depths were generally in the range 25 to 50 m with water transparencies which correspond to Jerlov's types 2 to 4. This situation was changed markedly by the third week in May when mixed layer depths of less than 5 m and Jerlov water-types 8 and 9 were found over large areas (including Station 1) due to the start of the annual Fraser River freshet. The average light intensity for the mixed layer at Station 7 was, at all times up to May, considerably greater than the average light intensity available in the water column at Station 1. Furthermore, by a comparison of these values with the compensation light intensities shown in Table II, it is apparent that conditions were suitable for a net positive production at Station 7 throughout the period February to May, but that at Station 1 a net increase in production was only possible during one period at the end of March, and again towards the end of April and beginning of May. In Table I the compensation light intensities of natural crops are shown for four different days during the investigation. The difference between these values on the basis of a 4 h light incubation are appreciable. Thus, phytoplankton present in February had about half the compensation light intensity of those in May. If, however, these values are compared after weighting the results for seasonal changes in day length, then there is considerably less difference between the values for each month and some similarity with other values in the literature (see Strickland, 1960, for discussion). Differences in the compensation light intensities given in Table I may have been sufficient, however, to bring about in part the phytoplankton species succession which is described by Parsons *et al.* (1969).

In Table III the daily primary production within the plume shows a maximum during the period March 14th to April 9th when it was approximately ten times that in February and two to three times that in May. The total production for the period February to May was approximately 50 g C/m².

From Table III it is apparent that the levels of primary production described above were independent of nutrient depletion, at least as indicated by the amount of nitrate remaining in surface samples; while nitrate was the only nutrient measured during the course of this work, it has been found from two years of data collected in the same general area that large blooms of phytoplankton do not occur in the area after exhaustion of nitrate (Bishop, *et al.*, 1966; Fulton, *et al.*, 1967).

While we do not intend to review previous work on estuarine productivity, it is pertinent to make certain comparisons with the productivity of other estuarine areas. It has been generally recognized, for example, that the primary production of estuaries is greater than that of surrounding waters (*e.g.* Riley, 1937; Hobson, 1966). The two reasons for this phenomenon most often cited are that rivers contribute nutrients to the surrounding waters and that the increased stability of estuarine waters permits an earlier and greater primary production than in adjacent areas. There is no question in the present study that the latter factor was a major influence on the primary production of the Fraser River plume from February to May. The question of whether the river waters contribute significantly to the nutrient enrichment of the plume is less clear. The level of production at Station 7 given in Table II does require, however, nitrate to be continuously added to the surface layers during the course of the investigation. This could have occurred either as a result of eutrophication of the plume waters from a domestic outfall or from entrainment of nutrient-rich deep water into the surface layers. Since the general ecology of the area appeared from our work to be the same as that described by Hutchinson, Lucas & McPhail (1929), the installation, in 1962, of a sewer outfall five miles northeast of Station 7 would not appear to have contributed significantly to qualitative changes in the production of the area. Certainly our results do not indicate that the area was a polluted estuary as compared with many others (*e.g.* Jeffries, 1962, reports estuarine nitrate concentrations in Raritan Bay, N.J. up to 96 μg - at/l). While the final analysis of this phenomenon must await a physical explanation, it is tentatively believed that nutrient-rich deep water was entrained into the euphotic zone by the surface flow of the Fraser River. A similar entrainment of nutrient-rich deep water has been discussed by Stefansson & Richards (1963) with respect to the Columbia River plume and a cursory examination of this problem was made in our own waters by LeBrasseur (1955).

REFERENCES

BISHOP, S. O., J. D. FULTON, O. D. KENNEDY & K. STEPHENS, 1966. Physical chemical and biological data, Strait of Georgia, March to October 1965. *Manuscr. Rep. Fish. Res. Bd Can.*, No. 211, 171 pp.

CUSHING, D. H. & H. F. NICHOLSON, 1966. Method of estimating algal production rates at sea. *Nature, Lond.*, Vol. 212, pp. 310–311.

FULTON, J. D., O. D. KENNEDY, K. STEPHENS & J. SKELDING, 1967. Physical chemical and biological data, Strait of Georgia, 1966. *Manuscr. Rep. Fish. Res. Bd Can.*, No. 915, 145 pp.

FULTON, J. D., O. D. KENNEDY, K. STEPHENS & J. SKELDING, 1968. Physical chemical and biological data, Strait of Georgia, 1967. *Manuscr. Rep. Fish. Res. Bd Can.* No. 968, 150 pp.

HOBSON, L. A., 1966. Some influences of the Columbia River effluent on marine phytoplankton during January, 1961. *Limnol. Oceanogr.*, Vol. 11, pp. 223–234.

HUTCHINSON, A. H., C. C. LUCAS & M. McPHAIL, 1929. Seasonal variations in the chemical and physical properties of the water of the Strait of Georgia in relation to phytoplankton. *Trans. R. Soc. Can.*, 3rd Ser., Vol. 23, pp. 177–183.

JEFFRIES, H. P., 1962. Environmental characteristics of Raritan Bay, a polluted estuary. *Limnol. Oceanogr.*, Vol. 7, pp. 21–31.

JERLOV, N. G., 1957. Optical studies of ocean waters. *Rep. Swed. deep Sea Exped.*, Vol. 3, pp. 1–59.

JITTS, H. R., C. D. McALLISTER, K. STEPHENS & J. D. H. STRICKLAND, 1964. The cell division rates of some marine phytoplankters as a function of light and temperatures. *J. Fish. Res. Bd Can.*, Vol. 21, pp. 139–157.

LEBRASSEUR, R. J., 1955. Oceanography of British Columbia mainland inlets. VI. Plankton distribution. *Prog. Rep. Pacif. Coast Stn, No.* 103, *Fish. Res. Bd Can.*, pp. 19–21.

LORENZEN, C. J., 1966. A method for the continuous measurement of *in vivo* chlorophyll concentration. *Deep Sea Res.*, Vol. 13, pp. 223–227.

McALLISTER, C. D., N. SHAH & J. D. H. STRICKLAND, 1964. Marine phytoplankton photosynthesis as a function of light intensity: a comparison of methods. *J. Fish. Res. Bd Can.*, Vol. 21, pp. 159–181.

PARSONS, T. R., R. J. LEBRASSEUR, J. D. FULTON & O. D. KENNEDY, 1969. Production studies in the Strait of Georgia. II. Secondary production under the Fraser River plume, February to May, 1967. *J. exp. mar. Biol. Ecol.*, Vol. 3, pp. 39–50.

RILEY, G. A., 1937. The significance of the Mississippi River drainage for biological conditions in the northern Gulf of Mexico. *J. mar. Res.*, Vol. 1, pp. 60–74.

RILEY, G. A., 1957. Phytoplankton of the north central Sargasso Sea, 1950–52. *Limnol. Oceanogr.*, Vol. 2, pp. 252–270.

SHELDON, R. W. & T. R. PARSONS, 1967. *A practical manual on the use of the Coulter Counter in marine research.* Coulter Electronics Sales Co., Canada, 66 pp.

STEEMANN NIELSEN, E. & V. K. HANSEN, 1959. Measurements with the carbon-14 technique of the respiration rates in natural populations of phytoplankton. *Deep Sea Res.*, Vol. 5, pp. 222–233.

STEFANSSON, U. & F. A. RICHARDS, 1963. Processes contributing to the nutrient distributions off the Columbia River and Strait of Juan de Fuca. *Limnol. Oceanogr.*, Vol. 8, pp. 394–410.

STEPHENS, K., 1967. Continuous measurement of turbidity. *Deep Sea Res.*, Vol. 14, pp. 465–467.

STRICKLAND, J. D. H., 1960. Measuring the production of marine phytoplankton. *Bull. Fish. Res. Bd Can.*, No. 122. 172 pp.

STRICKLAND, J. D. H. & T. R. PARSONS, 1965. A manual of sea water analysis. *Bull. Fish. Res. Bd Can.*, No. 125, 2nd ed., 203 pp.

TULLY, J. P. & A. J. DODIMEAD, 1957. Properties of the water in the Strait of Georgia, British Columbia, and influencing factors. *J. Fish. Res. Bd Can.*, Vol. 14, 241–319.

VON ARX, W. S., 1962. *An introduction to physical oceanography.* Addison-Wesley Publishing Co. Inc., London, 135 pp.

WALDICHUK, M., 1957. Physical oceanography of the Strait of Georgia, British Columbia. *J. Fish. Res. Bd Can.*, Vol. 14, pp. 321–486.

PRODUCTION STUDIES IN THE STRAIT OF GEORGIA.
PART II. SECONDARY PRODUCTION UNDER THE FRASER RIVER PLUME, FEBRUARY TO MAY, 1967

T. R. Parsons, R. J. LeBrasseur, J. D. Fulton and O. D. Kennedy

Fisheries Research Board of Canada, Pacific Oceanographic Group, Nanaimo, B.C.

Abstract: Secondary productivity measurements have been carried out during the period February to May, 1967 in the general vicinity of the Fraser River estuary. The quantity of phytoplankton grazed by zooplankton in the area was measured in incubated samples and the results show that both the quantity and type phytoplankton caused considerable changes in the amount of food eaten by the zooplankton. Grazing was found to start at threshold phytoplankton levels of between $\sim$ 50–190 μg C/l and maximum grazing was found to occur at phytoplankton concentrations of $\sim$ 400 μg C/l. The larger phytoplankton species were generally grazed in greater quantities than the smaller species.

From *in situ* observations the quantity of *Calanus plumchrus* produced during the investigation approximated to the amounts predicted from estimates of the growth rate of this organism. These results indicate that the large standing stock of zooplankton which developed in the vicinity of the Fraser River estuary resulted from the presence of adequate concentrations of the right size and shape of food organisms and from a lack of severe predation on the early growth stages of this copepod.

Introduction

The following account is concerned with the same area and time as already described (Parsons, Stephens & LeBrasseur, 1969).

The general intent was to determine quantitative relationships between the amount and type of phytoplankton available, and the quantities grazed by secondary producers. In this respect, the attempt was to quantify associations of predators and prey rather than to account for their exact distribution which was largely dependent on local advective effects. Thus, the extent of maximum zooplankton standing stock has already been shown (Parsons, *et al.*, *loc. cit.*) to be located about midway between the region of high primary production within the Fraser River plume and the region of low primary production outside the plume. Elucidation of the reasons for this is not a part of this work, but it may be pointed out that in general at least two factors are involved: first, that there is a greater depth of water away from the immediate influence of the plume so that since the principal zooplankton species, *Calanus plumchrus* Marukawa, winters in deep water, the arrival of copepod nauplii at the surface in the early spring is to some extent dependent on the depth of the water column; secondly, although the primary production was considerably less outside the plume than inside, a general surface transport of water from the river mouth could entrain an allochthonous supply of food and carry it toward the area of low primary production, so that for physical reasons, a distribution of zooplankton was produced which was not immediately associated with the trophic relationships to be described.

67

Reprinted from J. Exp. Mar. Biol. Ecol. 3: 39–50 (1969).

The extent to which the early development of zooplankton occurs in this region compared with the rest of the Strait of Georgia is illustrated in Fig. 1 using data which have been reported by Fulton, *et al.*, (1967). For stations in the vicinity of the Fraser estuary it may be seen that by March to April the standing stock of zooplankton is some five to ten times that of other areas. By May, when the maximum

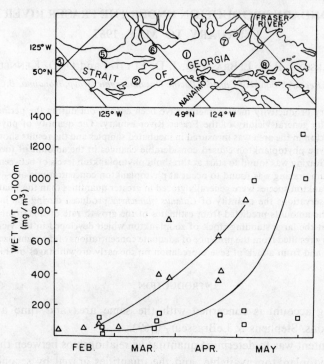

Fig. 1. Increase in zooplankton standing stock in the Strait of Georgia, February-May 1966: inset shows station positions ($\triangle$, data from Stations 1, 7, and 8; $\square$, data from Stations 2, 3, 5, and 6).

flooding of the Fraser River occurs, the zooplankton standing stock is disseminated northwest and southeast of the estuary, causing sudden and large increases in the zooplankton at more distant stations. We shall attempt to account for the growth of zooplankton in the vicinity of the Fraser River in terms of the availability of different phytoplankton to grazing by the zooplankton.

METHODS

ZOOPLANKTON GRAZING

Zooplankton grazing experiments were carried out in 1-litre glass jars. Successively greater dilutions of natural populations of phytoplankton were added to a series of jars and equal numbers of zooplankton were pipetted into each of the phytoplankton suspensions. The jars were fastened to a large wheel which rotated in a water trough once every three minutes to provide gentle mixing. The water in the trough was pumped from ~ 2 m below the sea surface (5–7° C) and flowed constantly during the period of incubation. Experiments were carried out with a minimum illumination by allowing the top of the chamber containing the rotating samples to be open a few inches. The quantity of light allowed in under these conditions was well below that required for photosynthesis so that the loss of plants by grazing represented the only appreciable change during the incubation period.

Zooplankton grazing was measured with a Coulter Counter as described previously by Parsons, LeBrasseur & Fulton (1967) and Sheldon & Parsons (1967a). In brief, these measurements were made by determining the particle size spectrum of material before and after grazing by secondary producers. The difference between these two values, representing the quantity grazed, was converted to the amount of organic carbon eaten. For the latter conversion, a regression equation was used which was determined during the investigation using phytoplankton carbon and Coulter volume data collected in the same area.

For purposes of comparing the effect of zooplankton grazing on different phytoplankton blooms, certain parameters have been employed where possible. These are contained in the following expression:

$$r = R[1 - e^{k(p_0 - p)}], \tag{1}$$

where r is the ration grazed, R is the maximum ration, p is the prey density, p_0 is the prey density at which grazing starts and k is a constant. Although this expression does not account for inhibition of grazing at high phytoplankton concentrations, or for the possible occurrence of superfluous feeding, its use has been found satisfactory for a large number of experiments performed in our laboratory. The derivation of the above equation has been reported by Ivlev (1961) and the modified form given above has been previously discussed by Parsons, LeBrasseur & Fulton (1967).

ZOOPLANKTON SAMPLES

Samples of zooplankton were collected from 0–20 m with a 0.25 m^2 black nylon net (330 μ mesh diameter) hauled at 1 m/sec. For smaller zooplankton, the samples were collected in 7-litre Van Dorn bottles and filtered through 35 μ mesh diameter nylon. All samples were preserved and examined ashore. Live copepods for grazing experiments were collected by towing a net obliquely from ~ 10 m. Samples were diluted with sea water and live animals selected and withdrawn with a pipette. For vertical and horizontal distributional studies, zooplankton were collected either with a series of Miller samplers placed at 1 m intervals on a wire or with a Hardy-Longhurst recorder (Longhurst, *et al.*. 1966).

GROWTH RATE OF *Calanus plumchrus*

The growth rate of *C. plumchrus* as %/day was determined from the average weight of individual stages of this organism and the time of their occurrence. The method was similar to that described by Cushing (1964) using the following expression:

$$\%\Delta W = 100[10^{1/t(\log W_2 - \log W_1)} - 1], \tag{2}$$

where $\%\Delta W$ is the percentage weight change per day, W_1 and W_2 are the wet weights of two stages of *C. plumchrus* and t is the time interval observed for their development. The time interval was determined from the position of the peak heights of relative numbers of nauplii, copepodid stages I, III and V plotted against time. The weight of individual stages was obtained from earlier data (*e.g.* Stephens, *et al.*, 1967).

RESULTS

Some effects of copepods grazing on different blooms of phytoplankton are shown in Figs 2 and 3.

In Fig. 2 representative size spectra of particulate material are shown for successive

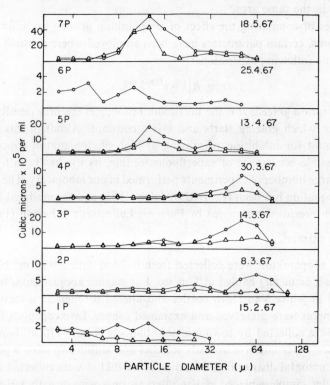

Fig. 2. Size spectra (○) and grazing (△) of some representative zooplankton in the area. Cruise 1-P, *Pseudocalanus minutus* grazing on μ-flagellates and *Skeletonema costatum*; Cruise 2-P, *Calanus pacificus* grazing on *Thalassiosira* spp.; Cruise 3-P, *Calanus pacificus* and *C. plumchrus* IV grazing on *Thalassiosira* spp.; Cruise 4-P, *Calanus pacificus* grazing on *Thalassiosira* spp.; Cruise 5-P, *Calanus plumchrus* III and IV grazing on *Skeletonema costatum* and μ-flagellates; Cruise 6-P, μ-flagellates, no grazing; Cruise 7-P, *Pseudocalanus minutus* and *Oithona* sp. grazing on *Skeletonema costatum* and μ-flagellates.

cruises (1–7) together with the results of some grazing experiments. During the first cruise (February) the particulate material consisted of μ-flagellates and *Skeletonema costatum* Greville, both at very low densities. For most of March the predominant phytoplankton organisms were *Thalassiosira* spp. chiefly *T. nordenskioldii* Cleve (?) with appreciable numbers of *T. rotula* Meunier during the first week of March. During April and May, *Skeletonema costatum* again occurred in various amounts together with unidentified μ-flagellates. Measurable quantities of phytoplankton were grazed by the zooplankton present on all cruises except on the sixth, on which no significant grazing was detected.

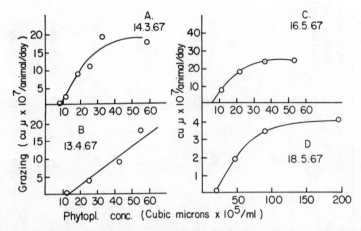

Fig. 3. Zooplankton grazing at different concentrations of phytoplankton. A. *Calanus pacificus* grazing on *Thalassiosira* spp. B. *Calanus plumchrus* III and IV grazing on *Skeletonema costatum* and μ-flagellates. C. *Calanus plumchrus* V. grazing on *Skeletonema costatum* and μ-flagellates. D. *Pseudocalanus minutus* and *Oithona* grazing on *Skeletonema costatum* and μ-flagellates.

Some results for zooplankton grazing on successive dilutions of the naturally occurring phytoplankton blooms are shown in Fig. 3, A to D. In a number of experiments, the asymptotic value R (maximum ration) was not reached (*e.g.*, Fig. 3B) and this was due to the fact that the natural plankton blooms could not be concentrated without altering the size spectrum of material available. A number of values for the prey density at maximum ration in Table I have, therefore, been entered as being greater than the highest concentration of phytoplankton naturally available. The values for the phytoplankton concentration at which grazing started have been taken from Fig. 3, A to D, and entered in Table I together with other data from figures not shown here. Data on the maximum and minimum concentrations of phytoplankton grazed, together with data on the quantity consumed per animal, have been converted from cubic microns, as measured with a Coulter Counter, to carbon using the following regression equation: Coulter vol. (ppm by vol) = 10.5 ± 2.1 carbon (ppm by wt) -0.02 ± 0.19, where the upper and lower limits of the slope and the intercept are the 95 % probable limits of these values. The equation reported here

TABLE I

Summary of zooplankton grazing experiments.

Date	Predominant zooplankton	Modal food size diam.(μ)	Predominant phytoplankter	Food density (μg C/l) min.	Food density (μg C/l) max.	Max. ration obtained (mg C/animal/day)	Mean zooplankter body weight (mg C)	Ration (% body wt/ day)
15.2.67	Pseudocalanus minutus Krøyer	8 and 14	Skeletonema costatum and μ-flagellates	81	>163	0.0004	0.010	4.0
9.3.67	Calanus pacificus Brodsky	90	Thalassiosira spp.	142	>305	0.0168	0.100	16.8
14.3.67	Calanus pacificus	57	Thalassiosira spp.	85	380	0.0184	0.100	18.4
30.3.67	Calanus pacificus and Calanus plumchrus IV	45	Thalassiosira spp.	48	>205	0.0202	0.100	20.2
10.4.67	Calanus plumchrus III and IV	5 and 9	μ-flagellates	39	>108	0.0017	0.030	5.7
13.4.67	Calanus plumchrus III and IV	16	Skeletonema costatum and μ-flagellates	119	>520	0.0181	0.030	60
16.5.67	Calanus plumchrus V	14	Skeletonema costatum and μ-flagellates	62	285	0.0260	0.175	14.8
18.5.67	Pseudocalanus minutus and Oithona sp.	14	Skeletonema costatum and μ-flagellates	190	760	0.0045	0.010	45

differs from one used previously (Sheldon & Parsons, 1967b) and this may be attributed to two factors: first, the correlation reported here is for Coulter volumes and plankton carbon as opposed to total carbon; and secondly, as has already been suggested (Sheldon & Parsons, 1967a) a regression equation of this type should be determined for each environment since ratios of carbon to cell volume may vary with a number of conditions (*e.g.* species, cell size, nutrient level).

Table I gives a summary of the data collected from the zooplankton grazing experiments. Values of the amount of food consumed have been expressed as a percentage of the animal's body weight (mg carbon). The individual body weights of animals and their carbon content were obtained during the course of the investigation and from previous data collected in the same area (*e.g.* Stephens, *et al.* 1967). In some cases, the sizes of the zooplankton organisms were so similar that two species or stages could not be separated and have been treated together.

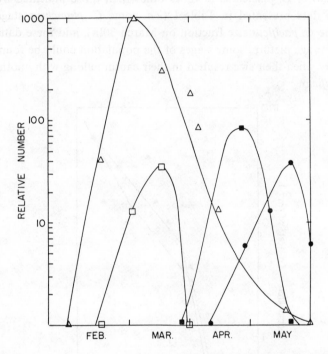

Fig. 4. Relative number of nauplii and copepodid stages of *Calanus plumchrus*, February-May 1967, (△ nauplii, □ copepodid I, ■ copepodid III, ● copepodid V).

Fig. 4 shows the relative number of different stages of *Calanus plumchrus* as they occurred. The predominant feature of these data is the continued presence of nauplii; in contrast, the occurrence of Stage III copepodids was confined to a single observation of very large numbers at the end of April. The date on which the maximum relative number of each stage occurred is given in Table II together with the wet

TABLE II

The *in situ* growth rate of *Calanus plumchrus*.

	Nauplii I-VI	I	Copepodids stage III	V
Date of maximum numbers	2.3.67	16.3.67	27.4.67	18.5.67
Time interval (days)		14	44	21
Wet wt (mg)	0.025	0.15	0.60	3.5
Growth (% per day)		14.0	3.5	8.8

weights of the stages. The former data may only approximate the true date of a maximum within several days. From these data, however, the growth rates in %/day have been calculated and are given in Table II for three time intervals. The occurrence of various stages of *C. plumchrus* at times other than those indicated by Fig. 4 is apparent from data presented in Table I (*e.g.* some *C. plumchrus* stage IV were separated in the *C. pacificus* size fraction on March 30th), and these data therefore represent an average picture; some stages of the population could be found at other times, especially when their size resulted in their capture along with another species, such as *C. pacificus*.

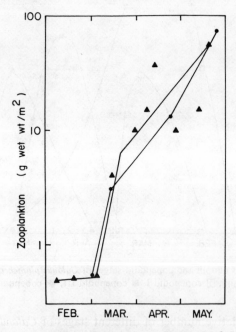

Fig. 5. The *in situ* (▲) and predicted (●) standing stock of *Calanus plumchrus*.

The standing stock of *C. plumchrus* from February to May, as determined from 0–20 m vertical net hauls, is shown in Fig. 5 and the approximate changes in the biomass of *C. plumchrus* are indicated by a line for these months. Starting with a biomass of 0.5 g wet wt/m^2 at the end of February and using the growth rates reported in Table II, a second line of predicted standing stock in terms of potential biomass is also given in Fig. 5.

DISCUSSION

The data shown in Fig. 2 indicate that there was a considerable change in the size spectrum as well as in the total quantity of particulate material available. The principal diatom species *Skeletonema costatum, Thalassiosira nordenskioldii* (?) and *T. rotula* were all grazed extensively. At times μ-flagellates augmented the total biomass and were sometimes effectively grazed (*e.g.* Cruise 1, February). At other times (*e.g.* Cruise 7, May, the left hand portion of the *Skeletonema costatum* and μ-flagellate peak) they were selectively avoided when diatom concentrations were adequate. During one cruise (Cruise 6, end of April) no significant grazing could be measured and this coincided with the occurrence of the smallest particles at the lowest concentrations found during the entire twelve week period.

Data on zooplankton grazing at different concentrations of phytoplankton are summarized in Table I from which it may be seen that the ration as a % of the body weight of the animals (both expressed in terms of carbon) was generally greater than 10 % and at times as high as 60 %. Some caution may be necessary in extrapolating these values, which were obtained in an incubator, to those actually found in nature. The results contrast with those of Parsons, LeBrasseur & Fulton (1967) who showed that the diatoms *Chaetoceros debilis* and *C. socialis* are poor food sources for three zooplankton organisms. In general, therefore, the phytoplankton occurring under the Fraser River plume at this time of year appears to have been an excellent food source for the copepods present. Differences were, however, found both in the amount of particular phytoplankton organisms consumed and in the concentration at which grazing started, as well as the concentration at which the animals reached maximum ration.

Limitations on growth due to restrictions on the availability of food were apparent for *Calanus plumchrus*, stages III and IV (Table I, 10.4.67) feeding on a small flagellate at low concentrations. The growth rate of *C. plumchrus* averaged about 7 %/day which would require a food intake of at least 14 %/day assuming a high assimilation efficiency. This type of restriction, imposed by a low concentration and a particularly small prey size, became apparent *in situ*, both in the predicted and actual increase in biomass (Fig. 5) during the period at the end of April when there was a marked slowing in the growth rate of the population. From Fig. 2 (25.4.67) it may be seen that this was almost certainly due to a lack of an adequate food supply which, at this time, consisted only of small flagellates at very low concentrations. With the addition of an appreciable quantity of *Skeletonema costatum* to the water, the experiment reported on 13.4.67 shows that the same organism obtained a ration of 60 % of its body weight per day. This would have allowed them to grow at ∼ 7 %/day, or faster, even assuming a low assimilation efficiency.

Three experiments on the grazing of *Calanus pacificus*, the results of which are given in Table I indicate that the adult organism obtained between 15 and 20 % of its body weight per day from a bloom of *Thalassiosira* spp. and that in terms of prey concentration feeding started at 50 to 150 μg C/l and reached a maximum at ∼ 400 μg C/l. The main activity of this organism during the early part of the period was egg laying. The ration which the animal obtained for this purpose (15–20 % of the body weight) is similar to the ration taken by *Euphausia pacifica* during egg laying as reported previously (Parsons, LeBrasseur & Fulton, 1967).

Pseudocalanus minutus was observed to graze on a mixture of *Skeletonema costatum* and μ-flagellates during two cruises. On the first of these (15.2.67) the total density was too low for the animal to obtain food at more than a subsistence level. On the second occasion, however, the animal got a ration of 45 % of its body weight from a similar bloom, present at a much higher density. In the latter circumstance, *Pseudocalanus minutus* consumed only the *Skeletonema costatum* fraction of the crop which was in contrast to its grazing on both the μ-flagellates and *S. costatum* at low crop density (*cf.* Fig. 2, Cruises 1 and 7). The presence of *Oithona* sp. is recorded in the data reported in Table I (18.5.67) but due to its comparatively small size, it was believed to constitute only about 10 % of the zooplankton biomass in this experiment.

From Fig. 4, it may be seen that *Calanus plumchrus* nauplii were found in relative concentrations of an order of magnitude greater than the copepodid stages. This suggests that a very high natural mortality occurred between the nauplii and copepodid stages. Another reason for the removal of large numbers of nauplii might have been advection but this should have been equally apparent on successive copepodid stages, which from Fig. 4 does not, however, appear to have been the case.

Dates on which maxima occurred in the relative numbers of the different stages of *C. plumchrus* are shown in Table II. From these dates and the wet weights of the different stages, the growth per day has been calculated as ranging from 3.5 to 14 %, with an overall growth rate of 6.5% /day. From these results it is apparent that the predicted increase in the standing stock of *C. plumchrus* shown in Fig. 5 is comparable with the observed increase in the biomass of zooplankton found in the plume during this period. This comparison has been made on a semi-log plot which tends to obscure arithmetic differences in the predicted and actual increase in biomass. However, the important feature of Fig. 5 is that if at the beginning of March there had been a 90 % predation of copepodid stage I, then the remaining 10 % growing at an average rate of 6.5 %/day could have reached a standing stock of stage V animals of only about 10 g wet wt/m^2 by the end of May, whereas the standing stock reached by this time was closer to 100 g wet wt/m^2. Thus, although the production of large amounts of *C. plumchrus* in the area of the plume was in part due to the availability of the adequate phytoplankton as discussed above, the lack of severe predation on the early stages of *C. plumchrus*, is an important factor.

In view of the lack of observations in the literature on the minimum concentration of particulate organic material at which animals begin grazing, the data of Table I are particularly interesting. Cushing (1964), for example, in his work on models of zooplankton grazing, has referred to the mechanical volume swept clear by a copepod in *barren* water. This appears to us from the data presented here and earlier (Parsons, *et al.*, 1967) to be a hypothetical value which the animals never reach because grazing ceases at some low prey density. From Table I, this value varies from about 50 to 190 µg of particulate carbon per litre. Quite recently, it appears from work by Adams & Steele (1966) that these authors have also found a level of particulate carbon at which the filtering rate declines (150 µg C/l), although they did not attach significance to the value at which grazing apparently ceases ($\sim$ 70 µg C/l from Adams & Steele, 1966, Fig. 4). If the cessation of feeding at low particle concentrations as reported by us occurs in nature, then it emphasizes their conclusion that the growth of phytoplankton at low population densities is made possible by a decreased filtering rate of zooplankton at these densities. It further indicates that grazing by zooplankton may not generally occur under the euphotic zone where particulate carbon concentrations may be as low as 20 µg C/l (Menzel, 1967). In some surface waters (*e.g.* those of the Sargasso Sea) the consideration of a minimum particle concentration for grazing emphasizes the importance of contagious particle distributions in which the local concentration of particles may be very much higher than that in the surrounding water.

Acknowledgements

The authors would like to their express gratitude to Mr. C. D. McAllister for discussions and assistance in the preparation of this manuscript and to Misses J. Tuttle and D. Kerr for technical assistance in the preparation of data.

References

Adams, J. A. & J. H. Steele, 1966. Shipboard experiments on the feeding of *Calanus finmarchicus* (Gunnerus). In, *Some contemporary studies in marine science*, edited by H. Barnes, George Allen and Unwin Ltd., London, pp. 19–35.

Cushing, D. H., 1964. The work of grazing in the sea. In, *Grazing in terrestrial and marine environments*, edited by D. J. Crisp, Blackwells Sci. Publ., London, pp. 207–225.

Fulton, J. D., O. D. Kennedy, K. Stephens & J. Skelding, 1967. Data record, physical, chemical and biological data, Strait of Georgia, 1966. *Manuscr. Rep. Fish. Res. Bd Can.*, No. 915, 145 pp.

Ivlev, V. S. 1961. *Experimental ecology of the feeding of fishes*. Translated by D. Scott, Yale Univ. Press, New Haven, pp. 20–21.

Longhurst, A. R., A. D. Reith, R. E. Bowen & D. L. R. Siebert, 1966. A new system for the collection of multiple serial plankton samples. *Deep Sea Res.*, Vol. 13, pp. 213–222.

Menzel, D. W., 1967. Particulate organic carbon in the deep sea. *Deep Sea Res.*, Vol. 14, pp. 229–238.

Parsons, T. R., R. J. LeBrasseur & J. D. Fulton, 1967. Some observations on the dependence of zooplankton grazing on the cell size and concentration of phytoplankton blooms. *J. oceanogr. Soc. Japan*, Vol. 23, pp. 10–17.

Parsons, T. R., K. Stephens & R. J. LeBrasseur, 1969. Production studies in the Strait of Georgia. I. Primary production under the Fraser River plume, February to May, 1967. *J. exp. mar. Biol. Ecol.*, Vol. 3, pp. 27–38.

SHELDON, R. W. & T. R. PARSONS, 1967a. *A practical manual on the use of the Coulter Counter in marine research*. Coulter Electronics Sales Co., Canada, 66 pp.

SHELDON, R. W. & T. R. PARSONS, 1967b. A continuous size spectrum for particulate matter in the sea. J. *Fish. Res. Bd Can.*, Vol. 24, pp. 909–915.

STEPHENS, K., J. D. FULTON, O. D. KENNEDY & A. K. PEASE, 1967. Biological, chemical and physical observations in Saanich Inlet, Vancouver Island, British Columbia. *Manusc. Rep. Fish. Res. Bd Can.*, No. 912, 119 pp.

Photosynthesis and Fish Production in the Sea

The production of organic matter and its conversion to
higher forms of life vary throughout the world ocean.

John H. Ryther

Numerous attempts have been made
to estimate the production in the sea of
fish and other organisms of existing or
potential food value to man (1–4).
These exercises, for the most part, are
based on estimates of primary (photosynthetic) organic production rates in
the ocean (5) and various assumed
trophic-dynamic relationships between
the photosynthetic producers and the
organisms of interest to man. Included
in the latter are the number of steps
or links in the food chains and the
efficiency of conversion of organic matter from each trophic level or link in
the food chain to the next. Different
estimates result from different choices
in the number of trophic levels and in
the efficiencies, as illustrated in Table 1
(2).

Implicit in the above approach is the
concept of the ocean as a single ecosystem in which the same food chains
involving the same number of links
and efficiencies apply throughout. However, the rate of primary production is
known to be highly variable, differing
by at least two full orders of magnitude
from the richest to the most impoverished regions. This in itself would be
expected to result in a highly irregular
pattern of food production. In addition,
the ecological conditions which determine the trophic dynamics of marine
food chains also vary widely and in
direct relationship to the absolute level
of primary organic production. As is
shown below, the two sets of variables—primary production and the associated food chain dynamics—may act
additively to produce differences in fish
production which are far more pronounced and dramatic than the observed variability of the individual
causative factors.

Primary Productivity

Our knowledge of the primary organic productivity of the ocean began
with the development of the C^{14}-tracer
technique for in situ measurement of
photosynthesis by marine plankton
algae (6) and the application of the
method on the 1950–52 Galathea expedition around the world (5). Despite
obvious deficiencies in the coverage of
the ocean by Galathea (the expedition
made 194 observations, or an average
of about one every 2 million square
kilometers, most of which were made
in the tropics or semitropics), our concept of the total productivity of the
world ocean has changed little in the
intervening years.

While there have been no more expeditions comparable to the Galathea,
there have been numerous local or regional studies of productivity in many
parts of the world. Most of these have
been brought together by a group of
Soviet scientists to provide up-to-date
world coverage consisting of over 7000
productivity observations (7). The result has been modification of the estimate of primary production in the
world ocean from 1.2 to 1.5×10^{10} tons
of carbon fixed per year (5) to a new
figure, 1.5 to 1.8×10^{10} tons.

Attempts have also been made by
Steemann Nielsen and Jensen (5), Ryther (8), and Koblentz-Mishke et al. (7)
to assign specific levels or ranges of
productivity to different parts of the
ocean. Although the approach was
somewhat different in each case, in
general the agreement between the
three was good and, with appropriate
condensation and combination, permit
the following conclusions.

1) Annual primary production in the
open sea varies, for the most part, between 25 and 75 grams of carbon fixed
per square meter and averages about 50
grams of carbon per square meter per
year. This is true for roughly 90 percent of the ocean, an area of 326×10^6
square kilometers.

2) Higher levels of primary production occur in shallow coastal waters,
defined here as the area within the
100-fathom (180-meter) depth contour. The mean value for this region
may be considered to be 100 grams of
carbon fixed per square meter per year,
and the area, according to Menard and
Smith (9), is 7.5 percent of the total
world ocean. In addition, certain offshore waters are influenced by divergences, fronts, and other hydrographic
features which bring nutrient-rich subsurface water into the euphotic zone.
The equatorial divergences are examples of such regions. The productivity of these offshore areas is comparable to that of the coastal zone.
Their total area is difficult to assess,
but is considered here to be 2.5 percent
of the total ocean. Thus, the coastal
zone and the offshore regions of comparably high productivity together represent 10 percent of the total area of the

The author is a member of the staff of the
Woods Hole Oceanographic Institution, Woods
Hole, Massachusetts.

Reprinted from Science 166: 72–76 (1969).

oceans, or 36×10^6 square kilometers.

3) In a few restricted areas of the world, particularly along the west coasts of continents at subtropical latitudes where there are prevailing offshore winds and strong eastern boundary currents, surface waters are diverted offshore and are replaced by nutrient-rich deeper water. Such areas of coastal upwelling are biologically the richest parts of the ocean. They exist off Peru, California, northwest and southwest Africa, Somalia, and the Arabian coast, and in other more localized situations. Extensive coastal upwelling also is known to occur in various places around the continent of Antarctica, although its exact location and extent have not been well documented. During periods of active upwelling, primary production normally exceeds 1.0 and may exceed 10.0 grams of carbon per square meter per day. Some of the high values which have been reported from these locations are 3.9 grams for the southwest coast of Africa (5), 6.4 for the Arabian Sea (10), and 11.2 off Peru (11). However, the upwelling of subsurface water does not persist throughout the year in many of these places—for example, in the Arabian Sea, where the process is seasonal and related to the monsoon winds. In the Antarctic, high production is limited by solar radiation during half the year. For all these areas of coastal upwelling throughout the year, it is probably safe, if somewhat conservative, to assign an annual value of 300 grams of carbon per square meter. Their total area in the world is again difficult to assess. On the assumption that their total cumulative area is no greater than 10 times the well-documented upwelling area off Peru, this would amount to some 3.6×10^5 square kilometers, or 0.1 percent of the world ocean. These conclusions are summarized in Table 2.

Food Chains

Let us next examine the three provinces of the ocean which have been designated according to their differing levels of primary productivity from the standpoint of other possible major differences. These will include, in particular, differences which relate to the food chains and to trophic efficiencies involved in the transfer of organic matter from the photosynthetic organisms to fish and invertebrate species large and abundant enough to be of importance to man.

The first factor to be considered in this context is the size of the photosynthetic or producer organisms. It is generally agreed that, as one moves from coastal to offshore oceanic waters, the character of these organisms changes from large "microplankton" (100 microns or more in diameter) to the much smaller "nannoplankton" cells 5 to 25 microns in their largest dimensions (12, 13).

Since the size of an organism is an essential criterion of its potential usefulness to man, we have the following relationship: the larger the plant cells at the beginning of the food chain, the fewer the trophic levels that are required to convert the organic matter to a useful form. The oceanic nannoplankton cannot be effectively filtered from the water by most of the common zooplankton crustacea. For example, the euphausid *Euphausia pacifica*, which may function as a herbivore in the rich subarctic coastal waters of the Pacific, must turn to a carnivorous habit in the offshore waters where the phytoplankton become too small to be captured (13).

Intermediate between the nannoplankton and the carnivorous zooplankton are a group of herbivores, the microzooplankton, whose ecological significance is a subject of considerable current interest (14, 15). Representatives of this group include protozoans such as Radiolaria, Foraminifera, and Tintinnidae, and larval nuplii of microcrustaceans. These organisms, which may occur in concentrations of tens of thousands per cubic meter, are the primary herbivores of the open sea.

Feeding upon these tiny animals is a great host of carnivorous zooplankton, many of which have long been thought of as herbivores. Only by careful study of the mouthparts and feeding habits were Anraku and Omori (16) able to show that many common copepods are facultative if not obligate carnivores. Some of these predatory copepods may be no more than a millimeter or two in length.

Again, it is in the offshore environment that these small carnivorous zooplankton predominate. Grice and Hart (17) showed that the percentage of carnivorous species in the zooplankton increased from 16 to 39 percent in a transect from the coastal waters of the northeastern United States to the Sargasso Sea. Of very considerable importance in this group are the Chaetognatha. In terms of biomass, this group of animals, predominantly carnivorous, represents, on the average, 30 percent of the weight of copepods in the open sea (17). With such a distribution, it is clear that virtually all the copepods, many of which are themselves carnivores, must be preyed upon by chaetognaths.

Table 1. Estimates of potential yields (per year) at various trophic levels, in metric tons. [After Schaeffer (2)]

| Trophic level | Ecological efficiency factor | | | | | |
| | 10 percent | | 15 percent | | 20 percent | |
	Carbon (tons)	Total weight (tons)	Carbon (tons)	Total weight (tons)	Carbon (tons)	Total weight (tons)
0. Phytoplankton (net particulate production)	1.9×10^{10}		1.9×10^{10}		1.9×10^{10}	
1. Herbivores	1.9×10^9	1.9×10^{10}	2.8×10^9	2.8×10^{10}	3.8×10^9	3.8×10^{10}
2. 1st stage carnivores	1.9×10^8	1.9×10^9	4.2×10^8	4.2×10^9	7.6×10^8	7.6×10^9
3. 2nd stage carnivores	1.9×10^7	1.9×10^8	6.4×10^7	6.4×10^8	15.2×10^7	15.2×10^8
4. 3rd stage carnivores	1.9×10^6	1.9×10^7	9.6×10^6	9.6×10^7	30.4×10^6	30.4×10^7

Table 2. Division of the ocean into provinces according to their level of primary organic production.

Province	Percentage of ocean	Area (km²)	Mean productivity (grams of carbon/m²/yr)	Total productivity (10⁹ tons of carbon/yr)
Open ocean	90	326×10^6	50	16.3
Coastal zone*	9.9	36×10^6	100	3.6
Upwelling areas	0.1	3.6×10^5	300	0.1
Total				20.0

* Includes offshore areas of high productivity.

Table 3. Estimated fish production in the three ocean provinces defined in Table 2.

Province	Primary production [tons (organic carbon)]	Trophic levels	Efficiency (%)	Fish production [tons (fresh wt.)]
Oceanic	16.3×10^9	5	10	16×10^5
Coastal	3.6×10^9	3	15	12×10^7
Upwelling	0.1×10^9	1½	20	12×10^7
Total				24×10^7

The oceanic food chain thus far described involves three to four trophic levels from the photosynthetic nannoplankton to animals no more than 1 to 2 centimeters long. How many additional steps may be required to produce organisms of conceivable use to man is difficult to say, largely because there are so few known oceanic species large enough and (through schooling habits) abundant enough to fit this category. Familiar species such as the tunas, dolphins, and squid are all top carnivores which feed on fishes or invertebrates at least one, and probably two, trophic levels beyond such zooplankton as the chaetognaths. A food chain consisting of five trophic levels between photosynthetic organisms and man would therefore seem reasonable for the oceanic province.

As for the coastal zone, it has already been pointed out that the phytoplankton are quite commonly large enough to be filtered and consumed directly by the common crustacean zooplankton such as copepods and euphausids. However, the presence, in coastal waters, of protozoans and other microzooplankton in larger numbers and of greater biomass than those found in offshore waters (15) attests to the fact that much of the primary production here, too, passes through several steps of a microscopic food chain before reaching the macrozooplankton.

The larger animals of the coastal province (that is, those directly useful to man) are certainly the most diverse with respect to feeding type. Some (mollusks and some fishes) are herbivores. Many others, including most of the pelagic clupeoid fishes, feed on zooplankton. Another large group, the demersal fishes, feed on bottom fauna which may be anywhere from one to several steps removed from the phytoplankton.

If the herbivorous clupeoid fishes are excluded (since these occur predominantly in the upwelling provinces and are therefore considered separately), it is probably safe to assume that the average food organism from coastal waters represents the end of at least a three-step food chain between phytoplankton and man.

It is in the upwelling areas of the world that food chains are the shortest, or—to put it another way—that the organisms are large enough to be directly utilizable by man from trophic levels very near the primary producers. This, again, is due to the large size of the phytoplankton, but it is due also to the fact that many of these species are colonial in habit, forming large gelatinous masses or long filaments. The eight most abundant species of phytoplankton in the upwelling region off Peru, in the spring of 1966, were *Chaetoceros socialis, C. debilis, C. lorenzianus, Skeletonema costatum, Nitzschia seriata, N. delicatissima, Schroederella delicatula,* and *Asterionella japonica* (11, 18). The first in this list, *C. socialis,* forms large gelatinous masses. The others all form long filamentous chains. *Thalossiosira subtilis,* another gelatinous colonial form like *Chaetoceros socialis,* occurs commonly off southwest Africa (19) and close to shore off the Azores (20). Hart (21) makes special mention of the colonial habit of all the most abundant species of phytoplankton in the Antarctic—*Fragiloriopsis antarctica, Encampia balaustrium, Rhizosalenia alata, R. antarctica, R. chunii, Thallosiothrix antarctica,* and *Phaeocystis brucei.*

Many of the above-mentioned species of phytoplankton form colonies several millimeters and, in some cases, several centimeters in diameter. Such aggregates of plant material can be readily eaten by large fishes without special feeding adaptation. In addition, however, many of the clupeoid fishes (sardines, anchovies, pilchards, menhaden, and so on) that are found most abundantly in upwelling areas and that make up the largest single component of the world's commercial fish landings, do have specially modified gill rakers for removing the larger species of phytoplankton from the water.

There seems little doubt that many of the fishes indigenous to upwelling regions are direct herbivores for at least most of their lives. There is some evidence that juveniles of the Peruvian anchovy (*Engraulis ringens*) may feed on zooplankton, but the adult is predominantly if not exclusively a herbivore (22). Small gobies (*Gobius bibarbatus*) found at mid-water in the coastal waters off southwest Africa had their stomachs filled with a large, chain-forming diatom of the genus *Fragilaria* (23). There is considerable interest at present in the possible commercial utilization of the large Antarctic krill, *Euphausia superba,* which feeds primarily on the colonial diatom *Fragilariopsis antarctica* (24).

In some of the upwelling regions of the world, such as the Arabian Sea, the species of fish are not well known, so it is not surprising that knowledge of their feeding habits and food chains is fragmentary. From what is known, however, the evidence would appear to be overwhelming that a one- or two-step food chain between phytoplankton and man is the rule. As a working compromise, let us assign the upwelling province a 1½-step food chain.

Efficiency

The growth (that is, the net organic production) of an organism is a function of the food assimilated less metabolic losses or respiration. This efficiency of growth or food utilization (the ratio of growth to assimilation) has been found, by a large number of investigators and with a great variety of organisms, to be about 30 percent in young, actively growing animals. The efficiency decreases as animals approach their full growth, and reaches zero in fully mature or senescent individuals (25). Thus a figure of 30 percent can be considered a biological potential which may be approached in nature, although the growth efficiency of a population of animals of mixed ages under steady-state conditions must be lower.

Since there must obviously be a "maintenance ration" which is just sufficient to accommodate an organism's basal metabolic requirement (26), it must also be true that growth efficiency is a function of the absolute rate of assimilation. The effects of this factor will be most pronounced at low feeding rates, near the "maintenance ration," and will tend to become negligible at high feeding rates. Food conversion (that is, growth efficiency) will therefore obviously be related to food availability, or to the concentration of prey organisms when the latter are sparsely distributed.

In addition, the more available the food and the greater the quantity consumed, the greater the amount of "internal work" the animal must perform to digest, assimilate, convert, and store the food. Conversely, the less available the food, the greater the amount of "external work" the animal must perform to hunt, locate, and capture its prey. These concepts are discussed in some detail by Ivlev (27) and reviewed by Ricker (28). The two metabolic costs thus work in opposite ways with respect to food availability, tending thereby toward a constant total effect. However, when food availability is low, the added costs of basal metabolism and external work relative to assimilation may have a pronounced effect on growth efficiency.

When one turns from consideration of the individual and its physiological growth efficiency to the "ecological efficiency" of food conversion from one trophic level to the next (2, 29), there are additional losses to be taken into account. Any of the food consumed but not assimilated would be included here, though it is possible that undigested organic matter may be reassimilated by members of the same trophic level (2). Any other nonassimilatory losses, such as losses due to natural death, sedimentation, and emigration, will, if not otherwise accounted for, appear as a loss in trophic efficiency. In addition, when one considers a specific or selected part of a trophic level, such as a population of fish of use to man, the consumption of food by any other hidden member of the same trophic level will appear as a loss in efficiency. For example, the role of such animals as salps, medusae, and ctenophores in marine food chains is not well understood and is seldom even considered. Yet these animals may occur sporadically or periodically in swarms so dense that they dominate the plankton completely. Whether they represent a dead end or side branch in the normal food chain of the sea is not known, but their effect can hardly be negligible when they occur in abundance.

Finally, a further loss which may occur at any trophic level but is, again, of unknown or unpredictable magnitude is that of dissolved organic matter lost through excretion or other physiological processes by plants and animals. This has received particular attention at the level of primary production, some investigators concluding that 50 percent or more of the photoassimilated carbon may be released by phytoplankton into the water as dissolved compounds (30). There appears to be general agreement that the loss of dissolved organic matter is indirectly proportional to the absolute rate of organic production and is therefore most serious in the oligotrophic regions of the open sea (11, 31).

All of the various factors discussed above will affect the efficiency or apparent efficiency of the transfer of organic matter between trophic levels. Since they cannot, in most cases, be quantitatively estimated individually, their total effect cannot be assessed. It is known only that the maximum potential growth efficiency is about 30 percent and that at least some of the factors which reduce this further are more pronounced in oligotrophic, low-productivity waters than in highly productive situations. Slobodkin (29) concludes that an ecological efficiency of about 10 percent is possible, and Schaeffer feels that the figure may be as high as 20 percent. Here, therefore, I assign efficiencies of 10, 15, and 20 percent, respectively, to the oceanic, the coastal, and the upwelling provinces, though it is quite possible that the actual values are considerably lower.

Conclusions and Discussion

With values assigned to the three marine provinces for primary productivity (Table 2), number of trophic levels, and efficiencies, it is now possible to calculate fish production in the three regions. The results are summarized in Table 3.

These calculations reveal several interesting features. The open sea—90 percent of the ocean and nearly three-fourths of the earth's surface—is essentially a biological desert. It produces a negligible fraction of the world's fish catch at present and has little or no potential for yielding more in the future.

Upwelling regions, totaling no more than about one-tenth of 1 percent of the ocean surface (an area roughly the size of California) produce about half the world's fish supply. The other half is produced in coastal waters and the few offshore regions of comparably high fertility.

One of the major uncertainties and possible sources of error in the calculation is the estimation of the areas of high, intermediate, and low productivity. This is particularly true of the upwelling area off the continent of Antarctica, an area which has never been well described or defined.

A figure of 360,000 square kilometers has been used for the total area of upwelling regions in the world (Table 2). If the upwelling regions off California, northwest and southwest Africa, and the Arabian Sea are of roughly the same area as that off the coast of Peru, these semitropical regions would total some 200,000 square kilometers. The remaining 160,000 square kilometers would represent about one-fourth the circumference of Antarctica seaward for a distance of 30 kilometers. This seems a not unreasonable inference. Certainly, the entire ocean south of the Antarctic Convergence is not highly productive, contrary to the estimates of El-Sayed (32). Extensive observations in this region by Saijo and Kawashima (33) yielded primary productivity values of 0.01 to 0.15 gram of carbon per square meter per day—a value no higher than the values used here for the open sea. Presumably, the discrepancy is the result of highly irregular, discontinuous, or "patchy" distribution of biological activity. In other words, the occurrence of extremely high productivity associated with upwelling conditions appears to be confined, in the Antarctic, as elsewhere, to restricted areas close to shore.

An area of 160,000 square kilometers of upwelling conditions with an annual productivity of 300 grams of carbon per square meter would result in the production of about 50×10^6 tons of "fish," if we follow the ground rules established above in making the estimate. Presumably these "fish" would consist for the most part of the Antarctic krill, which feeds directly upon phytoplankton, as noted above, and which is known to be extremely abundant in Antarctic waters. There have been numerous attempts to estimate the annual production of krill in the Antarctic, from the known number of whales at their peak of abundance and from various assumptions concerning their daily ration of krill. The evidence upon which such estimates are based is so tenuous that they are hardly worth discussing. It is interesting to note, however, that the more conservative of these estimates are rather close to figures derived independently by the method discussed here. For example, Moiseev (34) calculated krill production for 1967 to be 60.5×10^6 tons, while Kasahara (3) considered a range of 24 to 36×10^6 tons to be a minimal figure. I consider the figure 50×10^6 tons to be on the high side, as the estimated area of upwelling is probably generous, the average productivity value of 300 grams of carbon per square meter per year is high for a region where photosynthesis can occur during only half the year, and much of the primary production is probably diverted into smaller crustacean herbivores (35). Clearly, the Antarctic must receive much more intensive study before its productive capacity can be assessed with any accuracy.

In all, I estimate that some 240 million tons (fresh weight) of fish are produced annually in the sea. As this figure is rough and subject to numerous sources of error, it should not be considered significantly different from Schaeffer's (2) figure of 200 million tons.

Production, however, is not equivalent to potential harvest. In the first place, man must share the production with other top-level carnivores. It has been estimated, for example, that guano birds alone eat some 4 million tons of anchovies annually off the coast of Peru, while tunas, squid, sea lions, and other predators probably consume an equivalent amount (22, 36). This is nearly equal to the amount taken by man from this one highly productive fishery. In addition, man must take care to leave a large enough fraction of the annual production of fish to permit utilization of the resource at something close to its maximum sustainable yield, both to protect the fishery and to provide a sound economic basis for the industry.

When these various factors are taken into consideration, it seems unlikely that the potential sustained yield of fish to man is appreciably greater than 100 million tons. The total world fish landings for 1967 were just over 60 million tons (37), and this figure has been increasing at an average rate of about 8 percent per year for the past 25 years. It is clear that, while the yield can be still further increased, the resource is not vast. At the present rate, the industry can continue to expand for no more than a decade.

Most of the existing fisheries of the world are probably incapable of contributing significantly to this expansion. Many are already overexploited, and most of the rest are utilized at or near their maximum sustainable yield. Evidence of fishing pressure is usually determined directly from fishery statistics, but it is of some interest, in connection with the present discussion, to compare landings with fish production as estimated by the methods developed in this article. I will make this comparison for two quite dissimilar fisheries,

that of the continental shelf of the northwest Atlantic and that of the Peruvian coastal region.

According to Edwards (38), the continental shelf between Hudson Canyon and the southern end of the Nova Scotian shelf includes an area of 110,000 square miles (2.9×10^{11} square meters). From the information in Tables 2 and 3, it may be calculated that approximately 1 million tons of fish are produced annually in this region. Commercial landings from the same area were slightly in excess of 1 million tons per year for the 3-year period 1963 to 1965 before going into a decline. The decline has become more serious each year, until it is now proposed to regulate the landings of at least the more valuable species such as cod and haddock, now clearly overexploited.

The coastal upwelling associated with the Peru Coastal Current gives rise to the world's most productive fishery, an annual harvest of some 10^7 metric tons of anchovies. The maximum sustainable yield is estimated at, or slightly below, this figure (39), and the fishery is carefully regulated. As mentioned above, mortality from other causes (such as predation from guano birds, bonito, squid, and so on) probably accounts for an additional 10^7 tons. This prodigious fishery is concentrated in an area no larger than about 800×30 miles (36), or 6×10^{10} square meters. By the methods developed in this article, it is estimated that such an upwelling area can be expected to produce 2×10^7 tons of fish, almost precisely the commercial yield as now regulated plus the amount attributed to natural mortality.

These are but two of the many recognized examples of well-developed commercial fisheries now being utilized at or above their levels of maximum sustainable yield. Any appreciable continued increase in the world's fish landings must clearly come from unexploited species and, for the most part, from undeveloped new fishing areas. Much of the potential expansion must consist of new products from remote regions, such as the Antarctic krill, for which no harvesting technology and no market yet exist.

References and Notes

1. H. W. Graham and R. L. Edwards, in *Fish and Nutrition* (Fishing News, London, 1962), pp. 3–8; W. K. Schmitt, *Ann. N.Y. Acad. Sci.* 118, 645 (1965).
2. M. B. Schaeffer, *Trans. Amer. Fish. Soc.* 94, 123 (1965).
3. H. Kasahara, in *Proceedings, 7th International Congress of Nutrition, Hamburg* (Pergamon, New York, 1966), vol. 4, p. 958.
4. W. M. Chapman, "Potential Resources of the Ocean" (Serial Publication 89–21, 89th Congress, first session, 1965) (Government Printing Office, Washington, D.C., 1965), pp. 132–156.
5. E. Steemann Nielsen and E. A. Jensen, *Galathea Report*, F. Bruun et al., Eds. (Allen & Unwin, London, 1957), vol. 1, p. 49.
6. E. Steemann Nielsen, *J. Cons. Cons. Perma. Int. Explor. Mer* 18, 117 (1952).
7. O. I. Koblentz-Mishke, V. V. Volkovinsky, J. G. Kobanova, in *Scientific Exploration of the South Pacific*, W. Wooster, Ed. (National Academy of Sciences, Washington, D.C., in press).
8. J. H. Ryther, in *The Sea*, M. N. Hill, Ed. (Interscience, London, 1963), pp. 347–380.
9. H. W. Menard and S. M. Smith, *J. Geophys. Res.* 71, 4305 (1966).
10. J. H. Ryther and D. W. Menzel, *Deep-Sea Res.* 12, 199 (1965).
11. ——, E. M. Hulburt, C. J. Lorenzen, N. Corwin, "The Production and Utilization of Organic Matter in the Peru Coastal Current" (Texas A & M Univ. Press, College Station, in press).
12. C. D. McAllister, T. R. Parsons, J. D. H. Strickland, *J. Cons. Cons. Perma. Int. Explor. Mer* 25, 240 (1960); G. C. Anderson, *Limnol. Oceanogr.* 10, 477 (1965).
13. T. R. Parsons and R. J. Le Brasseur, in "Symposium Marine Food Chains, Aarhus (1968)."
14. E. Steemann Nielsen, *J. Cons. Cons. Perma. Int. Explor. Mer* 23, 178 (1958).
15. J. R. Beers and G. L. Stewart, *J. Fish. Res. Board Can.* 24, 2053 (1967).
16. M. Anraku and M. Omori, *Limnol. Oceanogr.* 8, 116 (1963).
17. G. D. Grice and H. D. Hart, *Ecol. Monogr.* 32, 287 (1962).
18. M. R. Reeve, in "Symposium Marine Food Chains, Aarhus (1968)."
19. Personal observation; T. J. Hart and R. I. Currie, *Discovery Rep.* 31, 123 (1960).
20. K. R. Gaarder, *Report on the Scientific Results of the "Michael Sars" North Atlantic Deep-Sea Expedition 1910* (Univ. of Bergen, Bergen, Norway).
21. T. J. Hart, *Discovery Rep.* 21, 261 (1942).
22. R. J. E. Sanchez, in *Proceedings of the 18th Annual Session, Gulf and Caribbean Fisheries Institute, University of Miami Institute of Marine Science, 1966*, J. B. Higman, Ed. (Univ. of Miami Press, Coral Gables, Fla., 1966), pp. 84–93.
23. R. T. Barber and R. L. Haedrich, *Deep-Sea Res.* 16, 415 (1952).
24. J. W. S. Marr, *Discovery Rep.* 32, 34 (1962).
25. S. D. Gerking, *Physiol. Zool.* 25, 358 (1952).
26. B. Dawes, *J. Mar. Biol. Ass. U.K.* 17, 102 (1930–31); *ibid.*, p. 877.
27. V. S. Ivlev, *Zool. Zh.* 18, 303 (1939).
28. W. E. Ricker, *Ecology* 16, 373 (1946).
29. L. B. Slobodkin, *Growth and Regulation of Animal Populations* (Holt, Rinehart & Winston, New York, 1961), chap. 12.
30. G. E. Fogg, C. Nalewajko, W. D. Watt, *Proc. Roy. Soc. Ser B Biol. Sci.* 162, 517 (1965).
31. G. E. Fogg and W. D. Watt, *Mem. Inst. Ital. Idrobiol. Dott. Marco de Marshi Pallanza Italy* 18, suppl., 165 (1965).
32. S. Z. El-Sayed, in *Biology of the Antarctic Seas III*, G. Llano and W. Schmitt, Eds. (American Geophysical Union, Washington, D.C., 1968), pp. 15–47.
33. Y. Saijo and T. Kawashima, *J. Oceanogr. Soc. Japan* 19, 190 (1964).
34. P. A. Moiseev, paper presented at the 2nd Symposium on Antarctic Ecology, Cambridge, England, 1968.
35. T. L. Hopkins, unpublished manuscript.
36. W. S. Wooster and J. L. Reid, Jr., in *The Sea*, M. N. Hill, Ed. (Interscience, London, 1963), vol. 2, p. 253.
37. *FAO Yearb. Fish. Statistics* 25 (1967).
38. R. L. Edwards, *Univ. Wash. Publ. Fish.* 4, 52 (1968).
39. R. J. E. Sanchez, in *Proceedings, 18th Annual Session, Gulf and Caribbean Fisheries Institute, University of Miami Institute of Marine Science* (Univ. of Miami Press, Coral Gables, 1966), p. 84.
40. The work discussed here was supported by the Atomic Energy Commission, contract No. AT(30-1)-3862, Ref. No. NYO-3862-20. This article is contribution No. 2327 from the Woods Hole Oceanographic Institution.

Grazing
and
Predation

Introduction

The transformation of energy fixed by primary producers from autotrophic tissue into herbivore and carnivore tissue is dependent upon herbivore grazing and carnivore predation. The rates of these processes depend upon various factors such as water temperature and food concentration. Such feeding has a profound effect not only upon the rate of growth of the organism involved, but also upon the population size of the organisms being fed upon, and the structure of the community. In this section we examine papers dealing with feeding dynamics at several levels of the trophic system. Much of the scientific work so far has dealt either with larger animals and been descriptive, examining stomach contents to determine trophic position, or with planktonic organisms, experimenting upon feeding rates at various concentrations and on various types of foods. Some experimental work has also been done on the effects of grazing or predation upon the structure of marine, particularly intertidal, communities.

Feeding mechanisms, obviously, vary with the type of food. Most of the herbivores of the ocean are filter feeders, creating a current that brings water-containing food particles to a sieve in which the particles are retained according to size and shape. Other animals feeding on suspended matter do so by carrying water-containing particles across ciliary surfaces where mucus entraps the particles and the cilia convey the food to the mouth. Jorgenson (1966) extensively reviews the habits of animals that feed on suspended material. The particles removed by these methods may vary widely in size, but generally the ciliary-mucus feeders retain smaller particles than the animals with structural sieves. Rates of filtering vary greatly depending upon the organism; in a day, a copepod may filter less than a hundred milliliters, while an oyster may pass 240 liters across its gill surfaces (Nicol, 1967). A wide variety of animals also feed herbiverously by ingesting multicellular algae in a variety of ways.

The description of predator-prey relationships has been very important to ecologists for many years. The Lotka-Volterra equations first described the relationship between prey and predator densities. The relationship between prey density and predator behavior has been of considerable theoretical and experimental interest. Ivlev (1961) described the feeding of planktiverous fishes as being minimal at very low prey densities, but increasing rapidly, then continuing at a relatively constant feeding rate at high prey densities. Holling (1965) also analyzed predator-prey relationships, and showed two types, one which he called the "invertebrate response" (Figure 1a), essentially like Ivlev's curve; the other, S-shaped curve (Figure 1b), was termed the "vertebrate response." This may have been an unfortunate choice of words, for his "vertebrate response" shows that there is a threshold concentration below which feeding is negligible when alternate prey is available, a phenomenon that has since been shown to exist in invertebrates as well as vertebrates. Nevertheless, these models of predator behavior in response to prey concentration have proven useful, as the papers on zooplankton feeding included here show.

Another aspect of feeding behavior is the

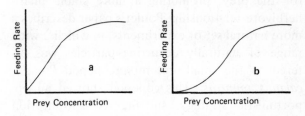

Figure 1. Holling's "invertebrate" (*a*) and "vertebrate" (*b*) responses to prey concentration.

selection of food. It has long been believed that suspension feeders indiscriminately remove particles from the water. More recent evidence shows that feeding strategies are employed to maximize food intake. Generally it would appear to be the best strategy to choose the largest packet of nutrition, because the energy used in capturing it will be less than capturing a nutritionally equivalent number of smaller packets. Ivlev (1961) showed how the selection of a particular food type could be inferred from an analysis of stomach contents and prey availability. His "electivity index" simply shows the relation between the food available and the food ingested. If any particular food type (or size) is preferred, the electivity index is positive; it is negative when the food type is selected against, and zero when no selection is taking place.

Prey populations are greatly affected by grazing and predation pressure. Selection of a particular species or a particular size can markedly alter the structure of a community, even when the pressure is not great. This phenomenon has been particularly well shown in freshwater planktonic ecosystems (Brooks and Dodson, 1965, Zaret and Paine, 1973) and in the marine intertidal regions (Paine, 1966). We examine these effects in the section entitled "Characterization."

The papers by Frost and by Poulet in this section deal with the grazing of copepods upon phytoplankton. Frost examined the dynamics of feeding over long time periods using several different sizes of phytoplankton species at varying concentrations. The decreased feeding rate at low food concentrations shown by Frost is important ecologically since it means that grazing pressure on species that are in low concentrations is lowered, providing a measure of security for the prey, promoting a more stable plant-herbivore relationship. Poulet's paper describes a more natural set of experiments, in which a wide range of naturally occurring particles was offered to the small, common copepod *Pseudocalanus minutus*. This copepod showed an opportunistic strategy, shifting its consumption from one cell size range to another depending upon relative particle concentration in order to maintain optimum food intake rates.

The paper by Hamner et al. provides a fascinating insight into the feeding behavior of gelatinous zooplankton and points out the vast differences between observations made in the laboratory and the field. O'Connell has examined differences in feeding behavior in anchovies, showing that they filter feed on small prey, but actively bite larger prey. A previous paper (Leong and O'Connell, 1969) had suggested that the anchovy could not meet daily food requirements by filter feeding alone. The paper included here shows that when there is a mixture of large and small prey organisms, daily food requirements can be met by a mixture of biting and filtering.

Whales are by far the largest animals in existence, and one would expect them to feed on large prey. Some do, but, as Nemoto has shown, the baleen whales feed on a wide variety of planktonic organisms, including krill (*Euphausia superba*), copepods, and small schooling fishes. This paper points out not only that baleen whales do not feed solely on krill, as has been supposed, but also that feeding must depend upon food concentration of the prey species in question. Many more questions are raised than are answered in this consideration of whale feeding patterns.

LITERATURE CITED

Brooks, J. L., and S. I. Dodson. 1965. Predation, body size and composition of plankton. Science 150:28–35.

Holling, C. S. 1965. The functional response of predators to prey density and its role in minicry and population regulation. Mem. Entomol. Soc. Can. 45. 60 pp.

Ivlev, V. S. 1961. Experimental ecology of the feeding of fishes. Yale University Press, New Haven, Connecticut.

Jorgenson, C. B. 1966. Biology of Suspension Feeding. Pergamon Press Inc., New York.

Leong, R. J. H., and C. P. O'Connell. 1969. A laboratory study of particulate and filter feeding of the northern anchovy (*Engraulis mordax*). J. Fish. Res. Board Can. 26:557–582.

Nicol, J. A. C. 1967. The biology of marine animals. 2nd Ed. Pitman Publishing Corporation, New York. 699 pp.

Paine, R. T. 1966. Food web complexity and species diversity. Amer. Naturalist 100:65–75.

Zaret, T. M., and R. T. Paine. 1973. Species introduction in a tropical lake. Science 182:449–455.

EFFECTS OF SIZE AND CONCENTRATION OF FOOD PARTICLES ON THE FEEDING BEHAVIOR OF THE MARINE PLANKTONIC COPEPOD *CALANUS PACIFICUS*[1]

B. W. Frost
Department of Oceanography, University of Washington, Seattle 98195

ABSTRACT

When adult females of *Calanus pacificus* are fed on monospecific cultures of centric diatoms which grow as single cells, a predictive relationship is found between feeding behavior of the copepods and size and concentration of food particles. Ingestion rate of copepods increases linearly with cell concentration up to a maximal rate. This maximal ingestion rate, expressed as carbon, is the same for copepods feeding on diatoms ranging in diameter from 11–87 μ. As the size of food particles increases, the carbon concentration at which this ingestion rate is achieved decreases. Thus females of *C. pacificus* can obtain their maximal daily ration at relatively low carbon concentrations of large cells.

INTRODUCTION

In oceanic food webs, calanoid copepods constitute prominent pathways for energy flow between primary producers and the larger predatory species of zooplankton and nekton. An understanding of quantitative trophic interactions between phytoplankton and herbivorous copepods is required to elucidate the nature of marine food webs in terms of rates. Much progress has been made in studies of species of *Calanus* and related genera.

The mechanics of filter feeding in copepods are well known (Marshall and Orr

1955). Unlike the more automatic, relatively unselective filter-feeding crustaceans such as *Artemia* (Reeve 1963) and *Daphnia* (McMahon and Rigler 1965; *but see also* Burns 1969), copepods apparently feed discontinuously and show considerable discrimination when presented with a choice of several food organisms. The latter behavior is known for copepods feeding on both mixed laboratory cultures of phytoplankton (Harvey 1937; Mullin 1963, 1966; Petipa 1965; Mullin and Brooks 1967) and on natural phytoplankton (Parsons et al. 1967, 1969; Parsons and LeBrasseur 1970; Hargrave and Geen 1970). Selective feeding may be based on quality of food but copepods also apparently tend to pick large-sized particles when given a choice (Mullin 1963; Richman and Rogers 1969).

It has not been clearly established whether size-selective feeding in *Calanus*

[1] Contribution No. 661 from the Department of Oceanography, University of Washington. This work was supported by National Science Foundation Grants GA-25385, GB-20182, and GA-31093, and U.S. Atomic Energy Commission Contract AEC AT(45-1)-2225, TA 26 (ref. RLO-2225-T26-6).

89

Reprinted from Limnol. Oceanogr. 17:805–815 (1972).

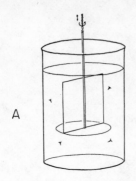

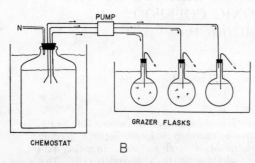

FIG. 1. A. Experimental container used in batch culture feeding experiments with *Calanus*. The Plexiglas stirrer, driven by a 1 rpm motor, rotates and oscillates up and down within the 4-liter beaker. B. Experimental apparatus used for continuous (chemostat) feeding experiments with *Calanus*. The seawater-nutrient mixture enters the 16-liter chemostat at N. Arrows indicate flow of *Thalassiosira fluviatilis* into and out of the 3-liter flasks.

reflects active hunting for larger cells (Richman and Rogers 1969) or is simply due to greater feeding efficiency of the animals on larger cells. Further, the relationship between cell concentration and feeding behavior has not been properly established in selective feeding experiments. In unialgal cultures the ingestion rate of *Calanus* increases with concentration of food up to some maximal rate, then decreases (Mullin 1963) or, more likely, remains constant with further increases in cell concentration (McAllister 1970, 1971). How ingestion curves vary with different sizes of food particles was usually confounded in previous investigations by the

use of algae of markedly different shapes or chemical composition.

In this paper I describe the feeding behavior of adult females of *Calanus pacificus* exposed to batch and continuous cultures of centric diatoms. This is the species described by Brodsky (1948, 1965) and is probably conspecific with the populations called *Calanus helgolandicus* by Mullin (1963), Mullin and Brooks (1967), Richman and Rogers (1969), and Paffenhöfer (1971). The purpose of this work is to demonstrate predictable relationships between feeding behavior of *Calanus* and both concentration of food and size of food particles.

I thank Dr. J. Lewin for providing cultures of diatoms and for advice on maintaining laboratory cultures of phytoplankton. B. Booth identified the diatoms isolated from Puget Sound and J. Vidal gave valuable technical assistance.

MATERIAL AND METHODS

Batch culture feeding experiments

Copepods feed at abnormally high rates during the first few hours of a grazing experiment, because of previous starvation or of handling during transfer (Mullin 1963; Hargrave and Geen 1970; McAllister 1970). Since this abnormal behavior may be evident for up to 12 hr, long-term (2–5 day) grazing experiments were used. Ten to thirty adult females of *Calanus* were placed in each of two 4-liter beakers containing 3.5 liters of twice glass-fiber filtered seawater; both the copepods and seawater were freshly collected at a depth of 50 m or more in Puget Sound (Washington). The contents of beakers were kept homogeneous by modified plunger-jar stirrers (Fig. 1A). The effect of food concentration on feeding behavior was assessed by adding a suspension of algae to the beakers and following the changes in concentration in the beakers as the animals grazed the suspension down. A single control beaker, containing algae but no grazers, was always used. Feeding behavior was monitored at short, irregular intervals (5–14 hr) by pipetting a 100–200-ml

TABLE 1. *Dimensions and estimated carbon content of diatoms used in grazing experiments. The two species of* Thalassiosira *were isolated from Long Island Sound by R. R. L. Guillard. The other three species were obtained from Puget Sound*

	Mean cell diam (μ)	Mean cell vol (μ^3)	Estimated carbon/cell (μg × 10^{-6}/cell)
Thalassiosira pseudonana	3.8	55	8
Thalassiosira fluviatilis	11	1,450	94
Coscinodiscus angstii	35	26,000	840
Coscinodiscus eccentricus	75	63,000	1,644
Centric sp.*	87	160,000	3,334

* Cf. *Coscinodiscus angstii* var. *granulomarginatus*.

sample from each beaker and making 6–8 cell counts with a model B Coulter counter. After counting, the remaining portions of the samples were returned to the beakers. The control was sampled and counted each time the beakers containing grazers were counted. On two occasions, at the end of long grazing experiments, copepods were removed from both grazer beakers and the growth rates of the algae determined after further incubation; in both cases no difference in algal growth rates was found between grazer beakers and the control beaker, indicating that the copepods were not affecting the growth of algae by remineralizing nutrients. All experiments were run at 12.5C in continuous dim light.

The effect of size of food particles on feeding behavior of *Calanus* was studied by using as food centric diatoms which grow as single cells in culture. Five species were used; all are of similar pill-box shape, but differ markedly in size (Table 1). Cultures of diatoms were maintained in medium "f" (Guillard and Ryther 1962). Cultures less than 6 days old, still in logarithmic growth, were used for grazing experiments. Average cell volumes of diatoms were determined with the Coulter counter, which was periodically calibrated with pollen grains of two different sizes. The carbon content of diatom cells was estimated from average cell volumes (Strathmann 1967). Only monospecific cultures of diatoms were fed to *Calanus*; no selection experiments are reported here.

Ingestion rates were calculated from cell counts of the control beaker and sepa-

rately for each beaker with grazers. The growth constant for algal growth, k, was calculated from

$$C_2 = C_1 e^{k(t_2 - t_1)},$$

where C_1 and C_2 are cell concentrations (cells/ml) in the control beaker at t_1 and t_2. For each beaker with grazers the grazing coefficient, g, was calculated from

$$C_2^* = C_1^* e^{(k-g)(t_2-t_1)},$$

where C_1^* and C_2^* are cell concentrations in a beaker with grazers at time t_1 and t_2. Using values of k and g the average cell concentration, $\langle C \rangle$, for each grazer beaker during a time interval $t_2 - t_1$ is:

$$\langle C \rangle = \frac{C_1^*[e^{(k-g)(t_2-t_1)} - 1]}{(t_2 - t_1)(k - g)}.$$

Although *Calanus* obtains its food by means of filtering maxillae, true filtering rate (volume of water passing through the maxillary filter per unit time) cannot be directly measured. The volume swept clear, F (Harvey 1937), is given by

$$F = V g/N \qquad \text{(ml copepod}^{-1} \text{ hr}^{-1}\text{),}$$

where V is the volume (ml) of the beaker and N is the number of copepods in the beaker. "Volume swept clear" is defined as the volume of ambient medium from which cells are completely removed by copepods to achieve the measured ingestion rate. Volume swept clear, as used here, is synonymous with the term "filtering rate" used in many studies of filter-feeding crustaceans and the term "grazing rate" used by Mullin (1963). Notice that filtering rate and volume swept clear are equivalent only if a copepod is 100% efficient at removing particles from the water passing through the maxillary filter. The ingestion rate, I, is then

$$I = \langle C \rangle \times F \quad \text{(cells eaten copepod}^{-1}\text{ hr}^{-1}\text{)}.$$

The effect of cell concentration on volume swept clear and ingestion rate is demonstrated by plotting the rates against the average cell concentration for each period of grazing.

Continuous culture feeding experiments

A continuous culture (chemostat) containing *Thalassiosira fluviatilis* was run with artificial seawater (Kester et al. 1967) enriched with solutions of nutrients, trace metals, and vitamins. The nutrient limiting algal growth was ammonium or silicate. The chemostat in steady state was used for four experiments by connecting separate outflow tubes from the chemostat through a peristaltic pump to 3-liter flasks (Fig. 1B) or 4-liter beakers. The contents of the containers were mixed with stirring bars and by air bubbling. Twenty to forty adult females of *C. pacificus* were used in feeding experiments. Diatom cells were counted periodically in the inflowing and outflowing medium of each copepod container. Both the chemostat and copepod containers were maintained in a water bath at 12.5C in a light–dark cycle (16L: 8D). The chemostat was illuminated directly by a fluorescent light bank (0.099 ly/min); fluorescent light to the grazer containers was first passed through a blue Plexiglas sheet and a 50% light screen, and the resultant flux was 0.019 ly/min.

In the large steady-state chemostat the growth rate of *Thalassiosira* was always kept low (about 0.6% per hour). About the same or a lower algal growth rate prevailed in the grazer containers since the illumination was a fifth as great as in the chemostat. Flow rates and numbers of copepods were adjusted so that the effect of these factors on cell concentration in the grazer containers was much greater than the effect of algal growth.

In only one experiment were the copepod containers run long enough to achieve steady state. For nonsteady-state conditions the growth rate of algae in the grazer flasks was assumed to be the same as the growth rate of algae in the chemostat (turnover rate of algal population in steady-state chemostat = turnover rate of liquid volume in chemostat). Ingestion rates were calculated from an equation that balanced, for each grazer flask, 1) inflow of cells, 2) growth of cells, 3) outflow of cells, and 4) removal of cells by grazers. In the one long steady-state experiment the number of grazers in the two flasks was slightly different (23 and 26 copepods/flask); the growth rate of algae was assumed to be identical in the two flasks containing copepods and ingestion rates were then calculated simply from the difference between cell concentrations in the outflows of the two flasks at steady state. Obviously the algal growth rate should be slightly higher in the flask containing more grazers if all copepods graze at the same rate and if a true steady state is reached; in this experimental design the result of assuming equal algal growth rates is that the ingestion rate of the copepods may be slightly underestimated.

At the end of all experiments the copepods were fixed in filtered seawater with 5% Formalin. After 2–3 weeks they were removed from the preservative, washed briefly with distilled water, dried at 60C on preweighed aluminum pans, and weighed on a Cahn electrobalance.

RESULTS

In three batch culture experiments, I could not measure ingestion rates when *Calanus* was fed *Thalassiosira pseudonana*, even though a few fecal pellets were produced. The average spacing of filtering setules on the seivelike second maxilla of adult *Calanus* is significantly greater than the cell diameter of the diatom, so that most cells probably pass between the setules.

Calanus displays a predictable feeding behavior when fed in batch cultures at

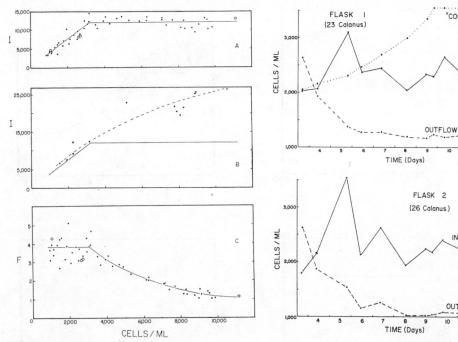

FIG. 2. Results of batch culture (dots) and continuous culture (circles) experiments with *Calanus* feeding on *Thalassiosira fluviatilis*. A. Effect of cell concentration on ingestion rate, *I* (cells eaten copepod^{-1} hr^{-1}). B. Ingestion rates of starved (dots and dashed line) and unstarved (solid line from part A) *Calanus* in batch cultures; starvation rates were those measured in the first 4–12 hr of a feeding experiment. C. Effect of cell concentration on volume swept clear, *F* (ml copepod^{-1} hr^{-1}).

FIG. 3. Results of experiments in which females of *Calanus* fed on *Thalassiosira fluviatilis* from a continuous culture (*see* Fig. 1B). Inflow: concentration of cells in medium flowing into grazer flask from the continuous culture (flow rate 96 ml/hr). Outflow: concentration of cells in medium flowing out of grazer flask. "Control" represents the concentration of cells in a flask containing no grazers. Average concentrations in the inflows to flask 1 (2,326 cells/ml) and flask 2 (2,288 cells/ml) are not statistically different.

different concentrations of *T. fluviatilis* (Fig. 2A). The ingestion rate increases with cell concentration up to a maximal rate that remains essentially unchanged with further increase in cell concentration. The cell concentration at which the maximal ingestion rate is first achieved will be referred to below as the critical concentration (McMahon and Rigler 1963). Figure 2A includes rates determined for unstarved copepods. An unstarved *Calanus* adjusts its feeding rate at high concentrations of cells (>4,000 cells/ml), since starved animals can ingest at unusually high rates (Fig. 2B). The adjustment in feeding behavior is also evident from the pattern of volume swept clear (*F*). Going from high

to low concentrations of *T. fluviatilis*, *F* increases curvilinearly to an average maximal rate about which it oscillates with considerable amplitude (Fig. 2C). While true filtering rate of *Calanus* cannot be measured, volume swept clear is useful for comparative purposes since it seems to represent, when it is measured at low food concentrations, the physiologically maximal rate at which a copepod can process a particular type of food. Thus, although at high cell concentrations (>4,000 cells /ml) a starved *Calanus* ingests at a higher rate than an unstarved *Calanus*, at cell densities below the critical concentration starved animals feed like unstarved animals (Fig. 2B).

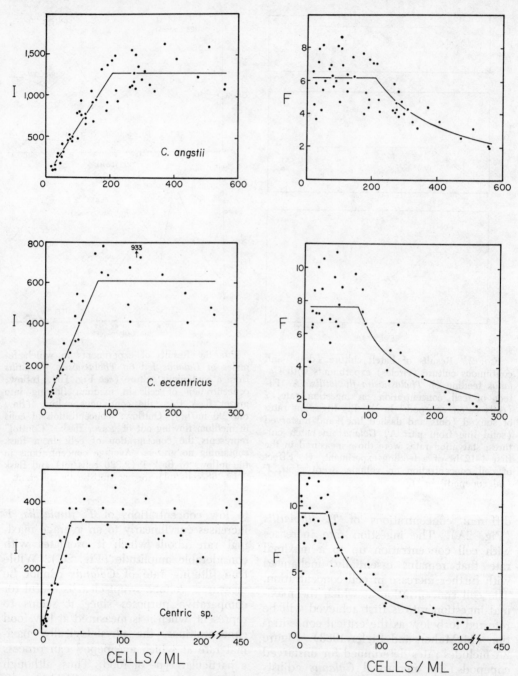

FIG. 4. Effect of cell concentration on ingestion rate, I, and volume swept clear, F, of adult females of *Calanus* feeding on *Coscinodiscus angstii* (top graphs), *Coscinodiscus eccentricus* (center graphs) and centric sp. (lower graphs).

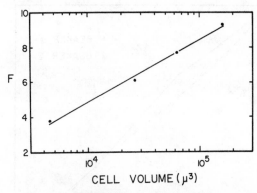

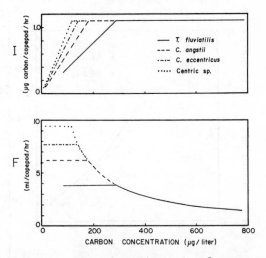

FIG. 5. Relationship between volume swept clear, *F*, for adult females of *Calanus* and mean cell volume of diatoms used as food. Values of *F* are means based on rates measured at cell densities below the critical concentration for each species of diatom (Figs. 2 and 4). *F* is predicted by the least-squares regression line $F = 2.61 (\log V) - 4.84$, where *V* is the cell volume (μ^3) of the centric diatom used as food. The correlation coefficient between log V and *F* is 0.79 ($N = 95$).

FIG. 6. Effect of size (species) and concentration (as carbon) of food particles on ingestion rate, *I*, and volume swept clear, *F*, of adult females of *Calanus*.

The curved part of the line in Fig. 2C represents the volume that must be swept clear by a *Calanus* female to obtain the maximal ingestion rate (12,066 cells copepod⁻¹ hr⁻¹ based on 26 points for cell concentrations exceeding 3,500 cells/ml). The straight part of the line in Fig. 2C represents the average volume swept clear (3.81 ml copepod⁻¹ hr⁻¹) at cell concentrations below 3,500 cells/ml. Thus I represent the relationship between ingestion rate of *Calanus* and cell concentration of *T. fluviatilis* as two intersecting straight lines (Fig. 2A). Note that the slope of the ascending part of the line in Fig. 2A is equivalent to the average maximal volume swept clear. A curvilinear function could be fitted to Fig. 2A (e.g. Parsons et al. 1967) but there is no a priori reason for doing so and my representation seems to me to provide clearer insight into the feeding behavior of *Calanus*.

The above results describe how adult females of *Calanus* might feed on a patch of *T. fluviatilis*. But the batch culture system is somewhat unreal in the sense that it is a completely closed system; the algal population often ceased growing in both the control and grazing beakers before the end of a long feeding experiment. The results of feeding experiments in continuous cultures of *T. fluviatilis* are therefore of interest because cell concentrations remain high and the cells are always in a logarithmic phase of growth. Ten days of results from a long steady-state experiment (Fig. 3) show that *Calanus* fed continuously and at a constant rate from day 8 onward. Ingestion rate (4,610 cells copepod⁻¹ hr⁻¹) and volume swept clear (4.3 ml copepod⁻¹ hr⁻¹) were calculated from the averaged data for days 8 to 13. These rates and those of three other experiments agree well with the results of batch culture experiments (Fig. 2, A and C).

When larger diatoms are fed to *Calanus* (Fig. 4) the effect of cell concentration on ingestion rates and volumes swept clear parallels that described above, with one important exception. Average maximal volumes swept clear increase with the size of cell, evidently in a predictable way (Fig. 5). The effects of cell concentration and size of cell on feeding behavior may be more clearly seen by replotting the data after converting cell concentrations and ingestion rates to carbon equivalents (Fig. 6). Above the critical concentrations of

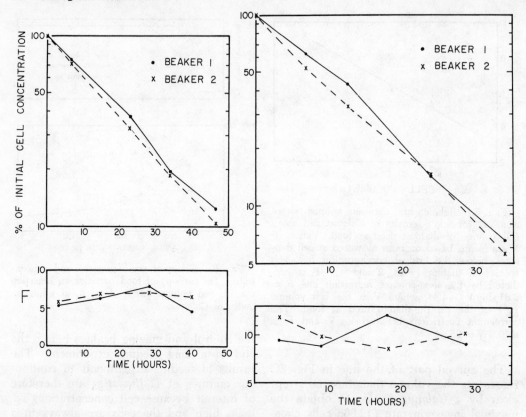

Fig. 7. Changes in cell concentration and volume swept clear, *F*, with time during experiments in which the initial cell concentration was below the critical concentration. Left graphs: *Calanus* feeding on *Coscinodiscus angstii* (initial cell concentrations: beaker 1—140 cells/ml; beaker 2—143 cells/ml). Right graphs: *Calanus* feeding on centric sp. (initial cell concentrations: beaker 1—47 cells/ml; beaker 2—32 cells/ml). Each beaker contained 30 *Calanus*.

plant carbon for all species of diatom, the copepods ingest and sweep clear water at the same rate regardless of the species of diatom used for food (Table 2). From this I conclude that *Calanus* does not distin-

TABLE 2. *Mean ingestion rate (A) (µg C cope-pod⁻¹ hr⁻¹) of adult females of* Calanus pacificus *feeding on diatoms of different sizes. Means are based only on rates measured at cell densities above the critical concentration for each species of diatom.* SE = *standard error;* N = *number of observations*

	A	SE	N
T. fluviatilis	1.13	0.031	26
C. angstii	1.07	0.038	18
C. eccentricus	1.01	0.072	14
Centric sp.	1.13	0.051	23

guish between these four species of diatom with respect to food quality, at least over the 3–5 day term of an experiment. Thus, the progressively higher carbon ingestion rates measured just at critical concentrations of larger cells (Fig. 6) most likely reflects differences in the efficiency with which *Calanus* handles and eats cells of different sizes. Clearly, females of *Calanus* can obtain their maximal daily ration at relatively low concentrations when feeding on the larger centric diatoms.

I suggest that below the critical cell densities of a particular diatom, *Calanus* ingests at a rate directly proportional to the amount of food available. This implies that volume swept clear is constant at low food concentrations. Feeding experiments initiated at cell densities just below critical food concentrations and run for 1.5–2 days demonstrate that this is essentially true. Cell concentrations decrease exponentially and volumes swept clear are independent of cell concentrations (Fig. 7).

Mean dry weight of the copepods ranged from 130–210 μg with no apparent systematic variation between experiments. If an average dry weight per copepod of 170 μg is used and the bodily carbon is taken as 40% of dry weight (Mullin 1969), then unstarved females of *Calanus* ingest daily an amount of carbon equivalent to about 39% of their bodily carbon when feeding at food concentrations above the critical concentration (Table 2).

DISCUSSION

When feeding on monospecific cultures of centric diatoms which grow as single cells, a female of *C. pacificus* displays an ingestion rate which is directly dependent on the concentration and size of cell at low food concentrations, but constant and apparently independent of concentration, size and type of cell at high food concentrations. A model of this behavior is simple: it assumes no preference for food quality (i.e. species of centric diatom), a maximal daily ration (~39% of the body weight at 12.5C) at high food concentrations, a critical food concentration (measured as carbon) that decreases with increasing cell volume, and an ingestion rate that (below the critical concentration) is linearly dependent on food concentration. Thus in Fig. 6 only four of a large family of ingestion lines are shown; lines for centric diatoms of other sizes could be found from Fig. 5 which gives the slope of the ingestion line for cell densities below the critical concentration. For now, I prefer this model because it is the simplest fit to the data in Figs. 2 and 4.

My model is not of the usual curvilinear form describing the feeding rate of predators in response to food concentration; however, Holling (1965) has pointed out that a rectilinear type of response curve may be typical of filter-feeding crusta-ceans (e.g. McMahon 1965). The response curve is characteristic of an animal which searches at random for prey and has a searching rate which is not affected by density of prey organisms. This applies to filter-feeding copepods, since food particles—diatoms at least—can be ingested only after falling from a feeding current onto the seivelike second maxillae (Conover 1966). Ingestion rate for such a feeder increases in direct proportion to increase in concentration of food up to a saturation point above which ingestion may be determined by the passage rate of food through the alimentary canal. Other mathematical expressions for feeding behavior of copepods (e.g. Parsons et al. 1967; Cushing 1968) were tried, but do not fit my data particularly well.

Below critical food concentrations volumes swept clear by *Calanus* increase as the size of food particles increases. For the range of cell volumes used in this study the relationship can be considered monotonic and linear when plotted as in Fig. 5. From this I conclude that *Calanus* probably feeds as a filterer on both large and small cells but handles and eats larger cells with greater effectiveness than it handles and eats small cells. The actual mechanism behind this phenomenon is not clear. Obviously there must be a maximal rate at which *Calanus* can filter water. For benthic filter-feeding molluscs, the maximal filtering rate is usually found for animals exposed to particle-free seawater or very dilute suspensions of food (e.g. Loosanoff and Engle 1947; Davids 1964). Cushing (1968) suggested that the maximal filtering rate of *Calanus* depends on swimming speed; however, a consequence of this is that he predicts the same volume swept clear by *Calanus* at low concentrations of different-sized cells, a feature not actually found in feeding experiments. Since calculation of volume swept clear is based on the number of cells eaten, it is possible that a few large cells might be manipulated and passed to the mouth more quickly than many small cells of equivalent total carbon content. If this is so, then *Calanus* could be filtering at the same rate on cells of all sizes, and the measure of volume swept clear could be indicative of how effectively *Calanus* handles and eats different-sized cells. Direct observations, designed after those of McMahon and Rigler (1963) and Burns (1968) on *Daphnia*, are needed.

Calanus may feed preferentially on larger cells when presented with a spectrum of particle sizes (*see* Richman and Rogers 1969). My results do not deny this possibility. However, consideration of the efficiency with which *Calanus* handles and eats cells of different sizes might shed light on some of the results of selective feeding experiments. Further, selective feeding experiments must be carefully designed with consideration of possible unwanted effects of cell concentration on feeding behavior. Adult females of *C. pacificus* feeding on a mixture of *T. fluviatilis* and *Coscinodiscus angstii* (each at a low concentration of carbon, i.e. <100 μg C/liter) ought to display a much higher ingestion rate, in terms of carbon, on the larger *Coscinodiscus*. When both diatoms are present at densities greater than the critical concentration for each, no selection by *Calanus* should be apparent. Richman and Rogers (1969) found that *Calanus* fed at significantly higher rates on paired cells than on single cells of the diatom *Ditylum brightwellii*. Inexplicably, the relationship in Fig. 5 qualitatively predicts this result but greatly overestimates the volume swept clear by *Calanus* when feeding on cultures of *Ditylum* dominated by single cells. This could be due to the difference in shape between a *Ditylum* cell and the cells used in my experiments. Alternatively, my study, utilizing unialgal cultures as food for *Calanus*, may possibly have missed other patterns of feeding behavior that emerge only when *Calanus* is exposed to heterogeneous mixtures of particle sizes or food species.

REFERENCES

BRODSKY, K. A. 1948. Free-living Copepoda of the Sea of Japan [transl. from Russian]. Izv. Tikhookean. Nauch.-Issled. Inst. Rybn. Khoz. i Okeanogr. **26**: 3–130.

———. 1965. Variability and systematics of the species of the genus *Calanus* (Copepoda) 1 [transl. from Russian]. Akad. Nauk SSSR Zool. Inst., Issled. Fauni Morei 3(11): 22–71.

BURNS, C. W. 1968. Direct observations of mechanisms regulating feeding behavior of *Daphnia* in lakewater. Int. Rev. Gesamten Hydrobiol. **53**: 83–100.

———. 1969. Particle size and sedimentation in the feeding behavior of two species of *Daphnia*. Limnol. Oceanogr. **14**: 392–402.

CONOVER, R. J. 1966. Feeding on large particles by *Calanus hyperboreus* (Kröyer), p. 187–194. *In* H. Barnes [ed.], Some contemporary studies in marine science. Allen and Unwin.

CUSHING, D. H. 1968. Grazing by herbivorous copepods in the sea. J. Cons., Cons. Perm. Int. Explor. Mer **32**: 70–82.

DAVIDS, C. 1964. The influence of suspensions of microorganisms of different concentrations on the pumping and retention of food by the mussel (*Mytilus edulis* L.). Neth. J. Sea Res. **2**: 233–249.

GUILLARD, R. R. L., AND J. H. RYTHER. 1962. Studies on marine planktonic diatoms 1. Can. J. Microbiol. **8**: 229–239.

HARGRAVE, B. T., AND G. H. GEEN. 1970. Effects of copepod grazing on two natural phytoplankton populations. J. Fish. Res. Bd. Can. **27**: 1395–1403.

HARVEY, W. H. 1937. Note on selective feeding by *Calanus*. J. Mar. Biol. Ass. U.K. **22**: 97–100.

HOLLING, C. S. 1965. The functional response of predators to prey density and its role in mimicry and population regulation. Mem. Entomol. Soc. Can. 45. 60 p.

KESTER, D. R., I. W. DUEDALL, D. N. CONNORS, AND R. M. PYTKOWICZ. 1967. Preparation of artificial seawater. Limnol. Oceanogr. **12**: 176–178.

LOOSANOFF, V. L., AND J. B. ENGLE. 1947. Effect of different concentrations of microorganisms on the feeding of oysters (*O. virginica*). Fish. Bull. **51**: 29–57.

MCALLISTER, C. D. 1970. Zooplankton rations, phytoplankton mortality and the estimation of marine production, p. 419–457. *In* J. H. Steele [ed.], Marine food chains. Univ. Calif.

———. 1971. Some aspects of nocturnal and continuous grazing by planktonic herbivores in relation to production studies. Fish. Res. Bd. Can. Tech. Rep. 248.

MCMAHON, J. W. 1965. Some physical factors influencing the feeding behavior of *Daphnia magna* Straus. Can. J. Zool. **43**: 603–611.

———, AND F. H. RIGLER. 1963. Mechanisms regulating the feeding rate of *Daphnia magna* Straus. Can. J. Zool. **41**: 321–332.

———, AND ———. 1965. Feeding rate of

Daphnia magna Straus in different foods labeled with radioactive phosphorus. Limnol. Oceanogr. **10**: 105–114.

MARSHALL, S. M., AND A. P. ORR. 1955. The biology of a marine copepod. Oliver and Boyd. 188 p.

MULLIN, M. M. 1963. Some factors affecting the feeding of marine copepods of the genus *Calanus*. Limnol. Oceanogr. **8**: 239–250.

———. 1966. Selective feeding by calanoid copepods from the Indian Ocean, p. 545–554. *In* H. Barnes [ed.], Some contemporary studies in marine science. Allen and Unwin.

———. 1969. Production of zooplankton in the ocean: the present status and problems. Oceanogr. Mar. Biol. Annu. Rev. **7**: 293–314.

———, AND E. R. BROOKS. 1967. Laboratory culture, growth rate, and feeding behavior of a planktonic marine copepod. Limnol. Oceanogr. **12**: 657–666.

PAFFENHÖFER, G. A. 1971. Grazing and ingestion rates of nauplii, copepodids and adults of the marine planktonic copepod *Calanus helgolandicus*. Mar. Biol. **11**: 286–298.

PARSONS, T. R., AND R. J. LeBRASSEUR. 1970. The availability of food to different trophic levels in the marine food chain, p. 325–343.

In J. H. Steele [ed.], Marine food chains. Univ. Calif.

———, ———, AND J. D. FULTON. 1967. Some observations on the dependence of zooplankton grazing on the cell size and concentration of phytoplankton blooms. J. Oceanogr. Soc. Jap. **23**: 10–17.

———, ———, AND O. D. KENNEDY. 1969. Production studies in the Strait of Georgia. Part 2. J. Exp. Mar. Biol. Ecol. **3**: 39–50.

PETIPA, T. S. 1965. The food selectivity of *Calanus helgolandicus* (Claus), p. 102–110. *In* Plankton investigations in the Black and Azov Seas [transl. from Russian]. Akad. Nauk Ukr. SSR.

REEVE, M. R. 1963. The filter-feeding of *Artemia*. 1. J. Exp. Biol. **40**: 195–221.

RICHMAN, S., AND J. N. ROGERS. 1969. The feeding of *Calanus helgolandicus* on synchronously growing populations of the marine diatom *Ditylum brightwellii*. Limnol. Oceanogr. **14**: 701–709.

STRATHMANN, R. R. 1967. Estimating the organic carbon content of phytoplankton from cell volume or plasma volume. Limnol. Oceanogr. **12**: 411–418.

GRAZING OF *PSEUDOCALANUS MINUTUS* ON NATURALLY OCCURRING PARTICULATE MATTER

S. A. *Poulet*

Fisheries Research Board of Canada, Marine Ecology Laboratory,
Bedford Institute of Oceanography, Dartmouth, Nova Scotia

ABSTRACT

The quantity and size of particulate matter consumed by *Pseudocalanus minutus* were studied in seawater samples collected from different depths and from closely spaced stations. The heterogeneity in particle distribution resulted from quantitative and qualitative fluctuations in the particle spectrum, although at times the total concentration was about the same. *Pseudocalanus minutus* consumed particles between 4 and 100 μ. An electivity index value was more often positive for 25.4–57.0-μ particles. On the average, particles <39 μ were more readily eaten than larger particles. The consumption by copepods at different locations was related not only to particle concentration but also to the pattern of the particle size spectrum. *Pseudocalanus* was able to shift its grazing pressure from small to large particles to compensate for a reduction in density of small particles.

Copepods feeding under natural conditions have to deal with a heterogeneous distribution of particles whose size and concentration change in space and time. In this dynamic system copepods will have to adapt to changes in quality and quantity of food. They generally show considerable discrimination in selecting food when a choice is available. This behavior has been described for copepods feeding on both mixed cultures of phytoplankton (Mullin 1963; Mullin et al. 1966; Petipa 1965; Mullin and Brooks 1967) and natural phytoplankton (Parsons et al. 1967; Hargrave and Geen 1970). It has also been shown that copepods tend to graze on large size particles when a choice is given (Mullin 1963; Richman and Rogers 1969).

In nature small particles (5–15-μ diam) account for a major portion of the total living material (Riley 1957; Sutcliffe 1972). The possible importance of these small organisms to selective feeding in copepods has not been established. It is not clear if the selectivity for larger particles is due to their being captured with greater efficiency or if it is caused by a decrease in the relative concentration of smaller particles in the environment so that it becomes more profitable for the copepods to catch a few larger particles, balancing feeding efficiency and "hunting" efficiency.

I report here observations on the food consumption and selectivity of a common copepod, *Pseudocalanus minutus*, fed on naturally occurring particles.

I am grateful to R. J. Conover, R. W. Sheldon, and M. Paranjape for their criticism and their assistance in the preparation of this manuscript. I would like to thank D. Porteous for the computing programs.

METHODS

Bedford Basin, Nova Scotia (Fig. 1), is a small basin 70 m deep separated from the ocean by a channel 10 km long and 20 m deep. The water in the basin is more often turbid than that of the extremity of the channel, because of freshwater discharge and untreated sewage effluent from the cities of Halifax and Dartmouth; it represents a highly enriched environment.

The water samples were collected with a 30-liter Niskin bottle at sta. 2 (0, 5, 15, 25, and 60 m) and at 5 m at sta. 1, 2, 3, 4, and 5 (Fig. 1). The copepods were collected by an oblique tow with a 0.75-m-diameter No. 6 net (239-μ mesh size). On each station, plankton tows were made from 20 m to surface. For each grazing experiment 50 adult female *P. minutus* were picked out and added to a liter beaker containing seawater screened through a 160-μ sieve to remove larger zooplankton.

100

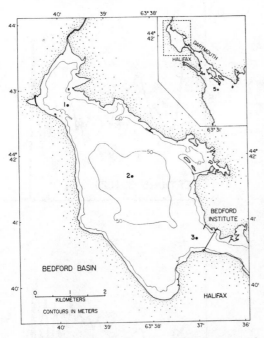

Fig. 1. Sampling stations in Bedford Basin and vicinity.

One beaker was prepared with water collected from each of the five selected stations or depths. Duplicate containers filled with screened seawater, without copepods, served as controls. The containers were placed on rotating trays and stirred gently with a fixed plastic spoon which served as a baffle. Dye studies and occasional spot checks of particle distribution failed to show any difference between or within the beakers. The experiments were carried out in the dark at a temperature of 4° C, close to the water temperature of the basin, for a period of 19 to 20 hr. The experiments were conducted three times during April and May 1971. Station 2 was visited three separate times from 20 April–6 May, and the five stations on the transect were sampled on three separate occasions, from 11–25 May all on a weekly basis.

At the end of the experiments the copepods were removed and the water samples were counted with a model T Coulter counter to produce a size frequency particle spectrum and to determine particle concentration. Aperture tubes of 50 and 280 μ were used to measure particle size distributions from 1.58–114-μ mean spherical diameter (Sheldon and Parsons 1967). The particle spectrum was divided into 40 size categories each with mean volume 2.5 times larger than the preceding (Sheldon 1969). The total particle spectrum was then split into five groups of eight particle sizes each: group 1—1.58 to 3.57 μ; group 2—4.0 to 8.97 μ; group 3—10.1 to 22.6 μ; group 4—25.4 to 57.0 μ; and group 5—64.0 to 144 μ. These size groups were chosen so that selectivity by the copepods could be studied. The size range 114–144 μ was included to help computer processing of the results. The number of particles in this size range was not measured but arbitrarily set at zero.

The quantity of particulate matter consumed by the copepods for the total spectrum and for each size group was calculated from the difference between the particle concentration in the controls and in the experimental containers. Then the electivity index E was calculated following Ivlev's (1955) equation:

$$E = (ri - pi)/(ri + pi),$$

where ri is the percentage of any particle size group in the ration, and pi the proportion of the same particle group in the water, expressed in terms of concentration. The whole index range lies between +1 and –1. In the case of a preference shown towards a particle group by the copepods the index E has a positive sign, and if there is a marked dislike it has a negative one.

RESULTS AND DISCUSSION

The vertical distribution of particles at sta. 2 is shown in Fig. 2 (control). The total particle concentration showed a considerable decrease from 0–60 m (4.82 ppm at surface to 0.82 at 60 m). The particle spectrum at 0 m had a bimodal distribution with a peak at small size range (5–10 μ) reaching to 1.2 ppm and a secondary peak at large size range (100 μ) reaching to 0.5 ppm. The same bimodal distributions were observed at 5 and 15 m, although the con-

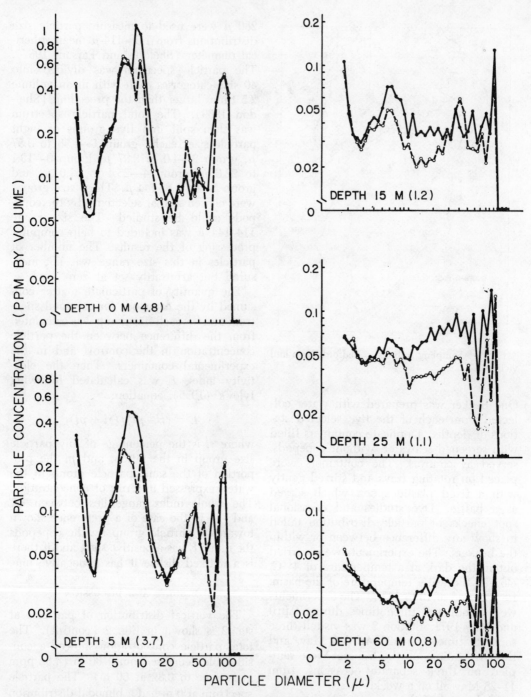

Fig. 2. Particle size spectra in the controls (●————●) and after grazing (○-----○) by *Pseudocalanus minutus* on water samples from 5 depths at sta. 2. Total particle concentrations (ppm) in the controls are in parentheses. Date of experiments: 27 April 1971.

centrations were lower. At 25- and 60-m depth the small particle concentration was less than 0.07 ppm but no such decrease in large particle concentration was noticed, and at these depths the spectrum remained mostly unchanged.

In the horizontal distribution (Fig. 3, control) total particle concentration at 5 m fluctuated between 2.3 and 2.8 ppm except at sta. 5 where it was 1.4. The particle spectra along these stations had the same basic pattern of distribution. The small particle peaks were higher inside than outside the basin, with maximum values reaching 0.4–0.5 ppm from sta. 1 to 3. The concentration of large particles also increased from sta. 1 to 4 (from 0.08–0.15 ppm). At sta. 5 concentrations of small and large particles were 0.150 and 0.09 ppm. It is obvious from the changes in total particle concentrations that the same factors do not affect the vertical distribution as the horizontal distribution. Using size spectrum to describe particles gives the best picture of both concentration and size distribution. It also indicates changes in either total concentration or in any component of the size spectrum. Modifications in concentration can easily be seen and result from either simultaneous or independent changes of these components.

The microscopic examination of freshly taken water samples from Bedford Basin showed that the peak in the small size range ($<10 \mu$) was dominated by a mixture of unidentified flagellates, while chain-forming diatoms such as *Thalassiosira* sp., *Chaetoceros* sp., and dinoflagellates *Ceratium* sp., and *Gymnodinium* sp. were abundant in larger size ranges.

The grazing behavior of copepods is also shown in Figs. 2 and 3. The copepods were able to consume particles from 1.58–114-μ diameter, but it was difficult to find any grazing within the 1.58–4-μ range. In certain size categories there was an increase in concentration of particles measured. This phenomenon frequently occurs in such experiments. It probably represents fecal materials, eggs, or aggregates at the large end of the spectrum, or breakage of large particles into smaller, as a result of copepod feeding, at the small end.

In these experiments particle size and relative concentration seem to play different roles in grazing. When the concentration of particles < 15-μ diameter was high (depth 0–15 m: Figs. 2 and 3), the main consumption by copepods took place in this range. But, when the concentration of this particle size range was lower, larger particles were consumed, as indicated at 15-, 25-, and 60-m depth as well as at sta. 3, 4, and 5. In nature, copepods swimming up and down in the water column may actually find such situations. It would be important for them to be able to catch particles from either large or small size ranges to obtain the amount of food they need. The tendency of the copepods to shift their consumption from one size range to another according to relative particle concentration seems to be an adaptation to maintain their food uptake.

The average number of particles consumed by *P. minutus* per hour in these experiments was 1,056 in group 2, 123 in group 3, 5.9 in group 4, and only 0.33 in group 5. On the assumption that the particles are phytoplankton, the carbon content of these size groups was calculated using the equation of Mullin et al. (1966). The corresponding carbon uptake becomes 24.4 $\times 10^3$ from group 2, 23.7 $\times 10^3$ from group

Fig. 3. Particle size spectra in the controls (●———●) and after grazing (○-----○) by *Pseudocalanus minutus* on water samples from five closely spaced stations. Total particle concentrations (ppm) in the controls are in parentheses. Date of experiments: 11 May 1971.

Fig. 4. Particle concentration (open bars), copepod grazing (closed bars), and electivity index *E*, calculated for each size group at different depths. Total particle concentrations (ppm) in the controls are in parentheses. Date of experiments: 20 April 1971 (11–15); 27 April 1971 (6–10); 6 May 1971 (1–5).

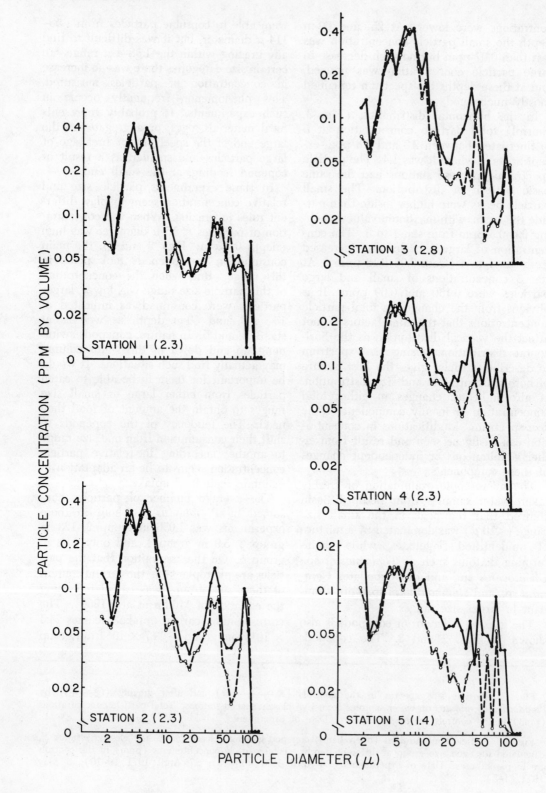

STATION 1 (2.3)

STATION 2 (2.3)

STATION 3 (2.8)

STATION 4 (2.3)

STATION 5 (1.4)

PARTICLE CONCENTRATION (PPM BY VOLUME)

PARTICLE DIAMETER (μ)

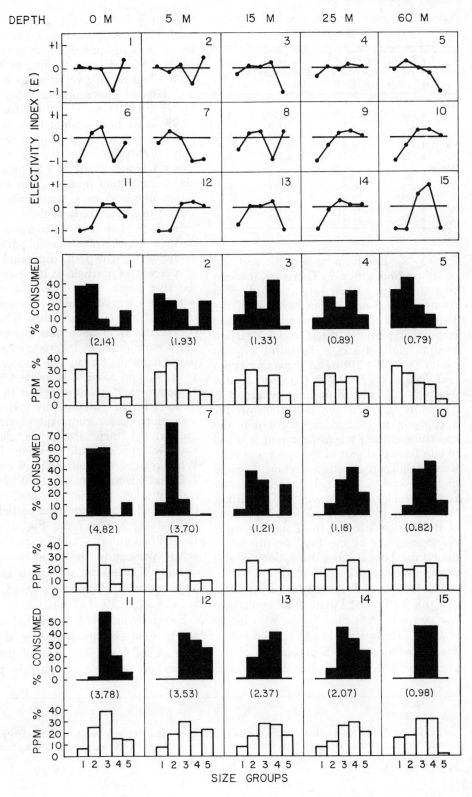

DEPTH   0 M   5 M   15 M   25 M   60 M

ELECTIVITY INDEX (E)

% CONSUMED

PPM %

SIZE GROUPS

Table 1. Summary of Pseudocalanus minutus grazing experiments

	Food density ($\mu^3 \times 10^6$)ml^{-1}	Ration ℓ^{-1} animal^{-1}day^{-1} ($\mu^3 \times 10^6$)	Ration animal^{-1}day^{-1} (mg C$\times 10^{-3}$)	Ration % body wt
Depth				
0	4.822	15.14	0.787	11.7
5	3.704	21.91	1.139	17.0
15	1.218	6.28	0.327	4.8
25	1.183	9.33	0.485	7.2
60	0.823	3.76	0.196	2.9
Sta.No.				
1	2.341	7.17	0.373	5.5
2	2.325	8.54	0.444	6.6
3	2.829	14.32	0.745	11.1
4	2.389	15.40	0.800	11.9
5	1.483	11.73	0.610	9.0

3, 9.3×10^3 from group 4, and 4.3×10^3 pg from group 5. Therefore, it seems that the main carbon ration would come from groups 2 and 3; but because of the large amount of carbon in large cells, taking large size particles could make copepod feeding more efficient. To fulfill a given ration (i.e. 23×10^3 pg of carbon) a copepod has to ingest 1,056 particles of size group 2, but only 15 particles of group 4. Therefore, it would be advantageous for a copepod to change the portion of the spectrum utilized when the ration it could obtain from that part of the spectrum, relative to the particle concentration, became more favorable. Such behavior has the further ecological advantage of shifting the grazing pressure to different portions of the particle size spectrum, making more economical use of the food supply while permitting heavily grazed locations of the spectrum to recover.

The feeding of *P. minutus* is summarized in Table 1. The total particle concentration was converted to carbon by use of a factor of 0.052 (Parsons et al. 1967). The ration has been expressed as a percentage of the average carbon content of an animal using the figure 0.0067 mg C animal^{-1} (Parsons et al. 1967). The maximum ration obtained by *P. minutus* per day was much higher when the animal was feeding at 0 and 5 m than when feeding on samples from 15 m and below. The ration per day obtained by *Pseudocalanus* increased from 5.5 to 11.9% of its body weight from sta. 1 to 3, although the increase in food density was only 0.52 ppm over the same stations. There was a decrease of 1.35 ppm in particle concentration from sta. 3 to 5 while the ration obtained as a percentage of body weight decreased only slightly.

In Figs. 4 and 5, histograms showing particle concentration and parallel consumption by copepods, expressed in terms of percent of particle volume in each of the five major size groups, are compared with the electivity index for each part of the two series of experiments. The particle concentration in each size category indicates considerable heterogeneity of the particulate matter in the water column (Fig. 4) and along the transect (Fig. 5). Nonetheless, *P. minutus* seems to adapt to these different situations by taking advantage of the most abundant particles, irrespective of their size, thus showing a strongly opportunistic feeding behavior. When relative concentrations of size groups 1, 2, and 3 were more than 50% of the total concentration, grazing occurred mainly in these small or medium sized particle ranges (Fig. 4: 1, 2, 5, 6, 7, 11; Fig. 5: 1, 7, 11, 12). When groups 4 and 5 increase slightly or are present in the same concentration as other size groups, grazing is greater on large particles (Fig. 4: 9, 10, 14, 15; Fig. 5: 4, 5, 8, 9, 10, 14, 15).

Electivity indices E calculated for different particle size groups show similar results (Figs. 4 and 5). Most of the time E values are negative for particle groups 1

→

Fig. 5. Particle concentration (open bars), copepod grazing (closed bars), and electivity index E, calculated for each size group at 5-m depth on the five stations. Total particle concentrations (ppm) in the controls are in parentheses. Date of experiments: 11 May 1971 (11–15); 18 May 1971 (6–10); 25 May 1971 (1–5).

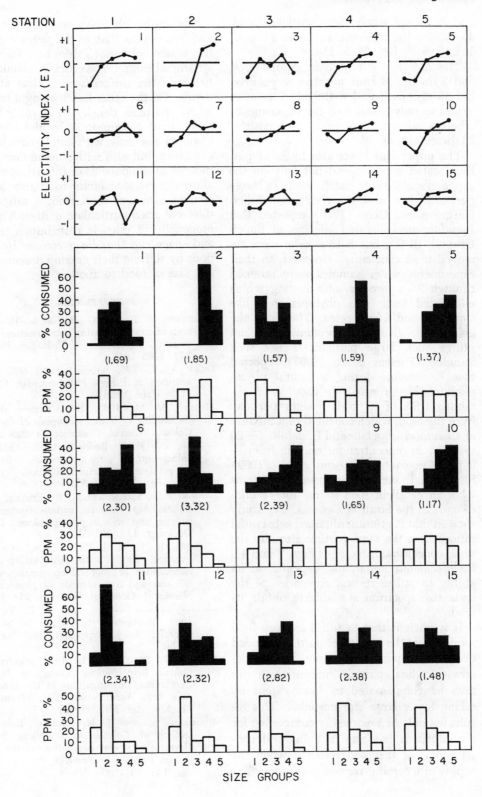

STATION

and 2, except when these particles reach 40–60% of the total concentration (Fig. 4: 1, 2, 5, 6, 7; Fig. 5: 3, 11).

The *E* values for particle groups 3, 4, and 5 fluctuated from negative to positive, but in groups 3 and 5 they had positive values in only about half the experiments, whereas values for size group 4 are positive 23 times out of 30.

The upper and lower size limits of particles eaten must depend not only on the copepod's ability to catch small or large particles, but also on their availability. Hargrave and Geen (1970) reported that *Pseudocalanus* ingested all sizes of flagellates (1–15 μ), but 5–15-μ cells were removed most efficiently. However, in their experiments water samples were strained through 35-μ-pore-diameter netting which eliminated large size phytoplankton like *Ceratium* and *Chaetoceros*. Therefore, the assumption that *Pseudocalanus* did not utilize such large particles is not well founded. Parsons et al. (1967) reported that *P. minutus* during a natural phytoplankton bloom was only capable of obtaining 0.8% of its body weight per day from the crop, which consisted of a mixture of *Chaetoceros socialis* and *C. debilis* ($\sim$ 20 to 40 μ) in concentrations ranging from 1–9 ppm; in contrast Parsons et al. (1969) found that such chain-forming diatoms were an excellent food source for *Pseudocalanus* in the Strait of Georgia. My study showed that *P. minutus* obtained substantial rations from the same particle size and the same concentration ranges (Figs. 2 and 4). *Pseudocalanus* apparently depends on its ability to utilize several categories of the wide size spectrum available to obtain its food.

It is obvious that to fulfill a given food requirement the number of particles needed decreases as particle size increases. A decrease in total particle concentration can thus be compensated to some extent by taking fewer large size particles. This results not only in a more efficient feeding for the copepods but also in a "natural conservation" of the food resource when the supply of a certain size becomes reduced.

The ration obtained was greater at 0- and 5-m depths than in the deepest part of the water column (Table 1). Therefore, an animal migrating vertically should feed better in the surface waters than at 60-m depth. On the other hand, along a transect where particle density was roughly the same, food consumption by the copepods changed not only with total particle concentration but also with fluctuations in the individual components of the size spectrum, allowing *Pseudocalanus* to derive a substantial ration of food from a mixture of different sized particulate matter. The heterogeneity of particle distribution in time and space can thus be overcome by copepods by shifting their grazing pressure from one size of food to another.

REFERENCES

HARGRAVE, B. T., AND G. H. GEEN. 1970. Effects of copepod grazing on two natural phytoplankton populations. J. Fish. Res. Bd. Can. **27**: 1395–1403.

IVLEV, V. S. 1955. Experimental ecology and nutrition of fishes. Pishchemizdat, Moscow. [Yale Univ. 1961.]

MULLIN, M. M. 1963. Some factors affecting the feeding of marine copepods of the genus *Calanus*. Limnol. Oceanogr. **8**: 239–250.

———, AND E. R. BROOKS. 1967. Laboratory culture, growth rate, and feeding behavior of a planktonic marine copepod. Limnol. Oceanogr. **12**: 657–666.

———, P. R. SLOAN, AND R. W. EPPLEY. 1966. Relationship between carbon content, cell volume, and area in phytoplankton. Limnol. Oceanogr. **11**: 307–311.

PARSONS, R. T., R. J. LeBRASSEUR, AND J. D. FULTON. 1967. Some observations on the dependence of zooplankton grazing on the cell size and concentration of phytoplankton blooms. J. Oceanogr. Soc. Jap. **23**: 10–17.

———, ———, ———, AND O. D. KENNEDY. 1969. Production studies in the Strait of Georgia. 2. J. Exp. Mar. Biol. Ecol. **3**: 39–50.

PETIPA, T. S. 1965. The food selectivity of *Calanus helgolandicus* (Claus), p. 102–110. *In* Plankton investigations in the Black and Azov Seas. Akad. Nauk SSSR. [Transl. N.S. 72. Min. Agr. Fish. Food.]

RICHMAN, S., AND J. N. ROGERS. 1969. The feeding of *Calanus helgolandicus* on synchronously growing populations of the marine diatom *Ditylum brightwellii*. Limnol. Oceanogr. **14**: 701–709.

RILEY, G. A. 1957. Phytoplankton of the north central Sargasso Sea, 1950–52. Limnol. Oceanogr. **2:** 252–270.

SHELDON, R. W. 1969. A universal grade scale for particulate materials. Proc. Geol. Soc. Lond. 1659, p. 293–295.

————, AND T. R. PARSONS. 1967. A practical manual on the use of the Coulter counter in marine science. Coulter Electronics Inc., Toronto. 66 p.

SUTCLIFFE, W. H., JR. 1972. Some relations of land drainage, nutrients, particulate material and fish catch in two eastern Canadian bays. J. Fish. Res. Bd. Can. **29:** 357–362.

Submitted: 17 January 1973
Accepted: 23 April 1973

LIMNOLOGY

AND

OCEANOGRAPHY

November 1975

Volume 20

Number 6

Underwater observations of gelatinous zooplankton: Sampling problems, feeding biology, and behavior[1]

W. M. Hamner,[2] L. P. Madin,[3] A. L. Alldredge,[4] R. W. Gilmer,[3] and P. P. Hamner[2]
Department of Zoology, University of California, Davis 95616

Abstract

Observations by SCUBA divers on the distribution and biology of gelatinous zooplankton have stimulated speculations about the structure of tropical oceanic ecosystems. The gelatinous group represents one of four apparent strategies for survival in pelagic animals. Conventional plankton collection probably does not sample these organisms accurately due to their patchy distribution, fragility, and escape responses. Many gelatinous plankters filter feed using mucous structures; these mechanisms are important because of their efficiency in collecting particulate material, and because mucus is a source of organic aggregates in the sea. Such aggregates are often large, irregularly distributed, and of complex composition; these properties are rarely discerned by conventional sampling gear. The aggregates contribute considerable spatial heterogeneity to a seemingly homogeneous environment. An entire category of pelagic animals lives in association with these floating substrates. The diversity and trophic complexity of epipelagic plankton communities have been underestimated by previous investigators.

During the past 5 years we have been collecting data on the natural history of various groups of gelatinous, transparent zooplankton. These observations on free-swimming, undisturbed animals were made by divers using SCUBA in the epipelagic blue water of several tropical water masses, the Florida Current, Sargasso Sea, North Pacific gyre, and Gulf of California. The logistics for this approach to biological oceanography are presented by Hamner (1975). A photographic survey of the remarkably beautiful animals is available (Hamner 1974) as are several papers concerned with the biology of particular groups (Alldredge 1972; Alldredge and Jones 1973; Bé and Gilmer in press; Madin 1974a; Swanberg 1974). Here we offer some general observations on the biology of gelatinous zooplankters and the physical and trophic structure of the pelagic community, and we speculate on the importance of these observations. We have been asked repeatedly by colleagues what we think about various subjects of traditional interest when viewed from our perspective. We hope that our comments here will answer some of these questions and, more important, stimulate additional research.

The groups which we call gelatinous zooplankton are the Hydromedusae, Siphonophora, Scyphomedusae, Ctenophora, Heteropoda, Pteropoda, Thaliacea, and Ap-

[1] This work was supported by grants from the National Science Foundation, the National Geographic Society, and the John Simon Guggenheim Foundation.

[2] Present address: Australian Institute of Marine Sciences, Townsville, Queensland.

[3] Present address: Woods Hole Oceanographic Institution, Woods Hole, Massachusetts 02543. Address reprint requests to L. P. Madin.

[4] Present address: Department of Biological Sciences, University of California, Santa Barbara 93106.

110

pendicularia, as well as many meroplanktonic larvae. We have been concerned with these forms primarily because they are relatively large (>1 mm) organisms that generally do not react to a diver unless physically disturbed. Consequently we have been able to observe undisturbed animals from all these groups at close range while diving in the upper sunlit regions of tropic seas. The fact that these organisms are large and do not avoid divers is convenient, but there are additional reasons to believe that we are dealing with a natural and important assemblage of tropical oceanic animals.

Gelatinous zooplankton comprises representatives of several phyla and spans two or three trophic levels. Notwithstanding this diversity, we consider this assemblage to be an ecological group on the basis of the predator-protection strategies which the animals have in common. The planktonic animals that inhabit the clear sunlit surface of tropic seas must avoid predators in an environment usually devoid of hiding places. They therefore possess a rather limited number of protective properties. We suggest that the organisms in our study area fall into one of four categories.

1. Animals that are too small to be seen by many predators (e.g. copepods).
2. Animals that are large but transparent and thereby invisible to many predators (e.g. ctenophores).
3. Animals that are large and readily visible but achieve protection either by aggregating in schools (e.g. herring) or by being very large, fast, and mean (e.g. sharks).
4. Animals that are nocturnal and seek cover by vertical diurnal migration (e.g. euphausids).

This scheme doubtless ignores many predator avoidance patterns, which operate within each of these assemblages, but we have found it useful and we believe it biologically reasonable to consider that the tropical, blue-water epipelagos has four basic predator avoidance systems.

We have directly observed each of these four assemblages. Members of the very small free-swimming zooplankton assemblage (<1 mm) are visible but hard to follow. The gelatinous animals are perhaps the easiest to observe underwater, although some practice is necessary to locate the most transparent forms. Once we learned how to look for them we found that they are almost always present and easily studied.

We see small schools of fishes while diving, but only when they are pursued by some larger fish. Presumably we are perceived as large predators by these schooling fishes, and they avoid us whenever possible. Large predatory fishes—sharks, marlin, sailfish, and dolphin fishes—were rarely seen in midocean but were common in the Florida Current. They are curious and unafraid, but do not react to the gelatinous animals in the water around them. Although blue-fin tuna are reported to eat salps (Reintjes and King 1953), we have the impression that transparent animals are either invisible or of no interest to the large, visual predators in the pelagic fish assemblage.

Thus, we favor an ecological taxonomy for the creatures of the surface of the sea that divides pelagic animals into the four major assemblages listed above. We believe that the gelatinous-transparent assemblage is important and coequal with the three other groups but is much less well known.

Equally striking to us in the course of our underwater investigations have been the abundance, size, and complexity of organic aggregates ("marine snow"). Many of these aggregations appear to originate from mucous secretions of gelatinous zooplankton. Like those organisms, the aggregates are fragile and not accurately sampled with conventional gear. We feel that they too need to be better understood.

Many people have given us support, advice, and encouragement during the course of these studies. We particularly wish to thank Cadet Hand and the staff of the Bodega Marine Laboratory, and Robert F. Mathewson and the staff of the Lerner Marine Laboratory. We are also indebted to the Woods Hole Oceanographic Institution (L.P.M., R.W.G.) and the Scripps Institution of Oceanography (W.M.H., L.P.M., A.L.A.) for the opportunity to dive from their ships and to discuss our work with their staffs. We thank M. Mullin, R. Strathmann, and an anonymous reviewer for useful criticism of several drafts of this paper.

Sampling

Among our most important motivations for studying gelatinous zooplankton by diving was the inadequacy of conventional plankton nets for collecting either intact specimens or reliable information on distribution and abundance. In the Florida Current, for example, we sampled regularly with meter nets for 15 months, but obtained only a few specimens of the common gelatinous forms and no specimens of others; yet we knew these animals were there because we saw them there almost every day while diving. Actually many were caught by the net, but then strained through the mesh and lost or damaged beyond recognition. Some of the gelatinous zooplankton are not only delicate but fast-swimming and can usually avoid nets. These observations convinced us that many previous estimates of the relative abundance of this fauna are incorrect.

By making direct observations in the sea we avoided many of these problems and at the same time saw some of the reasons why traditional methods of sampling fail for these large, gelatinous, and delicate organisms. For example, Gilmer (1974) has made in situ measurements of sinking and swimming rates of pteropods, which illustrate their capacity for avoiding nets. Pteropods are neutrally buoyant when they feed, but they are also extremely sensitive to turbulence and readily exhibit escape responses to local disturbances (Gilmer 1972, 1974). Pseudothecosomes (e.g. *Gleba*, *Corolla*, and *Cymbulia*) appear especially sensitive to turbulence; these pteropods are highly vulnerable to predators when feeding (Gilmer 1972), and they must be fast if they are to escape. They readily avoid capture by a diver, and it is therefore not

surprising to us that they have seldom been collected in large numbers (Chen and Bé 1964; Deevey 1971). Euthecosome species, on the other hand, are small, relatively slow, and encased in a hard shell; they are ideally suited to collection by nets.

The Gymnosomata, the shell-less opisthobranchs, also avoid nets. These fast swimming carnivores are found in all oceans. Although we rarely observed gymnosomes they were occasionally quite abundant, usually associated with the thecosomes on which they feed. Several genera, such as *Pneumodermopsis* and *Notobranchea*, are capable of swimming in excess of 1 m s⁻¹, or 2 knots (Gilmer unpublished observations); they too are seldom collected in nets.

Certainly most sampling techniques are biased; our own technique of collecting observational data misses things that are very small or very fast. But we believe also that in the case of the gelatinous zooplankton assemblage the sampling bias of traditional gear has been particularly misleading, and new approaches are necessary to understand the distribution and biology of these animals.

Filter feeding with mucous structures

Many aspects of animal behavior are dominated by trophic interactions; in the case of the gelatinous fauna we know very little about feeding biology. Of particular interest to us have been the filter-feeding mechanisms of pteropods, salps, and appendicularians. These animals collect, concentrate, and transport particulate organic matter using mucous sheets, nets, strands, and filters in conjunction with ciliated surfaces. The efficacy of these feeding mechanisms in collecting a broad size range of organic material is important to the organisms and to the ecosystem as a whole. While pteropods, salps, and appendicularians are particularly conspicuous mucous feeders, we have also observed mucous structures which appear to have a food collecting function in prosobranch veliger larvae (Gilmer unpublished observations) and polychaete larvae. The feeding mechanisms of some of these organisms have been described previously (*see* Jorgensen 1966 for general resume); we have added new information by direct observation (Gilmer 1972, 1974; Madin 1974*b*).

The mucous structures that we have seen may be classified as external or internal. One example of an external structure is the free-floating mucous webs produced by pseudothecosomatous pteropods (Gilmer 1972). These webs passively entrap suspended particles; the pteropods then draw portions of the mucous web into the mouth by ciliary action and ingest the material stuck on the mucus. Generally the mucous webs are much larger than the animals; we have seen webs more than 2 m across produced by an animal only 10 cm in diameter.

Appendicularian houses are doubtless the most complex and sophisticated external mucous structures; they are among the most complex external structures produced by any solitary animal. The structure and function of the house as a food collecting and concentrating apparatus was described by Lohmann (1909, 1933) and Körner (1952), but many details are still not clear. Field observations have enabled us to determine the morphology of several previously undescribed houses and to examine the functioning of the house in natural conditions (Alldredge unpublished observations).

An unusual external mucous feeding device is used by the larva of the polychaete *Poecilochaetus* sp. This worm secretes a 3-dimensional network of fine mucous strands. The larva is attached by its mouth to the threads and glides along them feeding on small adhering particles. In the field these threads ranged from 4 to 16 cm long; the worms are 0.5 to 1 cm long. *Poecilochaetus* larvae hand-collected in glass jars built threads in the laboratory, but only after the addition of a small amount of powdered milk to the water. *Poecilochaetus* occurred only rarely in the Florida Current, but animals were abundant when present. On one occasion about 58% of the larvae were feeding with their mucous threads (Alldredge unpublished observations).

Mucous secretions may serve a less direct feeding function. In the Gulf of California we observed another planktonic polychaete, the larva of *Loimia medusae*, which builds its own benthos about it. The larva secretes a clear, gelatinous barrel around itself and lies within the portable burrow with its prostomium and tentacles protruding from one of the open ends to filter particles. If disturbed it abandons the tube, swimming with rapid undulations. In the laboratory, one *Loimia* larva jettisoned its barrel and secreted another in 4 h. The completely transparent tube may provide the larva with some physical protection, but it is probably primarily a flotation device, allowing the larva to maintain its position while fully extending its tentacles to feed (P. Hamner and A. Alldredge unpublished observations).

Internal mucous feeding devices are best demonstrated by salps. In the Salpidae a conical mucous net within the cylindrical body strains particulate material from water which is actively pumped through the body. This water stream, with the aid of the ciliated tracts, moves the net continuously back from the oral opening, conveying trapped particles to the posterior esophagus. The mucus is ingested along with the collected food (Madin 1974*a*).

An important difference between collecting food with mucous structures and with rigid antennae or maxillae is that the size range of particles collected is often much greater for the mucous feeder. The lower size limit for setose filter feeders is determined by the minimum spacing between the rigid filter elements; the spacing sets a lower limit of 3–4 μ for most copepods (Gauld 1966; Poulet 1974). Mucous filters, however, can have extremely small pore sizes. The inner collecting filters of some appendicularian houses have a fibrillar structure with pores as small as 0.1 μm (Jorgensen 1966). Most appendicularians also have a coarse external filter that prevents large particles from entering and clogging the internal filter (Lohmann 1909). These animals can select a certain portion of the available spectrum of particle sizes by constructing mucous filters with specific pore sizes.

The external mucous webs produced by pteropods have pores 20 μm or more across, capable of physically filtering only larger particles. But very small particles, 1–10 μm, are also trapped in these webs, evidently by adhering to the mucus on contact; such particles compose up to 50% of the food ingested (Gilmer 1972, 1974). This adhesive quality of mucus may be quite important, enabling mucous structures to retain small particles without possessing a very fine ultrastructure, which would be difficult to create and maintain in large mucous nets or sheets. Adhesion may also be responsible for the ability of salps to retain cells as small as 0.7 μm.

Although their efficiency must more than compensate, the energetic cost of producing these mucous feeding devices seems remarkably high. For example, appendicularians expend considerable material and energy in building their house, yet discard it entirely when it becomes clogged and build another. *Oikopleura dioica* builds as many as six houses per day (G. Paffenhöfer personal communication). Pteropods must also realize a substantial energetic return on their investment of mucus. Although they ingest the mucus along with the particles, much of the web material is lost, and, if disturbed, pteropods abandon the entire gigantic structure (Gilmer 1972).

Salps are more conservative of their mucus. Since the filtering net is inside the body, it is not as liable to accidental loss and is completely ingested. This recycling of the mucus may reduce energy loss and enable salps to feed effectively with a filter that is small relative to their body size. Appendicularians and pteropods do not recycle mucus effectively, and their filters are quite large relative to their body size. Thus there may be two approaches to obtaining an adequate net energy gain, one involving a small, efficient, recycled filter, the other a much larger, but less efficient and partly expendable structure.

Structure, distribution, and origin of organic aggregates

The amount of mucus produced by animals such as pteropods, gastropod larvae, salps, appendicularians, annelids, fish, and reef corals is considerable. Our own observations were limited to the upper 30 m of the water column, but Neumann (personal communication) has made numerous dives in the submersible *Alvin* in the Florida Current and has seen similar mucous structures down to 400 m. Although the original description of marine snow by Nishizawa et al. (1954) was also made from a submersible, most attempts since then to study this aggregated organic material have been made from surface vessels using water samplers. Unfortunately, blind sampling from the surface can provide neither correct estimates of the size and shape of the aggregates nor insight into the fine-scale spatial relationships within the water column.

There is now a wealth of indirect information on the distribution, origin, and significance of particulate organic material (*see* Riley 1970), but our direct observations suggest that some earlier generalizations are incorrect. While particles collected on filters almost never exceed 100 μm in diameter (Gordon 1970), larger particles may actually be abundant in the sea. Johannes (1967) and Suzuki and Kato (1953) observed particles up to 3 cm in diameter. In the Gulf Stream, we frequently dove among amorphous aggregates up to 30 cm in diameter, at densities of 2–3 m^{-3}. These aggregates contained algae, foraminifera, detritus, and small particulates in a mucous matrix.

Aggregates are quite variable in size, composition, and distribution. Determination of the actual amount of particulate matter per cubic meter of seawater can be confounded by this patchiness, as recently demonstrated by Wangersky (1974). By visually selecting patches of marine snow, we have measured concentrations of particulate organic carbon in water samples that change more than fourfold over a distance of 10 cm. Sampling from the surface may provide extreme and average values of particulate carbon concentration if enough samples are taken and unusually high values are not discarded as contamination, but indirect sampling has two major weaknesses. Large aggregates are often widely dispersed and can be collected only in a large sample volume: sampling for a *Gleba* web with a Niskin bottle is like sampling for trees with a trowel. Large aggregates are also very fragile; they can often be dispersed by the wave of a diver's hand and may disintegrate even when collected by hand. Turbulence from most standard samplers certainly disturbs these aggregates and changes the apparent particle sizes.

The problems of sampling these large organic aggregates are considerable. Hand collection and quantitative visual or photographic estimates of particle density in the field are possible approaches. Collection of aggregates in situ by freezing (Schubel and Schiemer 1972) or filtering (Marshall et al. 1972) may give a more accurate picture of the structure of the aggregates.

A truly representative sample of large aggregates would be of considerable importance for accurate measurement of particulate carbon or C : N ratios. Coles and Strathmann (1973) found coral mucus flocs to contain ten times as much particulate carbon and twenty times as much particulate organic nitrogen as particles filtered from a liter of the surrounding water; Benson and Muscatine (1974) found wax esters and triglycerides in coral mucus and observed that reef fish "avidly ingest it."

Fecal material, detritus, bacteria, and adsorption of dissolved organics on surfaces are probably all important sources of particulate organic matter (Baylor and Sutcliffe 1973; Barber 1966; Paerl 1973), but the mucous products of the gelatinous organisms discussed here may be equally important. Alldredge (1972) observed abandoned appendicularian houses ranging in size from a few millimeters to 3 cm at densities of up to 600 per cubic meter. The feeding webs of the pteropod *Gleba cordata* are flat sheets of mucus up to 2 m in diameter; dispersion of these webs ranged from 10 m or more apart to a field of contiguous nets.

Not only is it produced in large amounts, but such mucus accumulates particles and smaller aggregates and may act as a surface for adsorption of dissolved substances. We do not know how much of the particulate organic material in the sea derives from mucus filter feeders, but the aggregates that we can visually identify (e.g. pteropod webs and appendicularian houses) have great significance in the organization of fine-scale community structure. Particulate organic carbon makes up only 1 to 10% of the total organic carbon present in the sea (Sharp 1973), but the small fraction tied up in aggregates may be of disproportionate importance as sources of food, surface microhabitats, or nuclei for the growth of larger aggregates. To understand the functional role of organic matter in the plankton community we must understand the origin, composition, and architecture of these macroscopic structures as well as of the microscopic particles. We feel that many previous discussions have missed the trophic and spatial importance of aggregates by focusing only on particles of 1–10 μm, a size range that is partially an artifact of the usual collection and filtration procedures (Strickland and Parsons 1968).

Utilization of surfaces in the plankton

Intuitively one conceives of the planktonic environment as a relatively homogeneous, fluid medium in which nutrients, phytoplankton, and most particles are invisibly small. Though zooplankters may attain macroscopic size, they too are thought of as adapted to a homogeneous fluid (Friedrich 1969).

Heterogeneity in this environment has been defined in terms of such physical parameters as microscale gradients (Gregg and Cox 1972; Gregg 1973), discrete oceanic or coastal water masses, current boundaries (Lovett 1968; Pingree et al. 1974), and thermo-, halo-, and pycnoclines (Harder 1968). However, the existence of actual physical heterogeneity in the plankton environment resulting in a variety of microhabitats like those of a tropical rain forest or a coral reef has been considered only with regard to a few conspicuous examples like *Sargassum*, tar lumps (Butler 1975), or bacterial colonization of detritus (Jannasch 1973; Paerl 1973).

This image of a homogeneous planktonic environment does not coincide with our direct field observations. Pteropod feeding webs, coral mucus flocs (Johannes 1967), abandoned appendicularian houses, decomposing gelatinous animals, and other macroscopic particulates can produce a complex 3-dimensional pattern of discrete organic aggregates in an environment often considered unstructured.

Our observations indicate that many "planktonic" organisms depend heavily on the presence of discrete surfaces, both living and nonliving, in the pelagic environment. We have frequently observed hyperiid amphipods, copepods, ostracods, euphausid larvae, polyclad flatworms, and crab megalops resting on the surfaces of organic aggregates, on pteropod feeding webs, and especially on abandoned appendicularian houses. Feeding on houses was demonstrated in the laboratory for *Oncaea mediterranea* (Alldredge 1972). Zooplankters treat these surfaces as benthic organisms might treat a rock, moving about on them and scraping off particles. Such zooplankters are often morphologically adapted to benthic feeding methods. For example, the head and first thoracic appendages of the genus *Oncaea* appear better adapted to the scraping of surfaces than to filter feeding or grasping (A. Fleminger personal communication).

The frequent occurrence of feeding on floating surfaces indicates that it may be important in the trophic structure of the plankton community. About 11% of the abandoned appendicularian houses observed over a 7-month period in the Florida Current had organisms visible on their surfaces. In the Gulf of California, the occurrence of organisms on the surfaces of ap-pendicularian houses approached 80%. These surfaces provide highly concentrated energy packets in an environment where food normally is thinly dispersed. Nannoplankton and microorganisms too small for most adult crustaceans to obtain by filter feeding are readily available on abandoned appendicularian house filters and organic aggregates.

The living bodies of various species serve as more or less stable habitats for other organisms. Bacteria and protozoa can also use most organic aggregates as microhabitats (Pomeroy and Johannes 1968; Paerl 1973). Three species of planktonic dinoflagellates utilize the pelagic foraminiferan *Hastigerina pelagica* as a surface habitat in the Florida Current (Alldredge and Jones 1973). Hyperiid amphipods use living surfaces (medusae, salps, siphonophores, ctenophores, and tornaria larvae) as both microhabitats and sources of food (Madin 1974*b*). Our observations so far confirm the hypothesis of Pirlot (1932) that hyperiid amphipods have conquered the pelagic space by attaching themselves as inquilines or parasites of the macroplankton.

Planktonic microhabitats are of particular interest as they relate to the origin and maintenance of diversity in planktonic animals. Attempts to resolve the "paradox of the plankton" (Hutchinson 1961) have taken several forms (e.g. *see* Harding and Tebble 1963; Richerson et al. 1970). The presence of heterogeneous microhabitats in the sea, with specific associated biota, may help.

Our direct field observations demonstrate that the planktonic environment contains considerable physical spatial heterogeneity in the form of macroscopic surfaces of a variety of sizes and shapes. We propose that the structure of this environment has two important characteristics not found in other habitats:

1. Constantly changing spatial relationships: Drifting aggregates are constantly changing relative to each other, due to the action of currents, microgradients of density, and the activity of animals. Organisms

remaining on specific surfaces are relatively unaffected by this motion, but organisms using these surfaces for feeding must develop suitable means of locating particles.

2. Small habitat size per unit animal: Many of the surfaces are only slightly larger than most of the organisms that utilize them, and thus the number of interacting animals per aggregate is probably much smaller than in other communities.

Predator-prey relationships

In situ observations and collections are particularly useful for investigating predation among planktonic animals. Events of predation and escape were seen in the sea and could often be recreated in aquaria with undamaged animals. Among the most active and effective predators we have observed are medusae, siphonophores, ctenophores, and heteropods.

The heteropod *Cardiapoda placenta* was observed on 38 occasions. On five of these, we saw *Cardiapoda* eating *Salpa cylindrica*; on seven other occasions the contents of the gut could be identified as *Salpa*. Time from ingestion to release of feces varies from 4.5 to 6.5 h, depending apparently on the relative size of predator and prey. In the laboratory, other prey were taken, including the pelagic nudibranch *Phyllirhoe atlanticum*, various fish larvae, and the heteropod *Firoloidea desmaresti*. The maximum time of digestion was never more than 7 h.

The attack behavior of *Cardiapoda* was observed once in the field and twice in the lab; it was initiated by the heteropod when the prey was as much as 60 cm away. In each case, the prey was above the heteropod (which swims on its back) suggesting that the silhouette of the prey can be seen by *Cardiapoda* from considerable distances. *Cardiapoda* swims rapidly to its prey (up to 40 cm s^{-1}), seizing it initially with the buccal cones. The entire prey is pulled into the esophagus by the combined action of the radula, buccal cones, and peristalsis of the proboscis. On one occasion, a heteropod attacked a salp twice its size, and after half was ingested severed the remainder

with its radula. Ingestion in all cases took about 10 min. Previous estimates of digestion times for net-collected specimens of *Carinaria japonica*, a close relative of *Cardiapoda* (Seapy 1970) averaged between 1.5 and 2.5 days, with some prey requiring up to 4 days to be digested. Compaction in the cod end of the net probably damages these organisms. Only a few percent of Seapy's specimens were healthy enough for observation in the lab, and the digestion times recorded were very likely affected by net damage.

Another heteropod, *Pterotrachaea coronuta*, was seen in the sea 27 times and on five occasions was eating physonect siphonophore bracts. No other prey items were observed underwater, nor would *Pterotrachea* feed on salps, ctenophores, or pteropods in the laboratory. Digestion times ranged from 6 to 8 h for the siphonophore bracts.

In the field (though not in the laboratory) these two heteropods swam actively only when attacking a prey item or avoiding capture by divers. Otherwise, they were rolled into a loose ball, with the posterior section of the tail in contact with the buccal region. The animal floats in this manner apparently as a neutrally buoyant mass; we could not detect any sinking.

We saw dramatic predation on salps by the hydromedusa *Aequorea* sp. Large *Aequorea*, with tentacles trailing 2 m or more, caught and ingested large salps by the dozen in one particularly dense swarm of *S. cylindrica*; the salps were apparently digested rapidly.

The tentaculate ctenophore *Lampetia* was generally no larger than its salp prey. On encountering a salp, the ctenophore opened its mouth and flattened itself over the surface of the prey (*see* Hyman 1940), eventually engulfing the salp.

The predatory behavior of *Beroe ovata* on other ctenophores, such as *Bolinopsis*, has been described by Swanberg (1974). *Beroe* maximizes the energetic gain in each encounter with its prey by eating as much of *Bolinopsis* as possible, either engulfing small ones whole or biting substantial portions out of larger ones with specialized ciliary "teeth." *Aequorea* also feeds on these ctenophores in aquaria. The tentacles of the medusae slice through the tissue of *Bolinopsis* and the individual pieces are eaten.

The pomocentrid fish *Abudefduf saxatilis* were seen in shallow waters of the Gulf of California feeding extensively on appendicularians in their houses (*see also* Shelbourne 1962). Nutritional utilization of the algae-clogged house as well as of the appendicularians would increase the caloric gain for these fish and shorten the food chain considerably, making nannoplankton directly available to upper trophic levels (Alldredge unpublished observations). Predation by fish on salps may also constitute a trophic shortcut from small primary producers to large consumers (Madin 1974a).

Our observations of the gelatinous zooplankton show that predation ranges from visually oriented and highly specific patterns, as in *P. coronuta*, to nonvisual and rather indiscriminate modes, as in *Aequorea*. Between these extremes many sorts of predatory feeding are undoubtedly important to the trophic and diversity structures of the zooplankton community. Most of these patterns will be discernible only by in situ observations.

The principal defense strategy against predation within this assemblage as we have defined it is transparency (Greze 1963). However, in addition, virtually all the animals we have seen exhibited behavior patterns, usually in response to the proximity of a diver, which seemed to be escape responses. Commonly the animals swam away in the normal or a different swimming mode when approached at a certain distance. Nonvisually oriented animals responded only to physical contact or local turbulence.

Heteropods abandon their enrolled postures and pteropods their feeding webs to swim away at remarkable speeds when approached. Some appendicularians have elaborate escape responses, retreating out through an escape hatch in the house when disturbed physically by a diver, or, as once observed, by a chaetognath. This hatch evidently serves as an exit port both when the house filters are clogged and to escape predation. Free-swimming appendicularians are extremely fast; we saw *Oikopleura longicauda* swimming at 30–50 cm s^{-1} over short distances. But this response is not universal. Other species remain within the house despite continued prodding by a diver, and their responses to an actual predator are unknown.

Salps respond to touch on the posterior end by swimming faster for a few seconds; they sometimes stop or reverse when touched on the anterior end. Siphonophores, spread in the fishing position, contract rapidly to a fraction of their extended length and swim away in response to turbulence or contact with a tentacle. Medusae respond to a touch on the bell by contracting the tentacles and swimming somewhat more rapidly.

Ctenophores are also responsive to touch. *Pleurobrachia* rapidly contracts its tentacles when touched and swims away. *Ocyropsis* swims powerfully with its muscular lobes when disturbed. *Cestum* switches from swimming as a rigid wing propelled by cilia to a fast undulating pattern that moves it at right angles to its previous course. Presumably these animals can differentiate between potential predators and prey items by the intensity or extent of the contact. In most cases the prey items are much smaller than the predators. Food or foe may also be identified chemically. In every case, rapid escape swimming persists only for a few seconds, moving the animal a meter or two away, and is followed by reversion to the normal speed or pattern.

A more complex response was seen in the ctenophore *Eurhamphea vexilligera* and several alciopid polychaetes that released reddish-brown streamers of pigment as they swam away; the secretion of *Eurhamphea* is also luminescent, producing a shower of green sparks in the water. This appears to be an "ink-cloud" escape behavior like that of cephalopods and bespeaks unexpected sophistication on the part of these plankters. These defenses may be effective only against visually oriented predators; the secretion of *Eurhamphea* seemed not to deter predation by the chemotactically oriented *Beroe* (Swanberg 1974). The diversity of predators and feeding behaviors among planktonic animals suggests that these instances are only a few of many escape responses that have arisen to deal with specific pressures.

We feel that our observations on the biology of the gelatinous zooplankton community amply demonstrate the utility of SCUBA diving in the open ocean as an investigative tool to complement conventional zooplankton methods. We hope that the thoughts presented here will stimulate additional study of the gelatinous macroplankton.

References

ALLDREDGE, A. L. 1972. Abandoned larvacean houses: A unique food source in the pelagic environment. Science **177**: 885–887.

————, AND B. M. JONES. 1973. *Hastigerina pelagica*. Foraminiferal habitat for planktonic dinoflagellates. Mar. Biol. (Berl.) **22**: 131–135.

BARBER, R. T. 1966. Interaction of bubbles and bacteria in the formation of organic aggregates in sea-water. Nature (Lond.) **211**: 257–258.

BAYLOR, E. R., AND W. H. SUTCLIFFE. 1963. Dissolved organic matter in seawater as a source of particulate food. Limnol. Oceanogr. **8**: 369–371.

BÉ, A. W. H., AND R. W. GILMER. In press. A zoogeographic and taxonomic review of euthecosomatous pteropoda. *In* A. T. Ramsay [ed.], Oceanic micropaleontology. Academic.

BENSON, A. A., AND L. MUSCATINE. 1974. Wax in coral mucus: Energy transfer from corals to reef fishes. Limnol. Oceanogr. **19**: 810–814.

BUTLER, J. N. 1975. Pelagic tar. Sci. Am. **232**: 90–97.

CHEN, C., AND A. W. H. BÉ. 1964. Seasonal distributions of euthecosomatous pteropods in the surface waters of five stations in the western North Atlantic. Bull. Mar. Sci. Gulf Caribb. **14**: 185–220.

COLES, S. L., AND R. STRATHMANN. 1973. Observations on coral mucus "flocs" and their potential trophic significance. Limnol. Oceanogr. **18**: 673–678.

DEEVEY, G. B. 1971. The annual cycle in quantity and composition of the zooplankton of the Sargasso Sea off Bermuda 1. The upper 500 m. Limnol. Oceanogr. **16**: 219–240.

FRIEDRICH, H. 1969. Marine biology. Univ. Wash.

GAULD, D. T. 1966. The swimming and feeding of planktonic copepods, p. 313–334. *In* H. Barnes [ed.], Some contemporary studies in marine science. Hafner.

GILMER, R. W. 1972. Free floating mucus webs: A novel feeding adaptation in the open ocean. Science **176**: 1239–1240.

————. 1974. Some aspects of feeding in thecosomatous pteropod molluscs. J. Exp. Mar. Biol. Ecol. **15**: 127–144.

GORDON, D. C. 1970. A microscopic study of organic particles in the North Atlantic Ocean. Deep-Sea Res. Oceanogr. Abstr. **17**: 175–185.

GREGG, M. C. 1973. The microstructure of the ocean. Sci. Am. **228**: 64–77.

————, AND C. S. COX. 1972. The vertical microstructure of temperature and salinity. Deep-Sea Res. Oceanogr. Abstr. **19**: 355–376.

GREZE, V. N. 1963. Determination of the transparency of the bodies of plankton and its protective significance [in Russian]. Dokl. Akad. Nauk SSSR **151**: 435–438.

HAMNER, W. M. 1974. Blue-water plankton. Natl. Geogr. Mag. **146**: 530–545.

————. 1975. Underwater observations of blue-water plankton: Logistics techniques, and safety procedures for divers at sea. Limnol. Oceanogr. **20**: 1045–1051.

HARDER, W. 1968. Reactions of plankton organisms to water stratification. Limnol. Oceanogr. **13**: 156–168.

HARDING, J. P., AND N. TEBBLE [Eds.]. 1963. Speciation in the sea. Syst. Assoc. Publ. 5. 199 p.

HUTCHINSON, G. E. 1961. The paradox of the plankton. Am. Nat. **95**: 137–145.

HYMAN, L. 1940. The invertebrates, v. 1. Protozoa through Ctenophora. McGraw-Hill.

JANNASCH, H. 1973. Bacterial content of particulate matter in offshore surface waters. Limnol. Oceanogr. **18**: 340–342.

JOHANNES, R. E. 1967. Ecology of organic aggregates in the vicinity of a coral reef. Limnol. Oceanogr. **12**: 189–195.

JORGENSEN, C. B. 1966. The biology of suspension feeding. Pergamon.

KÖRNER, W. F. 1952. Untersuchungen über die Gehäusebildung bei Appendicularien (*Oikopleura dioica* Fol). Z. Morphol. Oekol. Tiere **41**: 1–53.

LOHMANN, H. 1909. Die Gehäuse und Gallertblasen der Appendicularien und ihre Bedeutung für die Erforschung des Lebens im Meer. Verh. Dtsch. Zool. Ges. **19**: 200–239.

————. 1933. Tunicata-Appendicularia, p. 1–202. *In* W. Kukenthal und T. Krumbach [eds.], Handbuch der Zoologie, v. 5 pt. 2.

LOVETT, J. R. 1968. Vertical temperature gradient variations related to current shear and turbulence. Limnol. Oceanogr. **13**: 127–142.

MADIN, L. P. 1974*a*. Field observations on the feeding behavior of salps (Tunicata: Thaliacea). Mar. Biol. (Berl.) **25**: 143–148.

———. 1974*b*. Field studies on the biology of salps. Ph.D. thesis, Univ. Calif., Davis. 208 p.

MARSHALL, N., J. PARRISH, AND A. L. GALL. 1972. A technique for collection of particulate organic matter in situ. Mar. Biol. (Berl.) **12**: 194–195.

NISHIZAWA, S., M. FUKUDA, AND N. INOUE. 1954. Photographic study of suspended matter and plankton in the sea. Bull. Fac. Fish. Hokkaido Univ. **5**: 36–40.

PAERL, H. W. 1973. Detritus in Lake Tahoe: Structural modification by attached microflora. Science **180**: 496–498.

PINGREE, R. D., G. R. FORSTER, AND G. K. MORRISON. 1974. Turbulent convergent tidal fronts. J. Mar. Biol. Assoc. U.K. **54**: 469–479.

PIRLOT, J. M. 1932. Introduction à l'étude des Amphipodes Hypérides. Ann. Inst. Oceanogr. **12**: 1–36.

POMEROY, L. R., AND R. E. JOHANNES. 1968. Occurrence and respiration of ultraplankton in the upper 500 meters of the ocean. Deep-Sea Res. Oceanogr. Abstr. **15**: 381–391.

POULET, S. A. 1974. Seasonal grazing of *Pseudocalanus minutus* on particles. Mar. Biol. (Berl.) **25**: 109–123.

REINTJES, J. W., AND J. E. KING. 1953. Food of yellowfin tuna in the central Pacific. Fish. Bull. **54**: 90–110.

RICHERSON, P., R. ARMSTRONG, AND C. R. GOLD-MAN. 1970. Contemporaneous disequilibrium, a new hypothesis to explain the "paradox of the plankton." Proc. Natl. Acad. Sci. U.S. **67**: 1710–1714.

RILEY, G. A. 1970. Particulate organic matter in sea water. Adv. Mar. Biol. **8**: 1–118.

SCHUBEL, J. R., AND E. W. SCHIEMER. 1972. A device for collecting in situ samples of suspended sediment for microscopic analysis. J. Mar. Res. **30**: 269–273.

SEAPY, R. R. 1970. Analysis of the distribution and feeding habits of the heteropod mollusc *Carinaria japonica* in waters off southern California. Ph.D. thesis, Univ. Southern Calif. 172 p.

SHARP, J. H. 1973. Size classes of organic carbon in seawater. Limnol. Oceanogr. **18**: 441–447.

SHELBOURNE, J. E. 1962. A predator-prey size relationship for plaice larvae feeding on *Oikopleura*. J. Mar. Biol. Assoc. U.K. **42**: 243–252.

STRICKLAND, J. D. H., AND T. R. PARSONS. 1968. A practical handbook of seawater analysis. Bull. Fish. Res. Bd. Can. 167. 311 p.

SUZUKI, N., AND K. KATO. 1953. Studies on suspended materials marine snow in the sea. 1. Sources of marine snow. Bull. Fac. Fish. Hokkaido Univ. **4**: 132–135.

SWANBERG, N. 1974. The feeding behavior of *Beroe ovata*. Mar. Biol. (Berl.) **24**: 69–74.

WANGERSKY, P. J. 1974. Particulate organic carbon: Sampling variability. Limnol. Oceanogr. **19**: 980–984.

Submitted: 20 February 1975
Accepted: 22 July 1975

The Interrelation of Biting and Filtering in the Feeding Activity of the Northern Anchovy (*Engraulis mordax*)

CHARLES P. O'CONNELL

National Oceanic and Atmospheric Administration
National Marine Fisheries Service, Fishery-Oceanography Center
La Jolla, Calif. 92037, USA

O'CONNELL, C. P. 1972. The interrelation of biting and filtering in the feeding activity of the northern anchovy (*Engraulis mordax*). J. Fish. Res. Bd. Canada 29: 285–293.

A previous study showed that the northern anchovy (*Engraulis mordax*) captures *Artemia* adults (3.7 mm long) by biting and *Artemia* nauplii (0.65 mm long) by filter feeding. This study shows that the ratio of biting to filtering activity in small schools varies with the relative concentration of *Artemia* adults and nauplii in the water. Activity was half biting and half filter feeding when *Artemia* adults were 2% by dry weight of the available biomass, but was entirely biting when *Artemia* adults exceeded about 7% of the biomass.

It is estimated that when feeding activity is half biting and half filtering, *Artemia* adults and nauplii would contribute equal dry weights to ingestion, and that the sum of the two would be the same as that possible if total activity had been filtering. When relative concentration of *Artemia* adults is high enough to induce total biting activity, the dry weight ingested per unit time would be double that possible by filtering alone on the nauplii present.

Ratios of large-to-small crustaceans are relatively high near the surface at night off southern California, which suggests that biting activity could often exceed 50% of total feeding activity. If the plankton does support a high percentage of biting activity, a large part of the area should usually provide the anchovy with its daily nutritional needs.

O'CONNELL, C. P. 1972. The interrelation of biting and filtering in the feeding activity of the northern anchovy (*Engraulis mordax*). J. Fish. Res. Bd. Canada 29: 285–293.

Une étude antérieure a démontré que les anchois (*Engraulis mordax*) capturent les *Artemia* adultes (3.7 mm) à coups de dents et qu'elles se nourrissent par filtration de nauplii d'*Artemia* (0.65 mm). La présente étude révèle que le rapport morsure–filtration dans les bancs de taille restreinte varie en fonction de l'abondance relative des *Artemia* adultes et des nauplii dans le milieu. L'alimentation se fait moitié par morsure, moitié par filtration, quand les *Artemia* adultes constituent 2% en poids sec de la biomasse disponible, mais entièrement par morsure quand les *Artemia* adultes dépassent environ 7% de la biomasse.

Nous estimons que lorsque l'alimentation se fait moitié par morsure, moitié par filtration, les quantités d'*Artemia* adultes ingérées sont les mêmes en poids sec que celles des nauplii, et que le total des deux équivaut au total possible par filtration seulement. Lorsque la concentration en *Artemia* est suffisante pour induire une activité entièrement de morsure, le poids sec ingéré par unité de temps serait deux fois plus grand que celui possible par filtration seulement des nauplii.

La proportion des gros crustacés par rapport aux petits est relativement élevée près de la surface, la nuit, au large de la Californie méridionale. La morsure constituerait donc plus de 50% de l'activité alimentaire totale. Si, en réalité, le plancton est l'objet d'un fort pourcentage de morsure, une grande partie de la région pourrait généralement répondre aux besoins alimentaires quotidiens de l'anchois.

Received July 20, 1971

LEONG and O'Connell (1969) showed that the northern anchovy (*Engraulis mordax*) captured *Artemia* nauplii by filtering, but captured the much larger *Artemia* adults individually by directed biting (particulate feeding) when each food was presented separately. Food biomass was accumulated more rapidly by biting than by filtering. They hypothesized that the anchovy could not sustain its daily food requirement by filtering alone off southern California, except in limited areas of heavy plankton concentration. Their study suggested that the role of biting activity could not be satisfactorily appraised until the response of the fish to mixed size assemblages of food organisms was better understood.

121

The ability of the anchovy to capture larger but less abundant organisms by biting in addition to capturing the more numerous small organisms by filtering could be advantageous for two reasons. First, larger crustaceans with good escape responses might avoid capture by filtration. The directed nature of biting probably overcomes the escape response. Second, a greater volume of water can be searched in biting than can be strained in filtering. Either of these factors could raise the average amount of food eaten above what it would be by filtering alone.

The two modes of feeding suggest that the anchovy is a selective feeder. A number of authors have suggested on the basis of field studies that clupeoid fishes feed selectively and that food-organism size is a major determinant. Hiatt (1951), for example, showed that the nehu (*Stolephorus purpureus*) was larger and tended to have more food in its stomach in areas where large crustaceans were more available, and appeared to select these organisms. Brooks and Dodson (1965) concluded that the alewife (*Alosa pseudoharengus*) tended to select the larger crustaceans. Shen (1969) showed that zooplankters were generally more numerous than phytoplankters in the stomachs of *Engraulis japonicus*, and that the larger organisms, up to 5 mm long (Euphausiacea, Ostracoda, *Paracalanus*, and Rotifera), were commonly present in the stomachs, though only occasionally in relatively high abundance. Loukashkin (1970) showed that zooplankters were usually far more abundant than phytoplankters in the stomachs of *E. mordax*, and concluded that large copepods and euphausiids were the most important food items.

These studies imply that larger crustaceans are taken in preference to smaller organisms as their relative abundance permits. They suggest the hypothesis that ingestion rate will vary in respect to biomass concentration of plankton as the proportion of biting to filtering activity varies in some way with the relative abundance of large and small organisms in the concentration. This study attempts to describe the way in which these two kinds of feeding activity are used by small schools of anchovies in the laboratory when they are presented with different density combinations of *Artemia* adults and nauplii. The implications of these results are discussed in relation to anchovy feeding in the sea.

Procedure

Feeding trials were carried out in an arrangement of two plastic pools with connecting trough and gates (Leong and O'Connell 1969). The pools were supplied with a continuous flow of filtered sea water and each contained 3.65 m³ of water at a depth of 0.5 m. The pools were under a 12-hr day–12-hr night cycle of illumination, but all trials were during the day. Temperature ranged from 16 to 19 C.

For each trial, water flow was turned off and 16–20 fish were diverted from a large reserve school in one pool to the trough and then admitted to a prepared food situation in the other pool. Three observers scored feeding activity for a period of 8 min, after which the fish were returned to the reserve school. All trials were preceded by at least 24 hr without feeding and samples from the reserve school always showed empty digestive tracts.

The food situation was established for each trial by placing a known number of *Artemia* adults averaging 3.7 mm long (0.5 mg dry wt) and an approximately known number of Artemia nauplii averaging 0.65 mm long (0.00145 mg dry wt) in the experimental pool and dispersing them by stirring gently with wide mesh dipnets. *Artemia* adults were precounted for all but a few high density trials, for which they were estimated by wet weight. Post-trial estimates of nauplii density were made from samples taken from the pool just before the fish were admitted.

The basic units of observation over the 8-min feeding period were 10-sec intervals marked by an audible timing device. Each of the three observers recorded the activity of a single fish as biting (B), filtering (F), or none (0) for successive intervals. The two kinds of activity, which differ by a factor of at least 10 in duration of mouth opening (Leong and O'Connell 1969), were easily distinguished. Instances of one fish displaying both kinds of activity in the same 10-sec interval were rare. Prior to each trial, each observer was assigned a different part of the school — front, middle, or rear — and chose fish in these approximate positions at the start of each interval. Observer differences will be considered in conjunction with the analysis of position differences in activity.

A pilot study indicated that the proportion of biting to filtering activity in the school might vary in relation to the relative concentration of the two sizes of *Artemia* in the water, only when *Artemia* adults were at fractions of less than about 10% of the dry weight of total food concentration. Feeding activity appeared to be directed almost entirely at larger organisms when they constituted greater fractions than this of the available food. Groups of six fish allowed to feed for 5 min averaged 56% of adult *Artemia* in dry weight of digestive tract contents, when these larger organisms were 5% or less by dry weight of the food in the water (four trials), but averaged 95% or more adults in digestive tract contents when adults were between 7 and 99% of the food in the water. The experiment was therefore limited to density combinations in which *Artemia* adults were a small fraction of the total dry weight of available food. The fractions varied from 0.4 to 11% of total dry weight, with over half of the trials having 3% or less and another one-third having between 3 and 7%. The density ranges of each food type were extended as far as practicable within this limitation.

The pilot study also suggested that a behavioral measurement would probably be a more reliable indicator of the response to the food situation existing at the start of each trial than would digestive tract contents. Though both activities were evident in the trials that showed approximately equal dry weight fractions of the two food types in digestive tracts, and biting was distinctly dominant in the other trials, activity did appear to shift towards less biting and more filtering in the last few minutes in a few cases. Such change probably signified elimination of all or most of the larger organisms as feeding progressed.

Effect of *Artemia* Adult and Nauplii Densities on the Proportion of Biting to Filtering Activity

The response of the school to food concentrations in which *Artemia* adults were a small fraction of total dry weight at the start of the trial is described from the feeding activity occurring in the first 3 min of 46 trials. Though response was measured for 8 min, biting activity tended to be constant and maximal only during the first 3, as indicated by activity averaged for all trials (front bars in Fig. 1). Filtering activity was not maximal in the first 3 min, but rather started low and increased steadily over the feeding period. Schools exposed to nauplii alone (rear bars in Fig. 1) showed the same kind of steady increase in filtering activity.

Though interpretation of the first minute is uncertain, the response patterns in general suggest that where *Artemia* adults are present along with nauplii, biting tends to replace filtering rather than be in addition to it.

Figures 2 and 3 show the relation of biting and filtering activity in the first 3 min of feeding to the starting densities of *Artemia* adults and nauplii. Biting tended to increase with adult density and filtering tended to increase with nauplii density, but in each case the relation of response to density appeared to be lower where densities of the alternative food

were higher. This suggests that the school shows a ratio of biting to filtering activity that varies with the densities of both groups of *Artemia*.

Table 1 indicates the effect of each type of food on the relative amounts of biting and filtering, which was calculated as percent biting (the ratio of 10-sec time units scored as biting to the total scored as biting and filtering in the first 3 min). The table is based on the 24 trials with *Artemia* adults between 4 and 50/m³ and *Artemia* nauplii between 100,000 and 500,000/m³. The distribution of these trials permitted separation into four equal groups representing low and high density treatments of comparable range for each food factor. The food density means and ranges are given along with the average percent biting response for each of the food treatment combinations. Percent biting increased with *Artemia* adult density at both levels of nauplii density and decreased

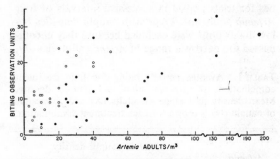

FIG. 2. The relation of biting observations to density of *Artemia* adults for 46 trials in which both adults and nauplii were present. Density of *Artemia* nauplii: ○, < 400,000/m³; ●, > 400,000/m³.

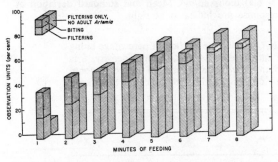

FIG. 1. Percentages of filtering and biting activity for each minute of the trial period. Front bars represent averages by minute for 42 trials in which fish were exposed to combinations of *Artemia* adults and nauplii. Rear bars represent averages by minute for 6 trials in which fish were exposed to nauplii alone. Average nauplii density was higher for the latter series of trials.

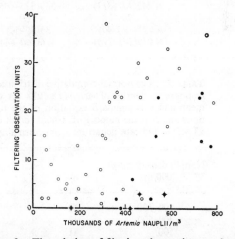

FIG. 3. The relation of filtering observations to density of *Artemia* nauplii for 46 trials in which both adults and nauplii were present. Density of *Artemia* adults: ○, < 40/m³; ●, > 40/m³; ✛, > 100/m³.

with nauplii density at both levels of adult density. An analysis of variance showed that only the two main effects were important, and that percent biting was largely associated with the density of *Artemia* adults (F = 11.0, df 1/20), but nauplii did have a discernible effect (F = 1.58, df 1/20), which would probably have been stronger if the average density of *Artemia* adults had been more nearly alike for the two treatment combinations in which adults were high. In any event, percent biting will be described as a function of *Artemia* adult density for broad ranges of nauplii density.

Trends of relative biting and filtering are shown by normal probability plots of percent biting (PB) on *Artemia* adult density for three different ranges of nauplii density (Fig. 4). Probit analysis was used to estimate the parameters of the fitted lines for suitably limited ranges, i.e., for values excluding 0 and 100% response. Individual points represent PB values for trials pooled in successive intervals of four *Artemia* adults/m³. Trials with nauplii densities below 100,000/m³ were excluded because they encompassed too narrow a range of *Artemia* adult density,

and trials with adult densities above 100/m³ were excluded because a few would have produced responses of 100%, a nonexistent value on the probit scale.

The data for the highest range of nauplii density were judged to be inconclusive because the PB values obviously did not show a significant trend. The only distinguishable increase in response was the PB value for the highest *Artemia* adult density. The arrays for the other two ranges, however, do show good evidence of linear trends, and regressions for each can therefore be taken to represent cumulative normal distributions (Bliss 1967) with parameters $\hat{\mu}$ and σ, where $\hat{\mu}$ is the mean density of adults at 50% response and $\hat{\sigma}$ is the standard deviation. Values for these parameters are given in Table 2. The χ^2 values show that the array of PB values for the lowest

TABLE 1. Average percent biting for four treatment combinations of *Artemia* adult and nauplii density. Mean density and range of adults (A) and thousands of nauplii (N) is shown for each treatment combination.

| *Artemia* adult density | *Artemia* nauplii density | |
	Low	High
Low	42	24
	A 10(4–15)	11(6–18)
	N 242(107–311)	389(325–465)
High	79	69
	A 27(20–40)	35(20–50)
	N 219(141–298)	416(360–484)

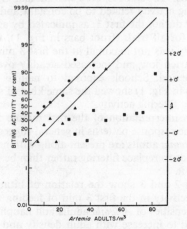

FIG. 4. Probit regressions of percent biting response on density of *Artemia* adults for three categories of *Artemia* nauplii density. *Artemia* nauplii density: ●, 100–300 thousand/m³; ▲, 300–500 thousand/m³; ■, > 500 thousand/m³. Mean and standard deviation of response are indicated at right.

TABLE 2. The normal cumulative distributions that describe the relation of percentage biting response to density of *Artemia* adults for three categories of *Artemia* nauplii density. $\hat{\mu}$ is mean and $\hat{\sigma}$ is standard deviation of the distributions. n is the number of 10-sec response units and N is the number of *Artemia* adult density classes in which they occurred. Significant χ^2 values indicate departure from normality.

Nauplii density range '000/m³	n	N	μ	$\hat{\sigma}$	χ^2
100–300	214	7	11.78±2.78	17.10±3.44	4.08
300–500	342	8	26.70±1.34	17.00±1.76	16.21[a]
500+	381	9	–	–	27.58[b]

[a]$P = 0.05.$
[b]$P = 0.01.$

range of nauplii density is acceptably normal, while that for the high range is not, and that for the middle range is uncertain. In view of the graphic evidence, the deviation from normality indicated for the middle range is probably attributable to inflation of variability, a not unexpected result where the experiment occurred over a long time and the results from individual trials were combined irrespective of the time element for analysis. Use of the uncorrected estimates of $\hat{\mu}$ and $\hat{\sigma}$ means only that the standard errors for each should be considered conservative.

The regression lines for the low and middle ranges of nauplii density, which have the same slope (0.59), and which represent normal cumulative functions that differ only in mean ($\hat{\mu}$), show that the proportion of feeding activity devoted to biting rather than filtering by small schools of anchovies increased as density of *Artemia* adults increased, and/or as density of nauplii decreased. The expected percent of biting activity can be estimated directly from Fig. 4 for any density of *Artemia* adults with *Artemia* nauplii assumed to be at a density of 200,000/m^3 or 400,000/m^3.

The above distributions define percent biting as an average for the school, but the average is based on summation over three positions in the school; and analysis by position indicates that there is a gradation of feeding response within the school. To evaluate the significance of position differences, however, it is necessary to test the hypothesis that variation in feeding activity with position is not associated with variation between observers. Since the positions

were not occupied an equal number of times by each observer among all trials, the hypothesis was tested for 24 trials in which each observer was at each position eight times. The numbers of 10-sec intervals scored as biting (B), filtering (F), and zero (0) activity over the first 3 min is shown in Table 3 by position for all of these trials, and by observer for each position. Feeding activity differs markedly between positions ($\chi^2 = 74.6$), with biting higher than filtering at the front and filtering higher than biting at the middle and rear. Though not grouped as such in the table, differences between position for each observer show the same marked pattern. χ^2 values are 43, 44, and 52 (df = 4) for observers A, B, and C, respectively.

The χ^2 values for the three positions, as given in the table, show, on the other hand, that all three observers scored essentially the same distribution of biting, filtering, and zero observations for the front and rear positions. There is significant deviation from expected at the middle position, although all three observers did score more filtering than biting. Inequities in the average food situation represented by the eight trials for each observer were greater at the middle than at the front and rear positions, and may have been partly responsible for the higher deviations in this set. Variation at the middle position notwithstanding, it can be concluded from these tests that differences between positions are far greater than differences between observers.

The nature of the gradient in feeding activity is evident in probit regressions for each of the three school positions — front, middle, and rear — for

TABLE 3. Contingency tables and χ^2 values for relation of the number of 10-sec intervals scored as biting (B), filtering (F), and zero (0) feeding activity in 3 min to position of observation in the school, and to observer at each position. n = number of feeding trials in which each position was observed. df = degrees of freedom. Expected distributions are shown in parentheses.

Position	Observer	n	B	F	0	χ^2	df
Front	All	24	132	51	249	74.6***	4
Middle	All	24	59	174	199		
Rear	All	24	74	137	221		
			(88.3)	(120.7)	(223.0)		
Front	A	8	51	19	74	8.32	4
	B	8	42	10	92		
	C	8	39	22	83		
			(44.0)	(17.0)	(83.0)		
Middle	A	8	29	56	59	19.3***	4
	B	8	18	45	81		
	C	8	12	73	59		
			(19.7)	(58.0)	(66.3)		
Rear	A	8	28	50	66	7.14	4
	B	8	20	52	72		
	C	8	26	35	83		
			(24.7)	(45.7)	(73.7)		

trials in the 100,000–300,000 nauplii/m³ range (Fig. 5A) and for trials in the 300,000–500,000 nauplii/m³ range (Fig. 5B). The regressions were fitted to PB values calculated for each position for the same combinations of *Artemia* adult and nauplii densities for which the average PB values were calculated. For the lower nauplii range (Fig. 5A) no regression was calculated for the front position because five of the PB values were 100%. These and the two values shown for the front position, nevertheless, demonstrate that fish in the lead position consistently showed far higher percentages of biting than fish in following positions. For the higher range of nauplii density (Fig. 5B) the same phenomenon is evident but less pronounced. Here, however, the line for the middle position indicates that percent biting response became as strong in the middle as in the front position when *Artemia* adult density was high. The general pattern is clearly a gradation from higher percent biting at the front to lower percent biting at the rear with variations in the strength and intensity of gradient dependent on the relative abundance of the two sizes of organism in the water.

Contributions of Biting and Filtering to Ingestion Rate

Average ingestion rate for individuals in the school, and the contribution of *Artemia* adults and nauplii to this average, must depend on the rates at which the two sizes of organism are captured as well as on the ratios for the two associated feeding activities in the school. The relation of numbers captured to density for nauplii is suggested by equation 5 in Leong and O'Connell (1969). Those authors also described feeding rate for *Artemia* adults, but only for densities above 1000/m³. Feeding rate for low densities of adults can be estimated for simulated daylight conditions from the pilot study described under methods above. In 10 of the trials nauplii were essentially ignored by the fish, and the relation between *Artemia* adults in the water and in digestive tracts was

No./m³	6	6	8	8	12	18	100	100	1000	10,000
No./fish	1	1.5	2	3	4.7	8.3	19	25	95	78

The correlation is obvious, except that average number consumed was no higher for a density of 10,000/m³ than for a density of 1000/m³. This is consistent with the finding by Leong and O'Connell (1969) that number of *Artemia* adults consumed by the anchovy was not related to density over the range 1000/m³–25,000/m³. If the 10,000/m³ entry is excluded on the assumption that it is far above some inflexion point where feeding rate ceases to increase with density, the relation for the remaining nine pairs of values ($r = 0.97$) is described by the line

$$\text{Log } N = 0.82 \text{ Log } D - 0.36 \qquad (1)$$

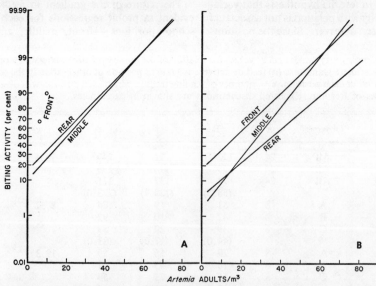

FIG. 5. Probit regressions of percent biting response on density of *Artemia* adults for front, middle, and rear positions in the school for trials with A, *Artemia* nauplii in the 100–300 thousand/m³ range, (No regression line is shown for the front position because 5 of the 7 points were off scale at 100%), and B, *Artemia* nauplii in the 300–500 thousand/m³ range.

where N = average number of *Artemia* adults consumed in 5 min, and D = average number of *Artemia* adults/m³ present in the water at the start of feeding.

Under the assumption that basic rates derived from equation 1 above and equation 5 in Leong and O'Connell (1969) are effective in proportion to the ratios for the two kinds of feeding activity in the school, the products of these basic rates and ratio values for the densities present should provide a partial rate for each size of organism. The sum of the partial rates, expressed in comparable units, will be the average total rate for the density combination. Basic, partial, and total rates are shown for four density combinations in Table 4. The rates are expressed as the milligram dry weight that would be consumed in 1 min by a fish weighing 5 g (average weight of fish in pilot study). The four density combinations are based on the midpoints of the nauplii density ranges for which probit regressions were calculated and on the adult densities indicated by those regressions (Fig. 4) for 50 and 99% biting activity.

The values in Table 4 indicate that when dry weight concentration of *Artemia* adults in the water is 2% of the total concentration, the basic rates are similar for the two sizes of organism and particulate and filter feeding activity are equal. Under these circumstances the ingestion rate for both sizes together, or total rate, is no greater than would be possible by filtration alone on the nauplii. However, when the dry weight concentration of *Artemia* adults is in the order of 5.5–8% of the dry weight of adults and nauplii together, the basic rate for *Artemia* adults is approximately double that for nauplii. Under these circumstances total ingestion rate is approximately double that possible by filtration alone, and filter feeding contributes little to total ingestion.

The possibilities of ingestion rate, of course, are not limited to the combinations illustrated. Total ingestion rate would presumably be intermediate between those for 50 and 99% biting activity if density of larger organisms were intermediate between those that support such levels of biting. Ingestion rate might also be more than double the filtration potential if the density of larger organisms were markedly greater than necessary for the 99% level of biting activity. Also, ingestion rates might be somewhat lower, or the concentrations of large organisms necessary to sustain them somewhat higher than indicated, if escapement of larger crustaceans in the sea is appreciable. Though biting may tend to counteract escapement, the calculations are based on *Artemia* adults, which appear to have no effective escape response.

Interrelation of Biting and Filtering in the Sea

Although the anchovy probably relies on filter feeding for a steady supply of food in the sea, this supply may often fail to sustain the daily nutritional requirement, as already hypothesized. The experiments described here suggest that biting could be an augmenting activity of some consequence, depending on the size composition of the plankton. O'Connell (1971) showed that plankton pump samples from surface waters off southern California vary greatly in relative abundance of large and small crustaceans. Of the constituents described, small copepods, with an average dry weight equivalent of 0.0025 mg, and euphausiids, with average dry weight equivalents of 0.042 mg for day samples and 0.293 mg for night samples, most closely approximate the *Artemia* nauplii and adults used in the laboratory

TABLE 4. Estimated ingestion rates for density combinations of *Artemia* adults and nauplii at which biting activity is 50 and 99% of total feeding activity.

Percent biting activity Organism type	Low nauplii density				High nauplii density			
	50		99		50		99	
	Nauplii	Adults	Nauplii	Adults	Nauplii	Adults	Nauplii	Adult
Density no./m³	200,000	12	200,000	52	400,000	27	400,000	66
Dry wt mg/m³	290	6	290	26	580	13.5	580	33
Adult % of total dry wt		2.0		8.2		2.3		5.4
Ingestion rates (mg dry wt/min) for above densities								
Basic	0.49	0.34	0.49	1.1	0.73	0.65	0.73	1.4
Partial	0.24	0.17	0.01	1.1	0.37	0.33	0.01	1.4
Total		0.41		1.1		0.70		1.4

study. They also constituted the bulk of the dry weight of the samples. Averages for five cruises indicated that euphausiids were 3–5% of the dry weight of both groups during the day and 21–66% at night. These percentages suggest that anchovy feeding in surface waters would be chiefly filtering in the daytime and chiefly biting at night, with ingestion rate far exceeding the filtration potential in the latter period.

However, the small laboratory school can scarcely represent more than the leading edge of many schools in the sea. Since percentage of biting diminished noticeably in following positions in a small school, it is possible that diminution extends to a complete lack of opportunity for biting activity for most of the fish in large, compact schools. This would result in a lower capture rate of the larger crustaceans and a lesser elevation of total digestion rate, on the average, than might be expected for a small school exposed to the same relative concentrations.

There is circumstantial evidence to suggest, on the other hand, that the anchovy schools more loosely when larger food organisms are abundantly available. In the laboratory experiments the fish schooled under all trial circumstances, but in the presence of the higher densities of *Artemia* adults, where initial activity was largely biting, they appeared to swim faster and be more widely separated than in circumstances where most of the activity was filtering. Hobson (1968), on the basis of extensive field observations, found that the flatiron herring, *Harengula thrissina*, remained inshore in dense schools during the day and moved offshore, possibly in looser aggregations, to feed at night when shrimp-like crustaceans were most abundant in midwaters. He cited other clupeids that have been observed to break up into smaller, more loosely associated feed-ing groups at night. Such a behavior pattern, which might well apply to the northern anchovy, would operate to improve the average rate of encounter with larger crustaceans, and thereby raise the average rate of ingestion in respect to the filtration potential, although perhaps still not as much as might be expected for a small laboratory school.

Whatever the daily feeding cycle of the anchovy, it appears that opportunities for biting activity would be greatest in dim light — either at depth during the day, or near the surface at night. Because of strong vertical migration, high densities of euphausiids are common only under these conditions, and O'Connell (1971) has shown that the larger copepods, which may also elicit biting, have a vertical migration pattern similar to that of the euphausiids. Small copepods, on the other hand, showed evidence of only a slight nighttime increase at the surface.

Feeding by biting, because it is so obviously dependent on visual perception of individual organisms, may not seem compatible with low nighttime levels of illumination, but there is some evidence to the contrary. Hobson (1968) commented that some authors greatly underestimate the extent to which vision can be used at night by both predators and prey. Hunter (1968) showed that juvenile jack mackerel (*Trachurus symmetricus*) obtained 50% of the usual daytime ration when feeding by biting on *Artemia* adults at a light intensity as low as would occur near the surface on a full moonlit night. He noted, also, that more visible energy may exist in the sea than his calculations indicate. Blaxter (1964) listed a number of "visual" feeders that stopped feeding at similarly low light intensities. In regard to the northern anchovy, O'Connell (1963) showed on anatomical grounds that this species has a retina that is probably better adapted

TABLE 5. Hours of feeding necessary for an anchovy weighing 5 g to obtain its daily nutritional requirement (74 mg dry wt) from observed nighttime plankton concentrations for four different percentages of particulate feeding activity. Rules separate the values at 12 and 24 hr.

Zooplankton concentration mg/m³ dry wt	% samples	Hr of feeding			
		Filtering only	50% biting	75% biting	99% biting
10	27	84	84	56	42
20	17	42	42	28	21
30	41	28	28	19	14
40	10	21	21	14	11
50	5	17	17	11	9

than that of the jack mackerel for dim-light vision.

From all of the above evidence it seems highly probable that the anchovy can feed by biting at night, and by so doing effectively exceed the limitations of filtering. Loukashkin's (1970) finding that euphausiids and large copepods were the most important items in stomach contents of the anchovy tends to confirm this view.

Leong and O'Connell (1969) evaluated the food potential of zooplankton concentrations in the sea for the anchovy by estimating the hours of filtering necessary to obtain the daily nutritional requirement. Table 5 suggests the way in which different percentages of biting activity might effect this index of food potential, where the different percentages of biting imply that a given concentration is composed of different though undefined proportions of large and small organisms. The percentage distribution for dry weight concentrations is based on the night samples from five plankton pump cruises (O'Connell 1971). Hours of filtering for each dry weight class were calculated on the assumptions that a 5 g anchovy requires 74 mg dry weight of food per day and filters water at the rate of 1.46 liters/minute (Leong and O'Connell 1969). Hours of feeding for percentages of biting were calculated on the basis of the relations suggested in Table 5. That is, ingestion rate is no different for plankton that elicits 50% biting than for plankton that elicits filtering only, but varies proportionately from a factor of one at 50% biting to a factor of two at 99% biting. Feeding duration is inversely proportional to feeding rate.

The calculated values, which are separated by lines at 12 and 24 hr in the table, indicate that with low biting activity only 15% of the area sampled would have provided the anchovy with its nutritional requirement in one day. The same fraction of the area would have provided the daily requirement in less than 12 hr if the plankton concentrations supported relatively high average percentages of biting. If feeding extended over more than 12 hr with a high percentage of biting most of the time, more than half of the area could have provided the required nutrition. The anchovy may capture large organisms less effectively in relation to density at

night than indicated earlier for simulated daylight conditions, but given the high ratios of euphausiid to small copepod dry weight in night samples, some average percentage of biting well above 50% seems entirely possible. Availability of the daily nutritional requirement would also tend to be widened if the anchovy utilizes phytoplankton along with zooplankton. These considerations suggest that plankton is usually adequate for the nutritional needs of the anchovy in a large part of the surface waters off southern California.

BLAXTER, J. H. S. 1964. Effect of change of light intensity on fish. Int. Comm. Northwest Atl. Fish. Spec. Publ. 6: 647–661.

BLISS, C. I. 1967. Statistics in biology. Vol. I. McGraw-Hill Book Co., New York, N.Y. 558 p.

BROOKS, J. L., AND S. I. DODSON. 1965. Predation, body size and composition of plankton. Science 150: 28–35.

HIATT, R. W. 1951. Food and feeding habits of the nehu, *Stolephorus purpureus* Fowler. Pac. Sci. 5: 347–358.

HOBSON, E. S. 1968. Predatory behavior of some shore fishes in the Gulf of California. U.S. Fish Wildl. Serv. Res. Rep. 73: 92 p.

HUNTER, J. R. 1968. Effects of light on schooling and feeding of jack mackerel, *Trachurus symmetricus*. J. Fish. Res. Bd. Canada 25: 393–407.

LEONG, R. J. H., AND C. P. O'CONNELL. 1969. A laboratory study of particulate and filter feeding of the northern anchovy (*Engraulis mordax*). J. Fish. Res. Bd. Canada 26: 557–582.

LOUKASHKIN, A. S. 1970. On the diet and feeding behavior of the northern anchovy, *Engraulis mordax* (Girard). Proc. Calif. Acad. Sci. 37: 419–458.

O'CONNELL, C. P. 1963. The structure of the eye of *Sardinops caerulea*, *Engraulis mordax*, and four other pelagic marine teleosts. J. Morphol. 113: 287–330.

——— 1971. Variability of near-surface zooplankton off southern California, as shown by towed pump sampling. U.S. Fish Wildl. Serv. Fish. Bull. 69: 681–697.

SHEN, S.-C. 1969. Comparative study of the gill structure and feeding habits of the anchovy, *Engraulis japonica* (Hout.). Bull. Inst. Zool. Acad. Sinica (Taipei) 8: 21–38.

Feeding pattern of baleen whales in the ocean

Takahisa Nemoto
Ocean Research Institute
University of Tokyo
Japan

ABSTRACT. A study of the main feeding structures of baleen whales—baleen plates, ventral groove extension, tongues and the shape of heads—prove that there are three feeding types, blue whale (blue, fin, humpback Bryde's), right whale (right, Greenland) and grey whale types. These types are divided into "skimming", "swallowing" and a combination of these two. Food species are described both for the southern hemisphere and northern Pacific, and a clear selection of foods is observed. The swallowing type whales demand heavy patches of food organisms such as euphausiids and gregarious fish, but skimming whales depend on more sparse plankton patches. Feeding activity is high in the morning and becomes low in daytime. Food quantities in stomachs are considered to coincide with the probable caloric intake estimated from the huge weight and big proportion of the blubber tissue in whales. Two main food chains involving whales in the Antarctic and a more complex one in the northern hemisphere are discussed. Seasonal feeding migration to seek heavy patches of zooplankton is suggested chiefly for swallowing type whales. The feeding range and areas are also restricted by the biological strength (number, size, school, time of migration) and interspecific interferences.

INTRODUCTION

The status of baleen whales in marine food chains is very interesting considering their huge body sizes, in conjunction with their consumption of comparatively minute foods. Baleen whales had been considered to feed generally on small zooplankton, but recent studies show that the habits and selection of food vary considerably from species to species according to their feeding apparatus (Tomilin, 1954; Nemoto, 1959; Kulmov, 1961).

For a study of feeding patterns eleven baleen whales belonging to three families and six genera are considered here.

130

Reprinted from Marine Food Chains ed. by J. H. Steele, pp. 241–252 (1970). Univ. Cal. Press, Berkeley, Ca.

TABLE 1

Balaenidae		
Balaena mysticetus	(Greenland right whale)	Northern Polar Seas
Eubalaena glacialis	(Right whale)	Northern and Southern hemisphere
Caperea marginata	(Pygmy right whale)	Southern hemisphere
Eschrichtiidae		
Eschrichtius gibbosus	(Grey whale)	North Pacific
Balaenopteridae		
Balaenoptera musculus	(Blue whale)	Northern and Southern hemisphere
Balaenoptera physalus	(Fin whale)	Northern and Southern hemisphere
Balaenoptera edeni	(Bryde's whale)	Northern and Southern hemisphere
Balaenoptera borealis	(Sei whale)	Northern and Southern hemisphere
Balaenoptera acutorostrata	(Little piked whale, Minke whale)	Northern and Southern hemisphere
Balaenoptera bonaerensis	(New Zealand piked whale)	Southern hemisphere
Megaptera novaeangliae	(Humpback whale)	Northern and Southern hemisphere

Feeding apparatus

The main structures affecting the feeding of baleen whales are baleen plates, head structures including mouth opening and tongue, and ventral grooves in the abdominal part of the body. The characteristics of baleen plates are summarized in Table 2. The shape of plates in right, Greenland and pygmy right whales is slender and elastic, and the fringes along the inner margin of plates are very fine and numerous. The plates of Balaenopteridae whales (blue, fin, humpback, etc.) are short and tough, and have rather rough baleen fringes. Sei whales have somewhat slender plates and sei and minke whales have finer baleen fringes. Grey whales have thick, short plates with coarse, short fringes. The younger whales generally have finer baleen fringes than the adults, and local differences in the character of baleen plates in the same species are also observed in many species if they feed regularly on different kinds of foods.

TABLE 2. Characteristics of baleen plates of whales in the North Pacific

	Blue	Fin	Bryde's	Sei	Minke	Hump-back	Right	Grey
Mean number in one side	360	355	300	340	280	330	245	160
Mean diameter of baleen fringe (mm)	1·1	0·8	0·6	0·2	0·3	0·7	0·2	1·0
Number of baleen fringes per 1 cm	10-30	10-35	15-35	35-60	15-25	10-35	35-70	10-15
Shape and quality	tough	tough	tough	tough	somewhat elastic	tough	slender elastic	tough

The arrangement of baleen plates indicates three types. Balaenidae whales lack the plates or smaller tuft at the tip of the upper palate. Grey whales also lack the plates at the tip but the filtering area of baleen plates is limited as they have short plates. *Balaenoptera* whales have smaller plates or tufts at the tip. The skull is curved in Balaenidae whales and rather straight in *Balaenoptera* whales, although sei whales show a somewhat different curve from other *Balaenoptera* whales. Right and Greenland whales have panels in both sides of the lower jaw, which grows upwards to fit the lower edge of the curved arch of the skull.

Although the tongue of right and grey whales is rather tough, in *Balaenoptera* whales it is very flabby. The former is effective in conducting water which contains planktonic food along the inner plane of the rows of baleen plates in continuous feeding; the latter works in such a way as to make the mouth cavity expand for swallowing large amounts of water with the food.

Balaenidae whales lack the ventral groove. Grey whales have only two to four furrows in the abdominal part instead of grooves. The grooves of humpback whales are wide, and their number (about 18 to 24) is far less than other *Balaenoptera* whales (about 52-82). However, the extension is the same with blue and fin whales ranging from 55 to 58% of the body length. Sei and minke whales have a shorter extension of grooves ranging from 45 to 47% of the body length.

TABLE 3.   Feeding apparatus types of baleen whales in the North Pacific*

Whale species	Baleen plates			Head, mouth and tongue	Ventral grooves	Apparatus type
	Shape	Fringe	Row			
Blue	Blue	Blue	Blue	Blue	Blue	Blue
Fin	Blue	Blue	Blue	Blue	Blue	Blue
Bryde's	Blue	Blue	Blue	Blue	Blue	Blue
Sei	Blue	Right	Blue	Blue (Sei)	Sei	Blue (Sei)
Minke	Blue	(Blue)	Blue	Blue	Sei	Blue (Sei)
Humpback	Blue	Blue	Blue	Blue	Blue	Blue
Grey	Grey	Grey	Grey	Grey	Grey	Grey
Right	Right	Right	Right	Right	Right	Right
Greenland	Right	Right	Right	Right	Right	Right

* Modified from the list by Nemoto (1959)

Those characteristics suggest three main types of feeding apparatus, namely blue, grey and right whale types (Table 3), although some parts of sei whales and sometimes minke whales demonstrate a separate sei whale type.

Food species and selection of food

Antarctic. Euphausia superba had been considered as the only important food for baleen whales in the Antarctic; however, recent investigations (Nemoto, 1962; 1968) have described many other planktonic animals as food of baleen whales (Table 4). Blue and fin whales feed mainly on euphausiids and the species they take are considered to form dense swarms in the sea.

TABLE 4. Stomach contents of baleen whales caught by Japanese pelagic catch from 1961 to 1965 in the Antarctic*

Food species	Whale species				
	Blue†	Fin	Sei	Humpback	Minke
Euphausiids	517	16158	5936	7	88
Euphausiids and others	4	18	4	—	—
Copepods	2	—	2472	—	—
Amphipods	6	9	1514	—	—
Munida decapods	—	—	75	—	—
Fish	—	76	31	—	—
Squids	—	—	5	—	—
Vacant	674	18878	16145	2	10
No. of whales examined	1203	35139	26182	9	98

* Sei whales include 1966 season.
† Mainly subspecies *Balaenoptera musculus brevicanda* distributed in the lower Antarctic.

Euphausia crystallorophias is often observed in a heavy patch along the Antarctic continental shelf and is fed upon by blue and minke whales (Marr, 1956). *Euphausia superba, Thysanoessa macrura* and *T. vicina* are found within the Antarctic convergence. There are other euphausiids in the Antarctic waters, such as *Euphausia triacantha, E. longirostris, E. lucens, E. fringida,* (John, 1936; Baker, 1965) which have not been described as the main food of baleen whales owing to their non-swarming habit in the ocean (Baker, 1959, for *E. triacantha*). *Euphausia vallentini* is mainly found along the Antarctic convergence (Nemoto, 1962). The subspecies of blue whales, pigmy blue whales, are confined to those waters of lower latitudes where *E. vallentini* is abundant, where they are separated from the larger normal form of blue whales.

Sei whales in the southern hemisphere feed on various plankton, i.e. animals such as carnivorous amphipods (*Parathemisto gaudichaudi*) and herbivorous copepods (*Drepanopus pectinatus*, etc). These are mostly found along the Antarctic and subtropical convergences, and this is the reason why sei whales feed in the comparatively lower latitudes of the Antarctic and subantarctic region. There are also large stocks of zooplankton, including the planktonic decapod *Munida gregaria* (or lobster krill), in the coastal waters of Patagonia.

Minke and Antarctic minke (*Balaenoptera bonaerensis*) whales feed on *Euphausia superba* or *E. crystallorophias* in the pack ice region and offshore feeding is rarely observed.

North Pacific. The food in the north Pacific (Table 5) varies considerably (Nemoto, 1959). Blue whales feed almost entirely on euphausiids. Fin whales feed on both euphausiids and copepods (*Calanus cristatus* and *Calanus plumchrus*). Shoaling fish, herring (*Clupea pallasii*), Alaskan pollack (*Theragra chalcogramma*), capelin (*Mallotus catervarius*) also form part of their diet. Sei whales on the other hand feed mainly on copepods (*Calanus*

TABLE 5. Stomach contents of baleen whales caught by Japanese pelagic catch from 1952 to 1965 in the North Pacific

Food species	Whale species				
	Blue	Fin	Sei	Humpback	Right
Euphausiids	455	4818	85	238	—
Euphausiids and Copepods	5	321	2	2	—
Euphausiids and others	—	34	—	12	—
Sergestes shrimp	1	—	—	—	—
Copepods	6	1877	1459	2	9
Copepods and others	—	3	12	—	—
Fish	—	469	36	53	—
Fish and others	—	—	1	—	—
Squids	—	51	21	1	—
Vacant	504	8794	3565	150	0
No. of whales observed	971	16367	5181	458	9

plumchrus) and those feeding on euphausiids are rather scarce. Humpback whales appear to feed mainly on euphausiids and gregarious fish, although the number of observations is low. Right whales feed only on copepods such as *Calanus plumchrus* and *C. cristatus*. Bryde's whales feed in comparatively warmer waters both in the northern and southern hemispheres (Nemoto, 1959). In the north Pacific they sometimes feed on oceanic micronekton such as lantern fish, *Yarrella microcephala* and *Myctophum asperum*.

Species of euphausiids important as food of baleen whales in the northern Pacific are *Euphausia pacifica, Thysanoessa inermis, T. longipes, T. spinifera* and *T. raschii*, which form extensive swarms in the sea (Nemoto, 1959). In the southern waters of the north Pacific, *E. pacifica, E. similis, E. recurva* and *E. nana* are important as food for fin whales. The copepods *Calanus cristatus, C. plumchrus* and *Metridia lucens* are important as food. *Eucalanus* and other non-swarming copepods never form part of the food supply of baleen whales.

Thus, the order of selection of the food in baleen whales is as follows (= shows equivalence and > shows the dominance to the left).

Blue whale	Euphausiids
Fin whale	Euphausiids = Copepods (large) = Gregarious fish > Copepods (small) > Squids
Bryde's whale	Euphausiids = Gregarious fish > Copepods (small)
Sei whale	Copepods ≥ Amphipods ≥ Euphausiids = Swarming fish = Squids
Humpback whale	Euphausiids = Gregarious fish
Minke whale	Euphausiids = Swarming fish > Copepods
Right whale	Copepods > Euphausiids

Feeding types

Two basic feeding types, defined as swallowing and skimming have been proposed (Nemoto, 1959; 1968). *Balaenoptera* whales (except *B. borealis,*

B. acutorostrata) show swallowing type feeding. Sei whales show both types according to food species present (skimming was observed by Ingebrigtsen, 1929) and *Balaenidae* whales show skimming type feeding. Swallowing type whales (blue, fin, Bryde's and humpback whales) swallow the food found in the patch or swarm, along with water, then discharge the sea water through the baleen plates while the food remains in the mouth cavity. They never take scattered plankton nor carry out feeding by swimming with their mouths open. Skimming type whales (right,* Greenland whales) take their food by swimming with their mouths open and the food retained in the mouth cavity is gulped. They may sometimes take a sparse or smaller patch of copepods and amphipods.

The exact position of grey whales is still not clear. However, they possibly combine these two types since field observations have shown them to feed intensively on swarms of zooplankton and also to skim off mud in shallow water in the Bering Sea to feed on benthic amphipods.

The feeding types can be classified as follows:

> Swallowing type
>> blue whale
>> fin whale
>> Bryde's whale
>> humpback whale
>> minke whale
>> Antarctic minke whale
> Skimming type
>> right whale
>> Greenland whale
>> pygmy right whale
> Skimming and swallowing type
>> sei whale
>> grey whale

Diurnal and seasonal variations

The feeding activity in the open ocean is high in the early morning, becomes lower later in the day and increases again late in the evening (Nemoto, 1959). Other marine mammals such as fur seals in the north Pacific also show the same tendency in feeding on squid and myctophid fish (Tayler, Fujinaga and Wilke, 1955). The variation in the feeding activity is due to the diurnal migration of food species. In the shallow water region on the continental shelf of Anadyle gulf in the north Pacific, fin whales still feed very actively on euphausiids *Thysanoessa raschii*, and capelin (*Mallotus catervarious*) in the daytime. When fin whales are feeding on copepods, *Calanus cristatus* and

---

* Field observations show two right whales swimming along the current rip with their mouths open in order to take a sparse patch of copepods which were so scattered as not to be recognizable from the airplane above (W. Schevill, personal communication on a film shown at Harvard University in 1962).

C. plumchrus, they do not exhibit such a clear diurnal variation in their feeding activity, the reason for which may be due to less intense migration of copepods in the summer.

The seasonal vertical migration of plankton also affects the feeding of baleen whales. The copepod *Calanus cristatus* in the north Pacific spends spring and summer in the upper layers, especially as copepodite stage V. They feed to store oils in their bodies and at this stage are fed upon by fin whales. After summer, *Calanus cristatus* goes deeper than 500 metres (Heinrich, 1957) and then fin whales cannot take them as the diving range of large fin whales is limited to about 300 metres (Scholander, 1940). Adults of both *C. plumchrus* and *C. cristatus* have not been observed in the stomachs of baleen whales (Nemoto, 1963).

TABLE 6. Examples of weights of foods in the stomachs of baleen whales in the North Pacific

Food and whale species	Food species	Amount of foods (kg)
Euphausiids		
Fin whale (18-19 m)	*Thysanoessa* and *Calanus*	425 (Ponomareva, 1949)
Fin whale (18 m)	*Thysanoessa inermis*	113 (Nemoto, 1959)
Bryde's whale (12 m)	*Euphausia similis*	204 (Nemoto, 1959)
Fin whale (?)	?	364 (Betesheva, 1954)
Fin whale (18-19 m)	*Thysanoessa raschii*	340 (Ponomareva, 1949)
Copepods		
Fin whale (18-19 m)	*Calanus plumchrus*	255 (Ponomareva, 1949)
Fin whale (19 m)	*Calanus cristatus*	107 (Nemoto, 1963)
Fin whale (18 m)	*Calanus plumchrus*	72 (Nemoto, 1963)
Fin whale (18 m)	*Metridia lucens*	80 (Nemoto, 1963)
Sei whale (?)	*Calanus plumchrus*	370 (Betesheva, 1954)
Fish		
Fin whale (17 m)	*Theragra chalcogramma*	759 (Nemoto, 1959)
Fin whale (?)	*Corolabis saira*	464 (Betesheva, 1954)
Squids		
Fin whale (?)	*Todarodes pacificus*	560 (Betesheva, 1954)
Sei whale (?)	*Todarodes pacificus*	600 (Betesheva, 1954)

More than 1000 kg of *Euphausia superba* were found in the stomachs of baleen whales in Antarctic waters (Marr, 1962). For the north Pacific, data on the quantity taken by baleen whales are shown in Table 6. The fresh condition of food suggests it is usually recently taken. There is no information about the weight of amphipods in the north Pacific, but sei whales took 200 kg amphipods in the southern hemisphere (Kawamura, 1968).

Food chain through baleen whales

Namoto (1968) gives two main flows through baleen whales in the Antarctic. One is "phytoplankton→herbivorous euphausiids *Euphausia superba*→blue and fin whales" which is possibly the shortest one in the ocean (Mackintosh, 1965). The other is "small phytoplankton→protozoa, larvae of zooplankton and small copepods→carnivorous amphipods (*Parathemisto gaudichaudi*)→

sei whales". In the northern hemisphere, the relationships are far more complicated as there are many different feeding patterns observed.

The feeding patterns and feeding mechanisms of euphausiids (Nemoto, 1968) show that the gregarious euphausiids feed mainly on phytoplankton or on small zooplankton. True carnivorous euphausiids however are never considered to form heavy swarms in the sea.

The relation between food concentrations and the filtering volumes of the baleen whales may be the key in solving the question of the feeding of these large animals. Large right whales have about 2·70 cm baleen plates in the centre of the baleen plate row. Left and right rows form about 13·5 m² of a

TABLE 7. Approximate filtering area of baleen whales formed by baleen plate row

Species	Body length (m)	Filtering area (m²)	Maximum length of baleen (cm)
Blue	27	4·6	85
Fin	24	4·0	80
Sei	16	2·2	70
Bryde's	15	1·7	60
Humpback	15	2·6	70
Black right	17	13·5	270
Fin*	18-19	2·4-2·8	62-70

* Estimated by Kulmov (1961).

filtering curtain which skims small zooplankton in continuous feeding as a maximum estimate. If the right whales swim with the mouth open, the mouth opening may be about 8·9 m² as a maximum. The filtering areas of other baleen whales by baleen plates are shown in Table 7. If right whales filter the sea water by swimming at a speed of 6 km per hour (Nishiwaki, 1965), then the amount of food taken by filtering is shown in Table 8. These values give the maximum, as the mouth opening may be less than the filtering area, and the values with 8·9 m² are also given. They show us that they can take suitable amounts of Calanoid copepods (*Calanus plumchrus* and *Calanus*

TABLE 8. Amounts of food taken according to different food plankton concentration in the sea by skimming method (kg per hour)

Speed in feeding (km)	Food concentration (mg/m³)				
	100	500	1000	2000	4000
13·5 m² filtering area					
1	1·4	6·8	13·5	27·0	54·0
2	2·7	13·5	27·0	54·0	108·0
4	5·4	27·0	54·0	108·0	216·0
6	8·1	41·1	81·0	162·0	324·0
8·9 m² filtering area					
1	0·9	4·5	8·9	17·8	35·6
2	1·8	8·9	17·8	35·6	71·2
4	3·6	17·8	35·6	71·2	142·4
6	5·3	26·7	53·4	106·8	213·6

cristatus), as the feeding activity of right whales is still high even in the day-time. For swallowing type whales (blue, fin, humpback and others) the volume of the mouth cavity is calculated as 4·5 m³ (Fraser, in Marshall and Orr, 1955) and 6·0 m³ by Kulmov (1966). The stretch of the ventral grooves (Matsuura, 1943) when feeding suggests that the cavity volume may be slightly greater than these values. If we use these volumes the amount of food taken by a swallowing action is shown in Table 9. Tables 8 and 9 show the filtering by baleen plates in skimming food is effective for right and Green-land whales which have broader filtering areas of baleen plates and curved skulls. For the swallowing types it is difficult to explain the quantity of food found in the stomachs from the average zooplankton biomass. In the feeding grounds of baleen whales in the north Pacific, Bogorov and Vinogradov (1956) describe the maximum biomass slightly exceeding 2000 mg/m³. Ponomareva (1966) also described the abundance of euphausiids as slightly more than 1/m³ both in the Antarctic and Arctic. Thus the standing stock is 100 to 200 mg/m³ in most parts of the feeding area of the north Pacific and 0·75 to 1·5 g/m³ in the Antarctic.

TABLE 9. Amount of food taken according to different food concentration in the sea and mouth cavity volume swallowed by baleen whales (g per swallowing)

Swallowing volume one action (m³)	Food concentration (g/m³)								
	0·1	0·5	1	10	100	500	1000	5000	10000
4·5	0·45	2·25	4·5	45	450	2250	4500	22500	45000
6·0	0·6	3	6	60	600	3000	6000	30000	60000
10·0	1	5	10	100	1000	5000	10000	50000	100000
15·0	1·5	7·5	15	150	1500	7500	15000	75000	150000

It is very clear that food organisms, plankton, fish and squid must swarm in very heavy concentrations for swallowing type whales to be able to feed. The study of the actual concentration of plankton in patches is very important for the study of these food chains in the future. If the concentration of euphausiids is considered according to the maximum values shown in the literature (Boden, 1952; Marr, 1962; Ponomareva, 1966; Nemoto, 1968), then swallowing whales might take sufficient food. The calorimetric balance in baleen whales is discussed by Kulmov (1966). The basal metabolism measured by Benedict (in Nakaya, 1961) and Sherman (1949) gives us insight into the problem. Kulmov (1966) has already drawn attention to the agree-ment of calories in the food with metabolic rate. Data from different sources for fin and sei whales are given in Table 10, which also confirm the consider-able amount of food which must be taken. The basal metabolism in baleen whales as a fraction of their weight must be very low. The high proportion of blubber shown in Table 11, which is very peculiar to baleen whales, also reduces the basal metabolism even further and improves flotation.

TABLE 10. Tentative kilo-calories of standing stocks of food of fin (18 m, 35 ton) and sei (16 m, 18 ton) whales in the North Pacific per 1 kg flesh of the body*

FIN WHALE

Food species	Food quantity (kg)					
	300	400	500	600	700	800
Euphausiids	9	13	**16**	19	22	25
Copepods	12	15	**19**	23	27	31
Allaska pollack	7	9	11	13	15	**18**
Squids	7	10	12	**14**	17	19
Saury	14	19	**23**	28	32	37

SEI WHALE

Food species	Food quantity (kg)					
	100	200	300	400	500	600
Mackerel	6	13	19	25	32	38
Saury	9	18	27	**36**	45	54
Anchovy	7	14	21	28	35	42
Squid	5	9	14	19	23	**28**
Copepods	8	15	23	**30**	38	45
Amphipods†	10	20	30	40	50	60

* Field observed food amounts are in bold figures
† Based on Kulomov's calorie value (1961).

TABLE 11. Body weight of baleen whales

Species	Length (m)	Weight (ton)	Blubber weight (ton)	Locality
Blue	29·5	108·3	20·9 (19·3%)	Antarctic
	27·8	96·5	18·7 (19·4%)	Antarctic
	26·2	81·2	15·8 (19·5%)	Antarctic
Fin	26·2	60·8	14·2 (23·4%)	Antarctic
	24·6	54·0	12·4 (23·0%)	Antarctic
	23·0	45·6	10·5 (23·0%)	Antarctic
	19·7	36·3	8·3 (22·9%)	Antarctic
	18·2	30·0	7·0 (22·9%)	Antarctic
Sei	18·0	22·8	3·9 (17·1%)	North Pacific
	16·4	18·2	3·1 (17·0%)	North Pacific
	14·7	14·1	2·5 (17·7%)	North Pacific
Right	16·4	78·5	29·2 (37·2%)	North Pacific
	15·1	55·3	23·0 (41·6%)	North Pacific
	14·1	47·6	20·0 (42·0%)	North Pacific

Figures concerning examples for sei and fin whales in the north Pacific are given, underlined in Table 10. If whales can take this amount of food twice a day in the summer, they may easily store the surplus calories as storage oil in their bodies for their winter migration to the south (Kulmov, 1961).

The distribution of baleen whales and zooplankton standing stocks in the ocean show good agreement. Skimming type whales such as right whales

(*Eubalaena glacialis*) are distributed mainly in the waters where the copepods are dominant. An example of this occurs in the sea off Japan where *Calanus plumchrus* occupy two-thirds of the total zooplankton standing stock in the summer season (Nakai, 1942), and many right whales have been caught there. Right whales were also distributed between the Antarctic and subantarctic convergences in the southern hemisphere, and in the northern part of the north Pacific, and give evidence of a short migration from winter to summer season (Townsend, 1935). Greenland whales are confined to the polar seas. On the other hand, swallowing type baleen whales show a typical seasonal migration from the high latitudes where the swarm of zooplankton (euphausiids and copepods) is available, to poorer waters in winter for breeding (Mackintosh, 1965). The feeding range of baleen whales in the ocean is also restricted by inter-specific competition (Nemoto, 1959; Sergent, 1968). In Antarctic waters, blue whales had occupied the pack ice region in former years, but in recent years fin whales have penetrated more into higher latitudes because of the decrease in the stock of blue whales. The fact that sei whales are also often observed south of the Antarctic convergence in recent years may be attributable to the decrease of fin and blue whales in those areas (Nemoto, 1962). In the north Pacific, the main areas of feeding in blue, fin and sei whales often show discrepancies. Of course few sei whales feed on euphausiids but they compete with fin whales for *Calanus* copepods in the feeding areas.

The number of baleen whales in the Antarctic has been estimated by the International Whaling Commission and the national scientific groups concerned. More than 400 000 blue, fin and humpback whales were estimated as a virgin stock in the Antarctic before 1930 (Mackintosh, 1965) and the total weight of baleen whales would have been 25·76 million tons (Crisp, 1962). At that time they must have consumed more than 772 800 tons of *Euphausia superba* per day, as the caloric value of *E. superba* is 1000 cal/kg and 1 kg flesh of baleen whales needs 30 calories per day. The summer feeding season may be about 100 days in the Antarctic and so the *Euphausia superba* consumed amounted to 77·3 million tons a year, which is somewhat larger than the value given by Marr (1962, about 37·8 million tons). This is only speculation on the food chains through the Antarctic whales. However, it suggests that at present a vast stock of *E. superba* remains unused because of the decrease of baleen whales in the Antarctic. The decrease of an ocean harvester, such as baleen whales, may thus cause very large changes in marine food chains and the consequences of this are still a problem to be solved both in the Antarctic and in the northern hemisphere.

REFERENCES

BAKER, A. de C. 1959. Distribution and life history of *Euphausia triachantha*, Holt and Tattersall. *Discovery Rep.*, **29**, 309-40.
BAKER, A. de C. 1965. The latitudinal distribution of *Euphausia* species in the surface waters of the Indian Ocean. *Discovery Rep.*, **33**, 309-34.

BETESHEVA, E. I. 1954. Data on the feeding of baleen whales in the Kurile region. *Trudȳ inst. Okeanol.*, **2**, 238-45.

BODEN, B. P. 1952. Plankton and sonic scattering. *Rapp. P.-v. Réun. Cons. perm. int. Explor. Mer*, **153**, 171-76.

BOGOROV, B. G., and VINOGRADOV, M. E. 1956. Some essential features of zooplankton distribution in the North-Western Pacific. *Trudȳ Inst. Okeanol.*, **18**, 60-84.

CRISP, D. T. 1962. The tonnages of whales taken by Antarctic pelagic operations during twenty seasons, and an examination of the blue whale unit. *Norsk Hvalfangsttid.*, **51** (10), 389-93.

HEINRICH, A. K. 1957. The propagation and the development of the common copepods in the Bering sea. *Trudȳ vses. gidrobiol. Obshch.*, **8**, 143-62.

INGEBRIGTSEN, A. 1929. Whales caught in the North Atlantic and other seas. *Rapp. P.-v. Réun. Cons. perm. int. Explor. Mer*, **56**, 1-26.

JOHN, D. D. 1936. The southern species of the genus *Euphausia*. *Discovery Rep.*, **16**, 193-324.

KAWAMURA, A. 1968. Quantity of foods in baleen whales in the Antarctic. *Geiken-Tsushin*, (201) (in press).

KULMOV, S. K. 1966. Plankton and the feeding of the whalebone whales (Mystacoceti). *Trudȳ Inst. Okeanol.*, **51**, 142-56.

MACKINTOSH, N. A. 1965. *The stocks of whales.* London, Fishing News (Books) Ltd. 232 pp.

MARR, J. W. S. 1956. *Euphausia superba* and the Antarctic surface currents. *Norsk Hvalfangsttid.*, **45** (3), 127-34.

MARR, J. W. S. 1962. The natural history and geography of the Antarctic Krill (*Euphausia superba* Dana). *Discovery Rep.*, **32**, 33-464.

MARSHALL, S. M., and ORR, A. P. 1955. *The biology of a marine copepod.* Edinburgh, Oliver and Boyd. 388 pp.

MATSUURA, Y. 1943. (Kujira satsuroku) Miscellaneous note on whales, 3, *Dobutsu-Shokubutsu (Animals and plants)*, **11** (12), 1001-3.

NAKAI, J. 1942. The chemical composition, volume, weight and size of the important marine plankton. *J. oceanogr. Soc. Japan*, **1** (1), 45-55.

NAKAYA S. 1961. (Eiyogaku) *Nutrition* 2, Tokyo, Japan Women. University Press.

NEMOTO, T. 1959. Food of baleen whales with reference to whale movements. *Scient. Rep. Whales Res. Inst., Tokyo*, **14**, 149-290.

NEMOTO, T. 1962. Food of baleen whales collected in recent Japanese Antarctic whaling expeditions. *Scient. Rep. Whales Res. Inst., Tokyo*, **16**, 89-103.

NEMOTO, T. 1963. Some aspects of the distribution of *Calanus cristatus* and *C. plumchrus* in the Bering and its neighbouring waters, with reference to the feeding of baleen whales. *Scient. Rep. Whales Res. Inst., Tokyo*, **17**, 157-70.

NEMOTO, T. 1968. *Feeding of baleen whales and krill, and the value of krill as a marine resource in the Antarctic.* Presented at the Symposium on Antarctic Oceanography at Sanchiago in 1966. (In press).

NISHIWAKI, M. 1965. *Whales and pinnepeds.* Tokyo, University of Tokyo Press.

PONOMAREVA, L. A. 1949. On the nourishment of the plankton eating whale in the Bering sea. *Dokl. Akad. Nauk SSSR*, **18** (2).

PONOMAREVA, L. A. 1966. Quantitative distribution of Euphausiids in the Pacific Ocean. *Oceanology*, **6** (4), 690-92.

SCHOLANDER, P. F. 1940. Experimental investigations on the respiratory function in diving mammals. *Hvalråd. Skr.*, **22**, 1-131.

SERGENT, P. F. 1968. *Feeding ecology of marine mammals. Symposium on diseases and husbandry of aquatic mammals at Florida, Feb. 21 and 22, 1968.*

SHERMAN, H. 1949. *Chemistry of food and nutrition.* New York, Macmillan and Co.

TAYLER, R. J. F., FUJINAGA, M., and WILKE, F. 1955. Distribution and food habits of the fur seals of the north Pacific Ocean. *Rep. Coop. Invest. Canada, Japan and U.S.A., Feb-July, 1952.*

TOMILIN, A. G. 1954. Adaptive types in the order Cetacea (The problem of an ecological classification of Cetacea). *Zool. Zh.*, **33** (3), 677-92.

TOWNSEND, C. H. 1935. The distribution of certain whales, shown by logbook records of American whaleships. *Zoologica, N.Y.*, **18** (1), 1-50.

Detritus, Decomposition, and Regeneration of Nutrients

Introduction

The preceding section dealt with feeding strategies in which plant matter is rearranged into animal protoplasm at one or more levels. Organic matter breaks down through microbial action, making nutrients available for photosynthesis, growth, and metabolism in the next generations. The rates are slow and the circuits are not always closed. For example, minerals may be stored in the sediments or detritus cast ashore and dried up, delaying this regeneration.

On a global and geological scale, elements moving through living and nonliving components of the ecosystem are in balance, but in an isolated locality, diversions may appear. For example, nitrogen from protoplasm is released as excretory urea or ammonia that may be assimilated by plants as ammonia or its oxidized product, nitrate, and converted to amino acids that are assembled as proteins for new tissue or as enzymes. Some nitrogen may be diverted from the system through denitrification and lost to the atmosphere, then returned, through fixation of molecular nitrogen, by lightning or reduction processes in bluegreen algae and bacteria.

Detritus is nonliving organic matter, derived from plants and animals. It is the foodstuff for microorganisms such as bacteria and fungi, and results from maceration by inorganic forces, or from consumer organisms. The papers selected for this unit describe steps on the circular route that break down organic matter and release minerals. Decomposition is slower in the sea than on land, but the process is speeded up by animals that reduce particle size and thereby increase surface area available to bacteria and fungi. Several papers showing this concept are included. In the next section, Nixon and Oviatt describe how the grass shrimp *Palaemonetes* not only breaks down plant matter, but also enriches the substratum by adding its own excrement. This understanding was drawn from the work of Welsh (1975) who partitioned the community and contrasted nutrient cycling with noncyclic energetic relationships. The studies of Fenchel demonstrate that the direct relationship between amphipod activity and the rate of turnover for turtle grass, *Thalassia,* is explained by the increased surface area made available to microbes. The paper by Gerber and Marshall is included to make two points: that detritus is consumed by copepods and that detritus can be made available by herbivorous fish that pass algae macerated but undigested through their systems. The activity in a tropical lagoon is compared with that in a temperate coastal estuary.

Dr. Sieburth and his students at the University of Rhode Island have used the scanning electron microscope to examine a variety of plant, animal, and inorganic surfaces. The paper included shows that microbiota colonized every surface investigated, and that the pattern of colonization differs according to the substratum and the environment in which it is found. In addition, it shows that surfaces of plants and animals have a great many irregularities (crevices, pits) in which microbes concentrate.

There is active interest in determining the rate of decomposition. Some of the work of investigators at two major laboratories show somewhat different but not incompatible viewpoints. Jannasch et al. describe what happened to the contents of lunch boxes aboard a submersible vessel sunk for ten months at 1,540 m and to experimental food materials maintained at 5,000 m for two to five months. Sieburth and Dietz built upon their conclusions by sinking unsealed lunches at both deep sea and inshore stations. Concurrently, Jannasch and Wirsen exposed sterile enrichments to deep sea microflora for a year. This set of three papers aptly demonstrates how scientists in one institution stimulate those in another institution to test ideas and build up an understanding of ongoing processes. The general conclusion from these observations has been extrapolated to what must be happening or, more appropriately, what is not happening to the compacted garbage that man disposes in the sea.

The study by Pomeroy compares mineral cycling in a marine grass, a coral reef, and a planktonic community. It relates these processes to diversity, stability, and disturbance from a viewpoint not previously expressed. It also examines successional patterns brought on by nutrient status of the medium surrounding the organisms. This paper was selected over others that deal with nutrient uptake partly because of the interesting concepts discussed but mostly because it does not require the reader to have a background in enzyme kinetics.

The experiments described in the papers in this unit deal with the conversion of plant matter to detritus by amphipods (Fenchel) and fish (Gerber and Marshall), of organic matter to microbial biomass (Jannasch et al., Jannasch and Wirsen, Sieburth et al., Sieburth and Dietz), and mineralization within communities (Pomeroy). Algae and seagrasses take up the nutrients once they are mineralized and either pass them on to animal communities, return them as detritus, or excrete them into surrounding waters.

Despite the number of scientists working in marine ecology, few have seriously concerned themselves with the critical step of returning nutrients to living systems. Interest in this topic is increasing, however, and the study of decomposition of organic matter in the sea with the subsequent renewal of resources for phytoplankton and seaweed promises to be an active one in the future.

LITERATURE CITED

Welsh, B. L. 1975. The role of grass shrimp, *Palaemonetes pugio,* in a tidal marsh ecosystem. Ecology 56: 513–530.

STUDIES ON THE DECOMPOSITION OF ORGANIC DETRITUS DERIVED FROM THE TURTLE GRASS *THALASSIA TESTUDINUM*[1]

Tom Fenchel[2]

Institute of Marine Sciences, University of Miami, Miami, Florida 33149

ABSTRACT

A study was made of the quantitative composition of the microbial communities living on detrital particles derived from the turtle grass *Thalassia testudinum*. The number of organisms on and the rate of oxygen consumption of the detritus are approximately proportional to the total surface area. Field samples of detritus harbored about 3×10^9 bacteria, 5×10^7 flagellates, 5×10^4 ciliates, and 2×10^7 diatoms and consumed from 0.7 to 1.4 mg O_2/hr per g dry wt.

The detritus-consuming amphipod *Parhyalella whelpleyi* feeds on detrital particles and on its own fecal pellets but it only uses the microorganisms; the dead plant residue passes undigested through the intestine. After a few days the microbial communities living on fecal pellets are qualitatively and quantitatively comparable to those living on other detrital particles.

The amphipods decrease the particle size of the detritus thereby increasing its total surface and thus the microbial activity. In less than 4 days, the mechanical activity of the amphipods may increase the detrital O_2 uptake by 110% of their own metabolic rate. Measurements of respiratory rates of the amphipods will therefore give a much too low estimate of their total role in the ecosystem.

INTRODUCTION

Much recent literature (e.g., Burkholder and Bornside 1957; Darnell 1967; Odum and de la Cruz 1967) has treated the trophic significance of organic detritus in the estuarine ecosystem. It is generally agreed that detritus, mainly derived from the benthic macrovegetation, is of major importance as a link between primary and secondary production in shallow-water areas. Many animals have been shown to feed on detritus (e.g., Muus 1967; Newell 1965; Odum 1968).

Organic detritus is not clearly defined and is known to be of heterogeneous nature and to harbor rich microbial communities. The composition and quantitative importance of these communities have not been treated in detail with the exception of the work of Rodina (1963) on detritus in lakes. The work of Newell (1965) and of Fenchel (1969) has indicated that microorganisms constitute the real food source for the detritus consumers, rather than the nitrogen-poor residues of the macrovegetation.

The macrofauna may play a role in the mechanical breakdown of organic detritus (Darnell 1967), but no quantitative evaluation of this has been carried out in the aquatic environment. In the terrestrial soil ecosystem, Edwards and Heath (1963) found that although the macrofauna contribute but little to community respiration through their own metabolism, their mechanical activity is important to subsequent decomposition by microorganisms.

This paper describes the microbial community associated with detritus derived from the turtle grass *Thalassia testudinum* under aerobic conditions. Previous work (Moore 1963; Odum 1957; Zieman 1968) has shown that *Thalassia* is an important primary producer in shallow-water areas off the Florida coasts. The greater part of

[1] Contribution No. 1118, Inst. Marine Sci., Univ. Miami.

[2] Present address: Marinbiologisk Laboratorium, DK-3000 Helsingør, Denmark.

My stay at the Division of Functional Biology, Institute of Marine Sciences, was supported by a grant from the Nordic Council of Marine Biology. My sincere gratitude is due to Prof. E. J. F. Wood for help in many ways during my stay and also to W. E. Odum and J. C. Zieman for help and discussions. Dr. E. L. Bousfield, Ottawa, identified the amphipods.

147

Reprinted from Limnol. Oceanogr. 15:14–20 (1970).

the detritus samples collected during my study were derived from this plant.

The significance of mechanical breakdown to subsequent decomposition was studied as well as the use of detritus. The decomposition of detritus by a detritus-consuming amphipod, *Parhyalella whelpleyi* Shoem, common in the study area, was also studied.

MATERIALS AND METHODS

Samples of detritus were collected at Bear Cut on Key Biscayne at low tide in 0- to 0.5-m depth. The detritus is found on the sand in layers 1–5 mm thick, especially between the sand ripples; it was collected with a pipette.

Different size fractions of natural detritus were obtained by wet sieving in Millipore-filtered seawater through sieves with mesh sizes of 1, 0.45, 0.295, 0.149, and 0.072 mm. Pure *Thalassia* detritus was made from dead, nearly transparent leaves that were floating around or lying on the bottom at the sampling locality. My observations and the work of Zieman (1968) indicate that these leaves have been detached for at least several months and that all easily decomposed substances have been lost. The leaves were dried overnight at 100C, crushed in a mortar, and sieved as above to obtain different size fractions. Four to six days after this artificial detritus has been placed in seawater at about 24C (avg temp in the sea at this time), the normal microbial community is almost fully restored.

Bacteria, flagellates, and diatoms were counted under a fluorescence microscope after vital staining with Acridine Orange (Wood 1955). All living organisms show a clear green fluorescence; the chloroplasts of photosynthetic organisms also show a red fluorescence. Dead bacteria become orange and can be recognized in the feeding vacuoles of protozoa. A very few detrital particles (not those derived from *Thalassia*) also show a green or orange fluorescence.

Ten fields with an area of 66×66 μ were counted with a 100× objective (total

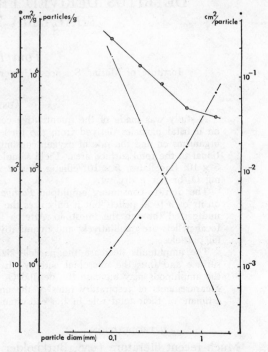

FIG. 1.   The relationship between diameter and the surface of the particles, the number of particles per gram dry weight, and the total surface per gram dry weight of detritus particles derived from *Thalassia*. The average diameter of each fraction is considered to be the geometrical mean of the mesh sizes in between which the fraction in question is retained.

magnification, 1000×) for each sample. Field samples were stored in plastic bottles and were always counted within 2 hr after collection. The numbers per gram dry weight of detritus could be calculated from the surface:weight ratio of the detritus (Fig. 1 and below). Ciliates were counted directly under the microscope (magnification, 100×) as numbers per particle.

Specimens of the abundant *P. whelpleyi* were picked from the detritus samples in the laboratory or were sampled directly in the field with a pipette.

Respiration of the detritus and of the amphipods was measured in containers with 32 ml of Millipore-filtered seawater at 24C with a polarographic oxygen electrode for 30 and 60 min respectively. The water was stirred constantly with a magnetic stirrer.

RESULTS

Composition of the detritus

To evaluate the relative importance of various sources of organic material, a number (100–200) of particles of different size fractions of three different detritus samples was identified under the microscope.

In one sample the fraction > 1 mm consisted of 87.1% *Thalassia* remains, 2.1% other marine flowering plants, 4.6% algae, 0.4% remains of animals (crustacean cuticles, sponge spicules, etc.), 3.3% mangrove leaves, and 2.5% other terrestrial material (leaves, pieces of wood; pollen grains were found in finer fractions).

The finer fractions had a similar composition although an increasing number of particles in the finest fractions could not be identified. In the subsieve fraction (<17 μ), around 50% of the particles consisted of calcium carbonate or other minerals.

The second sample had a similar composition; in the third sample, remains of mangrove leaves played a much larger role (i.e., 36.4% of the particles), while *Thalassia* remains constituted only 46.3%.

Since *Thalassia* remains form the most important part of the detritus in the locality studied, and since detritus particles derived from this plant are regular in shape, nearly transparent and thus more easily studied under the microscope, the *Thalassia* fraction of the detritus was chosen as the main object of the study.

Surface to weight ratio of the detritus particles

Large *Thalassia* particles tend to break perpendicular to leaf surfaces and the smaller particles tend to break more irregularly; the surface to weight ratio of each fraction therefore had to be calculated separately. The surface of a number of particles of each size fraction was calculated individually from microscopical measurements, and dried samples with a known number of particles were weighed (Fig. 1).

Since each *Thalassia* particle consists mainly of the walls of empty cells, the

Fig. 2. The microbial community of a detrital particle derived from a *Thalassia* leaf. Scale: 0.1 mm.

graph in Fig. 1 gives only the external surface; the total surface increases less with decreasing particle size than is indicated in Fig. 1. However, only the distal parts of the "internal surface" (generally the outermost or the two outermost cell layers) are generally populated by microbial communities.

Composition of the microbial communities of natural Thalassia detritus

The microbial community living in the detritus collected in the field consists mainly of bacteria, small zooflagellates and diatoms, and, to a lesser extent, other unicellular algae and ciliates (Fig. 2). Among metazoans, only nematodes seem to be directly associated with individual particles. Mycelia of fungi and actinomycetes were also regularly observed but were not counted.

Counts from intact dead leaves and from particles of different sizes constantly gave values of the same magnitude when calculated as numbers per surface area.

Counts of bacteria gave values between 1 and 9×10^6/cm^2, values around 3×10^6 being most common. This corresponds to around 6×10^9/g dry wt of detritus for the finest size fraction (72–149 μ) and around 10^9/g for the coarsest size fraction (1–2 mm).

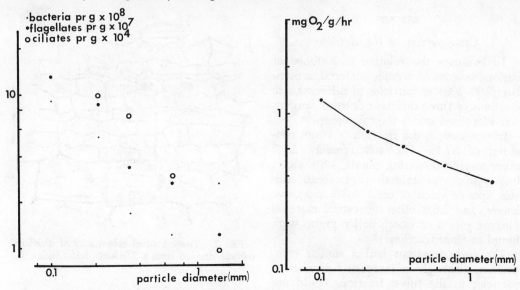

Fɪɢ. 3.   The numbers of bacteria, flagellates, and ciliates (left) and the oxygen uptake (right) of different size fractions of detrital particles.

The most numerous protozoans are the zooflagellates (3.6–8.6×10^4/cm^2); that is, around 10^8 for the finest and around 2×10^7 for the coarsest fractions per gram dry weight of detritus. The most frequently found flagellate type is a species of *Monas*; other common genera are *Oikomonas*, *Tropidoscyphus*, *Entosiphon*, *Bodo* and *Rhynchomonas*. All of these contained bacterial cells and are therefore consumers of bacteria.

Ciliates are also common, though they occur in much lower numbers than the flagellates; on intact dead leaves of *Thalassia* an average of 65/cm^2 of leaf surface was found. This corresponds well with counts on individual detritus particles of between 10^4 and 10^5 cells per gram of detritus. The dominant ciliates were *Euplotes*, which feed on bacteria and zooflagellates, and *Aspidisca*, *Uronema marina*, and *Cyclidium*, all of which feed exclusively on bacteria. *Diophrys*, which feed on diatoms, were also observed regularly as was *Litonotus duplostriatus*, which preys on smaller ciliates (*see* Fenchel 1968). More rarely occurring ciliates on detritus are *Holosticha*, *Dysteria*, a small species of

Sonderia, peritrich ciliates, and a few others.

Some amoebae were observed, but they were too few to count.

Diatoms were frequent in the detritus and often filled out the empty *Thalassia* cells completely. Counts ranged from 1.4×10^4 to 1.5×10^5/cm^2 with values around 2.5×10^4 predominating. Other unicellular algae, probably blue-greens, were also encountered. In one sample, 4.5×10^4 cells/cm^2 were found; usually there were fewer.

Diatoms, as well as other algae, were significantly less common in artificial *Thalassia* detritus than in field samples; presumably the light conditions of the laboratory were too poor to sustain a rich growth of these organisms.

The detritus studied had a preponderance of large particles; the finer particles probably were kept in suspension by tide and wave action. From sieving data of the particle size distribution the numbers of microorganisms per gram of natural detritus samples can be calculated. Thus 1 g dry wt of detritus was found to harbor around 3×10^9 bacteria, 5×10^7 flagellates, 5×10^4 ciliates, and 2×10^7 diatoms.

Large numbers of small metazoans were found among the detritus: nematodes, ostracods, rotifers, turbellarians, and harpacticoids. Except for the nematodes, which often occur inside large detritus particles, metazoans were not counted. Around 7.5 nematodes/cm² were found on intact dead *Thalassia* leaves.

Oxygen uptake of the different size fractions of detritus

Oxygen uptake in the dark at 24C was measured for the 5 size fractions of artificial *Thalassia* detritus that had been kept in seawater for 5 days. Results are shown in Fig. 3 together with the results of counts of the corresponding microbial populations. As expected, respiration increased with decreasing particle size. Measurements of the oxygen uptake of natural detritus samples gave values between 0.75 and 1.4 mg O_2 g^{-1} hr^{-1}, of the same magnitude, or a little higher, than could be expected from the oxygen uptake of artificial detritus.

These findings accord with the measurements of oxygen uptake of detritus particles derived from the dead *Spartina* leaves of Odum and de la Cruz (1967).

Activity of detritus-feeding amphipods

Besides the very abundant *P. whelpleyi*, another amphipod, *Lembopsis* sp., occurred commonly, though in smaller numbers, in the detritus. It apparently has a biology similar to that of *Parhyalella*, but it was not studied in detail.

When specimens of *Parhyalella* were placed in dishes together with detritus, their feeding behavior could be observed under the dissecting microscope. The amphipods held detritus particles with their appendages while they tore bits of the particles with their mouthparts. Some of these bits were swallowed while others were spread around.

Microscopical observation of fecal pellets up to 0.4 mm long showed that the bits of detritus were completely undigested; often bits of intact *Thalassia* tissue several hundred microns long were seen. Freshly formed fecal pellets are almost devoid of

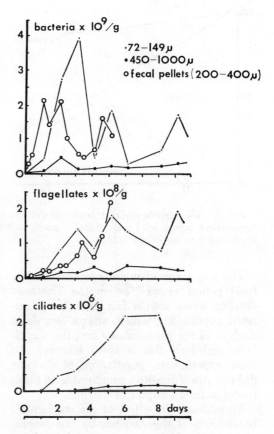

FIG. 4. The growth of bacterial, flagellate, and ciliate populations on fresh fecal pellets of amphipods and on previously oven-dried *Thalassia* detrital particles of two size fractions (larger and smaller than the fecal pellets). Ciliates were not counted on the fecal pellets.

microorganisms. When fecal pellets were kept in Millipore-filtered seawater, a microbial community typical of the detritus was restored after about 4 days (Fig. 4).

The development of the microbial community of the fecal pellets is comparable with that developing on *Thalassia* detritus, and the amphipods would, after some time, eat their own fecal pellets as frequently as other detritus particles. However, the initial development of bacteria is somewhat faster in the fecal pellets; this may be due to the presence of some mucus on the surface of the pellets.

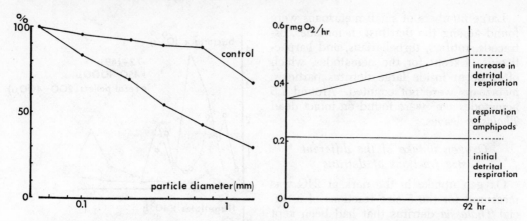

Fig. 5.   The particle size distribution of detritus samples kept for 92 hr with and without amphipods (cumulative curves, left) and (right) a graph showing the increase in detrital oxygen uptake after 92 hr exposure to amphipod activity.

The succession of microorganisms on the fecal pellets as well as on the *Thalassia* detritus starts with a fast growth of bacterial populations, which, after a few days, decline as the protozoans (first the flagellates and later the ciliates) increase. In these experiments, populations of diatoms did not attain densities comparable to those found in field samples of detritus.

To evaluate the effect of the mechanical breakdown of the detritus by the amphipods, I placed equal amounts (about 0.5 g dry wt) of field detritus in 2 petri dishes with Millipore-filtered seawater. About 20 amphipods (dry wt, about 0.05 g) were added to one dish. At the end of an experiment the particle size composition of the detritus in each dish, or its respiration, was measured, as was the respiration of the amphipods, after which both were dried and weighed. Different experiments were terminated after 72 to 144 hr. The results of all were comparable to those shown in Fig. 5. In the left panel, the size distribution at the end of the experiment is shown as per cent dry weight, graphed cumulatively. A significant increase in the amount of fine particles is evident; this corresponds to an increase in external surface by a factor of 2.3.

The effect on the detrital oxygen uptake is shown in the right panel (Fig. 5). The respiration of the amphipods was 2.95 mg O_2 (g dry wt)$^{-1}$ hr^{-1}. In this example, the oxygen uptake was calculated from Fig. 3. In a similar experiment, also extending for 92 hr, the detrital oxygen uptake was measured directly; it gave results of the same magnitude.

The mechanical activity of the amphipods increased the detrital oxygen consumption by a factor of nearly 2 in less than 4 days. This increase corresponds to about 110% of the respiratory rate of the amphipods.

DISCUSSION

This study investigated the general composition of the microbial communities associated with organic detritus under aerobic conditions. The plant residue is decomposed by bacteria and fungi. The former are consumed by flagellates and ciliates, which are again consumed by other ciliates. Another food chain is initiated by fungi which are eaten by nematodes, as has been demonstrated by Meyers and Hopper (1966).

The relatively large number of diatoms in the detritus indicates that it may, in addition to forming the basis for a heterotrophic community of microorganisms, also be the site of significant primary production. This aspect was not studied.

There are many reports demonstrating that animals consuming detritus are common, and animals eating fecal pellets also seem to be frequent (Frankenberg and Smith 1967; Newell 1965).

My findings indicate that in at least some cases, the real food for these animals is the microbial population although the plant residue cannot be used by them. These results accord with those of Newell (1965) who found evidence that the gastropod *Hydrobia* can only use the nitrogen-containing organics in detritus and that the initially nitrogen-poor fecal pellets of the gastropod become enriched with nitrogen when isolated in seawater for some time, presumably the result of microbial activity.

At the same time as the amphipods use the microorganisms, they increase the rate of production of these organisms (i.e., increase the rate of decomposition) by decreasing the particle size of the detritus, thus increasing its biologically active surface.

It is hard to evaluate precisely the significance of the activity of the amphipods for the rate of decomposition in the field. Five core samples of the sediments showed that there are about 9 amphipods/cm^2 corresponding to 0.0037 g dry wt/cm^2 and 0.045 g dry wt detritus/cm^2 (excluding particles smaller than 72 μ) that is, about the same proportion between amphipods and detritus used in the experiments. The significance of the effect will, however, be dependent on other factors that were not studied, such as the possibility that the amphipods prefer certain particle sizes and the transport rate of detrital particles of different sizes to and from the area studied. Nevertheless the effect is large and the mechanical breakdown of the detritus contributes more to the community respiration than does the metabolic activity of the amphipods themselves.

In ecological studies, the metabolic rates of the various components of the ecosystem are usually accepted as measures of the importance of the component in question. This study demonstrates that measurements of metabolic rates do not always give a complete picture of the role played by a specific component.

REFERENCES

BURKHOLDER, P. R., AND G. M. BORNSIDE. 1957. Decomposition of marsh grass by aerobic marine bacteria. Bull. Torrey Botan. Club **84**: 366–383.

DARNELL, R. M. 1967. Organic detritus in relation to the estuarine ecosystem, p. 376–382. *In* G. H. Lauff [ed.], Estuaries. Publ. Am. Assoc. Advan. Sci. 83.

EDWARDS, C. A., AND G. W. HEATH. 1963. The role of soil animals in breakdown of leaf material, p. 76–84. *In* J. Docksen and J. V. Der Drift [eds.], Soil organisms. North-Holland.

FENCHEL, T. 1968. The ecology of marine microbenthos. II. The food of marine benthic ciliates. Ophelia **5**: 73–121.

——. 1969. The ecology of marine microbenthos. IV. Structure and function of the benthic ecosystem, its chemical and physical factors and the microfauna communities with special reference to the ciliated protozoa. Ophelia **6**: 1–182.

FRANKENBERG, D., AND K. L. SMITH, JR. 1967. Coprophagy in marine animals. Limnol. Oceanog. **12**: 443–449.

MEYERS, S. P., AND B. E. HOPPER. 1966. Attraction of the marine nematode, *Methoncholaimus* sp., to fungal substrates. Bull. Marine Sci. Gulf Caribbean **16**: 143–150.

MOORE, D. R. 1963. Distribution of the sea grass, *Thalassia*, in the U.S. Bull. Marine Sci. Gulf Caribbean **13**: 329–342.

MUUS, B. 1967. The fauna of Danish estuaries and lagoons. Medd. Komm. Dan. Fisk.-Havunders. **5**: 1–316.

NEWELL, R. 1965. The role of detritus in the nutrition of two marine deposit feeders, the prosobranch *Hydrobia ulvae* and the bivalve *Macoma Balthica*. Proc. Zool. Soc. London **144**: 25–45.

ODUM, E. P., AND A. A. DE LA CRUZ. 1967. Particulate organic detritus in a Georgia salt marsh-estuarine ecosystem, p. 333–388. *In* G. H. Lauff [ed.], Estuaries. Publ. Am. Assoc. Advan. Sci. 83.

ODUM, H. T. 1957. Primary production in eleven Florida springs and a marine turtle-grass community. Limnol. Oceanog. **2**: 85–97.

ODUM, W. E. 1968. The ecological significance of fine selection by the striped mullet *Mugil cephalus*. Limnol. Oceanog. **13**: 92–98.

RODINA, A. G. 1963. Microbiology of detritus of lakes. Limnol. Oceanog. **8**: 388–393.

WOOD, E. J. F. 1955. Fluorescent microscopy in marine microbiology. J. Conseil, Conseil Perm. Intern. Exploration Mer **21**: 6–7.

ZIEMAN, J. C. 1968. A study of the growth and decomposition of the sea-grass, *Thalassia testudinum*. M.S. thesis, Inst. Marine Sci., Univ. Miami (Fla.). 50 p.

Ingestion of detritus by the lagoon pelagic community at Eniwetok Atoll

Ray P. Gerber and Nelson Marshall

Graduate School of Oceanography, University of Rhode Island, Kingston 02881

Abstract

The gut contents of *Undinula vulgaris* (a calanoid copepod) collected from Eniwetok lagoon consisted of about 95% detritus with only 2% of the gut material fluorescing as chlorophyll. That of *Oikopleura longicaudata* (a larvacean) consisted of about 89% detritus with only 6% fluorescing as chlorophyll. By contrast, the gut contents of the calanoid copepod, *Acartia tonsa*, from Narragansett Bay, Rhode Island, consisted of about 34% detritus with 36% of the gut material fluorescing as chlorophyll. The remaining material in the gut of all these organisms included various microorganisms and diatom frustules.

Plankton-feeding fishes from behind a reef and island were found to have consumed both zooplankton and detrital algal fragments.

The higher levels of particulate carbon and nitrogen in the lagoon, which had a lower C:N ratio than found in the incoming oceanic water, indicated that reef detritus enriches the lagoon environment.

Little is known of the productivity and of the food chains in lagoon waters behind coral reefs. Extremely low primary production in typical coral atoll lagoons is indicated by low chlorophyll values, by approximations of phytoplankton abundance, and by data on ^{14}C uptake (e.g. Johnson 1954; Gilmartin 1958; Jeffrey 1968; Michel 1969). Recently, various workers have been stressing the potential of organic matter exported from the reef as an additional primary food source available in lagoons. Reef export studies (e.g. Marshall 1965; Klim 1969; Qasim and Sankaranarayanan 1970; and our work in progress) indicate an abundance of particulate organic matter in the form of benthic algal fragments, fecal pellets, coral mucus, and aggregated organic matter flowing into the lagoon. Such export might support the zooplankton concentrations known to be consistently higher in lagoons than in the surrounding oceanic waters (e.g. Russell 1934; Motoda 1938). Furthermore our own underwater observations, supported by those of Johannes and Randall (personal communications) on the feeding habits of certain plankton-feeding fishes, suggested that reef detritus was being consumed in addition to zooplankton.

No one has tested the suggestion that organic export from the reef is consumed by the basic consumer populations in the lagoons. To provide evidence on this matter, we have analyzed the stomach contents of two zooplankters collected from lagoon stations: a tropical filter-feeding calanoid copepod, *Undinula vulgaris*, predominately neritic and a lagoon form in these environments (Johnson 1954; E. C. Jones personal communication), and the larvacean, *Oikopleura longicaudata*. Also, several species of plankton-feeding fishes from selected patch reef or coral knoll stations were collected for gut analysis. For comparison, gut analyses were performed on the typically neritic copepod, *Acartia tonsa*, from Narragansett Bay, Rhode Island, a productive temperate area.

The relative abundance in the lagoon of reef-derived algal fragments, total particulate carbon and nitrogen, and chlorophyll were analyzed to determine the relationship between stomach contents and food available from the same stations.

We are especially indebted to P. Helfrich for his cooperation and support at Eniwetok and to J. E. Randall for his assistance in collecting the fishes. We would also like to thank M. E. Q. Pilson, T. A. Napora, S. V. Smith, and M. Fine for critical reading of the manuscript.

Materials and methods

Four stations were sampled (Fig. 1) at Eniwetok Atoll, Marshall Islands, from 31

154

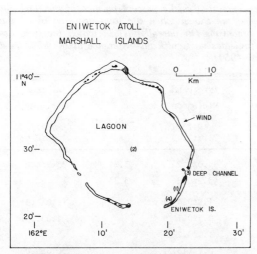

Fig. 1. Locations of sampling stations (shown in parentheses) at Eniwetok Atoll, Marshall Islands. Stippled areas indicate islands; clear outlined areas indicate reef tracts.

December 1971 to 8 February 1972. Station 1 was located along an interisland reef in the zone of larger coral heads about 200 m behind the reef crest. This station was about 8 m deep and characterized by a continuous unidirectional flow of water from in front of the reef, created by the heavy surf breaking across the reef crest. Station 2 was located about 8–10 km northwest of Eniwetok Island well into the lagoon, in about 50 m of water. Station 3 was in the southeastern pass, or deep channel, with an average depth of 31 m. A pinnacle reef or knoll on the northern side of this pass, harboring many plankton-feeding fishes, served as a collection source for these forms. Water from outside the atoll enters the lagoon via this pass. Station 4 was situated in about 6 m of water in scattered patch reefs about 200 m on the lagoon side of Eniwetok Island.

Plankton was collected with a 0.5-m No. 6 nylon plankton net (mesh 0.239 mm), having an internally mounted flowmeter, by oblique tows made from about 30 m in the deeper water and 4 m in shallow areas. Since the shallow stations, 1 and 4, were largely devoid of *U. vulgaris* and *O. longicaudata*, zooplankton gut analyses were

confined to samples from stations 2 and 3. Plankton samples used for gut analysis were preserved in 5% glutaraldehyde in seawater, kept on ice and in the dark as recommended by the National Research Council, Committee on Oceanography (1969). *Acartia tonsa* from Narragansett Bay were collected and treated in the same manner.

The entire stomach contents of *O. longicaudata* and only the foregut material of the copepods were analyzed. Only specimens with full or nearly full stomachs were used. After a given specimen was teased apart, the various components of the gut contents were identified and measured under a microscope with an ocular micrometer. Then, with fluorescence microscopy (Wood 1962) and a Whipple micrometer disk, the percent by area of fluorescing (red and green) material was determined for the contents of each gut. The technique of quantifying the contents by area, as flattened on a slide, indicates the relative amounts of the various food particles ingested and potentially available for assimilation.

Spear-fishing, rotenone poisoning, and netting were used to collect the plankton-feeding fishes, which were also preserved in 5% gluteraldehyde and kept on ice and in the dark. Components of the gut material were identified under a dissecting microscope and their settled volume estimated with a small graduated cylinder.

Total particulate carbon was determined by filtering 9–10 liters of seawater through a pad of two Gelman type A glass-fiber filters, precombusted at 450°C for 3 hr. The filters were stored under vacuum desiccation and, using a CHN analyzer, combusted at 725°C, at which temperature carbonate interference is minimal (Telek and Marshall 1974). The top filter was assumed to collect both particulate and adsorbed organic matter whereas the lower was assumed to collect only adsorbed, and its value was deducted to give a corrected particulate value.

Chlorophyll *a* and total pheo-pigments

Table 1. *Results of the zooplankton gut analyses, expressed as the percent by area of the total gut content area. (±) 95% confidence limits of the mean are shown when the item occurred in all the guts; otherwise, only the mean and the number of organisms (in parentheses) in which the item occurred are shown. Underscoring of adjacent means indicates no significant difference at the 95% level (Duncan's new multiple range test: Steel and Torrie 1960).*

	U. vulgaris Midlagoon Sta 2 N=35 2.56±0.12	*U. vulgaris* Deep channel Sta 3 N-23 2.66±0.11	*O. longicaudata* Midlagoon Sta 2 N=23 7.17±0.49	F-values*	*A. tonsa* Narr. Bay,R.I. N=23 1.40±0.03
Amorphic, non-fluor.	91.8±1.7	85.5±8.6	85.0±0.2	7.74	33.9±8.8
Granular, non-fluor.	4.8±1.2	2.6±0.5	3.9±0.8	3.23	-
Large cells	1.5 (8)	1.8 (5)	1.8±0.4		1.0 (3)
Small cells	1.0 (12)	1.2 (5)	0.7±0.1		1.0 (13)
Microflag.	0.5 (3)	0.5 (4)	1.0±0.2		
Diatoms/frag.	0.5 (6)	0.5 (4)	0.7 (13)		36.0±5.0
Radiolaria/frag.	2.4 (18)	10.5 (9)	-		
Red fluor.	1.3±0.3	1.9±0.8	6.3±1.2	58.21	30.5±5.4
Green fluor.	0.9 (13)	1.8 (11)	1.2 (16)		-

* Original data were transformed to ARCSIN functions for analysis of variance and range test (Snedecor and Cochran 1967).

were determined by filtering 3–4 liters of seawater onto HA Millipore filters of 0.45-μ pore diameter. The filters were stored frozen and later analyzed by the method of Strickland and Parsons (1968).

Results

The results of the zooplankton analyses are summarized in Table 1. These data show a predominance of the nonfluorescing amorphic material in the guts of specimens from Eniwetok. This material appeared as a conglomerate of undifferentiated particles lacking observable structural remnants of tissues or cells. Under white light it was mostly translucent and varied from light to very dark brown. Even the gut contents of *A. tonsa* from Narragansett Bay contained a substantial amount of this amorphic material. A small undetermined amount of clear mucus, perhaps secreted by the gut lining, was occasionally noted.

Nonfluorescing granular material, occurring only in the Eniwetok specimens, did not appear to be the remains of skeletal fragments of microorganisms since the in-

dividual granules were smooth, irregular in size and shape, and lacked surface features such as pits or grooves characteristic of skeletal fragments. Some may have been small bits of calcium carbonate such as Johannes (1967) observed in the mucus flocs streaming off corals at Eniwetok.

Generally, large nonfluorescing cells (12.5–28.0 μ) without skeletal structure constituted < 2% of the gut area. Only in *O. longicaudata* did they occur regularly. Microscopic examinations of these cells suggest that some, at least, may be dinoflagellates and protozoans.

The gut contents of *U. vulgaris* and *A. tonsa* were largely devoid of small naked nonfluorescing cells (4.0–10.0 μ) and microflagellates, yet they occurred in almost every *O. longicaudata* gut analyzed. In the case of the Eniwetok specimens, these small cells were primarily coccolithophores, and perhaps some bacteria.

Though these foregoing cell types did not fluoresce red, it is possible that some had contained chlorophyll which was decomposed in the gut.

Table 2. The gut content analyses of plankton-feeding fishes, expressed as the percent by volume. (±) 95% confidence limits of the mean are shown when N ≥ 4; otherwise only the mean is given.

	Chromis caeruleus			Dascyllus aruanus		D. reticulatus			Abudefduf curacao		Pomacentrus vaiuli		Pomacentrus pavos	Acanthiurus thompsoni
	Sta 1* N=11 52.1±7.0	Sta 3 N=8 48.6±4.9	Sta 4 N=20 41.6±5.1	Sta 1 N=9 32.3±10.7	Sta 4 N=5 21.2±4.6	Sta 1 N=4 67.3±12.9	Sta 3 N=4 55.5±5.0	Sta 4 N=11 19.8±4.1	Sta 1 N=1 77.0	Sta 3 N=1 89.0	Sta 1 N=2 54.0	Sta 3 N=3 43.0	Sta 4 N=2 58.0	Sta 3 N=2 139.0
Calanoid copepods	12.8±3.5	15.3±4.4	9.7±5.3	10.8±5.7	0.6±1.0	21.5±16.6	6.3±2.9	3.6±2.1	11.0	2.0	1.0	–	8.0	21.0
Cyclopoid copepods	23.3±6.4	17.4±3.8	13.5±3.9	29.3±9.1	10.0±4.6	27.0±18.7	14.5±5.8	41.1±3.3	21.0	1.0	3.5	–	31.5	14.5
Harpacticoid cop.	17.5±9.1	3.6±1.7	34.7±6.7	17.0±6.5	39.2±4.1	7.5±6.5	–	41.1±3.2	17.0	–	4.0	14.3	3.0	0.5
Decapod larvae	3.4±3.6	3.1±2.0	1.7±1.0	0.7±1.0	0.4±1.0	4.3±2.9	4.3±6.5	2.0±1.8	–	–	0.5	–	2.0	–
Branchyura zoea	–	–	0.4±0.5	–	–	–	–	–	–	–	–	–	–	–
Isopods	–	–	–	–	–	–	–	–	–	–	–	–	–	3.5
Gammarid amph.	–	–	–	–	–	–	–	–	11.0	–	–	–	–	11.5
Ostracods	–	0.9±1.2	1.7±1.3	0.1±0.3	1.2±2.0	–	0.7±2.2	1.9±1.9	–	–	–	1.0	–	–
Crustacea frag.	1.0±0.9	–	–	–	–	–	–	2.5±1.7	–	–	–	–	–	1.0
Gastropod larvae	–	1.9±1.7	2.1±1.3	0.7±0.8	–	–	–	–	–	–	–	–	0.5	–
Pteropods	–	–	–	–	–	–	–	–	6.0	–	–	–	1.5	3.5
Round worms+	2.1±1.7	–	0.3±0.2	–	–	1.3±2.2	–	–	–	2.0	–	–	–	–
Polychaete larvae	–	54.6±9.9	2.8±4.2	2.7±2.6	1.6±2.6	–	71.5±10.8	0.5±1.1	6.0	–	1.5	–	–	6.5
Fish eggs	2.2±2.6	–	18.7±8.8	2.0±1.6	–	0.8±1.4	–	–	–	–	–	–	2.0	–
Fish larvae	0.2±0.6	–	–	–	–	1.2±3.6	–	–	–	89.0	–	–	–	–
Larvaceans	17.8±13.5	3.3±2.3	4.0±1.5	1.6±0.5	2.8±1.0	1.3±0.7	1.5±2.2	1.5±2.2	17.0	–	2.0	2.3	1.5	4.0
Foraminifera	2.1±0.7	–	–	–	38.2±8.7	37.0±18.7	2.5±2.2	31.4±5.7	3.0	5.0	–	–	–	3.5
Algal frag.	–	–	–	–	–	–	–	–	8.0	1.0	87.5	80.7	50.0	16.0
Fecal pellets	17.5±6.4	–	10.9±5.6	33.7±15.6	4.8±6.6	–	–	–	–	–	–	–	–	–

*Station locations: Sta 1 - near reef; sta 3 - deep channel; sta 4 - near island.
+These may be parasitic.

More than half of the *O. longicaudata* and less than a fifth of the *U. vulgaris* from Eniwetok contained pennate diatoms but, by area, these amounted to less than 1% of the gut contents in which they occurred. The gut contents of *A. tonsa* from Narragansett Bay, on the other hand, included, by area, about 36% diatom frustules and fragments of *Skeletonema* sp., *Rhizosolenia* sp., and *Thalassiosira* sp.

Radiolarian spines and body fragments occurred in the gut contents of about half the *U. vulgaris*, yet none was observed in gut contents of *O. longicaudata*.

The amount of red fluorescence (which may include detrital chlorophyll) in the gut contents of the Eniwetok specimens was very low compared to the amount in *A. tonsa* from Narragansett Bay. The size range of the red fluorescing particles, which may include individual chloroplasts in addition to nannoplankton and small diatoms, was 5.0–25.0 μ in diameter in *U. vulgaris* and 2.5–25.0 μ in *O. longicaudata*. The red fluorescing particles in *A. tonsa* were not measured.

Green fluorescing cells from 7.5–30.0 μ in diameter occurred in at least a third of the guts of *U. vulgaris* and in more than half of the *O. longicaudata*. The exact nature of these cells was not determined; however, small (2–10 μ) green fluorescing naked flagellates were observed by Pomeroy and Johannes (1968) in the Gulf Stream and Sargasso Sea. They suggest that these cells are autotrophic and that some accessory pigment masks the red fluorescence of chlorophyll. No green fluorescing cells were found in the gut contents of *A. tonsa* from Narragansett Bay.

Table 2 summarizes the results of the gut analyses on plankton-feeding fishes. This table indicates that, in addition to their diet of zooplankton, the fish are ingesting a substantial amount of detrital algal fragments. These algal fragments were readily distinguishable filamentous types, varying in size from a few cells to large fragments about a centimeter long. The amount of algal fragments in the gut of *Chromis caeruleus*, *Dascyllus reticulatus*, and *Dascyllus aruanus* from the near-reef station 1 and the near-island station 4 were not significantly different ($P > 0.45$, "Student's" t-test). Except for *Pomacentrus vaiuli*, the quantity of algal fragments in the guts of fishes from the coral knoll at station 3 was exceedingly small.

The abundance of calanoid and cyclopoid copepods and decapod larvae was greatest in the guts of fishes from station 1 and over the knoll at station 3, and lowest at station 4; the reverse was true for harpacticoid copepods. This trend may reflect the influx into the lagoon of oceanic water, which contains many more calanoids and cyclopoids than harpacticoids. Ostracods, gastropod larvae, polychaete larvae, fish larvae, and larvaceans occurred infrequently and only in the guts of fishes from stations 1 and 3. The greatest abundance of fish eggs was found in the guts of fishes from station 3, with little obvious differences between stations 1 and 4. Some fish contained fairly long coiled "round worms" in their gut, which appeared to be parasitic rather than planktonic.

In addition to those listed in Table 2, gut analyses were performed on the following plankton-feeding fishes: *Caesio caeruleureux*, *Caesio* sp., *Pterocaesio tile*, *Pterocaesio marri*, *Mirolabrichthys tuka*, *Amblyglyphidon curacae*, *Chromis ternatensis*, *Chromis agilis*, *Chromis atripectoralis*, *Chromis lepidoleptris*, *Chromis* sp., *Cirrhilabrus* sp., *Pseudocoris* sp., and family Branchiostegidae (two species). The gut contents of these fishes consisted almost exclusively of zooplankton; no reef detritus was found.

Underwater observations behind the reef at station 1 indicated a greater abundance of algal fragments than at the near-island station 4. Day net-plankton samples confirmed this observation (Table 3) and showed that remnants of *Calothrix* spp., *Asparagopsis taxiformis*, and *Dictyota* spp. made up the bulk of the samples, the zooplankton biomass (wet weight) being negligible.

Table 3. The average biomass of net plankton in mg m⁻³ (wet weight). Number of samples in parentheses.

	Sta 1 Near reef (8)	Sta 4 Near island (2)	Sta 2 Midlagoon (8)	Sta 3 Deep channel (3)
Algal frag.	23.5	15.5	-0-	-0-
Zooplankton	5.5	2.0	239.4	139.6

At the midlagoon station 2 and in the deep-channel station 3, algal fragments were not seen in underwater observations nor were any taken in plankton nets. These day samples consisted exclusively of zooplankton, of which there was more in the midlagoon than in the incoming oceanic water and the shallow areas behind the reef and island.

The data for particulate organic carbon (POC) and particulate organic nitrogen (PON) are summarized in Table 4 and treated statistically in Table 5. The concentration of POC at the near-reef station 1 was significantly higher than the deep-channel station 3, and the concentration of PON at station 1 was significantly higher than at the midlagoon station 2 and sta-

tion 3. The ratio of particulate carbon to nitrogen increased from station 1 to station 3.

Table 6 indicates extremely low values for chlorophyll *a* and pheo-pigment at all stations at Eniwetok Atoll. The mean and 95% confidence limits of the combined lagoon chlorophyll *a* and total pheo-pigment samples are 0.10 ± 0.03 and 0.09 ± 0.06 mg m⁻³ respectively.

Discussion

Zooplankton nutrition

Eniwetok Atoll is located in the southern half of the North Equatorial Current in an unproductive oceanic area (Taniguchi 1972). Production in Eniwetok Lagoon and the North Equatorial Current is compared in Table 7. The slightly higher concentration of chlorophyll *a* and productivity in the lagoon than in the surrounding waters may largely reflect the input of reef algal detritus into the lagoon. Dissolved nutrients further indicate a minimal phytoplankton food base. Phosphate concentrations in the lagoon are similar to those in the North Equatorial Current and the ni-

Table 4. A comparison of the concentrations and ratios of particulate organic carbon and nitrogen in mg m⁻³. ⟨±⟩ 95% confidence limits of the mean shown at bottom of table along with the mean C:N ratio. Stations 1 and 2 sampled approximately every 2 days; station 3 sampled when weather permitted.

Jan 1972	Sta 1 - Near reef			Sta 2 - Midlagoon			Sta 3 - Deep channel		
	Carbon	Nitrogen	C:N	Carbon	Nitrogen	C:N	Carbon	Nitrogen.	C:N
1	20.8	2.2	9:1	25.3	2.9	9:1	-	-	-
2	21.0	3.3	6:1	28.4	2.3	12:1	16.1	1.5	11:1
4	17.9	2.3	8:1	22.7	2.1	11:1	-	-	-
6	38.3	5.1	7:1	11.7	1.5	8:1	22.1	2.0	11:1
8	20.6	2.4	9:1	11.1	1.8	6:1	-	-	-
10	18.3	2.4	8:1	15.8	1.9	8:1	-	-	-
13	27.3	2.6	10:1	15.9	1.9	8:1	15.4	1.2	13:1
15	24.6	3.8	7:1	14.9	1.7	9:1	11.4	1.0	11:1
23	19.9	2.3	9:1	22.0	2.7	8:1	-	-	-
25	30.3	3.0	10:1	24.3	2.1	12:1	-	-	-
28	26.3	3.2	8:1	22.9	2.5	9:1	17.8	1.5	12:1
29	34.4	3.2	11:1	28.1	2.3	12:1	-	-	-
31	35.0	4.7	7:1	22.1	1.8	12:1	-	-	-
	25.75	3.12	8:1	20.40	2.11	10:1	16.56	1.44	12:1
	±4.15	±0.56		±3.54	±0.25		±4.82	±0.47	

Table 5. *Statistical analysis of the particulate organic carbon (POC) and particulate organic nitrogen (PON) data in mg m^{-3}. Underscored adjacent means = no significant difference at 95% level (Duncan's new multiple range test: Steel and Torrie 1960).*

	Sta 1 Near reef	Sta 2 Midlagoon	Sta 3 Deep channel	*F-values from ANOVA
POC	25.75	20.40	16.56	4.89
PON	3.12	2.11	1.44	9.50

*Degrees of freedom - 28.

Table 6. *The concentrations of chlorophyll a and total pheopigments in mg m^{-3}. Stations 1 and 2 were sampled weekly; station 3 when weather permitted.*

Jan 1972	Sta 1 Near reef		Sta 2 Midlagoon		Sta 3 Deep channel	
	Chl a	Pheo.	Chl a	Pheo.	Chl a	Pheo.
2	0.172	0.110	0.023	0.260	0.080	0.040
8	0.055	0.031	0.133	0.174	-	-
13	0.020	0.138	0.099	0.035	0.135	0.034
23	0.144	0.021	0.115	0.003	-	-

trate values recorded are less inside the atoll than in the ocean.

Hardy and Gunther (1935) suggested that radiolarians and foraminifera, which harbor symbiotic algae, assume the role of phytoplankton as primary producers and food for zooplankton in tropical oceans. This suggestion does not seem to apply to Eniwetok Lagoon where radiolarians, though the most abundant "photosynthetic" organisms found in the plankton samples, rarely occurred in the guts of *U. vulgaris* and not at all in *O. longicaudata*. The lagoon zooplankton, however, which consists of increased concentrations of oceanic forms as well as endemic types and larvae of reef fauna, is much richer than in the North Equatorial Current. The lagoon zooplankton must thus depend on some food source in addition to phytoplankton, and, as noted at the outset, the suggested source is reef detritus. This alternate food source

is reflected in the higher levels of particulate organic carbon and nitrogen behind the reef and in the lagoon than in the incoming oceanic water. Typically, open ocean values for particulate organic carbon average about 10 mg m^{-3} (unpublished data), less than half that found in the lagoon.

The low carbon-to-nitrogen ratio of the suspended particulate organic matter in the lagoon also supports the concept of reef export as an alternate food source. Johannes et al. (1972) found that the C:N ratio of the suspended particulates changed from 15:1 in oceanic waters to 7:1 in the downstream end of the reef transect; they concluded that nitrogen fixation must be occurring at a high rate.

Recently Johannes (personal communication) has indicated that the blue-green algae *Calothrix*, remnants of which were abundant in our plankton samples behind the reef, is the major nitrogen fixer on the windward reefs. He believes it is a major

Table 7. *Average values of parameters relating to primary production for Eniwetok Lagoon and the North Equatorial Current.*

Area	Chl a (mg m^{-3})	Phytopl. (cell m^{-3})	Production (mg C m^{-3} hr^{-1})	Zoopl.* (mg m^{-3})	PO_4-P (mg atom m^{-3})	NO_3-N (mg atom m^{-3})
Eniwetok Lagoon+	0.10 0.23	1.4×10^3	0.57	239	0.13	0.17
North Equatorial C‡	0.06	3.7×10^5	0.13	9	0.12	0.41

*Wet weights.
+First Chl a value and zooplankton wet weight, this study; second Chl a value and production, Doty and Capurro 1961; phytoplankton cells, Johnson 1954; PO_4-P value, Pilson and Betzer 1973; NO_3-N value, Johannes et al. 1972.
‡From Taniguchi 1972.

contributer, directly and through the food chain, to lowered C : N ratios in the lagoon.

We too found a low C : N ratio, and a ratio of 8:1 at the near-reef station 1. A ratio of 12:1 for the deep-channel station 3 is evidence that nitrogen enrichment affects the waters of the reef passes as well as those behind the reef proper. This may explain why we found no significant differences between the concentrations of POC and PON for stations 2 and 3. The midlagoon C : N ratio at station 2 is about halfway between the ratios at stations 1 and 3, perhaps due to mixing of the latter two sources or to a conversion of some of this detrital nitrogen to animal protein and excreta by consumer organisms.

Finally, gut analyses suggest that detritus is the most important food component of the lagoon zooplankton. Detrital material in the gut contents of the Eniwetok specimens may be composed partly of reef particulate organic matter from a wide range of possible sources, such as decomposed benthic algal fragments and fecal matter, which form the bulk of the reef detrital load (Johannes and Gerber in press), or of coral mucus flocs as suggested in Johannes (1967). It may even consist of particulate organic matter produced in the pelagic environment itself: fecal matter, or aggregated organic matter formed by mechanical and microbial processes or bacterioplankton (Baylor and Sutcliffe 1963; Sieburth 1968; DiSalvo 1970; Sorokin 1971).

Detritus is an important food source for pelagic zooplankton in other areas of the ocean. Detritus feeding is well documented for deep-sea forms and also occurs in various epipelagic open ocean areas in the tropical Pacific (Geinrikh 1958; Pomeroy and Johannes 1968: Arashkevich and Timonen 1970).

Our work suggests that detritus is also utilized by A. tonsa in a productive temperate zone estuary. It should be added that these copepods were collected in early November, when the phytoplankton standing crop was minimal (Smayda personal communication) and when detritus utiliza-

tion would, theoretically, be at a maximum (Riley 1963).

The semiclosed circulation patterns of atoll lagoons, as interpreted from von Arx (1948), help retain endemic fauna including various larvae as well as oceanic zooplankton swept into the lagoon (Johnson 1954). The steady input to the lagoon of enriched particulate organic matter from the surrounding reefs, and to a much lesser extent from the ocean, seems to form the basis of this pelagic food web.

Plankton-feeding fishes

We have demonstrated that certain plankton-feeding fishes in reef areas consume suspended algal fragments in addition to zooplankton and show a diversified feeding habit that reflects their locality at the time of capture. For example, algal fragments were not recorded in the gut of C. caeruleus from stations 1 and 4 by Hiatt and Strasburg (1960). The precise location of their collection is not given, but it was presumably in an area relatively free of suspended algal fragments, as was station 3 in this study. Alternately, the lack of algal fragments could be due to a greater abundance of fish eggs, etc. in the plankton at the time of capture; these were preferred over other foods. Smith and Tyler (1973) suggested that such opportunistic feeding in reef fishes is important in maintaining community stability.

The gut contents of D. aruanus and D. reticulatus are remarkably similar to those of C. caeruleus, perhaps because all three species are territorial and generally feed close to the coral heads. The more or less solitary pomacentrid, Pomacentrus pavo, had about 50% zooplankton and 50% algal fragments in its gut. It was observed feeding on suspended particulates from the interstices of the coral branches and from midwater suspended particles. On the other hand, the gut contents of Pomacentrus vaiuli consisted almost exclusively of algal fragments and epibenthic harpacticoid copepods, suggesting a grazing feeding habit rather than planktonic. The gut contents of *Abudefduf curacao* and *Acanthurus thompsoni* were dominated by pelagic zooplankton, with only a small amount of detrital algal fragments in *A. curacao* and none in *A. thompsoni*. However, the gut of *A. thompsoni* contained fecallike detrital material, perhaps of reef origin.

These findings further indicate that reef detritus, in the form of algal fragments, is consumed by certain lagoon fishes. The decomposition of this material by digestion within the fish may be a first step in its transformation to a form that can be readily consumed by lagoon zooplankton.

References

ARASHKEVICH, YE. G., AND A. G. TIMONEN. 1970. Copepod feeding in the tropical Pacific. Dokl. Akad. Nauk USSR 191: 935–938.

BAYLOR, E. R., AND W. H. SUTCLIFFE. 1963. Dissolved organic matter in seawater as a source of particulate food. Limnol. Oceanogr. 8: 369–371.

DiSALVO, L. H. 1970. Regenerative functions and microbial ecology of coral reefs. Ph.D. thesis, Univ. N. Carolina. 289 p.

DOTY, M. S., AND L. R. A. CAPURRO. 1961. Productivity measurements in the world oceans, part 1. IGY Oceanogr. Rep. 4: 72–83.

GEINRIKH, A. K. 1958. On the nutrition of marine copepods in the tropical region. Dokl. Akad. Nauk USSR 119: 229–232.

GILMARTIN, A. 1958. Some observations on the lagoon plankton of Eniwetok Atoll. Pac. Sci. 12: 313–316.

HARDY, A. C., AND E. R. GUNTHER. 1935. The plankton of the South Georgia whaling grounds and adjacent waters, 1926–1927. Discovery Rep. 11: 1–456.

HIATT, R. W., AND D. W. STRASBURG. 1960. Ecological relationships of the fish fauna on coral reefs of the Marshall Islands. Ecol. Monogr. 30: 65–127.

JEFFREY, S. W. 1968. Photosynthetic pigments of the phytoplankton of some coral reef waters. Limnol. Oceanogr. 13: 350–355.

JOHANNES, R. E. 1967. Ecology of organic aggregates in the vicinity of a coral reef. Limnol. Oceanogr. 12: 189–195.

——, AND R. P. GERBER. In press. Import and export of net plankton by an Eniwetok coral reef community. Proc. 2nd Int. Symp. Coral Reefs, Australia, 1973.

——, AND PROJECT SYMBIOS TEAM. 1972. The metabolism of some coral reef communities. BioScience 22: 541–543.

JOHNSON, M. W. 1954. Plankton of the North-ern Marshall Islands. U.S. Geol. Surv. Prof. Pap. 260-F, p. 301–314.

KLIM, G. D. 1969. Interactions between seawater and coral reefs in Kaneohe Bay, Oahu, Hawaii. Hawaii Inst. Geophys., Univ. Hawaii HIG 69-19. 56 p.

MARSHALL, N. 1965. Detritus over the reef and its potential contribution to adjacent waters of Eniwetok Atoll. Ecology 46: 343–344.

MICHEL, A. 1969. Plankton des lagoon et des abords exterieurs de l'atoll de Mururoa. Cah. Pac. 13: 81–132.

MOTODA, S. 1938. Quantitative studies on the macroplankton off coral reefs of Palao Port. Trans. Sapporo Nat. Soc. 15: 242–246.

NATIONAL RESEARCH COUNCIL. 1969. Recommended procedures for measuring the productivity of plankton standing stock and related oceanic properties. Comm. Oceanogr., Biol. Methods Panel, NAS-NRC. 59 p.

PILSON, M. E. Q., AND S. B. BETZER. 1973. Phosphorus flux across a coral reef. Ecology 54: 581–588.

POMEROY, L. R., AND R. E. JOHANNES. 1968. Occurrence and respiration of ultraplankton in the upper 500 meters of the ocean. Deep-Sea Res. 15: 381–391.

QASIM, S. Z., AND V. N. SANKARANARAYANAN. 1970. Production of particulate organic matter by the reef on Kavaratti Atoll (Laccadives). Limnol. Oceanogr. 15: 574–578.

RILEY, G. A. 1963. Organic aggregates in seawater and the dynamics of their formation and utilization. Limnol. Oceanogr. 8: 372–381.

RUSSELL, F. S. 1934. The zooplankton. 3. A comparison of the abundance of zooplankton in the Barrier Reef Lagoon with that of some regions in northern European waters. Great Barrier Reef Exped. Sci. Rep. 2: 159–201.

SIEBURTH, J. McN. 1968. Observations on bacteria planktonic in Narragansett Bay, Rhode Island; a resume. Bull. Misaki Mar. Biol. Inst., Kyoto Univ. 12: 49–64.

SMITH, C. L., AND J. C. TYLER. 1973. Population ecology of a Bahamian suprabenthic shore fish assemblage. Am. Mus. Nov. 2528, p. 1–38.

SNEDECOR, G. W., AND W. H. COCHRAN. 1967. Statistical methods. Iowa State.

SOROKIN, Y. I. 1971. Trophic role of microflora in the coral reef biocenosis. Dokl. Akad. Nauk USSR 199: 490–493.

STEEL, R. G. D., AND J. H. TORRIE. 1960. Principles and procedures of statistics with special reference to biological sciences. McGraw-Hill.

STRICKLAND, J. D. H., AND T. R. PARSONS. 1968. A practical handbook of seawater analysis. Bull. Fish. Res. Bd. Can. 167. 311 p.

TANIGUCHI, A. 1972. Geographical variation of primary production in the western Pacific

Ocean and adjacent seas with reference to the inter-relations between various parameters of primary production. Mem. Fac. Fish., Hokkaido Univ. **19**: 1–33.

TELEK, G., AND N. MARSHALL. 1974. Using a CHN analyzer to reduce carbonate interference in particulate organic carbon analyses. Mar. Biol. **24**: 219–221.

VON ARX, W. S. 1948. The circulation systems of Bikini and Rongelap Lagoons. Trans. Am. Geophys. Union **29**: 861–870.

WOOD, E. J. F. 1962. A method for phytoplankton study. Limnol. Oceanogr. **7**: 32–35.

Submitted: 18 February 1974
Accepted: 25 July 1974

MICROBIAL COLONIZATION OF MARINE PLANT SURFACES AS OBSERVED BY SCANNING ELECTRON MICROSCOPY

JOHN McN. SIEBURTH, RICHARD D. BROOKS, ROBERT V. GESSNER, CYNTHIA D. THOMAS, and J. LAWTON TOOTLE

Narragansett Marine Laboratory
Graduate School of Oceanography
University of Rhode Island

Scanning electron microscopy was used to examine the microbiota which develop on the submerged surfaces of macroscopic plants. This paper summarizes the first year's observations with selected micrographs which show the kinds of microbial assemblages that occur and the apparent differences between substrates and seasons. Driftwood and polypropylene strips were chosen as non-living surfaces to act as controls. The oak driftwood from an exposed rocky point was free of bacteria and algae and showed a definite sequence of fungal development. In contrast, polypropylene strips tied to seaweed fronds were colonized by algal and bacteria-like filaments and pennate diatoms after only 2 weeks immersion.

The grasses also showed markedly different patterns of colonization. The shaded and periodically immersed internodal area of cord grass, *Spartina alterniflora*, showed an initial colonization by the mycelium and hyphopodial appendages of the fungus *Sphaerulina pedicellata*. Sexual stages of this and other fungi developed as this grass matured and senesced during late summer and fall. In contrast, the emerging and submerged surfaces of eelgrass, *Zostera marina*, were colonized mainly by the pennate diatom *Cocconeis scutellum*

164

Reprinted from Effect of Ocean Environment on Microbial Activities
ed. by R. D. Colwell and R. Y. Morita. University Park Press pp. 418–432 (1974).

which formed a unialgal mat. As broken frustules and detritus adhered, a crust was formed which then permitted non-selective colonization.

The seaweeds, unlike the grasses and driftwood, appeared to support lesser population densities but greater diversity. The brown alga *Ascophyllum nodosum*, which is rich in inhibitory polyphenols, was relatively clean during its active growth periods in spring and early summer. During the winter microcolonies of diatoms, yeasts, and the filamentous bacterium *Leucothrix mucor* increased in density. The fine red alga *Polysiphonia lanosa*, which is an epiphyte on *Ascophyllum nodosum*, supports a readily detectable summer epiflora which becomes dense at protected bifurcations. Pennate diatoms, yeasts, and filamentous bacteria increased in density during the winter.

During periods of active growth, marine plants appear to limit the diversity and density of the populations that can develop on their surfaces. Colonization approaches that of inanimate surfaces during periods of dormancy or senescence and especially when the fouling crust isolates the fouling surface from the host.

The major limiting factor in the application of scanning electron microscopy to marine materials is the production of slime layers which obscure the cell surface. In the case of the more gummy seaweeds, like *Chrondrus crispus*, the surfaces are always amorphous and so littered with debris and microorganisms that the micrographs are extremely difficult to interpret. In the case of bacteria, the slime layer sometimes makes their resolution difficult and identification all but impossible.

INTRODUCTION

Heterotrophic and facultatively heterotrophic microorganisms colonize any surface where organic matter is produced or adsorbed. The primary sources of organic matter in the sea are the marine plants. The most obvious habitat for marine microbiologists to study would appear to be any surface where organic matter and microorganisms come in contact. Despite the accepted importance of surfaces (ZoBell, 1946) we microbiologists continue to be mainly preoccupied with studying microorganisms "free" in the water column, which occur in low numbers. This "unattached" or loosely dissociable microflora may be nothing more than transients liberated from a richer substrate which are starving to death while waiting for another surface to colonize.

The microbiota of plant surfaces have been mainly studied by phycologists and, to a lesser extent, by zoologists including protozoologists. The few studies by microbiologists have been mainly concerned with the role of bacteria in fouling. Three main approaches have been taken to study the

Aufwuchs or periphyton of surfaces. One has been to scrape the surface under study and to count and identify the microorganisms in the resulting suspension by light microscopy. This destroys microcolonies and the micro-zonal relationships between microorganisms. In an attempt to avoid this problem, the second method has been the examination of glass or plastic slides which have been immersed for varying periods of time. This is a variation of the buried-slide technique (Cholodny, 1930) which has been applied to water by a number of investigators including Henrici (1936), ZoBell and Allen (1935), Skerman (1956), and Kriss (1963). Although the glass slide gives an idea of what happens on an inanimate surface, the results can only be applied to discarded glass surfaces such as bottles. The third has been to limit oneself to fine filamentous forms and structures which can be examined in wet mounts (Brock, 1966; Johnson *et al.,* 1971).

Until the development of the scanning electron microscope some 5 years ago, there was no practical way to examine thick, opaque, and three-dimensional preparations. Anything that can be freeze dried, attached to a 12.5-mm diameter aluminum stub, and plated with a palladium-gold alloy can now be examined. Hard solid surfaces are best, while those that are soft and porous, and especially those that have thin, fine structures protruding, can give problems. The instrument permits the resolution of microorganisms as small as 0.2 μm (the limit of the light microscope) and the differentiation of anything with a characteristic form. A major advantage for the microbiologist who is really a bacteriologist is that it forces him to put his microorganisms in perspective with those of his fellow microbiologists who call themselves phycologists, protozoologists, and mycologists.

Much of the material presented in this paper has arisen from student research problems in Sieburth's course on marine microbiology at the University of Rhode Island. Individual papers on the colonization of driftwood (Brooks *et al.,* 1972), *Spartina alterniflora* (Gessner *et al.,* 1972) and *Zostera marina* (Sieburth and Thomas, 1973) are in press. A preliminary form of this paper was presented by Sieburth *et al.* (1972). The purpose of this presentation is to collate and synthesize our observations on the colonization of marine plant surfaces and to show the applications and limitations of scanning electron microscopy for the study of the microflora of natural marine surfaces.

MATERIALS AND METHODS

The driftwood specimens were remnants of oak lobster pots lodged at low tide on the exposed shore of Point Judith, Rhode Island, and were collected in October 1971. *Spartina alterniflora* was collected from southern Rhode

Island salt marshes in the fall of 1971 while *Zostera marina* was collected from the Pattaquamscutt River, Rhode Island, during the fall of 1971 and spring of 1972. Seaweeds were obtained from August 1971 to August 1972 from a densely populated cove adjacent to Camp Varnum some 7 km south of the laboratory. Strips of polypropylene were attached to *Ascophyllum* fronds at the above location in July 1972.

The freshly collected specimens were either fixed in the field or rushed to the laboratory. Fixation in 4% gluteraldehyde ranged from a 10-sec dip to 30-min immersion. After rinsing well with distilled water, representative areas were quick-frozen in isopentane with liquid nitrogen or directly with liquid nitrogen. The frozen sections were freeze dried below 60°C with an Edwards-Pearse tissue dryer (Edwards High Vacuum Ltd., Sussex, England), mounted on SEM stubs with Duco cement containing conductive silver paint, either single or double plated with palladium-gold depending on the specimen, and examined with a Cambridge S-4 scanning electron microscope. The positive micrographs arising from the Polaroid PN Type 55 packs were kept as data records while the 4 by 5 negatives were used with Agfa Brovira I-3 paper to produce figures. (See note added in proof.)

RESULTS

Driftwood which has become lodged in a fixed position is an ideal natural substrate for observing the colonization and decomposition of cellulosic materials in the marine environment. A simplified record of the microscopic appearance of hard, semisoft, and softened wood is shown in Figs. 1 *A, B,* and *C,* respectively. While the wood is in a hard condition, conidia from a number of species can occur in the same microzonal area. By the time the wood is in an intermediate stage of decomposition, conidia of a single species can dominate. When the wood has become soft, the mycelia and conidia are absent and only ascocarps remain. In the rocky and turbulent waters from which these specimens were obtained, diatoms and bacteria were not observed. Another non-living substrate examined has been polypropylene strips tied to fronds of seaweeds. After only 2 weeks immersion, bacteria-like filaments and pennate diatoms (Fig. 1*D*) as well as tufts of green algal filaments (Fig. 1*E*) were well established.

The development of microorganisms on the surfaces of living grasses can vary greatly with the substrate and the environmental conditions. In the shaded and moist internodal area of *Spartina,* mycelia with hyphopodial appendages belonging to the fungus *Sphaerulina pedicellata* develop on the growing plant (Fig. 2*A*). Dense mycelial mats (Fig. 2*C*) and the sexual spores of *Sphaerulina pedicellata* (Fig. 2*B*) and other fungi develop as the plants

mature and senesce. Bacteria and diatoms do not seem to play a major role until the grass falls, becomes broken, and is submerged. *Zostera*, on the other hand, is initially colonized by a few bacteria and pennate diatoms dominated by *Cocconeis scutellum* (Fig. 2D). An essentially unialgal mat develops (Fig. 2E) upon which broken frustules and detritus adhere to form a crust (Fig. 2F). This crust now supports nonselective colonization by pennate diatoms, filamentous algae, bacteria, and fungi, in addition to nematodes and protozoa. The crust can equal or exceed the biomass of the supporting grass.

The seaweeds do not support such a density of colonization. The diversity of colonization is also restricted and quite repititious. Fig. 3A shows the clean surface of *Ascophyllum nodosum* during its active growing period in early spring. Microzonal areas at this time can also have surface slime with adhering microbial cells (Fig. 3B) and short filaments of *Leucothrix mucor* (Fig. 3G). During winter, when the plant is presumably dormant, pennate diatoms (Figs. 3C,D), yeasts (Figs. 3 E,F) and *Leucothrix mucor* (Fig. 3H) can form dense microzonal areas.

The red alga *Polysiphonia lanosa* is a common epiphyte on *Ascophyllum nodosum*. Since it shares the same habitat it was chosen to compare its seasonal colonization. Although summer specimens of its fine filaments appear barren at first glance (Fig. 4A), closer examination of the bifurcations (Fig. 4B) and the area between the ridges shows a variety of bacteria-sized cells, filaments of *Leucothrix mucor*, yeasts, and pennate diatoms. In winter there is a marked increase in the density of yeast colonies, diatoms, and webs of *Leucothrix mucor* filaments.

Magnifications sufficient to see or identify some microorganisms limit the view to such a small area that one loses perspective. In order to visualize what a larger piece of algal filament would look like in winter, a panorama of a 0.8 mm length of *Polysiphonia lanosa* is shown in Fig. 5. The 2-μm thick filaments of *Leucothrix mucor* (barely visible) have largely coalesced to form thick bundles. The few recurring species of pennate diatoms are quite visible.

Fig. 1. Colonization of non-living surfaces showing a sequence of fungal deterioration of lodged driftwood, and the heterogeneous colonization of polypropylene. *A*, The surface of freshly collected oak driftwood showing mycelium and conidia of multiple species (*Cirrenalia macrocephala* and *Dictosporium pelagica*) in a hardened condition; *B*, domination by a single species (*Cirrenalia macrocephala*) in an intermediate condition; *C*, the absence of mycelia and conidia with the presence of ascocarps of *Leptosphaeria oraemaris* in the softened condition; *D*, a polypropylene strip which had been tied to a frond of *Ascophyllum nodosum* for 2 weeks showing filamentous microorganisms and the pennate diatoms *Cocconeis scutellum* and possibly an *Acanthes* spp.; *E*, tufts of green algal filaments with a scattering of pennate diatoms. Marker bars on *C* and *E* equal 100 μm; all others equal 10 μm.

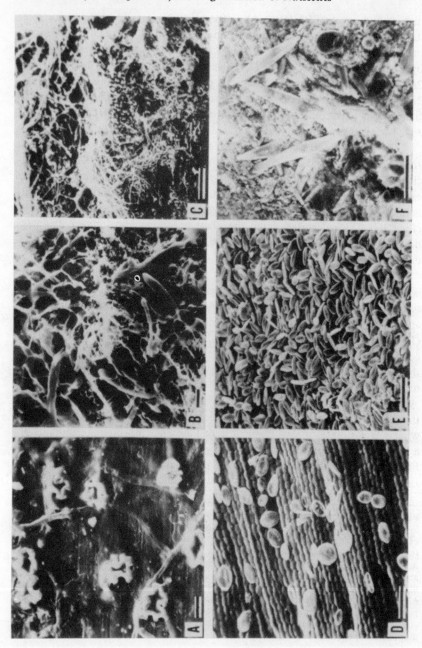

Individual yeast and bacterial cells are difficult or impossible to detect at this magnification.

DISCUSSION

The cellulosic substrates of driftwood and *Spartina alterniflora* showed a definite sequence of fungal development to the exclusion of other microorganisms. Since a presumably non-nutrient surface like polypropylene is readily colonized by a variety of microorganisms, the refractory nature of the cellulosic substrates does not explain the absence of other microorganisms. It is possible that the dominant fungal microflora is inhibitory and excludes other microorganisms. Another grass, *Zostera marina,* showed a completely different picture of colonization. A unialgal covering of a pennate diatom served as a base for a crust which became heavily colonized by a great variety of microorganisms. It is amazing that this plant can thrive with its central portion completely sheathed. Photosynthesis must be mainly limited to the emerged portion of the blade which is not colonized. It is hard to imagine an adaptive advantage to the host except protection from herbivores. The seaweeds, on the other hand, may be thought of as a photosynthetic root which both adsorbs nutrients and produces "photosynthesate". Heavy colonization like that on *Zostera* would surely interfere with both nutrient adsorption and photosynthesis and severely interfere with the metabolism of the plant. Some seaweeds, like *Ascophyllum* which belongs to the Fucales, are rich in polyphenols. They seem to function as plant-protective substances in two ways. The polyphenols are stored in physodes or fucosan bodies in the outer layers of cells. The ability of these polyphenols to tan polysaccharides as well as proteins may protect tissue damage by water turbulence and grazing animals such as *Littorina littorea* by forming a protective scab. These polyphenols, which are readily released, are also inhibitory to microorganisms (Sieburth, 1968) and may be responsible for the relatively clean surfaces

---

Fig. 2. Comparison of the microbial colonization of submerged portions of cord grass (*Spartina alterniflora* Loisel) with eelgrass (*Zostera marina* Loisel). The green plants of *Spartina* are initially colonized by the mycelia of *Sphaerulina pedicellata* with their "parasitic" hyphopodia (*A*), later by their ascospores (*B*), and mycelial mats (*C*). The cobblestone surface of *Zostera* after 2 weeks emergence (*D*) becomes colonized by *Cocconeis scutellum* and occasionally by other pennates such as *Licmophora* spp.; later it becomes dominated by *Cocconeis scutellum* (*E*) to form a crust (*F*) which is non-selectively colonized by a number of diatom taxa including *Cocconeis, Navicula, Pleurosigma, Amphora* and possibly *Nitzschia*. Marked bars on *A*, *B*, and *C* equal 10 μm and on *D*, *E*, and *F* equal 50 μm.

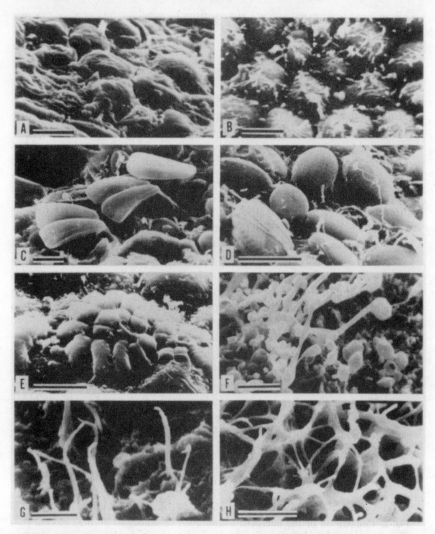

Fig. 3. The microcolonization of the surface of the brown alga *Ascophyllum nodosum* (Loisel) LeJol during the winter. *A,* The clean algal surface (November); *B,* mucoid surface with small microbial cells (November); *C,* the pennate diatom *Rhoicosphenia curvata* (February); *D,* the pennate diatom *Cocconeis scutellum* (February); *E,* a yeast microcolony (January); *F,* pseudomycelium and budding cells of the genus *Candida* (February)—note the dark bud scars on two cells, lower right; *G,* filaments of *Leucothrix mucor* showing individual cells and terminal gonidia (November); *H,* a web of *Leucothrix* filaments (February). All marker bars equal 10 μm.

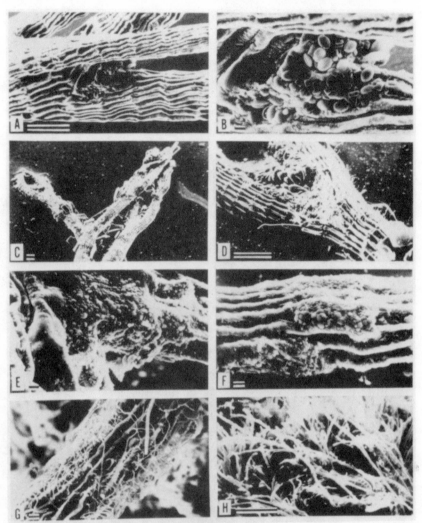

Fig. 4. The filamentous red alga *Polysiphonia lanosa* appears relatively free of an epiflora in summer (*A*) but on close examination shows a mixture of diatoms, yeast and bacteria-like cells (*B*). This plant supports colonization by diatoms in winter (*C* and *D*), yeasts in summer (*E*) and winter (*F*), and dense *Leucothrix mucor* infestation in winter (*G* and *H*). Marker bars on *A, C,* and *D* equal 100 μm; all others equal 10 μm.

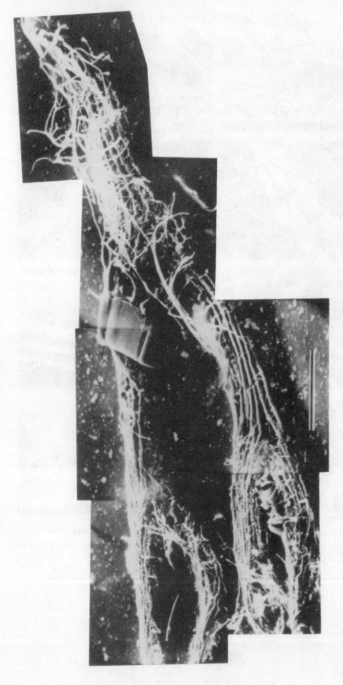

Fig. 5. A panorama of an 0.8 mm length of a filament of the red alga *Polysiphonia lanosa* (January specimen) showing coarser bundles and fine individual filaments of the bacterium *Leucothrix mucor* (upper right) and an assortment of pennate diatoms (in the lower left there is the avoid *Cocconeis scutellum*, wedge-shaped *Rhoicosphenia curvata* and possibly *Fragillaria pennata*, while the large central diatom is *Rhabdonema adriaticum*). Marker bar equals 100 μm.

observed on *Ascophyllum nodosum* during its active growing period. The exudation rate for *Ascophyllum nodosum* in spring was some 40 mg C/100 g tissue dry wt/hr. The comparable figure for *Polysiphonia harveyi* was some 0.4 mg C/100 g dry wt/hr (Sieburth, 1969). This negligible value may be due to its dense filamentous structure and its greater degree of spring-summer colonization.

Living-plant surfaces appear to have a markedly reduced diversity in comparison with the epibiota developing on inanimate surfaces such as glass slides. The density of colonization also appears much reduced during periods of active growth. During dormancy and senescence the density and diversity increase but still appear to be limited. An exception is the immersed portion of *Zostera* which supports an epibiota which equals its own biomass. Of what consequence is such a knowledge of the microbial colonization of marine plants?

Macroscopic marine plants play a vital role in the production of organic matter in inshore waters. The microorganisms which produce an Aufwuchs or periphyton may be an important part of this production. In the marshes, the cord grass *Spartina alterniflora* outproduces hybrid corn in productivity. In shallow protected waters, eelgrass *Zostera marina* produces dense stands of growth. The epiflora growing on it can equal or exceed the phytoplankton biomass of the water (Smayda, 1962). In the rocky intertidal and subtidal areas the seaweeds produce biomasses of several kilograms per square meter of area (Blinks, 1955). Both the flowering aquatic plants (Wetzel, 1969) and the seaweeds (Sieburth, 1969) have been shown to release some 30% of their "photosynthesate". Allen (1971) has observed that in ponds this material is immediately taken up by microorganisms in the Aufwuchs or periphyton. Sieburth has examined flowering aquatics *in situ* by snorkeling and scuba. In the proper light a translucent sheath which surrounds these plants can be seen. This is marked in small ponds and protected areas of large ponds where wind-induced turbulence is minimal. Certain areas of marshes and estuaries may also have quiet enough waters for such loose and large assemblages to form. In contrast, most seaweeds require a certain velocity of water movement to supply their nutrient needs (Conover, 1968). It seems highly likely that most of the organic substances released by seaweeds are carried away by the surging waters. The uptake of such materials by suspended particles and their microbiota, and their role in the food chain is another story.

Spartina is harvested by fall and winter winds and falls into the estuaries where it becomes broken and heavily colonized and enters the food chain (Welsh and Carney, 1971). Eelgrass forms a staple of migrating water fowl in the fall. Seaweeds are directly grazed by snails such as *Littorina littorea* and gammarid amphipods. Microbial enrichment of these plants and their remnants may be an important factor in their nutritional value.

This paper attempts to draw together observations made on the microbial epibiota of plants by scanning electron microscopy over a 1-year period. Its applications to describing the microbial habitats of surface films, suspended particles, animal and plant surfaces, and sediments must be obvious. Its limitations may not be. Fungi, yeast, diatoms, and the larger bacterial forms are readily recognized, some even to species. Bacteria can be detected but they are not the neat little bags we expect from light and transmission electron microscopy. A very large microflora consisting of single and paired cocci some 0.2 μm in diameter is missed by scanning surfaces of 500 X and 1000 X magnifications. They can only be detected at magnifications of 2000 X and higher where only a very small area can be seen at a time and where each field has to be refocused. Larger bacteria often appear as loosely stuffed pillows, since we are looking at their outer slime layer and not their more distinct cell wall. Bacteria suspended in water and lying between sand grains are often in microcolonies which are extremely difficult to recognize as such. The application of scanning electron microscopy to natural populations of non-filamentous coccobacillary forms at present is difficult.

ACKNOWLEDGMENTS

The National Science Foundation which not only supported this conference but the acquisition of our scanning electron microscope (GA28903) and the research itself (GB18000) is gratefully acknowledged for making this study and paper possible. We should like also to think Judith Murphy who helped unfold the mysteries of "SEMantha", our new love, and Don Scales who feeds and cares for her.

LITERATURE CITED

Allen, H. L. 1971. Primary productivity, chemo-organotrophy and nutritional interactions of epiphytic algae and bacteria on macrophytes in the littoral of a lake. Ecol. Monogr. 41:97–127.

Blinks, L. R. 1955. Photosynthesis and productivity of littoral marine algae. J. Mar. Res. 14:363–373.

Brock, T. D. 1966. The habitat of *Leucothrix mucor,* a widespread marine microorganism. Limnol. Oceanogr. 11:303–307.

Brooks, R. D., R. D. Goos, and J. McN. Sieburth. 1972. Fungal infestation of the surface and interior vessels of freshly collected driftwood. Mar. Biol. 16:274–278.

Cholodny, N. 1930. Ueber eine neue Methode zur Untersuchung der Boden-mikroflora. Arch. Mikrobiol. 1:620–652.

Cole, G. T., and H. C. Aldrich. 1971. Scanning and transmission electron microscopy and freeze-etching techniques used in ultrastructural studies of hyphomycetes. *In* B. Kendrick (ed.), Taxonomy of *Fungi Imperfecti,* pp. 292–300. Univ. Toronto Press, Toronto.

Conover, J. T. 1968. The importance of natural diffusion gradients and transport of substances related to benthic marine plant metabolism. Bot. Mar. 11:1–9.

Erlandsen, S. L., A. Thomas, and G. Wendelschafer. 1973. A simple technique for correlating SEM with TEM on biological tissue originally embedded in epoxy resin for TEM. *In* Scanning Electron Microscopy 1973, Part 3, pp. 349–356. ITT Research Institute, Chicago.

Gessner, R. V., R. D. Goss, and J. McN. Sieburth. 1972. The fungal microcosm of the internodes of *Spartina alterniflora* Loisel. Mar. Biol. 16:269–273.

Hanic, L. A., and J. S. Craigie. 1969. Studies on algal cuticle. J. Phycol. 5:89–102.

Henrici, A. T. 1936. Studies on fresh water bacteria. III. Quantitative aspects of the direct microscopic method. J. Bacteriol. 32:265–280.

Johnson, P. W., J. McN. Sieburth, A. Sastry, C. R. Arnold, and M. S. Doty. 1971. *Leucothrix mucor* infestation of benthic crustacea, fish eggs and tropical algae. Limnol. Oceanogr. 16:962–696.

Kriss, A. E. 1963. Marine Microbiology (Deep Sea). (Trans. by J. M. Shewan and Z. Kabata.) Oliver and Boyd, London.

Sieburth, J. McN. 1968. The influence of algal antibiosis on the ecology of marine microorganisms. *In* M. R. Droop and E. J. Ferguson Wood (eds), Advances in Microbiology of the Sea, pp. 63–74. Academic Press, New York.

Sieburth, J. McN. 1969. Studies on algal substances in the sea. III. The production of extracellular organic matter by littoral marine algae. J. Exp. Mar. Biol. Ecol. 3:290–309.

Sieburth, J. McN. and C. D. Thomas. 1973. Fouling on eelgrass (*Zostera marina* L.). J. Phycol. 9:46–50.

Sieburth, J. McN., C. D. Thomas, and J. L. Tootle. 1972. Microbial fouling of marine plants. *In* Abst. Annu. Meet. Amer. Soc. Microbiol., p. 76.

Skerman, T. M. 1956. The nature and development of primary films on surfaces submerged in the sea. N. Z. J. Sci. Technol. Sect. B. 38:44–57.

Smayda, T. J. 1962. Some quantitative aspects of primary production in a R.I. coastal salt pond. Proc. 1st Nat. Coastal Shallow Water Res. Conf. Tallahassee Fla., pp. 123–125.

Welsh, B. L., and E. J. Carney. 1971. Integration of computer modeling techniques with laboratory experiments. Univ. Rhode Island Sea Grant Reprint (1)1–5.

Wetzel, R. G. 1969. Excretion of dissolved organic compounds by aquatic macrophytes. BioScience 19:539–540.

ZoBell, C. E. 1946. Marine Microbiology. Chronica Botanica, Waltham, Massachusetts.

ZoBell, C. E., and E. C. Allen. 1935. The significance of marine bacteria in the fouling of submerged surfaces. J. Bacteriol. 29:239.

Note added in proof. Better preparative techniques, including the use of osmium tetroxide, glutaraldehyde fixation, dehydration in a graded ethanol series, and transfer to iso-amyl acetate before CO_2 critical point drying, have vastly improved preparations and the resolution and magnifications obtained. The problems of detecting imbedded microorganisms and confirming external morphological structure by internal ultrastructure has been overcome by combining the techniques of scanning electron microscopy with examination of thick and thin sections by light and transmission electron microscopy, respectively (Cole and Aldrich, 1971; Erlandsen *et al.,* 1973). Another mechanism for limiting the degree of fouling on seaweed surfaces appears to be the sloughing off of the surface layer of the proteinaceous cuticle (Hanic and Craigie, 1969) and the accompanying epibiota to expose a clean surface (J. L. Toottle, M.Sc. Thesis, Univ. Rhode Island, in preparation).

Microbial Degradation of Organic Matter in the Deep Sea

Abstract. *Food materials from the sunken and recovered research submarine* Alvin *were found to be in a strikingly well-preserved state after exposure for more than 10 months to deep-sea conditions. Subsequent experiments substantiated this observation and indicated that rates of microbial degradation were 10 to 100 times slower in the deep sea than in controls under comparable temperatures.*

On 16 October 1968, the research submersible *Alvin* of the Woods Hole Oceanographic Institution sank in about 1540 m of water, 135 miles southeast of Woods Hole, Massachusetts. The accident occurred when, because of a broken cable, the vessel dropped into the sea with an open hatch and sank after the crew of three escaped safely. A photograph taken on 13 June 1969 by U.S.N.S. *Mizar* prior to the retrieval operations showed the position of the vessel on the sea floor, the hatch still being open (Fig. 1). On 1 September 1969, *Alvin* was brought to the surface (*1*). Among the items recovered was the crew's lunch consisting of two thermos bottles filled with bouillon and a plastic box containing sandwiches and apples. From general appearance, taste, smell, consistency, and preliminary bacteriological and biochemical assays, these food materials were strikingly well-preserved. When kept under refrigeration at 3°C, the starchy and proteinaceous materials spoiled in a few weeks.

Possible implications of this unexpected finding led us to make some additional observations. The environmental conditions at a depth of 1500

179

Reprinted from Science 171: 672–675. (1971).

m are assumed to be fairly constant at about 3° to 4°C and 150 atm of pressure. There was no evidence of reducing conditions nor was there a noticeable lack of dissolved oxygen either in the pressure hull of the vessel or in the box containing the food materials. In addition, there was no evidence for the presence or the possible leakage of a soluble material that could have acted as a preservative. The plastic lids of the stainless steel thermos bottles were crushed by pressure, and some seawater must have penetrated and mixed with the contents.

Besides being soaked with seawater, the six sandwiches wrapped in waxed paper (Fig. 2a) appeared fresh by taste and smell. When pieces of the bread were streaked on seawater agar, bacteria and molds grew profusely. Placed in tubes with sterile seawater and kept at 3°C, the bread decayed with slight gas production (floating to the surface) within 6 weeks. The slices of meat (bologna) were grayish on the outside but still pink in the center. Submerged in sterile seawater, the meat spoiled with a putrefactive smell within 4 weeks at 3°C and within 5 days at 30°C.

The two apples found in the lunch box had a pickled appearance (Fig. 2b) but showed no sign of obvious decay. The pH of the tissue was the same pH (3.2), and the tyrosinase activity (2) was about half that of a fresh apple tested. The soup, originally prepared with hot (not boiling) water from canned meat extract, was perfectly palatable in hot and cold condition. Samples of this broth showed a maximum turbidity caused by bacterial growth in 22 days when incubated at 3°C, and in 5 days when incubated at 30°C. Sporeforming bacteria were observed while the majority of bacteria were represented by Gram-negative rods that grew well on seawater media.

In conclusion, the food materials recovered from *Alvin* after 10 months of exposure to deep-sea conditions exhibited a degree of preservation that, in the case of fruit, equaled that of careful storage and, in the case of starch and proteinaceous materials, appeared to surpass by far that of normal refrigeration.

The implications of this finding, if generally true, are of theoretical and practical interest. Viewing the ocean as the ultimate sink of inorganic as

Fig. 1. *Alvin* resting at a depth of 1540 m with open hatch as photographed about 8 months after her sinking.

well as organic materials, we have virtually no knowledge of qualitative and quantitative microbial decomposition processes. While the absolute amounts of nonliving organic matter calculated for all oceans by far exceeds that of the landmasses, the actual concentrations in seawater are extremely low. In fact, in the larger part of the oceans the concentration of dissolved organic carbon is too low for a direct measurement of oxygen consumption or any other parameter as an indicator of degradation processes. The constancy of organic carbon concentrations with depth in the sea suggests little or no microbial activity (3). On the other hand, results of experimental work on the effect of low temperature and high hydrostatic pressure (4) do not exclude considerable microbial activities in the deep sea if suitable energy sources and nutrients are available.

Research in this laboratory has been

Fig. 2 Food materials recovered from *Alvin* after exposure to seawater at a depth of 1540 m for 10 months.

directed toward measuring in situ rates of growth and biochemical activities of marine bacteria as measured by chemostat systems fed with natural seawater (5). This approach has been limited to the richer surface waters. The *Alvin* accident stimulated a direct experimental study of microbial activities as affected by deep-sea conditions.

Various experimental approaches were readily conceivable in order to confirm the observed phenomenon on a more general basis. In cooperation with the Department of Physical Oceanography a program was designed in which specially designed sample racks were attached to deep-sea moorings about 10 m above the sea floor at depths of about 5000 m and recovered by an acoustic release mechanism after exposure for 2 to 5 months.

The sample racks held about 50 bottles of 120-ml volume each and 20 plastic syringes containing liquid media of various types and concentrations. In experiments with ^{14}C-labeled substrates, the bottles were filled with seawater from 200 m collected at the site of launching. The substrates were added just before the samples were submersed. The serum stoppers used for sealing permitted pressure equalization. Parallel controls were kept under refrigeration at 3°C in the laboratory at 1 atm in the dark. In other experiments, bottles were inoculated with mixed microbial populations of heavily contaminated surface water (Eel Pond, Woods Hole). Some bottles were equipped with a simple device that provided for self-inoculation by hydrostatic pressure at depths from 150 m down to the sea floor. Other samples were inoculated with pure cultures of specific isolates and submerged in plastic syringes of 50-ml volume containing air in addition to the liquid media. The mechanical behavior of the bottles and syringes at increased hydrostatic pressure was tested in special pressure chambers equipped with viewing ports (6).

For the data presented in this account, the following brief indication of analytical methods may suffice. ^{14}C-Labeled substrates and metabolic products were counted in a Packard (Tri-Carb, model 3380) scintillation spectrometer in 10 ml of Bray's solution. The efficiency of all counts was corrected for quenching from a prepared external standard ratio curve. Ammonia released from nitrogenous substrates was determined by micro-

Table 1. Microbial degradation of four substrates exposed for 8 weeks at a depth of 5300 m (location 33°58′N, 70°W) as compared to controls kept at 3°C for 6 weeks. Percent values are corrected for the unequal exposure time. The microgram values are calculated from counts of ^{14}C radioactivity and are given for total volume of sample (120 ml). Cold Difco casamino acids were added to a mixture of 14 uniformly labeled ^{14}C-amino acids. The inoculum consisted of about 120 ml of seawater sampled separately at a depth of 200 m at the site of launching.

Substrate added (μg)	Substrate in particulate fraction (μg)		Sample control (%)
	Control	Sample	
Acetate			
3600	88.9	3.58	3.0
1200	146.1	2.08	1.07
600	138.7	0.29	0.15
240	16.2	0.073	0.34
Mannitol			
3600	166.6	3.45	1.55
1200	46.0	1.06	1.7
600	41.1	0.60	1.1
240	40.1	0.13	0.24
Sodium glutamate			
3600	252.5	6.50	1.9
1200	130.6	1.66	0.95
600	59.6	2.20	2.77
240	43.8	0.50	0.86
Casamino acids			
3600	406.7	48.80	9.0
1200	336.0	14.40	3.2
600	123.6	17.10	10.4
240	49.9	8.60	12.9

Kjeldahl distillation. Residual carbohydrates and sugars were determined by the phenol-sulfuric acid method (7). For the determination of bacterial growth, colonies were counted on seawater agar containing the particular substrate studied.

Table 1 represents data of an experiment with ^{14}C-labeled substrates in concentrations of 2 to 30 μg/ml. The total recovery of added ^{14}C activity in the three fractions—residual substrate, CO_2, and particulate carbon—ranged from 95 to 99 percent. The ratio of the amount of labeled CO_2 to the amount of particulate carbon in the laboratory controls ranged from 1.5 to 3.4. In the deep-sea samples, however, the amount of labeled CO_2 was too small for significant measurements and very low relative to the amount of labeled carbon in the particulate fraction. For this reason, only the data for the conversion of substrate into particulate carbon are given in Table 1. It might be assumed that dissolved products other than CO_2 were formed by fermentative interconversions. However, there was no indication of anaerobic or reducing conditions in any of the samples.

Table 1 shows that the amount of

substrate converted into the particulate fraction in the deep-sea samples ranged from 0.15 to 12.9 when expressed as the percentage of the corresponding conversion in the laboratory controls (in the calculation of these percentage figures, the values of columns 2 and 3 have been corrected for the unequal exposure time). In other words, in these two extreme cases the substrate decomposed 666 to 8.2 times more slowly in the deep-sea samples than in the refrigerated laboratory controls. The corresponding average figures for the two carbohydrates are 88 times, and for glutamate and casamino acids 62 and 11 respectively. With the exception of the casamino acids, these rates appear to decrease with increasing concentration of the particular substrate. In addition, on the basis of the turnover of organic carbon, in the deep-sea samples the carbohydrates decomposed two to four times more slowly than the nitrogenous substrates.

In another experiment at the same location, chemical analyses were used. For the sake of analytical accuracy, the substrate concentrations were chosen to be five to ten times higher (Table 2). The controls were checked after an incubation period of 6 weeks, at which time the degradation was clearly completed. Therefore, the ratios between the amount of substrate utilized in the controls (corrected for an incubation time of 19 weeks) to that metabolized in the deep-sea samples represent maximum values.

There was no perceptible quantitative difference in the rates of decomposition per bacterial cell when rich surface water or offshore seawater collected at 200 m was used as an inoculum. In pure culture experiments, we selected mesophilic and psychrophilic strains that had been isolated from various depths in the open ocean. Only an obligately psychrophilic bacterium produced a small but significant amount of ammonia in a peptone-yeast extract medium (Table 3). In no instance did any of the liquid media incubated in the deep sea give rise to turbid cell suspensions.

From this study it appears that the degree of preservation of the food materials recovered from *Alvin* is no chance observation, although our experiments were carried out at greater depths than those where the *Alvin* accident occurred.

The surprisingly large difference be-

Table 2. Microbial degradation of four substrates in 50 ml syringes (10 ml of liquid medium, 20 ml of air) exposed for 19 weeks at a depth of 5300 m (location: 33°58′N, 69°58′W) as compared to controls kept at 3°C for 6 weeks. Percent values are corrected for the unequal exposure time. The substrate concentrations are given as micrograms of starch, galactose, or ammonia nitrogen per milliliter, respectively. The inoculum was 5 ml of surface water from Eel Pond, Woods Hole.

Initial concentration (μg/ml)	Change in concentration (μg/ml)		Sample control (%)
	Control	Sample	
Starch			
1850	1330	260	6.2
	1290	170	4.2
Galactose			
1800	1680	220	4.15
	1580	280	5.6
Peptone			
57	263	9	1.1
	258	20	2.4
Albumin			
57	172	15	2.8
	173	17	3.1

tween rates of degradation in samples exposed to deep-sea conditions and those in controls appears to be real. The data support the notion of a general slow-down of life processes in the deep sea. No obvious explanation is readily conceivable except for some clues derived from an apparent temperature-pressure relation in microorganisms indicated by some of our data.

The experiment with pure cultures (Table 3) included, in addition to typical mesophilic bacteria, several psychrophilic strains that all grew readily at −1°C (not identical with the minimal growth temperature) in the laboratory. At deep-sea conditions, however, only the culture with the

Table 3. Microbial degradation of a complex nitrogenous medium (0.02 percent yeast extract, 0.2 percent peptone, and an initial ammonia nitrogen concentration of 39.0 μg/ml) in 50 ml syringes (10 ml of medium, 20 ml air) exposed for 18 weeks at a depth of 4300 m (location: 28°N, 70°W) as compared to controls kept at 3°C for 6 weeks. The ammonia nitrogen values have a standard deviation of ± 0.5 μg/ml. The inoculum was six strains of mesophilic and psychrophilic bacteria isolated in a preceding study (9).

Strain	Temperature range for growth (°C)	Change in concentration *	
		Control	Sample
44	17 to 36	3	0.1
36	8 to 36	1.5	−0.7
7	−1 to 36	130	−1.1
20	−1 to 27	128	+0.4
60	−1 to 23	87	−0.1
58	−1 to 17	15	+2.2

* Micrograms of ammonia nitrogen per milliliter.

lowest maximal growth temperature (strain 58) caused detectable biochemical changes of the substrate within the given exposure time. But even in this case, the rate is strongly reduced as compared to that in the laboratory controls.

These data suggest that, superimposed on a quantitative reduction of the rate of biochemical activity, the increased hydrostatic pressure may exert an effect on the cells, raising the minimal growth temperature. When this increase exceeds the environmental temperature, the cells will become inactive. This effect would be similar to, but not necessarily biochemically linked to, the observed increase in temperature tolerance of bacteria (4) and of isolated enzymes (8) when exposed to similar increases of pressure.

We now propose the hypothesis that, in an environment of low temperature, an increasing pressure will eliminate growth and biochemical activity of bacterial types successively as their minimal growth temperatures are shifted toward, and ultimately surpass, the environmental temperature. Thus, psychrophilism of our isolates at normal pressure may be defined as an expression of adaptability to the combined effect of high pressure and low temperature. Or, in other words, psychrophilic bacteria would not necessarily react as psychrophiles in the deep sea. Laboratory experiments in this direction are under way.

Our hypothesis may be further supported by the fact that in marine sediments from depths of 1300 and 2600 m extremely obligate psychrophilic bacteria that exhibited maximal growth temperatures between 8° and 15°C have been isolated. These types are not found in shallower waters where obligate psychrophiles with maximal growth temperatures between 17° to 24°C are present (9). Strain 58 belongs to the latter group but appears to have the potential of being biochemically active at 2° to 3°C at a depth of 4300 m.

In seawater collected at a depth of 200 m (17.6°C), mesophilic bacteria were predominant while obligate psychrophilic bacteria were absent (9). This may explain the low absolute rates of degradation in these samples when exposed to deep-sea conditions.

One obvious implication of our findings concerns the use of the deep sea as a dumping site for organic wastes.

The relatively low rates of microbial activity at deep-water conditions appear to render this way of waste disposal very inefficient compared to the degradation of organic wastes in land-disposal sites or in treatment plants. Accumulations of waste materials or intermediate decomposition products in the deep sea appear rather uncontrollable. Bruun and Wolff (10) mention the common recovery of waterlogged wood materials from deep-sea dredgings even far from land.

Normally, few solid organic materials, produced on land or in the sea, can be expected to reach the deep sea without passing surface waters or shallow-water sediments where considerable degradation occurs. If this step during offshore disposal were eliminated, it seems possible to trap substantial amounts of nutrients in solid form in the deep sea, and thereby remove them from natural or technically enhanced recycling processes. The notion of fertilizing the sea with manmade wastes might not be applicable with regard to deep-sea dumping.

Although neither microbial population collected from surface or deep waters showed appreciable activities when exposed to deep-sea conditions, our data do not entirely disprove the possibility of long-term enrichments in deep-sea sediments. Whether or not adaptive processes occur, the rates of oxygen supply and microbial degradation activities will determine the extent to which anaerobic conditions will arise, with possible elimination of the benthic nonmicrobial fauna.

HOLGER W. JANNASCH
*Woods Hole Oceanographic Institution,
Woods Hole, Massachusetts 02543*

KJELL EIMHJELLEN
*Department of Biochemistry,
The Technical University of Norway,
Trondheim*

CARL O. WIRSEN
Woods Hole Oceanographic Institution

A. FARMANFARMAIAN
*Department of Physiology and
Biochemistry, Rutgers University,
New Brunswick, New Jersey 08903*

References and Notes

1. W. O. Rainnie and C. L. Buchanan, *Ocean Ind.* 4, 61 (1969).
2. Y. M. Chen and W. Charin, *Anal. Biochem.* 13, 234 (1965).
3. D. W. Menzel and J. H. Ryther, *Deep-Sea Res.* 15, 327 (1968).
4. C. E. ZoBell, *Bull. Misaki Mar. Biol. Inst. Kyoto Univ.* 12, 77 (1968); W. Harder and H. Veldkamp, *Arch. Microbiol.* 59, 123 (1967).
5. H. W. Jannasch, *J. Bacteriol.* 99, 156 (1969).
6. Benthos Inc., North Falmouth, Mass. 02556.
7. M. Dubois, K. A. Gilles, J. K. Hamilton, P. A. Rebers, F. Smith, *Anal. Chem.* 28, 350 (1956).
8. R. Y. Morita and R. D. Haight, *J. Bacteriol.* 83, 1314 (1962).
9. K. Eimhjellen, unpublished.
10. A. F. Bruun and T. Wolff, in *Oceanography*, M. Sears, Ed. (AAAS, Publ. No. 67, Washington, D.C., 1961), p. 391.
11. We thank B.-A. Collins for technical assistance and G. H. Volkman and R. H. Heinmiller for performing the technical operations at sea. Supported by NSF grant BO 20956. Contribution No. 2573 of the Woods Hole Oceanographic Institution.

12 September 1970; revised 13 November 1970 ∎

BIODETERIORATION IN THE
SEA AND ITS INHIBITION

JOHN McN. SIEBURTH and ALLAN S. DIETZ

Narragansett Marine Laboratory
University of Rhode Island
Graduate School of Oceanography

At 1–3°C food materials were consumed or decayed when held in perforated
double containers at a depth of 2 m in inshore waters and at 5200 m in the
deep sea for 2 and 10 weeks respectively. When identical food materials were
held in triple enclosures which minimized water exchange and the passage of
omnivorous scavenging animals, they were in a remarkable state of preservation. The latter conditions were apparently present to a very high degree in
the well-preserved box lunches recovered from the research submersible
Alvin. Such inhibition of biodeterioration must also occur when dumped
organic wastes are compacted or allowed to accumulate in the sea, regardless
of depth.

INTRODUCTION

Microorganisms are widely distributed in the deep sea (ZoBell, 1946; Kriss,
1963). Bacteria-free environments are relatively rare (Sieburth, 1961; Watson
and Waterbury, 1969). Although bacteria can be inhibited or killed at
pressures exceeding 200 atm (ZoBell and Oppenheimer, 1950) and enzymic
reactions have been shown to be pressure sensitive (Morita, 1967), bacteria
surviving ascent from depths of 10,000 m can grow at hydrostatic pressures
of 1000 atm (ZoBell and Morita, 1957). Obligate marine psychrophiles

183

Reprinted from Effect of Ocean Environment on Microbial Activities
ed. by R. D. Colwell and R. Y. Morita. University Park Press pp. 318–326 (1974).

readily develop at temperatures below 10°C in inshore waters (Sieburth, 1967) and occur in the deep sea (Sieburth, 1971). Scavenging omnivores are ubiquitous in the sea and are present even in the less productive abysses at depths of 5300 m (Hessler *et al.,* 1972). Biodeterioration of organic materials in the sea is a result of both consumption by scavenging omnivores such as the gammarid amphipods, which tear and triturate the materials thereby increasing surface area and the penetration of oxygen, and the activities of microorganisms which are limited by surface area and the degree of oxygenation. The food ingested by the macrofauna is also subject to the microbial activities of the gut flora during its passage of the gut as well as in the feces.

Since both scavengers and microorganisms are ubiquitous, one might assume, *a priori,* that at similar temperatures the biodeterioration of organic matter in the deep sea would occur in a manner and rate not too dissimilar from that in shallow waters. The report by Jannasch *et al.* (1971) of the remarkable state of preservation of the box lunch accidentally submerged with the research submersible *Alvin* for 10 months at 1540 m is paradoxically at odds with what one might expect. In order to see if the conditions of the deep sea do slow the rates of microbial activity, Jannasch *et al.* (1971) used microorganisms from shallow water and a depth of 200 m to inoculate soluble substrates held in syringes and bottles sealed with serum stoppers which were incubated at the *in situ* temperatures and pressures of the deep sea for 2 to 5 months. They found that these samples decomposed 666 to 8.6 times more slowly than the refrigerated controls. They concluded that the degree of preservation of the foodstuffs recovered from *Alvin* was no chance observation and that there was a general slow-down of life processes in the deep sea, presumably due to an interaction of low temperature and high pressure.

One thing shared in common by the box lunch in *Alvin* and the experiments reported by Jannasch *et al.* (1971) is that the materials were isolated from a free interchange with the waters of the deep sea and their biota. The lunch in *Alvin* was in a covered plastic box. It seemed possible to us that the protection offered by the enclosure might be sufficient to interfere with the natural biodeterioration near the ocean floor. To see if this is the case, experiments were designed in a manner similar to those of Payne (1965) in which scavenging organisms would be excluded.

MATERIALS AND METHODS

The box lunch in *Alvin* contained bologna sandwiches. Similar solid food-stuffs were selected to simplify the design and construction of the inexpensive "lunch box" test packages (Fig. 1*B*) used to hold the test materials under

Fig. 1. The condition of the exposed, protected, and control test materials and the nature of the test packages. *A,* Current meter array with its test packages being recovered after 74 days immersion; *B,* a "lunch box" test package containing test materials protected in a triple enclosure (left) and exposed in a double perforated enclosure (right); *C,* the duplicate set of test materials in perforated containers from Station II (upper); *D–F;* a comparison of test materials from Station II (lower) showing frozen control (*D*), exposed (*E*), and protected (*F*) boxes.

both protected and exposed conditions. Polyethylene refrigerator boxes with a snug lid were used to hold 2.5 cm cubes of lean beef, suet, whole-wheat roll, 3% agar on a styrofoam core, a fresh crab claw from a Jonah crab, and corrugated cardboard which were strung and separated by sailmaker's Dacron twine to avoid protection of the surfaces of the foodstuff by the containers (Fig. 1D); the latter point is a critical feature of our experiments. Duplicate boxes were placed in a plastic lunch pail which was bolted to the inside of a plastic dishpan and enclosed with a matching dishpan held tightly lip to lip with bolts at each corner. These triple-enclosed test materials were intended to simulate the protected lunches in *Alvin*. In order to expose a replicate set of test materials to the environment, a second lunch pail and its enclosed food boxes were well perforated with 5-mm holes (Fig. 1C) and bolted to the outside of the dishpan enclosure. Plastic license-plate bolts were used throughout. Each completed test package was enclosed in a Dacron fish net bag (Fig. 1B) and frozen until used. On station, the bags were attached to the current meter arrays (Fig. 1A) which were submerged at the locations and depths and for the periods shown in Table 1. Test packages were also immersed in shallow water (5 m) from a dock in Narragansett Bay, Rhode Island, at a depth of 2 m for 2-week periods in February (1–2°C) and August (20–21°C). Frozen controls were kept for purposes of comparison. Upon recovery of the current meter arrays, the test packages were removed, the contents of the food boxes photographed to show gross changes and the materials were carefully examined to record their condition. Representative samples of the test materials were prepared for scanning electron microscopy (Gessner *et al.*, 1972) and kept frozen with Dry Ice.

RESULTS AND DISCUSSION

The results of the deep-sea tests are given in Table 1. The smallest differences in the condition of the test materials in the two types of containers were obtained with the upper packages held 1500 m off the ocean floor at depths of 3715–3875 m. Although all test materials were present in both types of containers, there was a detectably poorer quality in the color, texture, and feel of the more labile materials remaining in the perforated containers. The differences in the packages held some 5 m off the bottom were very striking as shown in Figs. 1D–F. Materials in the triple enclosure (Fig. 1F) were intact and of a good appearance while all readily decomposible or edible materials such as the beef, crabmeat, and bread, were absent from the perforated container (Fig. 1E). Several lysianassid gammarid amphipods were present in the perforated enclosures at Station II-lower. The crab claws in the perforated

Table 1. Station locations, incubation conditions, and state of the recovered "lunch box" test packages

	Station I		Station II				Station III			
Location	29°49.9'N, 70°22.3'W		28°54.3'N, 69°33.0'W				29°50.3'N, 60°34.2'W			
Date dropped	10/26/71		10/27/71				10/28/71			
Days incubation	73		74				72			
Incubation temp (°C)	2–3		2–3				2–3			
Package position	Upper[a]		Upper		Lower		Upper		Lower	
Depth (m)	3875		3815		5315		3715		5215	
Lunch box	Closed	Perf.[b]	Closed	Perf.	Closed	Perf.[c]	Closed	Perf.	Closed	Perf.[d]
Condition of food[e]										
Beef	G	P	G	P	G	Gone	G	P	G	Gone
Crabmeat	G	P	G	P	G	Gone	G	P	G	Gone
Cardboard	NC	NC	NC	NC	NC	NC	NC	NC	NC	NC
Agar	G	P	G	P	G	Gone	G	P	NC	F
Bread	NC	F	F	NC	NC	F	F	Gone	NC	Gone
Fat	NC	F	NC	NC	F	F	NC	F	F	F

[a] Upper, 1500 m above ocean floor; lower, 5 m above ocean floor.
[b] Perf., perforated.
[c] Lysianassid gammarid amphipods present.
[d] Sediment present.
[e] G, good appearance, intact, some slime; P, poorer appearance, some sediment present, copious slime; NC, no apparent change; F, fragmented.

enclosures had a tendency to be brittle or soft while those in the protected enclosures were hard and difficult to crush.

Striking differences were also obtained with the shallow-water test packages held in Narragansett Bay at $1-2°C$ for 2 weeks. The triple-enclosed materials were in an unbelievable state of preservation with the beef still red inside while the partially consumed materials remaining in the perforated container were discolored, slimy, and highly putrid. After 2 weeks of immersion at $20-21°C$ in Narragansett Bay, differences in the rate of biodeterioration of test materials in the two types of containers were much less, but still detectable. This was the only test package fouled by pennate diatoms and barnacles. The readily decomposible food stuffs were absent from the perforated container over which gammarid amphipods were swarming, while in the triple-enclosed containers the cube of beef and the crabmeat were intact although in a highly putrid state. The exposed crab claw was soft and friable while the protected one was in a hard state.

The test materials in the triple enclosures, although intact and of good appearance, were apparently subject to microbial attack as the enclosed water was turbid and contained in excess of 10^7 bacteria per milliliter as determined by direct microscopic examination. The original purpose in fixing and freezing the test materials was to characterize microbial colonization by scanning electron microscopy. Results obtained with most of the materials were difficult to interpret. An exception was the surface of the corrugated cardboard. The cellulose fibers, which were free of apparent microorganisms at the time of immersion, became covered with forms suggestive of both bacteria and fungi in both the exposed and the protected packages. If microorganisms are present and active in the triple enclosures, why was not biodeterioration greater? Biodeterioration is not normally restricted to one class of organisms. The decomposition of dead flesh is greatly expedited by the disruption of tissues by carrion-feeding insects (Payne, 1965). At the shore's edge, kelp-fly larvae play a similar role in hastening the decomposition of storm-tossed seaweed through trituration and aeration (Bunt, 1955). In the sea, gammarid amphipods and other omnivores have the role of the fly larvae on land. In the deep sea, as well as in Narragansett Bay, gammarid amphipods were present in the perforated containers in which the readily edible and decomposible foodstuffs were absent.

The differences observed in the test materials do not seem to be due to just the presence of animals. In the deep-sea packages held 1500 m above the ocean floor, where gammarids were not trapped in the perforated containers and all materials were intact, there was still a visibly detectable difference between materials held in the two types of containers. Also the softening of the crab claws in only the perforated containers must have been due to chitinoclastic bacteria. The restriction of water flow may have a detectable effect on microbial activity. Microbial activities are greatly decreased by oxygen depletion. The strong putrefactive odors of our food packages indicated this was the case. This possibility was recognized but rejected by Jannasch *et al.* (1971) as there was no evidence in the *Alvin* lunch of reducing conditions or the lack of dissolved oxygen.

The carefully controlled and executed experiments of Jannasch *et al.* (1971) were designed to obtain quantitative data fundamental to an understanding of the effect of temperature and pressure on bacteria. Despite the qualitative and subjective nature of our observations, we present them here as we feel they indicate that labile foodstuffs held near the ocean floor may readily undergo biodeterioration. Although the metabolism of deep-sea microorganisms and scavenging animals are apparently much reduced in the deep sea, their accumulative activities may not be as slow as indicated by observations on individual components or in exposure experiments in which the bulk of the test material is protected by its enclosure. The process of biodeterioration appears to be easily inhibited by small changes such as a protective covering which shields a labile substrate from scavenging animals and dissolved oxygen.

As pointed out by Jannasch *et al.* (1971), the rates of biodeterioration on the deep-ocean floor have obvious implications with regard to the dumping of organic wastes in the deep sea. Municipal wastes are largely composed of paper, which is slow to degrade, and a small content of food wastes. In tests for ocean dumping, this material is being compacted into large cubes of approximately 1 m^3 which will sink. Laboratory studies have shown that biodeterioration is impeded within solid waste deposits in seawater after oxygen has been consumed and hydrogen sulfide has been produced (S. D. Pratt, personal communication). Such observations, as well as ours, emphasize the fact that biodegradation is a surface phenomenon involving the total benthic community. Conditions which enclose organic matter or lead to passive deposits, thereby excluding animals and microbial oxidation, are, therefore, conducive to storage rather than to the oxidation and mineralization of organic matter.

SUMMARY

1. Readily consumable foodstuffs such as lean beef, crabmeat in a claw and bread held in perforated double containers were consumed or decayed at 1–3°C within 2 weeks in inshore waters and within 10 weeks in the deep sea.
2. Triple enclosures did not prevent bacterial development, but the foodstuffs were in an excellent state of preservation.

3. The exclusive of omnivorous scavengers in the sea has exactly the same delaying effect on decomposition as the exclusion of carrion-feeding insects on land.
4. Enclosures apparently prevented animal consumption and microbial decay of the well-preserved box lunches recovered from the research submersible *Alvin*.

ACKNOWLEDGMENTS

The scientific parties of TR 104 and 109 who helped to drop and recover the test packages and T. A. Napora who identified the amphipods are gratefully acknowledged. This study was supported in part by NSF Grants GB 18000 and GA 28903.

LITERATURE CITED

Bunt, J. S. 1955. The importance of bacteria and other microorganisms in the seawater at MacQuarie Island, Austr. J. Freshw. Res. 6:60–65.

Gessner, R. V., R. D. Goos, and J. McN. Sieburth. 1972. The fungal microcosm of the internodes of *Spartina alterniflora*. Mar. Biol. 16(4):269–273.

Hessler, R. R., J. D. Isaacs, and L. Mills. 1972. Giant amphipod from the abyssal Pacific Ocean. Science 175:636–637.

Jannasch, H. W., K. Eimhjellen, C. O. Wirsen, and A. Farmanfarmaian. 1971. Microbial degradation of organic matter in the deep sea. Science 171:672–675.

Jannasch, H. W., and C. O. Wirsen. 1972. *Alvin* and the sandwich. Oceanus 16 (Dec.):20–22.

Jannasch, H. W., and C. O. Wirsen. 1973. Deep-sea microorganisms: *in situ* response to nutrient enrichment. Science 180:641–643.

Kriss, A. E. 1963. Marine Microbiology (Deep Sea). (Trans. by J. M. Shewan and Z. Kabata.) Oliver and Boyd, London.

Morita, R. Y. 1967. Effects of hydrostatic pressure on marine microorganisms. Oceanogr. Mar. Biol. Annu. Rev. 5:187–203.

Payne, J. A. 1965. A summer carrion study of the baby pig *Sus scrofa* Linnaeus. Ecology 46(5):592–602.

Sieburth, J. McN. 1961. Antibiotic properties of acrylic acid, a factor in the gastrointestinal antibiosis of polar marine animals. J. Bacteriol. 82:72–79.

Sieburth, J. McN. 1967. Seasonal selection of estuarine bacteria by water temperature. J. Exp. Mar. Biol. Ecol. 1:98–121.

Sieburth, J. McN. 1971. Distribution and activity of oceanic bacteria. Deep-Sea Res. 18:1111–1121.

Watson, S. W., and J. B. Waterbury. 1969. The sterile hot brines of the Red Sea. *In* E. T. Degens and D. A. Ross (eds), Hot Brines and Recent Heavy Metal Deposits in the Red Sea, pp. 272–281. Springer-Verlag, New York.

ZoBell, C. E. 1946. Marine Microbiology. Chronica Botanica, Waltham, Massachusetts.

ZoBell, C. E., and R. Y. Morita. 1957. Barophilic bacteria in some deep sea sediments. J. Bacteriol. 73:563–568.

ZoBell, C. E., and C. H. Oppenheimer. 1950. Some effects of hydrostatic pressure on the multiplication and morphology of marine bacteria. J. Bacteriol. 60:771–781.

Note added in proof. In discussing our paper with H. W. Jannasch and C. O. Wirsen, it was learned that similar experiments were conducted 2 years earlier as part of the original study by Jannasch *et al.* (1971), but results were not included as they were not quantitative. The subsequent description of the "BIO-PACK" experiment (Jannasch and Wirsen, 1972) indicates an entirely different design and objective of the experiments. The paper by Jannasch and Wirsen (1973) ably demonstrates that the *in situ* sediment microflora is also slow to convert organic matter, but points out the possible importance of animals and their microflora, the major premise of this paper.

Deep-Sea Microorganisms: In situ Response to Nutrient Enrichment

Abstract. *After inoculation of sterile organic materials on the deep-sea floor and in situ incubation for 1 year, relatively minute rates of microbial transformation were recorded. This extremely slow conversion rate, as well as the type and quantity of organic matter normally reaching the ocean floor, appear to characterize microbial life in the deep sea.*

We have reported (*1*) that microbial conversion of a number of organic substrates was considerably retarded when laboratory cultures and mixed populations of surface-born marine bacteria were incubated in the deep sea. The assumption was made that the indigenous microbial flora of the deep water or sediment may respond differently.

To check this possibility, a housing for sterilized sample bottles was devised that permitted inoculation directly on the deep-sea floor. A rack holding 20 120-ml bottles was enclosed in a pressure-tight aluminum cylinder (Fig. 1). As in earlier experiments, the bottles contained the media in concentrated form in quantities of 1 to 10 ml and were equipped with punctured serum caps for self-inoculation (*1*). After closing, the cylinder was evacuated (i) to lower the concentration of gases that dissolve in the samples under elevated hydrostatic pressure and (ii) to give the lid a tight fit during the submersion operation.

This sample-housing vessel was attached to the research submarine *Alvin*. Upon reaching the sea floor, a valve was operated by the mechanical arm to fill the cylinder and the sample bottles with water from the top sediment, including suspended sediment particles. When the pressure had equalized after about 2 minutes, the sample housing opened, and the rack with the sample bottles was removed and placed on the sea floor by the mechanical arm of the submarine. The empty cylinder was returned to the surface. The place of incubation was the permanent bottom station at 39°46′N, 70°41′W, 1830 m depth; the temperature was 4°C. The site was marked by sonar reflectors and revisited by *Alvin* several times during the summer seasons in 1971 and 1972 for placing and retrieval of instruments and samples.

After being filled at the site of incubation, the bottles contained 0.1 percent starch, 0.033 percent agar, or 0.1 percent gelatin. In addition, a duplicate set contained 0.01 percent KH_2PO_4 plus K_2HPO_4 and 0.025 percent NH_4Cl. After retrieval and poisoning with 1.0 ml of 0.2 percent $HgCl_2$, starch and agar concentrations were determined by the anthrone method, and gelatin by the protein determination technique of Lowry *et al.* (*2*). Sterile samples of some solid materials (bond paper, paper towels, balsa wood, beech wood, and thalli of the marine algae *Ulva*) were also included in separate bottles and

192

Reprinted from Science 180:641–643 (1973).

their decomposition determined by weight loss of dry material (samples were dried at 103°C to constant weight after washing for the removal of salt). The bottles were deposited on 22 June 1971 and retrieved on 12 June 1972 (51 weeks of in situ incubation). Sterile controls were kept in the laboratory for the same time period and at the approximate deep-sea temperature of 3° to 4°C.

By the same inoculation and incubation procedure, isotopically labeled mannitol, sodium acetate, sodium glutamate, and casamino acids (in concentrations of 30, 10, 5, and 2 $\mu g/ml$ were deposited at the same site from 12 June to 25 September 1972 (14 weeks). In a duplicate set, the bottles were filled during *Alvin*'s descent at a depth of 200 m, also to be incubated at 1830 m. Upon retrieval and poisoning the degree of substrate conversion into cell material and CO_2 was measured by liquid scintillation spectrophotometry. Quenching corrections were made by the channel ratio method. The total recovery of labeled material averaged 97 percent.

Table 1 shows the percentage of substrate converted in 1 year on the deep-sea floor. The controls were parallel samples incubated at 4°C for 1 month in the laboratory. One set was inoculated with water from 1830-m depth, the other with surface water (Eel Pond, Woods Hole). The data for in situ in-

Page 642

cubation relative to controls are corrected for the difference in the times of incubation and indicate that the deep-water population converted the substrates 17.5 to 125 times faster when incubated in the laboratory (at deep-sea temperature) rather than in situ. In this response, this population hardly differs

from the microbial population of surface water.

The changes of dry weight of the paper, wood, and algal (*Ulva*) samples were statistically not significant (± 1 percent) and indicated no measurable degradation during the 1-year incubation. Parallel samples, incubated for 14 weeks at 1830 m, were sent for examination for evidence of bacterial and fungal decay to J. Kohlmeyer (Institute of Marine Sciences, University of North Carolina at Morehead City). None was found. Wood samples incubated for 1 year in open containers, not protected from animal attack as in the bottles, were attacked by boring mollusks (*3*) in the absence of visible microbial degradation.

In Table 2, conversion of ^{14}C-labeled substrates is given as percentage relative to control samples incubated in the laboratory at 4°C. Data for the 5300-m incubation were obtained in an earlier experiment; the same controls were used for these data and for the samples taken at 1830 and 200 m and incubated

Fig. 1. Two racks of 20 sample bottles (120 ml each) at the permanent bottom station at a depth of 1830 m. In the foreground are *Alvin*'s mechanical arm and two closed pressure-tight sample housings.

at 1830 m in the present experiment. Conversion data for the latter samples are essentially similar to those for samples inoculated and incubated at 1830 m, the differences probably being related to the size of the inoculated population.

In all experiments, the data indicate that the in situ microbial response to enrichment of deep-sea water and sediment with various organic substrates was between one to three orders of magnitude lower than in the controls. This was most recently supported by measurements of oxygen uptake in deep-sea sediments at the same site (4). Further, the response of microbial deep-sea populations was similar to that of surface-water populations incubated in the deep sea or in the laboratory at normal pressure and comparable temperature.

Two technical points have to be discussed. During the filling of the sample bottles, part of the inoculum underwent decompression, which must be assumed (until evidence to the contrary is available) to affect the viability of adapted deep-sea microorganisms. ZoBell (5) discussed this problem, giving experimental proof that successive compression and decompression of strains of marine bacterial isolates had little or no effect on their viability. Before the present study, however, no work had been reported on deep-sea bacteria that did not undergo previous decompression. In our experiments, this possible

Table 1. Percentage of substrate utilized after incubation with and without added nutrients (0.01 percent KH_2PO_4 plus K_2HPO_4 and 0.025 percent NH_4Cl). In situ samples (A) were inoculated with water at 1830 m and incubated for 12 months in situ at 1830 m. Controls were inoculated with water from 1830 m (B) or with surface water (C) and incubated for 1 month in the laboratory at 4°C. Columns A/B and A/C show in situ conversion as a percentage of control conversion, calculated on the basis of equal incubation periods.

Substrate	Added nutrients	In situ (A) (duplicates)		Controls		A/B (%)	A/C (%)
				B	C		
Starch, 1.0 mg/ml	Yes	7.7	3.9		28.0		1.7
	No	5.5	16.5	16.0	9.0	5.7	10.2
Agar, 0.33 mg/ml	Yes	1.3	1.7	13.0	26.5	0.9	0.5
	No	0.0	0.0	2.5	1.5		
Gelatin, 1.0 mg/ml	Yes	0.0	9.7	50.3	84.9	0.8	0.5
	No	0.0	1.8		48.5		0.2

Table 2. Microbial conversion of four ^{14}C-labeled substrates inoculated and incubated in the deep sea. Conversion in samples incubated in situ is expressed as percentage relative to control samples incubated in the laboratory at 4°C for an equal period. Data are given for incorporated substrate (particulate) and metabolized substrate (CO_2 plus particulate); I, incorporated; M, metabolized.

^{14}C-labeled substrate	Taken at 200 m; incubated at 5300 m		Taken at 200 m; incubated at 1830 m		Taken at 1830 m; incubated at 1830 m	
	I	M	I	M	I	M
Acetate	1.14	2.83	0.38	0.37	4.08	2.91
Mannitol	1.14	1.68	7.12	7.02	0.26	0.98
Glutamate	1.62	0.67	0.72	0.81	0.26	0.50
Casamino acids	8.87	7.71	11.74	30.86	1.70	3.08

effect is minimized because decompression lasted only seconds and did not affect the entire inoculum. In the absence of studies on pure cultures, no conclusive statement on the barophobic or barotolerant behavior of these populations can be made.

Another point of discussion is the increased solubility of oxygen at elevated hydrostatic pressure and the possible inhibitory effect of high oxygen concentrations (6). However, when the valve of the evacuated housing vessel was opened on the sea floor at about 183 atm, the critical pressure forcing the remaining air in solution was not reached in the vessel until most sample bottles were filled. In addition, undissolved air was observed to escape when the lid of the housing vessel opened following pressure equilibration. Experimental tests in pressure chambers showed that no gas is trapped in upright bottles with punctured serum caps. Therefore, we assume that little of the oxygen remaining in the housing vessel had dissolved in the sampled water, and that the critical concentration could not have been reached (6). As an additional check, samples first flushed with nitrogen gas have been deposited for retrieval in the summer of 1973.

It is unlikely that, after a 1-year incubation, the low rate of microbial activity is explainable by too small an inoculum (about 120 ml of surface sediment slurry). Thus, it appears that enrichments of deep-sea sediment with various organic substrates in different amounts do not result in any substantial activity of the indigenous microflora within a 1-year in situ incubation. The assumption may be made that more than 1 year is required to complete possible adaptive processes that may ultimately result in faster metabolic rates.

The fact that not only surface-born bacteria (1) but also microorganisms collected at the deep-sea floor are exhibiting extremely slow metabolic rates when incubated in situ may be interpreted in two ways.

Life processes in general may be slower at deep-sea conditions than at surface pressures and temperatures for reasons other than the low nutrient supply. Studies on the deep-sea benthic fauna (7), have reported (i) extreme diversity of species, (ii) small brood size, (iii) preponderance of adult individuals in most species, and (iv) abundant cases of endemism. These characteristics suggest slow growth and long life of the individual animal and could be the result of a relative retardation of certain critical metabolic processes.

The slow metabolic rate may also lead to the argument that an active, adapted microflora does not exist in the deep-sea sediment. The high colony counts usually found when deep-sea sediment samples are streaked on nutrient agar (5) may originate solely from surviving and viable cells that reached the ocean floor with sedimenting detritus particles. Particulate organic matter readily available for microbial decomposition will hardly reach the deep ocean. It will largely be degraded during the slow sedimentation, estimated to take from several weeks to more than a year per 1000 m of depth (8). The particulate organic matter in deep waters was shown to be "refractory" (9), that is, no degradability could be demonstrated. The total amount of larger particulate material (for instance, animal carcasses) reaching the deep sea undegraded will probably be very small, although of considerable significance for the highly diverse but scanty fauna of benthic scavengers. Nothing is known about the quantity of nonrefractory organic matter reaching the deep-sea floor with the relatively fast-sinking fecal pellets of zooplankton (8).

Thus, the top sediment being virtually void of nonrefractory organic matter readily available for degradation, the activity of microorganisms in the deep sea may be largely confined to intestinal tracts of animals, where the enriched nutrient milieu will enable microorganisms to decompose refractory materials (chitin, cellulose, and so forth) in an endosymbiotic fashion. This notion is supported by the finding of an enlarged gut in deep-sea mollusks (10). According to this hypothesis, the role played by microorganisms in the turnover of organic matter in the deep-sea sediments appears to be fundamentally different from that in shallow-water sediments, or, for that matter, in soil. Experiments on incubations of solidified organic materials (agar, starch, gelatin) on the deep-sea floor in open containers (11) show that after 1 year of exposure, marks of animal feeding appeared to be almost the only sign of disintegration. No work on the intestinal flora of deep-sea invertebrates has yet been done. Complementing our earlier

work (1), the data reported in this study confirm the conclusion that the deep sea must be considered extremely inefficient with respect to recycling of organic wastes.

HOLGER W. JANNASCH
CARL O. WIRSEN
Woods Hole Oceanographic Institution,
Woods Hole, Massachusetts 02543

References and Notes

1. H. W. Jannasch, K. Eimhjellen, C. O. Wirsen, A. Farmanfarmaian, *Science* 171, 672 (1972).
2. O. H. Lowry, N. J. Rosebrough, A. L. Farr, R. J. Randall, *J. Biol. Chem.* 193, 265 (1951).
3. Identified by R. Turner (Museum of Comparative Zoology, Harvard University).
4. K. L. Smith and J. M. Teal, *Science* 179, 282 (1973).
5. C. E. ZoBell, *Bull. Misaki Mar. Biol. Inst. Kyoto Univ.* 12, 77 (1968).
6. ——— and L. L. Hittle, *Can. J. Microbiol.* 13, 1311 (1967).
7. J. F. Grassle and H. L. Sanders, *Deep-Sea Res.*, in press.
8. T. J. Smayda, *Oceanogr. Mar. Biol. Annu. Rev.* 8, 353 (1970).
9. D. W. Menzel and J. H. Ryther, *Inst. Mar. Sci. Univ. Alaska Publ. No. 1* (1970), p. 31.
10. J. A. Allen and H. L. Sanders, *Deep-Sea Res.* 13, 1175 (1966).
11. H. W. Jannasch and C. O. Wirsen, in preparation.
12. We thank P. Holmes for assistance in one of the *Alvin* dives; and J. M. Teal, J. F. Grassle, and K. L. Smith for a critical discussion of the manuscript. Research supported by NSF grant GA 33405. This is contribution No. 2987 of the Woods Hole Oceanographic Institution.

6 December 1972   ■

MINERAL CYCLING IN MARINE ECOSYSTEMS

Reprinted From
MINERAL CYCLING IN SOUTHEASTERN ECOSYSTEMS
F. G. Howell, J. B. Gentry and M. H. Smith, Editors
ERDA Symposium Series (CONF-740513)

LAWRENCE R. POMEROY
Department of Zoology, University of Georgia, Athens, Georgia

ABSTRACT

Synthesis of recent work on marine grass communities, coral reefs, and oceanic plankton leads to a reexamination of some concepts of the flux of essential elements through marine communities. Where they occur, storage compartments and inputs from outside the system have a marked effect on stability. In planktonic communities, which have minimal storage and input, nutrient demand must be met largely from recycling and successional shifts of dominant species in response to changing nutrient limitations. Diversity in the plankton results in part from the successional mosaic created by local changes in nutrient limitations and by the presence of biochemical niches. Coral reefs also develop successional mosaics, but the forces at work here include storm and predator damage as well as interspecies competition. Marine grass meadows, on the other hand, show less spatial and temporal diversity, little succession, and high stability. In the marine communities considered, stability appears to be more strongly related to availability of nutrients than to diversity. The most stable communities are the least diverse.

Populations and communities are self-reproducing configurations of chemical elements that maintain their integrity and continuity through continuous reconstruction of tissues and reproduction of individuals. These processes require a constant flux of elements and energy through both individual organisms and communities. Although individual organisms may sometimes concentrate one or more essential elements and recycle them internally, populations and communities cannot do this because loss of elements is inevitable through the death and degradation of individuals. A source of essential elements is necessary, and communities may be shaped in part by their availability. The interaction between communities and the elements moving through them can influence species composition, diversity, and stability. Concepts of limiting nutrients must now take into account a number of complexities that did not concern Liebig (1840).

197

Reprinted from ERDA Symposium Series CONF – 740513 (1975).

Much of the conceptualization of diversity and stability of communities has been related to food-web structure and particularly to interspecies interactions. Moreover, there has been a dichotomy between results with mathematical and laboratory model systems, on the one hand, and observations of natural communities, on the other. Up to the present time, most theoretical models consider one or two factors at a time, eliminating by definition all others. Modulations of nutrient supply are usually among the factors eliminated from models for the sake of simplicity. Recent studies of real-world communities permit us to consider specifically the influence of essential elements on community structure and stability and to compare this knowledge with the output from theoretical models. The synthesis presented in this paper will focus on four contrasting marine communities. Although they do not include all the possibilities to be found in the biosphere, they represent some extremes of community development quite well, and all of them are well described by recent research reports.

MARINE GRASS COMMUNITIES

Mineral cycling in two contrasting estuarine communities has been studied in some detail: the subtidal eelgrass (*Zostera marina*) meadows and the intertidal cordgrass (*Spartina alterniflora*) meadows. In the *Zostera* meadows of Izembek Lagoon in Alaska, phosphorus and nitrogen move from the sediments into roots and then into leaves. Nutrients fixed in protoplasm are remineralized into the water when the leaves are eaten by consumers or die and decay. However, nearly half the phosphate, ammonia, and nitrate that reaches the leaves from the roots washes out into the water before being utilized by the leaf tissue (McRoy and Barsdate, 1970; McRoy, Barsdate, and Nebert, 1972; McRoy and Goering, 1974). Some of the lost nutrients are utilized immediately by layers of specialized periphytic algae which tightly cover the leaves (Harlin, 1973; Sieburth and Thomas, 1973; McRoy and Goering, 1974). Epiphytes on eelgrass, and probably on other sea grasses, recover dissolved materials that are lost from the leaves in much the same manner that epiphytes and understory plants in a rain forest recover materials washed from trees of the canopy (Tukey, 1970).

In the *Spartina* marshes of the Georgia coast, half the phosphate brought into the leaves from the sediments is fixed in plant tissues and half is lost at high tide or during rains. Epiphytes appear to be less well developed on the intertidal *Spartina* than on the subtidal sea grasses. Phosphate is lost from healthy growing leaves throughout the growing season (Reimold, 1972). This is net loss of reactive phosphate, measured chemically, and not a result of a tracer study, which might misrepresent net flux.

Both McRoy, Barsdate, and Nebert (1972) and Reimold (1972) estimated the loss of molybdate-reactive phosphate from individual plants cleaned of epiphytes and then extrapolated from these measurements to estimate the loss of phosphate from the meadows. The estimate for *Zostera* in Alaska was a flux of 2 mg atoms P m^{-2} day^{-1}, while the estimate for *Spartina* in Georgia was 20 mg atoms P m^{-2} day^{-1}. Izembek Lagoon has about twice the area of the Duplin River in Georgia, which was studied by Reimold, but the two watersheds have about the same high-tide volume. In spite of the difference in estimated loss from the two grass populations, the mean phosphate concentration in the water of both Izembek Lagoon and the Duplin River is 1 mg atom P m^{-3}, and the maximum observed values are between 5 and 10. This suggests that phosphate is being recycled more actively in the Duplin River despite the better development of epiphytes on *Zostera*. Most likely the phosphate is being more actively sorbed by sediments since the sediments of the Duplin River are fine clays (kaolinite and montmorillonite) and the sediments of Izembek Lagoon are volcanic sands. Since much of the sediment of the Duplin River is intertidal, there is good contact of water with sediments. The sorptive equilibrium concentration of phosphate in the water has been shown experimentally to be 1 mg atom P m^{-3} (Pomeroy, Smith, and Grant, 1965).

McRoy, Barsdate, and Nebert (1972) postulated a continuous loss of phosphate from Izembek Lagoon and resupplying by degradation of the volcanic sands. On the basis of the experimentally observed sorptive capacity of the clays, Pomeroy et al. (1972) and Reimold (1972) postulated concentration in the sediments of the Duplin River. Neither group has produced good evidence in the field to support their suppositions, however. High concentrations of phosphate extend out to sea from the inlets of both systems. Whatever the long-term geochemical processes may be, both sea-grass systems are clearly endowed with a nutrient reserve in their sediments. The principal limits to growth probably are crowding and shading.

CORAL REEF COMMUNITIES

In contrast to the grass communities, coral reefs, like planktonic communities and some tropical rain forests, have very small abiotic storage compartments for nutrients. There is storage in the tissue of living organisms, and there is continuous input of low concentrations of essential elements as seawater, propelled by wind, surf, and tides, passes over the reef. Recent work has markedly changed our concept of both the food web and the mechanisms of mineral cycling on coral reefs.

The first detailed examination of the food web of coral reefs was by Yonge and his colleagues on the Great Barrier Reef Expedition (Yonge, 1930). The presence of symbiotic algae in the tissues of reef corals, tridacnid clams, and other organisms raised questions about the nature of the trophic relationships of these organisms and the impact of the symbiosis on the food web and nutrient flux. Yonge concluded that corals are basically carnivores, feeding on zooplankton that are washed over the reef.

A reconsideration of the food web of coral reefs began when Sargent and Austin (1949) measured the change in oxygen in the water as it passed across a reef of Rongelap Atoll in the Marshall Islands. Diurnal changes in dissolved oxygen in the water flowing across the reef indicated that the reef was autotrophic. Sargent and Austin suggested that algae on the reef were major primary producers. Odum and Odum (1955), in a detailed study of Japtan Reef at Eniwetok, found that the reef was autotrophic. They also showed that individual coral heads were net producers of oxygen. While these field studies were going on, laboratory work with corals revealed that soluble organic materials of low molecular weight, such as glycerol, were lost by the symbiotic dinoflagellates and assimilated by coral tissue (Muscatine, 1967). A quantitative field study in Bermuda (Johannes, Coles, and Kuenzel, 1970) showed that the supply of zooplankton was not sufficient to sustain metabolism and growth in the absence of some other food source. All these studies led to the erroneous assumption that most primary production in coral reefs is accomplished by dinoflagellates living symbiotically within coral tissues. Pomeroy and Kuenzler (1969) showed that corals conserve phosphorus much more efficiently than do other heterotrophic organisms of similar body size. This suggested that the mineral cycles of coral reefs were specialized and partly internal. What was not clear was how corals were related to the food web of the dense and diverse consumer populations of the reef. Release of mucus by corals was suggested as a possible energy link to the consumers (Marshall, 1965; 1968; Johannes, 1967). Such a link would be analogous to a detrital system, with bacterial degradation of mucus a probable intermediate step.

New insight into the food web and mineral cycling of the Eniwetok reefs was provided recently by the Symbios Expedition. An important discovery was that photosynthesis is more intense on bare reef rock and rubble than in areas populated by corals (Johannes et al., 1972). Algal mats on the rocks are grazed so intensively by fishes and invertebrates that they are hardly evident, but their photosynthetic rate is impressive. Although this finding does not negate any previously established pathways of energy and materials, it adds a new pathway of substantial magnitude.

In an ecosystem as complex as a coral reef, the flux of materials takes many pathways through highly specialized populations although there may be few major primary and secondary pathways. Delineation of major pathways remains difficult. The state of our ignorance is summarized in Fig. 1. An evident pathway is filamentous algae to grazing fishes and invertebrates to feces to deposit feeders. The density of holothurians on many reefs is impressive, but other deposit feeders are also abundant though nocturnal and less conspicuous. The quantitative significance of this pathway is in doubt since reef fishes account for only a small part of total reef respiration and, therefore, can hardly be major movers of essential elements (Johannes, 1974). This being the case, invertebrates are probably the major grazers.

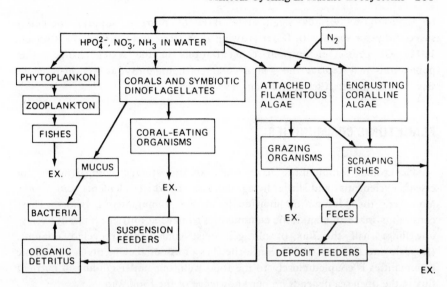

Fig. 1 A compartmental model of the flux of nitrogen and phosphorus through a coral-reef community which combines three postulates about the food web. EX. is the excretory flux of nitrogen and phosphorus. See text for discussion.

Another generally accepted pathway of materials is coral to mucus to bacteria to filter feeders. But, since there is little loss of phosphorus and probably a relatively small loss of nitrogen from the corals along this pathway, bacterial requirements for these two elements must come from phosphate and nitrate in reef water.

When we consider both these pathways, plus that through the symbiotic dinoflagellates in reef corals and other invertebrates, we are still unable to account quantitatively for the recycling of all the phosphorus and nitrogen known to be assimilated in primary production on the reef. Since many reefs are highly autotrophic, there may be a significant export of particulate materials that escape the many mouths and tentacles. Circumstantial evidence does not suggest this, however. Pilson and Betzer (1973) found no change in the concentration of molybdate-reactive phosphate in water as it passed over the windward reef of Eniwetok Atoll, and nitrate in the water actually increased as it crossed the reef (Johannes et al., 1972). These findings suggest a balanced uptake and release of nutrients by the community, with some nitrogen fixation by algal mats. Support for that view was found in the tracer studies of the Eniwetok reef community by Pomeroy, Pilson, and Wiebe (1974), which showed that, with the exception of the coral–dinoflagellate symbiosis, none of the reef populations showed any unusual retention of phosphate. Reef communities may be adjusted to various steady-state rates of nutrient supply, competing successfully for space with faster growing algae (Roy and Smith, 1971; Connell, 1974). Of course, if nutrient concentrations in the water change, other populations may replace the typical coral-reef community, as seen in Kaneohe Bay, Hawaii (Smith, Chave, and Kam, 1973).

PLANKTONIC COMMUNITIES

Planktonic communities have small abiotic storage compartments for essential elements and little living biomass in which elements can reside. Moreover, the lifetime of individuals in the community is short. These constraints make all planktonic communities somewhat nutrient limited. Even in upwellings, half the flux of nitrogen must come from recycled ammonia (Dugdale and Goering, 1967). Since the flux of essential elements in planktonic communities is coupled closely to the food web, our understanding of nutrient flux in the open sea depends on our knowledge of the food web.

Our concept of the planktonic food web is undergoing a radical change. The classical view of the food web in the open sea was a relatively simple chain from diatoms to copepods and other net plankton to fishes (compartments X_9 to X_{11} in Fig. 2). All theoretical constructs and models up to this time have been made on the basis of this food chain or fragments of it. Evidence that diatoms do not always dominate primary production in the sea began to appear long ago. Atkins (1945) pointed out that, since the depletion of silicate in the English Channel during the summer was not proportional to the depletion of phosphate, diatoms could be the major producers only if silicate were recycled more rapidly than phosphate. That seemed unlikely then, and we now know that it is not the case. Atkins was among the first to suggest that the nannoplankton (<60 μm) were significant primary producers. Recently it was suggested that the nannoplankton dominate the open sea and that large diatoms dominate the nutrient-rich regions, such as upwellings, the Antarctic Ocean, coastal waters, and estuaries (Ryther, 1969; Dugdale, 1972; Parsons and Takahashi, 1973). Ryther suggested that differences in the size of the primary producers led to a fundamentally different food web in the open sea, where nannoplankton must be consumed by other microorganisms, which results in an inefficient tertiary production of nekton. Most of the important fisheries were said to be in upwellings and coastal waters because the food chain from producers to nekton was shorter there. However, a number of studies of the relative rate of photosynthesis by net plankton and nannoplankton have shown that nannoplankton are responsible for over 90% of the photosynthesis in most marine waters most of the time, including upwellings, coastal waters, and estuaries (Pomeroy, 1974). Moreover, Semina (1972) reports that the mean size of phytoplankton is in fact smaller in the Peru Current than elsewhere in the Pacific Ocean.

If primary production is dominated by nannoplankton, the implication is that many of the primary consumers may also be very small. Empirical evidence suggests that this is so. Comparative studies of the respiratory rate of net plankton and microorganisms in the sea show that over 90% of the respiration is usually microbial (Pomeroy and Johannes, 1966). It is by no means certain, however, that the food web is a simple one of protozoans consuming small green flagellates. Andrews and Williams (1971) present evidence that a substantial fraction of the energy flux in the sea moves through a shunt of soluble organic compounds of low molecular weight which escape from the phytoplankton and are consumed by bacteria. Under the nutrient-deficient conditions of the open sea, the ratio of photosynthesis to respiration in the phytoplankton themselves is quite low and nearly half the carbon fixed by them is lost almost immediately as dissolved materials. In contrast, the photosynthesis-to-respiration ratio of phytoplankton in nutrient-rich upwellings is less than 10% of the carbon fixed (Thomas, 1971). Therefore, in nutrient-rich waters there are substantial numbers of phytoplankton of various sizes to be consumed directly, but, in the nutrient-depleted open sea, half or more of the flow of energy may be through the shunt of dissolved organic matter to heterotrophic microorganisms (Andrews and Williams, 1971; Pomeroy, 1974). This emerging view of the planktonic food web is far from clear and probably is still a minority view. Recent work on phytoplankton nutrition, however, tends to support the contention that the classical concept of the planktonic food web is oversimplified. Some of the alternate pathways are shown in Fig. 2, which is a highly simplistic representation of current speculation about the ocean's food web. The many roles of bacteria, for example, in the cycle of nitrogen, have not been delineated. We

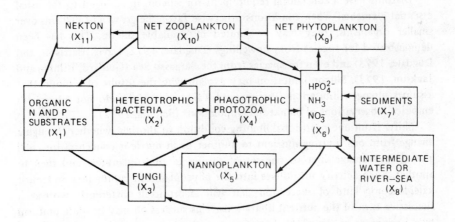

Fig. 2 A compartmental model of the flux of nitrogen and phosphorus through marine planktonic communities.

have long realized the complexity of the roles played by microorganisms, but we are just now beginning to appreciate the quantitative importance of total microbial metabolism.

BIOCHEMICAL NICHES

There is a long literature documenting experimentation and speculation on the factors permitting a diverse assemblage of phytoplankton species to persist, all presumably sharing one niche. One variable operating to preserve diversity is the periodic shift of nutrient limitation from one element to another. Nitrogen is now thought to be the most usual limiting nutrient in the sea although phosphorus is in nearly limiting supply. Even with a single limiting element, such as nitrogen, changes in nitrate and ammonia concentration shift the optimal growth rate from one phytoplankton species to another. There is experimental evidence that the half-saturation constants, K_s, and the growth constants, μ, of phytoplankters vary and are adapted to different regimes of each of several nutrients (Dugdale, 1967; Guillard, Kilhan, and Jackson, 1973). Moreover, since phytoplankton species have different daily regimes of nutrient uptake, some species are accumulating nutrients while others are not (Eppley et al., 1971). Induction of nutrient-specific enzymes or uptake sites may be inhibited either by the absence of a specific nutrient (NO_3^-) or by the presence of another (NH_3), depending on the algal species (Eppley and Rogers, 1970). Our understanding of these subtle interactions is imperfect, however, because it is based on batch-culture laboratory experiments at rather high nutrient concentrations. Metabolic activity or dormancy of populations can also be induced by changing temperature regimes (Wall and Dale, 1970).

Diatoms have a substantial requirement for silicon, in addition to the usual essential elements. When conditions otherwise favor the growth of diatoms over smaller organisms, they may be limited by available silicon. This has been demonstrated for species from upwellings (Dugdale, 1972; Davis, Harrison, and Dugdale, 1973) and even for species from the Sargasso Sea (Guillard, Kilhan, and Jackson, 1973). Silicon limitation may explain why the results of a model that assumes nitrogen limitation (Parsons and Takahashi, 1973) do not agree with empirical observations of planktonic populations (Semina, 1972).

Shifts from one nutritional limiting condition to another, whether through a change from one limiting nutrient to another or in nutrient concentration, will give the advantage first to one set of species of phytoplankters and then to another. Such activity introduces into the phytoplankton community an almost kaleidoscopic kind of succession that will assort itself differently with each successive turn of the nutrient limitations. This succession may be quite uniform over large areas of the ocean, but often it probably is not. Only occasionally are we able to recognize the time sequence of events, all parts of which usually appear at once as part of a spatial mosaic of species distribution. Margalef (1956) described a linear sequence of succession in the Ría de Vigo. The experimental production of a succession in phytoplankton has been demonstrated by Menzel, Hulbert, and Ryther (1963) and by Barber et al. (1971).

The relatively large diversity that is typical of oceanic plankton results at least in part from the presence of a mosaic of various stages of successional responses to nutrient limitation within a small distance, small, that is, in terms of the dimensions of the ocean. The complexity of the mosaic is enhanced by the differences in response of a set of phytoplankton species to constantly shifting concentrations of a set of limiting nutrients. Each species has evolved a distinct set of uptake constants and growth rates for the set of limiting nutrients, thereby finding for itself a biochemical niche in a nearly homogeneous environment.

RECYCLING

Nutrient sources for phytoplankton probably are somewhat more varied than we had realized. In the ocean two sources generally recognized are advection and recycling. Each appears to include a range of processes with varying rates. Advection operates especially well in upwellings, but the significance of such mechanisms as salt fingers in the permanently stratified tropical seas is not well understood. In coastal waters and estuaries, advection includes input from the bottom and from the land. Recycling, however, is a significant component of nutrient flux everywhere. In upwellings it supplies half the nitrogen (Dugdale and Goering, 1967), and some estuaries and coastal waters appear to function almost entirely on recycled nutrients (Pomeroy et al., 1972; Haines, 1974).

Recycling usually is attributed to bacterial metabolism, although experimental proof of this in natural waters is lacking. Many alleged experimental demonstrations of bacterial recycling of phosphate and ammonia do not discriminate between the activities of bacteria and other protists, and even zooplankton in some cases (Pomeroy, 1970). Attempts to isolate phosphatizing bacteria in seawater have been unsuccessful (Lear, 1965). Ammonifying bacteria do exist, but their quantitative significance in recycling has not been established in natural waters (Wiebe, 1974). An alternate pathway of recycling is excretion of metabolites and defecation of incompletely digested food by heterotrophic consumers of all kinds except bacteria.

Although there are a number of estimates of the roles of various consumer groups in excretory recycling, we still do not have enough information to evaluate fully its magnitude. The few existing estimates of the rate of recycling of nitrogen and phosphorus by zooplankton tend to suggest that it rarely supplies a significant fraction of photosynthetic demand (Pomeroy, 1970). In most marine ecosystems it is easy to calculate that nekton and other macrofauna cannot possibly be important in recycling (see Pomeroy and Bush, 1959;

Holdgate, 1967). Dugdale (1972) suggested that nekton are major recyclers in upwellings. Whitledge and Packard (1971), comparing the recycling rates of net zooplankton with those of anchovetas in the eastern tropical Pacific, concluded that anchovetas were the principal recyclers where they were abundant. The role of protists was not considered, however. A careful quantitative evaluation of all excretory processes seems desirable.

On the basis of relative respiratory rates, Pomeroy (1970) suggested that most recycling of phosphorus is accomplished by heterotrophic protists other than bacteria. Because of their relatively high concentration of phosphorus and their tendency to store polyphosphate, bacteria usually are net consumers of phosphorus, competing with phytoplankton for the available supply. Because of the difficulty of measuring both excretory and respiratory rates of specific kinds of marine protists, there is no experimental proof that they are in fact the major pathway of recycling in the sea. Microbial respiration exceeds that of net zooplankton and nekton by at least an order of magnitude (Pomeroy and Johannes, 1966; 1968). If excretion is correlated with respiration, microorganisms may be major agents in recycling. So far, however, no one has determined the relative rates of excretion of the various kinds of protists in the sea.

STABILITY, DIVERSITY, AND ESSENTIAL ELEMENTS

The response to limitation by essential elements at the population level was described in a general way by Liebig (1840). Populations may be limited absolutely by the availability of essential elements. At the community level the response to nutrient limitation is succession. The term "succession" is used here to mean a change in species composition of a community. The change usually represents a shift that serves to maintain community stability, but there is no implication here that it is part of an orderly or predictable sequence of events. The stimulus for succession may originate within the population structure, from the abiotic environment, from the impact of human activities, or from any combination of these factors. The organisms that will dominate a community at any given time are those with K_s and μ appropriate to the rate of supply of essential elements. To understand how mineral cycling affects communities, we must remember that community response to limitation is not the same as that of single-species populations.

Stability at the community or ecosystem level can be defined in various ways (Holling, 1973), but, in general terms, stability is the tendency of a community to persist and to return to an equilibrium state after being perturbed. The concept of Webster, Waide, and Patten (this volume) of two distinct components of relative stability, resistance and resilience, fits the observed responses of the four communities discussed here. The most stable of the four is the intertidal salt marsh. With a large reserve of nutrients in the sediment and a tidal energy subsidy to circulate them, the salt marsh is highly resistant to perturbation. It is also resilient when perturbed. Destruction of a salt marsh usually results in rapid leveling of the sediments by water currents and seeding in of new stands of *Spartina*. There is little succession other than microbial succession. Nearly as stable are the subtidal grass communities. They, too, have nutrient reserves and energy subsidies, they recover from mechanical damage by grazers or storms more slowly than do *Spartina* meadows.

Although coral reefs give the superficial impression of stability, they are rather easily damaged by mechanical insults from storms, predators, and man and are slow to recover. With a constant nutrient and energy input, coral reefs go through a complex and variable succession after disturbance (Connell, 1974). Their resistance is low, and their resilience is slow.

Least stable of the four are planktonic communities. They do not even have a substratum to support and conserve biomass. Nutrient reserves are severely limited, and there is constantly shifting species composition as the community, having little resistance, tracks the shifting nutrient regime. Even in upwellings, deficiencies develop and succession goes on. Stability in planktonic communities is achieved largely by the resilience provided by succession. In this case stability is survival—continuity of the community—and little more. In this sense all extant communities have absolute stability (Webster, Waide, and Patten, this volume) because they have not become extinct. In terms of relative stability, planktonic communities are unstable.

It follows that naturally enriched systems are least sensitive to pollution by nutrients. Enriched communities already have species with K_s and μ appropriate for high concentrations of nutrients, and there may be abiotic sinks to take up excess nutrients. Most sensitive to pollution will be the nutrient-impoverished systems with low K_s and μ and limited storage capacity. Their only recourse in the presence of pollution is succession, sometimes to an almost entirely different kind of community, as, for example, in the coral reefs in Kaneohe Bay (Smith, Chave, and Kam, 1973).

In theoretical ecology, stability of a community is usually related to its species diversity rather than to its supply of essential elements or the way it utilizes them. MacArthur (1955) developed a simple model in which energy flow was constant, nutrient limitation was not considered, and stability was defined as analogous to entropy. This model predicted that stability would always be a function of diversity. May (1971) showed, however, that other models may behave in the opposite way. Elton (1958) postulated that the stability of a community in the face of invasions by new species would be related to the number of branches in a food web. Population interaction theory was made more tractable and applicable to the real world by the treatise of MacArthur and Wilson (1967). They were concerned mostly with higher animals and with population interactions, however. Interactions between populations, communities, and their abiotic environment (except for its insularity) were not considered. Moreover, the effects of abiotic parameters, such as nutrients, light, and temperature, first show their influence on the lower trophic levels. These, in turn, set the stage for population interactions at the higher levels.

It has also been widely accepted in ecological theory that diversity increases as succession proceeds; this theory is based principally on the orderly, predictable successional sequences described for terrestrial plant communities. In the broader sense of the term, succession is not always orderly or predictable, as, for example, in the changes in phytoplankton communities. When we examine a broad spectrum of communities, it is evident that various factors may influence community diversity. These include population interactions of both the r and K types of MacArthur and Wilson (1967). Underlying these interactions are not only the effects of the populations on each other but also the effects of nutrient availability on the adaptive success of species populations. Also important in many communities is the effect of physical damage from storms, grazers and predators, and human impact, all of which tend to set the successional sequence back to its starting point. By their nature, however, these insults to communities affect only parts or patches at any time, thereby tending to increase diversity by the creation of a successional mosaic.

Both Hutchinson (1959) and Holling (1973) have drawn attention to the presence of an environmental mosaic underlying living communities. They point out that the abiotic environment is different enough from place to place to produce a mosaic of community response. In deciduous forests diversity is in part the result of disturbance and tends to diminish under the most protective management practices (Loucks, 1970). Probably disturbance is a significant cause of diversity in coral reefs, and, as with forests, the successional response is slow in the terms in which we measure time. Therefore, the coral reef we see is a successional mosaic. In planktonic communities adjacent patches of water in different nutrient-limitation states may also produce a successional mosaic that is particularly difficult to visualize since it is seen imperfectly by our sampling techniques. The net diversity of a community, therefore, is the result of population interactions taking place under temporally and spatially varying nutrient regimes and constantly interrupted by physical destruction of portions of the community which will have varying degrees of insularity with respect to repopulation. When seen in this light, the most diverse communities are not necessarily the most stable ones. Rather, diversified communities have moderate resistance and resilience.

Of the four communities we have considered, the intertidal *Spartina* meadows are the most stable and the least diverse. They can hardly be described as an early successional stage; in many cases they have persisted for thousands of years, giving way only to geological changes. Coral reefs are the most diverse of the four, and they seem to have intermediate stability. The planktonic communities are the most unstable, having moderate to high diversity. When we look across this spectrum of communities, it appears that, where essential elements are present in excess of needs, populations of high stability develop. Where nutrients are available, though not in great excess, communities of intermediate stability develop. Where nutrients are in short supply and where the specific limiting elements may change with time and space, communities are unstable. The relation between stability and species diversity in these communities seems less clear-cut than the relation between stability and the availability of essential elements.

ACKNOWLEDGMENTS

The research reported here was supported by the U. S. Atomic Energy Commission.

Preliminary drafts of this paper were reviewed by R. E. Johannes, Janet Pomeroy, J. E. Schindler, S. V. Smith, and W. J. Wiebe. Although they are in no way responsible for my speculations, their suggestions were most helpful.

REFERENCES

Andrews, P., and P. J. leB. Williams, 1971, Heterotrophic Utilization of Dissolved Organic Compounds in the Sea. III. Measurement of the Oxidation Rates and Concentrations of Glucose and Amino Acids in Sea Water, *J. Mar. Biol. Ass. U. K.*, **51**: 111-124.

Atkins, W. R. G., 1945, Autotrophic Flagellates as the Major Constituent of the Oceanic Phytoplankton, *Nature*, **156**: 446-447.

Barber, R. T., R. C. Dugdale, J. MacIsaac, and R. Smith, 1971, Variations in Phytoplankton Growth Associated with the Source and Conditioning of Upwelling Water, *Invest. Pesq.*, **35**: 171-193.

Connell, J. H., 1974, Population Ecology of Coral Reefs, in *The Geology and Biology of Coral Reefs*. O. A. Jones and R. Endean (Eds.), Academic Press, Inc., New York, in preparation.

Davis, C. O., P. J. Harrison, and R. C. Dugdale, 1973, Continuous Culture of Marine Diatoms Under Silicate Limitation. I. Synchronized Life Cycle of *Skeletonema costatum*, *J. Phycol.*, **9**: 175-180.

Dugdale, R. C., 1972, Chemical Oceanography and Primary Productivity in Upwelling Regions, *Geoforum*, **1972**(11): 47-61.

——, 1967, Nutrient Limitation in the Sea: Identification and Significance, *Limnol. Oceanogr.*, **12**: 685-695.

——, and J. J. Goering, 1967, Uptake of New and Regenerated Forms of Nitrogen in Primary Productivity, *Limnol. Oceanogr.*, **12**: 196-206.

Elton, C. S., 1958, *The Ecology of Invasions by Plants and Animals*, Methuen & Co. Ltd., London.

Eppley, R. W., and J. N. Rogers, 1970, Inorganic Nitrogen Assimilation of *Dilylum brightwellii*, a Marine Plankton Diatom, *J. Phycol.*, **6**: 344-351.

——, J. N. Rogers, J. J. McCarthy, and A. Sournia, 1971, Light/Dark Periodicity in Nitrogen Assimilation of the Marine Phytoplankters *Skeletonema costatum* and *Coccolithus huxleyi* in N-Limited Chemostat Culture, *J. Phycol.*, **7**: 150-154.

Guillard, R. R. L., P. Kilhan, and T. A. Jackson, 1973, Kinetics of Silicon-Limited Growth in the Marine Diatom *Thalassiosira pseudonana* Hasle and Heimdal (=*Cyclotella nana* Hustedt), *J. Phycol.*, **9**: 233-237.

Haines, E. B., 1974, Processes Affecting Production in Georgia Coastal Waters, Ph.D. Thesis, Duke University, Durham, N. C.

Harlin, M. M., 1973, Transfer of Products Between Epiphytic Marine Algae and Host Plants, *J. Phycol.*, **9**: 243-248.

Holdgate, M. W., 1967, The Antarctic Ecosystem, *Phil. Trans. Roy. Soc. London, Ser. B,* **252**: 363-383.

Holling, C. S., 1973, Resilience and Stability of Ecological Systems, *Ann. Rev. Ecol. Systemat.*, **4**: 1-23.

Hutchinson, G. E., 1959, Homage to Santa Rosalia or Why Are There so Many Kinds of Animals? *Amer. Natur.*. **93**: 145-159.

Johannes, R. E., 1974, Department of Zoology, University of Georgia, personal communication.

——, 1967, Ecology of Organic Aggregates in the Vicinity of a Coral Reef, *Limnol. Oceanogr.*, **12**: 189-195,

——, S. L. Coles, and N. T. Kuenzel, 1970, The Role of Zooplankton in the Nutrition of Some Scleractinian Corals, *Limnol. Oceanogr.*, **15**: 579-586.

——, et al., 1972, The Metabolism of Some Coral Reef Communities: A Team Study of Nutrient and Energy Flux at Eniwetok, *Bioscience*, **22**: 541-543.

Lear, D. W., 1965, Bacterial Participation in Phosphate Regeneration in Marine Ecosystems, Ph.D. Thesis, University of Rhode Island, Kingston.

Liebig, J., 1840, *Organic Chemistry and Its Applications to Agriculture and Physiology*, Taylor and Walton, London.

Loucks, O. L., 1970, Evolution of Diversity, Efficiency, and Community Stability, *Amer. Zool.*, **10**: 17-25.

MacArthur, R. H., 1955, Fluctuations of Animal Populations, and a Measure of Community Stability, *Ecology*, **36**: 533-536.

——, and E. O. Wilson, 1967, *The Theory of Island Biogeography*, Princeton University Press, Princeton, N. J.

McRoy, C. P., and R. J. Barsdate, 1970, Phosphate Absorption in Eel Grass, *Limnol. Oceanogr.*, **15**: 6-13.

——, R. J. Barsdate, and M. Nebert, 1972, Phosphorus Cycling in an Eelgrass (*Zostera marina* L.) Ecosystem, *Limnol. Oceanogr.*, **17**: 58-67.

——, and J. J. Goering, 1974, Nutrient Transfer Between the Seagrass *Zostera marina* and Its Epiphytes, *Nature*, **248**: 173-174.

Margalef, R., 1956, Estructura y dinámica de la "purga de mar" en la Ría de Vigo, *Invest. Pesq.*, **5**: 113-114.

Marshall, N., 1968, Observations on Organic Aggregates in the Vicinity of Coral Reefs, *Mar. Biol.*, **2**: 52-53.

——, 1965, Detritus over the Reef and Its Potential Contribution to Adjacent Waters of Eniwetok Atoll, *Ecology*, **46**: 343-344.

May, R. M., 1971, Stability in Multi-Species Community Models, *Math. Biosci.*, **12**: 59-79.

Menzel, D. W., E. M. Hulbert, and J. H. Ryther, 1963, The Effects of Enriching Sargasso Sea Water on the Production and Species Composition of the Phytoplankton, *Deep-Sea Res.*, **10**: 209-219.

Muscatine, L., 1967, Glycerol Excretion by Symbiotic Algae from Corals and *Tridacna* and Its Control by the Host, *Science*, **156**: 516-519.

Odum, H. T., and E. P. Odum, 1955, Trophic Structure and Productivity of a Windward Coral Reef Community on Eniwetok Atoll, *Ecol. Monogr.*, **25**: 291-320.

Parsons, T. R., and M. Takahashi, 1973, Environmental Control of Phytoplankton Cell Size, *Limnol. Oceanogr.*, **18**: 511-515.

Pilson, M. E. Q., and S. B. Betzer, 1973, Phosphorus Flux Across a Coral Reef, *Ecology*, **54**: 581-588.

Pomeroy, L. R., 1974, The Ocean's Food Web, a Changing Paradigm, *Bioscience*, **24**: 499-504.

____, 1970, The Strategy of Mineral Cycling, *Ann. Rev. Ecol. Systemat.*, **1**: 171-190.

____, and F. M. Bush, 1959, Regeneration of Phosphate by Marine Animals, International Oceanography Congress, Preprints, pp. 893-895, American Association for the Advancement of Science, Washington, D. C.

____, and R. E. Johannes, 1968, Occurrence and Respiration of Ultraplankton in the Upper 500 Meters of the Ocean, *Deep-Sea Res.*, **15**: 381-391.

____, and R. E. Johannes, 1966, Total Plankton Respiration, *Deep-Sea Res.*, **13**: 971-973.

____, and E. J. Kuenzler, 1969, Phosphorus Turnover by Coral Reef Animals, in *Symposium on Radioecology*, D. J. Nelson and F. C. Evans (Eds.), Proceedings of the Second National Symposium, Ann Arbor, Mich., May 15-17, 1967, USAEC Report CONF-670503, pp. 474-482, U. S. Atomic Energy Commission, The Ecological Society of America, and The University of Michigan.

____, M. E. Q. Pilson, and W. J. Wiebe, 1974, Tracer Studies of the Exchange of Phosphorus Between Reef Water and Organisms on the Windward Reef of Eniwetok Atoll. Proceedings of the Second International Symposium on Coral Reefs, in preparation.

____, L. R Shenton, R. D. H. Jones, and R. J. Reimold, 1972, Nutrient Flux in Estuaries, in *Nutrients and Eutrophication*, G. E. Likens (Ed.), American Society of Limnology and Oceanography, Special Symposium No. 1, pp. 274-291.

____, E. E. Smith, and C. M. Grant, 1965, The Exchange of Phosphate Between Estuarine Water and Sediment, *Limnol. Oceanogr.*, **10**: 167-172.

Reimold, R. J., 1972, The Movement of Phosphorus Through the Salt Marsh Cord Grass, *Spartina alterniflora* Loisel., *Limnol. Oceanogr.*, **17**: 606-611.

Roy, J. K., and S. V. Smith, 1971, Sedimentation and Coral Reef Development in Turbid Water: Fanning Lagoon, *Pac. Sci.*, **25**: 234-248.

Ryther, J. H., 1969, Photosynthesis and Fish Production in the Sea, *Science*, **166**: 72-76.

Sargent, M. C., and T. S. Austin, 1949, Organic Productivity of an Atoll, *Trans. Amer. Geophys. Union*, **30**: 245-249.

Semina, H. J., 1972, The Size of Phytoplankton Cells in the Pacific Ocean, *Int. Rev. Gesam. Hydrobiol.*, **57**: 177-205.

Sieburth, J. McN., and C. D. Thomas, 1973, Fouling on Eelgrass, *J. Phycol.*, **9**: 46-49.

Smith, S. V., K. E. Chave, and D. T. O. Kam, 1973, *Atlas of Kaneohe Bay: A Reef Ecosystem Under Stress*, Sea Grant Technical Report, TR-72-01, University of Hawaii Sea Grant Program.

Thomas, J. P., 1971, Release of Dissolved Organic Matter from Natural Populations of Marine Phytoplankton, *Mar. Biol.*, **11**: 311-323.

Tukey, H. B., 1970, Leaching of Metabolites from Foliage and Its Implication in the Tropical Rain Forest, in *A Tropical Rain Forest: A Study of Irradiation and Ecology at El Verde, Puerto Rico*, H. T. Odum and R. F. Pigeon (Eds.), USAEC Report TID-24270, pp. H-155—H-160, U. S. Atomic Energy Commission.

Wall, D., and B. Dale, 1970, Living Hystrichosphaerid Dinoflagellate Spores from Bermuda and Puerto Rico, *Micropaleontology*, **16**: 47-58.

Whitledge, T. E., and T. T. Packard, 1971, Nutrient Excretion by Anchovies and Zooplankton in Pacific Upwelling Regions, *Invest. Pesq.*, **35**: 243-250.

Wiebe, W. J., 1974, Department of Microbiology, University of Georgia, personal communication.

Yonge, C. M., 1930, Studies on the Physiology of Corals. I. Feeding Mechanisms and Food, *Sci. Rept. Great Barrier Reef Exped.*, **1**: 14-57.

Ecosystem Energy Flow and Modeling

Introduction

We now turn from the separate aspects of the functioning of marine ecosystems to a more holistic approach. The next three sections will be concerned with energy flow in these systems, with the structure of the communities, and with a novel and relatively recent type of experimental ecosystem analysis, the use of microcosms. The study of energy flow or nutrient cycling has lead some authors to an attempt to express their results in the form of models, and the papers included in this section give examples of several different types. Energy flow in ecosystems provides a unifying theme, for, as Steele (1974) points out in the introduction to his elegant little book, "The general structure of communities and the behavior of an individual when feeding, breeding or escaping from predators provide, indirectly, some idea of how species interact with one another. Between these two approaches is the study of energy flow through food webs where links between species—or groups of species—are the primary concern. This method appears relatively neutral, yet contains implicitly the assumption that energy is the unit by which other variables can be quantified."

Thus, the function of ecosystems is largely to be understood on the basis of the flow of materials or energy in the system of interest. Of the papers included in this section, E. Odum's provides a review of the concepts of energy movement through ecosystems. He points out that it is usually not satisfactory to fit individual species into certain trophic levels because an animal may operate at more than one level. This view is supported in the work of Nixon and Oviatt, who show that the grass shrimp *Paleomonetes* and the minnow *Fundulus* are essentially omniverous; and in the work of Miller, Mann, and Scarratt, who indicate that lobsters operate on both the primary and secondary carnivore level. One can study an entire ecosystem as Nixon and Oviatt have, or only a part of it, as was done by Miller, Mann, and Scarratt. In either case, estimates of the amount of energy coming in and going out, as well as transfers within the system, must be understood. The transfer of energy from one part of the system to another is never highly efficient; the usual figure quoted in 10%, and it is often less than that. It is, then, the transfer of energy between compartments and the storage of energy within compartments of the ecosystem that is of chief interest here.

Three papers in this section approach the study of energy flow in marine ecosystems from somewhat different angles. The first two concentrate on energy itself, the third on nutrient exchange. The first two papers on energy flow essentially deal with the benthic environment and the overlying watermass in very shallow water areas. These are considerably easier to deal with than the open sea, where it is often impossible to put boundaries on small areas of study. Nixon and Oviatt, as well as Miller, Mann, and Scarratt, present an ecosystem model in the energy circuit language developed by H.T. Odum (see, for instance, *Energy, Power and Society*, by Odum, 1971). In addition, Nixon and Oviatt present a simulation model of the diurnal changes in oxygen, and show, through use of the model, what changes might take place if perturbations such as a change in tidal exchange, increased sewage input, or an increase of temperature occurred.

Simulation models have obvious utility for the analysis of particular situations and the prediction of events should certain changes take place, but it must be emphatically pointed out that no model is better than the concepts and the data that go into it. Many models are verbal, but the ones referred to here are explicitly mathematical, designed to be analyzed with the aid of a computer. The investigator must reduce his hypotheses and discoveries to mathematical formulae that define the processes operating in the system. There can be no holes left in the definitions, no process undescribed, and no coefficients ignored, for the unforgiving computer does not make assumptions or allow minor missing details to pass by. Thus, one is forced to completely analyze the system or make simplifying assumptions when writing a model for it. The main aim of a model is, as Steele (1974) says, to determine where the model breaks down and use it to suggest further field or experimental work. If any part of the conceptual framework or the data base is misrepresented, an error in the prediction will occur. The magnitude of the error in the output depends upon the importance of the particular segment of the model in question. This important attribute of models can also be a blessing, for one can discover which parameters make the most difference in the functioning of the system. Parameters that make the greatest difference for the least change can be subjected to the most intense field study while those that are relatively insensitive need not be studied in such detail.

Walsh and Dugdale simulate an upwelling ecosystem in Peru in much the same way that oxygen was followed by Nixon and Oviatt, but adding spatial differences and advection between segments of the system. Many simulations owe their origins to the Lotka-Volterra (Volterra, 1928) equations for modeling two-species interactions, equations that were further developed for marine ecosystems by Gordon Riley in 1946 and 1947. Riley postulated the general equation $dP/dt = P(P_h-R) - G$, which, in words, reads: The rate of change of phytoplankton (P) equals the photosynthesis (P_h) of that population minus its respiration (R) and minus grazing (G) by herbivores. Into each one of these terms of the equation must come several factors. In order to know the photosynthetic rate (P_h) of a popula-

tion, for example, one must know such things as the amount of available light, the depth of the euphotic zone and the rate of mixing, the concentration of nutrients, and the uptake rate of nutrients. Riley wrote this part of the equation as $P_h = (P^{I_0}/kZ)(1-e^{-kZ})(1-N)(1-V)$, where $P =$ experimentally derived photosynthetic coefficient, $I_0 =$ surface light intensity, $Z =$ depth of the euphotic zone, $N =$ rate of nutrient depletion, $V =$ rate of vertical movement, $k =$ the extinction coefficient of the water, and $e =$ the base of natural logarithms. Many additions and corrections have been made and many more will be made as the knowledge of marine ecosystems increases, but much of the framework has been hung on these equations. The other terms, for respiration and grazing, can be expanded similarly (for further treatment see Parsons and Takahashi, 1973). Walsh and Dugdale's paper treats an admittedly preliminary simulation model and its application to a specific marine ecosystem of considerable economic importance. The model suggests that several environmental factors are uppermost in controlling phytoplankton standing crop, and others are less important. The fact that their model fits the data relatively well leads the authors to believe that "some of the processes are reasonably accurately postulated."

LITERATURE CITED

Odum, H. T. 1971. Energy, Power and Society. Wiley-Interscience, New York. 311 pp.

Parsons, T. and Takahashi, M. 1973. Biological Oceanographic Processes. Pergamon Press Inc., New York. 186 pp.

Riley, G. A. 1946. Factors controlling phytoplankton populations on Georges Bank. J. Mar. Res. 6:54–73.

Riley, G. A. 1947. Theoretical analysis of the zooplankton population of Georges Bank. J. Mar Res. 6: 104–113.

Steele, J. H. 1974. The Structure of Marine Ecosystems. Harvard University Press, Cambridge, Massachusetts. 128 pp.

Volterra, V. 1928. Variations and fluctuations of the number of individuals in animal species living together. J. Conseil 3:1–51.

Energy Flow in Ecosystems: A Historical Review

Eugene P. Odum

Institute of Ecology and Department of Zoology, University of Georgia, Athens, Georgia 30601

Synopsis. A generalized model of energy flow applicable both to individual populations and food chains is discussed. The basic ideas of energy flow and trophic levels are described, and it is emphasized that the concept of trophic level is not primarily applicable to individual species. The efficacy of rates of population energy flow as a measure of importance in community function is stressed, and the disadvantages associated with measures of density and biomass are pointed out. Finally, the historical development of energy-oriented thinking in ecology is traced in a series of ten steps dating from the late 19th century. The growing importance of systems analysis and the use of computer models to simulate ecological functions are recognized as major areas of emphasis during the next decade.

The study, understanding, and intelligent manipulation of our environment requires systematic investigation of the structure and function of ecological systems at various size levels ranging from simplified microcosms to the biosphere as a whole. In a recent article (Odum, 1962) I suggested that ecology could best be defined as the study of the relationships between structure and function in nature, and that the following breakdown might provide a simplification of the first order for purposes of study.

A. Structure.
  1. Composition of the biological community (species, numbers, biomass, life-history, dispersion, etc.).
  2. Quantity and distribution of abiotic materials (nutrients, water, etc.).
  3. Range or gradient of conditions of existence (temperature, light, etc.).

B. Function
  1. Range of energy flow through the system (eco-energetics).
  2. Rate of material cycling (eco-cycling).
  3. Regulation by physical environment and by organisms (eco-regulation).

Thus, we review in this refresher course one of the several fundamental approaches to ecology—one that lends itself especially to the teaching of the subject to student and citizen alike. My paper is concerned with a brief history of this approach as an introduction to the five papers which follow. Before going into history, I believe it would be well to delimit and define our subject.

The behavior of energy in ecosystems can be conveniently shorthanded as "energy flow" because energy transformations are directional in contrast to the cyclic behavior of materials. The potential and kinetic components of energy flow through an ecological system are lumped under the designations, production (P) and respiration (R), respectively. Consequently, energy flow (E) can be very broadly defined as the sum of P and R, or $E = P + R$. We need, of course, to further subdivide P and R into their ecologically significant subcompartments. Since terminologies have not been standardized and the usage of equation-symbols varies so widely as to be confusing, I believe the best way to establish communication, especially in the teaching arena, is to fall back on a graphic model that shows relationships in the form of an easily understood picture.

In Figure 1, I present what might be called a "universal" model of energy flow, one that is applicable to any living component whether it be plant, animal, microorganism—or individual, population, trophic group. Linked together, such graphic models can depict food-chains (as shown in

217

Reprinted from Am. Zoologist 8:11–10 (1968).

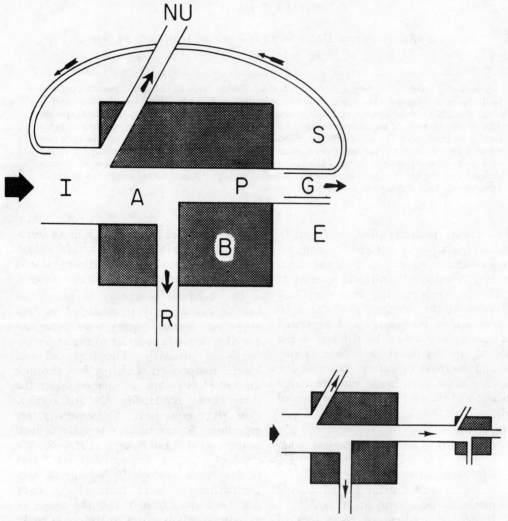

FIG. 1. **Components for a "universal" model of ecological energy flow. See text for explanation of symbols.**

the lower right of Fig. 1), or the bioenergetics of an entire ecosystem.

In Figure 1, the shaded box labelled "B" represents the living structure or "biomass" of the component. The designation, "standing crop," is a comparable term often used when we wish to speak of the total mass of a population or trophic grouping such as phytoplankton in a column of water or phytophagous insects in a grassland. Biomass is usually measured and expressed as some kind of weight: either total living (wet) weight, dry weight, or ash-free weight. Varying usages here and the lack of good

conversion factors often pose problems. From the standpoint of energetics it is also desirable to express biomass in terms of calories so that relationships between the rates of energy flow and the instantaneous or average standing-state biomass can be established. As will be briefly noted further along in this article, the ratios B/E, B/P, and B/R are of great theoretical interest with respect to community development and ecological succession.

The total energy input or intake is indicated by "I" in Figure 1. For strict autotrophs this is light, and for strict heterotrophs

trophs organic food. Some species of algae and bacteria can utilize both energy sources, and many may require both in certain proportions. A similar situation holds for invertebrate animals and lichens which contain mutualistic algae. In such cases the input flow in the energy flow diagram can be subdivided accordingly to show the different energy sources, or the biomass can be subdivided into separate boxes if one wishes to keep everything in the same box at the same energy level (*i.e.*, the same trophic level).

Such flexibility in usage can be confusing to the beginner. In teaching, I find that it is important to emphasize that the *concept of trophic level is not primarily intended for categorizing species.* Energy flows through the community in stepwise fashion due to the second law of thermodynamics, but a given population of a species may be (and very often is) involved in more than one step or trophic level. The universal model of energy flow illustrated in Figure 1 can be used in two ways. The model can represent a species-population, in which case the appropriate energy inputs and links with other species would be shown as a conventional species-oriented food-web diagram. Or the model can represent a discrete energy level, in which case the biomass and energy channels represent all or parts of many populations supported by the same energy source. Foxes, for example, usually obtain part of their food from eating plants (fruit, etc.) and part by eating herbivorous animals (rabbits, field mice, etc.). A single diagram of energy flow could be used to represent the species or the whole population of foxes if our objective is to stress intrapopulation energetics. On the other hand, two or more boxes (such as shown in the lower right of Fig. 1) would be employed should we wish to apportion the metabolism of the fox population into more than one trophic level. In this way we can place the fox population into the overall pattern of energy flow in the community. When an entire community is modeled one cannot mix these two usages unless all species happen to be restricted to single trophic levels (*e.g.*, a highly simplified blue grass-cow-man ecosystem).

So much for the problem of the source of the energy input. Not all of the input into the biomass is transformed; some of it may simply pass through the biological structure, as occurs when food is egested from the digestive tract without being metabolized, or when light passes through vegetation without being fixed. This energy component is indicated by "NU" ("not utilized"). That portion which is utilized or assimilated is indicated by "A" in the diagram. The ratio between these two components, *i.e.*, the efficiency of assimilation, varies widely. It may be very low, as in light-fixation by plants or food-assimilation in detritus-feeding animals, or very high as in the case of animals or bacteria feeding on high energy food such as sugars and amino acids.

In autotrophs the assimilated energy ("A") is known as "gross production" or "gross photosynthesis." Historically, the term, "gross production," has been used by some authors for the analogous component in heterotrophs. However, since the "A" component in heterotrophs represent food already "produced" somewhere else, the term "gross production" should be restricted to primary or autotrophic production. In higher animals, the term, "metabolized energy," is often used for the "A" component.

A key feature of the model is the separation of assimilated energy into the "P" and "R" components as previously described. That part of the fixed energy ("A") which is burned and lost as heat is designated as respiration ("R"), while that portion which is transformed to new or different organic matter is designated as production ("P"). This is the "net production" or "net photosynthesis" in green plants and simply "production" or "secondary production" in animals. It is important to point out that the "P" component is energy available to the next trophic level, as opposed to the "NU" component which is still available at the same trophic level.

The ratio between "P" and "R" varies widely and is of great ecological significance. In general, the proportion of energy going into respiration or maintenance is large in populations of large organisms, such as men and trees, and in mature (*i.e.*, "climax") communities. Conversely, the "P" component is relatively large in active populations of small organisms, such as bacteria or algae, and in the young or "bloom" stages of ecological succession.

Production may take a number of forms. Three subdivisions are shown in Figure 1: "G" refers to additions to the biomass or growth. "E" refers to assimilated organic matter which is excreted or secreted (*e.g.*, simple sugars, amino acids, urea, mucus, etc.). This "leakage" of organic matter, often in dissolved or gaseous form, may be appreciable but is often ignored because it is hard to measure. Finally, "S" refers to "storage," as in the accumulation of fat which may be reassimilated at some later time. The reverse "S" flow shown in Figure 1 may also be considered a "work loop" in that it depicts that portion of production which is necessary to insure a future input of new energy (*e.g.*, reserve energy used by a predator in the search for prey).

Figure 1 shows only a few of the ecologically useful subdivisions of the basic pattern of energy flow. In practice, we are often hampered by the difficulties of measurement, especially in field situations. A primary purpose of a model, of course, is to define components that we want to measure in order to stimulate research into methodology. Even if we are not yet able to chart all the flows, measurements of gross inputs and outputs alone may be revealing. Because energy is the ultimate limiting factor, the amounts available and actually utilized must be known if we are to evaluate the importance of other potentially limiting or regulating factors. Many of the controversies about food limitation, weather limitation, competition, and biological control could be resolved if we had accurate data on energy utilization by the populations in question.

Concepts of energy flow provide not only a means of comparing ecosystems with one another, but also a means of evaluating the

TABLE 1. *Density, biomass, and energy flow of different organisms.*

	Approximate Density (m²)	Biomass (g/m²)	Energy Flow (Kcal/m² /day)
Soil bacteria	10^{12}	0.001	1.0
Marine copepods (*Acartia*)	10^5	2.0	2.5
Intertidal snails (*Littorina*)	200	10.0	1.0
Salt marsh grasshoppers (*Orchelimum*)	10	1.0	0.4
Meadow mice (*Microtus*)	10^{-2}	0.6	0.7
Deer (*Odocoileus*)	10^{-6}	1.1	0.5
Insectivorous birds	10^{-3}	0.02	0.02
Blue grass (*Poa*)	40	200	20.0

relative importance of populations of diverse sizes and rates of metabolism. Table 1 lists estimates of density, biomass, and energy flow rates of eight populations differing widely in size of organism and trophic position. These estimates are based on densities and weight-specific metabolic rates (including an estimate for the "P" component) that one might expect to find in a temperate community of moderate nutrient fertility. For the first six populations listed, the estimated energy requirements are similar, even though the densities vary over 17 orders of magnitude and the biomass over five. This indicates that all six populations are functioning at approximately the same energy level (*i.e.*, as primary consumers or herbivores). The higher rates estimated for blue grass and the lower rates for insectivorous birds are indicative of their different trophic roles in the ecosystem. Because numbers overemphasize the importance of small organisms (*e.g.*, bacteria) and weight or biomass overemphasizes the importance of large organisms, we cannot use either as a reliable criterion for comparing the functional role of populations that differ widely in size-metabolism relationships. The rate of energy flow, however, provides a suitable index for comparing the importance, energetically speaking, of any and all components of an ecosystem.

The idea of looking at nature as an energy-flow system is deeply rooted in the early history of science. Interest in the "fires of life" goes back to antiquity. Many of the concepts that we now apply to the population and community level had their origin in the physical sciences and in the early history of physiology and medicine. For the purposes of this refresher course, I believe we can trace the recent history of ecological energetics in about ten steps. It should be emphasized that the following outline was prepared primarily as a chronology of ideas and not as a historical review of names and literature. Only a few samples of the latter are suggested for their value as background reading.

1. *Qualitative description of food webs.* The idea that organisms in nature are linked together in network fashion through food was expressed in various ways in the writings of 19th century naturalists. Stephen A. Forbes' classic essay on "The Lake as a Microcosm" (1887) is a good example, and provides an appropriate beginning for this historical review.

2. *Trophic levels and ecological pyramids.* In the 1920's, August Thienemann (1926) described trophic levels in terms of "producers" and "consumers," and Charles Elton (1927) wrote about the "ecological niche" and the "pyramid of numbers" in terms of organization of the food chain. As emphasized earlier in this paper, criteria of energy flow should replace numbers if the "Eltonian pyramid" is to remain a valid concept for all types of ecosystems.

3. *Application of thermodynamic principles.* The 1920's also saw the beginning of the influence that the second law of thermodynamics was to have on ecological theory. A. J. Lotka's book, "The Elements of Physical Biology" (1925), was a milestone. His concepts of the non-equilibrium steady state and the "law of maximum energy in biological systems" were forerunners of important ecological generalizations, *e.g.*, the theory of H. T. Odum and Pinkerton (1955) that nature's low efficiency of energy transfer is a consequence of the tendency for optimum efficiency for maximum power output to be less than maximum efficiency.

4. *Energy budgets and the concept of primary productivity.* Limnologists were among the first to develop these concepts, perhaps because lakes provide such convenient units for study and because heat and gaseous exchanges are more easily measured in standing bodies of water than in other ecosystems. The work of Birge and Juday in the 1930's comes to mind in this connection (see Juday, 1940). As instrumentation improved, it later became possible to deal with energy budgets of terrestrial environments, and even of the whole biosphere, as suggested in David Gates' little book, "Energy Exchange in the Biosphere" (1962).

To a considerable extent, progress in the study of primary productivity depended on the development of field methods. The following is a brief chronology of methods (together with one or more pioneer methodologists):

a. CO_2 uptake in terrestrial enclosures. Transeau (1926).

b. Dark and light bottle method (aquatic). Gaarder and Gran (1927).

c. The diurnal curve method (aquatic). Sargent and Austin (1949), H. T. Odum (1956).

d. Harvest methods. Penfound (1956), Ovington (1957), E. P. Odum (1960).

e. The pH method (aquatic). Verduin (1956), Beyers, and H. T. Odum (1959).

f. The light-chlorophyll method. Ryther and Yentsch (1957).

g. Infra-red gas analyzer measurement of CO_2 (enclosed terrestrial vegetation). Lemon (1960), Mooney and Billings (1961).

h. CO_2 vertical gradient method (unenclosed terrestrial vegetation). Monteith (1962).

5. *Trophic-dynamic concepts and energy flow by trophic levels.* Raymond Lindeman's classic paper on the "Trophic-dynamic Aspect of Ecology" (1942) ushered in the 1940's and did more than any other single contribution to bring concepts of energy flow to focus at the level of the ecosystem. One should not forget, how-

ever, the contributions to this synthesis by G. E. Hutchinson (1948), George Clarke (1946), Amyan Macfadyen (1949), and others.

6. *The energy-flow diagram and community metabolism.* The diagram of energy flow might be referred to by some as an "Odum" device (H. T. Odum, 1956, 1957; Odum and Odum, 1959; E. P. Odum, 1963), although flow-diagrams are routine in physics and engineering. The basic unit of the ecological energy-flow diagram has been described in this paper. In modified form, flow-diagrams have proved useful in emphasizing the fundamental partition of flow into grazing and detritus food-chains (E. P. Odum, 1962; 1963), and as a basis for models of electrical analogue circuits (H. T. Odum, 1960). The latter approach, which considers the energy channels to be the "invisible wires of nature," is especially appropriate for analogue computer manipulation (see item 10 below).

Studies during the 1950's demonstrated that important generalizations could be derived from measurements of the metabolism of whole communities without necessarily having detailed information on all component populations (Odum and Odum, 1955; H. T. Odum, 1957; Teal, 1957).

7. *Secondary production and energy flow in populations.* It is only natural that success in the study of primary production should be followed by increased interest in the energy flow of heterotrophs and concern with the utilization of net primary production. Here again, background experience in laboratory physiology was the basis for the first attempts at field measurements. I cite only a few examples of efforts to combine laboratory respirometry with field census procedures: Pearson (1954), Phillipson (1962), E. P. Odum and Smalley (1959), E. P. Odum, *et al.* (1962), Golley (1960). Ecologists are now seeking new methods which do not require "enclosement" (*i.e.,* confinement in cages or respirometers). Rates of uptake, elimination, and flux of radionuclide tracers provide the most exciting new tools (Crossley and Howden, 1961; E. P. Odum and F. B.

Golley, 1963; R. G. Wiegert, E. P. Odum, and J. H. Schnell, 1967; Reichle, 1967).

8. *Energetics of laboratory populations.* Just as some generalizations are best made from the study of "big nature" (such as a coral reef), so other useful generalizations come from the study of "little nature" in the laboratory. Ecologists are just beginning to take advantage of the precision, control, and experimental design-possibilities of the laboratory in studies that range from those focused on the energetics of populations of single species (Richman, 1958; Slobodkin, 1959) to those dealing with the community metabolism of self-sustaining micro-ecosystems (Beyers, 1963; Cooke, 1967). The latter will be of great interest in future attempts o design a regenerative ecosystem for man's space travel.

9. *The energetics of ecological succession.* H. T. Odum and R. C. Pinkerton (1955) were perhaps among the first to point out that ecological succession involves a fundamental change of the pattern of basic energy flows. As ecosystems develop toward maturity (*i.e.,* "climax"), the P/R ratio approaches one and the B/E (or B/P or B/R) ratio increases—the strategy being not to maximize efficiency of production (as is often desired by man), but to optimize the support of as large and complex a biomass structure as possible per unit of available energy flow. Margalef (1963*a*, 1963*b*) has recently documented and extended these basic ideas, and his papers should be required reading for all ecologists.

10. *Systems ecology.* The "input-output," "rate of change," and "flow chart" ways of thinking lead directly into "systems analysis," which is a state of mind as much as it is applied mathematics and computers (Watt, 1966). Because energy flow drives the complex cycles of materials that are the form, function, and diversity of life, "eco-energetics" is the core of "ecosystem analysis." It is not difficult to predict where the emphasis will be in the next decade!

EPILOGUE

Within our own exploding population,

there is increasing concern about "food ecology," which is essentially the same thing as "energy flow ecology" if we consider light as "plant food." The International Biological Program (IBP) now being planned worldwide around the theme, "the biological basis for productivity and human welfare," is but one indication of the recognition by scientist and citizen alike that a better understanding of the biosphere is urgent. The first phase of IBP, now well underway, involves the scheduling of symposia, planning of programs by national committees, and the preparation of manuals on methods.[1] The definitive phase will involve intensive, multidisciplinary studies of landscapes (such as forests or croplands), important processes (such as nitrogen fixation), and key problems (such as human adaptation). To achieve these goals, many more trained and motivated people than are now available will be needed. Hence, training is one of the primary concerns of IBP planners. In some small way we hope that this refresher course will prove a useful adjunct to the IBP.

---

[1] Bulletins outlining the U. S. National Program for IBP may be obtained by writing: U. S. National Committee for the IBP, National Academy of Sciences, 2101 Constitution Avenue, Washington, D. C. 20418.

REFERENCES

Beyers, R. J. 1963. The metabolism of 12 laboratory microecosystems. Ecol. Monographs 33:281-306.

Beyers, R. J., and H. T. Odum. 1959. The use of carbon dioxide to construct pH curves for the measurement of productivity. Limnol. Oceanog. 4:499-502.

Clarke, G. L. 1946. Dynamics of production in a marine area. Ecol. Monographs 16:321-335.

Cooke, G. D. 1967. The pattern of autotrophic succession in laboratory microcosms. BioScience 17:717-721.

Crossley, D. A., and H. F. Howden. 1961. Insect-vegetation relationships in an area contaminated by radioactive wastes. Ecology 42:302-317.

Elton, C. 1927. Animal ecology. The Macmillan Co., New York.

Forbes, C. A. 1887. The lake as a microcosm. Republished 1925. Illinois Nat. Hist. Survey Bull. 15:537-550.

Gaarder, T., and H. H. Gran. 1927. Investigations of the production of plankton in the Oslo Fjord. J. Cons. Intern. Explor. Mer 42:1-48.

Gates, D. 1962. Energy exchange in the biosphere. Harper and Row, New York

Golley, F. B. 1960. Energy dynamics of a food chain of an old-field community. Ecol. Monographs 30:187-206.

Hutchinson, G. E. 1948. Circular causal systems in ecology. Ann. New York Acad. Sci. 50:221-246.

Juday, C. 1940. The annual energy budget of an inland lake. Ecology 21:438-450.

Lemon, E. R. 1960. Photosynthesis under field conditions. Agron. J. 52:697.

Lindeman, R. L. 1942. The trophic-dynamic aspect of ecology. Ecology 23:399-418.

Lotka, A. J. 1925. Elements of physical biology. Williams and Wilkins, Baltimore.

Macfadyen, A. 1949. The meaning of productivity in biological systems. J. Animal Ecol. 17:75-80.

Margalef, R. 1963a. On certain unifying principles in ecology. Am. Nat. 97:357-374.

Margalef, R. 1963b. Succession of marine populations. Advancing Frontiers of Plant Science 2:137-188 (Institute for Advancement of Scientific Culture, New Delhi, India).

Monteith, J. L. 1962. Measurement and interpretation of carbon dioxide fluxes in the field. Netherlands J. Agr. Sci. 10:334-346.

Mooney, H. A., and W. D. Billings, 1961. Comparative physiological ecology of arctic and alpine populations of Oxyria digyna. Ecol. Monographs 31:1-29.

Odum, E. P. 1960. Organic production and turnover in old field succession. Ecology 41:34-49.

Odum, E. P. 1962. Relationships between structure and function in the ecosystem. Jap. J. Ecol. 12:108-118.

Odum, E. P. 1963. Ecology. Modern Biology Series. Holt, Rinehart and Winston, New York.

Odum, E. P., and A. E. Smalley. 1959. Comparison of population energy flow of a herbivorous and a deposit-feeding invertebrate in a salt marsh ecosystem. Proc. Natl. Acad. Sci. 45:617-622.

Odum, E. P., C. E. Connell, and L. B. Davenport. 1962. Population energy flow of three primary consumer components of old-field ecosystems. Ecology 43:88-96.

Odum, E. P., and F. B. Golley. 1963. Radioactive tracers as an aid to measurement of energy flow at the population level in nature, p. 403-410. In V. Schultz and A. W. Klement, Jr., [ed.], Radioecology. Reinhold Publ. Corp., New York.

Odum, H. T. 1956. Primary production in flowing waters. Limnol. Oceanog. 1:102-117.

Odum, H. T. 1957. Trophic structure and productivity of Silver Springs. Ecol. Monographs 27:55-112.

Odum, H. T. 1960. Ecological potential and analogue circuits for the ecosystem. Am. Scientist 48:1-8.

Odum, H. T., and E. P. Odum. 1955. Trophic structure and productivity of a windward coral reef community on Eniwetok Atoll. Ecol. Monographs 25:291-320.

Odum, H. T., and E. P. Odum. 1959. Principles and concepts pertaining to energy in ecosystems, p. 43-87. *In* E. P. Odum, Fundamentals of ecology. W. B. Saunders Co., Philadelphia.

Odum, H. T., and R. C. Pinkerton. 1955. Time's speed regulator: the optimum efficiency for maximum power output in physical and biological systems. Am. Scientist 43:331-343.

Ovington, J. D. 1957. Dry matter production by *Pinus sylvestris*. Ann. Bot. N. S. 21:287-314.

Pearson, O. P. 1954. The daily energy requirements of a wild Anna hummingbird. Condor 56:317-322.

Penfound, W. T. 1956. Primary production of vascular aquatic plants. Limnol. Oceanog. 1:92-101.

Phillipson, J. 1962. Respirometry and the study of energy turnover in natural systems. Oikos 13:311-322.

Reichle, D. E. 1967. Radioisotope turnover and energy flow in terrestrial isopod populations. Ecology 48:351-366.

Richman, S. 1958. The transformation of energy by *Daphnia pulex*. Ecol. Monographs 28:273-291.

Ryther, J. H., and C. S. Yentsch. 1957. The estimation of phytoplankton production in the ocean from chlorophyll and light data. Limnol. Oceanog. 2:281-286.

Sargent, M. C., and T. S. Austin. 1949. Organic productivity of an atoll. Am. Geophys. Union Trans. 30:245-249.

Slobodkin, L. B. 1959. Energetics in *Daphnia pulex* populations. Ecology 40:232-243.

Teal, J. M. 1957. Community metabolism in a temperate cold spring. Ecol. Monographs 27:283-302.

Thienemann, A. 1926. Der Nahrungskreislauf im Wasser. Verh. Deut. Zool. Ges. 31:29-79.

Transeau, E. N. 1926. The accumulation of energy by plants. Ohio J. Sci. 26:1-10.

Verduin, J. 1956. Primary production in lakes. Limnol. Oceanog. 1:85-91.

Watt, K. E. F. 1966. Systems analysis in ecology. Academic Press, New York.

Wiegert, R. G., E. P. Odum, and J. H. Schnell. 1967. Forb-arthropod food chains in a one-year experimental field. Ecology 48:75-83.

Production Potential of a Seaweed-Lobster Community in Eastern Canada[1,2]

R. J. MILLER,[3] K. H. MANN

Fisheries Research Board of Canada
Marine Ecology Laboratory, Bedford Institute, Dartmouth, N.S.

AND

D. J. SCARRATT

Fisheries Research Board of Canada
Biological Station, St. Andrews, N.B.

MILLER, R. J., K. H. MANN, AND D. J. SCARRATT. 1971. Production potential of a seaweed–lobster community in eastern Canada. J. Fish. Res. Bd. Canada 28: 1733–1738.

An energy-flow diagram for the bottom community in the seaweed zone of St. Margaret's Bay, Nova Scotia, has been constructed, using previous estimates of seaweed production and observations on animal biomass and respiration rates. Annual seaweed production exceeded the annual consumption by herbivores by a factor of more than 10 (7000 vs. 572 kcal/m² per year). This fact, plus observations of physical processes in the seaweed zone, indicate that most of the seaweed production is exported as particulate matter in suspension. Production rates of the American lobster (*Homarus americanus*) prey species exceed lobster ingestion rates, also by a factor of more than 10 (84 vs. 6 kcal/m² per year). The difference is presumably consumed by lobster competitors. The available information suggests lobster production in the seaweed zone could be increased by reducing predation and competition for food, together with an increase in suitable shelter.

MILLER, R. J., K. H. MANN, AND D. J. SCARRATT. 1971. Production potential of a seaweed–lobster community in eastern Canada. J. Fish. Res. Bd. Canada 28: 1733–1738.

Les auteurs ont construit un diagramme de transport énergétique pour la communauté benthique de la zone des algues de la baie Sainte-Marguerite, Nouvelle-Écosse, en utilisant des données antérieures de production d'algues et des observations relatives à la biomasse animale et aux taux respiratoires. La production annuelle d'algues est 10 fois supérieure à la consommation par les herbivores. Ceci, en plus des données physiques du milieu, indique que la majeure partie de la production d'algues est disséminée sous forme de matière en suspension. Le taux de production des proies du homard américain (*Homarus americanus*) est, lui aussi, 10 fois supérieur au taux d'ingestion par cette espèce. La différence est probablement utilisée par des espèces compétitrices. Les données actuelles suggèrent que la réduction des prédateurs et des compétiteurs, accompagnée d'une augmentation du nombre d'abris appropriés, favoriserait la production du homard dans la zone des algues.

Received April 8, 1971

A well-defined community on the rocky shores of eastern Canada is the sublittoral kelp zone, with its characteristic herbivores, such as sea urchins and periwinkles, and a variety of carnivores, of which the most valuable is the American lobster (*Homarus americanus*). A start has been made on the analysis of the trophodynamics of this system, with a view to determining the productivity of the seaweeds and the potential for increasing lobster production. This paper considers questions of economic importance, such as: Are the dynamics of the system such that lobster productivity is limited by productivity of the lower trophic levels? Could lobster production be enhanced? Could seaweeds be harvested without adversely affecting the yield of lobsters?

[1]Contribution to the International Biological Programme CCIBP 120.

[2]Bedford Institute Contribution BI 272.

[3]National Research Council of Canada Postdoctoral Fellow. Present address: Fisheries Research Board of Canada, Biological Station, St. John's, Nfld.

225

Animal population production may be limited by recruitment, food energy, food quality, predators, parasites, or space. Not all of these factors are easily quantified and not all are equally important. Food energy, expressed as production of prey, is frequently considered to be important; therefore, this is a logical start in studying what limits the size of a particular animal population. However, the usual method of estimating population production and consumption requires data on population dynamics, growth, metabolic rates, and assimilation efficiency and is therefore laborious. A recent paper by McNeill and Lawton (1970) demonstrated a correlation between population respiration and population production for a wide range of poikilotherms, so the opportunity has been taken to infer the productivity of the invertebrates in the food web from a combination of in situ measurements of their respiration and population biomass estimates. The confidence limits of such estimates are rather wide, but do not affect the major conclusions.

Methods

BIOMASS

Observations on the biomass of seaweed and of the major groups of invertebrates in the seaweed zone in St. Margaret's Bay, Nova Scotia, were made by scuba diving in the summer of 1968. Samples were taken on 24 transects laid out at right angles to the shore from randomly chosen points. Ten major seaweed zones were recognized (Mann 1971) and on each transect the zones that were well represented were sampled with $\frac{1}{2} \times \frac{1}{2}$-m quadrats. The average length of a transect was 369 m, the average maximum depth on a transect was 20 m, and the total number of quadrat samples was 165. The divers removed the macrophytes and macroinvertebrates from each quadrat, and identified, counted, and weighed them. The mean biomass per m^2 for each taxon was weighted according to the mean width of each seaweed zone to provide an estimate of the mean biomass per m^2 for the whole seaweed zone of the bay. The methods and results of the macrophyte sampling are given in more detail in Mann (1971).

The lobster biomass was estimated indirectly, since special techniques are required to sample lobster populations (Scarratt 1968). Two parameters were used: (1) the best estimate of average stock of legal-sized lobsters in this part of Nova Scotia, taken from Wilder (1965); (2) the ratio of legal-size stock (Wilder 1965) to total stock (D. J. Scarratt unpublished data) in Northumberland Strait.

Biomass was converted to calories using the data in Brawn et al. (1968) or by using ash-free dry weight/live weight ratios (Thorson 1957) times a constant of 5 kcal/g ash-free dry weight.

RESPIRATION RATES

The respiration rates (in ml O_2/animal per day) of six species of invertebrates important in the diet of lobsters were measured in situ in St. Margaret's Bay at depths of 2–4 m in August 1970. One or more specimens of a species were gently transferred to wide-mouthed bottles, which were capped and left on the bottom for about 6 hr; 5–8 bottles, each with a different mean size of animal, were used for each species. Rates of oxygen consumption were determined from the differences in dissolved oxygen concentration in experimental and control bottles.

In addition, the respiration of sea urchins was studied more intensively because of their importance in the seaweed community. Measurements were made in the laboratory at 2–month intervals at a temperature and salinity close to that in the field. Animals were held in the laboratory only a few days before the experiments. The procedure was similar to that used in the field. Lobster respiration rates were taken from McLeese (1964).

In extrapolating to field conditions, the values obtained in experimental bottles were taken to be fairly representative of the normal activity of the animals, all relatively sluggish in their habits.

To extrapolate from the temperature at which measurements were made to the seasonal range of temperatures we assumed a Q_{10} of 2.05. This value was obtained by fitting a regression line of respiration on temperature to published data for 14 species of marine temperate poikilotherms (R. J. Miller unpublished data). Seasonal temperatures in St. Margaret's Bay were taken from Sharaf El Din et al. (MS 1970).

Temperature was assumed constant for 2 months at a time. The annual respiration rate for each population (R_{ann}) was calculated by summing the 2-monthly values (R_i) as follows:

$$R_{ann} = \sum_{i=1}^{i=6} R_i$$

$$R_i = 61 R_d \times B \times 4.83 \times 10^{-3},$$

where R_d is ml O_2 consumed/g animal per day, derived from the observed respiration rate/g live weight for an animal of average size in the population, and corrected for the appropriate environmental temperature; B is population biomass in g/m^2, and 4.83×10^{-3} converts from ml O_2 to kcal.

PRODUCTION RATES

Invertebrate production was estimated from a modification of the regression of log population production on log population respiration given by McNeill and Lawton (1970). The modification was provided by using more recent data on fish populations in the River Thames (Mann et al. 1971). The new coordinates, $P_{ann} = 137$ and $R_{ann} = 671$, represent the combined populations of roach and bleak, and replace the points given in McNeill and Lawton (1970) for roach, bleak, dace, perch, and gudgeon. The revised regression equation is:

$$P_{ann} = 0.6440\, R_{ann}^{0.8517},$$

where P_{ann} is annual population production, and R_{ann} is annual population respiration, both in kcal/m² per year.

Seaweed production was estimated by punching holes in blades of *Laminaria* and *Agarum* and measuring the growth of new tissue from the base of the blade (Mann 1972). Growth in length was converted to growth in biomass by means of length/weight regressions and the annual value for the mean production/biomass ratio was applied to the biomass observed in the survey.

FOOD CONSUMPTION

Assimilation efficiency was assumed to be 80% for carnivores and 60% for herbivores. The former is the value given by Winberg (1956) as a reasonable average for all fish and the latter is an average over season and animal size for *Strongylocentrotus droebachiensis* feeding on *Laminaria longicruris* (R. J. Miller unpublished data). Food energy C was calculated from respiration R and production P (both in kcal/m² per year) as

$$C = \frac{P+R}{0.8} \text{ for carnivores and}$$

$$C = \frac{P+R}{0.6} \text{ for herbivores.}$$

SPECIES OF FOOD CONSUMED BY LOBSTERS

Stomach contents of lobsters collected in the Northumberland Strait were analyzed for food composition by volume (D. J. Scarratt unpublished data). Volumes of each species were converted to calories by assuming a specific gravity of 1 and using the calorific values given by Brawn et al. (1968).

Results

Sea urchins (*Strongylocentrotus droebachiensis*) clearly dominated the invertebrate biomass in the seaweed zone (Fig. 1). Periwinkles, primarily *Littorina littorea*, made a significant contribution to herbivore biomass, as did the mussels. The biomass of several groups not included in Fig. 1 was recorded in the survey of the seaweed zone, but none exceeded 1 g/m² and all totalled only about 3 g/m². Small animals, such as amphipods and polychaetes and the active rock crabs, may not have been sampled quantitatively. Note that "mussels" in Fig. 1 refers to both the blue mussel (*Mytilus edulis*) and the horse mussel (*Modiolus modiolus*).

A figure for probable density of commercial-size lobster stock in St. Margaret's Bay (interpolated from Wilder (1965)) is 1.8 g /m². The mean estimate

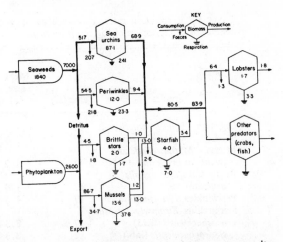

FIG. 1. Energy flow through the bottom community in the seaweed zone of St. Margaret's Bay, Nova Scotia. Units are kcal/m² per year, except for biomass, which is in kcal/m². Phytoplankton production figure from Platt (1971).

of total stock in Northumberland Strait off Richibucto, N.B., is 5.1 g/m², and in that area the commercial stock is thought to average 1.7 g/m². So the best estimate for the total stock of lobsters in St. Margaret's Bay is $1.8 \times \frac{5.1}{1.7} = 5.4$ g/m², or 1.7 kcal/m² (Fig. 1).

The results of respiration experiments are summarized in Table 1 and population respiration energy is given in Fig. 1 under the symbol suggested by Odum (1967) to indicate an energy sink. The value for sea urchin respiration in Fig. 1 is based on the bimonthly laboratory measurements.

Production and consumption energies have approximately the same ratio to biomass for each of the animal groups considered (Fig. 1), so the sea urchins have the largest values for these parameters as well. Consumption and production of the herbivores totals 663 and 93.5 kcal/m² per year, respectively. The consumption and production rates of the predators for which there is information are small relative to herbivore rates. Fish and crab predators are given a compartment in Fig. 1, since they are certainly important in energy transfer through this community, but no data are available.

The biomass estimate including all species of seaweed averages 4.1 kg/m². Paine and Vadas (1969) gave data on calorific value for 15 species of phaeophytes, which average 460 kcal/kg wet weight. Hence, the biomass at the time of the survey may be expressed as 1886 kcal/m². The ratio of production to biomass differs markedly according to species and depth, but the *Laminariales* con-

TABLE 1.   Respiration rates, expressed as oxygen consumption per gram per day, and correlation coefficient of regression of log respiration on log weight.

Species	Temp (C)	Mean size (g)	No. observations for regression	Correlation coefficient r	Oxygen consumption (ml/g per day)
Lobsters, *Homarus americanus*	12	450	14	–	0.514±0.095[a]
Starfish, *Asterias vulgaris*	17	5.0	5	0.99	1.14
Mussels, *Mytilus edulis* and					
Modiolus modiolus	17	2.2	8	0.59	0.876
Brittle stars, *Ophiopholis*					
aculeata	17	1.0	5	0.75	0.364
Periwinkles, *Littorina littorea*	17	1.4	6	0.95	0.600
Sea urchins, *Strongylocentrotus*					
droebachiensis (field)	17	8.3	6	0.99	0.464
S. droebachiensis (lab.)	2	8.3	18	0.98	0.284
	4	8.3	20	0.97	0.400
	11	8.3	18	0.99	0.531
	17	8.3	20	0.99	0.676
	8.5	8.3	18	0.99	0.417
	7.5	8.3	11	0.99	0.371

[a] From McLeese (1964).

stitute more than 80% of the biomass and their P/B ratios have been determined for a variety of environmental conditions. The problem of applying these to the bay as a whole has not yet been resolved, but a P/B ratio between 3 and 5 is a conservative estimate. Hence, production in the seaweed zone is likely to be at least 7000 kcal/m² per year[4].

The production of lobster prey species and probable rates of consumption of these species by lobsters are compared in Table 2. The energy contribution of each prey species identified in the stomachs of lobsters from the Richibucto area (estimated from the calorific value of each prey and its relative abundance) are included, since this (D. J. Scarratt unpublished data) is the most complete lobster food-habit study to date. However, there is evidence that lobster stomach contents reflect the relative abundance of prey species in the habitat. The percentage of lobster stomachs in which various species prey were found in a study off the Prince Edward Island coast (D. J. Scarratt unpublished data) are: mussels 67%, rock crabs 50%, polychaetes 42%, periwinkles 36%, sea urchins 23%, and starfish 5%. A study by Squires (1970) in western Newfoundland also includes percentage occurrence of prey; polychaetes 37%, rock crab 27%, periwinkles 22%, sea urchins 4%, mussels 4%, starfish 3%, and brittle stars 1%.

[4]Footnote added in proof: The production in the seaweed zone is now estimated at more than 17,000 kcal/m² per year (Mann 1972).

Because of the above evidence, the estimated consumption of each prey species by lobsters is given in the same proportion as production of these species (Table 2). In calculating these proportions the production of mussels is reduced by the energy ingested by starfish, since starfish are a competitor of lobsters. Comparison of columns c and d in Table 2 shows that prey production far exceeds lobster consumption for each prey species and for the column totals.

As an example of the degree of imprecision involved in the methods used here, confidence limits were accumulated through each major step of the calculations, from daily respiration rates of single animals to the annual production rate of a population (Table 3). The example used, the sea urchin population, includes the respiration rates at a single temperature in the field rather than the laboratory rates. The confidence limits are expressed as 2 standard errors of the mean (2 SE). Note that the data in column 2 are in log form. Since antilog (ln mean + ln 2 SE) = mean × (2 SE) and antilog (ln mean − ln 2 SE) = mean/2 SE, column 3 represents the value by which the mean would be multiplied and divided to get the mean ±2 SE. The upper confidence limit on the estimate of respiration rate at the temperature of measurement is 17% above the mean. The corresponding value for the annual respiration rate is 21% above the mean, the difference being derived from the confidence limits on the average Q_{10}. A further increase in upper limit to 60% above the mean is attributable to the

TABLE 2. (a) Percentage composition of the stomach contents of 550 lobsters from Richibucto, N.B.; (b) estimated percentage consumption of each type of prey by St. Margaret's Bay lobsters; (c) estimated consumption of each type of prey by a lobster population of density 5.4 g/m^2 (= 1.7 kcal/m^2); (d) production of each type of prey in St. Margaret's Bay.

Prey	(a) % of energy of stomach contents	(b) Estimated % consumption by lobsters	(c) Estimated consumption by lobsters (kcal/m^2 per year)	(d) Production by prey (kcal/m^2 per year)
Rock crab	28	a	–	a
Starfish	28	4	0.3	3.4
Sea urchin	10	83	5.2	68.9
Polychaete	10	a	–	a
Mussel	9	1	0.1	1.1
Periwinkle	–	11	0.1	9.5
Brittle star	–	1	0.1	1.0
Others	15	a	–	a
	100	100	6.4	83.9

aNot sampled quantitatively in St. Margaret's Bay.

TABLE 3. Accumulation of 2 standard errors (2 SE) from the measurement of individual sea urchin respiration to the estimate of population production.

Measurement	ln mean ± ln 2 SE	Antilog 2 SE
Respiration rate at temp of measurement	−0.9136±0.1616 ml O$_2$/g animal per day	1.17
Mean annual respiration rate	−1.6605±0.1926 ml O$_2$/g animal per day	1.21
Population respiration rate	4.6663±0.4778 kcal/m^2 per year	1.60
Population production	3.5344±1.6976 kcal/m^2 per year	5.46

confidence limits on population biomass estimates. The final step, introducing the production/respiration relation, brings the upper limit of confidence to 446% above the mean.

Discussion

Calculations of confidence limits on sea urchin production showed the production–respiration plot (modified from McNeill and Lawton 1970) to be the biggest contributor to imprecision. This is not surprising, considering the diversity of animals and habitats represented by the data and the variety of methods used. If future work is conducted using more uniform methods, such as those being developed in the International Biological Programme, some of the previous results may be shown to be unrepresentative of a taxon and may be eliminated from the regression.

The production of seaweeds (>7000 kcal/m^2 per year) far exceeds the ingestion by herbivores (663 kcal/m^2 per year). This figure is not derived entirely from seaweeds, since sea urchins and periwinkles make extensive use of periphyton and small macrophytes, and mussels and brittle stars feed primarily on phytoplankton. The biomass of other benthic herbivores was insignificant, and we know of no pelagic species that might take a measurable fraction of seaweed production. Therefore, it appears that most of the seaweed production is exported from the seaweed beds. In St. Margaret's Bay, at least, the export probably enters the water column as detritus eroded from the seaweed blades by wave action. Large pieces of seaweed are not often seen suspended in the water column or settled on the bottom outside the seaweed zone, and accumulations on the beach are small.

The existence of an excess of seaweed production over herbivore consumption does not mean that seaweed could be harvested without affecting the food web leading to lobsters. If the seaweed were harvested from a large area of bottom, the relatively immobile herbivores in that area would experience reduced food availability, which would probably limit their production.

The production of lobster prey species (less starfish ingestion) totals 83.9 kcal/m² per year (Fig. 1). Comparing this with the ingestion by lobsters, 6.4 kcal/m² per year, it appears that food production greatly exceeds that needed to support the lobster population.

There are several possible explanations of why, in spite of this large food production, lobster production is not greater. Not all of the production of prey species would be available to lobsters because some of it occurs in sizes and population densities that they cannot exploit, and because other predators compete for common prey. Moreover, food quality, predators, parasites, space, or recruitment might be limiting.

It is clear that much more research is needed to assess the importance of factors regulating lobster population production. However, as an alternative to waiting for more complete background information, an approach is suggested, based on available data and deductions from ecological principles, for increasing lobster production in a seminatural environment.

A natural area of lobster habitat would be surrounded by a barrier to exclude lobster predators and species competing with lobsters for food. The barrier should not impede water movements unduly since the productivity of seaweeds depends on free flow of water to transport nutrients and remove waste products. The design of the barrier would present problems. Extra lobster homes would be provided in the form of large rocks or pipes (Scarratt 1968; Chittleborough 1970). The area would then be stocked with lobsters at a density higher than is natural, and the resulting productivity determined. Questions requiring detailed study include: grazing efficiencies and productivity of lobsters at different stocking densities; competitive interactions, including cannibalism, between lobsters at high density; effect of removal of competing predators, including the possibility that some prey species not exploited by lobsters might undergo population explosion.

Acknowledgments

The authors are grateful to Drs L. M. Dickie and D. G. Wilder for helpful comments, and to Mr T. J. M. Webster for diving assistance in the field.

BRAWN, V. M., D. L. PEER, AND R. J. BENTLEY. 1968. Caloric content of the standing crop of benthic and epibenthic invertebrates of St. Margaret's Bay, Nova Scotia. J. Fish. Res. Bd. Canada 25: 1803–1811.

CHITTLEBOROUGH, R. G. 1970. Studies on recruitment in the western Australian rock lobster, *Panulirus longipes cyanus* George: density and natural mortality of juveniles. Aust. J. Mar. Freshwater Res. 21: 131–148.

MANN, K. H. 1971. Ecological energetics of the seaweed zone in a marine bay on the Atlantic coast of Canada. I. Zonation and biomass of seaweeds. Mar. Biol. (In press)

1972. Ecological energetics of the seaweed zone in a marine bay on the Atlantic coast of Canada. II. Productivity of the seaweeds. Mar. Biol. (Submitted for publication in 1971)

MANN, K. H., R. H. BRITTON, A. KOWALCZEWSKI, T. J. LACK, C. P. MATHEWS, AND I. McDONALD. 1971. Productivity and energy flow at all trophic levels in the River Thames, England. Proceedings of UNESCO/IBP Conference on Productivity Problems of Freshwaters. Pol. Acad. Sci. Warsaw. (In press)

McLEESE, D. W. 1964. Oxygen consumption of the lobster, *Homarus americanus* Milne-Edwards. Helgolaender Wiss. Meeresunters. 10: 7–18.

McNEILL, S., AND J. H. LAWTON. 1970. Annual production and respiration in animal populations. Nature (London) 225: 472–474.

ODUM, H. T. 1967. Biological circuits and the marine systems of Texas, p. 99–157. *In* T. A. Olson and F. J. Burgess [ed.] Pollution and marine ecology. Interscience Publishers, Inc., New York, N.Y.

PAINE, R. T., AND R. L. VADAS. 1969. Calorific values of benthic marine algae and their postulated relation to invertebrate food preferences. Mar. Biol. 4: 79–86.

PLATT, T. 1971. The annual production by phytoplankton in St. Margaret's Bay, Nova Scotia. J. Cons. Perma. Int. Explor. Mer. (In press)

SCARRATT, D. J. 1968. An artificial reef for lobsters (*Homarus americanus*). J. Fish. Res. Bd. Canada 25: 2683–2690.

SHARAF EL DIN, S. H., E. M. HASSAN, AND R. W. TRITES. MS 1970. The physical oceanography of St. Margaret's Bay. Fish. Res. Board Can. Tech. Rep. 219: 242 p.

SQUIRES, H. J. 1970. Lobster (*Homarus americanus*) fishery and ecology in Port au Port Bay, Newfoundland, 1960–1965. Proc. Nat. Shellfish. Ass. 60: 22–39.

THORSON, G. 1957. Bottom communities, p. 461–534. *In* J. W. Hedgpeth [ed.] Treatise on marine ecology and paleoecology. Vol. I. Ecology. Geological Society of America, New York, N.Y.

WILDER, D. G. 1965. Lobster conservation in Canada. Rapp. Proces-verbaux Reunions Cons. Perma. Int. Explor. Mer 156: 21–29.

WINBERG, G. G. 1956. Intensivnost obmena i pischevye petrebrosti ryb. (Rate of metabolism and food requirements of fish.) Nauch. Tr. Beloruss. Gos. Univ. Im. V. I. Lenina, Minsk. 253 p. (Transl. from Russian by Fish. Res. Board Can. Transl. Ser. No. 194, 1959)

ECOLOGY OF A NEW ENGLAND SALT MARSH[1]

Scott W. Nixon and Candace A. Oviatt

Graduate School of Oceanography
University of Rhode Island, Kingston, Rhode Island 02881

Table of Contents

Abstract

Measurements of the abundance of major populations, their metabolism, and the seasonal patterns of total system metabolism throughout a year were used to develop energy-flow diagrams for a New England salt-marsh embayment. The annual ecological energy budget for the embayment indicates that consumption exceeds production, so that the system must depend on inputs of organic detritus from marsh grasses. Gross production ranged from almost zero in winter to about 5 g O_2 m^{-2} day^{-1} in summer. Respiration values were similar, but slightly higher, with the maximum difference observed in fall. Populations of shrimp and fish were largest in fall, with a much smaller peak in spring. Few animals were present in the embayment from May to July, but fall populations of shrimp ranged from 250 to 800 m^{-2} and fish averaged over 10 m^{-2}. Birds were most abundant in winter and spring. In spite of high numbers, no evidence was found that the marsh embayment exported large amounts of shrimp or fish to the estuary. Production of aboveground emergent grasses on the marsh equaled 840 g m^{-2} for tall *Spartina alterniflora*, 432 g m^{-2} for short *S. alterniflora*, and 430 g m^{-2} for *S. patens*. These values are similar to those for New York marshes, but substantially lower than the southern marsh types. The efficiency of production of marsh grasses in the New England marsh was lower than reported for southern areas.

A simulation model based on the laboratory and field metabolism and biomass measurements of parts of the embayment system was developed to predict diurnal patterns of dissolved oxygen in the marsh. The model was verified with field measurements of diurnal oxygen curves. The model indicated the importance of the timing of high tides in determining oxygen levels and was used to explore simulated additions of sewage BOD and increases in temperature.

Key words: *New England, salt marsh, ecosystem, embayment, oxygen, model, Spartina, Palaemonetes, Fundulus, detritus, production, energy*

[1] Received August 29, 1972; accepted March 15, 1973.

231

Reprinted from Ecol. Monogr. 43:463–498 (1973).

INTRODUCTION

The extensive salt marshes of the Atlantic and Gulf coasts of the United States are among the most frequently studied ecological systems. Although marshes are present all along the coast, those in New England tend to be small and characterized by a heavy peat substrate, whereas the marshes south of New Jersey are better developed, with mineral sediments typical of a nonglaciated coastal plain (Chapman 1940, 1960). Most of our knowledge of the marsh ecosystem has come from studies of the latter type of salt marsh. The well-known synthetic treatment of many studies in the marshes of Sapelo Island, Georgia, by Teal (1962) has served for 10 years as a summary statement of ecological energy flow in this type of system. Drawing primarily on earlier work by Pomeroy (1959) on sediment algal production, Smalley (1960) on *Spartina* and insect populations, Kuenzler (1961) on mussels, his own work on crabs (Teal 1958, 1959), and his studies with Kanwisher (Teal and Kanwisher 1961) and Duff (Duff and Teal 1965) on gas exchange in marsh grasses and muds, Teal was able to illustrate the great productivity of the emergent grasses and the importance of detritus food chains on the marsh, in contrast to the grazing food chains of terrestrial grasslands.

Few, if any, such studies of ecological energetics have become available for the northern marsh types of New England (Cooper 1969). With the exception of primary productivity measurements on the marshes of Long Island by Udell et al. (1969) and estimates of primary production of sediment algae by Lytle (1969) in Rhode Island, almost all of the work in this region has been devoted to autecological studies of various component species of different marshes (Blum 1968) or to careful determinations of floristic composition, succession, and zonation (Chapman 1940, Miller and Egler 1950, Niering 1961). Because of clear zonation patterns in the northern marshes, the successional vegetation diagram of Miller and Egler may be cited as a structural analog of Teal's diagram of system function. Redfield (1972) has recently given a detailed description of the historical development of New England marshes.

In our efforts to describe the ecological energetics of a New England salt marsh, it also became apparent that little attention has been given anywhere along the coast to the associated networks of shallow tidal embayments and creeks that couple the emergent marsh with the larger estuary. When data are available, they are usually confined to individual species in separate marsh areas, such as studies by Wood (1967) on the physiology of grass shrimp, *Palaemonetes*, and by Schmelz (1964) on the com-

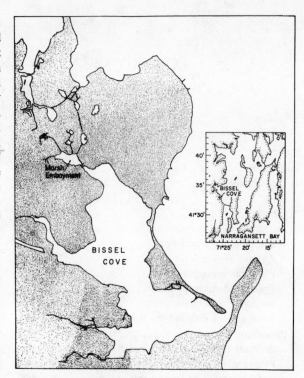

FIG. 1. Location of the marsh study area in Bissel Cove, Narragansett Bay, Rhode Island.

mon mummichog, *Fundulus heteroclitus*. Although some attention has been given to phytoplankton dynamics in the marsh creeks of Georgia (Ragotzkie and Pomeroy 1957, Ragotzkie 1959) and Long Island (Udell et al. 1969), few quantitative data are available on the overall patterns and magnitudes of metabolism of these systems or on the seasonal changes in the size and composition of their major populations. While it is understandable that the conspicuous stands of grass first captured the attention of coastal ecologists, it would seem equally important to understand the dynamics of this interface region of tidal creeks and embayments and to include them in an analysis of the marsh system.

Bissel Cove

Like a miniature of the well-known Barnstable Harbor on Cape Cod, Bissel Cove, the area chosen for this study, shows many of the "family" characteristics of New England marsh areas described by Ayers (1959). The cove and its associated marshes have developed from a coastal indentation that has been partially cut off from the West Passage of Narragansett Bay, R.I., by a barrier bar (Fig. 1). Freshwater inputs to the area from stream drainage and rainfall are small compared with tidal influences, so that the waters are characteristically marine, though great short-term variations in salinity occur. In the northern end of the cove the charac-

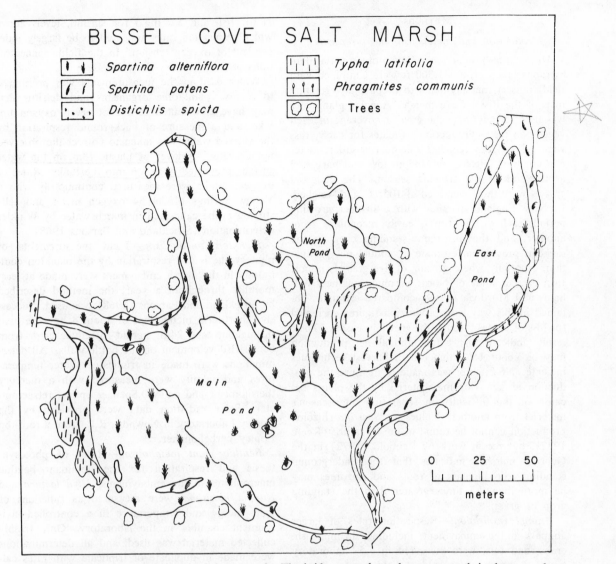

BISSEL COVE SALT MARSH

Spartina alterniflora	Typha latifolia
Spartina patens	Phragmites communis
Distichlis spicta	Trees

North Pond

East Pond

Main Pond

0   25   50

meters

FIG. 2. Vegetation map of the Bissel Cove marsh. The bridge over the embayment mouth is shown at the lower right.

teristic development of marsh and tidal creek networks has been modified somewhat by construction of a bridge over the main creek opening into the cove. As a result, the area beyond the bridge and its underlying culvert is a tidal embayment of 6,600 m² with a mean depth at low tide of about 0.25 m. Small ditches and creeks connect this embayment with a smaller saltwater pond to the east and two still smaller ponds in the north. These satellite ponds have areas of 1,600 m², 500 m², and 450 m² with average depths of about 0.15 m. An examination of old coast and geodetic survey maps indicates that the basic pattern of tidal ponds, emergent grasses, and bordering fringes of white oak shown in Fig. 2 has remained relatively unchanged for at least a hundred years.

The total area of the emergent marsh between the embayment and the trees is 16,800 m². This relatively small size of the combined embayment and emergent grass system, 23,400 m², made it possible to include many parameters in an almost synoptic sampling program and still maintain replication. The presence of a bridge and culvert across the mouth of the embayment instead of a large exposed tidal front provided an opportunity to regulate the tidal activity of the marsh and to get accurate estimates of the total input and output of organisms and detritus from the marsh. Morphological stability, floristic composition, small size, and susceptibility to experimental manipulation made Bissel Cove attractive for study as a model for energy flow in the New England marshes.

METHODS

Production and metabolism measurements

Marsh grass production.—Low altitude aerial photographs taken at 400–1,500 ft were combined with field transect measurements to prepare a vegetation map of the Bissel Cove marsh. A polar planimeter was used to determine the area of coverage of each vegetation type. Production estimates for each grass were taken from replicate one-quarter-square-meter clip quadrats selected by random toss and harvested at the end of the growing season. The collected grasses were then ovendried at 105° C and weighed. In New England marshes with a distinct growing period, each year's crop is easily recognized from another, and the maximum standing crop in the fall can provide an estimate of annual net production, especially where grazing of live grass is minor. In a careful study of North Carolina marshes, Williams and Murdoch (1969) compared estimates obtained in this way with production figures corrected for loss of blades through the growing season. Their results indicate that actual annual net production may be about 15% higher than harvest techniques indicate. No attempt was made to measure the production of new rhizome material by the plants because of the difficulty in separating one season's material from another. Values for *Spartina* rhizome production cannot be found in the literature (Keefe 1972), but recent work by Reimold (1972) in the Georgia marshes indicates that the underground standing crop biomass of roots and rhizomes may be more than 2.5 times greater than the standing crop of grass.

Animal respiration.—Respiration rates of larger animals in the embayment, including shrimp, fish, crabs, and eels, were measured in 5.5-l laboratory tanks of marsh water held at field temperature in the dark. The abundant shrimp and common mummichogs, which were present throughout the year, were measured at six temperatures from 3° to 30° C. Less abundant animals that were active only during the summer were measured at 20° C. All the animals were brought directly from the field, placed in the experimental tanks, and allowed to acclimate for 3–5 hr before determinations were begun. Each measurement involved 25–50 shrimp, 9–11 fish, and two to three crabs or eels. Only juvenile menhaden gave any evidence of hyperactivity or other crowding effects. Over 12 replicate measurements were made on random subsamples of field populations of the shrimp and common mummichog, instead of segregating animals on the basis of size, sex, or age. Four to six replicates were made with other species taken in a similar manner. Uptake rates were calculated from data taken over the first 2 or 3 hr, while oxygen concentrations remained above 2.0 mg

/l for fish and 5.0 mg/l for shrimp, levels above which respiration rates appeared to be largely independent of oxygen tension. In the field, concentrations often fell below these levels.

Water used in the measurements was prefiltered to remove planktonic organisms and detritus that may have contributed to the observed oxygen uptake, and a measure of background respiration by the filtered water was taken to correct the observed uptake rates. A layer of plastic film on the water surface prevented diffusion into the tanks. Rates of oxygen uptake were measured continuously with a Yellow Springs Model 54 oxygen meter and self-stirring probe calibrated in marsh water by Winkler determination (Strickland and Parsons 1968).

Plankton metabolism.—Field measurements of photosynthesis and respiration by the plankton community in the marsh embayment were made at least monthly throughout a year; the method described by Strickland and Parsons (1968) was used. Successive short-term incubations of about 4 hr were used throughout each 24-hr period to reduce error from bacterial development on the bottle walls. All determinations were made in triplicate. Water temperatures and salinity were measured with a mercury thermometer and an A.O.Spencer optical refractometer. Solar radiation data were provided by the Eppley Laboratory in Newport, R.I., from a roof-top Eppley pyrheliometer.

Benthic plant metabolism.—Rates of photosynthesis and respiration for the dominant benthic macrophytes in the embayments, *Ulva lactuca* and *Ruppia maritima*, were measured as functions of light energy and temperature in a controlled-environment chamber in the laboratory. Only freshly collected material was used, and all determinations were made in duplicate or triplicate with rates calculated from at least five points. Water taken from Bissel Cove was used and corrected for background metabolism as described for animal-respiration experiments. Oxygen changes were measured continuously with a Winkler calibrated Y.S.I. model 54 oxygen meter. On the basis of evidence presented by Hartman and Brown (1967) and others, that oxygen may be stored internally in the lacunar spaces of many vascular hydrophytes, Wetzel (1965) criticized the use of dissolved oxygen changes in short-term measurements of the metabolism of such plants. If this criticism is applicable to the measurements of *Ruppia* metabolism, the results would be an underestimate of photosynthesis and respiration. Sculthorpe (1967), however, does not make any mention of a well-developed lacunar system in *Ruppia*, and the use of continuous measurements eliminated the possibility of sharp transient changes in oxygen that may result from exchanges between the internal storages and the surrounding water (Nixon and Oviatt 1972). Radiation measurements in the laboratory were made with a Y.S.I. model 65 radiometer. After the determinations were completed, the plant materials were ovendried at 105°C and weighed.

On several occasions field measurements of the metabolism of *Ruppia* beds were made by enclosing areas of plants and sediment under 0.5-m-diameter plastic domes, one transparent and the other painted black and silvered to reduce heating. Again, changes in oxygen were monitored continuously with the self-stirring probe, and corrections were made for plankton metabolism and sediment respiration as determined from measurements of oxygen consumption by sediment cores in the laboratory and from field measurements under the blackened dome when placed over areas without benthic plants. Following the metabolism measurements, the plant material under the dome was harvested, dried, and weighed.

Sediment respiration.—Oxygen uptake by the top 5–10 cm of sediment from the marsh embayment was measured in replicate 4.5-cm-diameter cores across a range of temperature in the laboratory. In the spring and fall, measurements were taken over 2- to 4-hr periods with the self-stirring oxygen probe. Water over the core was agitated only during a 5-min period every half hour when readings were taken. In general, the method used here was similar to that described by Carey (1967) and others, except that a layer of thin plastic film was used to prevent exchange across the air-water interface. During the summer, measurements were made in the field with the darkened dome used for benthic plant-metabolism studies. Changes in dissolved oxygen in water overlying the sediments resulted from the combined actions of sediment infauna and bacteria, the respiration of organisms associated with large pieces of plant detritus, and from chemical oxidation reactions in the mud. Additional estimates of the metabolic activity of the total sediment community were obtained by subtracting the respiration of larger animals and plankton from measurements of total system metabolism in the embayment taken throughout a year.

Total system metabolism of the embayment.—The small size of the Bissel Cove embayment and the single opening of the embayment mouth for tidal exchange made it possible to regulate the flow of water into and out of the marsh. An inflatable rubber "pig" similar to those used in work on city sewage pipes was fashioned from a large truck tire inner tube on a heavy metal rim mounted between plywood disks. This plug was set in the culvert under the highway bridge crossing the embayment mouth and inflated with a SCUBA tank until it completely filled the passage and cut off all tidal flow. With a known water mass held in the marsh, the single diurnal curve method of Odum and Hoskin (1958) could be used to measure the total metabolism of the embayment.

Since water circulation is often important in regulating the metabolism of aquatic systems, confinement of water during the period of measurement may have introduced errors leading to an underestimate of both production and respiration (Nixon et al. 1971, Nixon and Oviatt 1972). These artifacts, however, would be small compared with similar artifacts long recognized and accepted in bottle experiments with plankton. Clear evidence of an effect of current speed on the metabolism of sediment is still lacking (Hargrave 1969), and it may be that diffusion effects within the sediment are much more important than the thickness of boundary layers in the overlying water. In fact, during a 24-hr period the diurnal curves showed no evidence of consistent change in the rate of production or consumption of dissolved oxygen associated with the amount of time the water had been held in the marsh.

Samples of water were taken in triplicate at least eight times during each 24-hr period (dawn, dusk, midnight, and at approximately 3-hr intervals during the day) and were analyzed for dissolved oxygen by the Winkler technique (Strickland and Parsons 1968). Water temperature and salinity were measured at each interval. Occasional continuous diurnal dissolved oxygen analyses with a Rustrak 190 oxygen-temperature meter indicated that our discrete sampling schedule gave an accurate picture of oxygen changes except during one or two days when the embayment became anoxic late at night. Diffusion rates across the water surface were measured directly with a floating plastic dome and oxygen meter (Copeland and Duffer, 1964, as modified by Hall, Day, and Odum, *unpublished data*), except during the winter when ice covered the embayment and no diffusion correction was used.

The diurnal curves were analyzed by plotting the rates of change in oxygen after correction for diffusion. The area under the positive portion of the rate-of-change curve represents apparent production during the day; the area of negative change is equal to night respiration. Linear extrapolation between the last negative rate-of-change point at dawn and the first at dusk provided a negative area that was attributed to daytime respiration. When added to apparent production, this value gave an approximation of gross production during the day. The procedure of extrapolating rates of dark respiration into the light probably provides a substantial underestimate of gross production, but more direct measures of this parameter have not yet been developed for field use (Odum, Nixon, and DiSalvo 1969).

Population estimates

Animals in the emergent marsh.—Rough estimates of the abundance of marsh snails, *Melampus bidentatus*, and marsh mussels, *Modiolus demissus*, were taken for preliminary comparison with other marsh areas from counts of twenty-six 0.5-m² quadrats positioned by random toss around the periphery of the embayment. All samples were collected on one day in July. An estimate of the population of fiddler crabs, *Uca pugnax*, was made in early fall by using twenty-two 1-m² quadrats on eight transects through the marsh. Small mammals were checked several times during the year with live traps. A study of the abundant and conspicuous bird populations of the marsh has been described by Lucid (1971), in which 2-hr observation periods were taken at random during the daylight hours twice each week for a 10-month period. With binoculars and a spotting scope, it was possible to observe the entire marsh and make a record of the numbers, species, and activities of all birds in the area every 5 min. This sampling schedule was selected after a continuous month of 2-hr observations made at random times twice each day indicated little day-to-day variation in results. Data collected in this way were supplemented with almost daily spot counts made on trips to the marsh. A study of feeding efficiency and energetics of the herring gull, *Larus argentatus*, a characteristic and abundant bird in the marsh during much of the year, was made on six captured adult birds over a 1-month period. The animals were kept in large cages and fed ad libitum on a diet of trash fish. The birds were weighed during the study, and a record kept of all food consumed and feces produced. Heats of combustion for food, feces, and birds were determined with a Phillipson microbomb calorimeter and a Parr bomb calorimeter, respectively.

Plankton in the embayment.—Counts of diatoms, larger flagellates, and zooplankton were made weekly throughout the summer and occasionally during other seasons. Estimates of phytoplankton were taken from single counts made with a Sedgwick-Rafter cell from 1 liter of freshly collected surface water. Zooplankton samples were collected at three stations in the embayment by pouring measured volumes of water through a #10-mesh plankton net. Single counts of each sample were made on formalin-preserved material.

Benthic plants and sediments.—During June, July, and August, quantitative surveys of benthic plant biomass were taken on four transects across the embayment. A benthic grab sample of 222 cm² area was used, and four to six samples were collected on each transect. The macro-plants were harvested and ovendried at 105° C. The extreme patchiness of benthic plant distributions made it difficult to arrive at a reliable estimate of plant biomass over the whole embayment, even though an attempt was made to do so by drawing contour maps of plant abundance and measuring them with a polar planimeter. The method probably gives results that are best interpreted as providing a range of plant densities in the area. Field estimates of the relative abundance of algae and submerged vascular plants were kept throughout the year by visually inspecting the embayment bottom through the clear, shallow water and making a weekly note of coverage and density relative to its appearance during the summer when measurements were being made.

A similar sampling program with the benthic grab was used to estimate the amount of macro-detritus (large pieces of dead grass and algae) that was in the surface layers of the embayment bottom. The distribution of this material was more uniform, and the method provided more consistent values of total detrital biomass. An estimate of total organic matter in the sediments of the embayment during summer was obtained by collecting 17 cores of sediment to a depth of 10 cm on transects across the marsh. These cores were divided into 1- or 2-cm-thick layers, dried at 105° C to constant weight, and then ashed at 500°–550° C for 4 hr. A check of the reliability of the ashing oven with eight replicates of a standard sample showed variations of only ± 1% of the mean (69.37% weight loss) regardless of sample position in the oven.

Infauna and epifauna.—Samples of infauna from the embayment sediments were collected weekly from February through July. Up to three grab samples of 173 cm² area each were taken to a depth of 10–15 cm and screened through a 0.5-mm-mesh sieve, and all animals were identified and counted. Dry weights were taken on the total sample for each week. The microbenthic infauna was sampled by Lavoie (1970) during early fall from six locations in the embayment. Single cores of 25 cm² cross-sectional area were removed from the sediment to a depth of 15–20 cm. The volume of each core was measured, and the material was washed through two screens of 1.9-mm mesh and 1-mm mesh. After a thorough washing, confirmed by microscopic examination of material retained on the screens, three 1-ml aliquots of the filtrate were examined for benthic copepods. Three other 1-ml subsamples of filtrate were stained with Rose Bengal to facilitate the counting and separation of nematodes from ostracods and ciliates.

Larger epifauna were sampled 1 or 2 days each

week throughout the year with four pull-up nets maintained around the marsh in a manner similar to that developed by Higer and Kolipinski (1967) for sampling shallow waters in the Everglades. Samples were not collected when thick ice was present during the winter and early spring. The nets used were 1.6 m on a side with a 6-mm mesh and could be pulled through the shallow water to the surface in a few seconds. When resting on the bottom, the nets sank into soft substrate and were nearly invisible from above. Shallow-water animals seem adapted to avoiding danger from above, and the rising nets captured a variety of organisms in great number with a relative abundance in agreement with seine samples.

Grass shrimp.—Population estimates for adults of the abundant grass shrimp, *Palaemonetes pugio*, were obtained in two ways. On four occasions, once during each season, a mark-and-recapture study was carried out following a design developed by Schnabel (DeLury 1951). Large numbers of shrimp were captured with dip nets and small seines, then immersed for about 15–30 sec in an aqueous solution of Alcian blue dye (64g l⁻¹), and counted directly or by volume displacement. Tagged animals were then released over the whole embayment. Checks on marked shrimp kept in the laboratory and in cages in the field showed little mortality due to the tag. In laboratory experiments with adult *Fundulus heteroclitus* and *F. majalis*, the fish would not feed on either marked or control shrimp. Gut analysis of fish in the field did not indicate any differential predation on tagged animals. The mark remained visible throughout the recapture period and often persisted through one molting. On successive days after the first tagging, shrimp were seined from all around the embayment, the number of recaptures noted, the remaining animals marked, and the entire sample returned over the whole embayment. During the winter, when ice prevented seining, the shrimp were dip-netted through holes cut in many places over the area.

A basic assumption of mark-and-recapture techniques is that there is no loss or gain of animals from the populations being measured (DeLury 1958, Cormack 1968). During each marking period a 1-m-diameter, 6-mm-mesh fyke net was placed across the mouth of the embayment, and stone weirs were positioned to channel migrating animals into the trap. The net was moved appropriately to capture flood- and ebb-tide movements. The total numbers of marked and unmarked shrimp entering and leaving the marsh on each tide during the sampling period were always very close, so that there was little net movement in either direction, and any excess in one direction was very small with respect to

the total population. The results of the shrimp census were analyzed according to the explicit formula developed by Schumacher and applied by DeLury (1958).

A plot of the shrimp data filled the necessary criterion that the proportion of recaptures in the tth sample be directly proportional to the total number of animals marked and released at the time the sample was taken.

A continuous check of the shrimp population was taken from the once- or twice-weekly counts of animals collected in the pull-up trap nets from four stations in the embayment. Although this method was probably not as reliable as the more elaborate seasonal estimates, the nets caught large numbers of shrimp and gave repeatable results. The movements of shrimp into and out of the marsh were monitored on two flood and two ebb tides weekly with the fyke net described earlier. The estimates of shrimp migration are probably low because of net avoidance and losses to predation in the net.

Fish.—Methods used for estimating populations of fish in the marsh embayment were similar to those used for the shrimp. Four seasonal mark-and-recapture studies were done on the most abundant fish, *Fundulus heteroclitus*, the common mummichog. Fish were captured throughout the embayment with an 18-m, 8-mm-mesh seine and anaesthetized with MS-222. During the winter fish were captured under the ice with baited traps. The anaesthetized fish were marked at the base of the dorsal fin or the caudal fin with a Panjet inoculator loaded with Alcian blue dye in aqueous solution at 64g l⁻¹ (Hart and Pitcher 1969). Mortality experiments in the laboratory and in field cages indicated that the larger fish were not affected by the procedure, but those under 35 mm suffered a high mark mortality. Accordingly, fish below this size were omitted from the marking program. It was assumed that differential predation on the tagged fish was not severe, since the small blue mark was not conspicuous when the fish were active underwater. On anaesthetized animals out of water, the tag was recognizable for 2 months or more. Because of the great amount of time and effort required to tag substantial numbers of fish, only an initial large sample was marked. On subsequent sampling days unmarked and recaptured animals were counted, and the entire sample was returned to the marsh. Recapture samples were taken for up to 4 weeks. Population estimates were calculated from the data with the standard Petersen formula (DeLury 1951).

During the recapture seining and trapping, counts were made of all other species of fish, so that a ratio of their abundance to that of the marked species could be developed. Assuming equal catch-

TABLE 1. Production of emergent plants in Bissel Cove Marsh

Zone	Area (m²)	Total solar radiation[a] (Kcal yr⁻¹)	Visible radiation[b] (Kcal yr⁻¹)	Aboveground production at the end of growing season				Efficiency[d] %
				g dry weight m⁻²	Kcal g⁻¹ [c]	Kcal m⁻² yr⁻¹	Kcal-yr⁻¹	
Spartina alterniflora (tall)	1.1×10^3	13.5×10^8	6.1×10^8	840	3.3	2.8×10^3	3.1×10^6	0.51
S. alterniflora (short)	12.1×10^3	14.9×10^9	6.7×10^9	432	2.6	1.1×10^3	13.3×10^6	0.20
S. patens	1.4×10^3	17.2×10^8	7.7×10^8	430	3.1	1.3×10^3	1.8×10^6	0.23
S. patens, Distichlis spicata mixture	1.8×10^3	22.0×10^8	9.9×10^8	680	2.8	1.9×10^3	3.4×10^6	0.34
Typha latifolia	3.6×10^2	4.43×10^8	2.0×10^8	693	4.3	3.0×10^3	1.1×10^6	0.54
Total	16.8×10^3	20.6×10^9	9.3×10^9				22.7×10^6	0.24

[a] Eppley pyrheliometer, Newport, R.I.: 12.3×10^5 Kcal m⁻² yr⁻¹.
[b] Calculated as 45% of total (Reifsynder and Lull 1965).
[c] Udell et al. (1969) and Boyd (1970).
[d] (Kcal yr⁻¹/visible radiation) (100).

ability, these ratios were used to extrapolate the results of the common mummie census to give a rough estimate of other, less abundant species.

Movement of fish in and out of the embayment was monitored once or twice weekly with the fyke net and sampling schedule described for the shrimp. Again, the error due to migration was minimal, since the numbers of marked fish entering and leaving the marsh were equal. No net movement of the other species was apparent during the recapture periods. Except during periods of thick ice, a weekly estimate of all fish populations was also taken from the four pull-up trap nets.

THE EMERGENT MARSH

Grass and detritus

The area between the embayment and a fringing wood of white oak trees consisted of 16,800 m² of characteristic marsh grasses, *Spartina alterniflora*, *S. patens*, and *Distichlis spicata*. The distribution of these species, as well as small clumps of cattail, *Typha latifolia*, and *Phragmites communis* is shown in Fig. 2. The *Phragmites* was confined to areas of freshwater input and marked the boundary of the salt-marsh system. As reported in other studies, growth per unit area of *S. alterniflora* was greatest in a 1-m-wide band of tall grass around the periphery of the embayment and along the creek banks and ditches. A shorter form of the same species covered a much greater area and thus contributed more to the overall production of the marsh. The amount of area covered by each plant species, its net production of aboveground parts at the end of the growing season, and its efficiency of production are summarized in Table 1. The 840 g m⁻² yr⁻¹ production of tall *Spartina* in the Bissel Cove marsh has been compared with a summary of production data for other marshes along the Atlantic coast

in Table 2. Apparently the decline in production at higher latitudes is neither as sharp nor as clear as suggested in previous reviews of marsh literature by Cooper (1969) and Keefe (1972) that lacked data for more northern areas.

Production values of the other grasses given in Table 1 agree with measurements reported by Udell et al. (1969) for a number of marshes on Long Island. The 693 g m⁻² biomass of *Typha* was about one-half to one-third of most of the values for this species reviewed by Keefe (1972), but Boyd (1970) has also reported almost identical low values for *Typha* marshes in South Carolina. Tall stands of *Phragmites* along the border of the marsh had 800–1,000 g dry weight m⁻², well within ranges cited by Keefe and reported in an extensive study by Björk (1967) for moderately eutrophic *Phragmites* biotopes.

The efficiency of net production for all grasses on the emergent marsh, 0.24%, is much smaller than the 1.1% annual efficiency of net grass production reported for the Georgia marshes by Teal (1962). Even the value for tall *S. alterniflora* is only about one-half that in Georgia, reflecting the more extreme environment and shorter growing season of the New England marsh. In his study of a Minnesota *Typha* marsh, Bray (1962) reported an aboveground net production efficiency of about 0.8% of visible radiation, a value over three times greater than that for the whole northern salt marsh, and almost 50% greater than for the *Typha* contribution to production on the marsh. As suggested by Teal, high respiratory rates in *Spartina* may reflect the energetic costs of success in the stressed marsh environment and lead to a lower net production efficiency than might be shown by similar communities at the same latitude. Even if Teal's value for *Spartina* respiration (70% of gross production) is ap-

TABLE 2. Biomass of *Spartina alterniflora* (tall and medium)[a] at the end of the growing season in some salt-marsh ecosystems

Marsh location	Biomass (g dry weight m^{-2})
Georgia (Teal 1962)[b]	1,290
North Carolina (Williams and Murdoch 1969)	1,100
Virginia (Wass and Wright 1969)[c]	1,332
Maryland (Johnson 1970)	1,207
Delaware (Morgan 1961)[d]	560
New Jersey (Good 1972)[e]	1,600
Long Island, N.Y. (Udell et al. 1969)	827
Rhode Island (this study)	840
Petpsewick Inlet, Nova Scotia (Mann 1972)[f]	580

[a] Aboveground portions only.
[b] Dry weight taken as 40% of fresh (Williams and Murdoch 1969).
[c] Cited in Keefe (1972).
[d] Size unspecified.
[e] Personal communication.
[f] Calculated from 290 g C m^{-2} yr^{-1}.

TABLE 3. Percentage of area covered by emergent grasses in some marine salt marshes

Marsh location	Species		
	Tall *Spartina alterniflora*	Other *S. alterniflora*	Mixed *S. patens* and *Distichlis*
Georgia (Teal 1962)	20	80	0
North Carolina (Stroud 1969)	6	54	—
Long Island, N.Y. (Udell et al. 1969)	10	67	23
Rhode Island (this study)	7	72	19

plied to the Bissel Cove marsh, the gross production efficiency is still 50% lower than the approximately 1.2% gross efficiency Bray calculated for the cattail marsh using respiration of 15% of gross production. Moreover, recent measurements by Lytle and Hull (*personal communication*) indicate that 70% may be an overestimate of *Spartina* respiration, at least in northern plants. There seems little question that the efficiency of production on the New England marsh is substantially lower than for southern marshes, and perhaps for similar but less stressed systems at higher latitudes as well. The relative coverage of the marsh by each vegetation type has been calculated in Table 3 for comparison with other marshes. A surprising regularity appears all along the coast, even for locations with quite different tidal regimes.

In late summer the *S. alterniflora* flowered and produced great quantities of seed which covered the water in windrows as it was carried back and forth by the tides. The grasses entered a dormant period in late fall and showed no growth until the following spring. This growth pattern is in contrast to that of the southern marsh, where some growth continues all year long. The dead grass remained standing until the thick ice of winter sheared it off and mechanically broke much of it up into smaller pieces during the spring thaw. In southern marshes a similar service may be provided by animals such as the abundant marsh crabs, *Sesarma*, and the fiddler crab, *Uca pugnax* (Crichton 1967). In Georgia marshes Wolf, Shanholtzer, and Reimold (1972) report a mean density in summer of 205 crabs m^{-2}, whereas only 2.7 ± 3.8 per m^2 were found during early fall in this study.

Major export of large pieces of detrital grass came in the spring with the first high tides that followed the ice melt. A smaller "burst" of dead grass export came in February during a brief midwinter thaw. Studies by Schultz and Quinn (*personal communication*) of the export from this system of fine detrital particles suspended in the water indicated a small but much more regular pattern of export throughout the year. Their values for filterable particulate matter in water ebbing from the Bissel Cove marsh were about a tenth of those found in Georgia by Odum and de la Cruz (1967).

Animals

Along with the fiddler crab discussed above, other ground animals were very scarce on the marsh. The values for our summer census of the 26 sample quadrats established around the embayment bank showed a range of 0–60 marsh mussels, *Modiolus demissus,* per square meter with a mean of 5.7 ± 13.2. The marsh snail, *Melampus bidentatus*, was clumped in shaded areas, with a range of 0–300 animals per square meter and a mean of 6.8 ± 60.8. In Georgia Kuenzler (1961) found an average for the marsh mussel of eight animals per square meter with densities four times greater in the most favorable areas along banks. In a study of Virginia marshes made during the fall, Kerwin (1972) reported mean *Melampus* densities of 7.2 animals per square meter. No significant populations of grasshoppers or other insects ever became apparent, in contrast to findings by Smalley (1960) in the Sapelo Island, Georgia, marsh. Occasional small mammals were sighted or trapped, including mice, voles, muskrats, and a raccoon, but they were never present in any number. Their impact on the marsh was probably slight, except for some cutting of *Spartina* for nest building by the muskrat.

THE TIDAL EMBAYMENT

General characteristics

As expected for a small shallow tidal embayment, almost all parameters measured showed wide diurnal

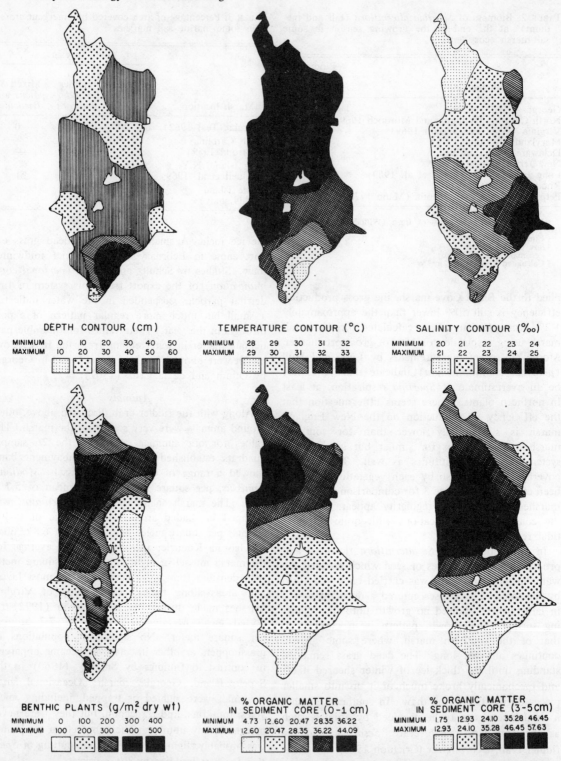

DEPTH CONTOUR (cm)

| MINIMUM | 0 | 10 | 20 | 30 | 40 | 50 |
| MAXIMUM | 10 | 20 | 30 | 40 | 50 | 60 |

TEMPERATURE CONTOUR (°C)

| MINIMUM | 28 | 29 | 30 | 31 | 32 |
| MAXIMUM | 29 | 30 | 31 | 32 | 33 |

SALINITY CONTOUR (‰)

| MINIMUM | 20 | 21 | 22 | 23 | 24 |
| MAXIMUM | 21 | 22 | 23 | 24 | 25 |

BENTHIC PLANTS (g/m² dry wt)

| MINIMUM | 0 | 100 | 200 | 300 | 400 |
| MAXIMUM | 100 | 200 | 300 | 400 | 500 |

% ORGANIC MATTER
IN SEDIMENT CORE (0–1 cm)

| MINIMUM | 4.73 | 12.60 | 20.47 | 28.35 | 36.22 |
| MAXIMUM | 12.60 | 20.47 | 28.35 | 36.22 | 44.09 |

% ORGANIC MATTER
IN SEDIMENT CORE (3–5 cm)

| MINIMUM | 1.75 | 12.93 | 24.10 | 35.28 | 46.45 |
| MAXIMUM | 12.93 | 24.10 | 35.28 | 46.45 | 57.63 |

FIG. 3. Some characteristic distributions of embayment parameters on a representative June day.

and seasonal variation. The range of variation became more extreme in the most northern and shallow sections of the embayment where small freshwater inflows were present. Over the year, water temperature ranged between –0.5° and 30.5° C with salinities between 1‰ and 28‰. Diurnal variations were least in the winter when daily changes in light and temperature were small compared with

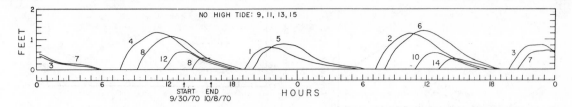

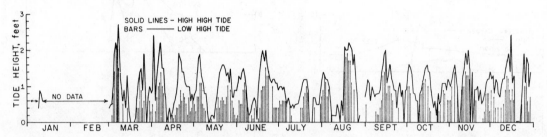

FIG. 4. Top: Daily pattern of the tides in the marsh embayment and the procession of 15 consecutive tides through a week in October. The sill effect completely eliminated high tides 9, 11, 13, 15. Bottom: Seasonal variation in the magnitude of the high tides in the marsh.

hot summer days and cool nights. During summer, ranges of 10°C were often measured over 24 hr. Contour maps of depth, temperature, salinity, sedimentary organic matter, and the biomass of benthic plants on a representative day in June are shown in Fig .3. Because of tidal exchange, ranges of 5°C and 5‰ at any one time were commonly found over the embayment. Benthic plants were distributed in patches. *Ruppia maritima* was most abundant in the lower salinity regions and was replaced by *Ulva lactuca* toward the mouth of the embayment. High densities of both species were often present in limited areas, with up to 500 g dry weight m⁻². Both plants went through several periods of rapid growth followed by death and decay and the export of large amounts of detritus. The very high levels of organic matter in the sediment were indicative of the peat-like structure of New England marshes and reflected the importance of the sediment as a storage of detrital material from the submerged plants as well as the emergent marsh grass.

Tidal patterns

The daily rise and fall of water in the Bissel Cove embayment regulates many features of the marsh. The floristic zonation pattern shown in Fig. 2 reflects to some extent the frequency and duration of submergence of the marsh grasses. Though ice prevented operation of a tide gauge during January and February, data for the rest of the year indicated an unusual tidal regime for the marsh. Fill used over the years for the construction and maintenance

of the bridge across the mouth of the embayment had formed a levee, so that tidal inputs to the system were limited to the upper one-third of the tidal cycle. This "sill effect" in the tidal signature is shown for an 8-day period in Fig. 4. The normal semidiurnal tide cycle was apparent only on days when both high tides were greater than the height of the sill. During the time of the study only about two-thirds of all days fell into this category. For such days the yearly average height of the high tide was 0.27 m above sill with a duration of 6.8 hr. The corresponding average for the low high tide was 0.21 m with a duration of 4.7 hr.

The difference in height between the two high tides increased during the spring and early summer, leading into a period from July through October in which there was often only one tide and flushing of the embayment was reduced. There seemed to be no clear seasonal differences in the maximum height of the tides, and the marsh was periodically flushed by several consecutive days of double tides throughout the year (Fig. 4). These modified tidal patterns did not seem to produce unusual features in the marsh, at least as reflected by an analysis of the distribution of indicator species of emergent grasses. The relative amount of area dominated by characteristic grasses in Bissel Cove is similar to values for various other marshes along the coast (Table 3).

Diurnal curves and system metabolism

The shallowness of Bissel Cove makes it susceptible to large and rapid diurnal and seasonal changes

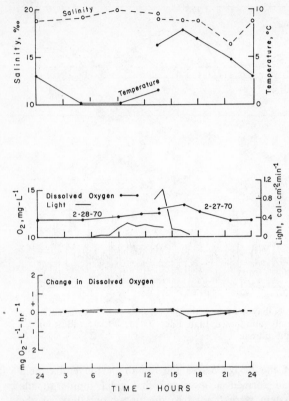

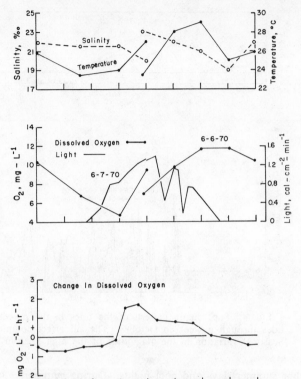

FIG. 5. Winter diurnal curves of salinity, temperature, solar radiation, and dissolved oxygen in the marsh embayment without tidal exchange.

FIG. 6. Summer diurnal curves of salinity, temperature, solar radiation, and dissolved oxygen in the marsh embayment without tidal exchange.

in dissolved oxygen as well as temperature and salinity that add to environmental stresses from the tides. For representative days in June and February, diurnal curves of these parameters, along with solar radiation, are shown in Fig. 5 and 6 with the embayment closed to tidal activity. Similar curves were collected at least once each month throughout the year to give a measure of total system apparent production and respiration. The magnitude of change increased dramatically between February and March, remained high throughout the summer, then declined again during the following November and December. During the winter the diurnal range was between 1 and 2 mg O_2 l^{-1}; comparable summer values were five times greater. Similar large diurnal oxygen changes have been described by Nichol (1935) for salt-marsh pools in England. Even with the culvert closed on warm summer nights followed by early morning fog, the cove seldom became anoxic, though oxygen values often dropped to less than 1 mg l^{-1}. Throughout the year oxygen curves tended to lag solar radiation by about 1 hr. Rates of apparent production by the entire marsh embayment system are plotted as a function of light energy in Fig. 7. Although the photosynthetic response of

individual plants to increasing light intensity is hyperbolic, a linear relationship provided the best fit to these data for the whole embayment, perhaps because the response of natural communities is complicated by the layering of photosynthetic pigments and the gearing of respiration to temperature and oxygen concentration. The fit is surprisingly good in spite of some scatter and indicates that primary production in the embayment is largely determined by light levels. Seasonal effects of population abundance and composition were less important. Similar findings were reported for sediment algae in the emergent marshes of Georgia by Pomeroy (1959). Data for the two smaller ponds fell within the range of values for the main embayment, indicating that their levels of metabolism were similar.

At night, when measurement of the respiration of the total embayment was possible, rates of oxygen consumption were an exponential function of water temperature (Fig. 8). The consumption of organic matter each night was also closely coupled with the production during the previous day, but with night respiration most often exceeding apparent production during the day. Fourteen points show a production-to-respiration ratio of less than 1, while only

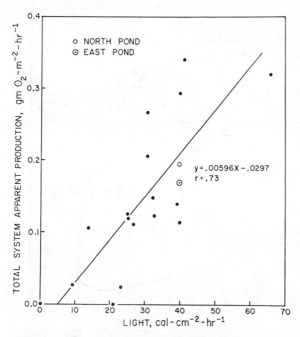

FIG. 7. Rates of apparent production during the day by the total embayment system as a function of solar radiation.

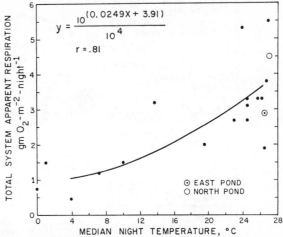

FIG. 8. Rates of night respiration by the total embayment system as a function of water temperature.

5 days had a *P*-to-*R* ratio greater than 1 (Fig. 9) indicating some net production over 24 hr. Values for total system apparent production during the day were less than 1 g O_2 m^{-2} day^{-1} October through February, then rose to a summer level of 2–3 g O_2 m^{-2} day^{-1} that remained relatively stable. Gross production ranged over the year from almost zero in midwinter to summer values of 5–6 g O_2 m^{-2} day^{-1}. Rates of calculated gross production reached 0.6 g O_2 m^{-2} hr^{-1}, but were usually between 0.2 and 0.45 g O_2 m^{-2} hr^{-1}. The seasonal pattern of total system metabolism, including gross production, apparent production, and respiration is summarized with data on light energy and water temperature in Fig. 10. The efficiency of gross production for the entire embayment community averaged 0.80% of visible solar radiation, with a low of 0.12% in November and a high of 1.56% in August. These values place the marsh embayment between very low production systems such as deserts, tundra, and subtropical ocean waters where efficiencies may be about 0.1%, and fertilized systems with high production and large storages, such as algal cultures, sugar cane, and water hyacinths, where gross efficiency may be 4% or more (Odum 1971). If Teal's (1962) value of 70% of *Spartina* gross production used in plant respiration is applied to the grass production at Bissel Cove, then a gross production efficiency of 0.8%, the same as for the embayment, results.

Plankton

Phytoplankton photosynthesis and respiration by phytoplankton and zooplankton in the embayment accounted for a relatively constant 10–20% of the total metabolism. Their importance increased during the winter when low temperature slowed sediment metabolism and the water entering the marsh from Narragansett Bay contained large plankton populations from the winter bloom characteristic of this region (Smayda 1957, Pratt 1965). Apparent production by the plankton community ranged from

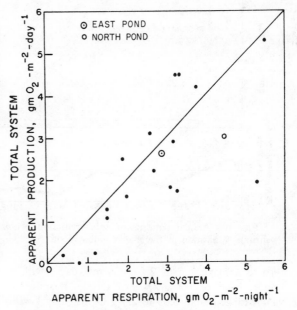

FIG. 9. The coupling of night respiration to production during the previous day in the marsh embayment. The diagonal line represents a *P*-to-*R* ratio of 1.

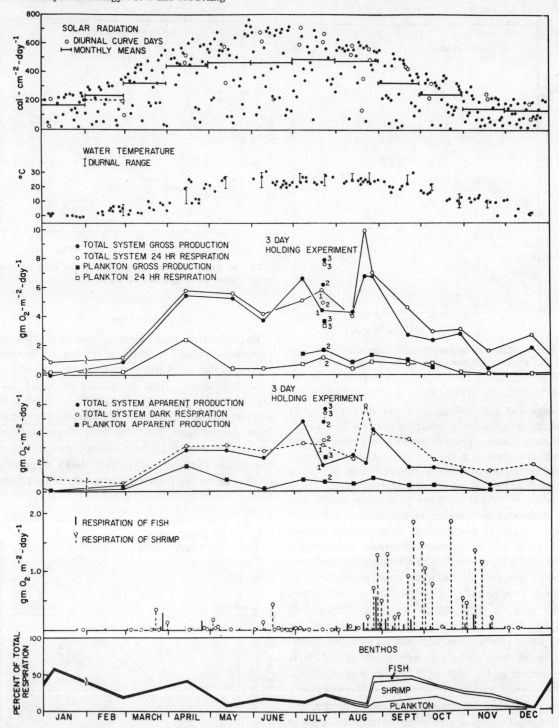

FIG. 10. Annual patterns of solar radiation, water temperature, total system metabolism, and fish and shrimp respiration in the marsh embayment.

zero to a maximum of 2.2 g O_2 m^{-2} day^{-1} with a yearly mean of 0.8 g O_2 m^{-2} day^{-1}. Respiration values ranged from 0.05 to 2.4 g O_2 m^{-2} day^{-1} with a mean of 0.6. Gross production and respiration by the plankton are shown as functions of light energy

and water temperature in Fig. 11. The substantial amount of scatter in the data reflects the erratic nature of plankton abundance in the marsh and indicates the importance of factors other than light and temperature in regulating plankton metabolism.

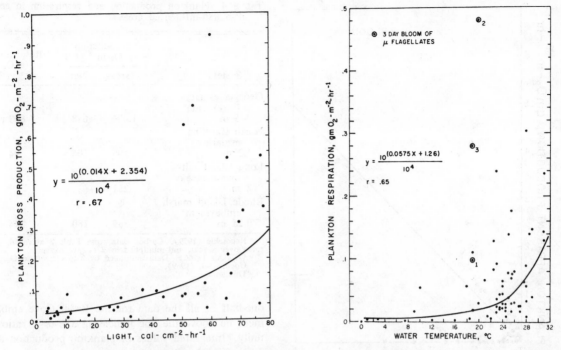

FIG. 11. Plankton gross production (left) and respiration (right) as functions of solar radiation and water temperature. The scatter reflects large and rapid changes in the plankton populations throughout the year, as well as variation in factors such as nutrient levels and species composition.

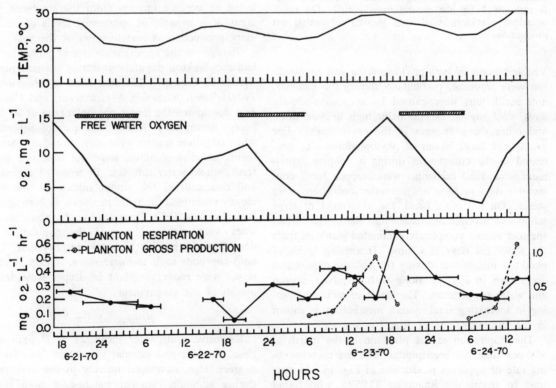

FIG. 12. Three-day holding experiment when tidal exchange was eliminated. Dissolved oxygen patterns remained relatively stable while water temperatures and plankton metabolism increased slightly. The bars on plankton metabolism indicate the period of measurement.

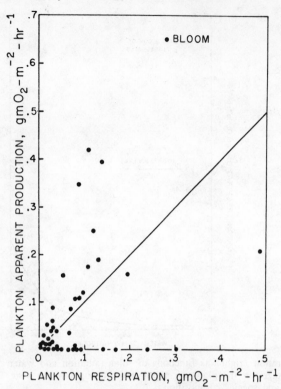

FIG. 13. The coupling of plankton respiration and production in light-dark bottle measurements in the marsh embayment. The diagonal line represents a P-to-R ratio of 1 for the measurement period. On many occasions plankton populations showed substantial net production.

TABLE 4. Plankton production and respiration in some salt-marsh-influenced areas

System	Production (g O_2 m^{-2} yr^{-1})		Respiration (g O_2 m^{-2} yr^{-1})
	Gross	Net	
Georgia estuary-Forks data[a]			
3–5 m	1,426	648	803
North Carolina estuaries[b]			
1.2 m	199	105	94
Long Island salt-marsh creeks[c]			
2 m	251		
Rhode Island marsh embayment[d]			
0.25 m	389	180	244

[a] Ragotzkie (1959). Carbon data from Table 2 averaged, prorated over a year, and multiplied by 2 to give oxygen.
[b] Williams (1966). Data multiplied by 2 to convert to O₂.
[c] Udell et al. (1969).
[d] This study.

one-half of all the data for the Bissel Cove embayment lie above the line indicating a P-to-R ratio of unity. Integrated values of plankton production and respiration in this system over the year are given in Table 4 for comparison with levels of plankton metabolism in other marsh areas. Our results indicate that while the plankton community of the salt marsh area appears to be a heterotrophic system living on organic imports from the emergent grass areas, it is capable of substantial levels of net primary production at certain times of the year.

Changes in the approximate size of phytoplankton and zooplankton populations during the summer are shown in Fig. 14. Diatoms such as *Asterionella*, *Thalassiosira*, *Nitzschia*, *Skeletonema*, and *Chaetocerus* dominated the larger phytoplankton at all times except during July, August, and part of September, when flagellate species were most abundant. Smaller forms of phytoplankton were not observed in the fresh, whole water samples. In terms of abundance and composition, the phytoplankton of the marsh closely resembled the patterns shown in Narragansett Bay (Smayda 1957, Pratt 1965). Zooplankton densities were usually low, and the population was dominated by benthic harpacticoid copepods. Calanoid copepods such as *Eurytemora*, *Acartia*, and *Oithona* were more abundant in deeper water at the mouth of the embayment.

Benthic plants

Extensive underwater meadows of *Ruppia maritima*, a submerged vascular plant, and *Ulva lactuca*, a green alga, developed quickly in the embayment during summer. Smaller patches of both species were present intermittently at other times of the year. Occasional blooms of *Enteromorpha* appeared

Visible blooms of both phytoplankton and zooplankton were frequent, particularly during the summer, and oscillations were caused by successive cloudy days, cool nights, exceptionally high or low tides, and other sharp changes in the environment. For example, a large bloom of phytoplankton was generated in the embayment during a holding experiment when tidal exchanges were stopped for 3 consecutive days in June when weather conditions were stable. On the other hand, the movement of large schools of juvenile menhaden into the marsh in late summer almost completely eliminated plankton from the water for days at a time. Increasing levels of plankton metabolism during the holding experiment are shown in Fig. 12 along with dissolved oxygen and water temperature. The effect of reduced flushing in increasing total system metabolism is shown in the summary data of Fig. 10.

The respiration of the plankton in the marsh at any one time has been plotted with the corresponding rate of apparent production in Fig. 13. In contrast to studies by Ragotzkie (1959), who found P-to-R ratios consistently less than 1 for the metabolism of plankton in Georgia marsh waters, over

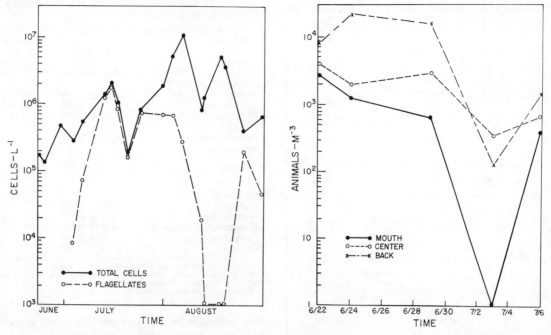

FIG. 14. Single-count estimates of phytoplankton and zooplankton during summer in the marsh embayment. Rapid short-term fluctuations in both populations occurred throughout the year.

during the spring and summer. The total density of benthic plants shown in Fig. 3 was highest in the rich organic sediments near the back of the embayment. The distribution of *Ruppia*, a brackish water species, reflected the input of a small stream in this area of the marsh. In June, a period of rapid growth, densities in the *Ruppia* bed ranged from 48 to 550 g dry weight m^{-2}, then peaked in July at 180–1,460 g dry weight m^{-2}, and declined in August. At the end of July, values of 80–160 g dry weight of dead *Ruppia* were found being carried out of the embayment on ebb tides. The growth of *Ulva* and *Enteromorpha* was confined to areas near the embayment mouth with higher and more constant salinities. At their peak, densities in the algae patches reached 260–600 g dry weight m^{-2}. Although production of submerged plants in the embayment was usually lower than that shown by the grasses on the emergent marsh, the maximum biomass of *Ruppia* per unit area was almost twice that of the fringing tall *Spartina*. The great abundance of these plants provided a substrate for a rich and diverse assemblage of epifauna and may serve as an important source of detritus for the estuary.

Photosynthetic rates in the plant beds were very high because of the large biomass and high light intensity that penetrated the shallow water. The results of laboratory measurements of apparent photosynthesis and respiration at varying levels of light and temperature for *Ruppia* and *Ulva* from the

marsh are shown in Fig. 15. *Ruppia* showed higher rates for each process, but it appears that under field conditions both species quickly become light saturated. Several field measurements in the plant beds under plastic domes showed rates of production and respiration that were very close to those found in the laboratory.

Sediments

The bottom of the Bissel Cove embayment consisted of very soft sediments containing large amounts of detrital organic matter. The organic content of sediment cores ranged from about 4% to over 50%. The highest values were found 3–5 cm below the surface and in the back of the embayment near the entrance of the small freshwater stream (Fig. 3). Lowest values for sediment organics were found near the mouth of the embayment, where fill from the bridge added a large amount of sand, and faster water movement from tides coming over the sill kept fine grain sediments from being deposited. Although submerged plants probably made a contribution to the organic detrital storage, examination of the sediments revealed large amounts of recognizable *Spartina* rhizomes distributed in small pieces throughout the cores. The standing crop of dead *Spartina alterniflora* detritus lying on the bottom of the embayment ranged from 100 to 900 g dry weight m^{-2} with an average of about 200 ± 86 g m^{-2}. Again, most of this detrital material appeared to be

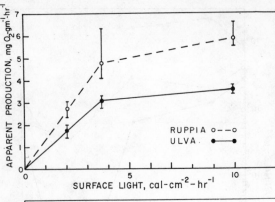

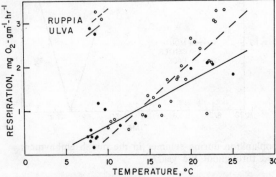

FIG. 15. Photosynthetic and respiratory rates for *Ruppia* and *Ulva*, the dominant macrophytes in the embayment. The bars in the top figure are ranges for three or more determinations. The bottom lines were fit by least-squares linear regression.

derived from grass roots and rhizomes, with little contribution from the leaves. No *S. patens* was found in the water or in the embayment sediment.

Measurements of oxygen uptake by the sediment were made in the spring and summer across a range of temperatures. Respiration rates near the middle of the embayment increased from 0.04 g O_2 m^{-2} hr^{-1} at 5°–10° C to 0.09 g O_2 m^{-2} hr^{-1} at 15°–20° C, whereas rates at the back of the embayment ranged from 0.08 g O_2 m^{-2} hr^{-1} at the lower temperatures to 0.13 g O_2 m^{-2} hr^{-1} at the higher. Maximum rates of uptake were found on four cores taken from the back region near the stream entrance, where the mean rate was 0.32 g O_2 m^{-2} hr^{-1}. The regression equation for spring and summer sediment respiration for all cores was:

$$R \text{ (g O}_2 \text{ m}^{-2} \text{ hr}^{-1}) = (1.57T + 9.7)\,10^{-3} \quad (r = 0.92),$$

where T = water temperature (°C).

This regression describes a line with a larger intercept and shallower slope than that derived by Hargrave (1969) in an extensive survey of benthic community respiration from a variety of marine and freshwater systems. The rates of oxygen uptake by the embayment sediments were similar to values for

TABLE 5. Number and biomass of frequent infauna in the Bissel Cove embayment[a]

Infauna and location	Number (m^{-2})	Biomass (g dry weight m^{-2})
Embayment mouth[b]		
Neanthes	63 ± 79	
Streblospio	79 ± 101	
Polydora	153 ± 223	
Capitella	394 ± 375	
Corophium	2,028 ± 2,765	
Polychaete	176 ± 207	
Total		5.9 ± 3.3
Back marsh[c]		
Capitella	788 ± 1,607	1.1
Nematodes	4.9 × 10^6	0.6
Harpacticoid copepods	4.6 × 10^4	0.05
Ostracods	2.7 × 10^4	
Ciliates	1.8 × 10^5	

[a] Mean ±1 standard deviation: N = 15 or more.
[b] Occasional species: *Mercenaria, Modiolus, Mya, Nereis, Scolecolepides, Euplana, Nematostella*, chironomid.
[c] Occasional species: polychaete.

other sediments in the emergent salt marsh reported by Duff and Teal (1965). These higher rates of metabolism may reflect the rich input of detrital fuels to salt-marsh sediments, in comparison with the plankton rain of sinking cells and fecal pellets that must feed the bottoms of lakes, rivers, and estuaries.

Infauna and epifauna

The soft sediments of the embayment provided a poor substrate for macrofauna, since no clams, large worms, or other animals were found except in the firm bottom areas near the entrance. The soft muds in the rest of the embayment were dominated by a small worm of the genus *Capitella* and a variety of nematodes, ciliates, ostracods, and harpacticoid copepods (Table 5). The most conspicuous group within the meiobenthos were the nematodes, whose numbers reached 10^7 m^{-2} in some areas. Estimates of nematode populations were similar to values reported by Wieser and Kanwisher (1961) for the emergent marsh sediments of Woods Hole, and by Teal and Wieser (1966) for Sapelo Island, except that maximum values here were five times greater than the maximum in Massachusetts, while the weight of organisms was four or five times smaller. Nematodes in the embayment muds here appear to be substantially smaller than those in the Woods Hole marsh. The great abundance of these organisms in marsh sediment-detritus systems in comparison with other areas is apparent from the summary of nematode population levels in various communities in Table 6.

Larger epifauna included abundant amphipods living in the detritus mat on the surface of the

TABLE 6. Number and biomass of nematodes in some intertidal and subtidal environments[a]

System	Number (m^{-2})	Biomass[b] (g dry weight m^{-2})	Source
Sandy Beach, Denmark	0.10 – 2.5 (10^5)		Fenchel (1969)
Narrow River, R. I.	0.15 – 12.0 (10^5)	3.3 – 11.8	Tietjen (1966)
Off Plymouth, England	0.5 – 1.8 (10^5)	0.23	Mare (1942)
Off Fladen, England	0.75 – 3.0 (10^6)	0.09 – 0.64	McIntyre (1964)
Off Martha's Vineyard, Mass.	0.5 – 6.8 (10^5)	0.04 – 0.51	Wigley and McIntyre (1964)
Buzzards Bay, Mass.	0.15 – 1.8 (10^6)	0.08 – 0.48	Wieser (1960)
Woods Hole marsh, Mass.	1.4 – 2.1 (10^6)	2.4 – 4.6	Wieser and Kanwisher (1961)
Sapelo Island marsh, Georgia	0.98 – 16.3 (10^6)	0.05 – 1.9	Teal and Wieser (1966)
Bissel Cove marsh embayment, R. I.	1.6 – 10.0 (10^6)	0.21 – 1.0	This study

[a] Modified and extended from Lavoie (1970).
[b] Calculated as 25% of wet weight (Wieser 1960, McIntyre 1964).

bottom with densities up to 0.5 g dry weight m^{-2}. Occasional green crabs, *Carcinides maenas*, appeared in the pull-up nets during summer, but were never abundant. Toward the end of summer juvenile blue crabs, *Callinectes sapidus*, began to appear regularly in densities of about 0.6 animals m^{-2}. They reached maximum numbers in mid-September with over 2.5 crabs m^{-2}; carapace widths ranged between 20 and 50 mm. The respiratory rate of juvenile crabs at summer temperature of 20° C averaged 0.95 mg O$_2$ g dry weight^{-1} hr^{-1} (Fig. 16).

Eels became abundant at various times and locations in the embayment during the summer. Maximum densities of five to six large animals per square meter were occasionally found, but values were usually 0.25 m^{-2} or less. Respiratory rates of eels at summer temperatures averaged 0.80 mg O$_2$ g dry weight^{-1} hr^{-1} (Fig. 16). Gut analysis of eels throughout the summer indicated that their diet

consisted largely of the small common mummichog, *Fundulus heteroclitus*, and grass shrimp, *Palaemonetes pugio*. Occasionally other fish and lumps of grass detritus were found. The eels either left the embayment or burrowed into the mud in the fall and did not appear until the following spring. Blue crabs congregated during the winter in deeper holes near the mouth of the embayment, where over 100 animals were lifted with the dip net at one time in January. The crabs remained a part of the community throughout the following summer.

Grass shrimp

The results of mark-and-recapture studies and pull-up net estimates of shrimp population size throughout the year established that from August through November, and again in early spring, grass shrimp were one of the most conspicuous elements of the embayment (Fig. 17). For the entire embayment, mark-and-recapture estimates with 95% confidence limits were:

	Lower limit	Mean	Upper limit
January	50,000	73,000	139,000
March	831,000	923,000	1,009,000
July	11,000	13,000	17,000
October	1,250,000	1,400,000	1,600,000

Coefficients of variation for the pull-up quadrat nets ranged from 0.43 to 1.95, with a yearly mean of 1.03. The highest variability was found during summer when the animals were scarce. When coupled with their great numbers, high shrimp respiration rates in the warm water (Fig. 16) amounted to about 1.9 g O$_2$ m^{-2} day^{-1}, or almost 30% of the total system fall metabolism. The shrimp made a substantial contribution to the excess of respiration over consumption in the embayment at the end of summer (Fig. 10). The low estimates for the summer population do not include the abundant juvenile shrimp which were too small for the nets. For larger, post-juvenile animals the size-frequency distribution throughout the year is shown in Fig. 18.

FIG. 16. Respiratory rates for yearly and seasonally abundant larger animals in the marsh embayment. Mean values are shown ± 1 standard deviation.

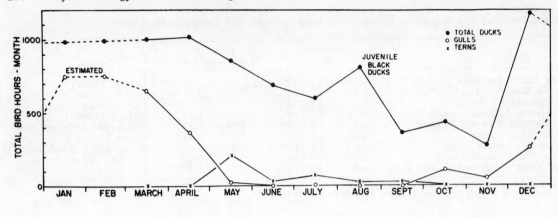

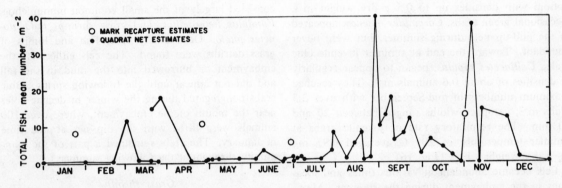

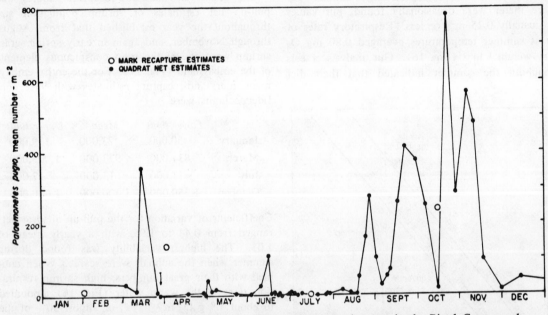

FIG. 17. The abundance of birds, fish, and shrimp throughout the year in the Bissel Cove marsh.

Females averaged 14% larger than males in winter, and up to 30% larger in spring and summer. Rapid growth in both sexes took place from mid-July to mid-October. There was little or no growth from this time through winter until the end of April, when

a second growth spurt, especially evident in the females, began.

The movement of shrimp in and out of the marsh largely reflected the abundance of animals in the embayment. Records of shrimp migration throughout the year are shown as part of Fig. 19. The massive movements of the fall season corresponded to maximum population sizes and spring tides. With the highest tides, up to 40% of the population moved during each cycle. Even during this period, it appeared from tagged animals in the mark-and-recapture study that the population within the embayment was a real entity that moved in and out of the area. Movement in both directions seemed coupled during the fall, but in the rest of the year, movement into the area almost always substantially exceeded emigration. Though some massive mortalities did occur during heavy ice periods in the winter, many of the shrimp moving into the marsh, and most of the animals produced in the embayment itself, probably were consumed by fish and birds in the marsh. There is no evidence to indicate that this marsh area served as an important source of shrimp for the Narragansett Bay estuary.

In laboratory aquaria the grass shrimp appeared to be omnivorous, with a nutritional pattern like that of the similar small estuarine shrimp, *Crangon* (Wilcox 1972). Adults of *Palaemonetes* from the embayment survived well on frozen fish and brine shrimp. In both the laboratory and in the field the shrimp were seen picking pieces of *Ulva*, *Ruppia*, and *Spartina* detritus. This feeding activity was probably directed more toward films of microflora, detritus, and bacteria growing on the plants than it was toward the plant tissue itself. Broad (1957) has shown that larvae of the shrimp lived through metamorphosis when fed a diet of plant and animal materials, but died if only given one or the other.

The interaction of the grass shrimp with marsh detritus is more complex than suggested by a simple feeding relationship. Detailed studies of shrimp ecology at this laboratory by Welsh (1973) indicate that the association of shrimp with detritus accelerates the breakdown and decomposition of the plant material. Scanning electron microscope studies of detritus exposed to the shrimp show structural changes in the grass tissue and the presence of well-developed diatom films growing in emptied cell space consisting only of the cellulose cell walls. Control microcosms with marsh detritus but no shrimp did not show these changes, and water over the control detritus was much lower in nitrogen, phosphorus, dissolved organic carbon, and particulate matter. Excretion and fecal pellets produced by the shrimp, along with their mechanical processing and break-up of detritus, may raise the level of nutrients available for development of bacteria and algae on the nitrogen and phosphorus-poor cellulose substrate of dead grass. Work by Ustach (1969) has shown that small additions of nitrogen and phosphorus can increase the consumption of *Spartina* detritus by

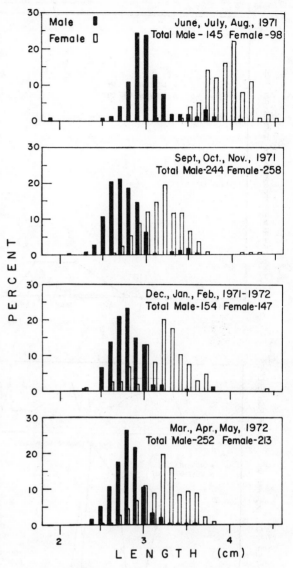

FIG. 18. Seasonal length-frequency distribution of grass shrimp, *Palaemonetes pugio*, in the marsh embayment.

heterotrophs, and Hargrave (1970) has found that changes in the density of a benthic amphipod, perhaps performing a role similar to that of *Palaemonetes*, may accelerate the metabolism of sediment microflora.

Although the shrimp population in the embayment consisted almost entirely of *Palaemonetes pugio*, other small shrimp, such as species of *Crangon* and other *Palaemonetes*, are often common in shallow coastal water (Table 7). One reason for the large, single species populations of *P. pugio* in marsh waters appears to be their tolerance for very low oxygen concentrations that other shrimp cannot survive (Welsh 1973). In waters with higher exchange rates and lower metabolism, dissolved oxygen levels

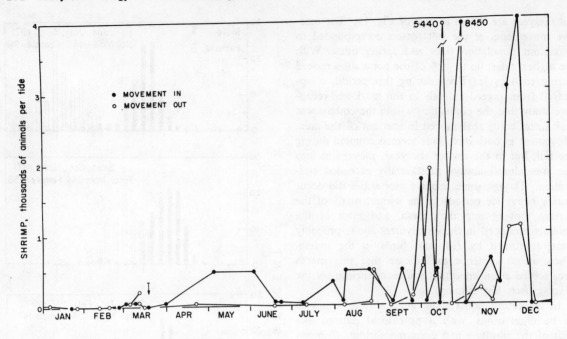

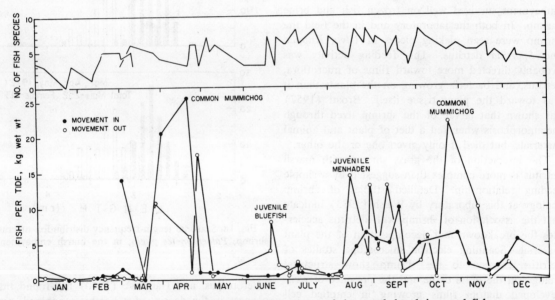

FIG. 19. Seasonal patterns of movement in and out of the marsh by shrimp and fish.

remain higher and other species may successfully compete. Gut analysis in net hauls indicated that shrimp served as a favorite food for fishes with a standard length of 60 mm or more while they were in the embayment, but fish in this size range generally moved out of the embayment during low tides. Since high tide periods averaged only about 6.5 hr each day, and about half of these occur during the dark when many fish cease or reduce feeding, the shrimp escaped intense predation over 75% of the time. Under conditions such as those

in the Bissel Cove embayment, where stresses in the system eliminate competition, predation is minimized, and a positive reward loop is maintained through nutrient exchanges between the animals and their detrital food supply, the biomass of shrimp in the natural marsh system may rival or exceed the shrimp farms of aquaculture (Table 7).

Fish

The shallow water of the marsh embayment contained large numbers of small fish, especially during

TABLE 7. Biomass of shrimp in some aquatic ecosystems

System	Biomass (g dry weight m⁻²)
Turtle grass community, Texas[a]	
Early spring maximum	1.75 (*Palaemonetes pugio*)
Spring maximum	2.55 (Penaeid shrimp)
Commercial shrimp culture, Panama[b]	5.5 – 7.3 (Penaeid shrimp)
Saltwater pond farming, U.S.[c]	14.0 (Penaeid shrimp)
Experimental estuarine pond, N.C.[d]	
Control pond, summer	0.3 (*Palaemonetes pugio* and *P. vulgaris*)
fall	0.2
Sewage pond, summer	2.9 (*Palaemonetes pugio*)
fall	4.1
max. value	9.1
Bissel Cove marsh embayment, R.I.[e]	
Summer	0.1
Fall	15.3 (*Palaemonetes pugio*)
Winter	2.7
Spring	9.0

[a] Hoese and Jones (1963).
[b] Smitherman and Moss (1970) in Sick, Andrews, and White (1972).
[c] Lunz (1967).
[d] Beeston (1971).
[e] This study.

the spring and fall. The yearly abundance of fish from the pull-up nets and the mark-recapture studies is very similar to the abundance pattern of the shrimp (Fig. 17). Because of the greater swimming speed and evasive capability of fish, the two techniques for measuring abundance do not agree as well as they did for shrimp, with the pull-up nets yielding an underestimate of population size. Coefficients of variation for fish in the nets were similar to those for shrimp, ranging from 0.11 to 2.01, with a mean for the year of 1.07. Again, variability was lowest with large fall populations. For the whole embayment the Petersen mark-recapture estimates of fish population sizes, with 95% confidence limits, were:

	Lower limit	Mean	Upper limit
January	49,540	57,000	67,900
April	67,300	75,500	85,900
June	33,000	37,700	43,800
October	79,700	93,000	112,000

During the summer large numbers of juvenile common mummichog were present that were not included in the pull-up nets or tag study for that period.

Twenty species of fish were found in the embayment over the year: seven species in winter, nine in spring, 12 in summer, and 14 in the fall. At all times the common mummichog was the dominant species. Only in the summer and fall, when striped mummichogs and sheepshead minnows made up 20% of the total fish, and in August, when large schools of juvenile menhaden moved into the embayment, were other species conspicuous. At the time of their maximum abundance over 40,000 juvenile menhaden were present in the marsh, as shown by

TABLE 8. Relative abundance of some fish species at each season in the Bissel Cove marsh embayment

Species	Winter	Spring	Summer	Fall
Common mummichog	1.000	1.000	1.000	1.000
Striped mummichog	0.026	0.149	0.247	0.051
Silverside	0.004	–	0.003	0.003
Sheepshead	–	0.021	0.002	0.230
Threespine stickleback	–	0.028	–	–
Mullet			0.001	
Menhaden	–	–	–	0.004
Winter flounder	–	–	–	0.001
Alewife	–	–	–	0.001

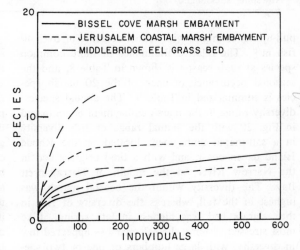

FIG. 20. Range in diversity of fish species from the marsh embayment, an open coastal marsh, and a tidal river eelgrass bed (after Sanders (1968) rarefaction method).

TABLE 9. Seasonal occurrence of fish species in the Bissel Cove marsh embayment

Species	Month											
	1	2	3	4	5	6	7	8	9	10	11	12
Fundulus heteroclitus (L.) Common mummichog	×	×	×	×	×	×	×	×	×	×	×	×
Fundulus majalis (Walbaum) Striped mummichog	×	×	×	×	×	×	×	×	×	×	×	×
Menidia menidia (L.) Silverside	×	×	×	×	×	×	×	×	×	×	×	×
Cyprinodon variegatus (Lacepede) Sheepshead	×	×	×	×	×	×	×	×	×	×	×	×
Anguilla rostrata (LeSueur) American eel			×	×	×	×	×	×	×	×		
Pomlobus pseudoharengus (Wilson) Alewife			×	×	×	×	×	×	×	×		
Mugil cephalus (L.) Mullet						×	×	×	×	×	×	
Brevoortia tyrannus (Latrobe) Atlantic menhaden						×	×	×	×	×		
Roccus americanus (Gmelin) White perch						×	×	×	×		×	
Pomatomus saltatrix (L.) Blue fish						×	×	×	×	×	×	
Tautoga onitis (L.) Tautog						×	×	×				
Clupea harengus (L.) Sea herring			×	×	×							
Stenotomus chrysops (L.) Scup									×	×	×	
Pseudopleuronectes americanus (Walbaum) Winter flounder									×	×	×	
Scophthalmus aquosus (Mitchill) Window pane flounder									×	×	×	
Hyporhamphus unifasciatus (Ranzani) Halfbreak									×	×	×	
Lucania parva (L.) Rainwater fish						×	×	×				
Gasterosteus aculeatus (L.) Threespine stickleback	×	×	×	×	×							×
Apeltes quadracus (Mitchill) Fourspine stickleback	×	×	×	×	×							×
Pungitius pungitius (L.) Ninespine stickleback	×	×										×

pull-up-net estimates with school densities of over 40 fish m⁻². The relative abundance of more common species at each season is shown in Table 8, and the seasonal occurrence of each of the 20 marsh species is summarized in Table 9. The annual species-diversity range in the marsh embayment is compared in Fig. 20 with the annual range of fish diversity in a salt-marsh embayment located on the Rhode Island ocean coast, and with a tidal eelgrass bed in the Narrow River, at the mouth of Narragansett Bay. The diversity within the coastal marsh was highest in the fall, whereas the diversity of fish in the eelgrass bed was highest in late spring. Like most stressed systems, the salt marshes appeared low in diversity, with large numbers of one or two species. These results suggest that the marsh is not used directly by large numbers of outside species as a nursery or feeding area.

Large common mummies were most abundant in spring, when spawning began in late April and continued through June. Size-frequency distributions for the common mummichog at each season indicate that a greater proportion of small fish enter the population in summer (Fig. 21). During winter the young-of-the-year fish were the most dominant group. Gut analysis of the small common mummichog indicated that their rapid growth was supported by a diet that consisted of harpacticoid copepods, amphipods, benthic diatoms, and unidentified bottom detritus, in order of abundance. Miscellaneous items, including fish eggs, worms, isopods, and ostracods, also appeared in some samples. Larger fish of both *Fundulus* species fed largely on the shrimp, *Palaemonetes*, juvenile fish of their own and other species, and detritus. None of the marked shrimp or fish were ever found in the gut analysis. Bio-

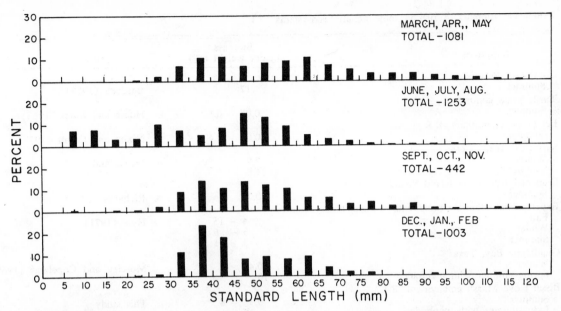

Fig. 21. Seasonal length-frequency distribution of common mummichog, *Fundulus heteroclitus*, in the marsh embayment.

chemical studies of feeding in both *Fundulus* species in the Bissel Cove embayment by Jeffries (1972) indicated that their diets consisted of five parts detritus to one part marine invertebrates. Eels and white perch were voracious feeders on shrimp as well as juveniles and adults of all the other fish species. As mentioned earlier, plankton feeding by the juvenile menhaden was so intense that water in the embayment became almost completely clear. Most, if not all, of their consumption was probably used in maintenance metabolism. Their respiratory rate (Fig. 16) was higher than that of other marsh animals, and their guts were almost always empty. The menhaden did not appear to grow at all during their stay in the marsh, and the heat of combustion for the Bissel Cove juvenile menhaden tissue (Table 14) is significantly lower than the 5.1 Kcal g^{-1} reported by Thayer et al. (1973) for postlarval and adult menhaden in North Carolina. With such poor nutritional conditions, the menhaden probably remained in the embayment only to escape intense predation by large blue fish that were abundant in the bay.

Respiration rates as a function of water temperature for the two species of mummies are shown in Fig. 16. During the fall, with menhaden present, total fish respiration in the embayment was about 0.05 g O_2 m^{-2} day^{-1}, with a more representative fall value of 0.01 g O_2 m^{-2} day^{-1}. Values were much smaller during the rest of the year. At its peak, fish respiration never exceeded 8% of the total system metabolism in the embayment (Fig. 10). Length-weight regressions developed for calculating fish biomass when coupled with wet-dry weight conversions are summarized in Table 10 for the major species in the embayment.

Total fish movement in and out of the marsh (Fig. 19) was approximately equal over the year. During the large migrations of common mummichog imbalances did occur when great numbers of fish moved in with the spring, then emigrated in the fall. This activity by the very abundant common mummichog was responsible for the strongly bimodal pattern of fish activity compared with that of the shrimp. Other species, including striped mummi-

TABLE 10. Length-weight regressions for some salt-marsh embayment fish

Species	N	Regression[a]	r
Common mummichog[b]	296	$\log W = (0.024L + 0.313)(10^{-4})$	0.97
Striped mummichog	59	$\log W = (2.76 \log L - 1.36)(10^{-3})$	0.97
Menhaden	85	$\log W = (0.0112L + 4.01)(10^{-4})$	0.73
American eel	38	$\log W = (2.94L + 0.326)(10^{-3})$	0.96
Sheepshead	22	$\log W = (1.2 \log L(10^{-6}))$	0.99
Juvenile blue fish	104	$\log W = (0.0403L + 1.44)(10^{-3})$	0.81

[a] Where W = wet weight (g) and L = standard length (mm).
[b] Dry weight (g) = 0.21 W + 0.074.

TABLE 11. Biomass of fish in some aquatic ecosystems

System	Biomass (g dry weight m^{-2})	Source
Coral reef Bermuda		
Summer	12	Bardach (1959)
Turtle grass community, Texas		
Summer	0.08 – 0.5	Hoese and Jones (1963)
Eel Grass community, R.I.		
Summer	0.04 – 0.4	Nixon and Oviatt (1972)
Surf fish, Texas Coast		
Winter	2.9	McFarland (1963)
Summer	11.6	
Demersal fish, Long Island Sound		
Summer	0.76	Richards (1963)
Experimental estuarine pond with sewage, N.C.		
Fall	4 – 15	Hyle (1971)
Winter	3 – 6	
Summer	3 – 11	
Guadalupe Bay, Texas		
Winter	0.4	Moseley and Copeland (1969)
Spring-summer with menhaden	10	
Bissel Cove Marsh embayment, R.I.		
Summer	0.3 – 8	This study
Late summer with menhaden	28	
Fall	7 – 14	
Winter	5	

chogs, silversides, sticklebacks, eels, and sheepshead minnows were much less abundant (Table 8), and their migration patterns showed small transient peaks at irregular intervals throughout the year. These species also remained in the embayment at all seasons (Table 9). Fish such as the menhaden and juvenile blue fish that showed a strong seasonality also showed marked migration peaks (Fig. 19). Net movements of these less abundant species in one direction or the other were not obvious in the data, but it was not possible from the number of samples available to make a reliable budget because of their small numbers and irregular patterns of movement. Fish movement could not be related in a direct way to water temperature or to tides. Additional attempts to resolve movement in terms of temperature changes and temperature-tide interactions were also unrewarding. Although the biomass of small fish produced and maintained in the marsh embayment compared favorably with that of other systems, (Table 11), the lack of a large clear pulse of net export during the year suggests that the marsh embayment is not supplying great numbers of animals to feed larger fish in the bay.

Birds

A 10-month study of birds in the Bissel Cove marsh in cooperation with Lucid (1971) recorded 27 species during the year, with a high of 16 in July and a low of 5 in March (Table 12). Of this total perhaps only 20 are marsh or shorebirds; the remaining seven are best described as occasional visitors from bordering woodlands and fields.

The seasonal abundance of ducks, gulls, and terns, the three major types of birds using the marsh, is shown with the seasonal patterns of fish and shrimp in Fig. 17. Black ducks and mallards were the heaviest users of the area throughout the year, with mallards most active in spring and summer. Use by gulls and terns also showed a sharp seasonal separation. In winter up to 30 herring gulls at a time were often seen "fishing" through small holes in the ice, but their numbers declined in early spring as the common terns arrived and began wheeling over the marsh to dive for fish. Later in the summer the least tern replaced the common and stayed on into the fall when the herring gulls returned.

No correlation was apparent between use of the marsh by birds and the abundance of plant or animal food in the embayment. The number of birds in the marsh was lowest when fish and shrimp were most abundant (Fig. 17). When present in the embayment, each species devoted different amounts of time that varied with season to feeding activities (Table 13). In general, smaller birds, such as terns, spent a greater proportion of their time in these activities, though food gathering increased to almost 80% of residence time for black ducks in the 2-month period before the hatching of new ducklings in August. Field observation showed that both ducks fed largely on *Ruppia* and *Ulva*, though mallards also were adept at capture of common mummichogs and shrimp. Terns and gulls fed entirely on fish. Laboratory experiments by Lucid (1971) indicated a daily maintenance ration for adult herring gulls of 50.9 ± 3.6 g dry weight of fish per bird, or

TABLE 12. Seasonal occurrence of bird species in the Bissel Cove marsh embayment[a]

Species	Month									
	3	4	5	6	7	8	9	10	11	12
Anas rubripes Black duck	×	×	×	×	×	×	×	×	×	×
Anas platyrhynchos Mallard	×	×	×	×	×	×				
Larus argentatus Herring gull	×	×	×			×		×	×	×
Larus marinus Great black-backed gull		×								
Larus delawarensis Ring-billed gull		×						×	×	×
Larus atricilla Laughing gull						×				
Sterna hirundo Common tern			×	×						
Sterna albifrons Least tern				×	×	×				
Podilymbus podiceps Pied-billed grebe							×	×		
Falco sparverius Sparrow hawk		×								
Ardea herodias Great blue heron		×	×						×	
Leucophoyx thula Snowy egret				×	×	×				
Butorides virescens Green heron			×	×	×	×	×			
Nycticorax nycticorax Black-crowned night heron						×	×			
Cygnus olor Mute swan					×	×				
Rallus longirostris Clapper rail								×	×	×
Tringa solitaris Solitary sandpiper							×			
Totanus flavipes Lesser yellowlegs			×	×		×	×	×		
Megaceryle alcyon Belted kingfisher	×			×	×	×	×	×	×	×
Tyrannus tyrannus Eastern kingbird				×	×					
Sayornis phoebe Phoebe					×					
Hirundo rustica Barn swallow				×	×					
Corvus brachyrhynchos Crow			×							×
Turdus migratorius Robin			×							
Sturnis vulgaris Starling			×	×	×					
Agelaius phoeniceus Red-wing			×		×					
Quiscalus versicolor Bronzed grackle	×		×		×					

[a] Modified and expanded from Lucid (1971).

17.7% of their dry body weight. Excreta produced amounted to 17.1 ± 4 g dry weight per bird, giving a weight-based assimilation efficiency of 66.4%. By using the measured heats of combustion shown in Table 14, these same data gave an energy-based efficiency of 86%. Even during their maximum abundance, the herring gulls consumed less than 0.5% of the standing crop of fish. Judging from the large amounts of excreta that accumulated at times on winter ice, however, they may be more important

TABLE 13. Percentage of residence time spent in feeding by dominant birds in the Bissel Cove marsh embayment

Species	Month									
	M	A	M	J	J	A	S	O	N	D
Black duck	33	45	—	79	78	6	4	62	0	0
Mallard duck	37	24	25	16	—	—	—	—	—	—
Herring gull	15	13	—	—	—	—	—	25	22	56
Ring-billed gull	—	—	—	—	—	—	—	—	13	28
Common tern	—	—	49	—	—	—	—	—	—	—
Least tern	—	—	—	100	97	75	—	—	—	—

in contributing to algal and microbial dynamics by increasing the supply of available nutrients, a role attributed to gulls living around rock tide pools on the Baltic by Ganning and Wulff (1969).

COMMUNITY ENERGY FLOW

The salt marshes of New England are marked by sharp seasonal changes and short-term swings of temperature, light, and salinity that make the marsh a different place from week to week as species enter and leave throughout the year. A phenological summary of some important changes in the Bissel Cove marsh is given in Fig. 22. Energy-flow diagrams for the marsh system on representative days in summer and winter (Fig. 23) have been prepared from the data presented in previous sections, along with the bomb-calorimetry values for marsh organ-

TABLE 14. Heats of combustion of some component species in the Bissel Cove salt marsh

Material	Kcal g dry weight^{-1a}
Plants	
Ruppia maritima	3.24 ± 0.05
Ulva lactuca[b]	1.61
Spartina alterniflora[b]	
Tall–medium	3.22
Short	2.57
Spartina patens[b]	3.08
Distichlis spicata[b]	2.96
Submerged macrodetritus	2.49 ± 0.03
Animals	
Palaemonetes pugio (mixed adults)	4.61 ± 0.03
Fundulus heteroclitus (adult)	
Female with eggs	4.55 ± 0.15
Female without eggs	4.59 ± 0.10
Male in breeding color	4.31 ± 0.10
Fundulus majalis (adult)[c]	
Brevoortia tyrannus (juvenile)	4.37 ± 0.06
Callinectes sapidus (juvenile)	3.07 ± 0.25
Anguilla rostrata[d]	6.4
Mixed amphipods	3.69 ± 0.28
Larus argentatus[e]	5.36 ± 0.05
Gull excretion	2.06 ± 0.18

[a] Mean ± 1 standard deviation, this study.
[b] Udell et al. (1969).
[c] Thayer et al. (1970).
[d] Hunter (1972).
[e] Lucid (1971).

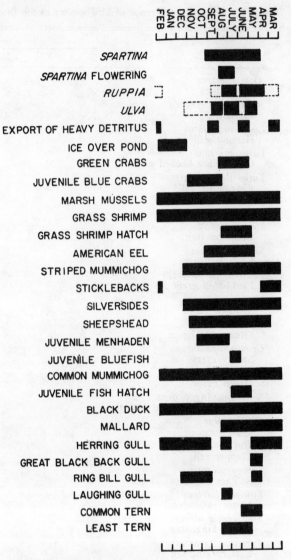

FIG. 22. Phenological summary showing seasonal changes of species composition in the Bissel Cove marsh during the study year.

isms in Table 14. While Teal's (1962) diagram of energy flow at Sapelo Island, Georgia, emphasized the emergent marsh, most of the effort here has been directed toward expanding details of the energy flow in the marsh creeks and embayments that are intimately coupled with the tall stands of grass and serve as tidal pathways linking the emergent marsh with larger estuaries and offshore waters. The diagrams emphasize the great importance of detrital food chains and large sedimentary organic storage in the embayment that agree with earlier findings by the Georgia group for the emergent marsh. They also indicate that large standing crops of plants and animals may be supported in the marsh-embayment complex, and that high levels of primary production

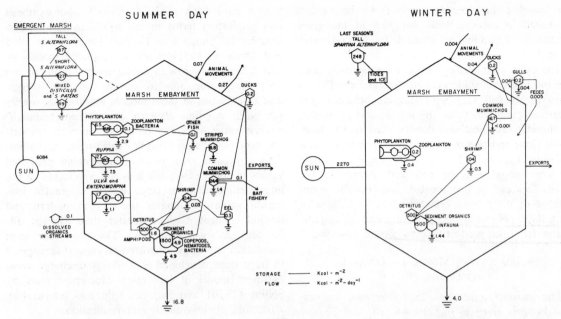

Fig. 23. Energy-flow diagrams for composite winter and summer days in the Bissel Cove marsh. Values were calculated with biomass estimates from the study, the heat of combustion measurements in Table 16, and an oxycalorific constant of 4 Kcal gm⁻¹ O₂ (Odum 1971). Symbols follow Odum (1967, 1971, 1972).

may result over short periods from submerged vascular plants, attached algae, and phytoplankton in the embayment. When metabolism, population sizes, and migration (Fig. 10, 17, and 19) are analyzed for the whole year, however, the marsh embayment emerges as a consumer system that must depend on the import of organic matter fixed on the emergent marsh by grasses and sediment algae. This result placed the marsh embayment with other systems, such as some highly polluted marine bays (Odum et al. 1963) and woodland streams with leaf litter (Hall 1970), where large inputs of organic matter are received and respiration exceeds photosynthesis.

An ecological energy budget for the marsh embayment over an annual cycle is given in Table 15. The excess of consumption over production of about 1.4×10^6 Kcal yr⁻¹ indicates that the embayment must depend on organic inputs to supplement production within the system. A possible second alternative, in which the system could be metabolizing past organic storages, is unlikely in an environment where low oxygen may limit respiration and organic matter is accumulating in the sediments. The value used for total organic input to the embayment is largely a function of the estimate of detritus entering from the emergent marsh. Measurements of the ice-shear zone in the Bissel Cove marsh indicated that only the fringing band of tall or creek-bank *Spartina* was carried into the water, at least as large pieces of grass. The budget in Table 15 has been calcu-

lated assuming an available grass input equal to the production of tall *S. alterniflora* only (Table 1), or 23% of the total emergent grass production. Estimates by Teal (1962) suggest that as much as 45% of the grass may be exported to the estuary via the tidal creeks and embayments. If this value is used, the excess organic matter available for storage in the embayment and export to the estuary would increase to 470 Kcal m⁻² yr⁻¹, or 4.3×10^6 Kcal yr⁻¹ from the entire area. The excess consumption of material in the embayment system amounted to about 14% of the total emergent marsh produc-

TABLE 15. Annual energy budget for the Bissel Cove marsh embayment

Item	Kcal m⁻² yr⁻¹	Kcal yr⁻¹
Production of organic matter within the embayment[a]	9.6×10^3	63.4×10^6
Consumption of organic matter within the embayment[a]	9.8×10^3	64.8×10^6
Excess of consumption over production	2.0×10^2	1.4×10^6
Imports of organic matter from streams[a]	15	1.0×10^5
from net immigration of fish and shrimp[b]	3.5	2.3×10^4
from emergent marsh[c]	2.4×10^2	1.6×10^6
Total imports	2.6×10^2	1.7×10^6
Excess organic matter available for storage and export	60	3.0×10^5

[a] Oxycalorific equivalent of 4 Kcal g⁻¹.
[b] From migration data and calorimetry (3 Kcal g⁻¹ for shrimp and 5 Kcal g⁻¹ for fish).
[c] Assuming input from 1-m-wide band of tall grass around the perimeter plus total area of the islands and 3.3 Kcal g⁻¹ (Udell et al. 1969).

tion, so that the effective transport to the estuary amounted to between 10% and 30% of the grass production, depending on which of the two estimates is used.

The exact relationships between emergent marshes and the interconnecting networks of tidal creeks and embayments will vary with flushing behavior, the shoreline development of the marsh, and a number of climatic and locational factors. The results from Bissel Cove indicate, however, that even in the less extensive marshes of New England, the development of large populations of fish and shrimp in the marsh area can be documented, and that the maintenance of the system necessary for the culture of these large popualtions depends on inputs of organic matter from the productive meadows of *Spartina*.

A SIMULATION MODEL OF DIURNAL DISSOLVED OXYGEN[2]

The seasonal summaries, flow diagrams, and annual budgets given in Fig. 10, 16, 18, and 22 and in Tables 1 and 15 help to synthesize many measurements of the diverse parts of the Bissel Cove marsh. It is more difficult, however, to give an impression of the ways in which these parts combine dynamically through time to produce characteristic patterns for the whole system; or, further, to determine if measurements of any one part are reasonable within the context of the larger system. In other studies, where the community of the emergent marsh was emphasized, this problem has led to the development of a compartment model for *Juncus* production in North Carolina (Williams and Murdoch 1972) and to a simulation of phosphorus flux in Georgia (Pomeroy et al. 1972). For Bissel Cove, where most of our effort was directed toward the marsh embayment, it was possible to incorporate biomass and metabolism measurements of dominant parts of the community into a mathematical model to simulate dissolved oxygen levels in the water.

Dissolved oxygen models

Dissolved oxygen has often been modeled with great success by sanitary engineers in water-quality studies of rivers and streams, but their efforts have usually been extensions of the basic oxygen-sag equation developed by Streeter and Phelps (1925). This approach can only be successful in situations approaching a steady state over short time intervals, and where the primary determining factors in the oxygen balance are high BOD loads, temperature, flow rates, and re-aeration. Although recent at-

[2] This section has been prepared in collaboration with James N. Kremer, Graduate School of Oceanography, University of Rhode Island.

tempts have been made to include photosynthesis and respiration terms in the equations (see discussion by O'Connell and Thomas 1965, Camp 1965 and Thomann 1971), most water-quality models either ignore the terms in actual computations or use very simple constants to add or subtract oxygen. The calculations in these models are also necessarily put on a long-time base to give a "representative" oxygen value each day or week, since the models do not realistically handle short-term variations. This approach may produce acceptable predicted values in some cases, but it is not applicable to the marsh embayment or other waters where the biological system dominates the oxygen pattern and diurnal changes may be greater than average differences in seasonal values. The only short-term simulation model of dissolved oxygen that appears to have been developed is a charge-discharge compartment model of laboratory blue-green mats by Sollins (1970) that includes light and temperature effects on photosynthesis and respiration.

The Bissel Cove model

Equations were developed relating the oxygen production and consumption by the major compartments of the embayment community to temperature and light energy. These were based on regression of laboratory data of individual metabolism and field biomass measurements discussed earlier (Fig. 11, 15, 16, and 17). For two of these compartments, gross production of sediment microflora and detritus respiration, it ·was necessary to use literature values (Odum and de la Cruz 1967, Pamatmat 1968). Since this was a diurnal model of oxygen flux, rather than a long-term growth model, the size of each compartment was input for each simulation day and held constant through that run. Simulations were made for 24-hr periods that coincided with each of the 17 days on which diurnal oxygen levels had been measured as part of the procedure to determine total system metabolism (Fig. 5, 6, 10). Compartment sizes, water temperatures, light values, and an initial oxygen concentration were taken from field measurements for the appropriate day. Calculations of flux due to diffusion used a measured diffusion constant of 0.1 hr⁻¹, with the saturation deficit defined as $SD = 1 - OX/OS$, where OX is the calculated instantaneous oxygen concentration, and OS is the saturation value for oxygen at each temperature and salinity calculated from an empirical regression developed by Truesdale, Bowning, and Lowden (1955). The advective contribution of tidal exchange was calculated as the product of flow and duration (tidal volume) times concentration, assuming saturation of incoming bay water. A function generator was used to approximate the

TABLE 16. Components of the Bissel Cove dissolved oxygen model—(H) hourly; (D) daily

Physical inputs	Biological inputs	Outputs
Solar radiation (H)	*Ruppia* biomass (D)	*Ruppia* metabolism (H)
Water temperature (H)	*Ulva* biomass (D)	*Ulva* metabolism (H)
Salinity (H)	Detritus biomass (D)	Detritus respiration (H)
Diffusion constant (D)	Shrimp biomass (D)	Shrimp respiration (H)
Depth ((H)	Fish biomass (D)	Fish respiration (H)
Tide height, duration (H)	BOD (H)	Plankton metabolism (H)
Initial O_2 concentration (D)		Sediment metabolism (H)
		Dissolved O_2 (H)

shape of the tide in the embayment as given by the recording tide gauge (Fig. 4).

The simulation program was written in G-level FORTRAN IV and processed by an IBM 360-50 computer. Numerical integration using Simpson's rule provided the predictions of dissolved oxygen at ¼₄-hr intervals throughout each 24-hr simulation run. A list of the equations used is included below. Units are g O_2 m^{-2} hr^{-1} unless otherwise defined.

Plankton gross production (Fig. 11):

$$PGP = 10^{(0.0138 \times SR - 1.655)}$$

where SR = solar radiation in cal cm^{-2} hr^{-1}.

Plankton respiration (Fig. 11):

$$PR = 10^{(0.058 \times TW - 2.74)}$$

where TW = water temperature in °C.

Benthic microflora gross production (Pamatmat 1968):

$$BMP = 1.79 \times 10^{-3} \times SR$$
$$+ 4.51 \times 10^{-3} \times TW - 0.0223.$$

Sediment community respiration (see text):

$$SDR = 1.57 \times 10^{-3} \times TW + 9.71 \times 10^{-3}.$$

Ulva respiration (Fig. 15):

$$UR = U(0.101 \times TW - 0.345).$$

Ulva gross production (Fig. 15):

$$UGP = UR + UAP(U)$$

where U = biomass of *Ulva* in mg m^{-2} and UAP = function generator based on Fig. 15.

Ruppia respiration (Fig. 15):

$$RR = R(0.172 \times TW - 1.34).$$

Ruppia gross production (Fig. 15):

$$RGP = RR + RAP(R)$$

where R = biomass of *Ruppia* in mg m^{-2} and RAP = function generator based on Fig. 15.

Detritus respiration (Odum and de la Cruz 1967):

$$DR = D(0.0259 \times TW + 0.097)$$

where D = biomass of detritus in mg m^{-2}.

Shrimp respiration (Fig. 16):

$$SR = S(0.123 \times TW + 0.917)$$

where S = biomass of shrimp in mg m^{-2}.

Fish respiration (Fig. 16):

$$FR = F(0.12 \times TW + 0.42)$$

where F = biomass of fish in mg m^{-2}.

With the above terms combined in an oxygen balance, the final equations for calculating changes in dissolved oxygen in the water appeared as follows:

$$\dot{O}_2 = PGP + BMP + UGP + RGP - PR - SDR$$
$$- UR - RR - DR - SR - FR \pm \text{diffusion}$$
$$\pm \text{tidal flux}$$

$$O_2 = \int (\dot{O}_2).$$

A summary of the components of the complete model is given in Table 16.

As with all such efforts, the amount of data available and the need for simplification placed limits on the embayment model. For example, while the correlation coefficients for the individual regressions (see individual sections) indicated that temperature and light were the dominant factors influencing metabolism in the embayment, neither the measurements nor the model included other factors such as salinity, nutrient levels, oxygen tension, or seasonal and daily behavioral rhythms that may have influenced metabolism. Since detailed measurements of benthic plant biomass were only taken during summer, when the plants were most abundant, the less reliable field estimates of coverage and relative abundance were combined with the detailed results of an earlier study of nearby Great Pond, Falmouth, Mass., by Conover (1958) to pro-rate values throughout the year. The metabolism of both the plankton and the sediment compartments was related to volume and area, respectively, rather than to numbers or biomass of any particular species. Although this is a common practice in ecological research, a substantial amount of detail has been missed. The error is probably greatest for plankton ($r = 0.65$), since the respiration of sediments appears to be due more to a ubiquitous and rich microbral flora rather than to larger infauna species that may fluctuate widely in numbers (Fenchel 1969). The respiration of less abundant animals, such as the eel and the blue crab, was not included in the model since preliminary calculations showed that their contribution to the diurnal oxygen budget was insignificant, even during their most conspicuous periods. With these limitations in mind, however, the model did serve as a useful method for resynthesizing measurements of individual parts to see if they were reasonable when combined and compared with independent measurements of the whole system.

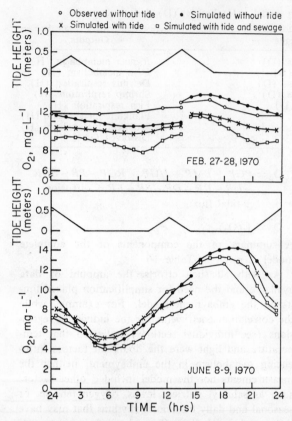

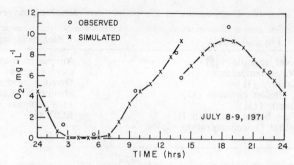

FIG. 25. Comparison of simulated and observed concentrations of dissolved oxygen in the marsh embayment when open to tidal exchanges.

FIG. 24. Comparison of simulated and observed concentrations of dissolved oxygen in the marsh embayment for representative winter and summer days without tidal exchange. The effects of appropriate tidal and sewage BOD inputs have also been simulated.

Simulation and verification

The model produced appropriate patterns and magnitudes of dissolved oxygen for each of the 17 days on which observed diurnal curves were available. Correlation coefficients between observed and simulated curves for each season are shown in Table 17, along with the correlation between observed data and two arbitrary reference functions, a straight line and a sine wave. The high values for the sine

function reflect the symmetry of the diurnal rate-of-change curves for this system, in which a slightly greater night respiration was closely coupled to apparent production during the day (Fig. 9). Representative diurnal curves with simulated and observed values are shown in Fig. 24 for winter and summer extremes, and in Fig. 25 for a day in July when the embayment was open to the tide.

Analysis of the simulation results indicated that phytoplankton contributed about 50% of total system production in the embayment model during winter, with about 40% from sediment microflora and 10% from macrophytes. The contributions of sediment algae remained at 30–40% throughout the year, while the importance of macrophytes increased to a peak of 55% in fall, as the phytoplankton contribution declined to 10%. Respiration results for the model system compartments were as shown in Fig. 10. Respiration by macrophytes, not shown separately, was lowest in winter with 1–2% of total oxygen consumption, then rose to 10–20% of the total during summer and fall. The relative importance of 10 major compartments in the model for a day in June is shown in Table 18. The ratio of total apparent production during the day to the night respiration of the model system has been calculated for each parameter level to give a measure of the

TABLE 17. Correlation coefficients between observed and simulated oxygen values, between observed values and a straight line[a], and between observed values and a fitted sine wave[b]

Date	Observed vs. simulated	Observed vs. straight line	Observed vs. sine wave
February 27–28	0.59	0.44	0.86
May 12–13	0.93	0.51	0.97
June 8–9	0.83	0.39	0.92
October 8–9	0.81	0.42	0.93
July 8–9 (with tide)	0.98	0.24	0.93

[a] Fit by the method of least squares.
[b] The sine wave for each day was fit by matching its period, amplitude, and time of maximum to the observed data.

TABLE 18. Response of the Bissel Cove oxygen model to changes in parameter values as indicated by the ratio of apparent production to night respiration (June 8–9, observed $P/R = 0.86$)

Test parameter (X)	Magnitude of parameter change						
	0.0X	0.5X	0.9X	1.0X	1.1X	1.5X	2.0X
Fish biomass	0.92	0.91	0.91	0.91	0.90	0.90	0.89
Shrimp biomass	0.95	0.93	0.91	0.91	0.90	0.88	0.86
Plankton respiration	1.32	1.10	0.94	0.91	0.87	0.73	0.59
Microbenthos respiration	1.34	1.10	0.94	0.91	0.87	0.74	0.60
Plankton production	0.42	0.65	0.86	0.91	0.96	1.17	1.43
Macrophyte biomass	0.27	0.57	0.84	0.91	0.97	1.20	1.43
Macrophyte respiration	1.70	1.25	0.97	0.91	0.85	0.64	0.41
Microbenthos respiration	0.20	0.50	0.82	0.91	0.99	1.33	1.76
Macrophyte production	0.0	0.37	0.78	0.91	1.03	1.54	2.17
Detritus biomass	2.87	1.62	1.02	0.91	0.80	0.48	0.02

relative sensitivity of the model to the magnitude of each compartment or flow. The center column with a *P*-to-*R* ratio of 0.91 resulted from the field and laboratory data as input, and closely approximates the observed ratio of 0.86 for that day.

Effect of the tide

Since small shallow embayments are often subjected to diking and filling operations that decrease or eliminate their exchange with the open estuary, the impact of tidal flows on diurnal oxygen patterns was explored with the model. Simulation runs were made with and without tidal inputs and with the phase moved through the 24-hr cycle. These trials indicated that tidal oxygen effects vary widely, depending on the timing of highs and lows. Results for a typical summer day are shown in Fig. 26 for a tidal excursion in agreement with the maximum observed in the cove. When high tides were simulated at noon and midnight, they had little effect on the oxygen pattern resulting from metabolism of the embayment. When the tides were shifted to morning and evening, however, large changes resulted that increased the diurnal oxygen range and produced almost anoxic conditions at dawn. Instead of damping cycles in the small storage of embayment water as one might first expect, the addition of water from the relatively constant oxygen reservoir in the bay may exaggerate diurnal oscillations. Since dissolved oxygen levels in the bay do not rise as high as those in the embayment, the addition of bay water at the end of the day may sharply lower the concentration of oxygen in the embayment. When imposed on these relatively low end-of-the-day oxygen concentrations produced by the tide, high rates of dark respiration may drive the levels in the marsh to zero during the night. The anoxic production of H_2S occasionally associated unfavorably with shallow coves and marsh embayments may result as much from tidal rhythms as from imports of organic matter injected into an otherwise balanced metabolism.

Simulation of sewage input

Although the Bissel Cove embayment did not have houses built directly by it, marsh waters are frequently used as receiving basins for domestic sewage from surrounding coastal housing. Accordingly, the model was used to determine the effect of sewage additions to the background metabolism of the embayment. On the basis of nearby housing patterns, about 30 houses containing an average of four people each might reasonably be developed in the area. With a typical BOD load of 77.2 g per person per day (C.P. Poon, University of Rhode Island Department of Environmental Engineering,

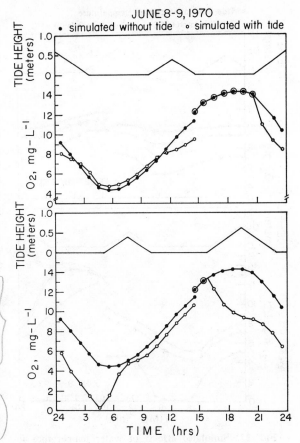

FIG. 26. Simulated effects of different times of high tide on the diurnal oxygen pattern in the marsh embayment on a summer day.

personal communication) the total input to the embayment would be on the order of 9.2 kg BOD day^{-1}. A step function was used to distribute this input so that 25% of the daily total was added between 0800 and 1200, 25% between 1600 and 2000, and the remaining 50% evenly throughout the remaining hours (Poon, *personal communication*). The model assumed a steady state for sewage and neglected probable effects of nutrients from the sewage in increasing algal biomass or photosynthetic rates. Results of the BOD addition on a winter and summer day with appropriate tide conditions are shown in Fig. 24. The general effect at this level of addition was a reduction in dissolved oxygen of about 1 mg l^{-1} with somewhat greater depressions at times of peak load. When the high tides came at dawn and dusk, however, the additional oxygen demand would produce anoxic conditions in the marsh for several hours.

Simulated temperature increases

The proposed construction of an electric power generating plant nearby on Narragansett Bay

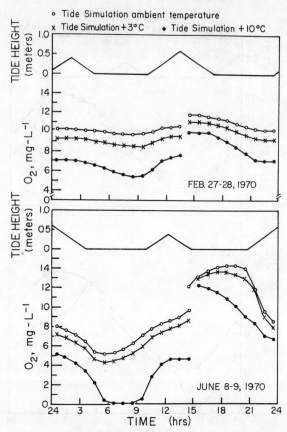

FIG. 27. Simulated effects of water temperature increases on the diurnal dissolved patterns of the marsh embayment.

prompted the simulation of an additional environmental modification, the imposition of thermal loads on the metabolism of the system. The model was run for 1 day each month throughout the year with temperature increases in the embayment for tidal water of 3° C and 10° C. Results for the winter and summer days are shown in Fig. 27. An increase of 3° C depressed the oxygen level somewhat less than the addition of 9.2 kg BOD, but a 10° C increase had a marked effect, driving the embayment anaerobic for four or five early morning hours in the summer. Subtle questions of long-term thermal influences on physiology and behavior remain to be answered, but the model is useful in showing that the more realistic small temperature increases that may be associated with thermal shadows from power plant cooling systems may be expected to produce measurable, but slight, lowerings of dissolved oxygen levels in coastal embayments like Bissel Cove.

ACKNOWLEDGMENTS

According to the time, task, and season, many friends, students, and colleagues helped us with the study of the

marsh. We are particularly grateful to B. Welsh, C. Rogers, K. Taylor, W. Macy, C. Hempstead, A. Myers, H. Donaldson, R. Wilcox, D. Lavoie, A. Forbes, B. Gonet, and V. Lucid for their help with the extensive field work. Solar radiation data were kindly provided by the Eppley Laboratories in Newport, Rhode Island. Nelson Marshall and H. P. Jeffries shared our enthusiasm for the marshes and criticized the manuscript. The research was supported by a grant from the Office of Sea Grant Programs, NOAA.

LITERATURE CITED

Ayers, J. C. 1959. The hydrography of Barnstable Harbor, Massachusetts. Limnol. Oceanogr. **4**: 448–462.

Bardach, J. E. 1959. The summer standing crop of fish on a shallow Bermuda reef. Limnol. Oceanogr. **4**: 77–85.

Beeston, M. D. 1971. Decapod crustacean and fish populations in experimental marine ponds receiving treated sewage wastes, p. 182–204. *In* E. J. Kuenzler and A. F. Chestnut [ed.] Structure and functioning of estuarine ecosystems exposed to treated sewage wastes. Sea Grant Rep., Inst. Mar. Sci., Univ. North Carolina, Chapel Hill, N.C.

Björk, S. 1967. Ecologic investigations of *Phragmites communis*. Folia Limnol. Scand. 218 p.

Blum, J. L. 1968. Salt marsh *Spartinas* and associated algae. Ecol. Monogr. **38**: 199–222.

Boyd, C. E. 1970. Production, mineral accumulation and pigment concentrations in *Typha latifolia* and *Scirpus americanus*. Ecology **51**: 285–290.

Bray, J. R. 1962. Estimates of energy budget for a *Typha* (cattail) marsh. Science **136**: 1119–1120.

Broad, A. C. 1957. The relationship between diet and the larval development of *Palaemonetes*. Biol. Bull. **112**: 162–170.

Camp, T. R. 1965. Field estimates of oxygen balance parameters. J. Sanit. Eng. Div., Proc. Am. Soc. Civ. Eng. **91**(5):1–16.

Carey, A. G., Jr. 1967. Energetics of the benthos of Long Island Sound: I Oxygen utilization of sediment. Bull. Bingham Oceanogr. Collect. **19**: 136–144.

Chapman, V. J. 1940. Succession on the New England salt marshes. Ecology **21**: 279–282.

———. 1960. Salt marshes and salt deserts of the world. L. Hill, London. 392 p.

Conover, J. T. 1958. Seasonal growth of benthic marine plants as related to environmental factors in an estuary. Publ. Inst. Mar. Sci., Univ. Texas **5**: 97–147.

Cooper, A. W. 1969. Salt marshes, p. 567–611. *In* H. T. Odum, B. J. Copeland, and E. A. McMahan [ed.] Coastal ecological systems of the United States. F.W.P.C.A. report, Univ. North Carolina, Inst. Mar. Sci.

Copeland, B. J., and W. R. Duffer. 1964. Use of a clear plastic dome to measure gaseous diffusion rates in natural waters. Limnol. Oceanogr. **9**: 494–499.

Cormack, R. M. 1968. The statistics of capture-recapture methods, p. 455–506. *In* H. Barnes [ed.] Oceanogr. Mar. Biol. Ann. Rev. 6. Allen and Unwin, Ltd., London.

Crichton, O. 1967. Caloric studies of *Spartina* and the marsh crab *Sesarma reticulatum* (Say). Ann. Pittman-Robertson Rep. to the Delaware Bd. Game and Fish. Comm. 20 p.

DeLury, D. B. 1951. On the planning of experiments for the estimation of fish populations. J. Fish. Res. Bd. Canada **8**: 281–307.

————. 1958. The estimation of population size by a marking and recapture procedure. J. Fish. Res. Bd. Canada **15**: 19–25.

Duff, S., and J. M. Teal. 1965. Temperature change and gas exchange in Nova Scotia and Georgia salt-marsh muds. Limnol. Oceanogr. **10**: 67–73.

Fenchel, T. 1969. Ecology of marine microbenthos IV. Structure and function of the benthic ecosystem, its chemical and physical factors and the microfauna communities with special reference to the ciliated protozoa. Ophelia **6**: 1–182.

Ganning, B., and F. Wulff. 1969. The effects of bird droppings on chemical and biological dynamics in brackish water rock pools. Oikos **20**: 274–286.

Hall, C. S. 1970. Migration and metabolism in a stream ecosystem. Ph.D. Thesis. Univ. North Carolina, Chapel Hill, N.C. 375 p.

Hargrave, B. T. 1969. Similarity of oxygen uptake by benthic communities. Limnol. Oceanogr. **14**: 801–805.

————. 1970. The effect of deposit-feeding amphipods on the metabolism of benthic microflora. Limnol. Oceanogr. **15**: 21–30.

Hart, P. B., and T. J. Pitcher. 1969. Field trials of fish marking using a jet inoculator. J. Fish. Biol. **1**: 383–385.

Hartman, R. J., and D. L. Brown. 1967. Changes in internal atmosphere of submerged vascular hydrophytes in relation to photosynthesis. Ecology **48**: 252–258.

Higer, A. L., and M. C. Kolipinski. 1967. Pull-up trap: a quantitative device for sampling shallow-water animals. Ecology **48**: 1008–1009.

Hoese, H. D., and R. J. Jones. 1963. Seasonality of larger animals in a Texas turtle grass community. Publ. Inst. Mar. Sci. Univ. Texas **9**: 37–47.

Hunter, B. 1972. The metabolic expenditures of the American eel (*Anguilla rostrata*) on its migration to the Sargasso Sea. Ph. D. Thesis. Univ. Rhode Island, Kingston, R. I. 115 p.

Hyle, R. A. 1971. Fishes of pond and creek systems, p. 285–296. *In* E. J. Kuenzler and A. F. Chestnut [ed.] Structure and functioning of estuarine ecosystems exposed to treated sewage wastes. Sea Grant Rep., Inst. Mar. Sci., Univ. North Carolina, Chapel Hill, N. C.

Jeffries, H. P. 1972. Fatty acid ecology of a tidal marsh. Limnol. Oceanogr. **17**: 433–440.

Johnson, M. 1970. Preliminary report on species composition, chemical composition, biomass and production of marsh vegetation in the upper Patuxent estuary, Maryland, p. 164–178. *In* Chesapeake Biol. Lab. Rep., Ref. No. 70-130.

Keefe, C. W. 1972. Marsh production: a summary of the literature. Contrib. Mar. Sci., Univ. Texas **16**: 163–181.

Kerwin, J. A. 1972. Distribution of the salt marsh snail (*Melampus bidentatus* (Say)) in relation to marsh plants in the Poropotank River area, Virginia. Chesapeake Sci. **13**: 150–153.

Kuenzler, E. J. 1961. Structure and energy flow of a mussel population in a Georgia salt marsh. Limnol. Oceanogr. **6**: 191–204.

Lavoie, D. M. 1970. A survey of benthic fauna in Bissel Cove salt marsh. Unpubl. project report. G.S.O., Univ. Rhode Island, Kingston, R.I. 20 p.

Lucid, V. J. 1971. Utilization of Bissel Cove salt marsh by birds of the families Anatidae and Laridae. M.S. Thesis. Univ. Rhode Island, Kingston, R.I. 84 p.

Lunz, G. R. 1967. Farming the salt water marshes,

p. 172–177. *In* Proceedings of the marsh and estuary management symposium, L.S.U., Baton Rouge, La.

Lytle, R. W. 1969. Primary productivity of photosynthetic microflora on a tidal marsh in Rhode Island. M.S. Thesis. Univ. Rhode Island, Kingston, R.I. 62 p.

Mann, K. H. 1972. Macrophyte production and detritus food chains in coastal waters. Paper presented at the Symposium on detritus and its ecological role in aquatic ecosystems, Pallanza, Italy.

Mare, M. 1942. A study of a marine benthic community with special reference to the micro-organisms. J. Mar. Biol. Assoc. U.K. **25**: 517–554.

McFarland, W. M. 1963. Seasonal change in the number and the biomass of fishes from the surf at Mustang Island, Texas. Publ. Inst. Mar. Sci., Univ. Texas **9**: 91–105.

McIntyre, A. D. 1964. Meiobenthos of sublittoral muds. J. Mar. Biol. Assoc. U.K. **44**: 665–676.

Miller, W. R., and F. E. Egler. 1950. Vegetation of the Wequetequock-Pawcatuck tidal-marshes, Connecticut. Ecol. Monogr. **20**: 143–172.

Morgan, M. H. 1961. Annual angiosperm production on a salt marsh. M. S. Thesis. Univ. Delaware, Newark, Del. 34 p.

Moseley, F. N., and B. J. Copeland. 1969. A portable dropnet for representative sampling of nekton. Contrib. Mar. Sci., Univ. Texas **14**: 37–45.

Nichol, E. A. 1936. The ecology of a salt-marsh. J. Mar. Biol. Assoc. U.K. **20**: 203–261.

Niering, W. A. 1961. Tidal marshes, their use in scientific research. Conn. Arbor. Bull. **12**: 3–7.

Nixon, S. W., and C. A. Oviatt. 1972. Preliminary measurements of midsummer metabolism in beds of eelgrass, *Zostera marina*. Ecology **53**: 150–153.

Nixon, S. W., C. A. Oviatt, C. Rogers, and K. Taylor. 1971. Mass and metabolism of a mussel bed. Oecologia **8**: 21–30.

O'Connell, L., and N. A. Thomas. 1965. Effect of benthic algae on stream dissolved oxygen. J. Sanit. Eng. Div., Proc. Am. Soc. Civ. Eng. **91**(3): 1–16.

Odum, E. P., and A. A. de la Cruz. 1967. Particulate organic detritus in a Georgia salt marsh-estuarine ecosystem, p. 383–388. *In* G. H. Lauff [ed.] Estuaries. AAAS Publ. 83. Washington, D.C.

Odum, H. T. 1967. Work circuits and system stress, p. 81–138. *In* Symposium on primary productivity and mineral cycling in natural ecosystems, ESA, Univ. Maine Press, Orono, Me. 245 p.

————. 1971. Environment, power, and society. Wiley-Interscience, New York. 331 p.

————. 1972. An energy circuit language for ecological and social systems: its physical basis, p. 140–211. *In* B. C. Patten [ed.] Systems analysis and simulation in ecology. Vol. 2. Academic Press, New York.

Odum, H. T., R. P. Cuzon du Rest, R. J. Beyers, and C. Allbaugh. 1963. Diurnal metabolism, total phosphorus, Ohle anomaly, and zooplankton diversity of abnormal marine ecosystems of Texas. Publ. Inst. Mar. Sci., Univ. Texas **9**: 404–453.

Odum, H. T., and C. M. Hoskin. 1958. Comparative studies on the metabolism of marine waters. Publ. Inst. Mar. Sci., Univ. Texas **5**: 16–46.

Odum, H. T., S. W. Nixon, and L. H. DiSalvo. 1969. Adaptations for photoregenerative cycling, p. 1–29. *In* J. Cairns, Jr. [ed.] The structure and function of fresh-water microbial communities. Res. Monogr. 3, Va. Polytech. Inst., Blacksburg, Va.

Pamatmat, M. M. 1968. Ecology and metabolism of

a benthic community on an intertidal sandflat. Int. Rev. Ges. Hydrobiol. **53:** 211–298.

Pomeroy, L. R. 1959. Algal productivity in the salt marshes of Georgia. Limnol. Oceanogr. **4:** 367–386.

Pomeroy, L. R., L. R. Shenton, R. D. H. Jones, and R. J. Reimold. 1972. Nutrient flux in estuaries, p. 274–293. *In* G. E. Likens [ed.] Nutrients and eutrophication. Am. Soc. Limnol. Oceanogr., Spec. Symp. 1.

Pratt, D. M. 1965. The winter-spring diatom flowering in Narragansett Bay. Limnol. Oceanogr. **16:** 173–184.

Ragotzkie, R. A. 1959. Plankton productivity in estuarine waters of Georgia. Publ. Inst. Mar. Sci., Univ. Texas **6:** 146–158.

Ragotzkie, R. A., and L. R. Pomeroy. 1957. Life history of a dinoflagellate bloom. Limnol. Oceanogr. **2:** 62–69.

Redfield, A. C. 1972. Development of a New England salt marsh. Ecol. Monogr. **42:** 201–237.

Reifsnyder, W. E., and H. W. Lull. 1965. Radiant energy in relation to forests. U.S. Dep. Agric., For. Serv. Tech. Bull. 1344. 111 p.

Reimold, R. J. 1972. Salt marsh ecology: the effects on marine food webs of direct harvest of marsh grass by man, and the contribution of marsh grass to the food available to marine organisms. Sea Grant Rep., Univ. Georgia, Athens, Ga. 4 p.

Richards, S. W. 1963. The demersal fish population of Long Island Sound. Bull. Bingham Oceanogr. Collect. **18:** 1–101.

Sanders, H. L. 1968. Marine benthic diversity: a comparative study. Amer. Natur. **102:** 243–282.

Schmelz, G. W. 1964. A natural history study of the mummichog, *Fundulus heteroclitus* (L.), in Canary Creek marsh. M.S. Thesis. Univ. Delaware, Newark, Del. 65 p.

Sculthorpe, C. D. 1967. The biology of aquatic vascular plants. Edward Arnold Publ. Ltd., London. 610 p.

Sick, L. V., J. W. Andrews, and D. B. White. 1972. Preliminary studies of selected environmental and nutritional requirements for the culture of Penaeid shrimp. Sea Grant Report, Skidaway Inst. Oceanogr., Savanna, Ga. 23 p.

Smalley, A. E. 1960. Energy flow of a salt marsh grasshopper population. Ecology **41:** 672–677.

Smayda, T. J. 1957. Phytoplankton studies in lower Narragansett Bay. Limnol. Oceanogr. **2:** 342–359.

Sollins, P. 1970. Measurement and simulation of oxygen flows and storages in a laboratory blue-green algal mat system. M.A. Thesis. Univ. North Carolina, Chapel Hill, N.C. 186 p.

Streeter, H. W., and E. B. Phelps. 1925. A study of the pollution and natural purification of the Ohio River. III. Factors concerned in the phenomena of oxidation and reaeration. U. S. Public Health Serv., Public Health Bull. 146. 75 p.

Strickland, J. D. H., and T. R. Parsons. 1968. A practical handbook of sea water analysis. Fish. Res. Bd. Canada Bull. 167. 309 p.

Stroud, L. M. 1969. Color-infrared aerial photographic interpretation and net primary productivity of a regularly-flooded North Carolina salt marsh. M.S. Thesis. North Carolina State Univ. at Raleigh, N.C. 97 p.

Teal, J. M. 1958. Distribution of fiddler crabs in Georgia salt marshes. Ecology **39:** 185–193.

——. 1959. Respiration of crabs in Georgia salt marshes and its relation to their ecology. Physiol. Zool. **32:** 1–14.

——. 1962. Energy flow in the salt marsh ecosystem of Georgia. Ecology **43:** 614–624.

Teal, J. M., and J. Kanwisher. 1961. Gas exchange in a Georgia salt marsh. Limnol. Oceanogr. **6:** 388–399.

Teal, J. M., and W. Wieser. 1966. The distribution and ecology of nematodes in a Georgia salt marsh. Limnol. Oceanogr. **11:** 217–222.

Thayer, G. W., W. E. Schaaf, J. W. Angelovic, and M. W. LaCroix. 1973. Caloric measurements of some estuarine organisms. Fisheries Bull. **71:** (in press).

Thomann, R. V. 1971. Systems analysis and water quality management. Environ. Sci. Div. E.R.A., New York. 286 p.

Tietjen, J. H. 1966. The ecology of estuarine meiofauna with particular reference to class Nematoda. Ph.D. Thesis. Univ. Rhode Island, Kingston, R.I. 238 p.

Truesdale, G. A., A. L. Bowning, and G. F. Lowden. 1955. The solubility of oxygen in pure water and sea-water. J. Appl. Chem. **5:** 53–62.

Udell, H. F., J. Zarudsky, T. E. Doheny, and P. R. Burkholder. 1969. Productivity and nutrient values of plants growing in the salt marshes of the town of Hempstead, Long Island. Bull. Torrey Bot. Club **96:** 42–51.

Ustach, J. F. 1969. The decomposition of *Spartina alterniflora*. M.S. Thesis. North Carolina State Univ. at Raleigh, N.C. 26 p.

Welsh, B. L. 1973. The role of the grass shrimp, *Palaemonetes pugio*, in a tidal marsh ecosystem. Ph. D. Thesis. Univ. Rhode Island, Kingston, R.I. 90 p.

Wetzel, R. G. 1965. Techniques and problems of primary productivity measurements in higher · aquatic plants and periphyton, p. 255–267. *In* C. R. Goldmen [ed.] Primary productivity in aquatic environments. Men. Ist. Ital. Idrobiol., 18 Suppl., Univ. California Press, Berkeley, Calif.

Wieser, W. 1960. Benthic studies in Buzzards Bay. II. The meiofauna. Limnol. Oceanogr. **5:** 121–137.

Wieser, W., and J. Kanwisher. 1961. Ecological and physiological studies on marine nematodes from a small salt marsh near Woods Hole, Massachusetts. Limnol. Oceanogr. **6:** 262–270.

Wigley, R. L., and A. D. McIntyre. 1964. Some quantitative comparisons of offshore meiobenthos and macrobenthos south of Martha's Vineyard. Limnol. Oceanogr. **9:** 485–493.

Wilcox, J. R. 1972. Feeding habits of the sand shrimp, *Crangon septemspinosa*. Ph.D. Thesis. Univ. Rhode Island, Kingston, R.I. 135 p.

Williams, R. B. 1966. Annual phytoplankton production in a system of shallow temperature estuaries, p. 699–716. *In* H. Barns [ed.] Some contemporary studies in marine science. Allen and Unwin, Ltd., London.

Williams, R. B., and M. B. Murdoch. 1969. The potential importance of *Spartina alterniflora* in conveying zinc, manganese, and iron into estuarine food chains, p. 431–439. *In* D. J. Nelson and F. C. Evans [ed.] Proceedings of the Second National Symposium on Radioecology, AEC.

——. 1972. Compartmental analysis of the production of *Juncus roemerianus* in a North Carolina salt marsh. Chesapeake Sci. **13:** 69–79.

Wolf, P. L., S. F. Shanholtzer, and R. J. Reimold. 1972. Population estimates for *Uca pugnax* on the Dauplim estuary marsh. Paper presented at the 35th annual meeting of ASLO, Tallahassee, Fla.

Wood, C. C. 1967. Physioecology of the grass shrimp, *Palaemonetes pugio*, in the Galveston Bay estuarine system. Contrib. Mar. Sci., Univ. Texas **12:** 54–79.

A simulation model of the nitrogen flow
in the peruvian upwelling system [*]

by

JOHN J. WALSH [**] and RICHARD C. DUGDALE [**]

INTRODUCTION

The present simulation model of nitrogen flow through the Peruvian upwelling system is based on data from an area off Punta San Juan, Perú (WALSH, KELLEY, DUGDALE, and FROST, 1971). The upwelling area appeared to be in quasi-steady state over at least a three-week period in March and April 1969. In response to the northerly wind stress, water upwells within a 10-20 km band off the coast carrying a seed population of phytoplankton and high nutrients. As the upwelled water drifts offshore in a persistent plume, the phytoplankton biomass increases and nutrients are depleted. Energy is passed up the food chain in the Peru system to support the world's largest fishery (RYTHER, 1969).

The model assumes that there is a continuous, steady gradient of biological properties from the rich inshore upwelling areas to the impoverished offshore regions. The simulation involves numerical solution of a series of coupled differential equations describing the behavior with time of nutrients and phytoplankton down the plume. The standing crop results predicted by the simulation are then compared with the observed distribution of nutrients and phytoplankton down the plume.

We would like to acknowledge the help of Mr. Perkins Bass in conversion of the simulation program for the IBM 1130. Drs. JAMES O'BRIEN and ALYN DUXBURY made helpful suggestions on treatment of the physical variables. NSF Grant GB-8648 provided financial support.

 [*] Contribution 567, Department of Oceanography, University of- Washington, Seattle, Washington.
 [**] Department of Oceanography, University of Washington, Seattle, Washington 98105.

267

Reprinted from Inv. Pesq. 35:309–330 (1971).

METHODS

The model was developed for use on the IBM 1130 computer aboard the R/V *Thomas G. Thompson* for comparison of the simulation results with the actual incoming data at sea. Restricted by the 8 k core storage of the IBM 1130, only five spatial blocks are included in this two layer model. The upper layer consists of distinct blocks 11 km wide, 11 km

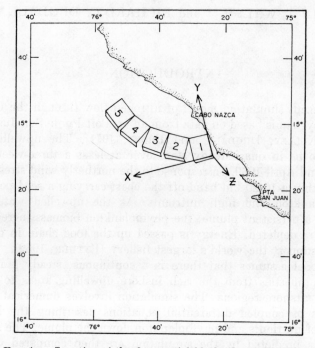

Fig. 1. — Location of the five spatial blocks down the plume.

long, and 10 m deep, which extend across the width of the plume and are distributed sequentially down the axis (figure 1). The lower layer extends below the plume from the 10 m depth to the bottom of the water column. At varying rates downstream, water upwells from the lower layer through the bottom face of each block into the upper layer (z-direction) and then flows downstream through the front and back faces (x-direction) of these blocks. Longshore water transport (y-direction) is considered to be negligible. For purposes of a simple coordinate system in the model, the plume is treated as a line perpendicular to the coast (x-direction) in the simulation calculations.

There are three biological compartments, nutrients, phytoplankton, and herbivores, within each spatial block. Figure 2 outlines in black box notations the inputs and outputs of each compartment within a block and the links to the upstream or downstream spatial blocks. N_1, P_1, and H_1 are the nutrient, phytoplankton, and herbivore compartments or

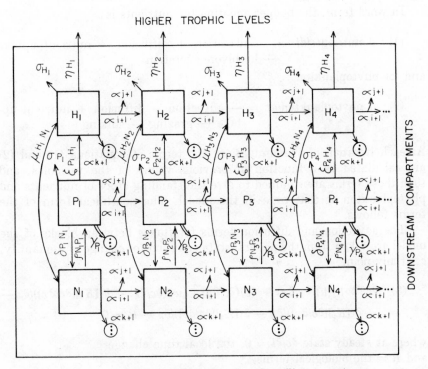

Fig. 2. — Compartment model of an upwelling ecosystem.

standing stocks in the first spatial block; N_2, P_2, and H_2 are the standing stocks in the second downstream block; and N_i, P_i, and H_i are the standing stocks in the ith downstream block. Fluxes between the compartments are nutrient uptake (ρ), grazing of the phytoplankton (ζ), and excretion of the phytoplankton (δ) and herbivores (μ); losses to the outside world are respiration (σ) and herbivore predation (ν); and fluxes between blocks are sinking of the phytoplankton (γ) and downstream, lateral, and vertical advection and diffusion ($\alpha_{i\ j\ k}$). The circles with three dots in the center indicate transport between the surface spatial block and lower levels of the water column.

The differential equations which describe the balance of fluxes controlling the standing crops of nutrients and phytoplankton at any point in the plume are the same for each of the five blocks, but individual values of the fluxes in each area depend on the position of the spatial block. Lack of data on herbivore biomass prevents inclusion of a budget equation for the herbivores in the present model, but herbivore interaction is simulated as an input term for the nutrients and a loss term for the phytoplankton.

In word form, the budget equation for nutrients is

(1) d nutrients/dt = — advection + diffusion — nutrient uptake + herbivore excretion

and for phytoplankton

(2) d phytoplankton/dt = — advection + diffusion + nutrient uptake — grazing — sinking

At each iteration of the simulation, the terms of equations (1) and (2) are calculated as a function of previous values of the variables, and then these terms are summed to give the standing crops of nutrients and phytoplankton at that time in the model. The non-linear form of the terms follows.

The advection and diffusion terms for fluxes between blocks of the model are taken from the general state equation for change of a quantity, c, at any point.

(3) $\partial c/\partial t + (u)(\partial c/\partial x) + (v)(\partial c/\partial y) + (w)(\partial c/\partial z) - \partial([K_x][\partial c/\partial x])/\partial x - \partial([K_y][\partial c/\partial y])/\partial y - \partial([K_z][\partial c/\partial z])/\partial z - R = 0$

where at steady state $\partial c/\partial t = 0$, the local time change and R = the biological terms.

The expression, $(w)(\partial c/\partial z)$, or $(w)(N_z)/dz$ in finite difference form for the nutrient flux, is the vertical advection term for the surface layer of the model, while $\partial([K_z][\partial c/\partial z])/\partial z$ is the vertical diffusion term. K_z is the vertical eddy coefficient and its contribution to the vertical flux is considered negligible compared to w, the upwelling velocity of the advection term. The vertical velocity, w, is a function of the wind stress and distance from shore and is calculated in the model from YOSHIDA's (1955) expression

(4) $w = - (k)(\tau_y)(e^{kx})/(\rho)(f)$

with

$$k = [f][(g)(z)(\Delta\rho)/\rho]^{-\frac{1}{2}}$$

where f = the Coriolis parameter $(2\omega \sin \theta)$; ω = angular velocity, θ =
= latitude

g = the gravitational field strength

z = depth of the nutrient compartment

$\Delta\rho$ = density difference between N_i and N_z

ρ = density of the N_i compartment

τ_y = the wind stress parallel to the coast, = $(\rho_{air})(C_D)(| U |)(U)$
where ρ_{air} = density of air

C_D = dimensionless drag coefficient, 0.0024 for winds
> 15 knots and 0.0015 for winds < 15 knots

U = the surface wind velocity

x = distance down the plume.

The other variables of the vertical advection term are dz = the depth of the block (10 m) and either N_z = the boundary layer nutrient concentration in the z direction (21.0 mg-at NO_3/m^3) or P_z = the boundary layer phytoplankton concentration in the z direction (2.5 mg-at particulate nitrogen/m³).

Using an average wind velocity of 4.6 m/sec for the Punta San Juan area (SMITH et al., 1969) and equation (4), one obtains an upwelling velocity, w, of 6×10^{-2} cm/sec at the shoreward boundary of the plume, i.e., $x=0$; 1×10^{-2} cm/sec in the middle of the first block ($x=5.5$ km) ; and 3×10^{-3} cm/sec at the interface between the first and second block ($x=11$ km). O'BRIEN (1971) has pointed out that decay of w offshore may occur in 5-10 km instead of the 35 km postulated by YOSHIDA (1955). However, O'Brien has suggested that vertical upward transport may still occur in offshore regions as a result of mixing rather than upwelling. For purposes of input fluxes to blocks in the model, mixing and upwelling transport are considered to be the same.

The plume curves along the coast rather than moving directly offshore, and one must integrate the values of w over the assumed homogeneous 11 km length of each block. Therefore a linear decay of w was assumed as a first-order approximation of YOSHIDA's (1955) exponential decay down the plume with 7×10^{-3} cm/sec in the center of the bottom face of the first spatial block, 5.6×10^{-3} cm/sec in the second block, 4.2×10^{-3} cm/sec in the third, 2.8×10^{-3} cm/sec in the fourth, and 0.0 in the fifth block.

The sinking loss of the phytoplankton is of the same form as the upwelling term but of opposite sign. $(w_s)(P_i)/dz$ is the sinking term where w_s = the sinking velocity as determined by SMAYDA and BOLEYN (1965, 1966 a, b) in the laboratory sinking experiments with diatom cultures. Their measured values of w_s ranged from 0.29 m/day to 0.73 m/day (0.3×10^{-3} cm/sec to 0.7×10^{-3} cm/sec) depending on the species and age of the culture.

The expression $(u)(\partial c/\partial x) - \partial([K_x][\partial c/\partial x])/\partial x$ of equation (3) is the downstream advection and diffusion term of the model and in finite difference form for the nutrient flux becomes

$$(\tfrac{1}{2})(u_{i-1})(N_{i-1})/dx - (\tfrac{1}{2})(u_i)(N_i)/dx - (K_x)(N_{i-1} + N_{i+1} - 2N_i)/(dx)(dx)$$

where u = dowstream velocity;

$\quad dx$ = the length of the block;

$\quad K_x$ = the downstream eddy diffusivity. It is determined by the scale length of the block according to PEARSON's (1956) $K_x = 0.01$ (length)$^{4/3}$ cm^2/sec. It is $= 1 \times 10^6$ for the present model, and the diffusion term is negligible compared to the advection term in the downstream direction.

The velocity in the x direction, u, increases with distance down the plume as an additive output at the downstream face of each block resulting from the upwelling velocity input at the bottom face of a

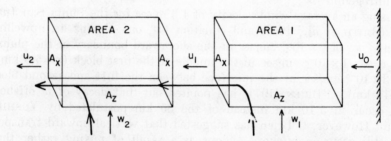

FIG. 3. — Water flow and continuity of mass in the first two spatial blocks.

block and the u input at the upstream face of a block (figure 3). The downstream u can be calculated from w of equation (4) and the upstream u through the mass continuity equation

(5) $$(u_i)(A_x) = (w_i)(A_z) + (u_{i-1})(A_x)$$

where $(u_i)(A_x)$ is the mass transport through the downstream face of the ith block, $(w_i)(A_z)$ is the mass transport upward through the bottom face of the ith block, and $(u_{i-1})(A_x)$ is the mass transport through the upstream face of the ith block. A_x and A_z are the areas of the x, z faces of the ith block. In figure 4, the downstream velocity from the second block away from the coast becomes

$$(u_2)(A_x) = (w_2)(A_z) + (u_1)(A_x)$$
$$u_2 = (w_2)(A_z)/(A_x) + (u_1)$$

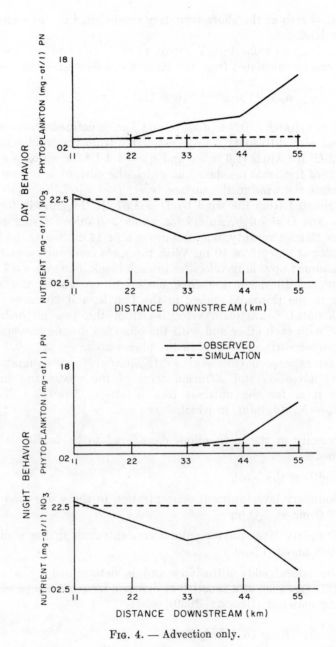

Fig. 4. — Advection only.

and the downstream velocity in the first block is

$$u_1 = (w_1)(A_z)/(A_x) + (u_0) = (w_1)(A_z)/(A_x)$$

because u_0 = zero at the shore boundary condition, i.e., no water flows out of the land.

Alternatively as a check to YOSHIDA's (1955) assumptions, the downstream u can be calculated from the EKMAN (1905) expression

(6) $$u_x = (Vo)(e^{-(\pi)(z/D)})(\cos [45° — (\pi)(z/D)])$$

where $Vo = \tau_y/[(\rho)(K_z)(f)]^{1/2}$ and τ_y, ρ, and f were defined previously. K_z, the vertical eddy diffusivity, can be estimated from the wind velocity by $K_z = 1.02 \ U^3$ for winds < 6 m/sec and $K_z = 4.3 \ U^2$ for winds > 6 m/sec. D is depth of frictional resistance at which the current is reversed from that cum sole the wind in the surface layer. $D = \pi/[(K_z)/(\rho)(1/2)(f)]^{1/2}$ and can be estimated from the wind by $D = 3.67 \ ([U^3]^{1/2})/[\sin \theta]^{1/2}$ for winds < 6 m/sec and $D = 7.6 \ U/[\sin \theta]^{1/2}$ for winds > 6 m/sec. Using equation (6) for the Ekman velocity, one obtains a u of 14 cm/sec at the surface and 9 cm/sec at a depth of 10 m. With the mass continuity equation (5) and the assumed upwelling velocities in each block, one gets $u = 7$ cm/sec at a depth of 10 m in the first block, 13 cm/sec in the second, 17 cm/sec in the third, 20 cm/sec in the fourth, and 20 cm/sec in the fifth. Calculated downstream velocities from the two methods agree fairly well with each other and with the observed drogue measurements of 12-24 cm/sec surface currents in the plume area.

The last expression $(v)(\partial c/\partial y) — \partial([K_y][\partial c/\partial y])/\partial y$ of equation (3) is the lateral advection and diffusion term of the model and in finite difference form for the nutrient flux becomes $(2)(v)(N_y — N_i)/dy + (2)(K_y)(N_y — N_i)/(dy)(dy)$, in which

v = velocity in the y direction, considered to be negligible in the model

dy = width of the block

N_y = boundary layer nutrient concentration in the y direction (3.0 mg-at NO_s/m^3)

P_y = boundary layer phytoplankton concentration in the y direction (2.5 mg-at PN/m^3)

K_y = the lateral eddy diffusivity and is determined by a variable 4/3 expression as a function of distance from the source of diffusing material (BROOKS, 1959), where

(7) $K_y = (K_0)(L/dy)^{4/3}$

with

$$K_0 = K_x \text{ at } x = 0.0, \text{ or } 1 \times 10^6 \text{ cm}^2/\text{sec}$$
$$L = (dy)(1 + (^2/_3)(J)(x/dy))^{^1/_2} \text{ and}$$
$$J = (12)(K_0)/(u)(dy)$$

The variable eddy coefficients in the y direction which were calculated from BROOKS' (1959) model are 0.36 km²/hr for the first spatial block, 0.59 km²/hr for the second, 0.87 km²/hr for the third, 1.20 km²/hr for the fourth, and 1.59 km²/hr for the fifth.

R of equation (3) represents the biological fluxes between compartments of the model. The poorly understood role of excretion in regeneration of nutrients (WHITLEDGE and PACKARD, 1970) and inadequate data on ammonia concentrations and uptake rates down the Peru plume require that an excretion term and its subsequent influence on the uptake term be implicit in the model without as yet an exact mathematical formulation.

The nutrient uptake term for nitrate utilization in the model, equation (8),

$$(8) \qquad \rho_{NO3} = (V_{\max})(N_i)(P_i)/(K_T + N_i)$$

is the Michaelis-Menten expression for nutrient uptake in a nutrient-limited system (DUGDALE, 1967). The term, $V_{\max}$, is the maximum uptake rate (hr^{-1}) at nutrient-saturated conditions. It is allowed to vary sinusoidally in the present model over 12 hours of daylight, with $V_{\max}$ assumed zero during the night. The full expression for $V_{\max}$ is $(1.43)(V_{\max})(\sin 0.2168\ t)$ where t is the cumulative time in the model up to each iteration and 1.43 adjusts observed $V_{\max}$ mean values to be mid-day maxima at the peak of the sinusoid. At dusk, $V_{\max}$ is set equal to zero so that nutrient uptake will not be a negative term as the sinusoid becomes negative in the night period. The term, K_T, is the nutrient concentration (1 mgAt NO$_3$/m³) at which the flux, ρ_{NO3} is half that at the maximum rate of uptake, when P_i is held constant.

In the model, the nutrient uptake term reflects both the distribution of nitrate down the plume and the estimated effect of the regenerated nutrient, ammonia, on nitrate uptake. $V_{\max}$ values of NO_3 uptake were entered in the model from field data collected by DUGDALE and MACISAAC (1971) down the Peru plume (table 1). With these values for each spatial block and equation (8), the nutrient uptake term of budget equation (1) was calculated as a loss from the nitrate concentration at each iteration.

The $V_{\max}$ of equation (8) was modified, however, for the uptake gain in budget equation (2) for the phytoplankton at each iteration. The implicit effect of excretion and regenerated nutrients is added in the

TABLE 1

Observed mean values of the variables in each of the five downstream spatial areas over 0-10 m depth.

	VARIABLES				
Area	NO_3 mgAt/m³	PNO_3 mgAt/m³/day	Chlorophyll a mg/m³	Phytoplankton particulate N mgAt/m³	V_{max} NO_3 hr⁻¹
1	22.37	1.018	2.30	2.76	0.032
2	16.00	0.710	2.01	3.28	0.019
3	12.20	1.355	3.32	6.03	0.020
4	14.42	2.126	5.01	7.71	0.017
5	5.99	1.639	13.13	14.39	0.011

model through a combined V_{max} which includes both the NO_3 and an estimate of the NH_3 contribution to total nitrogen uptake. A gradient of NH_3 V_{max} down the plume was estimated on the basis of possible grazing activities and NH_3 inhibition of NO_3 uptake.

Diel zooplankton data are inadequate for any firm interpretation of grazing patterns (WALSH, KELLEY, DUGDALE, and FROST, 1971), but there appears to be a diel variation in chlorophyll concentrations at the lower end of the plume and not at the upper. It is possible that higher grazing activity and resultant higher NH_3 excretion may occur in the downstream area as opposed to the upstream areas near the coast. Recent field experiments in the Mediterranean Sea (DUGDALE and MACISAAC, 1971) and laboratory studies of continuous cultures (CONWAY, personal communication) indicate that NH_3 is a preferential nitrogen source and in high concentrations inhibits NO_3 uptake. The downstream decline of NO_3 V_{max} in table 1 suggests that this hypothetical gradient of grazing, NH_3 concentration and uptake, and consequent inhibition of NO_3 uptake may actually occur down the plume. For purposes of this preliminary model, V_{max} values for NH_3 down the plume were assumed to be 0.001 in the first block, 0.005 in the second, 0.030 in the third, 0.040 in the fourth, and 0.050 in the fifth. The NH_3 values were then combined with NO_3 V_{max} values in each area and used to calculate the nutrient uptake for the phytoplankton at each iteration.

No data were available from the Peru area on grazing fluxes. If herbivore biomass figures were reliable, mathematical expression of a grazing flux could take the form, $G_{max}(H_i)(P_i - P^*)$ with $(P_i - P^*) \geq 0$, where G_{max} is the grazing rate, varying nocturnally as a cosine and zero during the day. G_{max} would be $(1.43)(G_{max})(\cos 0.2168\, t + 1.57)$ in analogy to the sinusoidal variation of V_{max} in equation (8), with t = the cumulative time interval. P^* = the threshold value below which herbivore grazing induces no change in the phytoplankton population, possibly

2.5 mg/m³ chlorophyll in the Peru plume (WALSH, KELLEY, DUGDALE, and FROST, 1971). The grazing loss in budget equation (2) for the phytoplankton was calculated at steady state, i.e., $dP_i/dt = 0$, by fitting the above grazing expression to the difference between observed and calculated night phytoplankton standing crop. Such a procedure is tenuous at best and independent measurements must be made in the future to provide data on the grazing flux.

The components of the model are then :

a)   constant wind along the coast
b)   upwelling velocity decreases down the plume
c)   downstream velocity increases down the plume
d)   eddy diffusivity in the y direction increases down the plume
e)   a sinusoidal NO_3 V_{max} decreases down the plume
f)   an adjusted NH_3 V_{max} increases down the plume
g)   a cosine grazing term increases down the plume.

Initial conditions of the model are 2.5 mgAt/m³ phytoplankton PN and 3.0 mgAt/m³ NO_3 in each block. Boundary conditions are 2.5 mgAt/m³ phytoplankton PN at the bottom, lateral, and downstream walls of the plume, 3.0 mg-at/m³ NO_3 at the lateral and downstream walls, and 21.0 mg-at/m³ NO_3 at the bottom wall. No fluxes occur through the upstream wall of the first block, i.e., the shore boundary. With the exception of the grazing simulation, the model was started at initial conditions and allowed to go to steady state during each simulation.

Differential equations (1) and (2) were solved with the Euler method of numerical integration utilizing a simulation program developed by BLEDSOE and OLSON (1968) and modified for the IBM 1130 computer. Simulations begin at 6 a.m. with the iterative time step equal to 1 hour, i.e., the losses and gains in equations (1) and (2) were summed every hour, and total simulated time was 10 days.

RESULTS

Simulations were run in a sequential fashion, adding one variable at a time, to see which fluxes were important in controlling the distributions of nutrient and phytoplankton concentrations down te plume. The output of each simulation was compared with the observed day gradients of nutrient and phytoplankton (table I) and with night gradients estimated from the underway map data presented by WALSH, KELLEY,

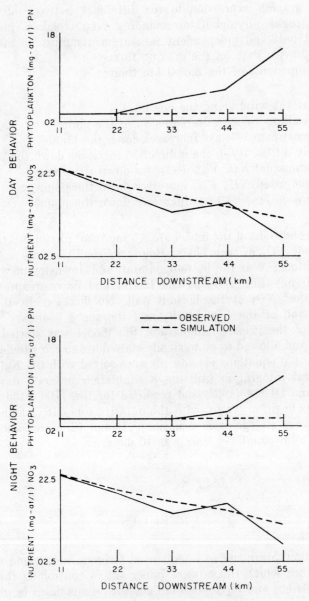

Fig. 5. — Advection and diffusion.

DUGDALE, and FROST (1971). The simulated day results in each block ($t = 228$ hours) are noon phytoplankton and nutrient values of the tenth day of the simulation, and night results ($t = 240$ hours) are midnight values of the variables during the same day. The importance of each flux was judged by how close the simulation results matched observed data.

Advection

Only the advective terms, w and u, were included in the first simulation. All other terms of equations (1) and (2) were set equal to zero. Figure 4 compares the simulation results (dotted line) and the observed data (solid line) down the plume. After 72 hours, the boundary concentration of 21.0 mg-at/m³ NO_3 in the upwelled water raised the initial concentration of 3.0 mg-at/m³ in each NO_3 compartment to the steady values of 21.0 mg-at/m³ NO_3. No losses were present in budget equation (1), and the upwelled input of nutrient simply filled up the NO_3 compartments during the transient state, i.e., $dN_i/dt \neq 0$.

Phytoplankton PN (Particulate Nitrogen) down the plume did not change during the simulation because both the initial and upwelled phytoplankton concentrations were the same, 2.5 mg-at/m³ PN, and no phytoplankton growth was allowed. As a result of the continuity equation, input = output for the phytoplankton mass transport, no transient state was observed, i.e., $dP_i/dt = 0$.

Advection and diffusion

When diffusion is added to the model (figure 5), simulated phytoplankton distribution is the same as in the advective case. There is still no nutrient uptake, and the gradient terms of equation (3) $(P_y - P_i)$ are equal to zero, i.e., no diffusion loss occurs. With no inputs, no losses, and the mass transport balanced there is no time change in the phytoplankton compartments.

The simulated nitrate distribution down the plume matches fairly closely that of the observed data. The nutrient term, $(N_y - N_i) \neq 0$, allows for a diffusion loss to be balanced by the upwelling input, and figure 5 shows the resultant steady state values of the NO_3 compartments down the plume. This simulation suggests that the purely physical diffusion flux away from a high nutrient source along the coast may be a very strong factor in contributing to the depletion of NO_3 down the plume.

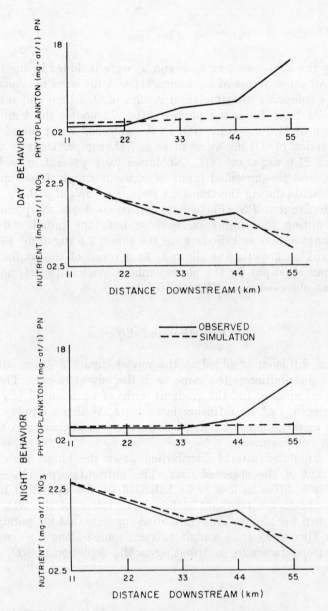

FIG. 6. — Advection, diffusion, and NO₃ uptake.

Advection, diffusion, and NO_3 uptake

If the phytoplankton are allowed to grow in the model at the observed NO_3 V_{max} uptake rates (figure 6), simulated NO_3 distribution converges on the observed data. Day and night differences in the steady state values of the NO_3 compartments are introduced by the sinusoidal input of equation (8). A balance of the upwelling input, diffusion loss, and uptake loss in budget equation (1) for NO_3 matches very well the observed surface NO_3 gradient down the plume and suggests that these are the major factors controlling NO_3 standing stocks in the Peru plume.

The simulated phytoplankton biomass of figure 6 does not come very close to the observed data, however. These results suggest that while the nitrate loss of equation (1) is well estimated, the uptake gain of the phytoplankton in equation (2) is not. The growth of the phytoplankton in the model must be supplemented from another nitrogen source.

Advection, diffusion, NO_3, and NH_3 uptake

With the hypothetical NH_3 uptake rates added to the model, a very close fit between the simulated and observed day phytoplankton distribution is obtained (figure 7). The simulated night phytoplankton distribution is higher than that postulated from the 10 underway night maps. Grazing has not been included in this simulation, however, and such a loss would lower the simulated night values. In contrast, if a constant 0.020 NH_3 V_{max} in each compartment is used down the plume for the supplemental nitrogen source (figure 8) instead of a gradient, the simulated steady state phytoplankton standing crop is higher than that observed.

The present fit of the gradient NH_3 model to observed data suggests that there may be a gradient in NH_3 concentration and utilization down the plume. The actual gradients may be higher than those used in the present model, however. If a day time grazing loss due either to zooplankton or fish is an important factor in the upwelling system, it could balance a higher gross nitrogen production than predicted by the model and still yield the same observed distribution of net phytoplankton PN standing crop.

Advection, diffusion, $NO_3 + NH_3$ uptake, and sinking

A sinking loss was then added to the model. SMAYDA and BOLEYN's (1965, 1966 a, b) smallest sinking rate, 0.29 m/day, produced lower

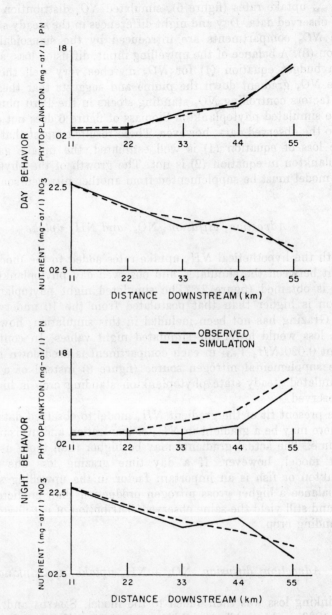

FIG. 7. — Advection, diffusion, NO₃, and gradient NH₃ uptake.

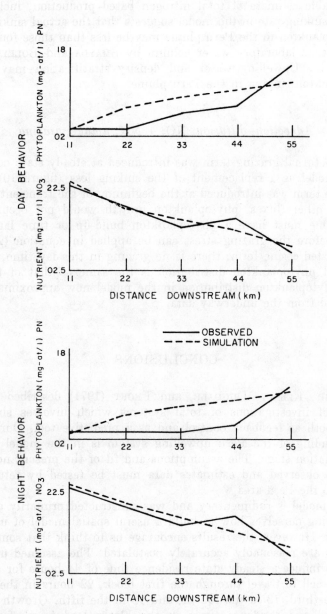

FIG. 8. — Advection, diffusion, NO₃, and constant NH₃ uptake.

simulated phytoplankton biomass than that observed (figure 9). The sinking term creates too large a loss flux in budget equation (2) and does not allow sufficient build-up of phytoplankton biomass in the model during the transient state. If the combined NO_3 and NH_3 uptake flux is a reasonable estimate of total nitrogen based production, inclusion of this low sinking rate in the model suggests that the actual sinking rates of phytoplankton in the Perú plume may be less than those found for a homogeneous laboratory water column by SMAYDA and BOLEYN (1965, 1966 a, b). Upwelling water and density stratification may impede phytoplankton sinking in the Peru plume.

Advection, diffusion, NO_3 + NH_3 uptake, grazing

A nocturnal grazing term was introduced at steady state conditions of the model as a replacement of the sinking loss (figure 10). If the herbivore term was introduced at the beginning of the transient state as were all other fluxes, phytoplankton growth would not occur in the model. One must have a phytoplankton build-up or time lag in the system before the grazing stress can be applied in equation (2). With this adjusted cosine term, there is no grazing in the day time, and the simulated phytoplankton distribution is the same as that of figure 7. Night phytoplankton distribution in the model now approximates that postulated from the underway data.

CONCLUSIONS

WALSH, KELLEY, DUGDALE, and FROST (1971) described a cyclic process of investigations of total systems which involves simulation models both as feedback control and as a predictive tool. Our present understanding of the Peru upwelling system is in the model building and validation stage. The assumptions and fit of the present non-linear model to observed and estimates data must be tested by return field studies to the Peru area.

This model is rudimentary and was constructed primarily as a tool for teaching ourselves how to build a useful spatial model of upwelling processes. However, the results encourage us to think that some of the processes are reasonably accurately postulated. The assumed upwelling velocities induce a steady state residence time of 44 hours for a phytoplankton cell to travel through the first block, 22 hours in the second, 18 in the third, 13 in the fourth, and 13 in the fifth. Growth rates in the model and residence times in the blocks lead to phytoplankton

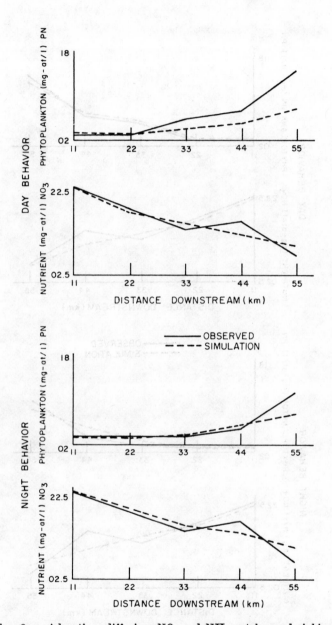

FIG. 9 — Advection, diffusion, NO₃ and NH₃ uptake, and sinking.

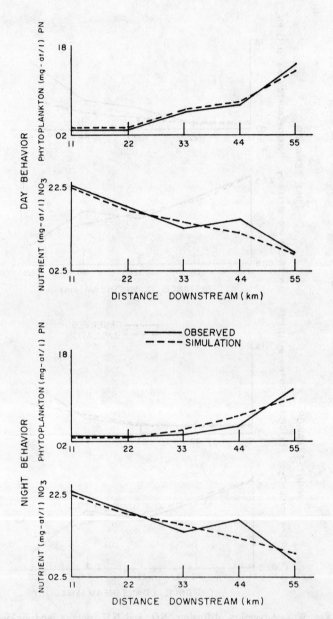

FIG. 10. — Advection, diffusion, NO$_3$ and NH$_3$ uptake, and grazing.

doubling times of 2 days at the upper end of the plume and ½ day at the lower end. These doubling rates are within the observed range of phyto-plankton growth in the Peru area (BARBER, personal communication).

In the last simulation (figure 10), however, an unknown input, regenerated nutrient as NH_3, is balancing an unknown output, grazing, and both are functions of each other. We must now validate the model with a future field study to see if at steady state a gradient of NH_3 concentration and utilization does exist down the plume and to see if the grazing flux of the model approximates the actual daily input to herbivore populations. Time series are also needed to document sugges-tions of nocturnal grazing and diel nutrient uptake.

When the present or second-generation model is validated, there will exist a working analog of the Peru system. Experiments can be run on the model as in figures 7 and 8 with two different sets of NH_3 V_{max} down the plume, or sensitivity analyses can be made on the components of the model, i.e., how sensitive is the phytoplankton biomass of the fifth compartment to a unit change of NO_3 concentration in the first compartment. Finite-difference approximation of continuous processes assumes partition of the process into finitely small discrete units. Five units 11 km long are a very crude approximation of a presumably continuous upwelling plume. We plan to expand the present model to as much as 1000 blocks to cover the same plume, and to consider species interactions, higher tropic levels, fluxes of other elements, and long-term variations of the system.

REFERENCES

BLEDSOE, L. J., and OLSON, J. S. — 1968. Comsys 1: A stepwise compartmental simulation program. ORNL TM (in preparation).

BROOKS, N. H. — 1959. Diffusion of sewage effluent in an ocean current, pp. 246-267. *In* E. A. Pearson (ed.), *Waste disposal in the marine environment*. Pergamon Press, New York.

DUGDALE, R. C. — 1967. Nutrient limitation in the sea: dynamics, identification, and significance. *Limnol. Oceanogr.*, 12(4): 685-695.

DUGDALE, R. C., and MacIsaac, J. J. — 1971. A computational model for the uptake of nitrate in the Peru upwelling region. *Inv. Pesq.*, 35: 299-308.

EKMAN, V. W. — 1905. On the influence of the earth's rotation on ocean currents. *Ark. f. Mat. Astr. och Fysik. K. Sv. Vet. Ak.*, Stockholm, 1905-06, v. 2, n. 11, 1905.

O'BRIEN, J. J. — 1971. A two-dimensional physical model of the North Pacific. *Inv. Pesq.*, 35: 331-349.

PEARSON, E. A. — 1956. An investigation of the efficacy of submarine outfall disposal of sewage and sludge. *State Water Pollution Control Board, Publ. 14*, Sacramento, California.

RYTHER, J. H. — 1969. Photosynthesis and fish production in the sea. *Science*, 166: 72-76.

SMAYDA, T. J., and BOLEYN, B. J. — 1965. Experimental observations on the flotation of marine diatoms. I. *Thalassiosira CF. nana, Thalassiosira rotula,* and *Nitzschia seriata. Limnol. Oceanogr.*, 10(4): 499-509.

SMAYDA, T. J., and BOLEYN, B. J.—1966a. Experimental observations on the flotation of marine diatoms. II. *Skeletonema costatum* and *Rhizosolenia setigera. Limnol. Oceanogr.*, 11(1): 18-34.

SMAYDA, T. J., and BOLEYN, F. J.—1966b. Experimental observations on the flotation of marine diatoms. III. *Bacteriastrum hyalinum* and *Chaetoceras lauderi. Limnol. Oceanogr.*, 11(1): 35-43.

SMITH, R. L.; MOOERS, C. N. K., and ENFIELD, D. B. — 1969. Mesoscale studies of the physical oceanography in two coastal upwelling regions: Oregon and Peru. *Paper presented at the International Conference on the Fertility of the Sea, Sao Paulo, Brazil, December 1969.*

WALSH, J. J.; KELLEY, J. C.; DUGDALE, R. C., and FROST, B. W. — 1971. Gross features of the Peruvian upwelling system with special reference to diel variation. *Inv. Pesq.*, 35: 25-42.

WHITLEDGE, T. E., and PACKARD, T. T. — 1971. Nutrient excretion by anchovies and zooplankton in Pacific upwelling regions. *Inv. Pesq.*, 35: 243-250.
zooplankton in Pacific upwelling regions. *Inv. Pesq.*

YOSHIDA, K. — 1955. Coastal upwelling off the California coast. *Rec. Oceanogr. Wkr.*, Japan, 2(2): 1-13.

Characterization

Introduction

The various approaches to the study of ecology embodied in the preceding sections are, in a way, smaller nibbles at a large chunk of a problem: How do marine communities function?

The concept of a biotic community has been stated in many ways, but perhaps, as Emlen (1973) points out, the simplest is best: A collection of organisms in their environment. Even to the most casual observer it must occur, however, that this "collection of organisms" is far from random. Species tend to occur together, giving a relative uniformity to the taxonomic composition. The assemblage of organisms functions in concert, allowing a metabolic pattern to be discerned and measured. The proper study of community ecology does not concern itself with the structure and function of the individual parts so much as of the whole, for the whole is more than the sum of the parts. Thus the marine ecologist involved in community studies asks questions concerning the abundance and diversity of species, the structure of the trophic web, and the pathways of nutrients and energy in the system. In this section we address the problem of how communities are structured: What species are present in what abundance and at what times of the year? What is the structure of the trophic web, and how is it maintained?

Between 1910 and 1920, in the shallow waters off the Danish Coast, C. G. J. Petersen took a large number of bottom grab samples in order to calculate the amount of food available to the demersal fish populations of the area. He was able not only to do this, but also to show that groups of species tended to occur together at certain depths and in particular sediment types. These were termed "level bottom communities" and were described by the numerically dominant, large, benthic invertebrates present. Considerably later, Thorson (1957) took the con-cept one step further to point out that "parallel" level bottom communities existed around the world, living on the same sediment types, at similar depths, with the same genera (but different species) represented. Since then it has been pointed out that the concept of parallel communities does not always hold, particularly in tropical waters. Additionally, much of the early description of community types was based on the macrobenthos component, ignoring all the organisms that passed through the 2 mm mesh of a sieve (Thorson, 1966). However, much of the impetus for the studies of marine community's structure comes from this early work.

Attempts at characterizing and quantifying planktonic as well as benthic community structure are now often made. These range from presenting species lists and identifying which species are numerically dominant, to complex statistical analyses that identify groups of organisms that tend to occur together and indices of the diversity of species.

Such analyses are interesting and aid greatly in understanding and comparing community structures. It is also important to ask why the communities are structured the way they are and why species diversity is high in one area and low in another. In this section three papers describe the plankton assemblage of one specific, well-known area on the east coast of the United States, Narragansett Bay in Rhode Island, one paper discusses a quasi-benthic community, the fauna found on Sargassum weed, and the rest then explore an hypothesis about the maintenance of species diversity in several marine ecosystems.

Studies on Narragansett Bay, a large, shallow, well-mixed estuary, have been continuing for many years. Pratt (1959) reported that in the yearly cycle, diatoms and flagellates alternate in dominating the phytoplankton assemblage of the bay. In the winter and early spring there is an enormous bloom of diatoms, primarily *Skeletonema costatum*, followed in the late spring and early summer by a dominance of flagellates, which grades into a late summer diatom dominance, then a fall flagellate dominance is seen. Pratt's paper in this volume examines the structure and control of the winter-spring flowering, emphasizing nutrient limitation, and Martin describes the zooplankton populations and how they may act to control phytoplankton species composition through grazing pressure. Recent research (Smayda, 1973, and A. Durbin, 1975) has shown that temperature, light, and grazing by animals other than copepods are also important in the regulation of succession of phytoplankton species in Narragansett Bay. E. Durbin et al. have pointed out the importance of various size fractions of phytoplankton in the bay, and their paper can be integrated with some of the readings presented earlier in this book. No comprehensive study of the benthic species assemblages of Narragansett Bay has been made, but Sanders (1958, 1960) investigated the benthos of Buzzards Bay, Massachusetts, which is quite comparable to Narragansett Bay. Oviatt and Nixon (1973) have described the demersal fish community of Narragansett Bay.

Data on the number of species present and the number of individuals of each species in any community allow the investigator to calculate the diversity of that assemblage. The species diversity of a community is a much-used parameter because it allows one to quantify the relative composition of individuals and species, to make statements about that community, and to compare it with others. Species diversity is measured in many ways; the best methods are designed to minimize a basic problem, that diversity tends to be dependent upon sample size. A commonly used measure, H', the information-theoretic expression of Shannon and Weaver (1963), is calculated by the equation $H' = \sum_{i}^{n} p_i(\log p_i)$ where p_i is the relative abundance of species i in the sample. Another technique, rarefaction (Sanders, 1968), has been applied to benthic assemblages with some success, and improvements have been suggested by Hurlbert (1971).

Comparing communities from different areas, as Sanders and Hessler have done in the paper on the ecology of deep-sea benthos included in this volume, quickly leads one to the realization that some communities are much more diverse than others. The questions of interest are: How has this come to be? What factors are important in determining the structure and diversity of a community? There are many different theories concerning the maintenance of species diversity (for a good review, see Emlen, 1973). Species diversity tends to be low in areas in which an unstable climate causes heavy decimation of species numbers; the same is true of heavy predation. On the other hand, where populations are held to a low level by climate variations, food may not be limiting, and there may be a tendency for trophic specialization, making sympatric speciation more likely and colonization by new species easier (Emlen, 1973). The stability-time hypothesis, on the other hand, postulates two intergrading types of communities, physically controlled and biologically accommodated. The physically controlled community, characterized by low species diversity, is one in which the organisms are subjected to severe physiological stress, and adaptations are primarily to the physical environment. The biologically accommodated community is one in which the physical conditions are constant, or at least predictable, for long periods of time. Here, adaptation to biological, rather than physical, factors leads to increased specialization, and a higher species diversity. Stable conditions over a long period of time do not, however, either produce or maintain a high species diversity; they simply are prerequisite.

Paine (1966), working with predator and prey populations in the rocky intertidal zone, has shown the importance of predation in determining species composition. His concept of a keystone predator, developed experimentally in an area where there was severe competition for a limited resource (space), postulates that species diversity is kept high by a predator efficiently removing the species that tend to monopolize limited environmental resources, thus preventing that species from outcompeting others. The papers by Paine and Vadas, Porter, and Estes and Palmisano presented here indicate how the structure of communities in three very different areas may be controlled. Breen and Mann (1976) have shown similar mechanisms at work in a seaweed-lobster community. Diversity in the deep sea is very high, despite very low food abundance. How this diversity is maintained is a subject of some debate, and two papers, by Dayton and Hessler and Grassle and Sanders, are included to serve as an example of how disagreements are argued in the literature. There is undoubtedly truth in both views, and only further research, such as that by Jumars (1975), will resolve the questions.

Finally, a word on diversity and ecosystems stability. It has often been remarked that the most stable ecosystems are the most diverse because they have the most interconnections in the food web, and the removal of one species might not greatly affect the structure of the community. More recent ecological thinking holds that diversity and stability probably are not related, at least in any direct sense. Ecosystem stability, according to Margalef (1975) is "the capacity of a system which has been disturbed by an external agent (force), to return itself to its original undisturbed state . . . " The paper by Pomeroy in an earlier section of this book addresses in part the question of the relationship between stability, diversity, and nutrients, and points out that the response of an ecosystem to external disturbance is often a decrease of diversity, followed by succession. The external disturbance (and, implicitly, the control of diversity) may be storms, removal of predators, influx of nutrients, or any one of a multitude of other external factors. The point to remember is that many factors contribute to the diversity and stability of an ecosystem.

LITERATURE CITED

Durbin, A. G. 1975. Ecological implications of the migratory behavior of the Atlantic menhaden *Brevevoortia tyrannus* and the alewife *Alosa pseudoharengus*. Ph.D. thesis, University of Rhode Island.

Emlen, J. M. 1973. Ecology: An Evolutionary Approach. Addison-Wesley Publishing Co. Inc., Reading, Massachusetts. 493 pp.

Hurlbert, S. H. 1971. The non-concept of species diversity: A critique, and alternative parameters. Ecology 52:577–586.

Jumars, P. A. 1975. Environmental grain and polycheate species diversity in a bathyl benthic community. Marine Biology 30:253–266.

Margalef, R. 1975. External factors and ecosystem stability. Schweiz. Zeitschr. Hydrologie 37:102–117.

Oviatt, C. A., and S. W. Nixon. 1973. The demersal fish of Narragansett Bay: An analysis of community structure, distribution and abundance. Est. Coastal Mar. Sci. 1:361–378.

Paine, R. T. 1966. Food web complexity and species diversity. Amer. Naturalist 100:65–75.

Pratt, D. M. 1959. The phytoplankton of Narragansett Bay. Limnol. Oceanogr. 4:425–440.

Sanders, H. L. 1958. Benthic studies in Buzzards Bay. 1. Animal-sediment relationships. Limnol. Oceanogr. 3:245–258.

Sanders, H. L. 1960. Benthic studies in Buzzards Bay. III. The structure of the soft bottom community. Limnol. Oceanogr. 5:138–153.

Sanders, H. L. 1968. Marine benthic diversity: A comparative study. Amer. Naturalist 102:243–282.

Shannon, C. E., and W. Weaver. 1963. The Mathematical Theory of Communication. University of Illinois Press, Urbana, Illinois. 117 pp.

Smayda, T. J. 1973. The growth of *Skeletonema costatum* during a winter-spring bloom in Narragansett Bay, Rhode Island, Norw. J. Bot. 20:219–247.

Thorson, G. 1957. Bottom communities (sublittoral or shallow shelf). *In* Geol. Soc. Amer. Memoir 67, 1:461–534. Chapter 17.

Thorson, G. 1966. Some factors influencing the recruitment and extablishment of marine benthic communities. Netherlands J. Sea Res. 3:267–293.

THE WINTER–SPRING DIATOM FLOWERING IN NARRAGANSETT BAY[1]

David M. Pratt

Graduate School of Oceanography, University of Rhode Island, Kingston

ABSTRACT

Weekly observations of phosphate, nitrate, silicate, and phytoplankton over a 4½-year period are analyzed to explain year-to-year variations in the time of inception and magnitude of the winter–spring diatom flowering in Narragansett Bay, Rhode Island. The population during this period (November–June) is dominated by *Skeletonema costatum*. The start of growth does not appear to be related to stability, temperature, or incident radiation; Martin (1965) has shown that it is triggered by the release of zooplankton grazing pressure. The diatom maximum is independent of temperature and incident radiation, and it is regulated by the concentrations of nitrate and silicate at the beginning of logarithmic growth, which also influences the magnitude of that part of the flowering that follows the winter maximum. In most years, considerable growth takes place after the exhaustion of nitrate, whereas growth was never observed to continue after silicate depletion. The relative importance and interplay of silicate and nitrogen as limiting factors are discussed. Since the flowering is greater the earlier it begins, its success is to some degree predetermined by zooplankton activity in the fall.

INTRODUCTION

The seasonal changes in abundance and species composition of the phytoplankton of Narragansett Bay, Rhode Island, have been described by Ferrara (1953), Smayda (1957), and Pratt (1959). The outstanding feature of the annual cycle is the winter–spring diatom flowering, which is extraordinary in its time of inception, intensity, and duration. Logarithmic growth begins usually in December, and after about a month terminates in a maximum sometimes exceeding 50,000 cells/ml; this is followed by a series of secondary peaks of diminishing amplitude, and the flowering period ends in late May or June. This "spring" flowering that begins in the fall, reaches a prodigious maximum in winter, and does not die away until nearly summer appears to be unique among the seasonal cycles so far described. Its unusual features are believed to be related to the moderate depth of the bay, adequate tidal mixing, and low flushing rate. The purpose of this paper is to discuss the roles and interactions of the environmental factors apparently responsible for year-to-year differences in the time of inception and magnitude of the flowering.

METHODS

Temperature and water samples were taken at the surface and the bottom at three stations in the west passage of Narragansett Bay, Station 1 near its head, Station 2 in the middle, and Station 3 at the mouth (respectively, 41°38'08" N lat, 71°22'17" W long, water depth 8 m; 41°34'07" N lat, 71°23'31" W long, depth 9 m; and 41°26'47" N lat, 71°25'09" W long, depth 19 m), weekly from 21 January 1959 to 25 June 1963. The samples, kept in polyethylene bottles, were analyzed within 4 hr for phytoplankton (identifications and counts on raw water as described in Pratt 1959), phosphate (Wattenberg 1937), nitrate (Mullin and Riley 1955) and silicate (Armstrong 1951). Chlorinity was determined by titration.

At somewhat longer intervals between January 1959 and October 1962, the zooplankton was sampled by oblique bottom-to-surface hauls with a Clarke-Bumpus sampler. During the portion of the year under consideration here, the sampler was equipped with a No. 12 mesh net. The samples from Stations 1 and 3 for the years 1959, 1960, and 1961 were analyzed by John H. Martin, to whom I am indebted for the zooplankton data presented below. I also wish to thank him, as well as Charles L.

[1] This work was supported by the U.S. Office of Naval Research, Contract Nonr-396(03).

295

Reprinted from Limnol. Oceanogr. 10:173–184 (1965).

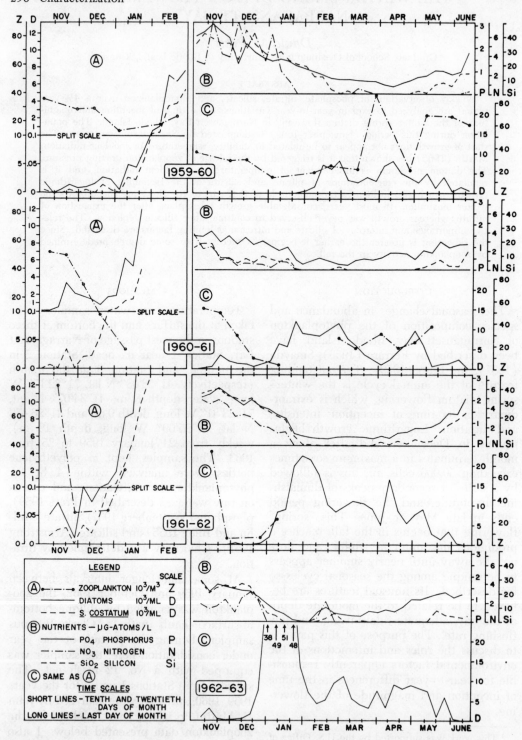

Fig. 1. Plankton and nutrients, Station 1. Panel A is an enlargement of the first part of panel B.

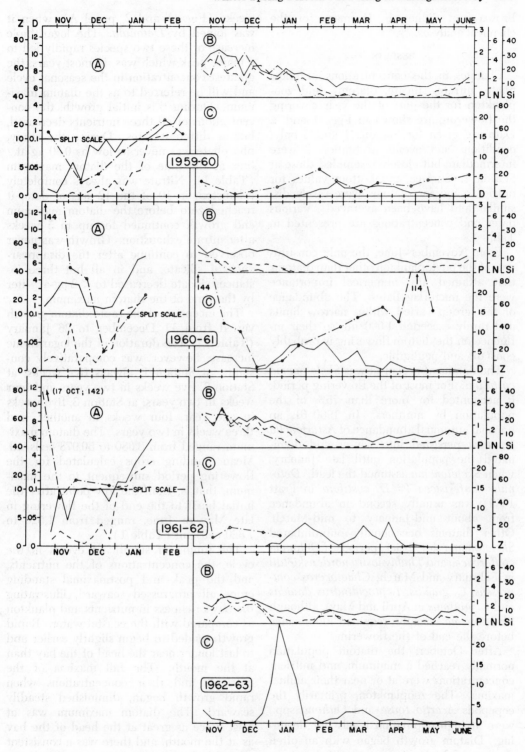

FIG. 2. Plankton and nutrients, Station 3. Legend as in Fig. 1.

Brown and Sidney S. Herman, for the chemical analyses.

RESULTS

Changes in the concentrations of phosphate, nitrate, silicate, diatoms, and zooplankton for the part of the year covering the flowering are shown in Figs. 1 and 2. Data are given for Stations 1 and 3 only; conditions and events at Station 2 were intermediate but closely resembled those at Station 1. Surface and bottom values for nutrients and for diatoms were rarely dissimilar and have been averaged. Various dates and concentrations are presented in Table 1.

From November–June, the only constituents of the phytoplankton other than diatoms that attained any numerical importance were the microflagellates. The abundance of this group varied within narrow limits and rarely exceeded 1,000/ml, so their influence on the diatom flowering is probably constant and negligible.

The dominant organism was *Skeletonema costatum*. For most of the flowering period, it accounted for more than 75% of the population by numbers. In 1960–61, an unusual autumnal abundance of *Asterionella japonica* persisted into winter and dominated the population until late January, when *Skeletonema* assumed the lead. *Detonula confervacea* (= *D. cystifera* in Pratt 1959) was usually second in abundance from about mid-January to mid-March. Other diatoms occasionally outnumbered *Skeletonema* for brief periods (*Detonula confervacea* and *Thalassiosira nordenskiöldii* in February and March; *Chaetoceros compressus*, *C. gracilis*, *Leptocylindrus danicus* and *L. minimus* in April and May), although *Skeletonema* usually regained dominance before the end of the flowering.

After October, the diatom population normally reached a minimum, and nutrient concentrations were at or near their annual maxima. The zooplankton, primarily the copepods *Acartia tonsa* and *Oithona* spp., were at high population levels but declining. Diatom growth began with an often slow and faltering increase in *Skeletonema* that usually gathered momentum in a few

weeks (Figs. 1 and 2, panel A), when it was joined by *Detonula*. The logarithmic increase of these two species rapidly led to a major peak which was, in most years, the greatest concentration in the seasonal cycle and will be referred to as the diatom maximum. During this initial growth, the concentrations of all three nutrients decreased, but at different rates. During the study, phosphate was never as low as 0.30 μg-at./liter at the time of the diatom maximum (Table 1). Nitrate was always completely exhausted; in 10 of the 12 sets of data, it reached zero before the diatom maximum and growth continued for up to 5 weeks after nitrate exhaustion. Growth was never observed to continue after the disappearance of silicate, and in all but three instances silicate decreased to < 2 μg-at./liter by the time of the diatom maximum.

The inception of the rapid diatom growth varied from 11 December to 26 January (Table 1). The duration of the logarithmic increase, however, was comparatively constant: at Station 1, always five weeks; at Station 2, five weeks in two years and four weeks in two years; at Station 3, five weeks in one year, four weeks in another, and three weeks in two years. The diatom maximum ranged from 3,680 to 50,978 cells/ml. Mean standing crops calculated for the flowering period subsequent to the maximum, that is, from the low point after the initial burst to the end of the flowering in late May or June, ranged from 1,520 to 7,620 cells/ml (Table 1).

The timing of significant events in the cycle, the concentrations of the nutrients, and the peak and postmaximal standing crops all progressed seaward, illustrating the bay's richness in nutrients and plankton as compared with the coastal water. Rapid growth tended to begin slightly earlier and to last longer near the head of the bay than at the mouth. The fall maxima of the nutrients and their concentrations when rapid growth began, diminished steadily seaward. The diatom maximum was at least twice as great at the head of the bay as at the mouth, and there was a consistent seaward decrease in mean standing crops for the remainder of the flowering. The

TABLE 1.   Diatoms and selected environmental variables during various stages in the winter–spring flowering.
(All concentrations and temperatures are means of surface and bottom values)

Year Station	Fall nutrient maxima (μg-at./liter)			Preceding rapid growth			At beginning of rapid growth				During growth		At diatom maximum					Postmaximal flowering	
	P	N	Si	Mean temp., C*	Mean langleys per day*	Last known zooplankton conc. (10³/m³)	Date†	P	N	Si	Mean temp., C	Mean langleys per day	Date	Diatoms/ml	P	N	Si	Dates	Mean diatoms/ml
1959–60																			
1	3.07	7.56	44.5	2.07	173	14.5	26 Jan	1.18	0	20.2	2.74	253	3 Mar	7,515	0.63	0	0.41	15 Mar– 7 June	3,680
2	2.89	6.95	41.0	2.13	173		26 Jan	1.15	0	18.9	2.66	236	23 Feb	8,495	0.52	0	2.80	15 Mar–24 May	2,730
3	2.07	4.63	27.0	4.28	173	8.0	26 Jan	1.07	0	13.9	3.41	236	23 Feb	3,680	0.47	0	1.50	3 Mar– 3 May	1,920
1960–61																			
1	1.66	3.75	18.3	3.98	157	9.5	20 Dec	1.12	0.69	5.3	1.61	184	24 Jan	12,100	0.40	0	1.10	14 Feb– 6 June	4,720
2	1.54	3.47	14.9	4.83	157		20 Dec	1.01	0.26	5.3	1.60	184	24 Jan	12,894	0.41	0	0	14 Feb– 1 June	3,210
3	1.20	1.58	9.0	7.72	157	3.4	20 Dec	0.94	0.22	6.5	3.98	184	24 Jan	6,985	0.51	0	0.80	7 Feb–11 Apr	3,360
1961–62																			
1	2.54	4.78	40.2	3.53	122	6.0	27 Dec	1.70	2.01	27.4	2.08	170	30 Jan	22,590	0.41	0	0.65	27 Feb–29 May	7,620
2	2.10	4.45	33.2	5.65	122		27 Dec	1.52	1.66	25.6	2.41	170	30 Jan	18,701	0.34	0	0.23	27 Feb–29 May	4,130
3	1.78	3.68	27.8	4.52	144	10.0	9 Jan	1.04	0.98	10.9	3.88	181	30 Jan	6,011	0.62	0	1.40	27 Feb– 8 May	2,450
1962–63																			
1	1.80	3.06	28.5	6.88	151		11 Dec	1.36	2.45	26.6	1.30	146	15 Jan	50,978	0.32	0	0	12 Feb–28 May	6,360
2	1.51	2.65	25.6	6.80	151		11 Dec	1.24	1.79	21.2	1.35	146	8 Jan	40,499	0.36	0	3.05	12 Feb–28 May	3,200
3	1.18	1.99	20.6	6.37	136		18 Dec	1.35	1.99	16.3	2.98	142	8 Jan	25,157	0.53	0	4.25	12 Feb–28 May	1,520

* Mean temperatures and langleys/day for 2-week period preceding beginning of rapid growth. One langley = one calorie/cm². Radiation data for Newport, R.I. from U.S. Department of Commerce.
† Dates of beginning of rapid growth from inspection of Figs. 1 and 2 (panel C).

seaward station (3) mirrored, in reduced amplitude, conditions at the other stations.

DISCUSSION

Smayda (1957) related the early inception of the winter–spring flowering to the bay's depth, tidal mixing, and flushing rate. These factors combine to render unnecessary, in Narragansett Bay, the degree of stability and incident radiation required for the development of the flowering in deeper and more exposed bodies of water in temperate regions, including areas similar to the bay, such as Long Island Sound (Riley 1959). When rapid growth begins in the fall or early winter, nutrient concentrations have passed their annual maxima and have been diminishing for a few weeks. The problem of the inception of growth is accounting for its delay.

There is considerable year-to-year variation in the timing of the flowering, and an explanation of this variation might identify the events or conditions responsible for the initiation of growth. Correlations of the onset of logarithmic diatom increase with various environmental factors have therefore been sought. The stability at this time of year is negligible, and its annual variations show no relation to the timing of the bloom. Mean water temperatures and solar radiation for the two weeks preceding the beginning of rapid growth, together with the last known zooplankton concentration, are given in Table 1. Neither the temperature nor the incident radiation shows any consistent relation to the date of inception of diatom growth. (It is true that in three of the four years light intensities *during* rapid growth were higher than in the preceding two-week period, possibly indicating a stimulus to photosynthesis. On the other hand, this was not the case in the last year, which produced the largest diatom population.)

Zooplankton concentrations, by contrast, yield a consistent picture. They were invariably low just before the diatoms bloomed; the growth of *Skeletonema*, and of the diatoms generally, always began after the zooplankton had decreased from an autumnal maximum to low values (Figs. 1 and 2A). Martin (1965), working with these same data, concludes that diatom growth in October is suppressed, under nutritionally favorable conditions, by the heavy grazing of a dense zooplankton population at moderate water temperatures. The population decrease in late fall, coupled with dropping temperatures, results in a marked reduction in grazing pressure, and the diatoms are allowed to increase. He argues that the grazing of the zooplankton dominants *Acartia tonsa* and *Oithona* spp. is especially severe for *Skeletonema*, and cites Curl and McLeod (1961), who state that this diatom is a particularly good food source for *Acartia tonsa*. This interpretation is here endorsed: the inception of the winter–spring flowering is evidently triggered by the release of grazing pressure.

The relatively constant length of the logarithmic growth period contrasts sharply with the range of maxima resulting from that growth. For example, at Station 1 this period lasted five weeks each year. The diatom standing crop at the beginning of growth varied only between 77 and 798 cells/ml, from year to year, but the subsequent maxima ranged from 7,515 to 50,978 cells/ml. It follows that the ultimate size of the population is determined neither by the initial seeding nor by the duration of growth but by the rate of growth.

The diatom maximum resulting from this logarithmic burst at the beginning of the flowering is compared with environmental factors that might affect its magnitude (Table 1): the fall maxima of phosphate, nitrate, and silicate, mean temperature and incident radiation during the period of rapid growth, and the concentrations of these nutrients at the time growth began. While there appears to be a slight negative correlation with temperature, the narrow range of the prevailing temperatures (2.68C) can scarcely have exerted an effect in any way commensurate with the spread of diatom maxima. The relation between diatom maxima and incident radiation during the period of logarithmic increase is actually inverse, and a causal connection at these low light intensities would appear to be impossible. The zooplankton data for this

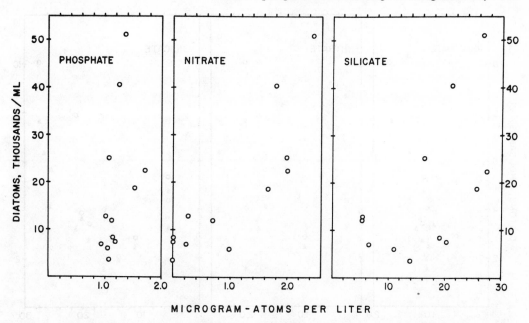

FIG. 3. Nutrient concentrations at beginning of logarithmic growth and subsequent diatom maxima. Correlation coefficients: phosphate, 0.507 (not significant); nitrate, 0.842 (significant at 1% level); silicate, 0.581 (5% level).

period are inadequate for quantitative comparison with the diatom maxima, but they do not suggest any consistent relationship (*see* Figs. 1 and 2).

It would be reasonable to assume that the size of the winter diatom maximum is controlled at least in part by nutrient supplies accumulated during autumn. The maximum concentrations of phosphate, nitrate, and silicate in the fall were therefore plotted against the subsequent diatom peaks. The resulting scatter of points was nearly random: no correlations were discernible. However, the logarithmic growth of the diatoms usually did not begin until several weeks after the fall nutrient maxima, and during this interval the concentrations of each of the nutrients diminished substantially.

In view of the delay in the start of growth, the influence of nutrients on the height of the diatom maximum was assessed by comparing the diatom maxima with the store of nutrients at the time that rapid growth began. Positive relationships of varying strength were found for all three nutrients (Fig. 3). The correlations between nu-

trient concentrations and diatom abundance extended beyond the initial peak. Fig. 4 shows the waning influence of the initial nutrient endowment persisting, in varying degrees, throughout the winter–spring flowering.

The involvement of all three nutrients in these relationships leads to questions of their possible influence in limiting diatom growth, as well as their relative importance. Nitrate was always exhausted during the initial burst of growth and silicate was nearly exhausted, although when the diatom maximum was reached and the population began to decline, phosphate concentrations ranged from 0.32 to 0.63 μg-at./liter (Table 1). The spring flowering in the English Channel begins at phosphate concentrations of approximately 0.45–0.70 μg-at./liter (as summarized in Raymont 1963, p. 148). Various authors (*see* Harvey 1945, p. 133; Riley 1946), citing the experimental work of Ketchum (1939*a*), give 0.55 μg-at./liter as the approximate concentration where phosphate values begin to limit phytoplankton growth, although Steele (1958) cites the figure of 0.40 μg-at./liter. It therefore seems

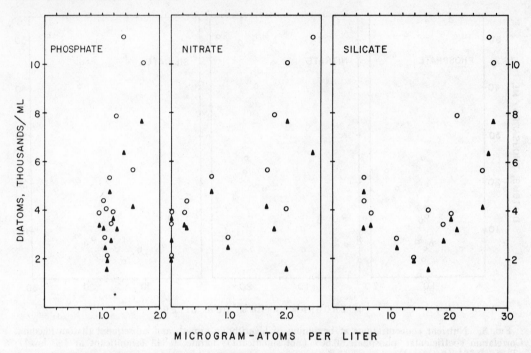

FIG. 4. Nutrient concentrations at beginning of logarithmic growth plotted against diatom abundance during flowering. Circles: mean of diatoms for entire flowering period. Correlation coefficients: phosphate, 0.736 (significant to 1% level); nitrate, 0.776 (1% level); silicate, 0.635 (5% level). Triangles: mean diatoms for postmaximal portion of flowering. Correlation coefficients: phosphate, 0.773 (1% level); nitrate, 0.479 (not significant); silicate, 0.491 (not significant).

unlikely that phosphate levels at the diatom maximum in Narragansett Bay limit further growth; in any event their effects must be negligible compared with those of nitrate and silicate. It will be suggested that the magnitude of the diatom maximum is determined by the concentrations of nitrate and silicate. If this is true, the most plausible explanation of the correlation between initial phosphate values and diatom maxima is that phosphate concentrations merely reflect the concentrations of the other nutrients. In years when rich supplies of nitrate and silicate are present at the onset of growth, phosphate values are correspondingly high because the concentrations of all three nutrients are governed by similar processes.

Evidently the nitrate supply is low in southern New England waters. Off the coast, Ketchum, Vaccaro, and Corwin (1958) report between 5 and 6 μg-at./liter in January and < 1 μg-at./liter in May, July, September, and November, with N : P ratios

varying seasonally from near zero to about 10 : 1 by atoms. Although Long Island Sound is better supplied in midwinter, with usual maxima of some 20 μg-at./liter and N : P ratios of about 8 : 1, nitrate concentrations are close to zero from the end of the spring flowering until September (Riley and Conover 1956). The scant supply of this nutrient probably imposes a significant limitation on phytoplankton growth in this general region.

The above limitation appears to be especially stringent in Narragansett Bay, where nitrate is scarcer than any other nutrient measured (including iron, Smayda 1957). In most years, it is not detectable except during the fall and early winter. Annual maxima rarely reach 5 μg-at./liter. Not only are the absolute concentrations unusually low but, in the 4½ years of weekly measurements, the N : P ratio was < 1 in 83% of the observations, and only for a 4-week period in December 1959–January 1960 did it exceed 4 : 1. The flowering's initial burst completely exhausts the nitrate supply, but growth continues for a few weeks thereafter. This indicates the production of nitrate deficient cells, a condition similar to that induced by Ketchum (1939*b*) in cultures of *Chlorella pyrenoidosa,* or a rate of nitrate supply just adequate to sustain continuing growth but not high enough to produce a measurable accumulation, or perhaps continued growth using ammonia or possibly dissolved organic nitrogen.

Since ammonia has not been measured in the bay, the information on its concentration in adjacent waters should be examined. Vaccaro (1963) has emphasized the necessity of accounting for ammonia, as well as nitrate (which, as measured, includes nitrite), in the total supply of inorganic nitrogen available for plant growth. He reports ammonia concentrations for August and January in the coastal waters of southern New England. During August, when nitrate in the photic zone is present only as traces, ammonia is the major source of available nitrogen. However, in January, before the main diatom flowering, nitrate is at its annual maximum and ammonia, at its seasonal minimum, is a minor fraction of the total available nitrogen. Harris (1959) described changes in nitrate, nitrite, ammonia, particulate nitrogen, and dissolved organic nitrogen from October 1954 to June 1955 in Long Island Sound. When the phytoplankton of the central part of the sound began its winter increase in December, more than half of the inorganic nitrogen was present as ammonia. Although the early stages of growth rapidly and preferentially diminished the ammonia, when the total inorganic nitrogen concentration fell to 0.6 μg-at./liter in February, ammonia again accounted for more than half of this low value.

The omission of ammonia measurements in the bay was based on the assumption that ammonia is preferentially absorbed by the phytoplankton, while nitrate is consumed only after ammonia is exhausted (Fogg 1953, p. 74). Harris' (1959) data suggest that this assumption may not be altogether justified. If it is valid, the omission of ammonia determinations simply means that some of the available nitrogen was not measured, and that the relationship of diatom growth to total nitrogen might be stronger than that observed with nitrate (+ nitrite) alone. If the assumption is not valid, ammonia persisting after nitrate depletion may have accounted for some of the additional diatom growth. In any event, nitrate appears to be important in regulating the amount of growth.

The possibility of a silicate limitation must also be examined. This nutrient has been measured over a long period of years in the English Channel (*see* Armstrong and Butler 1963), where maximal concentrations of only about 5 μg-at./liter occur in winter, followed by nearly complete exhaustion coincident with the spring diatom flowering. For the coastal waters of southern New England, Richards (1958) gives values of 0–20 μg-at./liter in May and July and suggests that in the latter month silicate limited phytoplankton growth.

In Narragansett Bay, silicate is usually undetectable during most of the winter and spring, reappears as the flowering wanes, attains an annual maximum of 30–60 μg-at./ liter (often in July), and is still present in considerable concentrations in the fall. In Fig. 5, there is a suggestion that the growth following nitrate exhaustion is influenced by the remaining silicate, but the correlation is poor. The silicate supply appears to be critical as the annual diatom maximum is reached. Not only is the magnitude of this maximum a function of the initial silicate concentration but, in the present study, growth was never observed to continue after silicate depletion, and in 7 of the 12 instances the population reached its maximum just as silicate approximated zero and decreased sharply the following week. Since observations were made only once a week, we cannot state that the maximum and the exhaustion of silicate were coincident, but a simple calculation suggests that these events may have been simultaneous. In the 10 instances in which growth apparently ceased in the presence of measurable silicate, extrapolation of the previous week's rates of population growth and utilization of silicate per population increment indi-

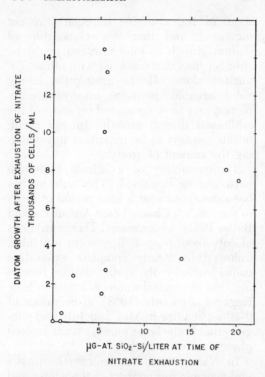

Fig. 5. Diatom growth after exhaustion of nitrate plotted against silicate concentration at the time of nitrate exhaustion, suggesting that the additional growth was a function of the remaining silicate.

cates that the remaining silicate would have been consumed in less than one week. It may be that this did happen, and that during that week the population declined from the actual (but unobserved) maximum to the level noted on the next sampling date.

Although culture conditions can be manipulated to induce some diatom species to develop without siliceous frustrules (*see* Hendey 1946; Jørgensen 1957), normally the silicon requirements of diatoms, while varying with the species, are relatively inflexible. According to Jørgensen (1953), cell division ceases in cultures of *Nitzschia palea* and *Bacillaria paradoxa* at silicate concentrations of 1.0–1.4 µg-at./liter, and silicate values in eutrophic Danish lakes decrease to this same level during the diatom maximum. Lund (1950) discusses the relatively high limiting concentration (about 8.3 µg-at./liter) that frequently stops the spring

increase of *Asterionella formosa* in the English Lake District. He demonstrates that the amount of silicate per unit area of cell in this species is nearly constant under all conditions, and explains the fact that this diatom often reduces the silicate concentration below 8.3 µg-at./liter by postulating that favorable illumination stimulates the cells to continued division in the absence of sufficient silicate for wall formation with death resulting.

Evidently the silicate requirement of the diatoms (predominantly *Skeletonema*) at the time of the winter maximum in Narragansett Bay differs from their nitrate requirement in that cell division cannot proceed in the absence of measurable silicate concentrations. The frequent coincidence of virtual exhaustion of silicate with cessation of growth indicates that the supply failure of this nutrient can at times produce a growth limitation more rigid than that imposed by the lack of nitrate.

It therefore appears that the magnitude of the flowering is determined by the concentrations of both nitrate and silicate at the beginning of rapid growth and, at the diatom maximum, there are reasons for believing that each of these nutrients limits further growth. The claim for silicate as the controlling nutrient appears to be more concrete than that for nitrate. It was noted that in every population the termination of growth may have completely exhausted the available silicate. While this would point to silicate as the critical factor, we cannot state positively that this occurred. It is also possible that the actual maximum for any given population was reached before the observed maximum or approximately coincided with it. Fig. 6 illustrates growth after nitrate exhaustion in relation to silicate concentration at the observed population maximum, i.e., at the apparent termination of growth. Assuming that the actual population maxima did not come after the observed maxima, the absence of points in the lower right part of the figure indicates that in no case did growth cease in the presence of a substantial concentration of silicate, except when considerable growth had occurred following the exhaustion of nitrate.

The upper right three points would then be interpreted as representing populations that stopped growing for want of nitrogen. Summarizing, if no substantial growth occurred after the observed maxima, and provided that Fig. 6 reflects the true pattern, the cessation of growth is not always caused solely by silicate depletion. These considerations make it likely that the availability of nitrogen in some form is involved, and the amount of growth after exhaustion of nitrate may be an index of the rate of nitrate supply or of the concentrations and rates of supply of other forms of nitrogen.

Factors responsible for the secondary peaks in diatom abundance in the postmaximal part of the flowering cannot be positively identified. However, in February 1963, following the extraordinary bloom of the previous month, diatom concentrations were lower at all stations than in any previous year before April. The consequent resurgence of nutrients in late February apparently produced the secondary peak of early March. Other minor bursts of growth may have been generated by temporary accelerations of nutrient supply that did not measurably increase nutrient concentrations. The effects of the animal population during winter and spring cannot be inferred from existing data although, as the water warms, the accelerated zooplankton metabolism doubtless combines with nutrient depletion to terminate the flowering.

The correlation of nutrient concentrations at the beginning of growth with the size of the postmaximal flowering population suggests that in the more successful years the larger amounts of nutrients captured by the initial bloom are retained and recycled in the bay's system of plankton, bacteria, detritus, solutes, and organic aggregates (Riley 1963), in large part accounting for the maintenance of relatively dense diatom populations in the postmaximal period. It is generally true that the earlier the growth begins, the greater the winter maximum and the mean standing crop for the remainder of the flowering period. This relationship involves the fact that, during the fall, while

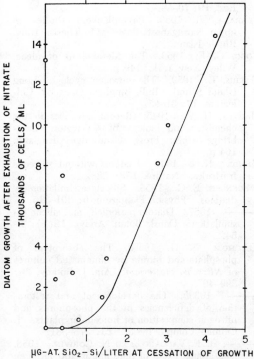

FIG. 6. Diatom growth after exhaustion of nitrate plotted against silicate remaining at the observed population maxima. If growth terminated before or at the observed maxima, the three points upper right appear to indicate nitrogen limitation (*see* text).

intense grazing holds the diatoms in check, nutrient stores are being expended without augmenting the standing crop. The release in grazing pressure not only sets the date for the inception of growth, but thereby checks the dwindling of nutrients. The earlier this happens, the greater will be the remaining nutrient endowment that governs the magnitude of the winter growth. In some measure, then, the autumnal zooplankton activity determines the success of the entire flowering.

REFERENCES

ARMSTRONG, F. A. J. 1951. The determination of silicate in sea water. J. Marine Biol. Assoc. U.K., **30**: 149–160.

———, AND E. I. BUTLER. 1963. Chemical changes in sea water off Plymouth in 1961. J. Marine Biol. Assoc. U.K., **43**: 75–78.

CURL, H., AND G. C. McLEOD. 1961. The physiological ecology of a marine diatom, *Skele-*

tonema costatum (Grev.) Cleve. J. Marine Res., **19**: 70–88.

FERRARA, R. 1953. Phytoplankton studies in upper Narragansett Bay. M.S. Thesis. Univ. Rhode Island.

FOGG, G. E. 1953. The Metabolism of algae. Wiley, New York. 149 p.

HARRIS, E. 1959. The nitrogen cycle of Long Island Sound. Bull. Bingham Oceanog. Collection, **17**: 31–65.

HARVEY, H. W. 1945. Recent advances in the chemistry and biology of sea water. Cambridge Univ. Press, Cambridge, England. 164 p.

HENDEY, N. I. 1946. Diatoms without siliceous frustrules. Nature, **158**: 588.

JØRGENSEN, E. G. 1953. Silicate assimilation by diatoms. Physiol. Plantarum, **6**: 301–315.

———. 1957. Diatom periodicity and silicon assimilation. Dansk Botan. Arkiv., **18**(1): 1–54.

KETCHUM, B. H. 1939*a*. The absorption of phosphate and nitrate by illuminated cultures of *Nitzschia closterium*. Am. J. Botany, **26**: 399–407.

———. 1939*b*. The development and restoration of deficiencies in the phosphorus and nitrogen composition of unicellular plants. J. Cellular Comp. Physiol., **13**: 373–381.

———, R. F. VACCARO, AND N. CORWIN. 1958. The annual cycle of phosphorus and nitrogen in New England coastal waters. J. Marine Res., **17**: 282–301.

LUND, J. W. G. 1950. Studies on *Asterionella formosa* Hass. II. Nutrient depletion and the spring maximum. J. Ecol., **38**(1): 1–35.

MARTIN, J. H. 1965. Phytoplankton-zooplankton relationships in Narragansett Bay. Limnol. Oceanog., **10**: 185–191.

MULLIN, J. B., AND J. P. RILEY. 1955. The spectrophotometric determination of nitrate in natural waters, with particular reference to sea-water. Anal. Chim. Acta, **12**: 464–480.

PRATT, D. M. 1959. The phytoplankton of Narragansett Bay. Limnol. Oceanog., **4**: 425–440.

RAYMONT, J. E. G. 1963. Plankton and productivity in the oceans. Macmillan, New York. 660 p.

RILEY, G. A. 1946. Factors controlling phytoplankton populations on Georges Bank. J. Marine Res., **6**: 54–73.

———. 1959. Environmental control of autumn and winter diatom flowerings in Long Island Sound, p. 850–851. *In* Mary Sears [ed.], Preprints Intern. Oceanog. Congr. New York 1959, AAAS.

———. 1963. Organic aggregates in sea water and the dynamics of their formation and utilization. Limnol. Oceanog., **8**: 372–381.

———, AND S. M. CONOVER. 1956. Oceanography of Long Island Sound, 1952–1954. III. Chemical oceanography. Bull. Bingham Oceanog. Collection, **15**: 47–61.

RICHARDS, F. A. 1958. Dissolved silicate and related properties of some western North Atlantic and Caribbean waters. J. Marine Res., **17**: 449–465.

SMAYDA, T. J. 1957. Phytoplankton studies in lower Narragansett Bay. Limnol. Oceanog., **2**: 342–359.

STEELE, J. H. 1958. Plant production in the northern North Sea. Scot. Home Dep. Marine Res. Ser., **7**: 3–36.

U.S. DEPARTMENT OF COMMERCE, WEATHER BUREAU. Climatological Data National Summary.

VACCARO, R. F. 1963. Available nitrogen and phosphorus and the biochemical cycle in the Atlantic off New England. J. Marine Res., **21**: 284–301.

WATTENBERG, H. 1937. Critical review of the methods used for determining nutrient salts and related consituents in salt water, 1. Rappt. Proces-Verbaux Reunions, Conseil Perm. Intern. Exploration Mer, **103**: 5–26.

PHYTOPLANKTON–ZOOPLANKTON RELATIONSHIPS IN NARRAGANSETT BAY[1]

John H. Martin

Graduate School of Oceanography, University of Rhode Island, Kingston

ABSTRACT

Zooplankton samples collected every other week in upper and lower Narraganset Bay, Rhode Island, were analyzed quantitatively and qualitatively. The seasonal occurrence and abundance of the various species identified and enumerated in the present study are presented in tabular form.

Analysis of the relative abundance of the principal grazers in the zooplankton population (*Acartia tonsa, A. clausi,* and *Oithona* spp.) and of the dominant phytoplankton species, *Skeletonema costatum,* upon which these copepods are believed to graze preferentially, shows that the seasonal changes in phytoplankton abundance are to a considerable extent attributable to differential grazing pressures.

Concentrations of plant nutrients (nitrate, phosphate, and silicate) were compared with zooplankton numbers. It is believed that the large summer–fall zooplankton population contributes to the phosphate maximum and possibly indirectly to the nitrate maximum as well.

INTRODUCTION

Zooplankton samples were collected over a three-year period in connection with a study of the phytoplankton of Narragansett Bay, Rhode Island, conducted by Pratt (1965). The collections were made near the head and the entrance of the west passage. Previous investigations have been limited to winter–summer observations of Williams (1906), and the one-year collections of Frolander (1955) and Faber (1959). Since both of the latter studies were restricted to single stations in the lower west passage, the annual cycle of the zooplankton in the upper parts of the bay has remained undescribed. The present study minimizes year-to-year variation and provides the first information on the less saline and phytoplankton-rich inner reaches of the bay. The objectives were to observe the seasonal cycle of the zooplankton species present at the two stations and to determine the effects of this population on the phytoplankton by grazing and possible enrichment of nutrients through excretion.

The author is indebted to Dr. David M. Pratt, Dr. Theodore A. Napora, and Dr. Milton Salomon for their assistance in this study.

[1] Based on a thesis submitted in partial fulfillment of the requirements for the M.S. degree, University of Rhode Island, 1964. This study was aided by Office of Naval Research Contract Nonr-396(03).

METHODS AND MATERIALS

The sampling points were two of Pratt's (1965) locations: Station 1 near the head of the west passage (41°38'08" N lat, 71°22'17" W long) and Station 3 at the entrance (41°26'47" N lat, 71°25'09" W long). Every other week from 28 January 1959 to 9 January 1962, zooplankton samples were collected with a Clarke-Bumpus sampler equipped with either a No. 2 or No. 12 net, in 4-min oblique tows. From these collections, counts of at least 1,000 specimens per sample were made on aliquots and converted to numbers per cubic meter, using a standard conversion factor of 4 liters per revolution of the sampler's flowmeter. The No. 12 net samples are believed to give a truer picture of the population than those taken by the No. 2 net, which did not retain quantitatively the smaller forms such as *Oithona* spp., *Microsetella norvegica,* copepod nauplii, early benthic larvae, and so on.

In an attempt to minimize the large variations arising from mechanical and natural causes, the concentrations of zooplankton collected in each month over the three-year period were averaged, yielding one estimate per month. In the calculation of this figure, the highest and lowest numbers for individual species or taxonomic groups were excluded and the remaining values were averaged. In the case of the smaller organisms not sampled adequately by the No. 2

307

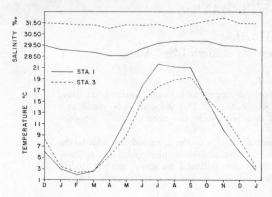

FIG. 1. Three-year averages of the salinity and temperature at Stations 1 and 3.

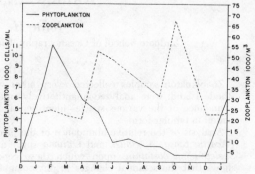

FIG. 2. Three-year average of the abundance of total phytoplankton and total zooplankton captured in No. 12 nets at Station 1. (No samples were taken with this size net in August.)

nets, averages were based on the existing No. 12 net collections. It is believed that this method yields the most reasonable estimate of the zooplankton population from the existing samples, since it minimizes errors due to plankton patchiness, year-to-year variations, atypical water masses, net clogging, and other procedural and natural deviations from normal. For comparison with the zooplankton, the data on temperature, salinity, phytoplankton, nutrients, and precipitation were treated in a similar manner.

The phytoplankton counts used in this study were made available by Dr. David M. Pratt. Chlorinity was measured by titration, inorganic phosphate by the Deniges-Atkins procedure (Wattenberg 1937), nitrate by the Mullin and Riley (1956) method, and silicate by the Armstrong (1951) method.

RESULTS

Total numbers of zooplankton (minus nauplii) were approximately equal at the two stations over the three-year average, as totals of the 12 "average" months were 259,000 m^{-3} yr^{-1} at Station 1 and 236,000 m^{-3} yr^{-1} at Station 3. The main periods of abundance in the headwaters of the bay were observed in May (three-year average 52,500/m^3) and October (67,500/m^3), while at the mouth of the bay, peaks occurred in February (24,000/m^3), April (44,000/m^3), and July (72,000/m^3). The major differ-

ences between the two areas were as follows: at Station 3 the copepod population always exceeded 70% of the total, while at Station 1 the copepods averaged 70% only from July through January, other holoplankton comprising at least 35% of the population from February through June. Larval numbers throughout the year were approximately equal in the two areas.

At the mouth of the bay (Station 3), 55 species and groups of species were recorded, including 26 copepods, 21 other holoplanktonic forms, and the larvae of 8 benthic taxa. At Station 1, 37 species and groups of species were found, including 16 copepods, 14 other holoplanktonic organisms, and 7 benthic forms.

The variation in the composition and abundance of the populations results from the occurrence and relative success of the different species existing at the salinities and temperatures of the two given areas (Fig. 1). The above-mentioned species together with their average monthly abundance are presented in Table 1. Seasonal changes in the total population are shown in Fig. 2 (Station 1) and Fig. 3 (Station 3).

DISCUSSION

Grazing

In an attempt to determine the effect of grazing on the bay's phytoplankton population, it is necessary to assess the importance

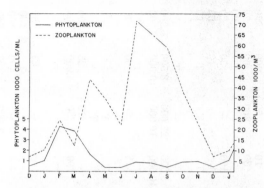

FIG. 3. Three-year average of the abundance of total phytoplankton and total zooplankton captured in No. 12 nets at Station 3. (No samples were taken with this size net in August.)

of the various means of phytoplankton removal. The three principal agencies are: 1) transport from the area by nontidal currents, 2) death and decomposition in the water column, and 3) grazing by the zooplankton. (Sinking to the bottom may be practically eliminated, since Narragansett Bay is a shallow, well-mixed body of water.) The first method, removal by currents, is important mainly at Station 3, where water leaving the bay enters the Rhode Island Sound current gyre. In the vicinity of the bay's entrance, the currents of this gyre set continuously southwest (Hicks 1959), and plankton thus leaving the bay is not returned. In contrast, the headwaters of the bay (Station 1) are relatively stable, and the phytoplankton, physically retained in this area, is removed mainly by death and grazing. For this reason, the discussion of grazing will be concentrated on the upper bay area.

In the winter at Station 1, the combination of dense phytoplankton and sparse zooplankton concentrations indicates that grazing pressure must be minimal. This condition may be attributed to the absence or limited abundance of the major bay grazers at the existing temperature-salinity combinations; that is, temperatures are too low for adequate production of *Acartia tonsa*, and *Oithona* spp., while salinities are suboptimal for the common colder-water forms such as *Pseudocalanus minutus*.

Acartia clausi, another important bay copepod, also occurs in limited abundance from February to May (Table 1). Although this form is tolerant of rather low salinities and temperatures, minimum bay temperatures may curtail winter–spring production, as indicated by the fact that Jeffries (1962) found no production of *A. clausi* in confined estuaries after the temperature fell below 4C. The winter–spring limitation of these forms is important because species of *Acartia* and *Pseudocalanus* have been shown to feed selectively on *Skeletonema costatum*, the dominant diatom of the winter–spring flowering. In describing experiments with *S. costatum*, Curl and McLeod (1961) state: "When plankton tows containing large quantities of *Skeletonema*, together with *Rhizosolenia*, *Nitzschia*, *Guinardia*, *Thalassionema*, and *Chaetoceros*, were set aside in the culture room, copepods (mainly *Acartia tonsa* and *Pseudocalanus* sp.) rapidly reduced the numbers of *Skeletonema*, leaving uneaten cells of *Rhizosolenia* which rapidly outgrew the other genera. *Skeletonema* seemed to be the favored food among the diatoms available, so far as these two species of copepods were concerned." Although it may not be assumed that *Acartia clausi* similarly selects *Skeletonema costatum*, it is known that this copepod will graze effectively on *Skeletonema* when it is abundant (R. Conover 1956). Since *Skeletonema* is by far the dominant bay phytoplankton organism from February through May, it seems logical to assume that any large *Acartia clausi* population present will utilize this diatom as its principal food source, and heavy grazing of *Skeletonema* will be inevitable. It is the sparcity of *Acartia clausi*, *Pseudocalanus minutus*, and the other grazers that permits *Skeletonema* production to proceed virtually uninhibited, resulting in the large population that is developed in the winter and maintained through the spring (Fig. 4).

Annual *Skeletonema* blooms of the magnitude and duration of those occurring in upper Narragansett Bay are extraordinary, if not unique. Other areas, for example, Long Island Sound, have only one short-

TABLE 1. Abundance of all organisms recorded at Stations 1 and 3 over the three-year period. T designates trace*

Organism	Station	Month											
		Jan	Feb	Mar	Apr	May	June	July	Aug	Sept	Oct	Nov	Dec
Copepods (adult only)													
Calanoid													
Acartia clausi	1	3,500	1,000	700	300	1,000	3,000	700	–	–	–	100	1,100
	3	300	200	300	200	400	3,500	4,000	T	T	–	–	200
Acartia longiremis	3	500	–	–	T	T	T	–	T	T	–	–	–
Acartia tonsa	1	T	–	–	–	–	–	9,500	6,000	1,500	3,500	9,500	1,100
	3	T	T	T	–	T	200	300	700	1,000	1,300	500	100
Calanus finmarchicus	3	T	T	T	T	T	T	T	T	–	–	T	T
Centropages hamatus	1	170	30	20	20	30	150	140	60	20	20	170	100
	3	50	100	–	70	270	310	500	530	200	110	130	50
Centropages typicus	3	10	–	–	–	–	–	T	T	30	60	260	80
Eurytemora herdmanni	1	–	–	–	–	–	120	70	10	–	–	–	–
	3	–	–	10	10	30	170	180	30	–	–	–	–
Paracalanus parvus	3	–	–	–	–	–	–	100	50	200	500	200	10
Pseudocalanus minutus	1	300	500	400	300	150	T	100	50	400	100	100	10
	3	500	1,700	1,300	1,900	1,700	1,300	1,900	200	10	50	200	150
Pseudodiaptomus coronatus	1	30	10	10	10	70	20	100	100	10	260	60	300
	3	T	–	–	–	–	–	100	100	30	50	200	20
Temora longicornis	1	60	30	50	30	30	20	T	–	100	–	T	20
	3	20	110	160	270	180	80	470	80	100	–	30	20
Tortanus discaudatus	1	20	10	20	50	60	40	10	–	–	T	10	50
	3	30	100	110	220	600	450	150	30	T	T	T	20
Cyclopoid													
Corycaeus spp.	3	T	–	–	–	–	–	–	–	–	–	T	T
Oithona spp.	1	1,000	500	500	400	100	200	5,000	?	8,000	22,000	8,500	5,200
	3	1,000	1,000	300	300	2,500	3,000	13,000	?	10,000	16,000	10,000	2,000
Oithona spinirostris	1	10	10	T	20	25	5	–	10	15	20	60	40
	3	–	–	–	–	110	–	–	–	–	–	–	–
Harpacticoid													
Harpacticus gracilis	1	–	T	T	T	T	–	–	–	T	T	T	–
	3	T	T	T	T	T	T	–	–	T	T	–	–
Microsetella norvegica	1	100	125	300	125	25	–	50	?	–	75	25	200
	3	700	900	850	1,000	300	250	300	?	100	750	5,200	800
Microsetella rosea	3	–	–	T	T	T	–	–	–	T	T	T	–
Temporary plankton													
Balanus larvae	1	750	300	100	100	100	–	50	40	T	T	25	50
	3	400	250	200	25	25	–	250	100	T	T	T	100
Bivalve larvae	1	T	–	T	100	–	20	250	250	50	1,000	400	100
	3	400	100	–	70	–	30	400	–	30	–	1,400	500
Bryozoan larvae	1	30	30	20	–	–	40	400	250	30	10	15	5
	3	–	–	–	–	10	–	–	–	–	–	–	–
Decapod larvae	1	–	–	–	–	–	–	–	–	–	–	–	–
	3	–	–	–	–	–	–	–	–	–	–	–	–

TABLE 1. (Continued)

Organism	Station	Jan	Feb	Mar	Apr	May	June	July	Aug	Sept	Oct	Nov	Dec
Temporary plankton (Continued)													
Fish eggs and larvae	1	–	T	T	T	T	T	T	T	T	–	–	–
	3	–	T	T	T	T	T	T	T	–	–	–	–
Gastropod larvae	1	–	–	–	–	T	1,500	2,200	400	100	100	100	–
	3	–	T	–	–	100	1,500	1,000	600	100	100	100	–
Littorina eggs	1	T	T	20	20	10	10	–	–	–	–	–	–
	3	10	160	200	40	10	150	50	25	25	500	–	–
Polychaete larvae	1	300	800	400	100	400	150	50	–	25	25	1,300	300
	3	100	600	300	200	100	25	25	25	25	25	250	50
Medusae													
Obelia spp.	1	–	–	T	135	80	–	–	–	T	20	T	–
	3	–	–	20	90	30	–	–	5	–	–	–	–
Rathkea octopunctata	1	–	–	5	125	70	5	20	–	T	–	–	–
	3	–	–	–	–	10	–	–	–	–	–	–	–
Sarsia tubulosa	1	–	T	T	T	T	–	–	–	–	–	–	–
Other Holoplankton													
Rotifers													
Asplanchna spp.	1	100	200	200	500	5,800	100	–	–	–	–	–	–
	3	–	200	300	–	1,000	–	–	–	–	–	–	–
Eucentrum spp.	1	200	4,800	9,400	1,000	100	–	–	–	–	–	–	–
	3	–	400	2,300	–	–	–	–	–	–	–	–	–
Synchaeta spp.	1	–	–	–	50	200	–	–	–	–	–	–	–
Cladocerans													
Evadni nordmanni	1	–	–	–	–	30	1,200	1,150	350	200	–	40	–
	3	–	–	–	–	–	400	800	230	200	100	T	30
Penillia avirostris	3	–	–	–	–	–	–	–	10	110	50	5	–
Podon polyphemoides	1	–	–	–	–	–	3,700	3,500	1,300	500	1,000	900	–
	3	–	–	–	–	–	100	500	100	1,000	100	50	–
Tunicates													
Oikopleura spp.	3	–	–	–	–	–	–	300	?	450	150	75	–
Salps													
Dolioletta gegenbouri	3	–	–	–	–	–	–	–	–	T	T	T	–
Chaetognaths													
Sagitta bipunctata	3	T	T	T	T	–	T	T	–	T	T	T	–
Sagitta elegans	1	25	20	20	25	10	–	–	–	–	–	–	5
	3	–	15	20	35	65	–	10	–	–	–	5	15
Mysids													
Neomysis americana	3	–	–	–	–	–	–	T	T	–	–	–	–

* A question mark is used in August when No. 12 net samples were unavailable for the smaller forms. The following organisms were found in traces once or twice at one or the other of the stations: the Calanoids, *Calanus helgolandicus, Candacea armata, Eucalanus elongatus, Labidocera aestiva, Meycynocera clausi, Metridia lucens, Rhincalanus nasutus*; the Harpacticoid, *Alteutha depressa*; the medusae, *Aurelia sp., Cosmetira pilosella, Cyanea capillata artica, Euphysa flamea, Hybocodon prolifer, Phialidium islandicum, Turritopsis nutricula*; the salp, *Muggsiaea sp.*; the Ostracods, *Conchoecia elegans, Conchoecia haddoni*; the polychaete, *Tomopteris helgolandica*; and the Ctenophore, *Mnemiopsis sp.*

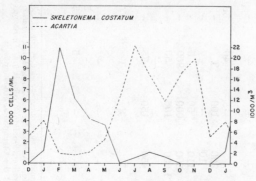

FIG. 4.   Three-year average of the abundance of *Skeletonema costatum* and of *Acartia* spp. (adult *A. tonsa, A. clausi,* and all *Acartia* copepodites).

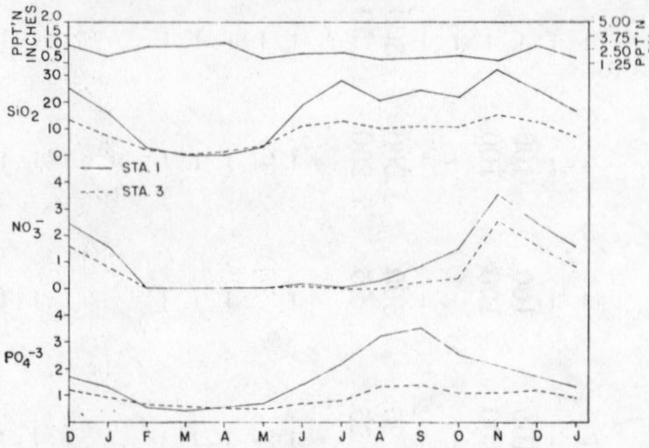

FIG. 5.   Three-year average of the concentrations of inorganic phosphate, silicate, and nitrate (μg-at./liter) at Stations 1 and 3 and the three-year average precipitation. (Precipitation data for Providence, Rhode Island from U.S. Department of Commerce, 1959, 1960, 1961.)

lived pulse of *Skeletonema* during the winter–spring period (S. Conover 1956). The differences between the *Skeletonema* populations of the two areas are probably due to many environmental variables, such as light, mixing rates, and so on, but it is interesting to note that R. Conover (1956) recorded much larger concentrations of *Acartia clausi* in Long Island Sound than those observed in Narragansett Bay during similar seasons. Perhaps, *A. clausi's* greater grazing contributes to the limitation of the *Skeletonema* population in the sound, thus preventing a sustained flowering such as that recurring annually in Narragansett Bay.

In Narragansett Bay, the winter–spring bloom is terminated when rising spring temperatures become favorable for reproduction by the zooplankton. At this time, the *Acartia* population (predominantly *A. clausi* with some immature *A. tonsa*) increases rapidly and, at the existing temperatures, can be assumed to be grazing at a high rate. Their increased grazing of *Skeletonema*, selective or otherwise, is thus primarily responsible for this diatom's sudden May–June decrease (Fig. 4).

Grazing pressure continues to increase during the hot summer months as the zooplankton population continues to expand. Consequently, the phytoplankton continues to decrease. As a result, competition for available food is maximal and perhaps contributes to the decline of the zooplankton

at Station 1 (Fig. 2). This may be true especially for *Acartia tonsa* which, according to R. Conover (1956), is an inefficient filter feeder whose numbers decrease when competition for available food becomes intense. (Its chief competitor at this time is *Oithona* spp.) *Skeletonema* continues in low abundance through the summer. However, short-lived blooms do occur, but they are rapidly consumed by the large zooplankton population.

At Station 1, annual phytoplankton minimum occurs during the fall months. While this may be due in part to reduced light intensities at relatively high temperatures, it would appear that in the case of *Skeletonema costatum*, the most important bay diatom, the principal factor is heavy grazing by the large zooplankton population. Curl and McLeod (1961, Table IV) reported the highest net production for this species in December, when light was at its minimum and the temperature was 12C. Since these conditions approximate those of the upper bay in the fall, particularly in November, it is assumed that *Skeletonema* should be quite productive and, with the abundance of nutrients (Fig. 5), a large standing crop would be expected. The absence of a considerable, sustained, autumn *Skeletonema* population is thus attributed to the intensive grazing of the large zooplankton population, including selective grazing by *Acartia tonsa* (Fig. 4). Consequently, the development of the annual *Skeletonema* bloom in Narragansett Bay is delayed until this grazing pressure is relieved. Because the waters cool more rapidly in the late fall and early winter, the zooplankton population (including *Acartia tonsa* and *Oithona* spp.) quickly diminishes, and the few remaining individuals (in part, *Acartia clausi*) graze at a reduced rate. With the decline in grazing, the phytoplankton (notably *Skeletonema costatum*) is favored by the existing environmental conditions and, in December, begins to increase to its yearly winter maximum.

At the mouth of the bay (Station 3), a similar seasonal progression occurs, with the exception that in the fall the zooplankton decreases markedly (Fig. 3), and the phytoplankton population becomes greater than that of the more heavily grazed Station 1. This is the only time in the year when the phytoplankton standing crop at the bay entrance exceeds that in the upper reaches. This fact, in conjunction with the simultaneous decline in zooplankton numbers at Station 3, points to the importance of zooplankton grazing in regulating phytoplankton abundance in Narragansett Bay.

Nutrient excretion

The zooplankton of Narragansett Bay must also have an important role in the production of nutrients, especially during the late summer and fall, when precipitation, and hence runoff, is relatively slight (Fig. 5). At this time, the heavily grazing, large zooplankton population probably provides large quantities of phosphate (Fig. 5) and ammonia (not measured in this study) that the phytoplankton, suppressed by grazing and now at its annual minimum, does not assimilate. The unutilized ammonia might subsequently be oxidized to appear as the nitrate that is measurable only in the fall and early winter (Fig. 5).

The Narragansett Bay zooplankton thus affects the phytoplankton population in two ways, which are of varying importance seasonally: by regulation of the standing crop by grazing, which increases from May to a maximum in autumn, and by augmentation of the nutrient supply, especially during the fall, thereby ultimately contributing to the success of the winter–spring flowering.

REFERENCES

ARMSTRONG, F. A. J. 1951. The determination of of silicate in sea-waters. J. Marine Biol. Assoc. U.K., **30**: 149–160.

CONOVER, R. J. 1956. Oceanography of Long Island Sound, 1952–1954. VI. The biology of *Acartia clausi* and *A. tonsa*. Bull. Bingham Oceanog. Collection, **15**: 156–233.

CONOVER, S. A. M. 1956. Oceanography of Long Island Sound, 1952–1954. IV. Phytoplankton. Bull. Bingham Oceanog. Collection, **15**: 62–112.

CURL, H. J., AND G. C. McLEOD. 1961. The physiological ecology of a marine diatom, *Skeletonema costatum* (Grev.) Cleve. J. Marine Res., **19**: 70–88.

FABER, D. J. 1959. Studies on the biology of the zooplankton of Narragansett Bay. M.S. Thesis, University of Rhode Island, Kingston. 131 p.

FROLANDER, H. T. 1955. The biology of the zooplankton of the Narragansett Bay Area. Ph.D. Thesis, Brown University, Providence, R.I. 94 p.

HICKS, S. D. 1959. The physical oceanography of Narragansett Bay. Limnol. Oceanog., **4**: 316–327.

JEFFRIES, H. P. 1962. Succession of two *Acartia* species in Estuaries. Limnol. Oceanog., **7**: 354–364.

MULLIN, J. B., AND J. P. RILEY. 1955. The spectrophotometric determination of nitrate in natural waters with particular reference to sea-water. Anal. Chim. Acta, **12**: 464–480.

PRATT, D. M. 1965. The winter–spring diatom flowering in Narragansett Bay. Limnol. Oceanog., **10**: 173–184.

U.S. DEPARTMENT OF COMMERCE. 1959, 1960, 1961. Local climatological data. Weather Bureau, Providence, R.I.

WATTENBERG, H. 1937. Critical review of the methods used for determining nutrient salts and relative constituents in salt water. Rappt. Proces-Verbaux Reunions Conseil, Conseil Perm. Intern. Exploration Mer, **103**: 5–26.

WILLIAMS, L. W. 1906. A list of the Rhode Island Copepoda, Phyllopoda, and Ostracoda with new species of Copepoda. Ann. Rept. R.I. Comm. Inland Fisheries, 37th, Spec. Paper **30**: 69–79.

Seasonal Studies on the Relative Importance of Different Size Fractions of Phytoplankton in Narragansett Bay (USA)

E. G. Durbin, R. W. Krawiec, and T. J. Smayda

Graduate School of Oceanography, University of Rhode Island; Kingston, Rhode Island, USA

Abstract

The composition and productivity of four different size-fractions (< 20, 20 to 60, 60 to 100, > 100 μm) of the phytoplankton of lower Narragansett Bay (USA) were followed over an annual cycle from November, 1972 to October, 1973. Diatoms dominated the population in the winter-spring bloom and in the fall; the summer population was dominated by flagellates. The nannoplankton (< 20 μm) were the most important, accounting for 46.6% of the annual biomass as chlorophyll a and 50.8% of the total production. The relative importance of the different fractions showed a marked seasonality. During the winter-spring and fall blooms the netplankton fractions (> 20 μm) were the most important. Nannoplankters dominated in the summer. The yearly mean assimilation numbers for the different fractions were not significantly different. During the winter-spring bloom, however, the assimilation numbers for the netplankters were significantly higher than those for the nannoplankton fraction. Temperature accounted for most of the variability in assimilation numbers; a marked nutrient stress was observed on only two occasions. Growth rates calculated from ^{14}C uptake and adenosine triphosphate (ATP)-cell carbon were generally quite high; maxima were > 1.90 doublings per day during blooms of a flagellate in the summer and of Skeletonema costatum in the fall. The series of short cycles observed in which the dominant species changed were related to changes in the physiological state of the population. Higher growth rates were generally observed at times of peak phytoplankton abundance while lower growth rates were observed between these peaks. The high growth rates and assimilation numbers usually found suggest that the phytoplankton in lower Narragansett Bay was not generally nutrient-limited between November, 1972 and October, 1973. Nutrient regeneration in this shallow estuary, therefore, must be very rapid when in situ nutrient levels are low.

Introduction

In the study of species succession in marine phytoplankton, floristic changes have traditionally been of primary concern. However, an understanding of the process of succession requires evaluation of the biological differences between species within the community (Smayda, 1973a). These biological differences will be reflected in differences in individual species' nutrient requirements and uptake kinetics, photosynthetic and respiration rates, sinking rates and in their suitability as food for grazers. Increasing evidence suggests that basic first-order differences in these factors among different species may be organized according to differences in cell size. Williams (1964) has

shown that small cells have higher intrinsic growth rates. In nutrient-uptake kinetics, cell-suface to volume considerations predict that large-celled species are less able to absorb nutrients from low-nutrient waters (Munk and Riley, 1952). This prediction has now been borne out in laboratory studies for the uptake of ammonium and nitrate (Eppley et al., 1969) and for vitamin B_{12} (Carlucci, unpublished, cited in Eppley, 1972). Furthermore, buoyancy is related to cell size (Smayda, 1970), as is grazing pressure (Mullin, 1963; Smayda, 1973a). Because of these differences it would be desirable to fractionate the natural phytoplankton population into different size-classes as a way of beginning to understand the relative importance of the different parts of the community,

314

Reprinted from Marine Biology 32: 271–287 (1975).

and to describe the behaviour of individual species for a given particular set of environmental conditions.

Previous size-fractionation studies have been concerned with sampling and describing the community; the problem of species succession was not a primary aim (Holmes, 1958; Yentsch and Ryther, 1959; Gilmartin, 1964; Anderson, 1965; Saijo and Takasue, 1965; Malone, 1971a, b). Holmes, Anderson, Saijo and Takasue and Malone (1971a) found that in a variety of oceanic environments nannoplankton (that fraction not retained by the fine-mesh net used, generally 20 to 60 μm) are often responsible for 80 to 99% of the observed production. In neritic waters in the Eastern Tropical Pacific and the Caribbean Sea Malone (1971a) found that netplankters (> 25 μm) were more important, contributing approximately 50% of the total production measured at these stations. Seasonal studies in neritic waters in Vineyard Sound (Yentsch and Ryther, 1959) and in the California Current off Monterey Bay (Malone, 1971b) showed the nannoplankton population to be the most important and relatively stable, while the netplankton showed marked seasonal trends, increasing when conditions became favourable for diatom growth during the winter-spring bloom or during periods of upwelling. At high latitudes, where the phytoplankton is made up mostly of chain-forming diatoms, the netplankton is also relatively more important (Digby, 1953). However, the relative importance of nannoplankton and netplankton in northern temperate and boreal coastal waters is not well documented, particularly with respect to any seasonal changes that may occur.

In this study the phytoplankton population in Narragansett Bay was fractionated into four size-classes and the cell numbers and species composition, chlorophyll a, ATP levels and rates of carbon assimilation were determined for each size class. This study is part of a program to more clearly define the successional changes in the composition of the different size classes, the rates at which these different fractions turn over, and to assess the role of individual species.

Materials and Methods

Around 09.00 hrs every second week from November 27, 1972 to October 15, 1973, a sample was collected from the bottom (9 m), mid-depth, and at the surface in 10-l Niskin bottles, at Station 2 (see Pratt, 1959) in lower Narragansett Bay.

Equal parts (6.25 l) from the three depths were combined to obtain a mixed sample. In the laboratory this sample was fractionated by filtering it successively through filter funnels with Nitex cloth of 100, 60 and 20-μm mesh size. Each filter-funnel consisted of a 13-cm length of 6.5 cm inner-diameter plexiglass tubing glued to a plexiglass base, to which Nitex cloth was permanently attached (Fig. 1). A base of flat plexiglass plate with a rubber tube leading from the centre of it to a collecting vessel was clamped to the filter-funnel; a rubber O-ring separated the two plates to make it water-tight. A clamp on the rubber outlet tube controlled the flow-rate through the filter. The whole rig was supported by a ring clamp placed under the filter base.

In carrying out the fractionation, water was passed through the 100-μm filter, taking care not to drain the filter completely dry. When about 50 ml of water remained above the filter the drain-tube was clamped off, filtered seawater from the same day's sample was added to the concentrated sample, and following gentle swirling to dislodge cells resting on the filter, the contents were poured into a measuring cylinder and made up to the desired volume with filtered seawater. The few milliliters of water remaining under the filter were drained into the vessel containing the filtrate. This procedure was repeated with the 60-μm and the 20-μm filters so that > 100, 60-100 and 20 to 60-μm fractions of known concentration and an unconcentrated < 20-μm fraction were obtained. Generally 2 l of sample were filtered and the concentrate made up to 500 ml, giving a concentration factor of 4X. Throughout the procedure flasks containing the samples were kept in a water bath to prevent warming.

Carbon-assimilation rate, chlorophyll concentration, ATP level and cell numbers were determined for each size fraction. Rates of carbon assimilation were estimated by the C-14 method (Steemann Nielsen, 1952). Samples (50 ml) were inoculated with about 1 μCi of bicarbonate C-14. These were placed in a tank containing running seawater located on the School of Oceanography dock, and incubated under ambient light from about noon to sundown on the day of sample collection. Following incubation, the samples were filtered on membrane filters, dried and counted on a Nuclear-Chicago Model 4338 planchet counting system. Chlorophyll a and phaeopigments were determined by fluorometry (Yentsch and Menzel, 1963) using a red-sensitive phototube (R-176) and the equations of

Lorenzen (1966). ATP was determined by the method of Holm-Hansen and Booth (1966) as modified by Cheer *et al.* (1974). Cell numbers were determined by counting a 1.0-ml sample in a Sedgwick-Rafter chamber. If samples were not counted immediately, Lugol's iodine solution was added as a preservative.

Results

Species Composition

Cell numbers for the different size fractions are shown in Fig. 2; the composition and abundance are given in Table 1. Because many species of diatoms form long chains or have long setae, or both, the cells retained by filters do not necessarily have dimensions greater than the filter mesh-size. This effect was particularly noticeable during the winter-spring bloom when chain-forming diatoms predominated, and when small cells such as *Skeletonema costatum, Detonula confervacea* or *Chaetoceros compressus* were found in the largest size fraction. However, the population was fractionated

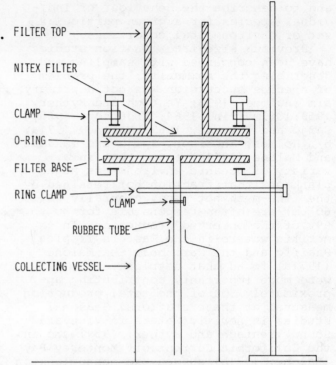

Fig. 1. Filter apparatus used to size fractionate natural phytoplankton population

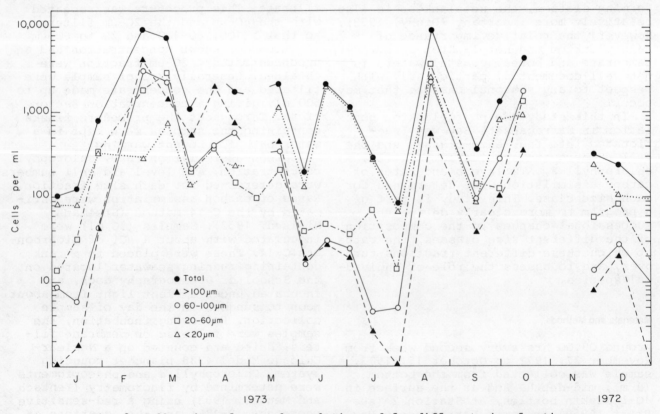

Fig. 2. Cell numbers for total population and for different size fractions

Table 1. Cell counts (cells/ml) for different size fractions (A: >100 μm; B: 60 to 100 μm; C: 20 to 60 μm; D: <20 μm); N.C.: counts not made

Species	27 November (1972)				14 December				2 January (1973)				15 January				29 January			
	A	B	C	D	A	B	C	D	A	B	C	D	A	B	C	D	A	B	C	D
Attheya decora			2.7				15.9													
Actinoptychus undulatus							0.1			0.5										
Asterionella japonica			3.7				0.6			3.2							8.2			
Cerataulina pelagica						0.1														
Chaetoceros affinis	3.6																			
C. decipiens																				
C. densus	0.3	0.4																		
C. diadema					2.4		0.7			1.2										
C. didymus					0.5															
C. laciniosus					0.7	0.8							1.0							
C. subsecundus	0.6																23.8			
Chaetoceros sp.			1.1				4.2								1.5				3.7	58.0
Corethron hystrix		0.1	0.2				0.5				0.1				0.3					
Cyclotella sp.			0.2	2.5												30.0				
Detonula confervacea																			16.0	19.3
Ditylum brightwellii			0.1																	
Leptocylindrus minimus				4.0		0.5										0.8				
Nitzschia closterium											0.2	0.2				0.3				2.4
N. reversa																			0.2	1.2
N. seriata			0.7																	
Paralia sulcata							2.6					1.0				0.8			6.6	
Rhizosolenia setigera										0.1	0.1		0.5						1.2	
Skeletonema costatum	8.5		24.5	94.0	14.0	24.4	51.1	52.0		1.9	6.6	1.0		4.3	19.3	38.0	9.0	18.0	222.0	168.0
Thalassiosira decipiens			0.1	5.7		6.5	4.6			0.2	3.7			1.0	2.5				1.8	
T. nordenskioeldii																			1.7	
Thalassiosira sp.	3.4	3.1	6.7	140.0	1.0	0.7	2.5	56.0	0.1	0.4	0.5	2.0			3.8	10.0	2.0		35.4	99.0
Thalassionema nitzschioides			2.0												1.0					
Thalassiothrix frauenfeldii											0.4				0.4					
Pennate	0.1	0.1	0.7		0.2	0.1	0.1	0.1						0.5	0.8					4.0
Peridinium sp.				2.0			2.0	4.0			0.1				0.8					4.0
Microflagellates				N.C.				N.C.				56.0				18.0				71.0
Total cells	8.1	14.2	59.2	240.0	20.4	27.9	82.4	114.0	0.1	8.6	13.4	80.0	1.8	5.8	31.9	80.0	44.0	42.0	352.0	337.0

Species	1973 26 February				12 March				26 March				9 April				23 April			
	A	B	C	D	A	B	C	D	A	B	C	D	A	B	C	D	A	B	C	D
Asterionella japonica					146.0	139.0			23.5				73.0	16.3	16.3		128.0	16.3		
Chaetoceros compressus									38.0				489.0	105.0	11.0		186.0	369.0	27.6	
C. decipiens					10.0	9.0			69.0	5.0			565.0	16.7			15.4	2.7		
C. diadema	54.3																			
C. didymus					9.0															
Chaetoceros sp.	9.0		8.7		9.0	40.0	99.0	23.2	3.6		1.4	0.5		14.5						
Detonula confervacea	2780.0	1939.0	700.0		451.0	1037.0	901.0	89.9		21.0	32.0	52.0			119.0	22.0	263.0	90.6	82.9	61.0
Leptocylindrus danicus				10.0	10.0															
L. minimus												12.0							9.5	43.5
Nitzschia closterium								2.9							3.6		1.8	11.0		
N. seriata			14.5							1.4			50.7	10.8	3.6		12.7			
Pennate sp.											3.0	0.72	25.4	5.4	18.0					
Rhizosolenia alata							19.0										0.9			
R. delicatula													87.0	20.0						
R. fragilissima						9.0			16.3	5.0	5.9			16.3			16.9	3.1		
R. imbricata var. shrubsolei																				
R. setigera	9.0				2.0	1.0			1.8								5.4	2.7		
Skeletonema costatum	692.0	931.0	719.0	103.0	471.0	892.0	1733.0	565.0		5.8	21.7	61.0	14.5	5.4	65.0		1.8	1.8	1.8	
Thalassiosira decipiens			29.0		339.0	125.0	51.0													
T. nordenskioeldii	558.0	46.4	5.8						391.0	177.6	128.0	8.0	714.0	186.7	76.0		10.8	26.3	19.4	
T. rotula		11.6																		
Thalassiosira sp.	87.0	52.2	282.0	162.0	9.0	23.0	23.0						25.0							
Thalassionema nitzschioides	3.6																			
Peridinium trochoideum								2.0												
Peridinium sp.			4.3									1.4								
Microflagellates				N.C.				N.C.				N.C.				N.C.			5.8	110.0
Choanoflagellate			7.0																	
Euglenid			1.0				55.1	14.5												
Dinobryon sp.					1.0															
Phaeocystis pouchetii					2.0															
Total cells	4193	3099	1799	283	1430	2272	2837	770	544	216	194	136.5	2046	404	367	N.C.	764	578	144	228.8

Table 1. (Continued)

319 Phytoplankton-Zooplankton Relationships in Narragansett Bay

Species	1973 7 May				21 May				4 June				18 June				2 July			
	A	B	C	D	A	B	C	D	A	B	C	D	A	B	C	D	A	B	C	D
Asterionella japonica			3.0		292.0	229.0	25.4				0.5	1.5					2.0			
Chaetoceros compressus					72.5	14.5			15.2	10.0		1.7								
C. diadema													0.7							
C. didymus									2.0											
Chaetoceros sp.			2.9										0.2	0.2			0.2			
Dactyliosolen sp.													0.2							
Ditylum brightwellii						1.2					0.2	1.0								
Leptocylindrus minimus			2.33		66.4	99.1	165.8	710.0		2.7	6.0	46.0	1.0	3.7	3.9	28.0				
Rhizosolenia alata							2.7		0.5											
R. delicatula													2.0	1.2	1.2	3.5				14.0
R. fragilissima			81.1	65.3			16.3		1.0			70.0						1.5	2.9	
Skeletonema costatum											8.0	16.0			3.1				1.3	28.0
Thalassiosira rotula			1.7						1.0	4.5			0.7	1.2	17.1					
Thalassiosira sp.																				
Cyclotella sp.																	0.7		4.6	70.0
Nitzschia closterium			2.3			2.4	2.4				0.5	6.0	10.2	2.7	0.2	0.4	0.5		1.3	14.0
N. seriata			15.7	4.8			1.8	14.5	1.5						2.7	0.8				
Thalassionema nitzschioides																			1.3	
Pennate			2.9	9.3						0.2		0.2								
Dinophysis sp.															3.5	3.5	0.2		0.6	7.0
Gymnodinium sp.													1.2	3.0	3.5	56.0				63.0
Prorocentrum redfieldii													4.5	2.2	3.1	126.0	0.2	0.6	0.6	49.0
Prorocentrum sp.																				
Peridinium sp.				23.3					0.5	0.5	2.2	2.0	0.2	1.6	1.6	28.0				7.0
Dinoflagellate				N.C.				362.0												
Microflagellates											2.7	40.0	5.5	2.7	2.7	280.0	13.5	21.0	27.4	35.0
Olisthodiscus luteus													3.7	8.7	8.5	1582.0				770.0
Total cells	N.C.	N.C.	95.5	97.8	450	349	213	1156	19.7	16.0	38.7	130.0	28.5	31.7	48.8	2100.0	19.2	25.1	37.2	1057.0

Species	1973 16 July				30 July				20 August				17 September				1 October				15 October			
	A	B	C	D	A	B	C	D	A	B	C	D	A	B	C	D	A	B	C	D	A	B	C	D
Asterionella japonica				3.2					78.7	471.0	267.0		3.5	1.4		5.8	22.1	156.7			217.0	532.0	42.0	
Chaetoceros compressus										15.7														
C. curvisetus									8.7	19.2	10.5						20.4	21.9	5.2			21.0		
C. decipiens									6.1	43.2	52.5						8.2				21.0			
C. didymus									3.5			2.3	15.8	12.6	9.3		4.2	7.9	10.5		45.5	14.0	5.25	28.0
Chaetoceros sp.	0.5	0.2																						
Corethron hystrix														0.7		1.2								
Coscinodiscus sp.									0.4															
Cyclotella sp.			1.0	5.6		0.2	5.0	5.0																
Dactyliosolen mediterraneus								0.2																
Ditylum brightwellii			0.5						1.3		3.5		0.8	4.9	2.3		0.6	3.5	4.3					
Eucampia zoodiacus														1.4			6.1		3.2			14.0	42.0	
Leptocylindrus danicus															5.8	11.6								
L. minimus													1.4	7.0					7.9				12.2	
Paralia sulcata							1.0																	
Rhizosolenia fragilissima	2.2	3.5	2.2	11.2		4.0	5.0	12.0					17.6	7.7	4.6	7.0								
R. hebetata f. *hemialis*					0.2																			
Skeletonema costatum									472.0	1967.0	1864.0	3108.0	3.5	49.0	54.8		5.8	7.8	14.0	14.0		1113.0	1249.0	928.0
Thalassiosira rotula																						7.0	9.3	
Thalassiosira sp.											2.3	70.0			14.0	70.0								23.0
Nitzschia closterium					0.5		2.0		0.4		1.7		3.5	2.1	19.8	14.0	36.1	79.6	39.5	779.0	45.5	14.0	12.2	1.7
N. seriata									4.8		3.5		8.7	10.5	2.4		3.5	0.9	0.8					
Thalassionema nitzschioides													1.7	4.9	16.3	46.6	6.4		41.1	23.3				18.6
Pennate							0.7	5.0					1.2			4.6				0.8				
Ceratium furca							0.5																	
Dinophysis sp.								1.0										0.8						
Gymnodinium sp.				14.0																				
Prorocentrum redfieldii																		0.8						
Prorocentrum sp.																								
Peridinium sp.											1.2	546.0		0.7										
Microflagellates		0.7	3.5	226.0			3.0	41.0																
Olisthodiscus luteus			0.5	22.4																				
Total cells	2.7	4.5	7.2	283.2	0.2	5.0	14.5	66.0	570.0	2534.0	2172.0	3724.0	38.4	97.3	145.8	88.4	115.4	292.0	128.2	816.0	329.0	1715.0	1364.0	928.0

to some extent, even though not strictly according to size. For example, during the winter-spring bloom when long chains of *Thalassiosira nordenskiöldii, S. costatum* and *D. confervacea* were numerous, *T. nordenskiöldii* was retained principally in the > 100-µm fraction, while the other two species were found in all three fractions > 20 µm, but were relatively more important in the 60 to 100-µm and 20 to 60-µm fractions.

During the minimum prior to the winter-spring bloom in February, small cells < 20 µm dominated the plankton, particularly *Skeletonema costatum*, a small solitary *Thalassiosira* sp., and microflagellates. Several diatom species were found in the larger fractions during this period, but were not important numerically. By February 26 the winter-spring bloom reached its initial peak. *Detonula confervacea, S. costatum, T. nordenskiöldii,* and the solitary *Thalassiosira* sp. dominated (Table 1). At this stage most of the cells were found in the two largest fractions (7292 cells out of a total of 9374 cells/ml). These species remained dominant in early March but, by March 26, *S. costatum* and *D. confervacea* had disappeared almost completely. *T. nordenskiöldii* remained the sole dominant in the > 100 and 60 to 100-µm fractions. At this time the total population had decreased from 12,370 cells/ml and 7310 cells/ml on the previous two sampling dates to 1090 cells/ml.

Following this decline, two successive peaks in abundance occurred: on April 9 and on May 7. In each of these the group of dominant species changed; *Chaetoceros compressus, C. decipiens* and *Thalassiosira nordenskiöldii* dominated on April 9. *T. nordenskiöldii* then dropped out, and the *Chaetoceros* spp. were joined by *Asterionella japonica, Leptocylindrus danicus* and *L. minimus*. During the entire winter-spring bloom the > 100-µm fraction was the most important numerically.

Large numbers of *Olisthodiscus luteus* were present during June, when the nannoplankton (< 20 µm) fraction became dominant as cell numbers. Flagellates completely dominated thereafter throughout the summer, and few diatoms were present until August 20.

A late-summer and early-fall bloom of *Skeletonema costatum* occurred, most of the cells were found in the < 20-µm and 20 to 60-µm fractions. Its initial peak on August 21 was followed by a period when several diatom species dominated the sparse population.

Standing Crop Measurements

Marked seasonal changes in the chlorophyll a content of all size fractions

occurred (Fig. 3, Table 2). On the average, the < 20-µm fraction was the most important; it comprised 46.6% of the annual biomass as chlorophyll a.

The nannoplankton (< 20 µm) fraction was most important during the mid-winter minimum (December-January) and during the summer (4 June-20 August). Over 75% of the chlorophyll a was then in this size class. Almost half (44%) of the total nannoplankton standing crop (as chlorophyll a) was found during the summer period.

During the spring and fall diatom-blooms, netplankters (> 20 µm) predominated, especially from 26 February - 7 May, when 77% of the total chlorophyll a for this period was contained in this size fraction.

Lowest chlorophyll a levels were recorded during the mid-winter minimum (< 1.3 mg chlorophyll a/m^3). Levels increased thereafter during the winter-spring bloom, characterized by considerable fluctuations and three successive peaks (Fig. 3). Over 50% of the total chlorophyll a at the time of these occurred in the > 100-µm fraction because of the predominance of long-chained diatoms (Fig. 2). During the initial peak (February 26), cell numbers and chlorophyll a were at a maximum for the bloom period (12,370 cells/ml and 9.78 mg chlorophyll a/m^3).

During the summer, chlorophyll a levels were generally quite high; a July (16) peak of 11.85 mg chlorophyll a/m^3 occurred. Cell numbers, however, were frequently low and dominated by flagellates < 20 µm. For example, on June 18, total cell numbers reached their summer maximum of 2209 cells/ml, of which 2000 cells/ml were in the < 20-µm fraction (Fig. 2; Tables 1, 2).

Diatoms dominated the fall peaks observed on August 20 and October 15 (Figs. 2, 3); the > 20-µm fractions then became relatively more important. Although cell numbers were then much higher (9000 and 4300 cells/ml, respectively) than in the summer, the chlorophyll a levels (4.31 and 8.04 mg chlorophyll a/m^3) were lower.

Throughout the year, the cell counts for the < 20-µm fraction were much lower than expected from the chlorophyll a levels, which suggests that the method used to count cells did not reliably estimate those < 20 µm. This discrepancy was most important during the summer when flagellates dominated the population. For example, on July 16 the chlorophyll a level in the < 20-µm fraction reached its annual maximum (10.96 mg chlorophyll a/m^3), but only 283 cells/ml were enumerated.

The ATP content of each size fraction was multiplied by a factor of 250 to obtain phytoplankton carbon content (Holm-Hansen, 1969, 1970); this value has been used by most other workers (e.g. Sutcliffe *et al.*, 1970; Eppley *et al.*, 1973). The carbon content of the different size fractions (Fig. 4, Table 2) generally followed the chlorophyll *a* trends, although occasionally values were higher than expected from the chlorophyll *a* levels. High C:chlorophyll *a* ratios then resulted (Table 2), especially on June 18 (125) and October 1 (149).

Production

Fig. 5 and Table 2 give the rates of carbon assimilation for each fraction, and as a percent of the assimilation of the total community. For the chlorophyll *a* content, the < 20-μm fraction was most important; it contributed 50.8% of the total annual production and 88% of that measured during the summer period (4 June - 20 August). Production in the netplankton fractions was greatest during the winter-spring and fall diatom-blooms.

Production was high throughout the year (> 5 mg C/m^3/h), except during the phytoplankton minimum prior to the winter-spring bloom. The maximum assimilation rate of 82.4 mg C/m^3/h and minimum of 1.34 mg C/m^3/h were measured on July 16 and January 15, respectively. The mean assimilation rate over the annual cycle was 18.4 mg C/m^3/h.

Specific Growth Rates of the Phytoplankton

Phytoplankton growth rates as doublings of carbon per day were determined, based on the rate of carbon assimilation per unit of living carbon from:

$$\mu = \left(\frac{1}{t}\right) \log_2 \left(\frac{C_1 + \Delta C}{C_1}\right) ,$$

in which ΔC is the daily increase in carbon due to photosynthesis, C_1 is the carbon content of the phytoplankton, and μ the specific growth rate as doublings of carbon per day ($t = 1$). The values for ΔC were obtained by multiplying the hourly assimilation rates by the number of daylight hours for that day. Values for C_1 were determined in two ways: Prior to the ATP measurements it was estimated from (chlorophyll *a*) (60), in which 60 is the assumed C: chlorophyll *a* ratio. Little information exists on the C: chlorophyll *a* ratios of the diatoms present in the winter-spring bloom in Narragansett Bay. A ratio of 67 was obtained for a natural, almost pure population of *Skeletonema costatum* growing at 12°C *in situ* within the 20 to 60-μm fraction which contained little detrital material. A ratio of 57 was obtained for *Thalassiosira nordenskiöldii* growing in dialysis sacs at 9°C. These observations suggest the ratio of 60 used. However, since a lower C:chlorophyll *a* ratio may characterize the beginning of the winter-spring bloom because of the lower temperatures and light intensities (Eppley, 1972), this value probably then overestimates the phytoplankton carbon and gives lower carbon doubling-rates. Accordingly, phytoplankton carbon doubling-rates were calculated only for the whole population during the winter-spring bloom (Fig. 6; Table 2).

C_1 was also estimated from ATP. These estimates of phytoplankton carbon are probably much more reliable then those based on the C:chlorophyll *a* ratio,

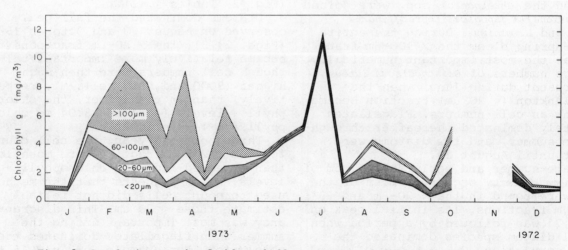

Fig. 3. Cumulative graph of chlorophyll *a* content of different size fractions

Table 2. Chlorophyll a (Chl a), phaeopigment (Phaeo) and ATP concentrations; carbon fixation and carbon doubling (K) rates and assimilation number (Assim. no.) for different size fractions during annual cycle. Prod.:production

Size fraction (µm)	Temperature (°C)	Chl a (mg/m³)	Phaeo (mg/m³)	Phaeo/Chl a	ATP (mg/m³)	Cell C (mg/m³)	C/Chl a	Prod. (mgC/m³) h	% total Prod.	Daily Prod. (mgC/m³) day	Assim. no.	K
27.XI.1972	7.0											
>100		0.20	0.10	0.50				0.25	4.1		1.25	
60-100		0.21	0.06	0.28				0.38	6.2		1.81	
20-60		0.81	0.18	0.22				1.28	20.8		1.58	
<20		1.52	0.90	0.59				4.23	68.9		2.78	
Total		2.74	1.24	0.45				6.14		65.2		
11.XII.1972	6.0											
>100		0.11	0.03	0.27				0.15	4.7		1.36	
60-100		0.17	0.03	0.18				0.19	5.9		1.12	
20-60		0.27	0.03	0.11				0.48	14.9		1.78	
<20		0.69	0.43	0.62				2.39	74.4		3.46	
Total		1.24	0.52	0.42				3.21		33.5		
2.I.1973	4.5											
>100		0.014	0.004	0.28				0.037			2.64	
60-100		0.034	0.009	0.26				0.051			1.5	
20-60		0.099	0.019	0.19								
<20		0.87	0.19	0.22								
Total		1.02	0.22	0.21								
16.I.1973	0.5											
>100		0.066	0.018	0.27				0.08	6.0		1.29	
60-100		0.042	0.009	0.21				0.04	3.0		0.97	
20-60		0.11	0.024	0.22				0.104	7.8		0.95	
<20		0.79	0.27	0.34				1.12	83.6		1.43	
Total		1.01	0.32	0.32				1.34		14.3		
29.I.1973	4.2											
>100		0.50	0.005	0.00				0.99	12.4		1.97	
60-100		0.56	0.03	0.05				0.88	10.0		1.42	
20-60		1.34	0.08	0.06				2.27	28.5		1.69	
<20		3.38	0.28	0.21				3.89	48.9		1.16	
Total		5.78	0.40	0.07				8.03		89.1		
26.II.1973	1.8											
>100		5.28	0.00	0.00				17.1	49.8		3.24	
60-100		1.71	0.00	0.00				6.91	20.8		4.04	
20-60		1.07	0.00	0.00				5.07	14.7		4.74	
<20		1.72	0.34	0.19				5.27	15.3		3.06	
Total		9.78	0.34	0.03		586[a]	60	34.3		343		0.65[a]
12.III.1973	5.5											
>100		3.49	0.05	0.01				7.46	34.1		2.13	
60-100		1.54	0.30	0.19				6.28	28.6		4.08	
20-60		1.42	0.42	0.29				5.64	25.7		3.98	
<20		1.70	1.07	0.63				2.52	11.5		1.48	
Total		8.35	1.74	0.20		501[a]	60	21.9		220		0.52[a]
26.III.1973	5.0											
>100		2.42	8.10	0.04				4.69	55.1		1.94	
60-100		0.70	0.02	0.03				1.62	19.0	*	2.32	
20-60		0.37	0.09	0.24				0.67	7.9		1.83	
<20		0.99	0.55	0.55				1.52	17.9		1.53	
Total		4.46	0.77	0.19		267[a]	60	8.50		85.2		0.39[a]
9.IV.1973	6.0											
>100		5.65	0.26	0.04				15.0	56.6		2.66	
60-100		1.30	0.04	0.03				5.21	19.6		4.01	
20-60		0.69	0.16	0.23				2.26	8.5		3.28	
<20		1.49	0.54	0.36				4.04	16.6		2.71	
Total		9.13	0.80	0.09		547[a]	60	26.5		265		0.57[a]
23.IV.1973	8.5											
>100		0.50	0.10	0.20	0.20	50	100	1.26	31.3	17.6	2.51	0.43
60-100		0.35	0.04	0.11	0.10	25	71.4	0.92	22.8	12.9	2.61	0.59
20-60		0.40	0.02	0.05	0.10	25	62.5	0.62	15.4	8.7	1.55	0.43
<20		0.76	0.49	0.64	0.45	112	148	1.22	30.3	17.1	1.60	0.20
Total		2.01	0.65	0.32	0.85	212	105	4.02		56.2		0.34
7.V.1973	11.4											
>100		3.23	0.29	0.09	0.86	215	66.5	16.1	60.8	241.0	4.97	1.08
60-100		0.91	0.16	0.17	0.24	60	65.9	4.37	16.5	65.5	4.80	1.06
20-60		1.05	0.15	0.14	0.24	60	57.1	2.64	9.9	39.6	2.52	0.73
<20		1.42	0.69	0.50	0.66	165	113	3.57	13.5	53.5	2.51	0.41
Total		6.61	1.28	0.19	2.00	500	75.6	26.7		400.0		0.85

Size fraction (µm)	Temperature (°C)	Chl a (mg/m³)	Phaeo (mg/m³)	Phaeo/Chl a	ATP (mg/m³)	Cell C (mg/m³)	C/Chl a	Prod. (mgC/m³)h	% total Prod.	Daily Prod. (mgC/m³)day	Assim. no.	K
21.V.1973	13.0											
>100		0.75	0.07	0.01	0.29	72.5	96.6	2.54	25.2	38.1	3.38	0.61
60–100		0.47	0.09	0.19	0.14	35	74.5	1.07	10.6	16.0	2.28	0.54
20–60		0.66	0.14	0.21	0.23	57.5	87	1.73	17.1	25.9	2.62	0.54
<20		1.94	0.97	0.29	0.86	215	112	4.76	47.1	71.4	2.45	0.41
Total		3.91	1.15	0.29	1.52	380	97.2	10.1		151.0		0.48
4.VI.1973	14.0											
>100		0.16	0.07	0.43				0.46	3.4	6.9	2.87	
60–100		0.12	0.06	0.50				0.36	2.6	5.6	3.00	
20–60		0.25	0.11	0.43				0.88	6.5	13.2	3.52	
<20		2.91	1.02	0.35				11.9	87.5	178.0	4.10	
Total		3.44	1.23	0.36				13.6		204.0		
18.VI.1973	16.5											
>100		0.13	0.05	0.38	0.16	40	307	0.44	2.1	6.9	3.37	0.23
60–100		0.11	0.04	0.37	0.09	22.5	204	0.42	2.0	6.6	3.79	0.37
20–60		0.18	0.06	0.33	0.14	35	194	0.95	4.4	15.0	5.29	0.51
<20		4.07	0.79	0.19	1.86	465	114	19.5	91.5	308.0	4.80	0.73
Total		4.49	0.84	0.19	2.25	562	125	21.3		336.0		0.68
2.VII.1973	19.0											
>100		0.13	0.04	0.31	0.14	35	269	0.17	1.9	2.5	1.31	0.10
60–100		0.12	0.04	0.33	0.05	12.5	104	0.22	2.5	3.3	1.83	0.34
20–60		0.18	0.09	0.50	0.09	22.5	125	0.40	4.6	6.6	2.22	0.37
<20		4.92	0.86	0.17	1.03	257	52	7.96	91.0	119.0	1.61	0.56
Total		5.35	1.03	0.21	1.31	327	61.1	8.75		131.0		0.48
16.VII.1973	21.0											
>100		0.28	0.15	0.55	0.19	47.5	169	2.88	3.6	42.5	10.3	0.92
60–100		0.29	0.15	0.53	0.07	17.5	60	1.66	2.1	24.9	5.8	1.27
20–60		0.32	0.27	0.59	0.06	15	46.8	2.66	3.3	39.2	8.3	1.85
<20		10.96	2.94	0.27	1.52	380	34.6	73.1	91.0	1078	6.7	1.94
Total		11.85	3.51	0.31	1.84	460	38.8	80.3		1184		1.84
30.VII.1973	19.5											
>100		0.08	0.10	1.25	0.16	40	333	0.79	5.9	11.3	6.58	0.36
60–100		0.11	0.12	1.10	0.03	7.5	68	0.58	4.4	8.3	5.27	1.07
20–60		0.14	0.14	1.00	0.04	10	71	0.94	7.1	13.5	6.71	1.23
<20		1.45	1.19	0.83	0.30	75	52	11.0	82.6	158.0	7.58	1.63
Total		1.78	1.54		0.54	135	758	13.3		190.0		1.27
21.VIII.1973	22.0											
>100		0.34	0.23	0.66	0.075	18.7	55	0.36	5.1	5.0	1.08	0.39
60–100		1.05	0.51	0.47	0.23	57.5	55	1.20	17.0	16.8	1.14	0.37
20–60		1.19	0.85	0.71	0.37	92.5	78	1.90	26.9	26.6	1.59	0.36
<20		1.77	1.43	0.83	0.56	140.0	79	3.60	51.0	50.4	2.03	0.44
Total		4.35	3.02	0.69	1.24	409.0	94	7.06		99.0		0.40
17.IX.1973	22.0											
>100		0.30	0.10	0.33	0.18	45.0	150	2.75	11.9	36.6	9.16	0.86
60–100		0.31	0.12	0.38	0.14	35.0	113	2.45	10.7	32.6	7.90	0.95
20–60		0.81	0.26	0.32	0.27	77.5	96	7.51	32.7	100.0	9.27	1.19
<20		1.12	0.31	0.28	0.39	97.5	87	10.25	44.6	136.0	9.15	1.26
Total		2.54	0.79	0.31	0.98	245.0	96	22.90		304.0		1.16
1.X.1973	17.0											
>100		0.22	0.03	0.14	0.097	24	111	1.53	10.9	17.2	6.95	0.77
60–100		0.44	0.09	0.22	0.21	52	119	3.62	25.7	40.7	8.23	0.83
20–60		0.44	0.29	0.67	0.21	52	119	3.55	25.2	39.9	8.07	0.82
<20		0.39	0.83	2.10	0.37	92	237	5.48	38.9	61.6	14.05	0.73
Total		1.49	1.22	0.81	0.89	222	149	14.20		159.7		0.85
15.X.1973	15.0											
>100		0.82	0.04	0.05	0.17	42	52	4.37	12.4	48.1	5.33	1.09
60–100		2.00	0.37	0.18	0.26	65	32	12.2	34.5	134.2	6.00	1.61
20–60		1.53	0.42	0.33	0.19	47	31	12.1	34.5	133.0	7.88	1.93
<20		0.96	0.95	0.98	0.26	65	68	6.67	18.9	73.4	6.95	1.08
Total		5.31	1.78	0.34	0.88	220	41	35.3		388.0		1.47

[a] Values calculated from C:chlorophyll a ratio of 60. Rest of numbers for cellular C (Cell C) calculated from ATP.

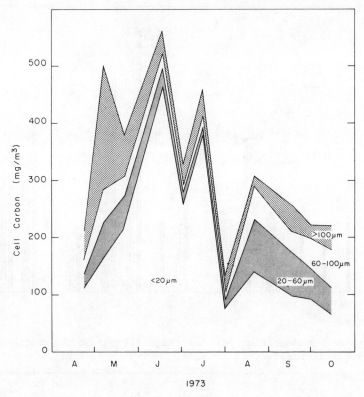

Fig. 4. Cumulative graph of carbon content of different size fractions, estimated from ATP concentration

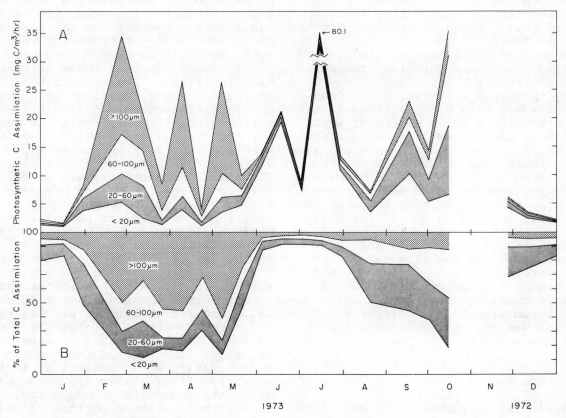

Fig. 5. (A) Cumulative graph of photosynthetic carbon assimilation by different size fractions of phytoplankton; (B) photosynthetic carbon assimilation by different size fractions as percentage of total population response

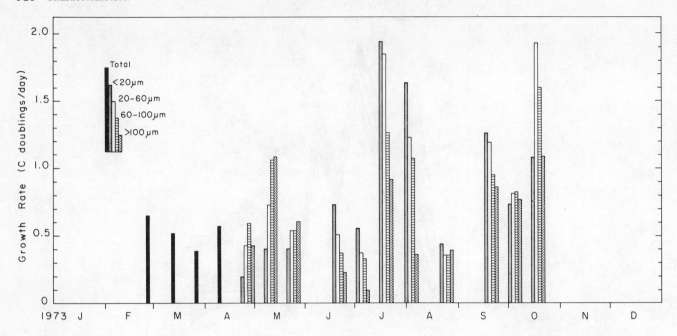

Fig. 6. Growth rates as carbon doublings per day for different size fractions and for total phyto-plankton population (see text for further details)

since ATP appears to show a much more constant relationship to cell carbon (Holm-Hansen, 1970). Carbon doubling-rates based on these estimates of C_1 were calculated for each size fraction sampled from the end of April to October (Fig. 6; Table 2).

During the winter-spring bloom, doubling rates were not high; the maximum was slightly > 1 per day on May 7. During the bloom's initial peak, the mean rate was only 0.65. Diatoms in the size fractions > 20 μm were most important then, and the smallest size fraction (< 20 μm) was growing at a significantly lower rate. However, when the latter fraction dominated (as biomass) during the summer, its doubling rates were also much higher than the > 20-μm fractions. On July 16 and 30 the daily carbon doubling-rates for the nanno-plankton fraction were 1.94 and 1.63, respectively. Low doubling rates (K = 0.40) were measured on August 21, but were higher thereafter. A maximum of almost 2 doublings/day was found for the 20 to 60-μm fraction on October 15. This fraction was entirely dominated by *Skeletonema costatum*.

Assimilation Numbers

The assimilation number (mg C/mg chloro-phyll a/h) may be used as an indicator of the growth rates of natural phyto-plankton populations (Ryther and Yentsch,

1957). Because of the intrinsically higher growth rates for smaller-sized cells, a higher assimilation number might be expected to characterize them. Howev-er, this was not the case; the mean assimilation numbers for the < 20, 20 to 60 and 60 to 100-μm fractions were 3.91, 3.88, and 3.72 mg C/mg chlorophyll a/h, respectively; none of these differences was significant. The assimilation numbers of the different fractions were signifi-cantly different only during the winter-spring bloom (Fig. 7), when the means for the < 20-μm and > 20-μm fractions were 2.24 and 3.18, respectively — a difference significant at the 5% level. The almost complete absence of cells > 20 μm during the summer preclude similar meaningful comparisons. In the fall there is no clear pattern, partly because *Skeletonema costatum* dominated the three smaller size fractions.

The low assimilation numbers for the < 20-μm fraction during the winter-spring bloom possibly indicates that this fraction contained a lot of dead cells and detrital material. The ratio of phaeopigments to chlorophyll a may be taken as an indicator of such materi-al, and in the < 20-μm fraction it was nearly always much higher than in the other fractions (Table 2). Associated with this phaeopigment would presumably be a much greater proportion of inactive chlorophyll.

Fig. 7 shows that higher assimilation numbers were found in the summer when

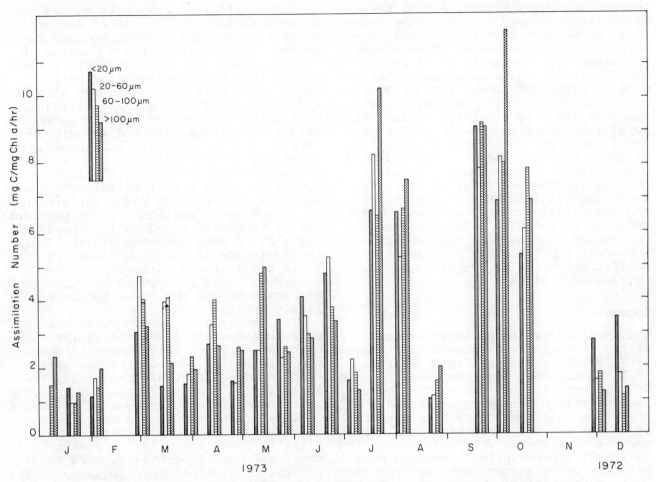

Fig. 7. Assimilation numbers (mg C/mg chlorophyll *a*/h) for different size fractions

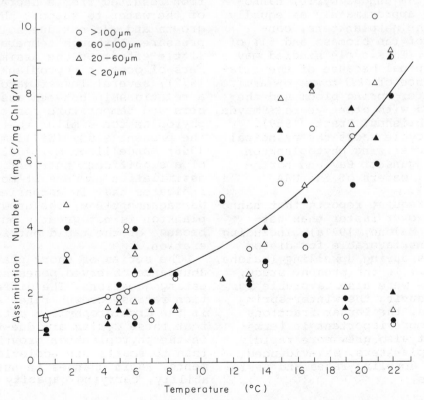

Fig. 8. Assimilation number versus temperature for different size fractions. Line is least-squares regression calculated from plot of the log of assimilation number versus temperature for all points except the two abnormally low groups at 19° and 22°C. Equation for the line is: log (assimilation

temperatures were higher. A plot of assimilation number against temperature (Fig. 8) shows a fairly good relationship, in which the assimilation numbers increased with temperature. On July 2 and August 18, the assimilation numbers were significantly lower (groups of points at 19° and 22°C in Fig. 8), and probably indicate a nutrient-deficient population (Curl and Small, 1965). On these two days the total concentrations of urea, NH_4, NO_2 and NO_3 were 0.46 and 0.40 µg-at/l, respectively. Excluding the highly aberrant data for these two days, the relation between the log of the assimilation number and the temperature was fitted to a least squares regression: log(mg C/mg chlorophyll a/h) = $0.0348T$ + 0.142, with a correlation coefficient of 0.79. The slope of the regression yielded a Q_{10} of 2.23.

Discussion

The spectacular and prolonged winter-spring bloom in Narragansett Bay (Smayda, 1957, 1973a, and unpublished; Pratt, 1959) is the most important event in the annual phytoplankton cycle. The domination of this bloom by long-chained diatoms suggests that the netplankters (< 20 µm) are more important than nannoplankters in the Bay. Indeed, during the 1972-1973 winter-spring bloom the netplankton was the most important fraction, contributing 77% of the biomass (as chlorophyll a) and 85% of the production. However, over the annual cycle, nannoplankters were approximately as equally important as the netplankters, contributing 47% of the biomass and 51% of the production. This cycle studied may perhaps be atypical because of the relative unimportance of Skeletonema costatum during the winter-spring bloom and the summer compared with other years (Smayda, 1957, and unpublished; Pratt, 1959). However, this cycle shows the principal features characterizing phytoplankton cycles in Narragansett Bay and northeastern coastal waters of the USA (Smayda, 1973b).

There are frequent reports that nannoplankters turn-over faster than netplankters (see Malone, 1971a), including under conditions favorable for diatom growth, such as during upwelling (Malone, 1971b). However, in the present study, when conditions were also favorable for diatom growth during the winter-spring and fall blooms, the larger fractions were not only more important in terms of biomass, but also grew more rapidly than the nannoplankters, as evidenced by both carbon doubling-rates and assimilation numbers.

These different responses may be due to several factors. The lower temperatures at which the diatoms bloomed may be one reason, since growth studies of certain small flagellates (Ukeles, 1961) showed that they did not grow well below 15°C. Nutrient uptake may also be a factor, since the half-saturation constants, K_s, and the maximum uptake rates, V_m, for nitrate and ammonia uptake vary with cell size and temperature (Dugdale, 1967; Eppley et al., 1969) with the larger cells having higher values of K_s and V_m. As a result of this, higher nutrient levels might favor the growth of larger cells while the lower nutrient levels generally found during the summer might favor nannoplankton growth. Thus, the dominance of the nannoplankters and their rapid growth rates during the summer would be favored by both the high temperatures and the lower nutrient levels present then.

The fairly good direct relationship between temperature and the assimilation numbers indicates that temperature was the main factor controlling the variation in these (Fig. 8). Abnormally low assimilation numbers, which may indicate nutrient deficiency or some other "water quality" factor (Curl and Small, 1965) were found on only two occasions. Slightly lower assimilation numbers and the lower mean growth rates were found for the population occurring between the peaks of the winter-spring bloom. These suggest that the decrease in the population resulted from a decreased capacity of the water to support phytoplankton growth and was not due solely to grazing pressure. Although temperature may have little effect on the assimilation numbers of oceanic phytoplankton (Eppley, 1972), several investigators have found a relationship between assimilation numbers and temperature for estuarine phytoplankton similar to that found in the present study (Williams and Murdoch, 1966; Mandelli et al., 1970). The lack of a significant nutrient effect on the assimilation numbers at most times indicates that in estuaries, including Narragansett Bay, the growth of phytoplankton is not greatly nutrient-limited, because of the rapid nutrient regeneration.

The series of short peaks in species' abundance observed poses some interesting questions. The different assimilation ratios, growth rates and changes in the C:chlorophyll a ratios indicate that these cycles are due to real changes in the phytoplankton growth rates rather than to scatter in ecological measurements. Rapid changes in nutrient availability, carrying capacity of the water, and its general suitability in supporting phytoplankton growth must occur. The low nutrient levels present during the peaks in phytoplankton abundance and the high growth rates measured at these times suggest that nutrient cycling is then very rapid in lower Narragansett Bay. For example, in order to sustain the growth rate measured in the > 100-μm fraction on May 7, 2.15 μg-at N/l/day are required, if a C:N ratio of 8 is assumed. The amount of NO_3, NO_2, NH_3 and urea measured *in situ* on this date was 0.30 μg-at N/l, which would supply only a small proportion of the nitrogen needs of the phytoplankton. This gives a regeneration time for the nitrogen in the Bay of 3.4 h. On July 16 the difference is even greater; 13.0 μg-at/l/day was required by the phytoplankton and only 1.62 μg-at N/l was present, giving a regeneration time of 3.0 h.

Size fractionation of phytoplankton provides a convenient way of getting at these problems, since generally the detrital material in the fractions >20 μm is very small compared to the phytoplankton. Possibly, reliable determinations of the phytoplankton C, N, and perhaps even Si content of these fractions can be made.

Acknowledgement. This study was supported by National Science Foundation Grant GA 31319X.

Literature Cited

Anderson, G.C.: Fractionation of phytoplankton communities off the Washington and Oregon coasts. Limnol. Oceanogr. *10*, 477-480 (1965)

Cheer, S., J.H. Gentile and C.S. Hegre: Improved methods for ATP analysis. Analyt. Biochem. *60*, 102-114 (1974)

Curl, H., Jr. and L.F. Small: Variations in photosynthetic assimilation ratios in natural, marine phytoplankton. Limnol. Oceanogr. *10* (Suppl.), R67-R73 (1965)

Digby, P.B.S.: Plankton production in Scoresby Sound, East Greenland. J. Anim. Ecol. *22*, 289-322 (1953)

Dugdale, R.C.: Nutrient limitation in the sea: dynamics, identification and significance. Limnol. Oceanogr. *12*, 685-695 (1967)

Eppley, R.W.: Temperature and phytoplankton growth in the sea. Fish. Bull. U.S. *70*, 1063-1085 (1972)

-, E.H. Renger and E.L. Venrick: A study of plankton dynamics and nutrient cycling in the central gyre of the North Pacific Ocean. Limnol. Oceanogr. *18*, 534-551 (1973)

-, J.N. Rogers and J.J. McCarthy: Half-saturation constants for uptake of nitrate and ammonium by marine phytoplankton. Limnol. Oceanogr. *14*, 912-920 (1969)

Gilmartin, M.: The primary production of a British Columbia fjord. J. Fish. Res. Bd Can. *21*, 505-538 (1964)

Holm-Hansen, O.: Determination of microbial biomass in ocean profiles. Limnol. Oceanogr. *14*, 740-747 (1969)

- ATP levels in algal cells as influenced by environmental conditions. Pl. Cell Physiol., Tokyo *11*, 689-700 (1970)

- and D.R. Booth: The measurement of adenosine triphosphate in the ocean and its ecological significance. Limnol. Oceanogr. *11*, 510-519 (1966)

Holmes, R.W.: Size fractionation of photosynthesizing phytoplankton. *In*: Physical, chemical and biological oceanographic observations obtained in expedition SCOPE in the Eastern Tropical Pacific, Nov.-Dec. 1956. Spec. scient. Rep. U.S. Fish Wildl. Serv. (Fish.) *279*, 169-171 (1958)

Lorenzen, C.J.: A method for the continuous measurement of *in vivo* chlorophyll concentration. Deep-Sea Res. *13*, 223-227 (1966)

Malone, T.C.: The relative importance of nannoplankton and netplankton as primary producers in tropical oceanic and neritic phytoplankton communities. Limnol. Oceanogr. *16*, 633-639 (1971a)

- The relative importance of nannoplankton and netplankton as primary producers in the California Current system. Fish. Bull. U.S. *69*, 799-820 (1971b)

Mandelli, E.F., P.R. Burkholder, T.E. Doheny and R. Brody: Studies of primary productivity in coastal waters of southern Long Island, New York. Mar. Biol. *7*, 153-160 (1970)

Mullin, M.M.: Some factors affecting the feeding of marine copepods of the genus *Calanus*. Limnol. Oceanogr. *8*, 239-250 (1963)

Munk, W.H. and G.A. Riley: Absorption of nutrients by aquatic plants. J. mar. Res. *11*, 215-240 (1952)

Pratt, D.M.: The phytoplankton of Narragansett Bay. Limnol. Oceanogr. *4*, 425-440 (1959)

Ryther, J.H. and C.S. Yentsch: The estimation of phytoplankton production in the ocean from chlorophyll and light data. Limnol. Oceanogr. *2*, 281-286 (1957)

Saijo, Y. and K. Takasue: Further studies on the size distribution of photosynthesizing phytoplankton in the Indian Ocean. J. oceanogr. Soc. Jap. *20*, 264-271 (1965)

Smayda, T.J.: Phytoplankton studies in lower Narragansett Bay. Limnol. Oceanogr. *2*, 342-359 (1957)

- The suspension and sinking of phytoplankton in the sea. Oceanogr. mar. Biol. A. Rev. *8*, 353-414 (1970)

- The growth of *Skeletonema costatum* during a winter-spring bloom in Narragansett Bay, Rhode Island. Norw. J. Bot. *20*, 219-247 (1973a)

- A survey of phytoplankton dynamics in the coastal waters from Cape Hatteras to Nantucket. *In*: Coastal and offshore environmental inventory, Cape Hatteras to Nantucket Shoals. Mar. Publ. Ser. Univ. Rhode Isl. *2*, 3-1, 3-100 (1973b)

Steemann Nielsen, E.: The use of radioactive carbon (C[14]) for measuring organic production in the sea. J. Cons. perm. int. Explor. Mer *18*, 117-140 (1952)

Sutcliffe Jr., W.H., R.W. Sheldon and A. Prakash: Certain aspects of production and standing stock of particulate matter in the surface waters of the Northwest Atlantic Ocean. J. Fish. Res. Bd Can. *27*, 1917-1926 (1970)

Ukeles, R.: The affects of temperature on the growth and survival of marine algal species. Biol. Bull. mar. biol. Lab., Woods Hole *120*, 255-264 (1961)

Williams, R.B.: Division rates of salt marsh diatoms in relation to salinity and cell size. Ecology *45*, 877-880 (1964)

- and M.B. Murdoch: Phytoplankton production and chlorophyll concentrations in the Beaufort Channel, North Carolina. Limnol. Oceanogr. *11*, 73-82 (1966)

Yentsch, C.S. and D.W. Menzel: A method for the determination of phytoplankton chlorophyll and phaeophytin by fluorescence. Deep-Sea Res. *10*, 221-231 (1963)

- and J.H. Ryther: Relative significance of the net phytoplankton and nannoplankton in the waters of Vineyard Sound. J. Cons. perm. int. Explor. Mer *24*, 231-238 (1959)

E.G. Durbin
University of Rhode Island
Graduate School of Oceanography
Kingston, Rhode Island 02881
USA

Date of final manuscript acceptance: May 20, 1975. Communicated by M.R. Tripp, Newark

Faunal variation on pelagic *Sargassum** **

M. L. FINE

Virginia Institute of Marine Science; Gloucester Point, Virginia, USA

Abstract

Pelagic *Sargassum* was collected in late summer, late winter, and early and late spring from inshore waters, the Gulf Stream and the Sargasso Sea of the Western North Atlantic Ocean. The noncolonial macrofauna was picked from the weed samples. The 34 samples contained 67 species and 11,234 individuals. The Shannon-Wiener index of diversity had a mean value of 2.419 ± 0.177 ($t_{.05}$ $s_{\bar{x}}$) and a statistical range between 1.401 and 3.437 ($t_{.05}$ s). Mean diversity values were not significantly different among the various sampling series, and diversity did not vary with raft volume. High diversity values were related to an equitable distribution of species resulting from a stable environment and an area low in productivity. Species composition of the *Sargassum* organisms varied seasonally and geographically. Animals were more abundant in the spring than in the fall samples. Samples collected on a transect in the Gulf Stream and Sargasso Sea maintained a similar faunal composition.

Introduction

The brown alga *Sargassum*, or gulf-weed, belongs to the order Fucales, which contains many species with vesicles or bladders for buoyancy. The presence of pelagic *Sargassum* with its attendant fauna is well known in the Sargasso Sea of the Atlantic Ocean, but the genus also occurs around Japan (IDA et al., 1967) and in the Red Sea (MARKKAVEEVA, 1965) with an associated fauna.

WINGE (1923) and DEACON (1942) have reviewed the early literature on *Sargassum*. KRÜMMEL (1891) attempted to fix the boundaries of the Sargasso Sea by studying the distribution of *Sargassum*. From records kept by German sea captains, he computed the number of times the weed was sighted in 1° squares and then, incorrectly, combined his results to give 10, 5, and 0.3% probability contours for 5° squares. WINGE (1923) collected *Sargassum* by plankton net and charted approximate boundaries of occurrence of the weed. PARR (1939) sampled extensive areas of the Sargasso Sea and the Gulf of Mexico and found that the sterile eupelagic species *Sargassum natans* and *S. fluitans* made up over

99% of the total pelagic vegetation in the Sargasso Sea, and that the 2 morphological types *natans* I and *fluitans* III composed between 88 and 99% of this total. Other forms of *S. natans* and *S. fluitans* were correspondingly rare, and species torn from littoral bottoms were insignificant. From a variety of evidence, PARR proved that attached coastal species, although occasionally encountered in the Gulf Stream, make no significant contribution to the flora of the Sargasso Sea proper.

PARR's (1939) work on vertical distribution of *Sargassum* demonstrated that only insignificant amounts of the weed are found below the surface. These results, buttressed by WOODCOCK's (1950) study of the extreme buoyancy of *Sargassum*, prove further that the weed is in its natural habitat on the high seas, and is not a coastal castaway with a short pelagic life.

Life associated with *Sargassum* divides into a myriad of forms including micro-, meio-, and macrofaunal components. CONOVER and SIEBURTH (1964) and SIEBURTH and CONOVER (1965) worked on the bacteriocidal effects of *Sargassum* tannins on vibrios and pseudomonads isolated from the alga. With few exceptions, the meiofauna is unstudied. THULIN (1942) found a tardigrade, *Styraconyx sargassi*, and YEATMAN (1962) investigated the copepods of gulf-weed, and hypothesized that the alga was the agent responsible for transplanting several American species to Europe. I filtered material from water in which the weed was agitated and found copepods, nematodes, amphipods, isopods, mites, and tardigrades.

Both sessile and motile forms compose the macrofauna. Many of the sessile species are colonial and, in the case of hydroids, often specific for different morphological types of *Sargassum* (WINGE, 1923; BURKEN-ROAD, in PARR, 1939; WEIS, 1968). HENTSCHEL (1922) found changes in presence or absence of sessile species on different samples and attempted to quantify these species by the number of colonies or the number of vertical branches of hydroid on 10 cm long *Sargassum* leaves. HENTSCHEL analyzed the guts of the important sessile forms (*Membranipora*, *Spirorbis*, *Lepas*, and *Diplosoma*) and discovered that these species subsisted largely on nannoplankton. Surprisingly, many of the

* Contribution No. 351 from the Virginia Institute of Marine Science, Gloucester Point, Virginia, USA.

** From a thesis submitted to the faculty of the School of Marine Science, The College of William and Mary, in partial fulfillment of the requirements for the degree of Master of Arts in Marine Science.

331

guts contained nematocysts from *Physalia* and un-identified coelenterates. He attributed absence of food contents in hydroids to regurgitation caused by their preservation in formalin. After looking at the gut contents of the nudibranch *Scyllaea pelagica* and the grapsid crab *Planes minutus*, he concluded that the sessile organisms were not an important component of their food. HENTSCHEL (1922) also discussed reproduction of the attached forms and described differences between the fauna of coastal and pelagic species.

Although THOMSON (1878) and MURRAY and HJORT (1912) mention weed animals they encountered during their cruises, TIMMERMANN (1932), a student of HENTSCHEL, has done the only extensive work concerning motile forms. Unfortunately, he attempted to cover the whole Sargasso Sea with 55 samples, many of which were small and sporadically distributed. TIMMERMANN stated that the free-living animals were saved in only some of the samples, but that the remainder sufficed, in general, to recognize the characteristic features of the geographical distribution. His species list appears to be low in numbers of individuals and numbers of species. I believe that his samples are unrepresentative, and I cannot accept his discussion of distribution and his observation of a decrease in fauna during the winter.

PRAT (1935) discussed some of the animals and algae he found on *Sargassum*, but gave no quantitative or station data. ADAMS (1960) described the postlarval development of the *Sargassum* fish *Histrio histrio*. Her paper ends with a discussion of the *Sargassum* complex from the literature and a rather large, although sourceless, list of species found on *Sargassum*. WEIS (1968) dipped 4 samples of gulf-weed from the Gulf Stream and identified the animals to genus. She found large numbers of the shallow water snails *Bittium* and *Rissoa* on the weed, but unfortunately chose to explain their presence by suggesting a benthonic origin for the *Sargassum*. Winds at times pile up great masses of weed on beaches of the Atlantic and Gulf coasts. A change in wind direction will carry the weed back out to sea along with any newly recruited species, even intertidal forms.

Available literature does not give more than a vague idea of the numerical distribution of organisms in the pelagic *Sargassum* community. My approach was to take a detailed look at that part of the *Sargassum* macrofauna which could be readily counted. Variations in time and space could then be charted with some confidence and indices of community ecology applied.

Materials and methods

Sargassum samples were dip-netted in the Atlantic Ocean at a number of stations. I took 18 late summer samples between 1 and 5 October, 1968 in 3 areas surrounding Cape Hatteras, North Carolina, USA. Four of the samples came from north of the Cape (I), 5 adjacent

to the Cape (II), and the remaining 9 to the south (III). All further samples were taken south of Hatteras. On a late winter cruise in March (RR), I managed to obtain only 1 small sprig of *Sargassum* in a plankton tow (33°27′ N, 76°56′ W, temperature 22.3 °C, volume 1.3 ml). Scientists in an airplane, looking for fish shoals, did not detect *Sargassum* north of Charleston, South Carolina. Nine early spring samples from 29 April of

Table 1. *Positions and surface temperatures where Sargassum samples were collected, together with respective raft volumes*

Sample	Latitude (N)	Longitude (W)	Temperature (°C)	Raft volume (ml)
I 4	36°55′	74°44′	21.9	15
I 5	36°38′	74°42′	22.6	255
I 5a	36°38′	74°42′	22.6	151
I 6	36°37′	74°44′	22.3	325
II 7	35°24′	75°23′	23.6	82
II 7a	35°24′	75°23′	23.6	74
II 7b	35°24′	75°23′	23.6	242
II 7c	35°24′	75°23′	23.6	202
II 8	35°18′	75°03′	25.0	322
III 1	34°35′	76°14′	27.4	708
III 1a	34°35′	76°14′	27.4	25
III 9	34°18′	75°37′	27.7	339
III 9a	34°18′	75°37′	27.7	387
III 10	34°14′	75°51′	26.6	562
III 10a	34°14′	75°51′	26.6	817
III 10b	34°14′	75°51′	26.6	1,327
III 10c	34°14′	75°51′	26.6	424
III 11	34°16′	76°17′	27.5	388
S 1	34°16′	75°48′	23.0	157
S 2	34°16′	75°48′	23.0	64
S 3	34°16′	75°48′	23.0	71
S 4	34°16′	75°48′	23.0	40
S 5	34°16′	75°48′	23.0	38
S 6	34°16′	75°48′	23.0	30
S 7	34°16′	75°48′	23.0	33
S 8	34°16′	75°48′	23.0	20
S 9	34°16′	75°48′	23.0	13
D 1	34°21′	75°36′	26.2	102
D 1a	34°21′	75°36′	26.2	127
D 2	33°56′	74°27′	21.6	92
D 3	33°32′	72°37′	21.8	124
D 4	33°26′	71°56′	22.1	134
D 5	33°15′	71°01′	22.2	269

the previous year (S) came from a limited area within the Gulf Stream. Late spring samples from 25 and 26 May, 1969 (D) were collected along a transect from the Gulf Stream into the Sargasso Sea. All samples were collected within a temperature range of 22 °C to 28 °C. Fig. 1 is a chart of the stations, and Table 1 lists the position, temperature, and raft volume for each sample. Samples were preserved in 10% buffered formalin and later picked for countable animals. All motile forms of approximately 1 mm and larger were selected as were the noncolonial sessile forms. The calcareous polychaete *Spirorbis* was not considered. Raft volumes were quantified by water displacement.

The organisms were identified to species when possible. Identification of portunid crabs in the late summer samples presented a problem because both megalopa and juveniles were present. The larval forms were designated by letter (Portunid A, B, etc.), but the juveniles were only partially separated, resulting in the lumped category of *Portunus* spp. Many of the juveniles had autotomized their chelae, a structure needed for identification. In addition, there was undoubtedly overlap between megalopa and juvenile forms. Statistical treatment of the portunids varied, and will be explained in each case.

Diversity was calculated from SHANNON's equation (1948) with the aid of tables provided by LLOYD et al.

$$\varepsilon = s'/s$$

where s′ = number of species conforming to MACARTHUR's model which would give the observed value for species diversity,

s = the number of species present in the sample.

When larval and juvenile portunids occurred in the same sample, individual categories of megalopa and juveniles were arbitrarily paired until the smaller category of the two was exhausted. For example, if 4 Portunid B megalopa, 4 *Portunus sayi* and 5 *P. anceps* occurred together, they would be treated as one species

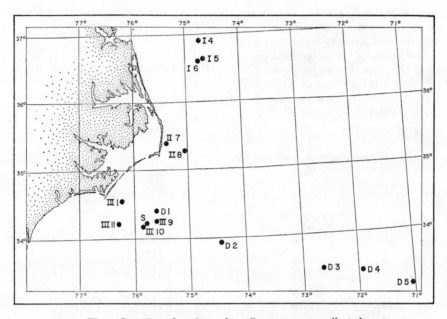

Fig. 1. Location of stations where *Sargassum* was collected

(1968). The diversity index (H′) is based upon the proportion of the number of individuals of each species to the total number of individuals in the sample:

$$H' = -\Sigma p_i \log_2 p_i$$
where $p_i = n_i/N$,

n_i = number of individuals in the ith species,

N = total individuals in the sample.

This index is sensitive to both numbers of species and their distribution.

Equitability (ε) (LLOYD and GHELARDI, 1964) specifically isolates the evenness of species diversity by comparing the number of species in a given sample to the number predicted by a hypothetical standard of species distribution, in this case MACARTHUR's (1957) model based on nonoverlapping niches:

of 8 organisms and a second species of 5 organisms to establish the number of species, diversity, and equitability of the sample.

In an attempt to define qualitative differences between various sets of samples, I calculated SANDERS' (1960) dominance-affinity index for all possible sample pairs. It was obtained by computing the percentage of the total sample represented by each species present in both samples, and then summing the smaller percentage for each species. High values of the index indicated faunal homogeneity or affinity between the samples being compared. Portunids were treated in their separate categories.

In order to examine the numerical dominance of species in a series of samples, I used the biological index described by SANDERS (1960). The species were ranked 1 to 7 in each sample and assigned values in reverse order of abundance so that a rank of 1 was given 7 points, a rank of 2, 6 points, etc. The bioindex value for each species was determined by adding the number of points it scored in all of the samples considered. For example, if a species occurred in 6 samples and ranked first in 4 and second in 2, its index value would be 40. This index prevents the obvious bias inherent in ranking species solely by total number of individuals, namely that a species occurring with a low frequency but in large numbers will be ranked above other species present in moderate numbers at most stations. The portunids were treated as a group in this analysis.

Results

Numbers of species and individuals and values for diversity and equitability are listed in Table 2. The values for diversity do not appear to contradict a normal distribution, and normality was assumed for statistical treatment of the data. The mean values 2.592, 2.571, 2.228, 2.432, and 2.447 for areas I, II, and III, and series S and D, respectively, gave a nonsignificant F-test after analysis of variance (F = 0.5413; 4,28 df).

Table 2. *Number of species and individuals, diversity and equitability of Sargassum samples*

Sample	Species	Individuals	H'	ε
I 4	7	60	2.0600	0.75
I 5	10	200	2.4565	0.83
I 5a	14	80	3.0517	0.84
I 6	8	82	2.8004	0.82
II 7	11	35	3.0927	1.10
II 7a	13	106	2.6315	0.66
II 7b	15	141	2.5409	0.53
II 7c	10	137	2.0110	0.53
II 8	18	395	2.5775	0.46
III 1	13	480	1.8588	0.36
III 1a	5	17	1.9903	1.05
III 9	10	285	1.6803	0.41
III 9a	8	98	2.4783	0.95
III 10	10	179	2.3723	0.70
III 10a	12	804	1.8964	0.40
III 10b	14	730	2.4941	0.55
III 10c	12	179	2.5322	0.66
III 11	18	546	2.7469	0.52
S 1	19	599	2.9178	0.56
S 2	19	266	3.3529	0.77
S 3	16	131	2.6495	0.54
S 4	16	187	3.1783	0.80
S 5	9	47	2.7553	1.04
S 6	6	301	1.0270	0.41
S 7	9	266	1.9866	0.58
S 8	10	80	2.1123	0.58
S 9	6	37	1.9064	0.82
D 1	15	364	2.4978	0.52
D 1a	18	513	2.9069	0.58
D 2	15	505	2.2531	0.43
D 3	12	562	1.8550	0.39
D 4	16	976	2.8374	0.62
D 5	15	1,709	2.3299	0.45

The regression of diversity on raft volume (Fig. 2) showed that, in addition to not changing with season or geographical area, the diversity index did not vary with sample volume. The mean for 33 samples was 2.419 ± 0.177 ($t_{.05} s_{\bar{x}}$) and the confidence interval on the individual data points ranged from 1.401 to 3.437 ($t_{.05} s$). Variation in calculated diversity values was

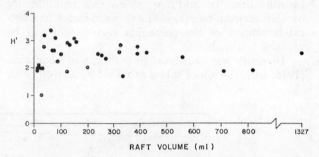

Fig. 2. Relationship of diversity (H') to raft volume for all *Sargassum* samples

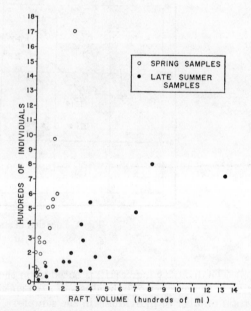

Fig. 3. Relationship of number of individuals to raft volume for *Sargassum* samples collected during spring and late summer

such that several samples in any one area are needed before a reliable estimate may be made.

Diversity is a function of the number of species, the number of individuals, and the distribution of the individuals among species, i.e. equitability. Numbers of species per sample did not change drastically during the year. Indeed, variation was as great within the fall samples as it was throughout the year. In general, within a given set of samples, larger rafts tended to hold more species.

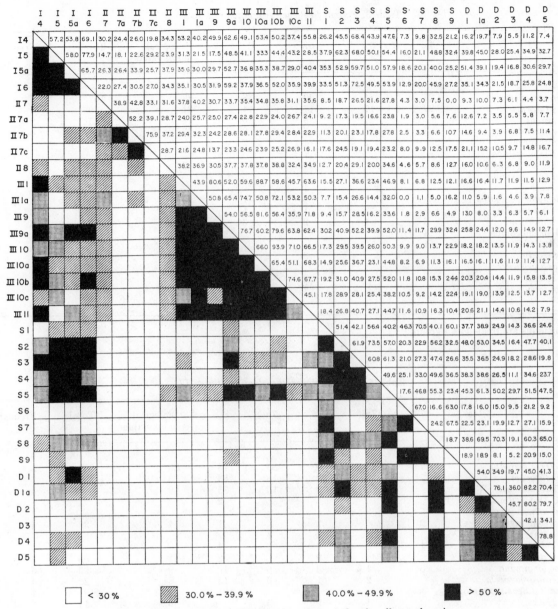

Fig. 4. Trellis diagram of dominance affinity index for all sample pairs

The number of individual animals in each sample fluctuated markedly through the year (Fig. 3). Samples from late summer were combined; except for the larger samples from area III, the points for the 3 areas were similar where they shared similar raft volumes. The regression for the spring samples (series S and D) has a higher slope than the regression for the late summer samples, indicating a more abundant fauna on smaller raft volumes. With one exception (sample Sl), the late spring samples had more organisms than early spring, but this is probably accounted for by the larger raft volumes of the D series and not by a change in faunal abundance.

Equitabilities were quite variable, ranging from 0.36 to 1.10. Twenty-two of the 33 values ranged between 0.50 and 1.00, with only 8 points below and 3 points above this range. These results indicate a high equitability.

The index of dominance affinity is shown on a trellis diagram (Fig. 4) arranged by groups of samples (I, II, III, S, and D). Such a diagram allows one to compare the affinities within an area and the affinities between areas. The mean affinity within areas I and III was 63.6 and 63.3, respectively. Such values indicate a homogeneous fauna (SANDERS, 1960). The mean affinity between samples in areas I and III dropped

Table 3. *Faunal frequency evaluation of area I*

	4	5	5a	6	Biological index value	Total	Frequency
Gnesioceros sargassicola	25	84	20	38	28	167	4
Latreutes fucorum	20	17	17	14	21	68	4
Janira minuta		27	8	10	14	45	3
Litiopa melanostoma	3	25	5	8	13	41	4
Portunid B megalopa		2	4	2			
Portunid D megalopa			1				
Portunid E megalopa			1		9	20	4[a]
Portunid spp. juvenile	1	3		5			
Cronius ruber			1				
Anoplodactylus petiolatus		11	14	1	8	26	3
Platynereis dumerilii		28	1	1	7	30	3
Leander tenuicornis	6	1	2	3	7	12	4
Dromiid sp. megalopa	4				4	4	1
Stephanolepis hispidus	1				2	1	1
Lucifer faxoni		3			1	3	1
Hoploplana grubei		2			0	2	1
Styliola subula			2		0	2	1
Ampithoe longimana		1			0	1	1

[a] Portunids are treated as a group.

Table 4. *Faunal frequency evaluation of area II*

	7	7a	7b	7c	8	Biological index value	Total	Frequency
Stephanolepis hispidus	7	23	70	85	18	29	203	5
Latreutes fucorum	7	13	22	11	83	28	136	5
Litiopa melanostoma		6	12	15	35	19	68	4
Portunid D megalopa			1		2			
Portunus anceps juvenile	5	3	1					
Portunus sayi juvenile	2		1		1	14	36	4[a]
Portunus spp. juvenile		6	2	4				
Cronius ruber juvenile	1		3		4			
Lucifer faxoni		44	9		2	11	55	3
Ampithoe longimana	5	1	3	6	10	9	25	5
Gnesioceros sargassicola		2	1	6	13	8	22	4
Dromidia antillensis megalopa					181	7	181	1
Atylus minikoi		1	9	5		6	15	3
Leander tenuicornis	3	2	2		9	5	16	4
Creseis virgula					23	4	23	1
Platynereis dumerilii	2		1		4	2	7	3
Janira minuta			2	4	1	1	7	3
Anadara ovalis	1	1				1	2	2
Cuthona sp.	1					1	1	1
Brachyura sp. A	1					1	1	1
Anoplodactylus petiolatus			2	3	2	0	7	3
Doridella obscura		2				0	2	1
Hemiaegina minuta				1		0	1	1
Hyperia galba			1			0	1	1
Gammarid sp. A					1	0	1	1
Gammarid sp. B					1	0	1	1
Brachyura sp. B		1				0	1	1
Tunicate sp.					1	0	1	1
Selar crumenophthalmus		1				0	1	1

[a] Portunids are treated as a group.

Table 5. *Faunal frequency evaluation of area III*

	1	2	9	9a	10	10a	10b	10c	11	Biological index value	Total	Frequency
Latreutes fucorum	309	4	180	24	49	476	187	23	212	58	1,464	9
Leander tenuicornis	46	7	63	17	47	118	172	24	125	51	619	9
Litiopa melanostoma	39	4	12	24	43	99	190	55	14	46	480	9
Platynereis dumerilii	44		3	4	24	56	93	52	36	34	312	8
Portunid A megalopa									3			
Portunid B megalopa						1						
Portunid C megalopa									5			
Portunus anceps juvenile	2											
Portunus ordwayi juvenile									1	28	116	9[a]
Portunus sayi juvenile	10	1	3	1	1		4	1	5			
Portunus spp. juvenile	5		15	6	1	7	10	4	28			
Cronius ruber	1					1						
Gnesioceros sargassicola	18		2	21	10	31	53	11	43	26	189	8
Stephanolepis hispidus	3	1		1		3	7	5	2	8	22	7
Probopyrus latreuticola			3			9	5	1	2	5	20	5
Dromidia antillensis meg.									17	2	17	1
Hoploplana grubei				2			5			2	7	2
Syngnathus pelagicus			2			1	1		1	2	5	4
Sagitta hispida									14	1	14	1
Leptochelia dubia				1						1	1	1
Seriola dumerilii			1					1		0	2	2
Anoplodactylus petiolatus	1						1			0	2	2
Planes minutus									2	0	2	1
Anemonia sargassensis							1			0	1	1
Glaucus atlanticus						1				0	1	1
Spurilla neapolitana								1		0	1	1
Lepas anatifera	1									0	1	1
Lepas fascicularis						1				0	1	1
Squilla sp. larva									1	0	1	1
Ampithoe longimana	1									0	1	1
Tunicate sp.									1	0	1	1
Abedefduf saxatilis				**1**						0	1	1
Abedefduf taurus								1		0	1	1
Caranx dentex							1			0	1	1
Caranx bartholomaei									1	0	1	1
Hyperglyphe bythites									1	0	1	1
Centrolophid sp.		1								0	1	1

[a] Portunids are treated as a group.

to 41.2, but still denotes a strong relationship between these areas. The affinity within area II is 40.8, and the affinities between areas I and II and areas II and

Table 6. *Faunal frequency evaluation of RR-20*

Species	Number of individuals
Lepas pectinata	91
Anemonia sargassensis	45
Gnesioceros sargassicola	2

III drop to 26.7 and 29.2. These values demonstrate greater variability in the samples taken offshore from Cape Hatteras, and a faunal change compared with the bordering regions. The mean within the S series was

43.6, but this is somewhat misleading because the samples appeared to fall into 2 groups. Stations 1 to 5 have higher affinities for each other, comparatively higher raft volumes, and higher diversities than the remaining 4 samples. A number of taxa, including *Gnesioceros*, *Litiopa*, the Nudibranchia, *Leander*, and *Latreutes*, are more conspicuously represented in the first 5 samples. The mean within the D series was 54.68 and compares reasonably with the S series ($\bar{x} = 30.70$). Comparisons of the spring and fall samples show some interesting trends:

S:I	$\bar{x} = 39.5$		D:I	$\bar{x} = 25.9$
S:II	$\bar{x} = 14.3$		D:II	$\bar{x} = 9.5$
S:III	$\bar{x} = 21.8$		D:III	$\bar{x} = 13.0$

Series S and area I had an amazingly high affinity considering the gulf in time and space that separated them. The early spring samples had higher affinities

Table 7. *Faunal frequency evaluation of series S*

	1	2	3	4	5	6	7	8	9	Biological index value	Total	Frequency
Gnesioceros sargassicola	113	82	57	55	7	22	23	23	6	52	388	9
Anemonia sargassensis	176	12	4	16		247	234		20	34	609	7
Hemiaegina minuta	44	40	3	18	9		6	37		32	157	7
Latreutes fucorum	20	21	31	22	12		3	2		28	111	7
Janira minuta	29	9	10	6	8	13	24	10		28	109	8
Lepas pectinata	132	2	3	14	1	5	69	1	3	26	229	9
Platynereis dumerilii	22	20	8	9	5	13	5	3		25	85	8
Litiopa melanostoma	20	15	3	1	3			1	6	15	49	7
Cuthona sp.	3	15		29						8	47	3
Anoplodactylus petiolatus	18	20		4					1	7	43	4
Hoploplana grubei	2	8	3	4			1		1	5	19	6
Portunus sayi	2	2						1		2	5	3
Hyperia galba			1			1				2	2	2
Portunid F megalopa								1		2	1	1
Leander tenuicornis	4	3	1	4	1					1	13	5
Tanystylum orbiculare	6	5								0	11	2
Doto sp.	2	5	2							0	9	3
Lucifer faxoni		2	2	1	1					0	6	4
Scyllaea pelagica		3		3						0	6	2
Lepas anatifera	2			1						0	3	2
Biancolina sp.	1		1							0	2	2
Spurilla neapolitana	2									0	2	1
Doridella obscura			1							0	1	1
Lepas anserifera							1			0	1	1
Cronius ruber		1								0	1	1
Ovalipes guadulpensis				1						0	1	1
Dromiid sp. megalopa								1		0	1	1
Cladoceran sp.		1								0	1	1
Cavolina longirostris	1									0	1	1
Allorchestes sp.				1						0	1	1

Table 8. *Faunal frequency evaluation of series D*

	1	1a	2	3	4	5	Biological index value	Total	Frequency
Janira minuta	32	81	120	301	206	348	35	1,088	6
Hemiaegina minuta	40	177	234	4	339	827	33	1,621	6
Sunamphitoe pelagica	14	49	61	165	109	78	22	476	6
Gnesioceros sargassicola	48	74	32	5	71	39	20	269	6
Anoplodactylus petiolatus	175	42	1		80	168	19	466	5
Platynereis dumerilii	6	34	31	36	36	44	14	187	6
Litiopa melanostoma	11	8	7	24	35	132	11	217	6
Latreutes fucorum	26	14	1	1	3	2	4	47	6
Planes minutus	1	3	2	16	6	6	3	34	6
Hoploplana grubei		1	1		14	40	2	56	4
Anemonia sargassensis					53		2	53	1
Cuthona sp.	1		7		4	17	2	29	4
Spurilla neapolitana				6			2	6	1
Biancolina sp.	5	14	4	1	1	4	1	29	6
Acerotisa sp.	1	2	1		13		0	17	4
Leander tenuicornis	2	5		1			0	8	3
Histrio histrio		1	2		1	1	0	5	4
Lepas pectinata					5		0	5	1
Portunus sayi	1	3					0	4	2
Scyllaea pelagica				2		2	0	4	2
Probopyrus latreuticola	1	2					0	3	2
Polyclad sp.		2	1				0	3	2
Doto sp.						1	0	1	1
Endeis spinosa		1					0	1	1

for the late summer samples than did the late spring samples, possibly indicating faunal changes are greater during the summer months than during the winter. In every set of comparisons in which it was involved, area II had the lowest value.

The species responsible for these affinities are listed in Tables 3 to 8 by decreasing bioindex, abundance and frequency. The one sample from March (RR-20) has not been treated statistically because of its small size and uniqueness.

Discussion

FAGER (1963) defined a community as a group of species which are often found together. Such a definition tacitly assumes the existence of communities, an assumption frequently made by marine biologists (FAGER, 1963; MARGALEF, 1967; JONES, 1969; MILLS, 1969). An opposite viewpoint holds that there are no communities, but rather randomly assembled collections of organisms whose ecological tolerance allows them to exist in a particular environment; each collection is an individual point on a continuum, and any grouping of them is, at best, artificial (FAGER, 1963). Since an individual *Sargassum* raft is discrete within the surrounding planktonic environment, and is populated by a sharply different fauna, it is best treated as a separate community.

Communities have often been named by dominant animals (biocenosis), substrate type (biotope), or by a combination of the two (JONES, 1969). Recently, ecologists have not felt the need for a specific name, which may be misleading, and have typified communities by groups of recurring organisms (FAGER, 1963; MARGALEF, 1967; JONES, 1969). However, I do not feel overly anachronistic in designating the weed complex as the *Sargassum* community. As well as being the substrate, the alga is the most obvious organism in the community.

A small raft of algae afloat on the Atlantic Ocean is a rather extreme habitat. One would expect relatively fewer species on these biotic islands than in the deep-sea benthos beneath them (SANDERS, 1968). This situation is reflected in the diversity, which averaged 2.419 bits of information per individual. Although comparisons of diversities of different communities and different habitats are extremely risky, I will attempt two such comparisons to give the reader a basic frame of reference. GRASSLE (1967) found diversities ranging from 4.023 to 5.083 from grabs on the North Carolina shelf and slope sieved to include meiobenthos. Diversity values for SANDERS' (1960) study of Buzzards Bay, Massachusetts, as given by GRASSLE (1967), varied between 1.558 and 3.466. Although the *Sargassum* community has a tropical affinity and a benthic origin somewhere in the distant past, it has a lower diversity than a tropical benthic habitat.

Considering the uniqueness of the habitat and the number of species encountered, the weed community is remarkably diverse. High diversities were supported by the equitable distribution of the fauna. LLOYD and GHELARDI (1964) hypothesized that the equitability component of diversity is sensitive to the stability of the physical conditions. Indeed, physical conditions are stable: Temperatures range between 22° and 28 °C, Sargasso Sea salinities are high and constant, and dissolved oxygen at the ocean surface approaches saturation. Another factor promoting high diversity is the low productivity of the Sargasso Sea, which has been frequently labeled an oceanic desert. MARGALEF (1968) indicated an inverse relationship between productivity and diversity, reasoning that rich conditions, such as those in a plankton bloom, will favor those few species maximally adapted to utilize the situation.

Among the many theories explaining high diversity, stability is the most widely accepted (PIANKA, 1966). Time by itself does not automatically permit a community to diversify, but it is certainly part of the stability theory. In this light, it is interesting to note that MARKKAVEEVA (1965) found 10 species on *Sargassum vulgare* afloat in the Red Sea which also occur on pelagic *Sargassum* in the Atlantic. This finding indicated that the floating community is old, probably extending back to the time when the Tethys Sea existed.

The dominance affinity index within the individual series of samples was remarkably high when one considers that the weed floats on the water surface, the most variable part of the sea. There is no doubt that the weed forms the basis of a community and not a haphazard congregation of individuals. The change in fauna evident in area II may have resulted from a prolonged residence within the area. The gyre, adjacent to Cape Hatteras but inshore from the Gulf Stream (HARRISON et al., 1967), may have trapped the *Sargassum* where it could be modified by the local fauna. Affinities within the late spring samples show a similarity between the *Sargassum* community in the Gulf Stream and in the Sargasso Sea.

Dominance varied among the samples, and I would consider only the polyclad *Gnesioceros sargassicola*, the polychaete *Platynereis dumerilii*, the snail *Litiopa melanostoma* and the shrimp *Latreutes fucorum* as having maintained dominant positions in each series of samples. Many of the species showed seasonal peaks of abundance. The anemone *Anemonia sargassensis* was only abundant in the late winter and early spring collections. By late spring it had disappeared in all but 1 sample. Nudibranchs were most abundant in the spring. The *Lepas* barnacles also had peak abundance in late winter and early spring. *Lepas pectinata* was the only abundant species; it did not occur in association with *L. anserifera* as reported by PILSBRY (1907). Amphipods exhibited several types of seasonal distribution. *Hemiaegina minuta*, the only caprellid found, was a dominant in both spring series. *Sunamphitoe pelagica* was a dominant in late spring, the only time it was collected, while *Biancolina* sp., a form which normally burrows into algae, had a maximum abundance in early spring although it was taken twice in late spring samples. *Ampithoe longimana* and *Atylus minikoi* were taken in late summer in the Hatteras area. The isopod *Janira minuta* was the dominant organism in late spring, but was also abundant in early spring and late summer in area I.

The pycnogonid *Anoplodactylus petiolatus* reached peak abundance in the late spring, but was present in every set of samples. TIMMERMANN (1932) found most of his *Anoplodactylus* in the central or eastern part of the Sargasso Sea. My observations show they can also be abundant in the western part of the sea and in the Gulf Stream. Although TIMMERMANN frequently encountered *Endeis spinosa*, I found only a single individual. HEDGPETH (1948) took *Tanystylum orbiculare* from gulf-weed cast ashore on the Gulf coast of Texas. I found only 11 individuals in 2 neighboring, early spring samples.

The portunids in the late summer samples were necessarily treated as a group, certainly elevating their position above that which an individual species could claim. Since most of the species were probably transients sharing similar niches, such treatment is not unjustified. Only *Portunus sayi* is commonly considered a resident of the community. The abundance of megalopa and juveniles (including dromiid megalopa) indicates that the weed might offer a protective advantage to the planktonic young. WILLIAMS (1965) lists the range of the portunid *Cronius ruber* as from South Carolina to Brazil. *Cronius* juveniles taken in the Virginian province probably represent a range extension for this species.

Planes minutus, a grapsid crab typically associated with *Sargassum*, was rare or absent except in late spring. The first 2 samples from the Gulf Stream contained 4 *P. minutus*, while the remaining 4 samples from the Sargasso Sea contained 30. Coincident with this was the disappearance of *P. sayi* from Sargasso Sea samples. Although both species occur in both localities, it is possible that *Planes* has a more pelagic distribution while *Portunus* remains closer to shore.

The shrimp *Leander tenuicornis* was dominant only in area III, although it was present in other series in low numbers.

Juvenile fishes were found chiefly in late summer in areas II and III. *Stephanolepis hispidus* was the dominant animal in area II. These juvenile filefishes lead a pelagic life, but associations with the weed remain transitory because the fishes leave for the bottom when between 50 and 100 mm in length (BERRY and VOGELE, 1961). Predation by these fishes in area II may have been partially responsible for the different faunal homogeneity. Seven of the 8 other species of juvenile fishes were found in area III, indicating a

tropical affinity. The pipefish *Syngnathus pelagicus* is a typical resident, but the other species were transients probably attracted to the weed for protection (GOODING and MAGNUSON, 1967). The *Sargassum* fish *Histrio histrio* was found only in late spring, although ADAMS (1960) took it year-round.

Regarding seasonal and local variation, this study has perhaps raised more questions than it has answered. I have no sure way of knowing if seasonal changes I observed were the result of real periodicity of the fauna or whether changes were due to variations within the great gyre of the Sargasso Sea. In other words, geographical variation within the gyre could be taken for seasonality because of sampling in one place at different times of the year. To rectify this situation and definitively establish spatial and temporal variation within the Western North Atlantic would require simultaneous sampling over many sections of the Sargasso Sea and the Gulf of Mexico, as well as repeated sampling over a several-year period at selected stations.

Summary

1. Informational diversity for the noncolonial macrofauna picked from pelagic *Sargassum* had a mean value of 2.419 ± 0.177 ($t_{.05} s_{\bar{x}}$) and a statistical range between 1.401 and 3.437 ($t_{.05} s$).

2. Mean diversity values were not significantly different among the various sampling series and diversity did not vary with raft volume.

3. High diversity values were related to an equitable distribution of species resulting from a stable environment and an area low in productivity.

4. Species' composition of *Sargassum* organisms varied seasonally and geographically.

5. Animals were more abundant in the spring than in the fall samples.

6. Samples collected on a transect in the Gulf Stream and Sargasso Sea maintained a similar faunal composition.

Acknowledgements. I wish to express my sincere appreciation to Drs. M. L. WASS, J. A. MUSICK, and G. C. GRANT who criticized the manuscript. In addition, R. G. SWARTZ, D. F. BOESCH and Dr. M. E. CHITTENDEN gave freely of their knowledge, often at critical times. Miss S. B. LEONARD and Mr. BOESCH, respectively, donated the early and late spring samples used in this study. The *Sargassum* was collected aboard R.V. "Eastward" of Duke University and the National Aeronautics and Space Administration vessel "Range Recoverer."

This project was immensely aided by many people who helped identify species from various animal taxa: A. R. LAWLER, Polycladida, Virginia Institute of Marine Science (VIMS); Dr. D. R. FRANZ, Nudibranchia, University of Connecticut, Storrs; Dr. J. C. McCAIN, Caprellidae, USNM, Smithsonian Institution, Washington, D.C.; J. K. LOWRY, Gammaridea and Hyperidae, University of Canterbury, Christchurch, New Zealand; Dr. M. H. ROBERTS, JR., decapod larvae, Providence College, Providence, Rhode Island; Dr. G. C. GRANT, Chaetognatha, VIMS; and J. D. McEACHRAN and Dr. J. A. MUSICK, Osteichthyes, VIMS.

Literature cited

ADAMS, J. A.: A contribution to the biology and postlarval development of the *Sargassum* fish, *Histrio histrio* (LINNAEUS), with a discussion of the *Sargassum* complex. Bull. mar. Sci. Gulf Caribb. **10**, 55—82 (1960).

BERRY, F. H. and L. E. VOGELE: Filefishes (Monocanthidae) of the Western North Atlantic. Fishery Bull. U.S. Fish Wildl. Serv. U.S. **61**, 61—109 (1961).

CONOVER, J. T. and J. McN. SIEBURTH: Effect of *Sargassum* distribution on its epibiota and antibacterial activity. Botanica mar. **6**, 147—157 (1964).

DEACON, G. E.: The Sargasso Sea. Geogrl. J. **99**, 16—28 (1942).

FAGER, E. W.: Communities of organisms. *In:* The sea, Vol. 2, pp 415—437. Ed. by M. N. HILL. New York: Interscience 1963.

GOODING, R. M. and J. J. MAGNUSON: Ecological significance of a drifting object to pelagic fishes. Pacif. Sci. **21**, 486—497 (1967).

GRASSLE, J. F.: Influence of environmental variation on species diversity in benthic communities of the continental shelf and slope. Unpubl. Ph.D. Dissert., Duke Univ., Durham, North Carolina 1967.

HARRISON, W., J. J. NORCROSS, N. A. PORE and E. M. STANLEY: Circulation of shelf waters off the Chesapeake Bight. Surface and bottom drift of continental shelf waters between Cape Henlopen, Delaware, and Cape Hatteras, North Carolina, June 1963 to December 1964. Prof. Pap. environ. Sci. Serv. Adm. **3**, 1—82 (1967).

HEDGPETH, J.: The Pycnogonida of the Western North Atlantic and the Caribbean. Proc. U.S. natn. Mus. **97**, 157—342 (1948).

HENTSCHEL, E.: Über den Bewuchs auf den treibenden Tangen der Sargassosee. Mitt. zool. Mus. Hamb. **38**, 1—26 (1922).

IDA, H., Y. HIYAMA and T. KUSAKA: Study on fishes gathering around floating seaweed. II. Behavior and feeding habit. Bull. Jap. Soc. scient. Fish. **33**, 930—936 (1967).

JONES, G. F.: The benthic macrofauna of the mainland shelf of Southern California. Allan Hancock Monogr. mar. Biol. **4**, 1—219 (1969).

KRÜMMEL, O.: Die nordatlantische Sargassosee. Petermanns geogr. Mitt. **37**, 129—141 (1891).

LLOYD, M. and R. J. GHELARDI: A table for calculating the equitability component of species diversity. J. Anim. Ecol. **33**, 217—225 (1964).

— J. H. ZAR and J. R. KARR: On the calculation of information-theorerical measures of diversity. Am. Midl. Nat. **79**, 257—272 (1968).

MACARTHUR, R. H.: On the relative abundance of bird species. Proc. natn. Acad. Sci. U.S.A. **43**, 293—295 (1957).

MARGALEF, R.: Some concepts relative to the organization of plankton. Oceanogr. mar. Biol. A. Rev. **5**, 257—289 (1967).

— Perspectives in ecological theory, 111 pp. Chicago: University of Chicago Press 1968.

MARKKAVEEVA, E. G.: The biocenosis of sargasso algae in the Red Sea. [Russ.] *In:* Bentos, pp 81—93. Kiev: Dumka Nauk 1965.

MILLS, E. L.: The community concept in marine zoology, with comments on continua and instability in some marine communities: a review. J. Fish. Res. Bd Can. **26**, 1415—1428 (1969).

MURRAY, J. and J. HJORT: The depths of the ocean, 821 pp. London: MacMillan and Co., Ltd. 1912.

PARR, A. E.: Quantitative observations on the pelagic *Sargassum* vegetation of the Western North Atlantic. Bull. Bingham oceanogr. Coll. **6**, 1—94 (1939).

PIANKA, E. R.: Latitudinal gradients in species diversity: a review of concepts. Am. Nat. **100**, 33—46 (1966).

PILSBRY, H. A.: The barnacles (Cirripedia) contained in the collections of the U.S. National Museum. Bull. U.S. natn. Mus. **60**, 1—122 (1907).

PRAT, H.: Remarques sur la faune et la flore associees aux Sargasses flottantes. Naturaliste Can. **62**, 120—129 (1935).

SANDERS, H. L.: Benthic studies in Buzzards Bay. III. The structure of the soft-bottom community. Limnol. Oceanogr. **5**, 138—153 (1960).

— Marine benthic diversity: a comparative study. Am. Nat. **102**, 243—282 (1968).

SHANNON, C. E.: A mathematical theory of communication. Bell Syst. tech. J. **27**, 379—423, 623—656 (1948).

SIEBURTH, J. McN. and J. T. CONOVER: *Sargassum* tannin, an antibiotic which retards fouling. Nature, Lond. **208**, 52—53 (1965).

THOMSON, C. W.: The voyage of the Challenger, the Atlantic, Vol. 2, 340 pp. New York: Harper and Bros 1878.

THULIN, G.: Ein neuer mariner Tardigrad. Göteborgs K. Vetensk.-o. VitterhSamh. Handl. **2** (5), 1—10 (1942).

TIMMERMANN, G.: Biogeographische Untersuchungen über die Lebensgemeinschaft des treibenden Golfkrautes. Z. Morph. Ökol. Tiere **25**, 288—335 (1932).

WEIS, J. S.: Fauna associated with pelagic *Sargassum* in the Gulf Stream. Am. Midl. Nat. **80**, 554—558 (1968).

WILLIAMS, A. B.: Marine decapod crustaceans of the Carolinas. Fishery Bull. U.S. Fish Wildl. Serv. U.S. **65**, 1—295 (1965).

WINGE, Ö.: The Sargasso Sea, its boundaries and vegetation. Rep. Danish oceanogr. Exped. Mediterr. **3** (Misc. Pap. No. 2), 1—34 (1923).

WOODCOCK, A. H.: Subsurface pelagic *Sargassum*. J. mar. Res. **9**, 77—92 (1950).

YEATMAN, H. C.: The problem of dispersal of marine littoral copepods in the Atlantic Ocean, including some redescriptions of species. Crustaceana **4**, 253—272 (1962).

Author's address: Mr. M. L. FINE
                  Graduate School of Oceanography
                  University of Rhode Island
                  Kingston, Rhode Island 02881, USA

Date of final manuscript acceptance: July 8, 1970. Communicated by G. L. VOSS, Miami

THE EFFECTS OF GRAZING BY SEA URCHINS, *STRONGYLOCENTROTUS* SPP., ON BENTHIC ALGAL POPULATIONS[1]

Robert T. Paine
Department of Zoology, University of Washington, Seattle 98105

and

Robert L. Vadas[2]
Department of Botany, University of Washington, Seattle 98105

ABSTRACT

A series of shallow intertidal pools at Mukkaw Bay, Washington, ranging in height from −0.3 to +0.6 m had the urchin *Strongylocentrotus purpuratus* removed from them. Subtidal rocks at Friday Harbor, −7.3 to −8.2 m, were either caged or had *Strongylocentrotus fransiscanus* removed at monthly intervals. Observations of the rate and pattern of algal succession for periods of up to three years showed that following an initial establishment of new species, brown algae began to dominate. The rate of domination is related to the area's tidal height, with succession most rapid in the lower intertidal areas or subtidally. After a variable period, the majority of the algal biomass was vested in a single perennial brown algal species, *Hedophyllum sessile* in the intertidal and *Laminaria complanata* or *Laminaria groenlandica* subtidally. These plants existed neither in the control areas throughout the study, nor in the experimental pools and rocks before urchin removal. Intermittent urchin browsing could make a major contribution to the variety of algae coexisting within limited areas on these rocky shores.

INTRODUCTION

In a trophic level of varied species composition exposed to fishing or browsing, certain trends in species composition should be predictable when the intensity of exploitation is varied. These should be especially observable when the limiting resource is space because effective use requires occupancy—permitting evaluation on a presence or absence basis. Marine rocky bottoms where space is of primary importance to potential competitors (Connell 1961; Paine 1966) should support communities in which the following relationships hold. If the exploiter is removed entirely, severe interspecific competition should eventually occur and a single or at least restricted number of dominant species should appear, their number probably

being directly related to habitat complexity (MacArthur and Levins 1964). On the other hand, continued overexploitation should lead to an impoverished species group characterized by more effective protective mechanisms, and eventually, under extreme grazing pressure, all inhabitants should be eliminated. Theoretically, then, a single winner should characterize interspecific competition for a simple resource, whereas a number of evolutionary pathways—protective chemicals, structures or behavior, relative unpalatability, or total predator avoidance—might permit the coexistence of many species under fairly severe exploitation. Between these extremes, a greater number of species should occur, their variety being strongly dependent on successional events and the size of the area being considered.

We are addressing ourselves to the following problem. If a section of fairly homogeneous shoreline is examined, varying numbers and species of marine benthic algae will be found. These will vary in size and density, and samples from the entire

[1] This work was supported by National Science Foundation Grant G20901 to the Friday Harbor Laboratories and GB341 and GB2950 to R. T. Paine. R. L. Vadas acknowledges the support of a N.D.E.A. Title IV Fellowship.

[2] Present address: Department of Botany and Zoology, University of Maine, Orono 04473.

342

Reprinted from Limnol. Oceanogr. 14:710–719 (1969).

area will be characterized by considerable internal variability in species composition. Although we recognize the basic importance of understanding broad patterns of diversity as well (Pianka 1966), we have concentrated our efforts on trying to determine the factors related to local species diversity. That is, by examining an ecological mechanism contributing to coexistence, we are seeking answers to the general question, "What permits more or different species to coexist under one set of conditions than under other similar ones?" This approach assumes, perhaps incorrectly, that a large-scale interpretation is ultimately derivable from the integration of many small-scale phenomena. Our paper gives the results of experiments designed to examine the degree to which voracious grazers, sea urchins of the genus *Strongylocentrotus*, might influence local algal species diversity.

We wish to thank Drs. J. M. Emlen, H. S. Horn, and A. J. Kohn for thoughtful comments on the manuscript; Drs. P. Dixon and R. Norris for identification of all the algae in 1967; S. Grant, B. Gregory, and K. P. Mauzey for assistance in diving operations; and Dr. R. L. Fernald for permission to use the facilities of the Friday Harbor Laboratories of the University of Washington.

MATERIALS AND METHODS

Urchins in the genus *Strongylocentrotus* are common both intertidally and subtidally, are primarily herbivorous, and feed on a variety of attached marine algae flourishing in the study areas. In this study, they constituted the predominant grazers, and the species whose numbers were experimentally manipulated.

Two study areas were chosen. One at Mukkaw Bay, just south of Cape Flattery on the western tip of the Olympic Peninsula, Washington, consisted initially of a series of three pairs of shallow pools ranging from about +0.6-m to −0.3-m tidal level (Table 1), the height being determined from tidal factors given for Aberdeen, Washington, adjusted for local factors. Each experimental pool was paired with

an adjacent control pool containing approximately equal densities of S. *purpuratus*. All urchins were removed from the experimental pool by handpicking; no toxic substances were used. The control pool remained unaltered. Experimental pool 4 held fewer urchins initially and was characterized by more algal species than its control. All pools were censused periodically, usually at monthly intervals, and the macroscopic species of algae (with the exception of encrusting *Lithothamnion* sp.) counted and occasionally measured. There was no larval urchin recruitment and minimal adult urchin migration into the experimental sites following their initial preparation. The average density of urchins (Table 1) was estimated on the basis of approximately 30 urchin counts made throughout the study on the control pools (with a 0.1-m^2 quadrat).

These three pairs of pools (2, 3, 4) all received at least two and probably most of three years of algal spore recruitment in the absence of urchin grazing pressure. A fourth pool (pool 1) was initiated in January 1965 at a height of +0.9 m but had only 70% of its urchin population removed. Differences between pool 1 and its control were minimal because of the height of the pool and the presence of urchins; therefore, we have omitted it from further consideration. In addition, three more pools were established in March–April 1966 in an attempt to replicate our earlier efforts in both time and space. Two pools, A and B, were cleared at a height comparable to pool 3. The other, pool 4P, originally the control for pool 4, was inadvertently cleared of urchins when the starfish *Pycnopodia helianthoides* moved in, consuming some urchins and stampeding the rest. These additional pools probably received algal spore recruitment for most of two spring seasons.

Sublittoral areas were established in the San Juan Islands, near the Friday Harbor Laboratories and observations were conducted with the aid of SCUBA. Experiments were designed to prevent grazers from invading selected test surfaces. Boul-

TABLE 1. *Relevant data on experimental sites*

Site		Tidal level (m)	Date of urchin removal	Date of terminal census	Surface area (m²)	Mean urchin density on control (No./m²)
Mukkaw Bay						
Pool 2	Expt	+0.6	May 1964	May 1967	0.27	
	Cont.				0.21	55
Pool 3	Expt	+0.3	Sep 1964	Jun 1967	0.36	
	Cont.				0.54	73
Pool A	Expt	+0.3	Apr 1966	Jun 1967	0.34	
	Cont.				0.21	60
Pool B	Expt	+0.3	Apr 1966	Jun 1967	0.63	
	Cont.				0.27	69
Pool 4	Expt	−0.3	Feb 1965	May 1967	0.33	
	Cont.				0.16	72
Pool 4P	Expt	−0.3	Mar 1966	May 1967	0.16	
	Cont.				—	—
Friday Harbor						
Cage 8	Expt	−7.3	Feb 1965	Apr 1966	0.80	
	Cont.	−7.3			0.82	6
Rock 8	Expt	−7.3	Apr 1964	Apr 1966	0.81	
	Cont.	−6.4			2.10	7
Rock 11	Expt	−8.2	Apr 1964	Apr 1966	0.90	
	Cont.	−4.6			4.70	6

ders (designated rock 8 and rock 11) were stripped of their flora and fauna, and subsequent algal colonization and succession were measured on a regular (monthly) basis. A limited number of adventitious grazers were removed at each visit. A second method consisted of placing denuded boulders inside a cage, one cubic meter in volume, made from plastic-coated wire. A large mesh size (ca. 5 × 6 cm) was selected to permit ample light to reach the boulder surface. The cages had to be brushed frequently to prevent diatom growth on the wire. Each cage had a swinging door to permit close inspection of the rock surfaces. Although we expected them to provide a more efficient means of limiting browsing, the cages proved unsatisfactory on two counts. In areas of very dense populations of S. *dröbachiensis* and S. *fransiscanus*, the average size of the former species was slightly smaller than the diagonal dimension of the mesh, so that

small individuals could enter and graze the experimental boulders. Other individuals climbed the cages and consumed the diatom film and plastic coating on the wire. The ensuing corrosion led to a weakening or total destruction of the wire in slightly over a year, and the cage studies were abandoned in April 1966. In addition, substantial algal growth on the boulders in the absence of urchins actually caused most of the cages to be lifted by currents and overturned; only one experimental cage (No. 8) and its control escaped such disturbance. Control situations were established either by adding 5 S. *fransiscanus* to a caged boulder or by allowing the usual complement of urchins to persist on uncaged boulders. Grazing by urchins continued throughout the study on control rocks. Data on all experimental and control sites are given in Table 1.

Wet weight, determined after blotting the organism to a damp-dry state, and length measurements were used to relate the size of the plants to an estimate of the wet weight per species at each site.

RESULTS

Urchin removal brought about an initial increase in the number of algal species represented in all plots with such species as *Spongomorpha* sp., *Codium fragile*, *Gigartina papillata*, *Delesseria decipiens*, and *Pterosiphonia* sp. attaining a recognizable size. This result was anticipated for two reasons. Experimental investigations on marine benthic herbivores have indicated that these, unlike their terrestrial counterparts, consistently overgraze and at times almost eliminate their food source (Stephenson and Searles 1960; Kitching and Ebling 1961; Southward 1964; Randall 1965; Jones and Kain 1967). The pools and boulders grazed by urchins, in contrast to adjacent ungrazed areas, showed a dearth of most algae, although, predictably, they were populated by a few species. Thus, the control areas for intertidal pools 2–4 began with a mixture of calcareous and small fleshy algae (*Corallina vancouveriensis*, *Corallina officinalis*, *Bossiella* spp., *Calliarthron tuberculosum*, *Lithothamnion* spp., *Polysiphonia* spp., and *Ulva rigida*) that was maintained throughout the study. In the subtidal cage with enclosed sea urchins and on the control boulders, the only macroscopic algae found continuously were *Lithothamnion* and *Ulva-Monostroma*. This pooling of various *Ulva* and *Monostroma* species was necessitated by their great morphological similarity and the logistic problems of adequate underwater sampling. We feel that such pooling can be justified on the grounds that these genera probably affect their environments in ecologically comparable ways, especially in their relative abilities to space-limit or shade other algal species. There was a brief seasonal establishment of both foliose and filamentous red algae and a few brown algae in the cage; none of these persisted for long.

Hence the immediate increase in species diversity as indicated by the length of the species list at all experimental sites was due to the successful establishment of certain species in the absence of significant browsing pressure. Algal colonization, however, also depends on the seasonal availability of spores. Thus, when present, *Prionitis lyalli* first appeared in November, *Desmarestia herbacea* and *Hedophyllum sessile* usually in April, and so on.

An analysis of the trends exhibited on the experimental plots through time has necessitated some arbitrary divisions of the algae into broad types. Obviously no species are identical either in their physiological tolerance to light intensity or exposure or in their eventual size, growth rate, or successional status. Thus, it is as difficult to discuss and compare the ecological roles of *Polysiphonia* and *Hedophyllum* as it would be those of grass and oak trees, for one is small and common, the other large and relatively rare. We will therefore consider two arbitrary algal types, understory and canopy-forming species.

Censuses taken in the intertidal habitats in May or June of 1965–1967, in which the standing crop wet weights of all algae were either estimated from frond length–wet weight relationships, or, in 1967, determined by cropping, indicate that on a damp-dry wet weight basis, the pools were dominated by a limited number of species (Table 2). Nine of these have been designated the canopy-forming species because they exerted considerable shading on other species in the pools and at one stage or another six of them (*Odonthallia*, *Desmarestia*, *Costaria*, *Alaria*, *Hedophyllum*, *Nereocystis*) comprised the dominant plant by weight. The remaining three canopy-forming species (*Laminaria*, *Egregia*, *Pleurophycus*), though never dominant, at times formed from 16–41% of the total biomass and were conspicuous elements of the flora. Unfortunately, no estimate of the error introduced by using wet weights is available, since certainly the point at which an alga is "damp-dry" is highly subjective. However, the maximum disparity between two measurements of water content for each of the canopy-forming species

TABLE 2. *Algal dominants, by per cent of the total wet weight of canopy-forming species in urchin-free areas*

Site	Date	Most common species	% Total wet wt	Second most common species	% Total wet wt
Pool 2	May 1965	Nereocystis luetkeana	24	Laminaria setchellii	16
	May 1966	Odonthallia floccosa	29	Laminaria setchellii	17
	May 1967	Hedophyllum sessile	47	Egregia menziesii	24
Pool 3	May 1965	Nereocystis luetkeana	54	Desmarestia herbacea;	16
				Costaria costata	16
	May 1966	Hedophyllum sessile	40	Laminaria setchellii	26
	Jun 1967	Hedophyllum sessile	48	Egregia menziesii	41
Pool A	Jun 1967	Hedophyllum sessile	48	Costaria costata	26
Pool B	Jun 1967	Costaria costata	45	Hedophyllum sessile	36
Pool 4	May 1965	Alaria nana;	27	Costaria costata	16
		Desmarestia herbacea	27		
	May 1966	Hedophyllum sessile	53	Laminaria setchellii	17
	May 1967	Hedophyllum sessile	80	Odonthallia floccosa	5
Pool 4P	May 1967	Hedophyllum sessile	57	Pleurophycus gardneri	24
Cage 8	May 1965	Nereocystis luetkeana	98	Alaria marginata	1
	Apr 1966	Laminaria groenlandica	96	Agarum fimbriatum	3
Rock 8	May 1964	Nereocystis luetkeana	86	Desmarestia herbacea	2
	May 1965	Laminaria groenlandica	63	Laminaria complanata	35
	May 1966	Laminaria complanata	60	Laminaria groenlandica	20
Rock 11	May 1964	Nereocystis luetkeana	97	Ulva-Monostroma	1
	May 1965	Laminaria groenlandica	65	Laminaria complanata	35
	May 1966	Laminaria groenlandica	53	Laminaria complanata	47

(except *Pleurophycus*, which was not measured) was 2.8%, suggesting some internal consistency in the technique. The range of water contents for the eight species was 77–90%, and for the six species that comprised the dominants, 81–90% (Paine and Vadas, unpublished). Thus measurements within 5–6% of one another are doubtfully distinguishable; differences in excess of this value have been assumed to indicate biologically distinguishable differences in dominance.

Inspection of Table 2 indicates a strong tendency with time for all intertidal areas to become dominated by *Hedophyllum sessile* and subtidal ones by *Laminaria* spp. The complex of intertidal canopy species has been further analyzed using an index of overlap (Horn 1966) in which the entries are the proportion of the total wet weight of all canopy species formed by any one species in May or June of 1965–1967. By calculating the extent of overlap between all pairs of samples, a matrix is developed (Fig. 1) which gives the appropriate values. Since the index varies from 0 (no species in common) to 1.0 (samples identical with respect to proportional species composition), the values in the matrix represent the relative similarity of the samples at different times and levels. The values have been shuffled, as is appropriate for trellis diagrams, to bring together the greatest number of entries showing the highest affinity. Each sample is indicated by a series of numbers or letters (i.e., 4M67 designates pool 4, May 1967 sample, etc.).

Three tendencies within the canopy-forming species are apparent (Fig. 1). There is relatively little overlap between the complex of these species existing in pools 2–4 in 1965 and the same pools in 1967. In other words, the local species composition has been radically changed due to the alteration of a single factor—sea urchin density. By 1967 all pools, with

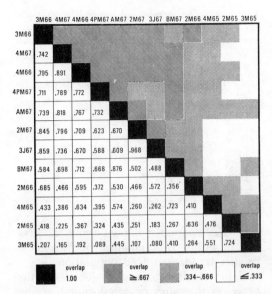

FIG. 1. Trellis diagram arrangement of overlap indices, following Horn (1966), of the nine canopy-forming species based on their wet weight.

the possible exception of B, show a high degree of overlap, as is indicated by their common occupancy of the upper left portion of the trellis diagram and overlap indices greater than 0.667. Coupled with other data (Table 2), this suggests that *Hedophyllum*, in the absence of urchins, would become the dominant alga between the –0.3-m and +0.6-m levels at Mukkaw Bay. Finally, the analysis shows that succession to the apparent climax association takes longer at higher than at lower levels. Thus, pool 4 in 1966, 15 months after urchin removal, is very similar (overlap of 0.89) to the 1967 condition, 27 months after. The overlap in 1967 between pools 4PM67 and 4M67 is 0.79, indicating the attainment by pool 4P of a closely comparable status to pool 4 within 14 months of urchin removal. Similarly the overlap between 4PM67 and 4M66, when the latter was 15 months old (Table 1) is high (0.77), again suggesting that succession to the final observed state takes a maximum of 14–15 months at the –0.3-m tidal level. At the +0.3-m level, pool 3 in 1966 and 1967 shows a 0.86 overlap, indicating a stabilizing of proportions after 20 urchin-

free months. Pools A and B, at the same level as 3, exhibit overlap values, after 14 urchin-free months, of 0.61 and 0.49 when compared with the 3J67 samples, while being internally consistent as indicated by comparisons between themselves (overlap 0.88). Apparently at this tidal height, between 14 and 20 months are needed before the relative proportions attain a fairly high level of overlap. Pool 2, on the other hand, although dominated by *Hedophyllum* in May 1967, shows an overlap of 0.47 with itself in May 1966, 24 months after urchin removal. Comparisons between 2M67 and 2M65, when this pool was 12 months old, show an overlap of only 0.25. The implication is that at the +0.6-m level between 2–3 years are needed before the dominant is expressed. The observations were not carried out long enough at this level to determine if the pattern had stabilized.

Another way of expressing the relationships between the canopy-forming intertidal algae is to show the per cent of the total wet weight of all the canopy-forming species formed by *Hedophyllum* as a function of the time (months) since urchins were removed (Fig. 2). The tendency for succession to be most rapid at the lowest level (pool 4) and slowest at the highest (pool 2) is obvious. Pools A, B, and 4P are similar to pool 4 in their patterns of domination by *Hedophyllum*, despite their different tidal heights (Table 1). We believe this indicates the importance exercised by the season in which the experiment was initiated. One might have expected pools A and B to be more similar to pool 3 than to pool 4. However, Widdowson (1965) observed that along the Strait of Juan de Fuca *Hedophyllum* produced spores from November to March and that the greatest number of young sporophytes appeared in July. If the same seasonal pattern applies at Mukkaw Bay, only a few kilometers distant, the patterns in Fig. 2 are explicable. Pools A and B, established in April and sampled 14 months later, were colonized by spores either before or immediately after clearance and contained *Hedophyllum* benefiting

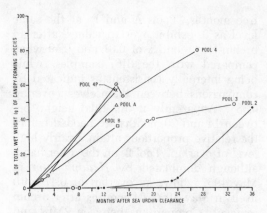

FIG. 2. The tendency toward domination by *Hedophyllum* in urchin-free intertidal pools.

TABLE 3. *The number of canopy-forming and understory species present in the major censuses*

Year	Pool					
	2	3	A	B	4	4P
Canopy-forming species (total of 9)						
1965	5	4			7	
1966	4	6			5	
1967	4	6	7	6	7	6
Understory species (total of 26)						
1965	10	5			7	
1966	4	5			9	
1967	11	5	9	10	9	13

from one full growing season. A similar trend is shown by pools 4 and 4P, also started just before the period of maximum spore settlement and sampled 13–14 months later. Pool 3, on the other hand, was initiated in September 1964 and did not receive *Hedophyllum* spores in significant enough numbers to be recorded until June 1965, in accord with Widdowson's phenological observations. Once established, however, the initial slope of the line appears comparable to that of all the other pools except pool 2. Another possibility is that, if latent spore settlement did occur in pool 3 in September 1964, reduced winter light intensities may have prevented gametophyte maturation and subsequent sporophyte growth until the following spring. A similar situation has been shown to influence other local kelp species (Vadas 1968). In pool 2, successful establishment took between 12–23 months, the initial *Hedophyllum* growth appears slower, but beyond a certain threshold, the rate of domination, indicated by the slope, is again comparable to that of the other areas.

The break in the slope of *Hedophyllum* domination in pool 3 after 20 months is due to the development of a significant mass of *Egregia*, represented by two plants. A comparable situation may have been developing at a reduced tempo in pool 2 (Table 2), in which, after 36 months, *Egregia* comprised 24% of the standing crop of the canopy-forming species. Not enough is known about the ecology of either *Hedophyllum* or *Egregia* to predict trends in mixed populations.

Table 3 illustrates a major distinction between the nine canopy-forming species and the understory. Although the abundance and size of the former varied somewhat, as reflected in the variable overlap indices, between 44–77% of the species were present at all times. The remaining 26 species or genera recorded in the major censuses—the understory—represent a much more heterogeneous assemblage, with some being found in almost all censuses (*Ulva*, *Spongomorpha*, *Iridaea*, *Polysiphonia*, *Corallina*) and others at unpredictable intervals but perhaps influenced by season or pool height (e.g., *Codium*, *Bryopsis*, *Prionitis*, *Rhodymenia*, *Laurencia*, *Opuntiella*, *Pterosiphonia*). It is these understory species which, while accounting for very little of the standing crop wet weight, make a significant contribution to both the variety of species present and their temporal and spatial variability.

On subtidal rocks protected from urchin browsing, events were remarkably uniform. Algal species abundance reached a peak in April or May of all three years (1964–1966). All three areas were dominated in the first year by *Nereocystis* (97, 86, and 98% of the standing crop biomass) and in the second by *Laminaria groenlandica*. However, in the third year on rocks 8 and 11, *Laminaria complanata* appears gradually to be replacing *L. groenlandica* and, if given enough time, might well become the dominant species (Table 2). Again, as in the lower intertidal, the apparent climax dominant is clearly expressed a little more than one year after ridding the area of urchins. These subtidal rocks depart significantly from the intertidal situation when the paucity of species in the understory is considered. This might be due to incidental but experimentally unavoidable herbivory or, more likely, to greatly reduced light intensities (Vadas 1968). Thus, in cage 8, which was exposed to the lowest level of browsing and hence is our best example of an urchin-free subtidal boulder, only seven species were found in the April 1966 census, and of these, four are canopy-forming.

DISCUSSION

Along the outer coast of Washington, *Hedophyllum* is one of the more conspicuous intertidal algae. Rigg and Miller (1949) found that it could occupy a –0.3-m to +1.5-m vertical distribution in the Neah Bay region, a few kilometers from Mukkaw Bay. We have observed it on exposed rock surfaces from the –0.3-m to +1.2-m levels, and our experiments have indicated that *Hedophyllum* can predominate in shallow pools, in the absence of sea urchins, from –0.3 to +0.6 m. Potentially then, in the absence of severe disturbance, *Hedophyllum* could become the most conspicuous alga between –0.3 and +1.5 m. That it does not, despite a large holdfast and high growth rate (Widdowson 1965), can be attributed to at least three factors: physical disturbance, topographic complexity, and herbivory. Similar conditions probably influence subtidal algae, for in our experiments, *Laminaria* spp. consistently become dominant in the absence of urchins, although in their presence, *Laminaria* are much less conspicuous members of the flora.

We have no data on the role that variations in topographic relief play in determining the presence or absence of algae, and thus their ability to coexist. Physical disturbance is obviously important, although our observations were not structured to measure it. We have observed relative plant stunting, marginal fraying of the fronds of the larger species due to whiplash effects, and severe bleaching and even killing of individual *Nereocystis* and *Laminaria,* all of which will influence the variety of benthic algae inhabiting any particular space.

Biological disturbance due to herbivory is certainly a major factor. Our data show that, depending on tidal height, urchin removal leads to significant changes in algal species composition within periods of less than a year. Urchin removal also occurs naturally, for the starfish *Pycnopodia* consumes them (personal observations; Mauzey, Birkeland, and Dayton 1968), often but unpredictably clearing a pool (e.g., pool 4P). Such events indicate that our experimental procedures were realistic and suggest a natural cause for the patchwork of areas in the lower littoral with and without urchins, and hence characterized by a great variety of algae.

The controls for all areas, both intertidal and subtidal, maintained a characteristic appearance throughout the study: high urchin density, a pavement of *Lithothamnion,* and only a few other algae. Intertidal urchin removal initiated a procession of events very much dependent on the season but also generalizable: 1) an immediate increase in the number of algal species present, with from 6–12 new species being added to the pool's flora; 2) the rapid establishment of a canopy of larger, usually brown, algae; 3) a succession within these so that after a suitable period the majority of the algal biomass was vested in the single perennial species, *H. sessile*; 4) the establishment under the canopy of an understory composed of a variety of smaller, usually red, algae, no one species of which was prevalent. These account, on a presence or absence basis, for over 60% of the species present within the experimental and control areas; but because of small size, frequent rarity, and unpredictable temporal and spatial patterns, they make analysis difficult.

Events in the subtidal are comparable

except that the perennials, *L. complanata* or *L. groenlandica* come to predominate and an extensive understory does not develop. We attribute the latter to a marked reduction in light intensity due both to the development of a dense canopy and to the natural extinction of light passing through water (Kitching 1941). Pool 3 in the intertidal may have exhibited a similar effect. In the June 1967 sample, only five understory species were recorded, about half as many as at any of the five other pools (Table 3). Pool 3 was characterized by the development of both *Egregia* and *Hedophyllum*, these species having a combined wet weight of 2,975 g, 2–3 times that of any other pool's entire flora on a square meter basis. Also suggestive of a shading effect was the reduced presence of *Lithothamnion* and the absence of *Corallina* and *Bossiella*, the calcareous red algae which characterize control pools but also persist in most of the experimental areas. If this trend in pool 3 is general rather than exceptional, it would indicate a drop in the variety of species inhabiting the area with time, as the canopy becomes denser in the absence of sea urchins.

In conclusion, the overall effects of urchin browsing on algal diversity can be viewed in two ways. First, considering only the size of a patch that might develop on urchin removal, there is a complex of species that then establishes itself locally. This complex contains many more species than areas inhabited by dense urchin populations. Hence sea urchin presence reduces diversity by preventing the development of combinations of algae capable of coexistence. This situation, contrary to the expectations suggested in the introduction, could be stable or could have resulted from an insufficient period of observation. Thus, more than three years' data seem to be necessary to determine if a monoculture could develop from currently mixed populations of *Hedophyllum* and *Egregia* in the intertidal and *Laminaria* spp. in the subtidal. We would predict, then, assuming the eventual resolution of this indeterminate situation, that one species would

clearly dominate the area, on a wet weight basis, at the expense of (with the local elimination of) the other canopy-forming species. Concomitant reductions in the list of understory species would be expected to follow, similar to those observed in pool 3 and the subtidal areas. The other general interpretation invokes a broader consideration of the local landscape. Urchin browsing, especially when the animals are dense, reduces the associated algal standing crop to a brief list. Urchin browsing that is intermittent, because of local destruction of the urchins or their movements, introduces a heterogeneity factor of space available for algal colonization that, coupled with asynchronous algal development in urchin-free areas, could lead to a substantial increase in the number of species "coexisting" within larger portions of these rocky areas.

REFERENCES

CONNELL, J. H. 1961. The influence of interspecific competition and other factors on the distribution of the barnacle *Chthamalus stellatus*. Ecology, **42**: 710–723.

HORN, H. 1966. Measurement of "overlap" in comparative ecological studies. Am. Naturalist, **100**: 419–424.

JONES, N. S., AND J. M. KAIN. 1967. Subtidal algal colonization following the removal of *Echinus*. Helgolaender Wiss. Meeresuntersuch., **15**: 460–466.

KITCHING, J. A. 1941. Studies in sublittoral ecology. III. *Laminaria* forest on the west coast of Scotland: a study of zonation in relation to wave action and illumination. Biol. Bull., **80**: 324–337.

———, AND F. J. EBLING. 1961. The ecology of Lough Ine. XI. The control of algae by *Paracentrotus lividus* (Echinoidea). J. Animal Ecol., **30**: 373–383.

MACARTHUR, R., AND R. LEVINS. 1964. Competition, habitat selection and character displacement in a patchy environment. Proc. Natl. Acad. Sci. U.S., **51**: 1207–1210.

MAUZEY, K., C. BIRKELAND, AND P. DAYTON. 1968. Feeding behavior of asteroids and escape responses of their prey in the Puget Sound region. Ecology, **49**: 603–619.

PAINE, R. T. 1966. Food web complexity and species diversity. Am. Naturalist, **100**: 65–75.

PIANKA, E. R. 1966. Latitudinal gradients in species diversity: a review of concepts. Am. Naturalist, **100**: 33–46.

RANDALL, J. E. 1965. Grazing effects on sea

grasses by herbivorous reef fishes in the West Indies. Ecology, **46**: 255–260.

RIGG, G., AND R. MILLER. 1949. Intertidal plant and animal zonation in the vicinity of Neah Bay, Washington. Proc. Calif. Acad. Sci., **26**: 323–351.

SOUTHWARD, A. J. 1964. Limpet grazing and the control of vegetation on rocky shores, p. 265–273. *In* D. J. Crisp [ed.], Grazing in terrestrial and marine environments. Blackwell.

STEPHENSON, W., AND R. B. SEARLES. 1960. Experimental studies on the ecology of intertidal environments at Heron Island. I. Exclusion of fish from Beach Rock. Australian J. Marine Freshwater Res., **11**: 241–267.

VADAS, R. 1968. The ecology of *Agarum* and the kelp bed community. Ph.D. Thesis, Univ. Wash., Seattle. 282 p.

WIDDOWSON, T. 1965. A taxonomic study of the genus *Hedophyllum setchell*. Can. J. Botany, **43**: 1409–1420.

James A. Estes and John F. Palmisano

Reprinted from Science 185:1058–1060 (1974).

Sea Otters: Their Role in Structuring Nearshore Communities

Abstract. A comparison of western Aleutian Islands with and without sea otter populations shows that this species is important in determining littoral and sublittoral community structure. Sea otters control herbivorous invertebrate populations. Removal of sea otters causes increased herbivory and ultimately results in the destruction of macrophyte associations. The observations suggest that sea otter reestablishment indirectly affects island fauna associated with macrophyte primary productivity.

Destruction of subtidal and intertidal kelp and sea grass beds because of overgrazing by dense populations of sea urchins has been observed over a wide geographical range (*1, 2*). Removal of sea urchins by experimental manipulations (*2*) and by accidental oil spills (*3*) has resulted in the rapid development of marine vegetation. Because community structure differs in the presence and absence of kelp beds (*4–6*) and prey density in marine communities can be significantly influenced by predation (*7*), the structure of a marine community could be determined by the intensity of herbivore predation (*8*).

Speculation regarding the interrelations of sea otters (*Enhydra lutris*) and marine invertebrates has generated controversy in California. However, only slight consideration has extended beyond economic and esthetic arguments by commercial abalone interests and groups concerned with the sea otters' welfare. The observations discussed in this report suggest that sea otters have a profound effect on the structure of marine communities.

Historically, the sea otter occupied a range from the northern Japanese archipelago, through the Aleutian Islands, and along the coast of North America as far south as Morro Hermoso, Baja California (*9*). At present, the sea otter occupies only remote portions of this original range in the Kuril, Commander, and Aleutian islands and parts of southeastern Alaska (*10*). There is an isolated population off the coast of central California, and recent transplants have reintroduced the sea otter into Oregon, Washington, and British Columbia. Continued expansion of the sea otters' range may be expected.

The sea otter population of Amchitka Island, in the Rat Island group (*11*) of the Aleutian archipelago, has been estimated to be 20 to 30 animals per square kilometer of habitat (*12*). The feeding habitat of the sea otter is limited to the intertidal and sublittoral regions within the 60-m depth contour (*10*). Adult, captive sea otters require 20 to 23 percent of their body weight daily in food, and in the natural environment forage species include benthic invertebrates and fish (*10, 13*). Considering the sea otters' average weight as about 23 kg (*10*), we conservatively estimate that 35,000 kg km^{-2} year^{-1} of animal biomass is consumed by foraging sea otters at Amchitka Island. Thus, a high-density sea otter population is an important member of the nearshore marine community.

Such high-density populations have existed in the Rat Island group for about 20 to 30 years, after almost complete annihilation by Russian fur traders during the 18th century. Apparently, the once abundant sea otter population of the Near Islands was extirpated by overexploitation. Until recently, immigrants from the densely populated Rat Islands have been unable to reach the Near Islands, which are located approximately 400 km west-northwest and are separated from the Rat Islands by wide, deep oceanic passes. Since 1959 there have been scattered reports of sea otters in the Near Islands (*10*), although no major population reestablishment has yet occurred.

We have studied the nearshore marine communities of Amchitka Island in the Rat Island group and Shemya Island in the Near Island group. Field observations were made at Amchitka at approximately bimonthly intervals from October 1970 to August 1973 and at Shemya for 1 week each in September 1971 and July 1972; observations were also made at Attu in the Near Islands for 4 days in July 1972.

We propose that the sea otter is the primary cause of the differences observed between the nearshore marine communities of the Rat Island and the Near Island groups. Sea urchins (*Strongylocentrotus* sp.) (*14*) are an important sea otter food and are known to be voracious algal grazers which can consume and destroy large quantities of kelp. Our hypothesis is that a dense population of sea otters reduces the sea urchins to a sparse population of small individuals by size-selective predation. The resultant release from grazing pressure permits a significant increase in the size of nearshore and intertidal kelp beds and associated communities.

Benthic macrophytes in the Rat Island group extend from the intertidal region and cover most of the surface of the rock substrate to depths of 20 to 25 m (Fig. 1). Major contributors to these plant communities are Phaeophyta (brown algae), *Alaria fistulosa*, *Laminaria longipes*, *L. groenlandica*, *L. yezoensis*, *L. dentigera*, *Agarum cribrosum*, *Thalassiophyllum clathrus*, *Desmarestia* sp., and various Rhodophyta (red algae). Sea urchins are generally not conspicuous in shallow areas (0 to 20 m). However, relatively high densities of sea urchins occur in microhabitats along more protected cracks and beneath holdfasts of macrophytic vegetation. Beginning at depths of 10 to 20 m, sea urchin densities increase with depth and vegetation coverage de-

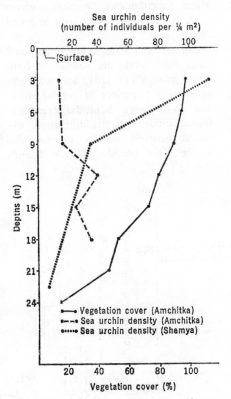

Fig. 1. Vegetation coverage and sea urchin density plotted against depth. The data for Amchitka Island and Shemya Island represent averages from four and three study areas, respectively. Vegetation cover at Shemya Island is coincident with the ordinate.

creases in areas of solid substrate (Fig. 1). Densities of sea urchins are highly variable at these depths, but range up to 680 m^{-2} (15). The majority of these sea urchins have test diameters of less than 32 mm (16). The increase in sea urchin density with depth is probably related to decreased predation by sea otters (and perhaps diving birds). Feeding on small sea urchins at these depths may be energetically infeasible for predators.

Conversely, the Near Island group is characterized by a distinct lack of macrophytic vegetation below the lower intertidal region. In many areas, sea urchins almost completely carpet the sublittoral immediately adjacent to the littoral, but densities decrease as a function of depth (Fig. 1). Differences in size class distribution and biomass between Near Island and Rat Island sea urchin populations are shown in Fig. 2. The larger size (age) classes of sea urchins are missing from the Rat Island group.

Despite the physical similarities and geographical proximity of the Rat Islands and the Near Islands, there are major floral and faunal differences between the marine communities of their lower intertidal rock platforms (benches). The Rat Islands have an almost complete mat of benthic marine brown algae (kelp), predominantly *Hedophyllum sessile* and *L. longipes*, covering these benches. Sessile, filter-feeding invertebrates—barnacles (*Balanus glandula* and *B. cariosus*) and mussels (*Mytilus edulis*)—and motile, herbivorous invertebrates—sea urchins and

chitons (*Katharina tunicata*)—are inconspicuous, small, and scarce. At the Near Islands, *H. sessile* and *L. longipes* are heavily grazed by dense populations of sea urchins and chitons, and there are extensive mussel beds and dense populations of barnacles. Less than 1 percent of the attached kelp examined at the Rat Islands was grazed (17). At the Near Islands all kelp overhanging channels and tide pools was grazed, and more than 75 percent of the *L. longipes* plots and 50 percent of the *H. sessile* plots sampled contained grazed plants (17). Barnacle and mussel densities, respectively, averaged 4.9 m^{-2} and 3.8 m^{-2} at the Rat Islands and 1215 m^{-2} and 722 m^{-2} at the Near Islands (17). Sea urchin and chiton densities, respectively, averaged 8 m^{-2} and less than 1 m^{-2} at the Rat Islands and 78 m^{-2} and 38 m^{-2} at the Near Islands (17).

Kelp beds at the Rat Islands shelter the shore from wave action to an appreciable extent. Populations of sessile intertidal invertebrates decline drastically at the Rat Islands since they cannot compete successfully with kelp for space and they are hampered by silt which accumulates because wave-induced turbulence has been reduced (18).

Climate, sea state, tidal ranges, and mean tidal levels are similar at both island groups (19, 20), and we compared only coastlines of similar structure (with wide intertidal benches). We conclude that the differences observed between benthic communities of the Near Islands and Rat Islands

are probably related to the presence or absence of sea otters. The otters effectively control sea urchin populations, and the absence of grazing pressure allows vegetational communities to flourish. Reducing the population of sea otters makes it possible for the sea urchin population to increase, and this leads to a significant reduction in the size of the kelp beds and associated communities.

More far-reaching consequences of these relations are suggested by comparing food webs and faunal distributions between the island groups. Benthic macrophytes are of considerable importance to nearshore productivity in temperate waters (21). Species whose food webs originate from macrophytic algal productivity would certainly be adversely affected by its removal. We believe that some faunal differences between the Near Islands and Rat Islands are related to the presence or absence of benthic macrophytes as a nutritional base. Rock greenling (*Hexagrammos lagocephalus*), harbor seals (*Phoca vitulina*), and bald eagles (*Haliaeetus leucocephalus*) are abundant in the Rat Islands but are scarce or absent in the Near Islands (19, 22). These species depend largely on nearshore marine productivity in the Aleutians (23). We propose that reduced populations of these (and perhaps other) species in the Near Islands may be related to reduced macrophyte productivity.

Our results suggest that reestablishment of sea otters along the Pacific coast of North America will have pro-

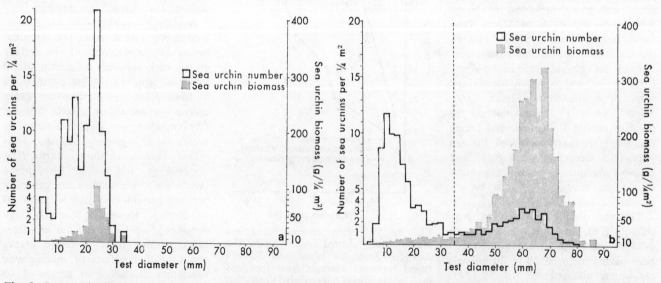

Fig. 2. Sea urchin size class distributions and associated biomass contributions. (a) Data collected from Amchitka Island (high-density sea otter populations). (b) Data collected from Shemya Island (sea otters absent). The dotted line represents the largest sea urchin size class observed at Amchitka Island.

found ecological effects. That this is currently happening is indicated by the sea otter–abalone controversy in California. A decrease in sport and commercial abalone fisheries has been reported following the influx of sea otters into areas of previously unoccupied habitat (24). Surveys conducted in 1967 by the California Department of Fish and Game revealed that throughout the sea otters' range preferred sea otter forage items were reduced in number and restricted to protected habitat as compared with habitat outside the range (25). Also, an increased diversity in sea otter forage items has been reported in areas long inhabited by sea otters. This is apparently the result of reduced availability of preferred sea otter forage items (24).

The sea otter may also be important in restoring kelp beds (and associated species of animals) in southern California. Sea otters in California completely remove large sea urchins (Strongylocentrotus franciscanus) from areas by predation, permitting luxuriant development of the Nereocystis-Pterygophora (brown algae) association (4). Recent increases in sea urchin populations are correlated with kelp bed reduction (5). Although kelp bed reductions are obviously related to phenomena more recent than the disappearance of sea otters (26), the reestablishment of sea otters should decrease invertebrate populations and increase vegetational biomass.

The sea otter is an important species in determining structures and dynamic relations within nearshore communities, and so fits Paine's (27) concept of a keystone species. Many changes have resulted from the near extinction of the sea otters in these communities during the 18th and 19th centuries. In modern biological studies of nearshore marine communities along the Pacific coast of North America the species' ecological importance has not been considered in sufficient detail. We believe that the sea otter is an evolutionary component essential to the integrity and stability of the ecosystem.

JAMES A. ESTES
Arizona Cooperative Wildlife Research Unit, University of Arizona, Tucson 85721
JOHN F. PALMISANO
College of Fisheries, University of Washington, Seattle 98195

References and Notes

1. N. S. Jones and J. M. Kain, *Helgol. Wiss. Meeresunters.* **15**, 460 (1967); J. H. Himmelman and D. H. Steele, *Mar. Biol.* **9**, 315 (1971); D. K. Camp, S. P. Cobb, J. F. VanBreedveld, *BioScience* **23**, 37 (1973); P. K. Dayton, R. J. Rosenthal, L. C. Mahan, *Antarct. J. U.S.* **8** (No. 2), 34 (1973); J. C. Ogden and R. A. Brown, *Science* **182**, 715 (1973).
2. R. T. Paine and R. L. Vadas, *Limnol. Oceanogr.* **14**, 710 (1969).
3. A. Nelson-Smith, in *The Biological Effects of Oil Pollution on Littoral Communities,* J. D. Carthy and D. R. Arthur, Eds. (Field Studies Council, London, 1968), vol. 2, supplement.
4. J. H. McLean, *Biol. Bull.* **122**, 95 (1962).
5. W. J. North, *Kelp Habitat Improvement Project, Annual Report for 1964–1965* (California Institute of Technology, Pasadena, 1965).
6. J. C. Quast, *Calif. Dep. Fish. Game Fish Bull.* **139**, 109 (1968).
7. R. T. Paine, *Am. Nat.* **100**, 65 (1966); J. W. Porter, *ibid.* **106**, 487 (1972).
8. R. L. Vadas, thesis, University of Washington (1968).
9. A. Ogden, *The California Sea Otter Trade 1784–1848* (Univ. of California Press, Berkeley, 1941); I. I. Barabash-Nikiforov, *Kalan* (Soviet Ministrov RSFSR, 1947), published in English as *The Sea Otter,* A. Birron and Z. S. Cole, Transl. (Israel Program for Scientific Translations, Jerusalem, 1962).
10. K. W. Kenyon, *The Sea Otter in the Eastern Pacific Ocean* (Government Printing Office, Washington, D.C., 1969).
11. The Rat Islands are located at approximately 52°N, 178°E.
12. J. A. Estes and N. S. Smith, *USAEC Res. Dev. Rep. NVO 520-1* (1973).
13. P. Morrison, M. Rosenmann, J. A. Estes, in preparation.
14. There is some doubt about the species identification of the green sea urchin in this area (that is, *S. drobachiensis* or *S. polyacanthus*).
15. L. Barr, *BioScience* **21**, 614 (1971).
16. Test diameter refers to a measurement of the external skeleton diameter, not including spines.
17. Data were collected from randomly selected ¼-m² plots (Rat Islands, $N = 171$; Near Islands, $N = 9$) and from 1/16-m² plots at intervals along transect lines (Rat Islands, $N = 32$; Near Islands, $N = 23$) [J. F. Palmisano and C. E. O'Clair, unpublished results; C. E. O'Clair and K. K. Chew, *BioScience* **21**, 661 (1971)].
18. The results of experiments that confirm these conclusions will be presented by J. F. Palmisano (in preparation).
19. J. A. Estes and J. F. Palmisano, personal observations.
20. U.S. Department of Commerce, Coast and Geodetic Survey, *Tide Tables, West Coast, North and South America, 1969* (Government Printing Office, Washington, D.C., 1968).
21. L. R. Blinks, *J. Mar. Res.* **14**, 363 (1955); K. H. Mann, *Mar. Biol* **14**, 199 (1972).
22. C. J. Lensink, thesis, Purdue University (1962); K. W. Kenyon and J. G. King, "Aerial survey of sea otters, other marine mammals and birds, Alaska Peninsula and Aleutian Islands, 19 April to 9 May 1965." Bureau of Sport Fisheries and Wildlife report, on file at the Fish and Wildlife Service, Department of Commerce, Washington, D.C. (1965).
23. T. H. Scheffer and C. C. Sperry, *J. Mammal.* **12**, 214 (1931); V. B. Scheffer and J. W. Slipp, *Am. Midl. Nat.* **32**, 373 (1944); C. M. White, W. B. Emison, F. S. L. Williamson, *BioScience* **21**, 623 (1971).
24. P. W. Wild, paper presented at the Conference of the American Association of Zoological Parks and Aquariums, Western Region, San Diego, California, 21 February 1973.
25. E. E. Ebert, *Underwater Nat.* **5**, 20 (1968).
26. *Sport Fish. Inst. Bull.* **238** (1972), p. 1.
27. R. T. Paine, *Am. Nat.* **103**, 91 (1969).
28. Supported by AEC contracts AT(26-1)-520 and AT(26-1)-171 through subcontract from Battelle Memorial Institute, Columbus, Ohio. We are indebted to S. Brown, R. Glinski, P. Lebednik, C. O'Clair, and N. Smith for field assistance. We thank P. Dayton and R. Paine for helpful comments in preparing the manuscript and J. Isakson for assistance with logistic problems. The U.S. Air Force and U.S. Coast Guard provided access to their facilities in the Near Islands.

21 January 1974; revised 16 April 1974

COVER

Captive sea otter at Amchitka Island, Alaska. The reestablishment of sea otter populations in the western Aleutian Islands has produced remarkable changes in the structure of nearshore communities after near extinction from overhunting.

Community Structure of Coral Reefs on Opposite
Sides of the Isthmus of Panama

James W. Porter

Community Structure of Coral Reefs on Opposite Sides of the Isthmus of Panama

Abstract. *Competition for space among reef corals includes interspecific destruction by extracoelenteric digestion, rapid growth, and overtopping. No Caribbean species excels in all strategies, and on western Caribbean coral reefs there is a positive correlation between coral abundance and diversity. On eastern Pacific coral reefs, however,* Pocillopora damicornis *excludes other corals, and on these reefs there is an inverse relation between coral abundance and diversity, except in areas where disturbances, such as* Acanthaster *predation, offset space monopolization.*

The eastern tropical Pacific and the western Caribbean have been separated for only 3 million to 5 million years (1), yet they are strikingly different in the taxonomic composition and arrangement of their faunas. In this report the community structure and composition of coral reefs on opposite sides of the Isthmus of Panama are compared. The effects of physical and biological processes, especially predation and competition, on the production and maintenance of these communities' structures are assessed.

On reefs where crowding occurs, corals are space-limited. As in other benthic environments (2), it can be expected that the resolution of this competition will be a major determinant of the community's structure. Due to their dependence on zooplankton for food (3) and on light for their symbiotic algae (4), the corals' growth form as well as growth rate is important in space competition. Some corals have overtopping morphologies, as found in plant species for growing over and shading out neighboring individuals (5). Another mode of competition is to eat neighboring corals. Lang (6) has shown that certain species, when they come into physical contact with other species, can extrude their mesenterial filaments, digest away the living tissue of the neighboring coral, and then overgrow the exposed skeleton. In the Caribbean (6) and the eastern Pacific (7), corals are arranged in a hierarchy such that species higher up the scale are capable of digesting all those beneath them (position on this scale is referred to here as digestive dominance). This is complicated only by occasional equalities where no digestive interaction is observed between the touching species. Normally the hierarchy is inflexible and does not depend on the physiological state or age of the interacting corals (6).

Despite conspicuous examples of biological interactions that occur on reefs, many of the fundamental controls over the location and intensity of reef development are purely physical. Factors such as temperature and salinity fluctuations, heavy sedimentation, low light penetration (8), and storm or wave damage (9) are important. In their extremes, these stresses can exclude certain species or prevent the accumulation of coral biomass in

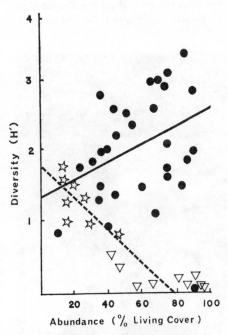

Fig. 1. For transects along Caribbean and eastern Pacific coral reefs, species diversity, H', is plotted against coral abundance (percentage of transect line covering living coral) (12). A positive correlation is found for Caribbean transects (closed circles); correlation coefficient $r = +.41$, significant at $P < .025$. The number of species under the transect also correlates positively with the abundance of coral under the transect; $r = +.56$ and $P < .01$. For eastern Pacific transects with (open stars) and without (open triangles) *Acanthaster* predation, $r = -.91$, significant at $P < .001$. The number of species under the transect also correlates negatively with abundance; $r = -.76$, with $P < .001$.

an area. Only in areas of moderated physical regimes where coral growth is favored do the complex biological controls influence the division of biomass among coral species.

Coral reefs of the eastern Pacific are composed predominantly of monospecific stands of *Pocillopora damicornis* Linn. (= *P. lacera* Verrill?) (10–12). In areas where coral coverage is high, the space is monopolized by this species; in areas with less coverage, space is progressively less monopolized; thus, the inverse relation between species diversity and the abundance of living coral is produced (Fig. 1). Several factors contribute to the high proportion of *Pocillopora*. In localities most favorable to reef development, *Pocillopora* has a maximum growth rate [up to 6 cm per branch per year (13)] an order of magnitude greater than that of any other eastern Pacific coral. This rapid growth rate is coupled with an overtopping morphology and high digestive dominance. Field studies (7, 14) place *Pocillopora* near the top of the digestive dominance hierarchy, followed by species in the genus *Pavona*, and finally species of *Porites*. Such a dominance interaction can be observed in nature (Fig. 2). Finally, *Pocillopora* is able to inhabit water depths near the low tide mark, even being awash at lowest spring tides. It is for these reasons that this coral occupies a high proportion of the substrate suitable for coral growth in the eastern Pacific.

Reversal of space monopolization is possible through a variety of disturbances (2, 15). Most interesting is predation by *Acanthaster planci*, the crown of thorns starfish (14). All of the transects with the highest diversity (starred points in Fig. 1) come from reefs with a high frequency of coral predation by *Acanthaster*; the transects with lower diversity come from reefs where *Acanthaster* is rare (Islas Secas) or absent (Islas Perlas). The same number of species is found on both kinds of reef; the increase in the number of species observed under the transect for the preyed-on reefs (Fig. 2) is due to an increase in abundance in these species, making them common enough to be counted by the transect method employed (12). Although the relative abundances of the coral species change markedly, for the most part, the order of abundance does not. On all reefs *Pocillopora* is most abundant, followed by *Pavona*, with very little coverage

357

Reprinted from Science 186:543–545 (1974).

contributed by the rest of the fauna. Totaling the amount of coral coverage on the two kinds of reef (Fig. 2) shows that the reefs with a high frequency of *Acanthaster* have almost one-third less coral biomass than those lacking the starfish. By cropping down the coral, the starfish leaves space where other species, in addition to *Pocillopora*, can (and do) colonize in areas once exclusively occupied by *P. damicornis*. Differential recolonization rates as well as differential predation can cause the observed shift in relative abundances. The Indo-Pacific character of the fauna indicates that eastern Pacific coral larvae are capable of colonizing reefs from long distances.

In addition to localized *Acanthaster*, there are a number of cosmopolitan corallivores capable of producing bare spaces for coral settlement. The feeding rates of a guild of corallivorous fishes, a gastropod, and two paguridean decapods are described by Glynn *et al.* (7); these corallivores consume approximately one-third of the new growth of

Pocillopora communities each year. Such predation is significant, especially when compounded by *Acanthaster* predation (16). Thus, there is a considerable biological component in the production and maintenance of eastern Pacific coral community structure. Because of the broad taxonomic base of this control, it can be said to be interphyletic in origin.

On the Caribbean side of Panama, areas most favorable to reef growth have coral densities comparable to those observed on the eastern Pacific reefs, but the areas of maximum coverage, instead of being monopolized by one species, have the greatest number of species and the highest species diversity (Fig. 1). Goreau and Goreau (17) noted that except for several acroporids, which calcify relatively rapidly, no consistent relation exists between a coral's calcification rate and its prevalence on a Jamaican reef. This is similar to the situation on the Atlantic side of Panama, but opposite to that on the Pacific side. A partial explanation for these inter-

oceanic differences is suggested by a comparison of the growth rates of Caribbean coral species with their dominance ranks. In the eastern Pacific, rapid growth and high digestive dominance are present in the same species. In the Caribbean, the opposite is true: The species that calcify and grow most rapidly (those in *Acropora*, *Porites*, and *Agaricia*) are the least capable of eating other species, while the species that grow most slowly (those in *Scolymia*, *Mussa*, *Isophyllia*, *Mycetophyllia*, *Isophyllastrea*, and *Meandrina*) are highest on the digestive dominance scale. Furthermore, none of the strong aggressors employs an overtopping morphology, as is found in the weak genera *Acropora* and *Agaricia*. As a speculation, high digestive dominance and rapid growth rate may represent physiological alternatives which, to date, have not evolved together in one Caribbean species. There appears to be a balance of abilities divided among the Caribbean corals such that no one species is competitively superior in acquiring and holding space. The effect of this balance of competitive abilities is to retard, even in high-density situations, the rapid competitive exclusion that takes place on undisturbed eastern Pacific reefs. In areas where physical factors allow for the accumulation of coral biomass, a large number of species coexist. Hence, for the Atlantic side of Panama there is a positive relation between diversity and abundance.

One exception occurs (bottom Caribbean point in Fig. 1) in extensive shallow-water reef flats composed almost exclusively of *Porites furcata*. In San Blas, off the Atlantic coast of Panama, this species was observed to survive a 1-hour exposure to air at midday without apparent morphological damage, as well as a continuous 3-hour flow of silt-laden river water with a salinity of 19 parts per thousand, which followed the first torrential rain of the 1972 wet season. *Porites furcata* is at the bottom of the digestive dominance hierarchy (6), but its physiological tolerance to desiccation and surface salinity fluctuations allows it to escape in shallow water the competitive interactions which, under normal circumstances, would prevent its monopolization of space on the reef.

The destruction by predation of at least one-third of new coral growth per year observed in the eastern Pacific appears to be absent from the Caribbean.

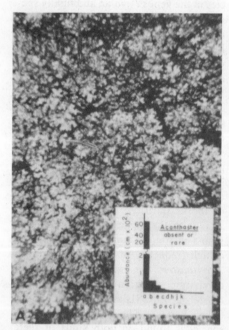

Fig. 2. Eastern Pacific *Pocillopora* reef crests. (A) Monospecific stand of *P. damicornis* where *Acanthaster* is absent (Islas Perlas); (B) crest where *Acanthaster* is common (Isla Uva). Note the black, killed area (arrow) on the margin of the white *Pavona clivosa* colony in the center of (B). This dead area is an outline of the neighboring branched *Pocillopora* colony whose mesenterial digestive filaments were seen extruded onto this *Pavona* colony at the region of contact. Note the lower coral cover but higher number of species on this reef section. The white tips of the *Pocillopora* colonies in (A) and (B) are due to predation by various corallivorous fish; each area is roughly 0.75 m². The number of centimeters of transect covering each species of coral on eastern Pacific reefs is shown in the bar graphs for reefs where *Acanthaster* is absent or rare [none to one individual per 50,000 m² in (A)] and where it is common [one or more individuals per 100 m² in (B)]. (a) *Pocillopora damicornis, Pocillopora elegans,* (c) *Pavona varians,* (d) *Millepora* sp., (e) *Porites californica,* (f) *Porites panamensis,* (g) *Pavona clivosa,* (h) *Psammocora brighami,* (i) *Pavona ponderosa,* (j) *Pavona gigantea,* and (k) *Tubastrea aurea.*

Although these observations are preliminary, the direct intraphyletic interaction of coral versus coral in the Caribbean seems less modified by predation pressure than on some eastern Pacific coral reefs (18). Space is produced in the Caribbean by slower means, such as the slump or collapse of larger coral heads due to intense boring by sponges and bivalves (19). This might be analogous to the creation of space in highly dense rain forests by tree fall (20) and seems sufficient to offset the slow competitive exclusion which would occur if the system were left totally undisturbed. Elsewhere in the Caribbean, weather disturbances such as hurricanes are probably of major importance to shallow-water coral community organization as well, but Panama does not have such severe disturbances.

In looking at species succession, Pielou (21) found that species diversity increased over time for some small plots of young temperate forest trees. She concluded that natural thinning by interspecific competition caused the decrease in segregation and dominance by single species, the result being an increase in the local diversity. While few data exist for reef succession, most phases of Caribbean coral succession may prove to be similar to this pattern, whereas, without large-scale physical or biological disruptions, eastern Pacific coral succession may prove to be the opposite. In support of this, Grigg and Maragos (22) have shown that, all else being equal, coral communities on the oldest lava flows in Hawaii have the lowest diversities and that the same negative relation between diversity and

abundance exists in Hawaii as on the eastern Pacific reefs. They suggest that interspecific competition of the kind outlined above might be the cause for this decrease in diversity over time. Additional support for this hypothesis comes from their observation that in areas exposed to heavy swell and periodic storm damage, abundance remains low and diversity high, regardless of the age of the colonized surface. In Hawaii, where coral fauna and reefs are similar to those of the eastern Pacific, storm and swell damage may act as a diversifying force in the same way as *Acanthaster* predation does in the eastern Pacific.

JAMES W. PORTER

School of Natural Resources,
University of Michigan,
Ann Arbor 48104

References and Notes

1. F. C. Whitmore and R. H. Stewart, *Science* 148, 180 (1965).
2. P. K. Dayton, G. A. Robilliard, R. T. Paine, L. B. Dayton, *Ecol. Monogr.* 44, 105 (1974).
3. J. W. Porter, in *The Second International Symposium on Coral Reefs*, G. R. Orme, Ed. (Great Barrier Reef Committee, Brisbane, Australia, in press).
4. T. F. Goreau, *Biol. Bull. (Woods Hole)* 116, 59 (1959); L. Franzisket, *Int. Rev. Hydrobiol.* 55, 1 (1970).
5. J. W. Porter, cited in H. S. Horn, *The Adaptive Geometry of Trees* (Princeton Univ. Press, Princeton, N.J., 1971), p. 129.
6. J. Lang, *Bull. Mar. Sci.* 23, 260 (1973).
7. P. W. Glynn, R. H. Stewart, J. E. McCosker, *Geol. Rundsch.* 61, 598 (1972).
8. J. W. Wells, *Mem. Geol. Soc. Am.* 67, 609 (1957). Physical control is especially prominent on the eastern Pacific side, which is by far the more physically stressed of the two reef environments. In the Bay of Panama (Pacific) tidal amplitude can exceed 7 m and water temperature ranges from 31° to 15°C, compared to 0.5 m and 30° to 26°C on the Atlantic side [I. Rubinoff, *Science* 161, 875 (1968)].
9. D. R. Stoddart, *Biol. Rev.* 44, 433 (1969).
10. The origin of the modern eastern Pacific coral fauna is unclear. Although they are generically similar to Miocene and Tethyan fossils from the Panama region, they have no species in common with the Caribbean fauna today. Species similarities with the present Indo-Pacific and remote Pacific island corals, as well as the presence of genera without prior fossil representation in the New World, indicate that some of the present eastern Pacific fauna immigrated from the Indo-Pacific recently (11).
11. J. W. Porter, *Bull. Biol. Soc. Wash.* 2, 89 (1972).
12. Lines 10 m long were laid parallel to reef depth contours. The number and relative abundances of living coral species under the transect line were used to calculate the Shannon species diversity (H') and the total abundances (percentage of line covering living coral) [J. W. Porter, *Ecology* 53, 744 (1972); see also Y. Loya, *Mar. Biol.* 13, 100 (1972)].
13. P. W. Glynn and R. H. Stewart, *Limnol. Oceanogr.* 18, 367 (1973).
14. J. W. Porter, *Am. Nat.* 106, 487 (1972); thesis, Yale University (1973).
15. R. T. Paine, *Am. Nat.* 100, 65 (1966).
16. P. W. Glynn, *Science* 180, 504 (1973).
17. T. F. Goreau and N. I. Goreau, *Biol. Bull. (Woods Hole)* 117, 239 (1959).
18. The outcome of connecting these two oceans with a sea-level canal is problematic, but what little predictive evidence exists is ominous: eastern Pacific *Acanthaster* are capable of eating all of the Caribbean coral species so far fed to them (11). Why should Caribbean species have any resistance to predators with which they have never come in contact? Further, while dominance experiments between eastern Pacific and Caribbean corals have not been tried, analogy with Indo-Pacific and Caribbean species—the Indo-Pacific species killed even the most aggressive Caribbean corals [J. C. Lang, thesis, Yale University (1970)]—is disturbing, considering the close systematic relationship between the eastern Pacific and the Indo-Pacific corals (10, 11).
19. T. F. Goreau and W. D. Hartman, *Science* 151, 343 (1966).
20. J. H. Vandermeer, unpublished data.
21. E. C. Pielou, *J. Theor. Biol.* 10, 370 (1966).
22. R. W. Grigg and J. E. Maragos, *Ecology* 55, 387 (1974).
23. I thank W. D. Hartman, G. E. Hutchinson, R. K. Trench, D. C. Rhoads, and P. W. Glynn for advice and criticism throughout this research. K. G. Porter, D. H. Janzen, S. T. Hubbell, L. Johnson, J. C. Lang, D. S. Wethey, and S. Ohlhorst have provided helpful comments. Supported by a Smithsonian predoctoral fellowship, NSF grant GA-30970, and a grant-in-aid of research from the Society of Sigma Xi.

13 December 1973; revised 4 March 1974

COVER

Plating *Montastrea annularis* (center) on Caribbean coral reefs off the Atlantic coast of Panama. In contrast to reefs on the Pacific coast of Panama, areas of dense coral growth in the Caribbean are also areas of high coral diversity.

Ecology of the Deep-Sea Benthos

More detailed recent sampling has altered our concepts about the animals living on the deep-ocean floor.

Howard L. Sanders and Robert R. Hessler

Marine benthic communities cover most of the earth's surface. At least 94 percent of the ocean bottom lies below the permanent thermocline, and, in its physical and chemical parameters, this region is remarkably stable and homogeneous. It is constantly dark, with bottom water of constant salinity, oxygen content, and low temperatures. Even the bottom sediments, largely derived from planktonic organisms, are unvarying for hundreds of square kilometers. Great pressures characterize the environment. Probably as a consequence of its separation both spatially and temporally from the primary organic production at the surface of the sea, there is a low rate of food supply to the deep-ocean floor.

There is little information about the kinds of animals living in this vast environment and about the relation of the high environmental stability to the ecology and physiology of the deep-sea fauna.

Because of the sparseness of animal life and the technical difficulties in sampling the deep-sea benthos, relatively few specimens were collected in the past century. Even nonquantitative trawls and dredges traversing appreciable distances on the deep-ocean floor have captured only a few animals. However, understanding of the deep-sea fauna requires that samples contain enough individuals to give statistical support to conclusions. We developed the large deep-sea *Anchor Dredge* (*1*) for quantitatively sampling the infauna (animals living in the bottom) and the *Epibenthic Sled* (*2*) to collect both epifauna (animals living on the bottom) and infauna in large quantities. From 1960 to 1966, we made a study of a transect of the ocean floor, between southern New England and Bermuda (the Gayhead-Bermuda transect, Fig. 1). Subsequently, we extended our study to the tropical Atlantic.

Faunal Composition

Initially, our sampling of the deep-sea benthos was done by the *Anchor Dredge*. These predominantly infaunal samples were dominated by Polychaeta, Crustacea, and Bivalvia. Polychaetes, comprising 40 to 80 percent of these samples by abundance and represented by numerous species, were the most abundant. Hartman (*3*) found from 65 to 77 polychaete species in each of the five quantitative samples which covered 0.5 to 1.0 square meter of bottom at upper slope depths. Crustaceans, the second most common group, formed 3 to almost 50 percent of the fauna and were represented by many isopod, amphipod, cumacean, and tanaid species. Other common faunal elements were glass sponges, sea anemones, pogonophores, sipunculids, echiurids, tunicates, priapulids, brittle stars, and starfish.

In the past most deep-sea collections were made with coarse-meshed trawls which collected predominately larger epifaunal organisms. To provide basic information on the neglected smaller epifaunal animals, as well as to obtain better samples of the total benthic macrofauna, we constructed the *Epibenthic Sled* (*2*). A door which closes the mouth of the net reduces the effect of winnowing of the sample as it is brought up through the long water column. A fine-meshed net in the sled and fine-meshed screens for processing the samples on board ship retain the abundant smaller animals which would otherwise be lost.

The effectiveness of the method is demonstrated by the following collection records.

Malletia abyssorum Verrill and Bush (bivalve); previous record, single specimen; sled samples, 3257 specimens from 19 samples.

Malletia polita Verrill and Bush (= *Malletia bermudiensis* Hass) (bivalve) both known from single empty valve;

sled samples, 191 specimens from ten samples.

Tindaria callistiformis Verrill and Bush (bivalve); previous records, two specimens; sled samples, 1708 specimens from 11 samples.

Serolis vemae Menzies (isopod); previous record, two specimens from South Atlantic; sled samples, 255 specimens from seven samples on transect and in tropical Atlantic.

Desmosoma insigne Hansen (isopod); previous records, six individuals from Davis Strait; sled samples, 294 specimens from four samples.

In contrast to the pronounced abundance of polychaetes in the infaunal *Anchor Dredge* samples, dominance in the sled samples is shared about equally among the crustaceans, polychaetes, bivalves, and brittle stars. When these four groups are compared on the basis of diversity, crustaceans invariably have the largest number of species, followed by polychaetes and bivalves; ophiuroids are always represented by few species. If 1000 individuals of each group are counted for every species of brittle star there are 4.1 bivalve, 9.7 polychaete, and 16.5 crustacean species [see (*4*) for method of rarefaction].

Diversity

Our most unexpected finding was a high faunal diversity in individual samples. Such diversity is far in excess of anything obtained in the past, in contrast to the belief that the deep sea harbors a qualitatively restricted fauna (*5–7*). The results of the first five sled samples taken during the summer of 1964 are given in Table 1; subsequent samplings demonstrated similar diversity.

Only two previous samples from depths greater than 1000 meters have yielded more than 100 species of benthic invertebrates. They are *Challenger* station 320, at 1096 meters, with 124 species and 496 individuals (*8*), and *Galathea* station 716, at 3570 meters, with about 2100 specimens divided among 132 species (*9*).

The greater diversity shown by our samples may be due to a more complete sampling of the total benthic fauna. Earlier samples contained few specimens, and the total number of species was therefore small. The restricted number of specimens in these samples made

Dr. Sanders is senior scientist in biology at the Woods Hole Oceanographic Institution, and Dr. Hessler is an associate professor at the University of California, San Diego.

Reprinted from Science 163: 1419–1424 (1969).

it impossible to determine the significance of apparently small taxonomic differences. When a specimen resembled a known species, it was usually included in that species. The resultant lumping helped to create the impression of relatively few, broadly distributed species. Our large samples allow us to conclude that most of the major taxa are characterized by numerous closely related species both at a specific locality and among spatially separated localities.

The data obtained from both the *Anchor Dredge* and *Epibenthic Sled* samples indicate that the deep-sea benthos is not impoverished but, instead, is represented by a remarkably diverse fauna. How, then, does the diversity in the deep sea compare with the diversities occurring in other regions of the world?

We answered this question by collecting benthic samples from boreal estuary, boreal shallow marine, tropical estuary, and tropical shallow marine environments. In all cases, the sediments were soft oozes and were therefore comparable in particle size. All sam-

Table 1. The number of species and individuals collected in five *Epibenthic Sled* samples.

Station	Depth (m)	Individuals (No.)	Species (No.)
73	1400	25,242+	365
62	2496	13,425+	257
72	2864	5,897+	208
64	2891	12,083+	310
70	4680	3,737+	196

ples were processed in a similar manner. The analysis is based on total fauna (2) and on the polychaete-bivalve fraction of the fauna (4). The polychaetes and bivalves comprise about 80 percent of the animals in most of the infaunal samples, and thus we can generalize from the results. Diversity in the deep sea, measured by a rarefaction method (4) which allows direct comparison of samples with differing numbers of specimens, is about the same as that in the physically stable, shallow, tropical marine environment and significantly greater than that of the other three environments (Fig. 2).

We believe that the constancy of phy-

sical conditions and the long past history of physical stability in the deep sea have permitted extensive biological interactions and accommodations among the benthic animals to yield the diverse fauna of this region. Such communities evolve wherever physical conditions remain constant and uniform for long periods. Other than the deep-sea, tropical shallow-water environments and tropical rain forests best approximate these conditions. Communities found in these environments are characterized by many species and can be termed biologically accommodated communities (4).

At the other end of the diversity spectrum are the physically unstable communities where physical conditions fluctuate widely and unpredictably, and thus the organisms are exposed to severe physiological stresses. Here the adaptations are primarily to the physical environment (4). Hypersaline bays and temporary ponds exemplify this state, and certain shallow boreal marine and estuarine environments approach such conditions. Physically unstable communities are characterized by a small number of species. A similar paucity of species is found in environments of recent past history, such as most freshwater lakes.

The term "diversity" as used above means "within-habitat" diversity, that is, the number of species in a specific habitat, and not "between-habitat" diversity or the total number of species for all habitats (10). (The habitat under study is that of soft, fine-grained sediments.) The pronounced homogeneity of the deep sea permits fewer habitats than do the shallow depths. Thus, although the animal diversity of the soft sediments in the deep sea is well above that of the equivalent inshore boreal habitat, the total between-habitat diversity may be lower.

Zonation and Zoogeography

How are the environmental stability and homogeneity on the deep-ocean floor reflected in faunal vertical zonation? Known abyssal and hadal records of vertical distribution for 1144 species of deep-sea benthic invertebrates, reviewed by Vinogradova (11), showed rapid decrease of species from 2000 to 6000 meters, with a much slower reduction at greater depths. (The bathyal region encompasses depths from 200 meters to 2000 or 3000 meters and includes the continental slope; the abyssal zone covers from 2000 or 3000 meters

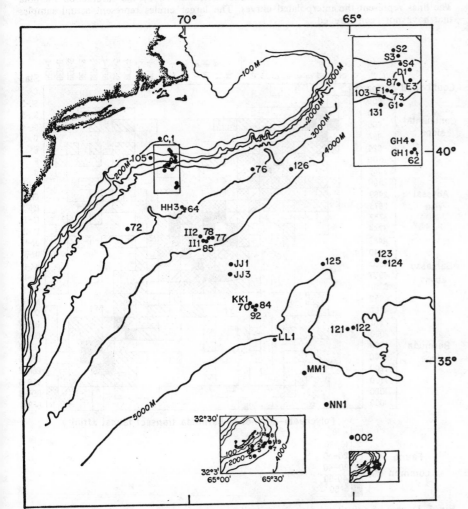

Fig. 1. Depth contours and locations of sampling stations of the Gayhead-Bermuda transect.

to, 6000 meters; the hadal zone is at depths greater than 6000 meters and includes the deep-sea trenches.) At about 3000 and 4500 meters, important changes occur in the taxonomic composition of the benthic fauna. Numerous species and higher taxa, broadly distributed on the slope and even shallower, disappear and are replaced by new species, genera, and families found only at greater depths. For these reasons, Vinogradova concluded that 3000 meters represents the true upper limit of the abyssal zone.

We did not find such abrupt boundaries at bathyal and abyssal depths. Our analysis is not dependent on criteria of presence or absence; we measured the percentage of fauna shared by each possible pair of stations on the transect. The data for polychaetes, extracted from Hartman (3) and based on 264 species and almost 14,000 specimens, are given in Fig. 3.

The stations are arranged sequentially in a north to south direction traversing, in order, the outer continental shelf, the continental slope, the abyssal rise, the Sargasso abyss, and the Bermuda slope. The diagonal pattern running from the upper left to the lower right corner means that highest faunal indices are always with neighboring stations, that is, there is a gradual and continuous faunal change with depth and distance along the transect.

The very small faunal index values shared by stations C (97 meters) or SL-2 (200 meters) with other stations on the transect marks this portion of the transect as a region of pronounced faunal change. In fact, it is the sharpest zoogeographical boundary encountered. This discontinuity is also true for groups other than the polychaetes. Among the bivalves, the transition of faunal change at the shelf-slope break is even more pronounced. Within the isopods, the typically deep-water subtribe Paraselloidea is absent from station C and extensive samplings in shallower waters. The group makes its appearance on the upper slope at station SL-2 and is a major constituent in all deeper samples.

We believe that this faunal break is related to temperature. In 98 meters at station C, the seasonal temperature change is 10.5°C; in 300 meters at slope station 3, it is 5.1°C; and in 487 meters at station D, only 1.4°C. Therefore, the boreal continental shelf, with highly variable seasonal temperatures, supports a qualitatively impoverished eurytopic (broad physical tolerances)

fauna, whereas the neighboring physically stable continental slope harbors a different stenotopic (narrow physical tolerances) benthic fauna of high diversity. Entirely analogous conditions

are found in ancient Lake Baikal in Siberia, similarly dominated at shallow depths by a continental boreal climate (12). We conclude, on the basis of taxonomy, diversity, and environmental

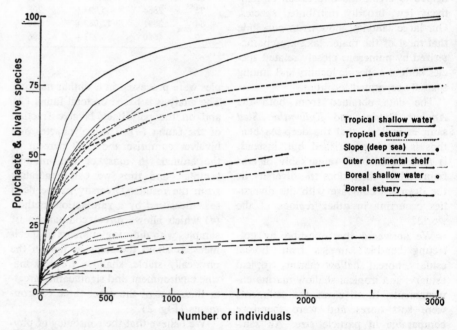

Fig. 2. Diversity values for different benthic environments by the rarefaction method. The lines represent the interpolated curves. The larger circles represent actual samples that have not been rarefied.

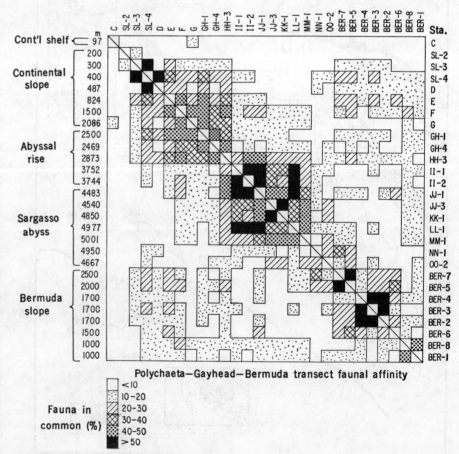

Fig. 3. Degree of polychaete faunal similarity among the stations (Sta.) of the Gayhead-Bermuda transect.

Table 2. Degree of bivalve faunal similarity and percentage of bivalve species shared per 500 specimens with station 64 on the Gayhead-Bermuda transect.

Station	Depth (m)	Species shared* (%)	Fauna shared† (%)	Depth deviations‡ (m)
105b	530	7.77	2.61	−2361
87	1102	25.79	4.80	−1789
73	1400	24.47	5.90	−1491
103	2022	22.87	5.80	− 869
131	2178	30.68	6.81	− 713
62	2496	46.41	28.71	− 395
72	2864	60.19	77.74	− 27
64	2891	100.00	100.00	0
126	3806	53.05	9.02	+ 915
85	3834	35.42	6.34	+ 943
70	4680	10.64	1.45	+1789
92	4694	11.65	1.36	+1803
84	4747	16.70	2.68	+1856
121	4800	7.57	1.03	+1909
125	4825	5.75	0.93	+1934
122	4833	13.98	1.38	+1942
123	4853	3.88	0.83	+1962
124	4862	3.88	0.83	+1971

* There are 25.75 species per 500 individuals at station 64.   † With station 64.   ‡ Minus (−) values mean that the station is the given number of meters shallower than station 64; plus (+) values mean that the station is that number of meters deeper than station 64. Station 64 is at a depth of 2891 meters.

factors, that the self-slope break within the depth range of 100 to 300 meters represents the true upper boundary of the deep-sea benthos in this part of the ocean.

Changes in species composition with depth also can be demonstrated by the use of a modification of the rarefaction method (Table 2). The criterion is the number of species shared between a given station and each of the other stations along the transect when equal-sized samples are compared. The results of such an analysis for the bivalves (excluding the taxonomically difficult family Thyasiridae) are presented in Table 2 for station 64 (2891 meters). The stations are arranged in sequential order by depth from the upper continental slope to the abyssal plain in the Sargasso Sea; the depths, the percentage of species per 500 individuals (by rarefaction), the percentage of the fauna that each station shares with station 64, and the deviation of each station in depth from station 64 are given.

In percentages of species and of fauna shared, there is a continuous decrease in similarity with departure in depth from station 64. The only small alterations in this pattern occur with stations that are widely separated in depth from station 64 (stations 103, 73, 87, and 122). The gradient of change is steeper, with percentage of fauna than with percentage of species shared. This difference in the steepness of the two gradients means that species that are numerically dominant at station 64 are relatively less important or even rare

faunal constituents at stations separated from station 64 by more than 800 meters. Station 126 is a case in point. With 53 percent of the species in common, it shares many species with station 64. Yet with only 9 percent of the fauna common to both stations, species that are numerically dominant at one station are not numerically dominant at the other.

The horizontal and vertical components of zonation can also be measured. Station 73 at 1400 meters on the Gayhead-Bermuda transect shares 47.9 percent of its bivalves with station 142 at 1700 meters off the coast of West Africa (10°30.0′N; 17°51.5′W). If we follow the contours of 1000 to 2000 meters in depth along northern North America, Greenland, Iceland, Europe, and Africa, these two stations are separated by about 16,500 kilometers. Yet, if the depth is altered as little as 800 meters either shallower or deeper from station 73 on the transect (a distance of 10 kilometers landward or 12 kilometers seaward), the number of species shared with station 73 falls below 48 percent. Thus a change of only 800 meters in a vertical direction on the transect is at least equivalent to a 16,500-kilometer change horizontally. The implication, at least for bivalves, is that faunal composition is far more sensitive to change in depth than to the effects of distance.

To assess the patchiness of the fauna, we took three samples (stations 72, 64, and 76) within the depth range of 2862 to 2891 meters over a horizontal dis-

tance of 366 kilometers. Another series of five samples (121, 122, 123, 124, and 125) was taken within the depth range of 4800 to 4862 meters, at three widely separated localities (Fig. 1). In both series, the faunal index of each pair of stations was very high and, with two minor exceptions, higher than with any station outside the series. These high faunal values imply that many components of the deep-sea benthic fauna at any given depth are uniformly and homogeneously distributed over extensive distances of the ocean floor.

Breeding Patterns

On land and in the shallow boreal and arctic marine regions, reproduction is usually coupled with some cyclical and often seasonal phenomenon such as day length, temperature, or rainfall. This synchrony allows the adult portion of the population to be ready for reproduction at the same time, and the resulting young or larvae are usually present when feeding conditions are best.

The monotonously constant conditions of the deep sea provide no such obvious cyclical environmental phenomena. Even seasonal pulses of primary production in the surface waters are probably dampened and dissipated long before the slow rain of these minute particles reaches abyssal depths.

George and Menzies (13) suggest that cyclic reproduction may be present among deep-sea isopods. Their data, derived from studies of the genus Storthyngura, are based on only six ovigerous specimens of three species taken at widely differing localities and depths. The statistical inadequacy of these data make their conclusion conjectural.

Schoener (14) studied the state of reproduction in two species of brittle stars, Ophiura ljungmani and Ophiomusium lymani, collected from the continental slope, and abyssal rise on the Gayhead-Bermuda transect. Her analysis of a large number of specimens obtained at different times of the year provides convincing evidence for reproductive periodicity for these deep-sea species.

We are now gradually accumulating information on other deep-sea groups. An August sample from the lower continental slope (station 73, 1400 meters) contained 71 females of an unidentified species of isopod belonging to the genus Ilyarchna. Twenty-four percent of the females had well-developed marsupial plates and were therefore in the brooding condition. (As with all peracarid crustaceans, the embryos of isopods are carried in a marsupium.) A December sample (station 131, 2178 meters) yielded 59 females of the same species, with 27 percent bearing well-developed marsupial plates.

Sections were made on 25 adult specimens of the bivalve *Nucula cancellata*, 2.4 to 2.8 millimeters in size, from each of three samples, station 103 (May, 2022 meters), station 73 (August, 1400 meters), and station 131 (December, 2178 meters) (15). The findings for each of the stations were essentially identical. At least part of the gonad contained ripe sexual products in almost every individual examined.

These results for *Ilyarchna* sp. and *Nucula cancellata* indicate that reproduction is constant and continuous in the two forms. Schoener's evidence of reproductive periodicity for the two brittle star species suggests that no pattern need be universal. Analysis of numerous additional species will indicate whether there is a characteristic pattern of reproduction in the deep sea.

Animal Abundance and Food

Deep-sea benthos show a decrease in animal life with increasing depth of water and distance from land (5, 6, 8, 16–20). Although certain departures from this pattern occur in the shallower water of the sublittoral and upper bathyal zones (1, 21), the lower bathyal and abyssal zones invariably support a smaller biomass or density than shallower depths. Island arc trenches are ideal traps for organic material, and, therefore, the high concentration of life occasionally reported from these hadal depths (22) are special cases.

In our series of *Anchor Dredge* samples taken along the transect, numerical density of animals increased from near shore (approximately 6000 per square meter) to the edge of the shelf (13,000 to 23,000 per square meter) (1). From this depth to the bottom of the continental slope, density dropped precipitously to approximately 500 animals per square meter, with a more gradual decrease on the continental rise, and then a leveling off on the abyssal plain under the Sargasso Sea (25 to 100 animals per square meter). Thus, from the edge of the continental shelf to the abyssal plain, there was a reduction in density by a factor of several hundreds.

The biomass of our samples has never been analyzed; the material was biologically too valuable for mechanical abuse. However, density is an adequate index of biomass in our samples, because the size of animals and the relative proportions of major animal groups do not change between localities.

The abundance of food is generally assumed to control density and biomass in the deep sea (5, 17, 19, 23, 24). Oxygen concentration, sediment type, and temperature either do not show correlated changes or never reach levels regarded as limiting. Hydrostatic pressure is depth-dependent, but although pressure could limit the kinds of species present, there is no apparent way in which it can regulate biomass.

Sources of food fall into two broad categories: either it is produced in the euphotic zone (the relatively narrow, lighted, upper water layer) and conveyed into the deep sea, or it is generated *in situ* in the deep-sea environment. The theory that food is of euphotic origin has until recently dominated our thinking (6, 23) because the distribution of biomass in the deep sea correlates with primary production. The highest productivity is in shallow coastal waters (25). This concentration of food is enhanced by the outwash from terrestrial sources: the farther from inshore waters of highest production, the lower the benthic biomass. Biomass also decreases with increasing depth, because during its transport the amount of food is progressively diminished by autolysis, bacterial decay, and scavengers.

The rapid decline in animal density on the continental slope, the region of greatest rate of depth change on the transect, labels depth as having more importance than distance from land in determining the amount of available food (26). Density decreases 25 times from 200 to 2500 meters of depth. Yet, these localities are so close (61 kilometers) that primary production at the surface barely changes. Stations at 4500 to 5000 meters have densities 50 to 390 times less than at 200 meters, and pro-

ductivity in surface waters is only half as great (25) (Table 3).

The mode of transport of food in the deep sea has long been regarded as a rain of dead plants and animals from the euphotic zone (6, 23, 27). This mechanism has often been criticized because settling velocities are so slow and intermediate attrition is so efficient that little of nutritive value would reach the bottom. Inefficiency alone is not a valid reason for rejecting this mechanism. The deep-sea biomass is so much smaller than that of surface waters that an inefficient food supply can account for it. Downward movement by means of diurnal migration would increase the efficiency of food transport (28). However, the chain of migrating animals postulated by Vinogradov might be less efficient than passive settling since it entails large energy losses in every change of trophic level.

Turbidity currents provide another possible means of rapid downward movements of organics (29). Yet, the infrequency and destructiveness of such phenomena suggest that they are of no more than minor importance in food transport. The presence of layers of carbon associated with turbidity current deposits (30) indicates that while organics have been moved, they were never made available to organisms.

Organic aggregates, with considerable surface activity, adsorb dissolved organics and convert them into particulate compounds. Thus, they may serve as an *in situ* source of nutrition for intermediate bacteria (31) or directly for metazoans in deep water. However, the concentration of dissolved organics and the amount of carbon bound up in organic aggregates are relatively constant in deep water (32) and therefore correlate poorly with the marked decrease in benthic biomass with depth. There is no quantitative data on deep-sea heterotrophic algae (33, 34); therefore this food source cannot be evaluated.

Extensive degradation of organics during the slow transport from the sur-

Table 3. A comparison of the range in amount of benthic faunal density to the range of the amount of organic carbon in the sediment and the annual primary productivity in surface waters.

Area of study	Bathyal-abyssal bottoms		Annual productivity in surface water (cg/m²)
	Density (No./m²)	Organic carbon* (%)	
Gayhead-Bermuda transect	21,263–33(1)	1.0–0.1(1) 2.5–0.8(20)	180–72(25)
Oregon	2,200–14(20)	3.5–0.1(39)	152–60(40)

* Excluding sand bottoms

face waters to the deep-sea benthos permits only the most refractory substances to reach the bottom. Here benthic microorganisms (bacteria and, possibly, fungi) are probably a critical intermediate in the food cycle (6, 7). The abundance of bacteria has been reported (35), and some metazoans show adaptations to utilize an intestinal flora (36). The true importance of bacteria and fungi must still be determined.

The amount of organic carbon in deep-sea sediments follows a variety of patterns with relation to depth and distance from land (1, 20, 28, 29, 37, 38, 39). Yet, the absolute range of variation is usually less than an order of magnitude, far less than the corresponding variation in animal density (Table 3). Therefore, organic carbon in the sediment is not a good index of available food. If bacteria are nutritional intermediates of major importance, then the ratio of bacterial biomass to metazoan biomass or density should remain constant from place to place, and changes in bacterial biomass would not be correlated with changes in organic carbon content in the sediment. Such a pattern, if true, would suggest that in deep water much of the deposited organic material is not readily available to bacteria.

Feeding Types

According to the principle of competitive exclusion, ecological niches of species living together must have areas of nonoverlap. The uniformity of the deep sea and the limited amount of food indicate that niche separation is especially critical with respect to feeding.

Our knowledge of feeding types in this environment is fragmentary, but interpretation of morphology, analogy with shallow-water relatives, and gut-content analyses show that deposit feeders dominate in our samples. At station 64 (2891 meters), for example, the most diverse groups contain the following percentage of detritus feeders: Polychaeta, 60 percent; Tanaidacea, greater than 90 percent; Isopoda, 90 percent (23); Amphipoda, greater than 50 percent (41); and Pelecypoda, 45 percent; totaling 47 percent of the species in the whole sample. Many species of other, less diverse groups (Holothuroidea, Sipunculida, Oligochaeta, and others) also eat detritus, and thus this feeding type includes well over half of the fauna.

The high diversity in the deep sea implies a degree of niche fractionation far greater than in equivalent shallow boreal communities. Yet, studies of zonation indicate that appreciable flexibility must exist in intraspecific interactions, for each species has a zone depth preference whose limits vary with the species. Therefore, each species must be associated with a continuously altering assemblage of animals within its depth range.

Intraspecific flexibility, in conjunction with the low density of the benthic fauna, may explain high faunal diversity coupled with environmental homogeneity. Since density is low, only a fraction of the total number of species lives in any one area, for example in a single square meter. An individual organism, because of its small size and presumably restricted mobility, will interact with a relatively small suite of competing species. This hypothesis permits high faunal diversity with an increased, but not necessarily extraordinary, degree of specialization.

Summary

The benthos of the deep sea in a region between southern New England and Bermuda can be characterized by low density but high within-habitat diversity. The low density is probably determined by the amount of food present and is correlated with depth and distance from land. Depth is the more critical variable. The high diversity, about the same order as that found in shallow tropical seas, can be related to the seasonal and geological stability of the deep-sea environment. An unexpectedly large number of deposit-feeding species are found in individual samples. The composition of the benthic fauna gradually and continuously changes with depth throughout the bathyal and abyssal regions, but an abrupt faunal discontinuity is found at the shelf-slope break in from 100 to 300 meters of water. In this region of our study, the fauna shallower than the zone of discontinuity is eurytopic, of low diversity, and taxonomically distinct from the stenotopic, highly diverse populations living below the discontinuity. In comparing faunal composition, a vertical change of a few hundred meters is equivalent to a change of thousands of kilometers horizontally. The few data on reproduction reveal that some elements of the deep-sea fauna breed continuously, whereas others are restricted to a limited period of the year.

References and Notes

1. H. L. Sanders, R. R. Hessler, G. R. Hampson, *Deep-Sea Res.* 12, 845 (1965).
2. R. R. Hessler and H. L. Sanders, *ibid.* 14, 65 (1967).
3. O. Hartman, "Deep-Water benthic polychaetous annelids off New England to Bermuda and other Atlantic areas," *Allen Hancock Found. Publ. Occas. Pap. No. 28* (1965), p. 1.
4. H. L. Sanders, *Amer. Natur.* 102, 243 (1968).
5. S. Ekman, *Zoogeography of the Sea* (Sidgwick and Jackson, London, 1953).
6. N. B. Marshall, *Aspects of Deep Sea Biology* (Hutchinson, London, 1954).
7. A. F. Bruun, *Geol. Soc. Amer. Mem. No. 67* (1957), vol. 1, p. 641.
8. J. Murray, *Challenger Rep.* 1895, 1608 (1895).
9. T. Wolff, *Galathea Rep.* 5, 129 (1961).
10. R. H. Whittaker, *Science* 147, 250 (1965).
11. N. G. Vinogradova, *Deep-Sea Res.* 8, 245 (1962).
12. M. Kohzov, *Monogr. Biol.* 11, 352 (1963).
13. R. Y. George and R. J. Menzies, *Nature* 215, 878 (1967).
14. A. Schoener, *Ecology* 49, 81 (1968).
15. R. Scheltema sectioned, stained, and interpreted the bivalve material.
16. R. J. Menzies, *Oceanogr. Mar. Biol. Annu. Rev.* 1, 195 (1965).
17. L. A. Zenkevitch and J. A. Birstein, *Deep-Sea Res.* 4, 54 (1956).
18. R. L. Wigley and A. D. McIntyre, *Limnol. Oceanogr.* 9, 485 (1964).
19. N. G. Vinogradova, *J. Oceanogr. Soc. Japan* 20, 724 (1962).
20. A. G. Carey Jr., *Trans. Joint Conf. Exhibit Ocean Sci. Ocean Eng.* 1, 100 (1965).
21. A. P. Kuznetsov, *Benthic Invertebrate Fauna of the Kamchatka Waters of the Pacific Ocean and the Northern Kurile Islands* (Moscow, 1963).
22. G. M. Belyaev, *Bottom fauna of the Ultraabyssal depths of the World Ocean* (Academy Nauk USSR Inst. Okeanol., Moscow, 1966).
23. R. J. Menzies, *Int. Rev. Ges. Hydrobiol.* 47, 339 (1962).
24. L. B. Slobodkin, F. E. Smith, N. G. Hairston, *Amer. Natur.* 101, 109 (1967).
25. J. H. Ryther, in *Geographic Variations in Productivity, the Sea,* M. N. Hill, Ed. [Interscience (Wiley), New York, 1963], vol. 2, chap. 17, p. 347.
26. This is in contrast to the findings of G. M. Belyaev [*Trudy Inst. Okeanologii* 34, 85 (1960)].
27. A. Agassiz, *Bull. Mus. Comp. Zool.* 14, 1 (1888).
28. M. Vinogradov, *Rapp. Process-Verbaux Reunions Cons. Perma. Int. Explor. Mer.* 153, 114 (1962).
29. R. C. Heezen, M. W. Ewing, R. J. Menzies, *Oikos* 6, 170 (1955).
30. D. B. Ericson, M. Ewing, G. Wollin, R. C. Heezen, *Bull. Geol. Soc. Amer.* 72, 193 (1961).
31. G. A. Riley, D. Van Hemert, P. J. Wangersky, *Limnol. Oceanogr.* 10, 354 (1965).
32. D. Menzel, *Deep-Sea Res.* 14, 229 (1967).
33. R. V. Fournier, *Science* 153, 1250 (1966).
34. J. F. Kimball, E. F. Corcoran, E. J. F. Wood, *Bull. Mar. Sci.* 13, 574 (1963).
35. C. E. ZoBell and R. Y. Morita, *Galathea Rep.* 1, 139 (1959).
36. J. A. Allen and H. L. Sanders, *Deep-Sea Res.* 13, 1175 (1966).
37. D. Ye. Garshanovich, *Okeanology* 5, 85 (in English translation) (1965).
38. F. A. Richards and A. C. Redfield, *Deep-Sea Res.* 1, 279 (1954).
39. M. G. Gross, *Int. J. Oceanogr. Limnol.* 1, 46 (1967).
40. G. C. Anderson, *Limnol. Oceanogr.* 9, 284 (1964).
41. E. L. Mills, personal communication.
42. We thank L. B. Slobodkin and E. L. Mills for reading the manuscript. Contribution No. 2171 from the Woods Hole Oceanographic Institution. Supported by NSF grants 6027 and 810.

Role of biological disturbance in maintaining diversity in the deep sea

P. K. DAYTON and R. R. HESSLER*†

(*Received* 12 *April* 1971; *in revised form* 7 *August* 1971; *accepted* 17 *August* 1971)

Abstract—This paper presents the hypothesis that the maintenance of high species diversity in the deep sea is more a result of continued biological disturbance than of highly specialized competitive niche diversification. Detrital food is the primary resource for most of the deep-sea species, but we suggest that in deposit feeding, most animals would consume available living particles as well as dead. We call this dominant life-style 'cropping'. Predictable cropping pressure on smaller animals reduces the probability of their competitive exclusion and allows a high overlap in the utilization of food resources. Since cropping pressure is in part proportional to the abundance of the prey, proliferations of individual species are unlikely.

Through time many species have accumulated in the deep sea because of speciation and immigration. Extinction rate is low because the biological and physical predictability of the environment has suppressed the possibility of population oscillations. Predictability in food supply for smaller deposit feeders is enhanced by the larger, mobile scavengers which consume and disperse large particles of food which fall to the ocean floor.

INTRODUCTION

THE DEEP sea has long been regarded as one of the most rigorous environments on the planet. The combination of low temperatures, high pressure, absence of light, and low rate of food supply influenced the conclusion that this is not a likely place to find a rich fauna. Until recently most benthic samples from the deep sea supported this view. Though supposedly representing large areas of the bottom, these samples characteristically were low in numbers of both individuals and species. With the development of an improved sampling device, the epibenthic sled, a continuing series of samples has been obtained which demonstrates that, contrary to classical belief, deep-sea benthic communities have an extraordinarily high species diversity (HESSLER and SANDERS, 1967).

This discovery led SANDERS (1968) to review current ideas on the factors that control within-habitat diversity and to propose a general model which he called the Stability-Time Hypothesis. Stated briefly, this model says that physical instability in an environment prevents the establishment of diverse communities. However, if physically stable conditions persist for a long period of time, speciation and immigration will cause species diversity to increase gradually as the member species become biologically accommodated to each other. Thus, high diversity in the deep sea is a result of the great long-term stability of that environment. Basic to his view is the idea that each species must occupy an increasingly narrow, specialized niche.

Because of the high physical homogeneity, both temporally and spatially, of the

---

*The order of authorship was determined by the flip of a coin.

†Scripps Institution of Oceanography, La Jolla, California 92037, U.S.A.

366

Reprinted from Deep-Sea Res. 19:199–200 (1972).

deep-sea environment, there is little chance for the kind of niche diversification that results from environmental heterogeneity. It is generally agreed that by far the most important potentially limiting resource in the deep sea is food. It follows that if there is a high degree of specialization, it must relate to the food resource, and that empirical evidence of life habits of deep-sea organisms should indicate a trend toward higher specialization with regard to this resource than is found in less diverse shallow-water communities.

Deposit feeding is the dominant trophic type in the deep-sea, soft bottom communities. SOKOLOVA (1965) has argued that suspension feeders dominate in the deep, central, oligotrophic portions of the oceans, but no other workers have reported similar findings. Hessler (in preparation) finds a typical high diversity of deposit feeders in one oligotrophic area of the North Pacific. This suggests that even if suspension feeders are important in such areas, deposit feeders are at least as diverse as elsewhere in the deep sea.

There is no disagreement on the dominance of deposit feeders in deep eutrophic areas (as defined by SOKOLOVA, 1965). SANDERS and HESSLER (1969) claim that deposit feeders constitute well over half of the fauna, while SOKOLOVA (1965) calculates that they constitute over 70%. The deposit-feeding habit is virtually universal in some taxa, such as isopods (MENZIES, 1962), that show high within-habitat diversity. Finally, increased within-taxon emphasis on deposit-feeding with increasing depth has been found to hold true even in groups thought of as not normally utilizing this mode of nutrition (ascidians for example: MONNIOT and MONNIOT, 1968; MILLAR, 1970). Thus, in a single sample from the epibenthic sled there may be hundreds of species of deposit feeders. As it has been stated to date (SANDERS, 1968, 1969; SLOBODKIN and SANDERS, 1969), the Stability–Time Hypothesis explains this high diversity in one trophic type by suggesting that each species has evolved a sufficiently high food specialization that there is no competitive overlap with other members of the community.

This degree of specialization seems highly unlikely, because it implies that deposited food can be categorized into a number of types equivalent to the large number of co-occuring deposit feeding species. To date, there has accumulated no direct evidence that the feeding habits of deep-sea species are any more specialized than those of species from shallow water. In the absence of such data, the idea that animals are evolving increased food specialization simultaneously with an emphasis on the deposit-feeding habit is difficult to accept.

While direct evidence on the specific types of detritus that deep-sea species consume is lacking for the present, SANDERS and HESSLER (1969) discuss two aspects of deep-sea communities that indirectly suggest that deposit feeders are flexible in their habits.

First, deep-sea species show strong depth zonation with the only major zonal break in community composition (defined as a peak in a depth-correlated graph of rate of species replacement) occurring at the continental shelf-slope transition band. Otherwise there is a continuous turnover in the species composition of the community as depth increases. This means that within its individual depth range, each species is associated with a continuously changing assemblage of species. This varying association with other species suggests that unless the competitively differentiated food niche changes over a depth gradient, the various detritus-feeding individuals are probably harvesting a spectrum of detrital material.

Second, although diversity for the community is high, density is so low that the average, small, non-vagile individual will be in direct interaction with far fewer species (SANDERS and HESSLER, 1969). Unless these communities are highly structured mosaics (which we have no reason to believe), different individuals of a species will be interacting with different suites of species. Again, under these conditions, food specialization is hard to accept, and it seems more likely that each species is indeed harvesting a variety of available food particles. Both of these points suggest that species must possess a certain flexibility in order to survive.

In considering the question of food specialization, we purposely ignore the fact that deposit feeders are constructed differently and behave differently, thus handling their food in diverse ways. In the present context, when a resource is limiting, it does not matter what method an organism uses to obtain it as long as it is no longer available to other members of the community. This means that degree of taxonomic affinity *per se* may have no bearing on whether two species are competing for the same resource. In the deep sea, all deposit feeders are potentially competing provided that they are utilizing the same kind of detritus from the same level in the bottom. We believe that the number of such food-space categories is low, far lower than the number of species they maintain.

If greater niche specialization is concomitant with higher diversity, an obvious place for this to occur would be in prey preference of predaceous forms. SOKOLOVA (1959) inspected decapods, asteroids, and ophiuroids from the Northwestern Pacific and adjacent areas, and SCHOENER (1969) studied two species of ophiuroids from the North Atlantic. Both studies show the same thing: rarely are the animals in the stomachs of these predators confined even to one phylum. Even when one phylum predominates, several species are represented. Sokolova finds that within a predaceous species, prey may vary with locality, suggesting that to some extent faunal composition of the food simply reflects composition of the community. Thus, contrary to the predictions of the Stability–Time Hypothesis, deep-water predators are apparently as generalized as those in shallow water, in spite of the fact that this is one area where a clear-cut opportunity for specialization exists.

In summary, we can find no evidence suggesting increased food-niche specialization in the deep sea, and in fact what little information there is on feeding habits suggests the opposite. These considerations lead us to believe that the Stability–Time Hypothesis in its present form is not adequate to explain the diversity in deep-sea communities.

General factors limiting the growth of populations

There are two classically understood ecological mechanisms by which the growth and distributions of populations are regulated. Populations may be limited when some requisite resource is in short supply, or they may be limited by some disturbance operating somewhat independently of the resources which otherwise potentially limit the populations. In the former situation, the population is said to be limited by either intra- or interspecific competition. In the latter situation the outside disturbance limits the growth of the population before any resource becomes limiting.

Probably all populations share some potentially limiting resources with other populations. As such resources become predictably more limiting, those populations potentially limited by these resources are exposed to increased competition. So long as a few scarce resources are limiting the growth of the competing populations, the intensity of competition will increase the probability of competitive extinction (HARDIN, 1960). If, on the other hand, there are sufficient independent resources, competition can cause adaptive changes in the competing species. Such character displacement can result in the coexistence of species. The degree of similarity allowable which still permits coexistence between such species in a competition based system has been carefully considered and built into a strong theoretical model which predicts that there can be no more species than there are independent resources within a given level of environmental stability (MACARTHUR and LEVINS, 1964; 1967). This is part of the theoretical framework which gives support to the Stability–Time Hypothesis.

Disturbance is an equally effective mechanism by which the sizes of populations can be regulated. An immediate effect of disturbance is that the resources become less limiting, and the probability of competitive exclusion is reduced. The importance of disturbance in the form of predation and fire is well known (DARWIN, 1859; HUTCHINSON, 1961; LOUCKS, 1970), and the general ecological significance of predation in this role has recently been experimentally demonstrated (SLOBODKIN, 1964; CONNELL, 1970; PAINE, 1966; DAYTON, 1971). There is ample evidence that under predictably disturbed conditions, the probability of competitive exclusion is reduced, more overlap in the utilization of the resources is tolerated, and potential competitors for these resources will have greater opportunity for coexistence. Under such conditions it would be possible for a community to support more species than there are independent resources.

Reduction of competitive exclusion by croppers

Physical factors are relatively constant in the deep sea and therefore are unimportant as disturbance agents, but the role of biogenic disturbance is probably significant. Since the general importance of predators as disturbers is well recognized in other communities, their effect must be considered in the deep sea as well. Here their role essentially merges with that of the deposit feeders, which will probably readily accept living animals as well as dead organic and inorganic material as they forage (SANDERS, GOUDSMIT, MILLS and HAMPSON, 1962). Therefore, any distinction between predators and deposit feeders in the deep-sea is probably somewhat arbitrary, and none will be made here. To describe this broad but unified life style, we will use the term ' cropper ' as applying to any animal that ingests living particles, whether exclusively or in combination with dead or inorganic materials.

We suggest the hypothesis that by preying on the populations of other, smaller deposit feeders, croppers such as holothurians, echinoids, ophiuroids, asteroids, cephalopods, and some polychaetes, decapods, and fish are largely responsible for the maintenance of the high species diversity of small deposit feeders observed by HESSLER and SANDERS (1967) by reducing the probablity of competitive exclusion. To a certain extent, our hypothesis merges and applies earlier hypotheses of WILLIAMSON (1957), PAINE (1966), and LEVIN (1970).

Since we do not here distinguish between predators and deposit feeders, the entire community (except perhaps suspension feeders and bacteria) can be regarded as consisting of croppers. Viewed simplistically each of these croppers should be able to eat everything that is sufficiently smaller than itself unless that food item employs avoidance. Even the smallest meiofaunal animal is cropping bacteria. There would be croppers available to harvest the complete resource spectrum, from sporadic, high-yield food to relatively constant, low-yield foods; fish and holothurians, respectively, exemplify these extremes among the croppers.

The larger croppers probably exhibit size selectivity in that their prey must be large enough to be worth interrupting their search processes (MACARTHUR and PIANKA, 1966). The more motile croppers should have become efficient at locating and cropping patches of prey. It is important to note that this efficiency need not imply food specialization. Indeed, with the supposed low densities of prey patches and surface-derived carrion, search time for these large, motile croppers is probably high, and they would be expected to be food generalists (MACARTHUR and PIANKA, 1966).

These feeding patterns suggest that the smaller animals have more species of predators than do the larger animals. Because the probability of being resource limited is inversely related to the intensity of the cropping pressure, the logical result of this situation in which the cropping pressure becomes increasingly severe on populations of decreasing-size animals is that the smaller animals will be less resource limited. The obvious corollary is that the larger the animal, the greater the probability of resource limitation and therefore the lower the diversity of populations of increasingly large animals. While data on the larger, more motile deep-sea croppers are scanty, this prediction may be supported, as there does seem to be an inverse relationship between size and diversity.

The predicted motile efficient searcher and food generalist fits the observed pattern. Lack of trophic specialization among larger invertebrates has already been discussed. The efficiency with which the larger croppers find and consume patches of food is dramatically demonstrated by the Monster Camera photographs taken by John Isaacs and colleagues in the Marine Life Research Group (Scripps Institution of Oceanography). Cans of bait (usually fish) are dropped on to the deep-sea bottom with a camera arranged to photograph the bait and surrounding area at regular intervals. Figures 1 to 4 show that even in extremely sterile areas motile animals representing many species of a number of phyla (fish, natantian decapods, octopods, amphipods, ophiuroids, holothurians, polychaetes, brachyurans) were quick to locate and consume the bait (ISAACS, 1969). The alignment of these croppers frequently suggests that they have utilized current-borne clues to find the bait (Fig. 3) This quick response to the bait by apparently widely scattered individuals reinforces the supposition of searching efficiency. Furthermore, the fact that several species of large croppers quickly consume the bait denies obligate feeding specialization and supports our predicted food generalization.

In summary, given the generalized diets and foraging efficiencies of croppers, it seems logical to suggest that the prey populations are maintained at a sufficiently low level that their food resources are rarely limiting. Since competition would then be relatively reduced, the diets of these prey species would be expected to overlap extensively because the best strategy in a habitat with a relatively low productivity would be to eat any food item that they encountered and could handle.

The present hypothesis does not deny the presence of some niche specialization in the deep sea. It is clear that in being adapted to swimming, ambulation, or burrowing, etc., different species are adapted to operating in levels or zones within or above the sea bottom. Animals of different sizes or morphologies will prefer particles of different sizes. In contrast to nonselective deposit feeders such as holothurians, some croppers may show some qualitative selectivity. However, there is no evidence that the tastes of deep-sea animals are any more refined than those of shallow water; indeed the high search time suggests an advantage to being less specialized (MACARTHUR and PIANKA, 1966).

Role of predictability

Besides being ultimately responsible for relatively more flexible strategies with the corresponding loss of efficiency (MACARTHUR and LEVINS, 1964), unpredictable oscillations in the physical environment usually result in corresponding oscillations in the size of the populations. This can occur either by affecting mortality directly or by setting up oscillations in a critical resource or level of predation. The indirect result of the latter is that resources will aperiodically become limiting, with the utilizers of the resources being thrown into direct competition. In addition, with predators over-exploiting their prey, further oscillations will result. Thus, extinctions can occur, lowering species diversity.

In the deep sea, the primary stimulus for biological oscillation is for the most part missing; that is, the physical conditions are highly stable, and the food supply possibly more predictable than in other environments (see below). Therefore, the oscillations considered above will rarely be initiated. The most important consequence of this is that the amount of available food and the amount of predation pressure are relatively predictable for each species. Instead of arguing, as does the Stability–Time Hypothesis, that this biological accommodation results in competitive niche differentiation, we suggest that the species remain generalists. Indeed, Sanders' idea of the highly buffered structure of biologically accommodated communities requires generalized feeding, because only in that way can buffering cross-linkages exist. High trophic specialization results in straight chain linkages. If food is a limiting resource, such chains would actually be unstable configurations.

Predictability of food

One of the basic tenets of most deep-sea theories is that the habitat lacks the various stimuli causing biological oscillation and that the extreme environmental predictability is one of the most important components of the deep-sea community (SLOBODKIN and SANDERS, 1969). However, the possibility of unpredictability in food supply has been ignored, and thus deserves special attention. It is not known how food actually arrives at the deep-sea floor, but there cannot be much dispute that in the open ocean it comes from above. Every condition of this food, from organic aggregates and copepod ecdyses to dead fish and whales, has been postulated (MENZIES, 1962; SANDERS and HESSLER, 1969).Given the variable and extremely slow sinking rates, the smaller particles should arrive at the bottom in a homogeneous pattern from which unpredictable pulses of primary productivity at the surface have been damped out, except perhaps in coastal waters. In the deep sea, assimilation by the various animals and bacteria in the water column render it unlikely that these smaller particles have directly utilizable food value when they arrive. If this is so, the most significant

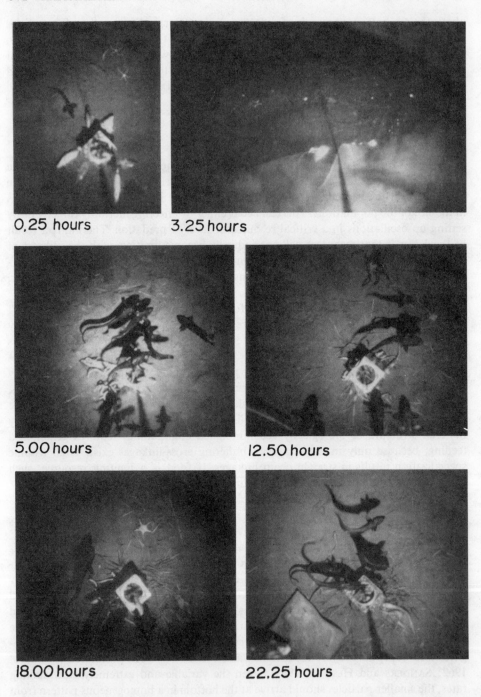

0.25 hours   3.25 hours

5.00 hours   12.50 hours

18.00 hours   22.25 hours

Fig. 1. Monster camera sequence from 2009 m depth in coastal waters off Baja California, 30° 53′N, 116° 45′W. The pictures show the variety of organisms that are attracted to bait (=large parcels of food): several species of fish, including *Antimora rostrata*, *Somniosus pacificus*, *Coryphaenoides acrolepis*, *Eptatretus stoutii*, *Lycodes* sp., and *Raja* sp.; asteroids, lithodid crabs, and the quill worm, *Hyalinoecia tubicola* (the rod-shaped objects). The camera's arrival on the bottom is the datum point for the time given under each picture.

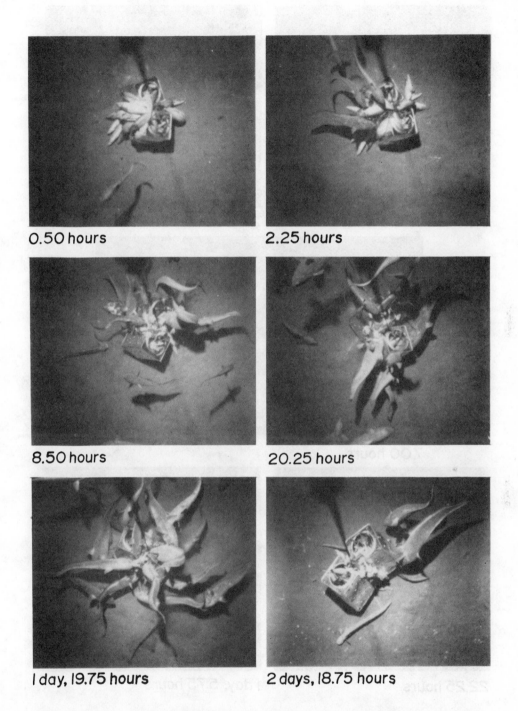

0.50 hours

2.25 hours

8.50 hours

20.25 hours

1 day, 19.75 hours

2 days, 18.75 hours

Fig. 2. Monster camera sequence from 3435 m depth in coastal waters off Baja California, 29°59'N, 118° 23'W. The diversity of mobile scavengers is distinctly lower than in Fig. 1. Note that the fish shown at 0·50 hr are actually the dead bait; this bait on the outside of the cans has been mostly removed by the last shot of the sequence (2 days, 18·75 hr).

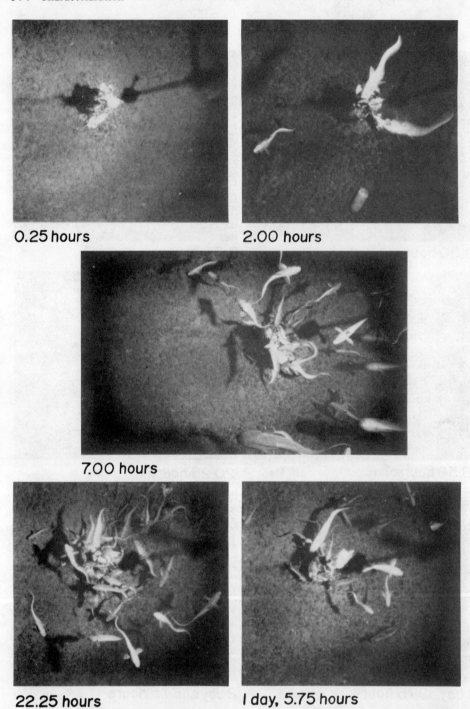

0.25 hours

2.00 hours

7.00 hours

22.25 hours

I day, 5.75 hours

Fig. 3.   Monster camera sequence from 5856 m depth in the northwestern Pacific, 34° 03′N, 163° 59′E. Except for the holothurian at 2 hr, all the scavengers are fish. In this area benthic standing crop is only moderately low (about 0·1 g wet wt. / m² according to Filatova, in ZENKEVICH, 1969), and the environment has been regarded as eutrophic by Sokolova (*ibid.*). Nevertheless, many more scavengers have been attracted than one would have expected in a manganese nodule environment from open-ocean waters so deep. The orientation of fish at 7 hr suggests an appreciable current may have done much to spread the odor of the bait.

0.50 hours

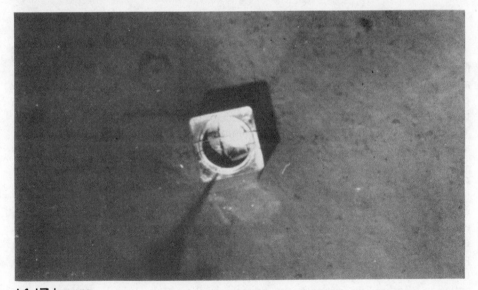

14.17 hours

Fig. 4.   Monster camera photographs from 5307 m depth in the north central Pacific, 23° 54·5′N, 144° 05′W. The two shots, one at the beginning of the sequence, the other, 13·67 hr later, demonstrate the efficiency with which the bait was dispersed. In the second shot, the can appears to be empty except for one voracious amphipod.

source of nutritive energy to the deep sea could be the faster falling, larger parcels of food such as various-sized dead euphausids, fish, squids, whales, etc. In terms of seasonality, the rate of supply of large particles to the deep sea should be more constant than with smaller organisms because the longer life span of larger animals will damp out many of the seasonal oscillations. Nevertheless, from such a source, food supply to any given small area in the deep sea would probably be more unpredictable than anywhere on the planet, and this could be a potent source of biological oscillation.

However, the Monster Camera data demonstrate that the deep benthic community contains a surprisingly large number of efficient scavengers which are widely dispersed but quick to concentrate on any such large parcel of food (Figs. 1–4). After they have consumed the food parcel, these scavengers again disperse. The important point is that such large patches of food are promptly homogenized and dispersed over the bottom as feces (or some other organic product) from the scavengers before a local population build-up of small detritus feeders can take place. That is, the majority of species in the community are never given an opportunity to take advantage of a large piece of food. Thus, a potential source of biological unpredictability is very much reduced. It is still an open question, of course, whether the feces from the scavengers furnish a predictable source of food. We think that in relation to the presumably very low turnover rates of the community, the organic by-products of the scavengers would be a rather reliable source of energy. That is, in terms of the generation time of the deposit feeders, the supply rate of such organics to any given small area may be reliable.

Although we feel that the possible importance of larger food particles has been under-rated in recent years (see also ISAACS, 1969), the present discussion is not meant to discredit organic aggregates or the rain of dead plankton as possible sources of food for the deep sea. What is important here is that regardless of their energy content, small particles are not likely to result in patchiness, either spatially or temporally.

DISCUSSION

It is becoming clear that some of the better known communities, such as forests (LOUCKS, 1970; JANZEN, 1970), lakes (BROOKS and DODSON, 1965; DODSON, 1970), and the intertidal (PAINE, 1966; CONNELL, 1970; DAYTON, 1971), are structured by patterns of natural disturbance that reduce the effects of competition. This is also seen to be the case with certain plant–insect associations (JANZEN, 1968; CANTLON, 1969). We envision that, in general, deep-sea communities are dominated by disturbance, as are these other communities. However, biological disturbance by cropping may play an even greater role here because of the environmental predictability and the fact that relatively few of the prey have refugia in size or space from their potential consumers.

Sanders' Stability–Time Hypothesis was developed to explain the high faunal diversity observed in the deep-sea. The essence of this hypothesis is that long term environmental predictability has allowed a highly refined competitive niche differentiation to result in a great many essentially non-overlapping niches. We agree that predictability is probably ultimately responsible for the high diversity; however, we do not agree that this increase in diversity need result from competitive niche differentiation. Rather, we propose the hypothesis that high diversity in this stable environment is allowed by predictable disturbance by the croppers, which in effect reduces the importance of competitive exclusion and thus allows the continued coexistence of many species which share the same resources.

An important criterion of any scientific hypothesis is that it generates testable predictions. The predictions discussed by SANDERS (1969) are designed to allow a test of the effect of environmental stability, but do not permit an evaluation of the mechanisms by which the high diversity is maintained. It is difficult to identify and contrast the roles of competition and disturbance in the deep-sea community. The precise experiments in the intertidal (DAYTON, 1971) and lake (DODSON, 1970) communities in which predators are experimentally excluded are clearly impossible in the deep sea. Neither can we extract convincing evidence such as that proposed by LOUCKS (1970) in his argument that fire is an important component of forest diversity.

Predictions or consequences which could be used to test our basic hypothesis include the following: (1) the smaller organisms such as polychaetes and small arthropods should be complete food generalists able to consume any appropriately small organic food particle which becomes available to them; (2) the intermediate-sized croppers such as holothurians, ophiuroids, echinoids, larger polychaetes, etc., should also be food generalists which do consume living animals smaller than them-selves as well as detritus; and (3) the larger, more motile croppers such as the largest crustaceans and fish should be extreme food generalists capable of locating and quickly consuming the large rapidly settling parcels of food. We have presented observations which support these predictions, but clearly the data are fragmentary, and much more needs to be done.

It might be objected that if we agree that this deep-sea community has an extremely low rate of food income, it is not possible to argue that most of the species are in fact not food limited considering that the community as a whole is clearly food limited. However, all communities are ultimately energy limited, and all communities, especially terrestrial communities, have many species which are limited by disturbance, especially predation and weather (SLOBODKIN, SMITH and HAIRSTON, 1967).

The deep-sea community differs only in that the trophic levels are almost com-pletely merged, so that the roles of most predators are not distinguishable from those of the decomposers. The situation that we envisage is that the smaller the organism, the more potential predators it will have and the less probability of its being food limited. Conversely, the larger the organism, the more it will have to search for sufficient food and the more likely it is that it will be food limited. The populations of some of the largest croppers seen in Isaac's deep-sea photographs are almost certainly food limited most of the time. Thus the HAIRSTON, SMITH and SLOBODKIN (1960), concept of decomposers being food limited is ultimately true also in the deep sea in that potential energy is not accumulating. However, in the deep sea, many of the predators which are limiting the growth of the populations of smaller organisms are also cropping the decomposers within the detritus.

It has been theoretically demonstrated that unselective predation can allow two otherwise competing species to coexist (SLOBODKIN, 1962; MURPHY, 1968); therefore the situation which we envision should not be unstable. In fact, the almost total intermeshing of the food webs might offer the community an ability to resist per-turbations (MACARTHUR, 1955). It is also germane to point out that predation is the active selection agent which differentially reduces the more abundant populations of prey species in the apostatic selection discussed by CLARKE (1962).

In this paper we have not been concerned with the possible sources of the species in the deep-sea community. Rather, we assume, with others (SLOBODKIN and SANDERS, 1969), that both speciation and immigration are occurring and act to enlarge the species list. Our only concern here has been to consider the question of what enables the species to coexist so that a high species diversity is maintained.

We are aware that we have taken a relatively extreme position in this paper, and we agree with LEVINS (1966) that the truth is usually the intersection of independent lies. However, we suggest that realistic deviations from our model such as less predictable food supply, the temporary formation of search images such as discussed by TINBERGEN (1960) and HOLLING (1965), various types of escape responses, nonpredator mortalities, etc., will all act to strengthen our main point: that it is unlikely that the patterns of coexistence in the deep sea depend entirely on competitive niche differentiation. In this context, future areas of deep-sea research should include a more sophisticated analysis of diet, determination of optimal predator foraging strategies, elucidation of the importance of formation and duration of search images (MURDOCH, 1969), evaluation of the cost and effectiveness of any existent defense mechanisms, both behavioral and chemical, and detailed analysis of microdistribution patterns of species within the community. With the improvement of deep-sea submersibles, experimental manipulation should be undertaken. This type of research will almost certainly demonstrate that community organization in the deep sea is much more complex than is currently envisioned in this paper and elsewhere.

Acknowledgments—We appreciate helpful and critical discussions with J. H. CONNELL, E. W. FAGER, J. F. GRASSLE, R. T. PAINE, J. A. McGOWAN, W. W. MURDOCH, H. L. SANDERS, D. SIMBERLOFF, M. WILLIAMSON and S. WOODIN. We are grateful to J. D. ISAACS for the use of his ' monster photographs '. This work was supported by Sea Grant GH-112 and NSF Grant GB 14488.

REFERENCES

BROOKS J. L. and S. I. DODSON (1965) Predation, body size, and composition of plankton. *Science*, 150, 28–35.

CANTLON J. E. (1969) The stability of natural populations and their sensitivity to technology. *Brookhaven Symp. Biol.*, 22, 197–203.

CLARKE B. (1962) Balanced polymorphism and the diversity of sympatric species. *Systematics Assoc. Publ.* No. 4., 47–70.

CONNELL J. H. (1970) A predator–prey system in the marine intertidal region. 1. *Balanus glandula* and several predatory species of *Thais. Ecol. Monogr.*, 40, 49–78.

DARWIN C. (1859) *On the origin of species.* Murray, London.

DAYTON P. K. (in press) Competition, disturbance, and community organization: the provision and subsequent utilization of space in a rocky intertidal community. *Ecol. Monogr.*, 41, (4), in press.

DODSON S. I. (1970) Complementary feeding niches sustained by size-selective predation. *Limnol. Oceanogr.*, 15, 131–137.

HAIRSTON N. G., F. E. SMITH and L. B. SLOBODKIN (1960) Community structure, population control, and competition. *Am. Nat.*, 94, 421–425.

HARDIN G. (1960) The competition exclusion principle. *Science*, 131, 1292–1297.

HESSLER R. R. and H. L. SANDERS (1967) Faunal diversity in the deep-sea. *Deep-Sea Res.*, 14, 65–78.

HOLLING C. S. (1965) The functional response of predators to prey density and its role in mimicry and population regulation. *Mem. Entomol. Soc. Can.*, 45, 1–60.

HUTCHINSON G. E. (1961) The paradox of the plankton. *Am. Nat.*, 95, 137–145.

ISAACS J. D. (1969) The nature of oceanic life. *Scient. Am.*, 221, 146–162.

JANZEN D. H. (1968) Host plants as islands in evolutionary and contemporary time. *Am. Nat.*, 102, 592–595.

JANZEN D. H. (1970) Herbivores and the number of tree species in tropical forests. *Am. Nat.*, **104**, 501–528.

LEVIN S. A. (1970) Community equilibria and stability, and an extension of the competitive exclusion principle. *Am Nat.*, **104**, 413–423.

LEVINS R. (1966) Strategy of model building in population biology. *Am. Scient.*, **54**, 421–431.

LOUCKS O. L. (1970) Evolution of diversity, efficiency, and community stability. *Am. Zool.*, **10**, 17–25.

MACARTHUR R. H. (1955) Fluctuations of animal populations and a measure of community stability. *Ecol.*, **36**, 533–536.

MACARTHUR R. H. and R. LEVINS (1964) Competition, habitat selection and character displacement in a patchy environment. *Proc. natn., Acad. Sci., U.S.A.,* **51**, 1207-1210.

MACARTHUR R. H. and R. LEVINS (1967) The limiting similarity, convergence and divergence of coexisting species. *Am. Nat.*, **101**, 377–385.

MACARTHUR R. H. and E. PIANKA (1966) On optimal use of a patchy environment. *Am. Nat.*, **100**, 603–609.

MENZIES R. J. (1962) On the food and feeding habits of abyssal organisms as exemplified by the Isopoda. *Int. Revue ges. Hydrobiol.*, **47**, 339–358.

MILLAR R. H. (1970) Ascidians including specimens from the deep sea, collected by the R.V. *Vema* and now in the American Museum of Natural History. *Zool. J. Linn. Soc.*, **49**, 99–159.

MONNIOT C. and F. MONNIOT (1968) Les Ascidies de grandes profondeurs récoltées par le navire océanographique americain *Atlantis II. Bull. Inst. océanogr. Monaco*, **67**, 3–48.

MURDOCH W. W. (1969) Switching in general predators: experiments on predator specificity and stability of prey populations. *Ecol. Monogr.*, **39**, 335–354.

MURPHY G. I. (1968) Pattern in life history and the environment. *Am. Nat.*, **102**, 391–403.

PAINE R. T. (1966) Food web complexity and species diversity. *Am. Nat.*, **100**, 65–75.

SANDERS H. L. (1968) Marine benthic diversity: a comparative study. *Am. Nat.*, **102**, 243–282.

SANDERS H. L. (1969) Benthic marine diversity and the stability–time hypothesis. *Brookhaven Symp. Biol.*, **22**, 71–80.

SANDERS H. L., E. M. GOUDSMIT, E. L. MILLS and G. R. HAMPSON (1962) A study of the intertidal fauna of Barnstable Harbor, Massachusetts. *Limnol. Oceanogr.*, **7**, 63–79.

SANDERS H. L. and R. R. HESSLER (1969) Ecology of the deep-sea benthos. *Science*, **163**, 1419–1424.

SCHOENER A. (1969) Ecological studies on some Atlantic ophiuroids. Ph. D. dissertation, Harvard University, 115 pp.

SLOBODKIN L. B. (1962) *Growth and regulation of animal populations.* Holt, Rinehart, and Winston, 172 pp.

SLOBODKIN L. B. (1964) Experimental populations of Hydrida. *J. Anim. Ecol.*, **33**, (*suppl.*), 131–148.

SLOBODKIN L. B. and H. L. SANDERS (1969) On the contribution of environmental predictability to species diversity. *Brookhaven Symp. Biol.*, **22**, 82–93.

SLOBODKIN L. B. and F. E. SMITH and N. G. HAIRSTON (1967) Regulation in terrestrial ecosystems, and the implied balance of nature. *Am. Nat.*, **101**, 109–124.

SOKOLOVA M. N. (1959) The feeding of some carnivorous deep-sea benthic invertebrates of the far Eastern Seas and the Northwest Pacific Ocean. (In Russion). In: *Marine biology.* B. N. NIKITIN, editor. *Trudy. Inst. Okeanol., Akad. Nauk. SSSR.*, **20**, 227–244. Transl. publ. by A.I.B.S.

SOKOLOVA M. N. (1965) Uneven distribution of bottom feeders in the deep-sea benthos as a consequence of uneven sedimentation. (In Russian). *Okeanologiia*, **5**, 498–506. English transl.; 85–92.

TINBERGEN L. (1960) The natural control of insects in pinewood—I. Factors influencing the intensity of predation by songbirds. *Arch. néerl. Zool.*, **13**, 265–343.

WILLIAMSON M. H. (1957) An elementary theory of interspecific competition. *Nature, Lond.*, **180**, 422–425.

ZENKEVICH L. A., editor (1969) *Deep-sea bottom fauna, pleuston.* (In Russian). *Biologiia Tikhogo Okeana Tikii Okeana*, V. G. KORT, Chief editor, Isdated Nauka. Moskva, **7**, (2), 3550 pp.

Life histories and the role of disturbance*

J. Frederick Grassle† and Howard L. Sanders†

(*Received* 19 *July* 1972; *in revised form* 14 *December* 1972; *accepted* 2 *February* 1973)

Abstract—An alternative to the stability–time hypothesis explaining the high benthic faunal diversities in the deep sea (SANDERS, 1968; SLOBODKIN and SANDERS, 1969) has been proposed by DAYTON and HESSLER (1972). According to Dayton and Hessler, nonselective predation reduces competition between species thereby allowing more species to coexist.

Much controversy relating to the concept of diversity and what it implies can be resolved by realizing that an increase of within-habitat diversity is achieved by two entirely different and un-related pathways. The resultant diversities are differentiated as follows: *short-term, non-equilibrium*, or *transient diversity*—induced by a low level or unpredictable physical or biological perturbation or stress resulting in biological 'undersaturation'. *Long-term* or *evolutionary diversity*—increase in diversity is the product of past biological interactions in physical benign and predictable environments. Although predation may play a role in the evolution of deep-sea species, the life histories indicate that it does not seem a likely means for control of population size, regardless of whether predation is selective or non-selective. The known life history characteristics of deep-sea animals—small brood size, age-class structure not dominated by younger stages, probable slow growth rates—are features that neither would be expected nor have high survival value in predator-controlled communities or any environment where short-term or transient diversity is important. Non-selective cropping proposed by Dayton and Hessler as a mechanism for controlling population size of prey species would result in rapid extinction of species with relatively low reproductive rates.

In addition to feeding behavior, niche diversification may be the product of biochemical specializa-tion, biotic relationships and microhabitat specialization. Niche diversification may also result from adaptation to different parts of a temporal mosaic. The stability–time hypothesis does *not* state that disturbance plays no role in predictable environments. The relative predictability of the environment enables species to survive with lower reproductive rates, lower mortality rates, and smaller population size. Rates of competitive exclusion are lower and species are able to become more specialized on both biotic and physical components of the environment. Control of population size is seldom the result of changes in the physical environment or any disturbance, including predation, so that we may say that the community is biologically accommodated rather than physically regulated.

INTRODUCTION

IN A THOUGHTFUL and provocative paper, DAYTON and HESSLER (1972) question the premise that niche diversification as reflected in feeding behavior or microhabitat specialization is the basis for the high benthic faunal diversities found in the deep sea. Instead, they argue that the environment is continuously disturbed by predators which non-selectively ingest prey species from the sediment in proportion to their abundance. Under intensive predation, no prey species is very abundant so that neither space nor food resources limit population size. Dayton and Hessler assume such resources are not limiting and therefore large numbers of species coexist in the absence of compe-tition.

Dayton and Hessler offer this theory as an alternative to the stability–time hypo-thesis proposed by SANDERS (1968) and SLOBODKIN and SANDERS (1969). We will

*Contribution No. 2914 of the Woods Hole Oceanographic Institution. This research was sup-ported by grants GB 6027X and GA 31105 of the National Science Foundation.
†Woods Hole Oceanographic Institution, Woods Hole, Mass. 02543, U.S.A.

380

attempt to answer their objections by showing that the life histories of animals evolving in relatively predictable environments such as the deep sea are consistent with the stability–time hypothesis and at variance with the Dayton–Hessler postulation. We intend to show that: (1) the available life history data indicate that a predation hypothesis to explain the high diversities is unlikely, (2) the Dayton–Hessler hypothesis cannot work with the assumptions of non-selective predation and uniform distributions, and (3) the stability–time hypothesis is consistent with the existing data.

LIFE HISTORIES

In a community that is intensively cropped (or suffers a sustained high mortality from any source), the population of prey species will be composed preponderantly of young stages. To withstand the considerable mortalities, the reproductive potential of the prey species must be high and selection should favor relatively early maturation. Let us, then, examine from the available information whether such features characterize the deep-sea benthic fauna.

Adequate data for testing this hypothesis must be based on samples containing numerous individuals belonging to single species. The other necessary requirement is the presence of the entire size spectrum in proportion to its actual abundance. The use of the deep-sea Anchor Dredge (SANDERS, HESSLER and HAMPSON, 1965) and, more particularly, the Epibenthic Sled (HESSLER and SANDERS, 1966; SANDERS and HESSLER, 1969), together with processing the samples through screens with 0·42 mm apertures, satisfy these criteria. A vast amount of information is being gathered from a number of geographically separated regions of the deep Atlantic, and, we have only begun to exploit this large source of fundamental knowledge.

A recent scholarly and monographic investigation of the deep-sea tanaid family, Neotanaideae, by GARDINER (1971) provides some relevant data. At WHOI Benthic Sta. 64 in 2886 m depth on the Gay Head–Bermuda transect, 74 individuals of an as yet unnamed species, temporarily designated as *Neotanais* No. 2, were measured and arranged into a length–frequency histogram. The interval chosen was 0·5 mm and the smallest specimen, belonging to the first free-living developmental stage, was 2·6 mm in length. This is well above the minimum length validly sampled by our screening procedures. Thus, the sample is an unbiased representation of the sizes available.

It is evident that the younger, smaller stages are not numerically predominant (Fig. 1). In fact, the first six size intervals, 2·5 to 5·5 mm in length, contain 33 individuals, while the remaining six intervals, 5·5 to 8·5 mm, are represented by 41 specimens. Preparatory males and females appear by the fifth interval (4·5–5·0 mm); copulatory males by the sixth interval (5·0–5·5 mm); and copulatory females by the ninth interval (6·5–7·0 mm). Essentially the same pattern can be seen for *Neotanais* No. 2 at WHOI Sta. 76 (2862 m, 94 intact specimens) and *Neotanais americanus* (WHOI Sta. 64, 40 intact individuals and WHOI Sta. 76, 37 specimens).

Another source of information is the elegant study by HESSLER (1970) of the deep-sea isopod family Desmosomatidae, found at stations on the Gay Head–Bermuda transect. Two species collected from WHOI Sta. 73 (1330–1470 m), *Eugerda tetarta* and *Chelator insignis*, were separated by Hessler according to developmental stage into size–frequency histograms. Each species was represented by about 230 individuals. We have transformed the histograms for *Eugerda tetarta* (HESSLER, 1970, Fig. 6) into

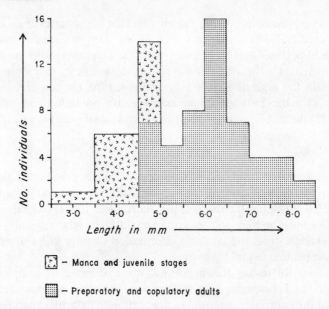

Fig. 1. Length–frequency distribution of an undescribed tanaid peracarid crustacean, *Neotanais* No. 2, collected from Sta. 64 under 2886 meters of water on the Gay Head–Bermuda transect (modified from GARDINER, unpublished Ph.D. thesis).

a life history diagram. The body length of the smallest stage (manca 1) was 1·5 mm, thus, each size category was sampled according to its abundance. As with the Neotanaidae, the smaller and younger stages are decidedly not the most common size groups (Fig. 2), suggesting that severe predation is an unlikely agent for the resultant size–distribution pattern. Length–frequency distributions for *Chelator insignis* (HESSLER, 1970, Fig. 7) again show that the younger and smaller sizes are not numerically predominant. None of the species of isopods or tanaids studied show dominance of younger stages as frequently found in shallow water.

The same pattern of non-dominance by smaller stages appears to be typical for the deep-sea Bivalvia as well (personal observations of more than 100 species). Each form

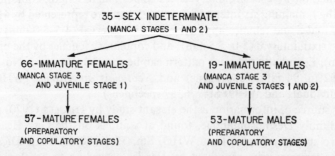

Fig. 2. The distribution of immature and mature males and females belonging to the desmosomatid isopod, *Eugerda tetarta*, collected from Sta. 73 in 1330–1470 meters of water on the Gay Head–Bermuda transect (modified from HESSLER, 1970, Fig. 6).

that has been carefully documented, *Silicula fragilis*, *Silicula* n. sp., and a species belonging to a new genus and family (ALLEN and SANDERS, in press), and two representatives of another new genus (SANDERS and ALLEN, in press) conform to this pattern.

For example, the length–frequency histogram for 38 specimens of the undescribed protobranch bivalve species belonging to a new genus and family, is shown in Fig. 3 (from ALLEN and SANDERS, in press). As in the other examples, the smallest individual, 1·70 mm long, is well above the minimum length validly sampled by our screening

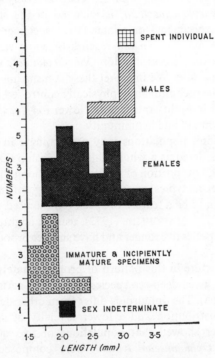

Fig. 3. Length–frequency distribution of specimens belonging to a protobranch bivalve representing a new species and genus and sorted according to sex and gonadal development. The sample was collected at Sta. 64 on the Gay Head–Bermuda transect in 2886 meters of water.

procedures. The smallest size category does not dominate numerically. Immature and incipiently mature specimens form only a small proportion of the sample. The presence of mature specimens in all but the smallest size category suggests that an individual can reproduce at least twice and possibly many times. The high proportion of mature specimens and small brood size of deep-sea species compared with shallow-water species (260 eggs/individual—ALLEN and SANDERS, in press) imply that high mortality, including predation, does not play a major role in determining the length–frequency composition for this species.

Five species belonging to the genus *Nucula* studied by SCHELTEMA (in press) and SCHELTEMA and SANDERS (in preparation), found off the northeastern United States from subtidal to abyssal depths, have a lecithotrophic mode of reproduction. The ratio of gonad to total body weight is used to indicate the portion of the food resources utilized for reproduction. When corrected for animal size, this proportion in the shelf forms (*Nucula proxima* and *Nucula annulata*) is about three times greater than in the slope and abyssal species (*Nucula granulosa*, *Nucula cancellata* and *Nucula verrilli*). If egg number rather than gonad volume or weight is compared, there are four times as many eggs per individual for the shelf species in contrast to their deep-sea congeners even when appropriate corrections are made for size differences.

Size–frequency histograms for both winter (9123 specimens) and summer (731 specimens) show that the vast majority of specimens of the shelf species, *Nucula annulata*, are contained in the smallest size categories. Alternatively, only a minor proportion of the slope–abyss form, *Nucula cancellata*, is included in the smaller-size groups in either the summer (352 specimens) or winter (1017 specimens) samples. Monthly measurements show that *Nucula annulata* reaches breeding size approximately one year after the larvae settle on the bottom. Both *Nucula annulata* and *Nucula cancellata* become mature at the same size. We interpret these contrasting patterns to mean that mortality rate is high in the predominantly physically controlled habitat of the inshore *Nucula annulata* while the mortality rate is much lower for *Nucula cancellata*, living in a largely biologically accommodated community.

At shallow depths, adult populations may predominate in localized areas. Yet, when the species are considered as a whole, this relationship does not hold. Alternatively, we interpret the high proportion of adults found among the bathyal and abyssal species cited in this paper as typical over the entire range of most deep-sea species. Small brood size is known to be a feature of species in the predictable portions of the tropics (MacArthur, 1972). Mauchline (1972), using a slightly different approach, shows that bathypelagic species live longer and have smaller brood sizes than epipelagic forms.

The one marked departure in this pattern of proportionately modest numbers of smaller size groups for a given deep-sea species is found in certain of the ophiuroids studied by Schoener (1968). The ophiuroids differ from other deep-sea taxa in having much broader vertical distribution and a lower diversity of species (Sanders and Grassle, 1971). Schoener showed that populations of two deep-sea brittle stars, *Ophiura ljungmani* and *Ophiomuseum lymani*, were composed almost entirely of small individuals in samples collected during the summers of 1964, 1965 and 1966 on the Gay Head–Bermuda transect. In winter and spring samples, taken at approximately the same depths, the numerical dominance of the small individuals was much less pronounced. On the basis of seasonal differences in size distribution and the observation that well-developed gonads were found only in winter samples for *Ophiura ljungmani*, Schoener (1968) concludes that reproduction is seasonal rather than continuous, the more typical condition for the deep-sea benthos (Sanders and Hessler, 1969; Scheltema, in press). *Nucula annulata* has seasonal reproduction, yet, the smaller size categories numerically dominate at all times of the year. The difference between the constant dominance in the case of *Nucula*, and seasonal predominance of smaller brittle stars for *Ophiura ljungmani*, might indicate a less intense mortality rate for deep-sea ophiuroids.

Microbial decomposers

One of the more profound insights in understanding the biological characteristics of the deep-sea biota came about by the recovery of the deep-sea submersible, *Alvin*.

This small submarine was lost in 1540 m of water, 135 miles southeast of Woods Hole, Massachusetts. Aboard were the *Alvin* lunches consisting of soup, bologna sandwiches, and apples. On September 1, 1969, ten and a half months later, the submersible was recovered with the lunches remarkably intact. In general appearance, their condition was little different than if they had been left in a refrigerator for two weeks (JANNASCH, EIMHJELLEN, WIRSEN and FARMANFARMAIAN, 1971). From taste, smell, and bacteriological and biochemical evidence, the food was excellently preserved.

These unexpected findings inspired Dr. Jannasch and his colleagues to delve further into the problem of bacterial breakdown of organic matter in the deep sea. They placed dissolved substrates in pressure-equalizing bottles and syringes on deep-sea moorings at depths of 4300 and 5300 m, where they were exposed to deep-sea conditions for periods of 2 to 5 months. As with the *Alvin* lunches, biodegradation was startlingly slow. For two carbohydrates, a mixture of amino acids, and a single amino acid (glutamic acid), biodegradation was approximately 10 to 100 times slower than in the controls kept under refrigeration in the laboratory at 3°C in total darkness. JANNASCH, EIMHJELLEN, WIRSEN and FARMANFARMAIAN (1971) conclude that "The data support the notion of a general slow-down of life processes in the deep sea".

As a result of the extensive degradation of small organic particles during the slow transport from the surface water to the deep ocean floor, only the more refractory components reach the bottom (MENZEL and RYTHER, 1968). Such organic materials cannot immediately be utilized by the bottom fauna and benthic microorganisms are the critical intermediaries. The recent findings reported by Jannasch and his associates can only be interpreted to mean that this refractory organic matter is being converted into more labile food at very slow rates. The bacterial decomposers are the pacemakers for the higher trophic levels. Since turnover rates for bacteria in the deep sea are apparently very slow, then the turnover rates for all of the other dependent trophic levels must be at least correspondingly slow. Although we are, at present, lacking the critical information on rates; the small broad size, and structure of the length–frequency histograms are consistent with such an interpretation. These are the features that would be selected against in a community controlled by predation.

Growth rates

Knowledge on growth rates of deep-sea animals is extremely meager. KOHLMEYER (1969) concludes, after examining an Ascomycete fungus found on test panels submerged off California between 1616 and 2073 m, that "development of fructifications of Ascomycetes in the deep sea is very slow and requires more than $1\frac{1}{2}$ years. . . ." Growth of the boring bivalve, *Xygophaga washingtona*, as measured by penetration into the wood in test panels off Oregon, was only 31·5% to 42·9% as fast at 1000 m as at 200 m depth (TIPPER, 1968, Fig. 40).

There is no knowledge on growth for any deep-sea infaunal or non-fouling epifaunal animals. Yet, despite the relatively small size of bivalves, polychaetes, peracarid crustaceans, and other groups in the deep sea, they may, on the average, be considerably longer-lived and slower-growing than their shallow water counterparts. Furthermore, metabolic rates are probably also very low. SMITH and TEAL (1973) found that oxygen consumption on the sea floor on the lower slope south of New England was about two orders of magnitude less than at shelf depths. However, animal and bacterial respirations are combined so that it is impossible to be certain that respiration per gramme of animal weight is significantly lower than in the shallow seas.

Scavengers

DAYTON and HESSLER (1972) argue that a significant fraction of the nutrition for deep-sea benthos could be in the form of large, fast falling food particles such as the carcasses of fish, whales, squid, euphausiids, etc. This food source is then immediately and directly used by large and motile scavengers that homogenize and disperse the food much more uniformly over the bottom as feces. To support this thesis, they cite data obtained from the Monster Camera. Photographs taken at regular intervals by the camera of bait dropped on to the deep-ocean floor showed that the bait was quickly located and consumed by a variety of motile fish and large invertebrates (ISAACS, 1969).

Certainly, the biomass of larger active scavengers must be small when compared with the rest of the benthos. They are seldom recorded in bottom photographs in the absence of bait and they are extremely rare in any of the benthic samples, although this may in part be a sampling problem. Since the scavengers are a proportionately small fraction of the total biomass and normal maintenance, growth and reproductive needs utilize the larger fraction of the consumed food, it would appear unlikely from an energetics point of view that much of the food for the rest of the benthic and pelago-benthic fauna would channel through the bodies of these highly motile forms.

Carcasses of large dead animals, and the feces of both scavengers and predators may well produce local environmental patchiness which DAYTON and HESSLER (1972) argue against. Scavengers are not predictable in the sense that they can be expected to pass over an area at regular intervals and thus distribute their feces uniformly over the bottom.

Predators

DAYTON and HESSLER (1972) have proposed that, with food in such short supply, high diversity is made possible by intense and uniform cropping by larger food generalists on smaller faunal constituents in proportion to their abundance. Prey species density stays sufficiently low and food does not become a limiting resource. As a result, numerous species (although in small numbers) are found together utilizing a food component such as detritus with little competitive interaction.

We feel that reduction of food resources in conditions of equilibrium restricts the number and size of individuals but not the number of species. Inherent in the Dayton–Hessler thesis is the idea that each predator (or 'cropper') is a broad generalist feeding on the smaller species with relatively little or no selectivity. Such a postulation implies that high diversity is the product of an underexploited resource, in this case, food. Through intensive cropping by predators and omnivores the environment is biologically 'undersaturated' in terms of individuals, thus making competitive exclusion largely inoperative. As a result, many potential competitors can coexist. However, the deep-sea benthos differs decisively from other habitats where this concept has been invoked (PAINE, 1966; CONNELL, 1970; DAYTON, 1971) for the highly predictable environment favors increased specialization and efficiency of utilization of resources.

To support their argument, Dayton and Hessler cite SOKOLOVA's (1959) and SCHOENER's (1969) studies on the food of predatory deep-sea benthic invertebrates.

Although *Amphiophiura bullata* and *Ophiura ljungmani* (SCHOENER, 1969) are, by definition, croppers, most of the stomach contents of these brittle stars were planktonic in origin (*Sargassum* and its associates, pteropods, heteropods, pelagic Foraminifera, Radiolaria, and algae). These organisms, together with the ubiquitous presence of bottom sediments in the stomachs, suggest that these species are not obligate predators on the macrofauna.

Sokolova, in contrast to Dayton and Hessler, concluded from her data that there is pronounced stenophagy (food specialization) in the diets of the carnivores she studied. The food of two crangonid shrimp differed considerably; *Nectocrangon dentata* fed predominantly on molluscs while *Sclerocrangon zenkewitchi* mostly consumed polychaetes. Similarly, the two ophiuroid carnivores selected contrasting faunal components; *Stegophiura* sp. ate mainly amphipods, and cumaceans were the sole animal remains present in the stomachs of *Ophiura leptoctenia*. The predominant prey group of two carnivore species from the same samples, the decapod, *Chionocetes angulatus*, and the asteroid, *Psilaster pectinatus*, were molluscs. Yet, the composition of the constituent food items diverged widely.

Clearly, Sokolova's conclusions do not support the Dayton–Hessler model. Deep-sea benthic carnivores are not complete generalists that prey upon smaller species with little or no selectivity. Whether they are more or less selective than their shallow-water counterparts is more difficult to determine. The very limited data for the deep sea, together with the general paucity of information on benthic carnivore feeding, render any statements speculative.

Life history consequences of non-selective predation

Even if we assume that food and space are not limiting on the deep ocean floor, there are serious difficulties in applying the Dayton–Hessler hypothesis to the deep sea. Their concept will work only if brood size or number of eggs produced is necessarily the same from species to species. Among a collection of deep-sea protobranch bivalves currently under investigation, a mature female may bear from 2 to 2000 eggs depending on the species. For five species of deep-sea brittle stars examined by SCHOENER (1972), those with the highest counts had from 12,000 to 93,000 eggs per specimen while the species with the lowest numbers varied from 210 to 360. With this in mind, let us consider a theoretical case.

Given two species of the same size and similar in all ways but for brood size; one species B produces twice the number of offspring per unit period of time as the other A, i.e. the reproductive potential is twice as great. The Dayton–Hessler theory requires that equilibrium numbers be maintained from generation to generation by non-selective predation. Let the combined number of individuals of these two species remain constant at 40 individuals per unit area. Initially, we will start with equal numbers of each species, 20 individuals, and assume non-overlapping generations. Within these constraints, the following pattern emerges (Table 1).

Thus, under the conditions that the Dayton–Hessler model sets forth, a species with but a slightly lower reproductive potential will be reduced in numbers and eliminated. The resultant qualitatively depauperate condition would be the complete antithesis of the actual highly diverse deep-sea benthic fauna. Furthermore, under the constraints set down by Dayton and Hessler, one would expect intense selection for a

Table 1. Hypothetical example of the effects of uniform, non-selective predation for two prey species with differing reproductive rates.

Generation	Species A	Species B	Abundance ratio
1 (40 ind.)	20	20	1:1
2 (40 ind.)	13·33	26·67	1:2
3 (40 ind.)	8	32	1:4
4	4·44	35·56	1:8
5			1:16
6			1:32
7			1:64
8			1:128
.			
.			
.			
25			1:1, 793, 216

high reproductive potential. All available evidence from the deep sea suggests that the reproductive potential is lower than in most other habitats.

Predation models in general, regardless of whether they involve selective or non-selective predation, can only hold in environments that have a significant component of unpredictability and this is where they have been demonstrated, the rocky intertidal (PAINE, 1966; CONNELL, 1970; DAYTON, 1971), high latitude terrestrial forests (LOUCKS, 1970) and early recovery stages under polluted conditions (SANDERS, GRASSLE and HAMPSON, 1972). In these environments, removal of a predator results in sharp increase in density for those prey species with high reproductive rates. The low reproductive rates of deep-sea species insure that a decreased rate of predation could not result in major changes in the community in a single generation. The predator's role in the deep sea is limited to the long-term evolutionary time scale rather than the short-term control of population size each generation. The role of herbivores in determining the distribution of rain forest trees is an example of the evolutionary importance of species functioning as 'predators' in a predictable environment (JANZEN, 1970).

Time and diversity

Much of the confusion relating to diversity and what it implies is attributable to the fact that an increase of within-habitat diversity can be achieved by two entirely different and unrelated pathways (SLOBODKIN and SANDERS, 1969). The resultant diversities are differentiated as follows:

Short-term, non-equilibrium, or transient high diversity—induced by unpredictable physical or biological perturbations or stress resulting in biological 'undersaturation' of the environment. Because there is at least a partial biological vacuum, more species can temporarily occupy the habitat until population sizes build up to densities where the species must interact. The effects are local and the time scale is always short—days, weeks, months, or years—and this type of diversity is manifested primarily in predominantly physically controlled habitats. The increase in diversity is rapid and is brought about by immigration of other species from the surrounding areas.

Long-term, equilibrium, or evolutionary high diversity. Increase in diversity is a product of past biological interactions in physically stable, benign and predictable environments. The time scale is geologic—at least thousands of years—and the resultant product is a biologically accommodated community. These assemblages are broadly distributed rather than sharply localized. Diversity increment is slow, the result of speciation and/or a low rate of immigration into the environment and the level of diversity is well above that achieved in those transient assemblages that are biologically undersaturated.

Of course, these two contrasting types of diversity cannot be set forth as simple alternatives. However, short-term diversity has a much greater probability of being an important feature in unpredictable environments and the likelihood of its contribution is progressively reduced along the gradient from physically controlled to biologically accommodated conditions.

Under extreme, variable conditions of low predictability a highly opportunistic strategy evolves. Life histories are keyed to the ephemeral and stressed nature of the habitat. Thus, opportunistic species have opted for high reproductive rates, short life spans, large population size, wide physiological tolerances, high mortality rates and broad, often cosmopolitan, distributions. They are unable to maintain themselves for extended periods of time under pressure from less opportunistic species.

'Holes' or undersaturated spaces become larger and more frequent as one approaches the physically controlled end of the community gradient because of progressively poorer coupling to an increasingly less predictable and more stressful environment. The more opportunistic species can settle in these 'holes'.

For example, pronounced opportunists such as the polychaete worm, *Capitella capitata* or the bivalve, *Mulinia lateralis*, are widely distributed in modest numbers. However, when they do occur in great concentrations, it is always under conditions of high stress or during the immediate aftermath of a catastrophe when the environment is temporarily denuded of much of its normal fauna. In the latter case 'holes' are magnified into 'voids'.

In the chemostatic, oligotrophic, and low temperature regimes of the deep sea and, to a lesser degree, the cave habitat (POULSON and WHITE, 1969), a different evolutionary strategy has been adopted. There has been selection for smaller population size, lower reproductive potential, longer life, narrower physiological tolerances and, probably, lower genetic variability. Both deep-sea and cave habitats are physically stable and oligotrophic, and the convergences in respective life histories are an evolutionary response to similar environmental stimuli.

POULSON (1963), comparing five species belonging to a small family of fishes, the Amblyopsidae, showed a sequential adaptation to life in caves. Obligate cave dwellers, in contrast to facultative and, even more so, non-cave dwelling species are long lived, late maturing, have low reproductive potentials, an age structure that is dominated by older age classes, low metabolic rates and smaller population sizes. POULSON and WHITE (1969) interpret these findings to mean "that when an organism invades the stable cave environment, selection no longer acts to maintain its ability to adjust ecologically and physiologically to variable conditions. The loss of adjustment on these levels may be associated with a decrease in genetic variability. There are also changes toward the lowering of population size, reproductive rate, and metabolic rate and toward the lengthening of life and development. These are strategies which do not lead to success in variable or unpredictable environments, where being an opportunist is important to survival". In discussing the similarities with other predictable habitats such as the deep sea and the lowland tropics, these authors point out that "competitive exclusion is rare, or difficult to reconcile with the high species diversities found within single habitats".

In comparing surface and cave populations of the characid fish *Astyanax mexicanus*, AVISE and SELANDER (1972) report high levels of genetic variability in surface populations and greatly reduced variability in populations in the relatively predictable cave environments. This reduction in variability is associated with the small size of the cave populations.

A fundamental difference between the views of Dayton and Hessler and ourselves is the interpretation of the importance of long-term stability in determining community structure. If disturbance occurs over a short-time scale, large population size, high reproductive and high mortality rates are favored. Under such circumstances, predation can be important in controlling population size. Reproductive rates and resistance to mortality are genetically determined. Regardless of the mechanisms involved in the maintenance of population size (be it resource limitation, predation, changes in the physical environment or competition), differences in these life history characteristics require explanation in terms of adaptive significance in predominantly biologically accommodated versus physically controlled environments.

A critical point to our thesis is the statement made by one of us (SANDERS, 1969) that "competition and predation may, indeed, be most severe in the physically controlled environments. Since the organisms must give adaptive priority to the physical regime, the biological interactions are not refined or mutually balanced. To the degree that species do not make subtle distinctions about the environment, as in the physically unpredictable or high stress habitats, competition or predation can be extreme. Thus, biological effects are often drastic or catastrophic". This comment serves as an adequate description of the "better known communities" cited by DAYTON and HESSLER (1972)—the rocky intertidal (PAINE, 1966; CONNELL, 1970; DAYTON, 1971), geologically young lakes (BROOKS and DODSON, 1965; DODSON, 1970) and boreal forests (LOUCKS, 1970) which "are structured by patterns of natural disturbance that reduce the effects of competition". We are not surprised that these studies are from environments that are primarily physically controlled.

Advantages of low reproductive rates, low mortality rates and small population size

Increased specialization resulting from low reproductive rates, low mortality rates and small population size, allows species to maintain maximum competitive ability with respect to neighboring species. DAYTON and HESSLER (1972) acknowledge that niche diversification results from increased specialization. We differ from them in considering more than feeding behavior as specialization. The amount of space occupied by a species depends on the size and density of individuals in the population. For a within-habitat comparison of taxa that are equally common in the differing areas studied, the relationship between species diversity and specialization becomes tautological. In collections of similar size, the average population size is smaller in more diverse areas. Small population size results in more efficient utilization of resources and enhances the ability of species to compete successfully with neighboring species. Specialization in terms of the space occupied by a species may be divided into three somewhat artificial categories: feeding behavior, relationships with other species, and microhabitat. Specialization has been shown to be a feature of tropical rain forest trees (ASHTON, 1969) and birds (HOWELL, 1971; KARR, 1971; MACARTHUR, 1972) and coral reef gastropods (KOHN, 1971).

In their discussion of feeding specialization, MACARTHUR and LEVINS (1964) do not simply say there can be only the same number of species as there are independent resources (DAYTON and HESSLER, 1972), but rather: "of this mixed resource type of species there can be as many as there are proportions of the resources which can be counted from season to season, i.e. very many in stable climates and fewer in unpredictable climates". Thus, in predictable environments, such as the deep sea, a number of species may coexist on relatively few kinds of food.

The ability to distinguish kinds of food before ingestion is not the only determinant of the number of separate food resources. Different species are likely to have alternative ways of utilizing the various organic molecules in the sediment. The number of alternative foods will depend on variations in the number and kinds of enzymes involved in digestion and metabolism. If biochemical specialization is involved, then it is not inconceivable "that deposited food can be categorized into a number of types equivalent to the large number of co-occurring deposit feeding species" (DAYTON and HESSLER, 1972).

MANGUM (1964) in a thorough study of distribution and competitive interaction of five sympatric species of polychaete worms found that two species of the genus *Clymenella* "appear to be making identical demands on the environment in terms of their substratum utilization". Although there were no differences in the sediment ingested, it was possible to demonstrate very different abilities to utilize a specific plant pigment. Conceivably, variations in the digestive capabilities, as well as presence or absence of specific enzymes, can account for considerably higher trophic diversity for deposit-feeders than can be based solely on mode of feeding.

Equally provocative are the observations made by WAVRE and BRINKHURST (1971) on the differential utilization of mud-dwelling bacteria by three morphologically similar, sediment-feeding tubificid oligochaete species from Toronto Harbor, Ontario. They conclude ". . . partitioning of the nutritional resources may provide a mechanism by which three or more unspecialized detritus feeders are able to coexist despite sharing of the same sediment".

Dayton and Hessler exclude microhabitat differentiation on the basis that large samples tend to collect similar faunas. In species with highly specialized microhabitat preferences, patchiness may develop on a much smaller scale than would be expected in shallow water. Disturbances that occur in the deep sea may also result in small-scale temporal mosaics. If a fish stirs up the bottom or mud slumps occur, a slow and highly localized succession will take place. We will return to this point in the subsequent discussion.

Advantages of large population size, high reproductive rates and high mortality rates

For sessile or slow-moving invertebrates the amount of environment occupied by a species will depend chiefly on population size. One of us (GRASSLE, in press) has suggested that large population size is directly related to the advantages resulting from the maintenance of a high degree of genetic variability—an adaption to unpredictable environmental change. Deviations that make environments less predictable differ greatly from the mean and occur sporadically and without autocorrelation within the life span of an individual. Those changes that are included within the physiological tolerance of every individual, i.e. that do not result in mortality or reduced viability of some portion of the population, do not affect predictability. Similarly, disturbances that non-selectively kill all individuals do not affect predictability. This means that for most species the variance and severity of sudden environmental change are sufficient to define differences in predictability (SLOBODKIN and SANDERS, 1969).

Large maximum population size, high reproductive rates, high mortality and short generation time are the hallmarks of the opportunist. These features of the life history, together with the ability to become widely dispersed, are the requisites for the maintenance of high genetic variability. A high level of genetic variability is insurance that some individuals in the population will be closely adapted to unpredictable events. In these species, high mortality is an essential component of the adaptation that enables a population to shift rapidly to a new adaptive peak through short-term selection. Large population size prevents loss of genetic variability through inbreeding. Shifts in the 'optimum' genotype, which occur from time to time and place to place, make certain that genetic variability will be maintained.

Species can be ranked on a continuous scale of relative opportunism according to the mortality rates sustained by the species. The most opportunistic species are the most likely to be preyed upon since they are the most abundant and the least likely to evolve mechanisms to avoid predation. The definition of relative opportunism in terms of mortality rates has the additional advantage of pinpointing the critical period of adaptation—the high mortality stages.

The more opportunistic species found in relatively unpredictable environments are characteristic of the first successional stages in recolonization. The presence of highly opportunistic species in such environments insures that succession and competitive exclusion occur rapidly.

In relatively predictable environments, such as coral reefs (JOHANNES, 1971; STODDART, 1969) and, probably, the deep sea, succession proceeds at a much slower rate. Not only are the species more likely to be separated by minor differences in the environment (the mountain passes are higher—JANZEN, 1967), but the life histories are such that immigration rates are lower.

The role of disturbance

Although we have presented data indicating that predation is unlikely to be the principal means for control of population size, we do not deny that disturbance occurs in the deep sea. Localized disturbances may be triggered by a variety of phenomena: mud slumps, the dead bodies of large animals such as whales, feces of scavengers, or activities of large motile deposit feeders. DAYTON and HESSLER (1972) believe that there is "little chance for the kind of niche diversification that results from environmental heterogeneity". We cannot accept the idea of a uniform, predictable disturbance. We believe disturbances in the deep sea are much less frequent, less severe, and of smaller spatial extent than in physically controlled communities.

We further believe disturbances that are likely to occur in the deep sea would lead to highly localized successions. The first successional stages would involve taxa better adapted to new unpredictable situations by virtue of their more effective dispersal and higher reproductive rates. One example of the relationship between a relatively opportunistic deep-sea species and an unpredictable disturbance would be the attack by boring molluscs on pieces of wood reaching the bottom. This has been documented in experiments at 1800 m by R. Turner (personal communication). As more species enter the area, the relatively specialized species gradually exclude the more opportunistic. However, competitive exclusion proceeds so slowly in the deep sea that a temporal mosaic results. In other words, a spatial mosaic emerges from local successional sequences that are out of phase. (This has been called the "contemporaneous disequilibrium hypothesis" by RICHERSON, ARMSTRONG and GOLDMAN, 1970.)

A small-scale temporal mosaic of the sort we envisage was observed by JOHNSON (1970) in a highly unpredictable intertidal area. He was able to rank the species on the basis of occurrence in a gradient of environmental predictability and diversity. Ranking into higher or lower order species corresponds to what we have referred to as the degree of opportunism. A mosaic of micro-succession stages is most easily observed in the many scales of patchiness that occur in coral reef communities (GRASSLE, in press). LANG (1970) and CONNELL (in press) have shown that slower growing coral species inhibit the growth of faster growing species by digesting the coral tissue of the competitor. Even though the slow-growing massive species may live for hundreds of years, they do not occupy all of the environment. Over a time scale of hundreds of years, disturbances such as tropical storms denude local areas of coral reef so that parts of the reef are never in equilibrium.

Such a system differs markedly from that proposed by Dayton and Hessler in that population size for most species is controlled by competition with other species. The existence of a temporal mosaic, with niche differentiation based on the different temporal adaptations of the component species, is fully consistent with earlier discussions on niche diversification (HUTCHINSON, 1965).

Our data on size frequencies and clutch size indicate differences between deep-sea and shallow-water benthic populations regarding fecundity, survivorship, and time to maturity. Yet, we realize we are not measuring rates. The prime obstacle in following populations through time in the seasonless deep sea is the absence of temporal markers. We plan to circumvent this difficulty by placing boxes of azoic sediment (frozen for a month and then thawed), using the deep-sea submersible *Alvin*, at the permanent bottom station beneath 1800 m of water on the continental slope of New England. We will put identical boxes of sediment at the R station site (SANDERS, 1960) in Buzzards Bay, Massachusetts in 20 m of water. We have already placed a sediment box at the permanent deep-sea station.

At both the deep-sea and shallow-water localities, colonization can be measured over different intervals of time (three months, one year, two years, etc.). Size–frequency histograms for each abundant species can then be compared with control samples taken from the immediately surrounding areas. The results should enable us to define growth rates, survivorship and time to maturity within narrow limits. The relative growth rates, survivorship and duration of time to maturity for the deep-sea and shallow-water benthos can be determined by comparing the experimental results from these two environments. The rate measurements from a number of deep-sea and shallow-water benthic species will provide us with ample information to test the conclusions we have drawn from the existing limited data.

CONCLUSIONS

(1) Much controversy relating to the concept of diversity and what it implies can be resolved by realizing that an increase of within-habitat diversity is achieved by two entirely different and unrelated pathways. The resultant diversities are differentiated as follows: *Short-term, non-equilibrium,* or *transient diversity*—induced by a low level or unpredictable physical or biological perturbation or stress resulting in biological 'undersaturation'. *Long-term or evolutionary diversity*—increase in diversity is the product of past biological interactions in physically benign and predictable environments.

(2) The known life history characteristics of deep-sea animals—small brood size, age–class structure not dominated by younger stages, probable slow growth rates—are features that neither would be expected nor have high survival value in predator controlled communities or any environment where short term or transient diversity is important.

(3) Although predation may play a role in the evolution of deep-sea species, the life histories indicate that it does not seem a likely means for control of population size, regardless of whether the predation is selective or non-selective.

(4) Non-selective deposit feeding as a mechanism for controlling population size of prey species would result in rapid extinction of species with relatively low reproductive rates.

(5) Even if predation were the principal mechanism for controlling population size, a different theory would have to account for actual differences in population size and other life history phenomena when comparing deep-sea forms with those from shallow water.

(6) In addition to feeding behavior, niche diversification may be the product of biochemical specialization, biotic relationships and microhabitat specialization. Niche diversification can also result from adaptation to different parts of a temporal mosaic.

(7) The stability–time hypothesis does *not* state that disturbance plays no role in predictable environments. The relative predictability of the environment enables species to survive with lower reproductive rates, lower mortality rates, and smaller population size. Rates of competitive exclusion are lower and species are able to become more specialized on both biotic and physical components of the environment. Control of population size is seldom the result of changes in the physical environment or any disturbance, including predation, so that we may say that the community is biologically accommodated rather than physically regulated.

Acknowledgements—We are most appreciative for the helpful comments made by R. R. HESSLER, L. B. SLOBODKIN, J. P. GRASSLE, Y. LOYA, J. A. ALLEN and V. L. GOODRICH, who carefully read this manuscript.

REFERENCES

ALLEN J. A. and H. L. SANDERS (in press) Deep-sea protobranchiate bivalves: the families Siliculidae and Lametilidae. *Bull. Mus. comp. Zool. Harv.*

ASHTON P. S. (1969) Speciation among tropical forest trees: some deductions in the light of recent evidence. *Biol. J. Linn. Soc.,* **1,** 155–196.

AVISE J. C. and R. K. SELANDER (1972) Evolutionary genetics of cave-dwelling fishes of the genus *Astyanax. Evolution,* **26,** 1–19.

BROOKS J. L. and S. I. DODSON (1965) Predation, body size, and composition of plankton. *Science*, **150**, 28–35.

CONNELL J. H. (1970) A predator–prey system in the marine intertidal region. 1. *Balanus glandula* and several predatory species of *Thais. Ecol. Monogr.*, **40**, 49–78.

CONNELL J. H. (in press) Population ecology of reef corals. In: *The geology and biology of coral reefs*, O. A. JONES and R. ENDEAN, editors, Academic Press.

DAYTON P. K. (1971) Competition, disturbance and community organization: the provision and subsequent utilization of space in a rocky intertidal community. *Ecol. Monogr.*, **41**, 351–389.

DAYTON P. K. and R. R. HESSLER (1972) The role of disturbance in the maintenance of deep-sea diversity. *Deep-Sea Res.*, **19**, 199–208.

DODSON S. I. (1970) Complementary feeding niches sustained by size-selective predation. *Limnol. Oceanogr.*, **15**, 131–137.

GARDINER L. F. (1971) The systematics, postmarsupial development, and ecology of the deep-sea family Neotanaidea (Crustacea Tanaidacea). Ph.D. Dissertation, Univ. of Rhode Island.

GRASSLE J. F. (in press) Species diversity, genetic variability and environmental uncertainty. In: *Fifth European symposium on marine biology. Archo Oceanogr. Limnol.*

GRASSLE J. F. (in press) Variety in coral reef communities. In: *The geology and biology of coral reefs*, O. A. JONES and R. ENDEAN, editors, Academic Press.

HESSLER R. R. (1970) The Desmosomatidae (Isopoda, Asellota) of the Gay Head–Bermuda transect. *Bull. Scripps Instn Oceanogr.*, **15**, 1–185.

HESSLER R. R. and H. L. SANDERS (1966) Faunal diversity in the deep-sea. *Deep-Sea Res.*, **14**, 65–78.

HOWELL T. R. (1971) An ecological study of the birds of the lowland pine savanna and adjacent rain forest in Northeastern Nicaragua. *The Living Bird*, **10**, 185–242.

HUTCHINSON G. E. (1965) *The ecological theater and the evolutionary play*. Yale Univ. Press, 139 pp.

ISAACS J. D. (1969) The nature of oceanic life. *Scient. Am.*, **221**, 146–162.

JANNASCH H. W., K. EIMHJELLEN, C. O. WIRSEN and A. FARMANFARMAIAN (1971) Microbial, degradation of organic matter in the deep sea. *Science*, **171**, 672–675.

JANZEN D. H. (1967) Why mountain passes are higher in the tropics. *Am. Nat.*, **101**, 233–249.

JANZEN D. H. (1970) Herbivores and the number of tree species in tropical forests. *Am. Nat.*, **104**, 501–528.

JOHANNES R. E. (1971) How to kill a coral reef. II. *Mar. Poll. Bull.*, **2**, 9–10.

JOHNSON R. G. (1970) Variations in diversity within benthic marine communities. *Am. Nat.*, **104**, 285–300.

KARR J. R. (1971) Structure of avian communities in selected Panama and Illinois habitats. *Ecol. Monogr.*, **41**, 207–229.

KOHLMEYER J. (1969) Deterioration of wood by marine fungi in the deep sea. Materials performance and the deep sea. *A.S.T.M. Stand.*, **445**, 20–30.

KOHN A. J. (1971) Diversity, utilization of resources and adaptive radiation in shallow-water marine invertebrates of tropical oceanic islands. *Limnol. Oceanogr.*, **16**, 332–348.

LANG J. C. (1970) Inter-specific aggression within the scleractinian reef corals. Ph.D. Thesis, Yale Univ.

LOUCKS O. L. (1970) Evolution of diversity, efficiency, and community stability. *Am. Zool.*, **10**, 17–25.

MACARTHUR R. H. (1972) *Geographical ecology*. Harper & Row, 269 pp.

MACARTHUR R. H. and R. LEVINS (1964) Competition, habitat selection and character displacement in a patchy environment. *Proc. natn. Acad. Sci. U.S.A.*, **51**, 1207–1210.

MANGUM C. P. (1964) Studies on speciation in maldanid polychaetes of the North American Atlantic coast. II. Distribution and competitive interaction of five sympatric species. *Limnol. Oceanogr.*, **9**, 12–26.

MAUCHLINE J. (1972) The biology of bathypelagic organisms, especially Crustacea. *Deep-Sea Res.*, **19**, 753–780.

MENZEL D. W. and J. H. RYTHER (1968) Organic carbon and the oxygen minimum in the South Atlantic Ocean. *Deep-Sea Res.*, **15**, 327–337.

PAINE R. T. (1966) Food web complexity and species diversity. *Am. Nat.*, **100**, 65–75.

POULSON T. L. (1963) Cave adaptation in amblyopsid fishes. *Am. Midl. Nat.*, **70**, 257–290.

POULSON T. L. and W. B. WHITE (1969) The cave environment. *Science*, **165**, 971–981.

RICHERSON P., R. ARMSTRONG and C. R. GOLDMAN (1970) Contemporaneous disequilibrium, a new hypothesis to explain the 'paradox of the plankton'. *Proc. natn. Acad. Sci. U.S.A.*, **67**, 1710–1714.

SANDERS H. L. (1960) Benthic studies in Buzzards Bay III. The structure of the soft-bottom community. *Limnol. Oceanogr.*, **3**, 245–258.

SANDERS H. L. (1968) Marine benthic diversity: a comparative study. *Am. Nat.*, **102**, 243–282.

SANDERS H. L. (1969) Benthic marine diversity and the stability–time hypothesis. *Brookhaven Symp. Biol.*, **22**, 71–80.

SANDERS H. L. and J. A. ALLEN (in press) Studies on deep sea protobranches. Prologue and the Pristiglomidae. *Bull. Mus. comp. Zool. Harv.*

SANDERS H. L. and J. F. GRASSLE (1971) The interactions of diversity, distribution and mode of reproduction among major groupings of the deep-sea benthos. *The world ocean, Proc. Joint Oceanogr. Assembly*, MICHITAKE UDA, editor, Tokyo, 1970, Japan Soc. Promotion Sci., pp. 260–262.

SANDERS H. L., J. F. GRASSLE and G. R. HAMPSON (1972) The West Falmouth oil spill. I. Biology. *Woods Hole oceanogr. Instn, Ref.* 72–20. (Unpublished manuscript.)

SANDERS H. L. and R. R. HESSLER (1969) Ecology of the deep-sea benthos. *Science*, **163**, 1419–1424.

SANDERS H. L., R. R. HESSLER and G. R. HAMPSON (1965) An introduction to the study of deep-sea benthic faunal assemblages along the Gay Head–Bermuda transect. *Deep-Sea Res.*, **12**, 845–867.

SCHELTEMA R. S. (in press) Reproduction and dispersal of bottom-dwelling deep-sea invertebrates: a speculative summary. *Proc. Workshop Conf.—High pressure aquarium systems as tools for the study of the biology of deep ocean fauna and associated biological problems.*

SCHOENER A. (1968) Evidence of reproductive periodicity in the deep sea. *Ecology*, **49**, 81–87.

SCHOENER A. (1969) Ecological studies on some Atlantic ophiuroids. Ph.D. dissertation, Harvard University, 115 pp.

SCHOENER A. (1972) Fecundity and possible mode of development of some deep-sea ophiuroids. *Limnol. Oceanogr.*, **17**, 193–199.

SLOBODKIN L. B. and H. L. SANDERS (1969) On the contribution of environmental predictability to species diversity. *Brookhaven Symp. Biol.*, **22**, 82–92.

SMITH K. L. and J. M. TEAL (1973) Deep-sea benthic community respiration: an *in situ* study at 1850 meters. *Science*, **179**, (4070), 282–283.

SOKOLOVA M. N. (1959) The feeding of some carnivorous deep-sea benthic invertebrates of the far eastern seas and the Northwest Pacific Ocean. In: *Marine biology*, B. N. NIKITIN, editor, *Trudy Okeanol. Akad. Nauk SSSR*, **20**, 227–244. Transl. publ. by A.I.B.S.

STODDART D. R. (1969) Ecology and morphology of recent coral reefs. *Biol. Rev.*, **44**, 433–498.

TIPPER R. C. (1968) Ecological aspects of two wood-boring molluscs from the continental terrace off Oregon. Unpublished Ph.D. thesis, Dept. of Oceanography, Oregon State Univ.

WAVRE M. and R. O. BRINKHURST (1971) Interactions between some tubificid oligochaetes and bacteria found in the sediments of Toronto Harbour, Ontario. *J. Fish. Res. Bd Can.*, **28**, 335–341.

Microcosms

Introduction

A microcosm is a device for studying ecosystems. It may take a variety of forms and sizes but it is always an isolated portion of a larger natural unit. The subunit may be brought into the laboratory or contained in the field. Because it is a simplified portion of the bigger world, it is easier to control and sample, to manipulate, and to replicate. Because it is more elaborate than the study of individual species in culture, it gives more information on the real ecosystem it represents. Additionally, the microecosystem is valuable because it can be subjected to stresses, including those induced by man's activities, such as pollution from domestic and industrial effluent, radiation, and siltation. It should therefore facilitate intelligent predictions on the future of a natural ecosystem.

Parameters such as light, temperature, waterflow, and species composition, along with basic shape and size, are determined by the questions being asked and the available budget and time. Obviously, the larger the unit the fewer replications possible, but the trade off comes with less interference from the artificiality of container walls, better circulation of water, and freer interaction among organisms. The kind of information that can be gained depends on how well the microecosystem mimics the larger unity it is designed to test.

Two large outdoor microcosms are described in the readings in this unit. One is a silo-like tank (3 m diameter and 10 m deep) erected at Scripps Institution of Oceanography in California. Diatoms and dinoflagellates were inoculated into these tanks. Growth and species composition of these populations were monitored as a response to light and nutrient regimes suggested by earlier laboratory experiments. Grazers were then added to these tanks to determine the efficiency of organic carbon transfer through the food chain. The test organisms were systematically harvested to learn what effect such pseudopredation would have on the phytoplankton. The deep tank was intended to allow natural vertical migrations and sampling at any depth, and to have the tremendous advantage over sampling off the coast, where water masses move entire populations of smaller organisms and migrating fish consume plankton without regard to the interests of the investigator. The second type of microcosm illustrated in this unit is a shallow (0.5 m deep) tank at the University of Rhode Island. This tank contained an estuarine benthic community dominated by polychaetes and mollusks. Growth and species composition were observed while temperature and salinity were controlled.

Other designs have been tried that are not described by papers in this section. They include a free-floating thin transparent sphere (6 m diameter) in British Columbia (Antia et al., 1963), and a cylindrical plastic bag (3 m diameter X 17 m deep) in Aberdeen, Scotland (Davies et al., 1975). Both these large systems were kept in the sea rather than on shore. Scientists from several programs have consolidated their approaches in a gigantic effort in Western Canada, Controlled Ecosystem Pollution Experiments (Chem. Eng. News, 1973: Takahashi et al., 1975). Plastic containers hold 70 tons of water and to date little difference has been observed between biological events measured in these containers and those in the water outside. Another interesting joint venture is one between ecologists at the Environmental Protection Agency and the University of Rhode Island now being conducted at the Water Quality Laboratory at Narragansett. Indoor barrels simulate Narragansett Bay, for which physical and biological parameters have been carefully selected (Nixon, Perez, Oviatt, personal communication).

Much smaller units have been operated in continuous series to test the sensitivity of different portions of an estuary to fresh water input (Cooper and Copeland, 1973). Laboratory aquaria are also being used to test effects of stresses on organisms. Examples are: siltation and nutrient load on pieces of coral, trace metals and DDT on algae, and oil on interstitial animal populations. Changes in temperature, nutrient load, silt, radiation, or chemical toxins can alter rates of such processes as photosynthesis, respiration, nutrient uptake, metabolism, and organic exudations. Although the physiological response of individual organisms to isolated stresses is important in studying a natural ecosystem, subtle interaction may not be realized until species are tested with one another. In a mixed-species microcosm one can see who has the competitive edge, an event that might be unpredictable from a single-species culture.

Microcosms may be as small as flasks, but in such vessels the volume of water per unit of organisms is limited, distorting the natural environment system. Where the container surface is highly exaggerated, the circulation of water and number of species is drastically reduced. A zooplankton community can, for example, quickly consume all the original phytoplankton in its container. Predators are heavily restricted and decomposition patterns altered. Water must be continuously enriched with nutrients and vitamins to support algal growth. The more components excluded from the basic unit, the more unnatural it becomes, but while not representative, it can still be very informative.

To the extent that a microcosm is synthetic, we can question what it tells us about a real ecosystem. For examples, Blake and Jeffries dealt with denser populations than would be found in the natural environment because higher trophic levels were not present. After four months, interference from the walls of the container was observed in the plastic bags in Scotland. Nonetheless, the microcosm is an invaluable approach to ecology because of its relatively small size, simple community structure, controllable parameters, reproducibility, and facility for experimental manipulation. It can teach us about natural events in an ecosystem and, in addition, can help us predict responses under specific stresses.

LITERATURE CITED

Antia, J. J., C. D. McAllister, T. R. Parsons, K. Stephens, and J. D. H. Strickland. 1963. Further measurements of primary production using a large-volume plastic sphere. Limnol. Oceanog. 8:166–183.

Chemical and Engineering News. 1973. Effects of pollutants on marine life probed. 17 December, pp. 17–23.

Cooper, D. C., and B. J. Copeland, 1973. Responses of continuous-series estuarine microecosystems to point-source input variations. Ecol. Monogr. 43:213–237.

Davies, J. M., J. C. Gamble, and J. H. Steele. 1975. Preliminary studies with a large plastic enclosure. *In* L. E. Cronin (ed.), Est. Research Vol. 1: Chemistry, Biology and the Estuarine System. Academic Press Inc., New York. pp. 251–264.

Takahashi, M., W. H. Thomas, D. L. R. Seibert, J. Beers, P. Koeller, and T. R. Parsons. 1975. The replication of biological events in enclosed water columns. Arch. Hydrobiol. 70:5–23.

THE USE OF A DEEP TANK IN PLANKTON ECOLOGY. I. STUDIES OF THE GROWTH AND COMPOSITION OF PHYTOPLANKTON CROPS AT LOW NUTRIENT LEVELS[1]

J. D. H. Strickland, O. Holm-Hansen, R. W. Eppley, and R. J. Linn

Institute of Marine Resources, University of California, San Diego, La Jolla 92037

ABSTRACT

A deep tank (3-m diam, 10 m deep), in which phytoplankton crops can be grown under simulated natural conditions of nutrient concentration and cell density, is described. Three growth experiments have been performed in the tank with *Ditylum brightwellii* (a centric diatom), *Cachonina niei* (an armored dinoflagellate), and with a mixed crop of the dinoflagellate *Gonyaulax polyedra* and *Phaeocystis* sp. (a colonial member of the Haptophyceae). Each experiment lasted about two weeks. In each case ammonia was assimilated first, then nitrate as the plants grew. Successful nitrogen budgets were achieved in the experiments with *D. brightwellii* and *C. niei*. The chemical composition of the tank-grown cultures of *D. brightwellii*, *C. niei*, and *G. polyedra* is reported for cell carbon, nitrogen, phosphorus, lipid, carbohydrate, DNA, ATP, chlorophyll *a*, and carotenoids. The composition of these cultures was very similar to that obtained in laboratory cultures grown with relatively high nutrient levels.

INTRODUCTION

The use of a large plastic sphere for the culturing of phytoplankton has already been described (Strickland and Terhune 1961). Such equipment was used to follow the growth of a mixed population of coastal phytoplankters and to study nutrient depletion, *in situ* photosynthesis, and nutrient regeneration (McAllister et al. 1961; Antia et al. 1963). Vessels of large volume are particularly useful for obtaining data on the composition of plankton grown at ecologically significant (low) levels of population density, as the sample volume available for analysis is virtually unlimited.

As discussed by Strickland and Terhune, submerged translucent plastic spheres or cylinders have advantages over land-based shallow tanks (e.g., Raymont and Miller 1962; Ansell et al. 1964), but practical considerations make their apparatus unsuitable when exposed to full ocean swells. *In situ* plastic vessels are also hard to provide with sophisticated monitoring equipment, and it is difficult to establish complete control of light and temperature regimes.

The construction of a deep-tank facility at the Hydraulics Laboratory of the Scripps Institution of Oceanography prompted us to examine the potentialities of a deep tank on land for the study of plankton ecology. Such an arrangement seems particularly suited where a considerable depth of water is required, as for example, in the study of vertical migrations.

The facility and its operation for culturing phytoplankton crops is described, and some analytical data for populations grown at ecologically realistic levels of cell concentration and nutrients are reported. We have also used the facilities for investigating the vertical migration of dinoflagellates, but this work is not described here.

Mr. C. R. Stearns constructed the light-measuring devices. We obtained valuable analytical assistance from Miss L. Solorzano, Miss H. Kobayashi, Miss G. Prevost, and Mr. J. L. Coatsworth. Mr. J. McCarthy prepared inocula for two of the experiments. Dr. R. C. Smith kindly measured the spectral distribution of the searchlight radiation.

CONSTRUCTION AND OPERATION
The deep tank

The tank (Fig. 1) is a cylinder 3 m in diameter by 10 m tall with a slightly

[1] Supported by U.S. Atomic Energy Commission Contract No. AT(11-1)GEN 10, P.A. 20.

401

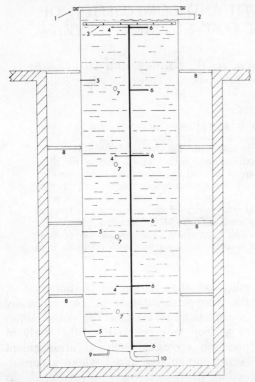

FIG. 1. Sectional view of the deep tank. 1. Transparent vinyl tank cover stretched on wooden frame. 2. Tank overflow to waste. 3. Transparent cooling coil suspended just beneath water surface. 4. Photocells. 5. Thermistor probe mounted through a 2-inch (5 cm) instrument port (24 ports available in tank wall). 6. Liquid sampling tubes. 7. 6-inch-diameter (15 cm) viewing ports. Two on each working level located 180° apart. 8. Work platforms. 9. Compressed air inlet. 10. Tank fill or drain line.

rounded bottom and is constructed of welded ¼-inch (0.6 cm) hot-rolled steel, coated on the inside with Laminar X 500, a black plastic (matte finish) which is inert toward growing phytoplankton. On the outside of the tank is 3.5 inches (9 cm) of polyurethane, a foamed plastic insulating material which reduces ambient heat flow to and from the tank to 0.001C/hr/1C differential between tank content and surroundings. A tank of the water a few degrees different in temperature from its surroundings maintains its temperature practically unchanged for many days.

A scaffolding and stairway around the tank enables the operator to move up and down and to look into any one of the eight 6-inch-diameter (15 cm) Plexiglas windows set opposite each other at four different levels down the tank. There are also 24 2-inch-diameter (5 cm) instrumentation ports 90° apart at six levels around the tank. Three of these, for our experiments, were fitted with thermistor thermometers.

Sampling tubes (⅜-inch-I.D. (0.9 cm) polyethylene tubing) were attached at six different depths to a central plastic pipe which extended from the base of the tank to the water surface. These six tubes were connected to a carboy which could be partially evacuated with a vacuum pump for drawing samples. This central plastic pipe also acts as a support for three photocells fitted with green glass filters and plastic cosine collectors. The cells are arranged, so as not to shade each other, from three arms which project 1.0 m from the center and which are 0.25, 4.25, and 8.25 m from the surface of the water. The photocells measure the attenuation of light by plankton as it grows. The whole centrally supported system of sampling tubes and photocells can be rotated.

Samples for the measurement of photosynthesis are suspended at known depths in bottles fastened to a nylon cord. The cord is fixed so that the bottles hang in substantially the same light field as that measured by the three photocells.

Before the tank is filled, its walls are scrubbed with detergent and rinsed well with freshwater. The central pole and sampling tubes and photocells are wiped with isopropyl alcohol and, immediately before filling, the tank and fixtures are irradiated by slowly lowering and raising a 1,200-w Hanovia Englehardt mercury arc tube (model 189A). The efficiency of this treatment was confirmed by tests with bacterial cultures; there is little doubt that all exposed surfaces are completely sterilized. It is not certain how worthwhile this operation may be, considering that the incoming water is not bacteria-free, but the operation is relatively simple and minimizes contamination.

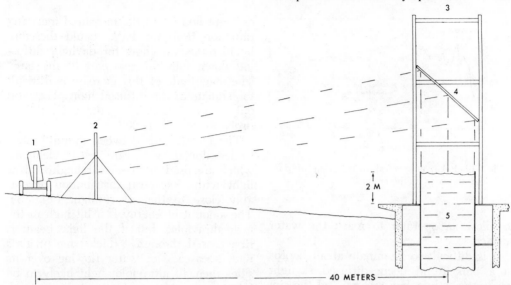

Fig. 2. Lighting arrangement for culturing of organisms in the deep tank. 1. 5-ft-diameter (152 cm) carbon arc searchlight. 2. Cooling window. ¼-inch (0.6 cm) plate glass in metal frame held in vertical position by guywires. 3. Steel service tower for deep tank. 4. Mirror mounted in tower. 5. Deep tank.

The tank is filled with freshwater or with seawater piped from just beyond the end of the Scripps Institution pier. Seawater is given a preliminary filtration through coarse sand. Such water would be unsuitable for ecological work without further filtration, and this is achieved by using a large swimming pool filter coated with fine diatomaceous earth. The second filtration removes all living phytoplankton and zooplankton and reduces the bacterial count in the water to only a small fraction of its initial value. The organic detritus is also reduced to an acceptable level (less than 15 μg of particulate carbon/liter). The plumbing is arranged to allow all lines and elbows to be preflushed thoroughly with filtered water before the tank is filled.

In addition to filtration, the water can be cooled so that the tank can be filled in about 24 hr with water at 5C and correspondingly faster if, as would be usual, warmer water were needed or the tank was to have a thermocline made in it. The temperature of the water in the tank can be raised about 1.5C/day by using three submerged 5-kw heating elements.

The possibility remains that the radiant heat from the searchlight and from daylight will slowly heat the surface water of the tank and produce an unwanted extra thermocline. This can be prevented by passing liquid from the main chiller through a flat helix of plastic tubing fastened at the surface of the tank. The need for this has proved marginal unless stirring is stopped for many days.

The water may be stirred by using a stream of air bubbles introduced at the bottom of the tank from a compressor. This soon moves the whole water column by forced convection and plankton is easily kept in uniform suspension. As large volumes of air are involved, it must be freed from organic matter and nitrogenous fumes in the atmosphere. We use a compressed air filter (model 130, Deltec Engineering, Inc., Newcastle, Del.).

The top of the tank is protected against dirt by a transparent plastic cover (thick clear vinyl) fastened over a frame which can be secured under tension by shock cording or suspended well clear of the rim

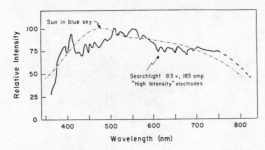

FIG. 3. Relative spectral distribution of daylight and the searchlight system.

to allow an operator to reach the water surface.

It is difficult to illuminate at all depths in a tank 10 m deep; diffused skylight penetrates only a few meters, and the sun is too unpredictable (cloud and fog) for a heliostat to be used. We have therefore used a 60-inch-diameter (152 cm) searchlight to provide illumination (Fig. 2). The searchlight, some 40 m from the tower, is directed at a plane mirror above the tower. The mirror is 4.4 m high by 3.4 m wide and consists of 12 panes of ³/₁₆-inch (0.48 cm) mirror glass; their backings are treated with uncut orange shellac and attached to a ¾-inch (1.9 cm) plywood base. The attachment is made with allowances for expansion and twisting of the board. The mirror assembly is on a steel frame that is hinged at the base and can be fixed vertically for storage or tipped forward and fastened at an angle of 47.5° to the horizontal.

The searchlight gives a collimated beam of light some 1.5 m in diameter which can be diverged slightly so that, when the light is situated some distance from the tank, the beam covers the top of the tank and near-parallel light passes down its entire length. The optics, although far from perfect, are adequate. Direct sunlight and as much as possible of the diffused daylight is shielded from the tank top by canvas screens. The residual light, which does not penetrate far into the water, can be measured by the top photocell assembly described later. For convenience the searchlight has been used mainly during

daylight hours and the measured incoming radiation from the light should therefore be increased to allow for daylight diffusing down into the top part of the tank. The magnitude of this increase is difficult to estimate as the diffused light penetrates only a few meters, but the correction is probably less than 20%.

The electric arc between graphite electrodes doped with rare earth elements, which are used in these searchlights, gives light with a spectral distribution acceptably close to that of sunlight (Fig. 3). The amount of energy is a bit high in the near-ultraviolet, but if the light beam is first passed through a thick pane of glass with a cascade of water flowing over its sides then all ultraviolet light likely to be injurious to plants and animals is removed. Nearly all long-wavelength heat is also absorbed. Some 70% of the resulting radiant energy is in the spectral range between 380 to 720 mμ. A very uniform illumination can be obtained with an average intensity over the entire water surface of 0.05 cal cm^{-2} min^{-1} of photosynthetically active light. This is about a tenth of the sea surface value to be expected from midday summer sun or equivalent to the intensity at a depth of about 10 m in moderately clear ocean water.

The searchlight working at 80 v and 175 amp consumed a pair of "high intensity" electrodes in about 90 min; the motor generator driving the searchlight needed 150 liters of gasoline for 12 hr of service.

Measurement of illumination

From a consideration of Fig. 3, we have assumed that the photosynthetically active radiation can be approximated to 70% of the reading of a thermopile radiometer (Stephens and Strickland 1962) which we then used to calibrate the three photocells carried on the central pipe. The searchlight was used at night for this purpose. The output of each photocell was put across a 50-ohm resistor and the potential drop measured with a millivolt recorder.

The average intensity of light over the entire 7 m² of water surface in the tower was estimated using an array of small selenium photocells, each glued to a 2-cm-diameter circular piece of frosted cosine collector. The sensitivity of each cell was adjusted to be the same by variable blackening of the surface of the collectors. Five of the cells were arranged along each of five transparent plastic arms radiating out from the center of the tank and suspended a few centimeters above the water surface. The cells were arranged in such a way that each of the resulting 25 units was situated in hypothetical segments of concentric annuli of identical area of tank surface. The circular central portion of the tank (with an area also equal to that of each hypothetical annulus) was monitored by a further 10 cells, making a total of 35 photocells in all. The output of each of these cells was calibrated against the radiometer using the searchlight beam, and hence the output of the whole assembly could be standardized by using a suitable multiplication factor. The output was additive and linear provided that the external resistance was kept below 10 ohms, but this still resulted in a voltage drop of 50–100 mv, sufficient to be easily measured by a recording voltmeter.

ANALYTICAL PROGRAMS

Samples from an aerated and well-mixed tank were taken at about 0900 hours each day. Samples for the analysis of soluble constituents were first filtered through washed Whatman GF/C glass filters. The methods used for the soluble constituents as well as for the particulate fractions have all been described in detail by Strickland and Parsons (1968) except for lipids, which was described by Holm-Hansen et al. (1967). A composite sample was generally obtained by collecting water in 20-liter carboy lots, running about one carboy to waste to flush the sampling tubes and collecting the second one with water pouring from all six sampling tubes at about the same velocity. Even after two weeks of intensive sampling less than 1% of the volume of the tank had been removed.

TABLE 1. *Initial concentrations of nutrients for three deep-tank experiments (including amounts added and amounts present in filtered seawater). In addition to the nutrients listed below, the following additions were also made to the 70,000 liters contained in the tower: vitamin B_{12}, 320 μg; thiamin, 4.0 mg; Fe (chelated with EDTA), 2.7 g; biotin, 320 μg*

| Nutrient | Expt | | |
| | 1 | 2 | 3 |
		(μg atoms/liter)	
Ammonium	0.9	4.60*	3.00
Nitrate	12.7	7.35	5.07
Nitrite	0.09	0.14	0.08
Phosphate	2.02	1.55	1.11
Silicate	35.4	6.92	2.60

* A value of 3.9 was found with the bis-pyrazolone method.

RESULTS AND DISCUSSION

Experiment No. 1: culture of Ditylum brightwellii

The tank was filled with filtered seawater of a salinity of 33.4‰, a small enrichment made with vitamins and chelated iron, and the inorganic nutrient concentrations adjusted to suitable values by the addition of potassium nitrate, potassium dihydrogen phosphate, and acidified sodium metasilicate (Table 1). After removing samples for the initial analyses the tank was inoculated in the evening with 130 liters of a culture of *Ditylum brightwellii* grown in artificial light to a chlorophyll *a* concentration of about 100 μg/liter. The contents of the tank were mixed all night and the searchlight turned on in the morning and the first samples taken. The light was used for 8 hr each day and samples were taken for analyses each morning at about 0900. The daily radiation varied between 34 and 44 cal/cm^2 with an average throughout the 14 days of the experiment of 38 cal/cm^2 or a total input of 37,000 kcal. This amount of incoming radiation produced 21 g of plant carbon. Thus the mean energetic efficiency of plant production was about 0.5% over the two weeks and nearly 1.6% during the last day, when most of the light was absorbed before it reached the bottom of the tank.

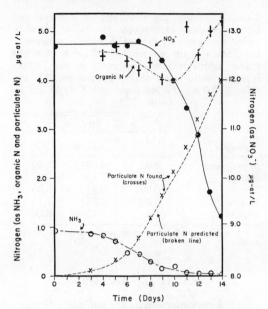

FIG. 4. Changes of dissolved and particulate nitrogen with time in the culture of *Ditylum brightwellii*.

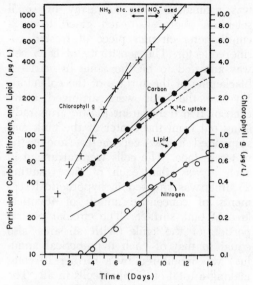

FIG. 5. Production of particulate matter with time in the culture of *Ditylum brightwellii*.

The decrease of plant nutrients in the seawater was determined daily. The assimilation of nitrogen and the incorporation of nitrogen into particulate matter are shown in Fig. 4. The most striking aspect was the use of nearly all ammonia before any nitrate was assimilated. The ammonia method used here measures some amino-acid nitrogen as well as ammonia so that by day 9 there may have been little or no ammonium ion left in the water. Direct assay of the cells for nitrate reductase (Eppley, Coatsworth, and Solorzano 1969) showed that this enzyme did not reach significant cellular concentrations until about day 8. A small concentration of nitrite initially present (0.09 μg-atoms N/liter) remained unchanged until the end of the experiment. There appeared to be a small amount of dissolved organic nitrogen initially used by the cells followed by a secretion of organic nitrogen back into the medium. The precision of the organic nitrogen data is marginal but a better nitrogen balance (loss of soluble versus production of particulate) is obtained if these changes of organic nitrogen are con-

sidered. The net loss of soluble nitrogen was close to the amount of nitrogen recovered in the particulate fraction (Fig. 4).

During the 14 days of growth, the inorganic phosphate concentration decreased from 2.0 to 1.6 μg-atoms P/liter, a decrease which agreed fairly well with the production of particulate phosphorus. The dissolved organic phosphorus changed little throughout the experiment and was present at a concentration of about 0.3 μg-atoms P/liter. Reactive silicate decreased from 35.4 to 31.0 μg-atoms Si/liter. This decrease has been used to compute the silicon content of the resulting diatom cells, as no such analysis was done directly.

The formation of particulate matter with time is illustrated for certain constituents in Fig. 5. For convenience of presentation, we have not subtracted the initial values for particulate matter in the tank so that specific growth rates cannot be calculated from the semilogarithmic plots in Fig. 5 except for chlorophyll *a* where little or no initial pigment was present. One sees, however, that there was a marked change of slopes at about the time the cells switched from ammonia to nitrate

TABLE 2. *Growth and composition of* Ditylum brightwellii. *The temperature of the tank-grown culture was 14.5C, and that of the laboratory culture 20C*

Constituents	Ammonia grown 3–8-day cells		Nitrate grown 12-day cells		Laboratory culture*	
	µg	pg/cell	µg	pg/cell	µg	pg/cell
Carbon	100	1,350	100	1,550	100	680
Nitrogen	22.5	300	20.5	320	18.0	120
Phosphorus	4.8	65	5.1	80	3.4	23
Protein (N × 6.25)	140	1,900	128	2,000	112	760
Lipid	23.5	320	38	590	—	—
Carbohydrate	13	180	11.5	180	18.5	125
Chlorophyll *a*	4.5	61	5.0	78	5.6	38
DNA	2.3	31	1.95	30	—	—
ATP	0.49	6.6	0.54	8.4	—	—
Silicon	58	780	47	730	28	190
Cells produced/liter	0.74×10^5		0.645×10^5		1.48×10^5	
k† (day^{-1}):C; Chl *a*; P	0.41		0.27		0.65 (Synchronous growth)	
k† (day^{-1}):cells; N	0.43		0.17			
Avg cell volume (μ^3)	Not measured				24,000	

* Eppley, Holmes, and Paasche (1967).
† Growth rate constants were calculated as $k = (1/\text{days}) \log_e (N_2/N_1)$. This $k/0.69$ gives doublings/day.

nutrition, in particular an increased rate of formation of lipid.

Table 2 shows the relative composition (carbon equals 100) and the actual composition of cells grown mainly with ammonia (days 3 to 8) and the composition of material formed by day 12 when the crop had doubled due to growth on nitrate. For comparison we show the composition of cells grown in a laboratory culture at high concentrations of silicate, nitrate, and phosphate (approx 250 µg-atoms Si/liter, 250 µg-atoms N/liter, and 25 µg-atoms P/liter) and exposed to cycles of 8-hr light and 16-hr dark (Eppley et al. 1967). These cells were noticeably smaller than those grown in the deep tank.

Cells growing on either ammonia or nitrate have substantially the same content of silicon, protein, carbohydrate, and DNA. There is, however, a significant increase in the chlorophyll, phosphorus, and ATP contents of the cells grown on nitrate, the cellular concentration of all three constituents increasing by 25% over the values found in ammonia-grown cells. The most striking difference in the two types of cell lies in the lipid content which increased nearly twofold and is responsible for most

of the increase of cell carbon. This is the first time, to our knowledge, that the nature of the source of nitrogen has been implicated as an agent affecting cellular composition. There was, of course, throughout the experiment, a slow decrease in the mean light intensity to which cells were exposed, but this does not seem to be causing the sudden change in cellular composition (*see* Fig. 5).

Some idea of the magnitude of reduction in light intensity due to growth can be had by comparing photosynthetic carbon assimilation rates per unit chlorophyll *a* measured in bottles suspended in the tank. For example, in bottles at the surface rates were 21, 20, and 9.9 g C/g chlorophyll *a* per day for days 1, 7, and 12, respectively. Corresponding rates were 9.5, 4.9, and 2.8 at 4 m and 7.1, 1.8, and 0.6 at 8 m. These shifts were gradual and progressive unlike the sharply defined and abrupt shift in nitrogen source and change in cell composition.

Direct comparison of cells produced in the deep tank with those grown in the laboratory is not possible as the latter were smaller. By considering composition relative to carbon, however, it is seen

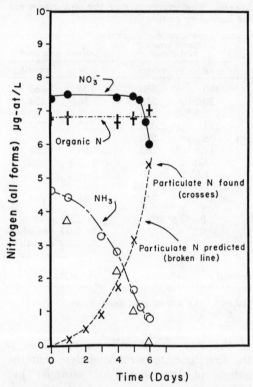

FIG. 6. Changes of dissolved and particulate nitrogen with time in the culture of *Cachonina niei* (circles show ammonia plus amino acids, triangles show ammonia only).

that the cells grown in the laboratory (on nitrate) have a similar nitrogen and pigment content but less silicon and phosphorus and considerably more carbohydrate. Lipid, DNA, and ATP data were not determined by Eppley et al. (1967), so these comparisons are impossible; even so, there is clearly no assurance that plants grown in the laboratory will be identical to those grown in the sea. In particular, it is important to note that cells of *D. brightwellii* in the deep tank at day 14 were almost completely neutrally buoyant. Stirring was stopped for three days and the tank sampled at various depths from 0.5 to 9.5 m. The chlorophyll content at all depths remained constant within experimental error. Only a fraction of the cells in any of our laboratory cultures have shown such buoyancy (cf. Eppley, Holmes, and Strickland 1968).

The estimation of photosynthesis using the radiocarbon method was carried out in bottles suspended at the surface, 4 m, and 8 m. A graphical integration of results is shown by the broken line in Fig. 5. Agreement with the measured particulate carbon is as good as could be expected. Direct analyses of the water for dissolved organic carbon indicated that any excretion of organic matter by the cells must have been less than 10% of the photosynthesate. Direct measurements of organic radiocarbon released from cells in 24-hr photosynthesis measurements averaged 5.3% of net photosynthesis.

Experiment No. 2: culture of Cachonina niei

This small armored dinoflagellate, isolated from the Salton Sea and kindly provided by Mr. A. Loeblich of the Scripps Institution, was chosen because of its vigorous growth and vertical migration.

A tank full of filtered seawater was suitably enriched (Table 1) and inoculated with 800 liters of a culture grown in diffuse daylight to a chlorophyll *a* concentration of about 60 µg/liter. It was sampled at 0900 each day that the contents were stirred (as shown in Fig. 7). Germanium dioxide was added to 1 µM in this experiment and in experiment 3 to prevent growth of diatoms (Werner 1967; Lewin 1966).

The balance of dissolved and particulate nitrogen was again satisfactory (Fig. 6). There appears to be no change in dissolved organic nitrogen. The true ammonia concentrations, determined by the bispyrazolone method (Johnston 1966) are about 0.7 µg-atoms N/liter less than those obtained by hypochlorite oxidation, indicating the presence of this much amino-N not available to the plants. As with *D. brightwellii*, nitrate was not utilized until nearly all the ammonia had been used up, at which time (day 6) nitrate reductase was first detected in the cells.

Because nearly all the dinoflagellate cells migrated to the top of an unstirred tank it was possible on day 6 to empty the

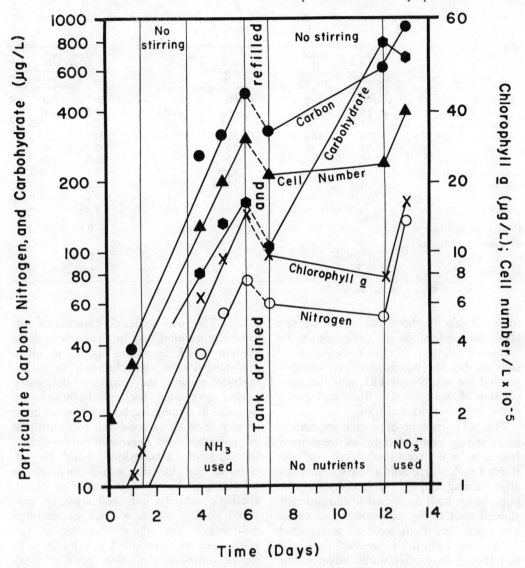

Fig. 7. Production of particulate matter with time in the culture of *Cachonina niei*.

tank almost completely and refill it with filtered nutrient-deficient seawater without losing much of the plant crop. Thus from day 7 to day 12 we could study cells in water containing less than 0.2 μg-atoms N/liter as ammonia, nitrite, or nitrate. These nitrogen-starved cells continued to migrate vertically but lost nearly all nitrate reductase. On day 12, 5.0 μg-atoms N/liter, as nitrate, and 0.5 μg-atoms P/liter, as phosphate, were added and the tank

stirred. Within a few hours nitrate reductase could be detected in the cells and after a day the crop had almost doubled and had consumed nearly all the added nitrate.

The formation of cells and most of the particulate material is shown in Fig. 7. These plots have been corrected for the presence of preinoculum particulate matter so that the slopes of these semilogarithmic plots indicate specific growth

TABLE 3. *Growth and composition of* Cachonina niei *at 20C. The cell volume of this organism was 1,300 μ³ at day 5*

Constituents	Ammonia grown 1–5-day cells		Nutrient starved 12-day cells		Nitrate grown 13-day cells	
	μg	pg/cell	μg	pg/cell	μg	pg/cell
Carbon	100	150	100	263	100	230
Nitrogen	16.2	24.5	8.2	21.5	14.5	34
Phosphorus	4.4	6.7	1.9	5.0	3.1	7.2
Protein (N × 6.25)	101	155	51	135	91	210
Lipid	57	86	42	110	35.5	83
Carbohydrate	35	53	127	335	73.5	170
Chlorophyll *a*	2.9	4.4	1.2	3.15	1.7	4.0
Carotenoids	4.2	6.4	1.95	5.15	2.3	5.35
DNA	4.0	6.1	2.9	7.6	3.2	7.5
ATP	0.45	0.68	0.24	0.63	0.29	0.68
Cells produced/liter	0.66×10^6		0.38×10^6		0.43×10^6	
k^* (day^{-1}): cells; N; P; Chl *a*	0.48		—		0.63	
k^* (day^{-1}): C	0.48		—		0.48	

* *See* footnote to Table 2 for explanation of rate constant.

rates. Table 3 shows the relative and absolute composition of cells grown on ammonia (days 1–5), the composition of cells on day 12 (which had been nitrogen starved for nearly a week), and the composition of the cells after they had grown (doubled) on added nitrate.

The ATP content of a cell appears to be relatively unaffected by all treatments. Nitrogen was conserved well in the starved cells; there was a marked increase after adding nitrate. Chlorophyll *a* and phosphorus both decreased in the nutrient-starved cells to the same fractional extent but much less than we had anticipated, which may indicate a peculiar ability for this species to survive under adverse conditions (cf. *Gonyaulax polyedra* later). There appeared to be no change in concentration of DNA during nitrogen starvation; following the addition of nitrate, the concentration of DNA increased at a rate just slightly less than the increase in cell number.

Although the lipid content of nitrogen-starved cells increased appreciably (on per cell basis) the values for lipid, chlorophyll, phosphorus, and ATP in cells grown on nitrate were much the same as those for cells grown on ammonia, which is quite unlike the behavior found with *D. bright-*

wellii. The most striking alteration of cellular composition was in the carbohydrate content which increased sixfold in nitrogen-starved cells. The increase in carbohydrate during the nitrogen deficiency period accounted for essentially all the increase in organic carbon. Upon addition of nitrate to these cells, the concentration of carbohydrate decreased appreciably; this loss of carbohydrate could be accounted for by the rapid increase in cellular protein. A high degree of photosynthetic activity and cell motility persisted during the five days of nitrogen deficiency. The slight increase in the carotenoid to chlorophyll *a* ratio in cells during nitrogen depletion gave a weak indication of the state of nutrition (cf. Yentsch and Vaccaro 1958). The chlorophyll *c* to chlorophyll *a* ratio (ca. 0.65) was largely unaffected.

Experiment No. 3: culture of Gonyaulax polyedra

This culture was contaminated with what we believe to be a *Phaeocystis* sp. (Kornmann 1955), but cell counts indicated that for the first five days the biomass of plant material in the tank was predominantly dinoflagellate, which enables us to estimate, after a first order

TABLE 4. *Growth and composition of* Gonyaulax *polyedra at 22C*

Constituents	Ammonia grown 1–5-day cells (μg)	Laboratory culture (μg)
Carbon	100	100
Nitrogen	13	14
Phosphorus	1.8	0.75
Protein (N × 6.25)	79	87
Lipid	72	63
Carbohydrate	32	43
Chlorophyll *a*	1.75	0.34
Carotenoids	2.60	—
DNA	2.4	1.7
ATP	0.30	0.30
Cells produced/liter	1.0×10^4	1.0×10^3
k^* (day^{-1}) : Chl *a*; C; cells	0.24	0.42
Cell volume (μ^3)	30,000	50,000

* *See* footnote to Table 2 for explanation of rate constant.

correction, the properties shown in Table 4. Unlike the *C. niei*, *G. polyedra* cultures lost motility and deteriorated rapidly in the absence of fixed nitrogen.

Again, nitrate was not assimilated until most of the ammonia had gone from the medium and it was again confirmed that nitrate reductase was induced in the cells only just prior to nitrate being used (Fig. 8). This preferential use of ammonia in competition with nitrate by marine phytoplankton thus may be fairly general. It has not, to our knowledge, been demonstrated before under near-natural conditions, although Grant, Madgwick, and DalPont (1967) recently showed a similar phenomenon with ammonia, urea, and nitrate in cultures of a pennate diatom. In their experiment, urea was used after the ammonia and before nitrate became incorporated in the cells.

We obtained a very poor nitrogen balance and also poor agreement between radiocarbon uptake data and particulate carbon production in this experiment. Sampling error was a likely cause of this trouble. *Phaeocystis* sp. colonies in the tank were up to 2–3 mm in diameter. As a result of shear developed during flow at sampling, most of these colonies were

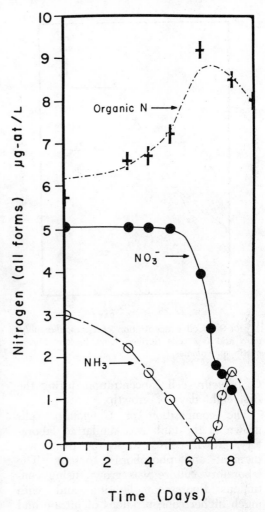

FIG. 8. Changes of dissolved nitrogen with time in the culture of *Gonyaulax polyedra*.

broken yielding many small cells 4–5 μ in diameter and much debris from the gelatinous matrix of the colonies. The fate of the latter material is unknown, but it probably accounts for the analytical disagreement mentioned above. Counts on the small cells over time indicated logarithmic growth of *Phaeocystis* sp. with $k = 0.59$ (compared with $k = 0.23$ for *G. polyedra*) until growth ceased as a result of nitrogen depletion (Fig. 9). The different growth rates resulted in an approximate 100-fold increase in *Phaeocystis* sp. as compared with an eightfold increase in

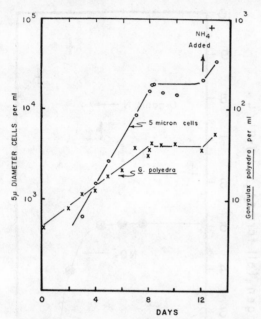

FIG. 9. Cell concentration of *Gonyaulax polyedra* and 5-μ cells derived from broken *Phaeocystis* sp. colonies.

G. polyedra cell concentration during the first eight days of growth.

The composition of *Gonyaulax* cells grown in the tank was similar to laboratory grown cells except for the lower pigment and phosphorus content. The laboratory culture was grown using continuous fluorescent lighting and with much higher concentrations of nitrate and phosphate.

The two dinoflagellates grown on ammonia resemble each other much more than the diatom when their major metabolites are compared on a normalized basis ($C = 100$). In particular, they are much higher in lipid and carbohydrate and lower in protein.

REFERENCES

ANSELL, A. D., J. COUGHLAN, K. F. LANDER, AND F. A. LOOSMORE. 1964. Studies on the mass culture of *Phaeodactylum*. IV. Production and nutrient utilization in outdoor mass culture. Limnol. Oceanog., 9: 334–342.

ANTIA, N. J., C. D. MCALLISTER, T. R. PARSONS, K. STEPHENS, AND J. D. H. STRICKLAND. 1963. Further measurements of primary production using a large-volume plastic sphere. Limnol. Oceanog., 8: 166–183.

EPPLEY, R. W., J. L. COATSWORTH, AND L. SOLORZANO. 1969. Studies of nitrate reductase in marine phytoplankton. Limnol. Oceanog., 14(2): (In press).

——, R. W. HOLMES, AND E. PAASCHE. 1967. Periodicity in cell division and physiological behaviour of *Ditylum brightwellii*, a marine diatom, during growth in light-dark cycles. Arch. Mikrobiol., 56: 305–323.

——, ——, AND J. D. H. STRICKLAND. 1968. Sinking rates of marine phytoplankton measured with a fluorometer. J. Exptl. Marine Biol. Ecol., 1: 191–208.

GRANT, B. R., J. MADGWICK, AND G. DALPONT. 1967. Growth of *Cylindrotheca closterium* var. Californica (Mereschk) Reimann and Lewin on nitrate, ammonia and urea. Australian J. Marine Freshwater Res., 18: 129–136.

HOLM-HANSEN, O., J. COOMBS, B. E. VOLCANI, AND P. M. WILLIAMS. 1967. Quantitative micro-determination of lipid carbon in micro-organisms. Anal. Biochem., 19: 561–568.

JOHNSTON, R. 1966. Determination of ammonia in sea water as rubazoic acid. Intern. Council Exploration Sea, Hydrograph. Comm., c.m. 10. 7 p.

KORNMANN, P. 1955. Beobachtungen an *Phaeocystis* Kulturen. Helgolaender Wiss. Meeresuntersuch., 5: 218–233.

LEWIN, J. 1966. Silicon metabolism in diatoms. V. Germanium dioxide, a specific inhibitor of diatom growth. Phycologia, 6: 1–12.

MCALLISTER, C. D., T. R. PARSONS, K. STEPHENS, AND J. D. H. STRICKLAND. 1961. Measurements of primary production in coastal sea water using a large-volume plastic sphere. Limnol. Oceanog., 6: 237–258.

RAYMONT, J. E. G., AND R. S. MILLER. 1962. Production of marine zooplankton with fertilization in an enclosed body of sea water. Intern. Rev. Ges. Hydrobiol., 47: 169–209.

STEPHENS, K. U., AND J. D. H. STRICKLAND. 1962. Use of a thermopile radiometer for measuring the attenuation of photosynthetically active radiation in the sea. Limnol. Oceanog., 7: 485–487.

STRICKLAND, J. D. H., AND T. R. PARSONS. 1968. A practical handbook of seawater analysis. Bull. Fisheries Res. Board Can. No. 167. 311 p.

——, AND L. D. B. TERHUNE. 1961. The study of in-situ marine photosynthesis using a large plastic bag. Limnol. Oceanog., 6: 93–96.

WERNER, D. 1967. Hemmung der Chlorophyllsynthese und der NADP$^+$-abhängigen glycerinaldehyde-3-phosphate—Dehydrogenase durch Germaniumsäure bei *Cyclotella cryptica*. Arch. Mikrobiol., 57: 51–60.

YENTSCH, C. S., AND R. F. VACCARO. 1958. Phytoplankton nitrogen in the oceans. Limnol. Oceanog., 3: 443–448.

The use of a deep tank in plankton ecology. 2. Efficiency of a planktonic food chain[1]

Michael M. Mullin and Pamela M. Evans

Institute of Marine Resources, University of California, San Diego, La Jolla   92037

Abstract

Populations of copepods and ctenophores were maintained in a quasi-steady state in a 70-m³ tank supplied with phytoplankton and harvested regularly for several months. *Paracalanus parvus* replaced *Acartia tonsa* as the dominant copepod; in the absence of ctenophores, the yield of copepods was 10.4–12.5% of the phytoplankton supplied as food. When the carnivorous ctenophores were present, their yield was 2.6% of the phytoplankton supplied but could have been as high as 3.2% had not copepods been harvested simultaneously.

Populations of marine zooplankton have been maintained in large outdoor tanks in attempts to measure such parameters as growth and reproductive rates or to examine succession of species, under conditions closer to nature than those usually established in laboratory containers or aquaria of a few liters in volume. Such processes are often difficult to study directly in nature because movement of water and patchiness of organisms complicate the interpretation of sequential samples from planktonic populations. Because of the importance of the vertical dimension to zooplankton, tanks in the form of upright cylinders are probably the most useful. Such a tank 2 m in diameter and 12 m deep was constructed at Göteborg (Pettersson et al. 1939*a, b*); it could be filled with nutrient-enriched seawater (which could be filtered and further sterilized with an ultraviolet lamp) of at least three different salinities, cooled by brine-containing coils in the tank, and illuminated from above. Although the published experiments concerned the growth of phytoplankton from an inoculum of unfiltered seawater, survival and phototaxis of copepods in the tank were also noted. Raymont and Miller (1962) studied the breeding and succession of coastal species of phytoplankton and zooplankton in two 20-m³ rectangular tanks containing unfiltered seawater enriched with nitrate and phosphate. McLaren (1969) placed polyethylene columns 1 m in diameter and 30.5 m long in a landlocked fjord in an atttempt to determine the effect of enrichment with nutrients on the population dynamics and production of the entrapped zooplankton populations. A column 3 m in diameter and 17 m deep was used by Davies et al. (in press) in a Scottish loch to examine whether, in such a container, standing stock and productivity of phytoplankton, and abundance and species composition of zooplankton, would agree with these same parameters in the adjacent, unenclosed water column. Similar studies are being conducted by investigators in the CEPEX program in Saanich Inlet on Vancouver Island, British Columbia (Chem. Eng. News 1973). In all of these cases, no attempt was made to select particular species from the natural assemblage for cultivation.

The use of a deep tank at Scripps Institution of Oceanography in the mass cultivation of phytoplankton has previously been described (Strickland et al. 1969); this tank provides an experimental container "between beakers and bays" in size (Strickland 1967). The purpose of the present experiment was to measure the efficiency of transfer of organic carbon in a simple food chain consisting of coastal species at approximately natural concentrations. Primary production was imitated by a measured rate of supply of phytoplankton to the tank, and the yield of zooplankton was

[1] Supported by U.S. Atomic Energy Commission Contract AT(11-1)GEN 10, P.A. 20., and National Science Foundation Grant GA-35507.

413

taken by net, imitating a predator which is relatively nonselective as to prey but can adjust the intensity of its predation. Since the same water column was sampled repeatedly for several months, the food chain efficiency in steady state (e.g. the ratio of the rate of yield from a herbivorous population to a predator to the rate of primary production) can be approximated, under conditions not too dissimilar from nature, by averaging over a long period of time to smooth out fluctuations in age frequency distribution of the herbivorous population and departures from the desired equilibrium between the biomass and primary production (= rate of supply) of phytoplankton and the biomass and grazing pressure of the herbivores. After measuring the food chain efficiency of grazing copepods, we added carnivorous ctenophores and determined the ratio of final yield to primary production for the resulting three-step food chain.

Food chain efficiency and ecological efficiency (the steady state ratio of the rate of yield from a population to its rate of ingestion) are thought to be less than 20% in natural systems (e.g. Slobodkin 1968; Kozlovsky 1968), and such values have been used to calculate potential yield of fish from measurements of primary production in the ocean (e.g. Ryther 1969). The yield of cultivated bivalves is less than 20% of the phytoplankton provided as food (Tenore et al. 1973). However, the relatively high gross efficiency of growth of marine zooplankton (e.g. Mullin and Brooks 1970, for two species of copepods; Reeve 1970, for a chaetognath; Hirota 1972, for a ctenophore) and the apparently minor role of uneaten plant detritus in the plankton, compared with ecosystems such as deciduous forest or salt marsh (see Steele 1972) suggest that marine planktonic food chains may operate with relatively high efficiency.

We established a linear food chain consisting of the chain-forming diatom *Skeletonema costatum*, the copepods *Acartia tonsa* and *Paracalanus parvus*, and ultimately the ctenophore *Pleurobrachia bachei*. Particle-grazing copepods of the genus *Acartia* are amenable to culture in the laboratory using phytoplankton as food (Zillioux and Wilson 1966; Zillioux and Lackie 1970), and population dynamics have been studied in both wild populations and laboratory populations subjected to various intensities of harvesting (Heinle 1966, 1970). The genus is a major constituent of estuarine and coastal plankton and is an important source of food for the ctenophore *Pleurobrachia* (Hirota 1974) and other coastal carnivores. Similar comments apply to the genus *Paracalanus*, species of which have also been cultured (Lawson and Grice 1973; E. R. Brooks personal communication).

The natural history, metabolism, and population dynamics of *P. bachei* in waters off La Jolla, California, have been thoroughly investigated by Hirota (1972, 1974). Further, Hirota (1972) showed that *Pleurobrachia* would survive and grow in the deep tank when natural zooplankton is available as food.

We thank J. D. Powell and his staff for assistance in operating the deep tank; J. Jordan for examining the phytoplankton cultures; J. Hirota, M. Pomerantz, and M. Schonzeit for occasional assistance; and especially C. Sapienza for major assistance in all phases of the experiment. J. R. Beers, M. R. Reeve, and J. H. Steele improved the manuscript with comments, and I. A. McLaren and D. R. Heinle were most helpful referees. We also acknowledge our intellectual indebtedness to J. D. H. Strickland, who pioneered in experiments of this sort in the marine environment.

Methods

The tank is a steel cylinder 10 m deep and 3 m in diameter, coated internally with an inert plastic and surrounded externally by an insulating layer of polyurethane foam. It contained about 70 m³ of seawater from the Scripps Institution of Oceanography seawater system, first filtered through fine diatomaceous earth. We did not use the searchlight or sampling tubes illustrated in Strickland et al. (1969) since it was necessary to darken the tank during our experiment.

Skeletonema costatum was grown in translucent polyethylene tubs, each containing 200 liters of medium, exposed to natural sunlight and cooled by dripping tapwater over their external surfaces, which were wrapped in cheesecloth to act as a wick and thus promote evaporation. The medium consisted of seawater which had been passed through a membrane filter, plus nitrate, phosphate, silicate, trace metals, and vitamins to give final concentrations as in half-strength IMR medium (Eppley et al. 1967). Bacteria-free cultures grown in sterilized IMR medium served as inocula for the tubs. Approximately every 2 weeks, each tub was drained, cleaned with isopropyl alcohol and freshwater, and reinoculated with *Skeletonema*. Twice each day a measured volume of culture was pumped from the tubs into the deep tank, and the concentration of chlorophyll in the culture was estimated from subsamples. Once a week the concentration of particulate organic carbon was also measured, so that the ratio of carbon to chlorophyll for the phytoplankton could be calculated. Filtered seawater and nutrients were added to the tubs each day to restore the culture to its original volume. Microscopic examination indicated little contamination of the culture by other photosynthetic phytoplankton, although heterotrophic flagellates were often present. Carbon: chlorophyll ratios indicated that these were only occasionally an important component of the biomass in the culture.

Chlorophyll *a* and pheophytin in phytoplankton and detritus were analyzed by fluorometry after filtration onto a glass fiber filter (Whatman GF/C) and extraction by grinding in 90% acetone with $MgCO_3$ (Strickland and Parsons 1968). Particulate organic carbon was analyzed by combustion of material retained on a glass fiber filter and measurement of the resulting CO_2 with a Hewlett-Packard C-H-N analyzer. The bodily carbon of ctenophores as a function of equatorial diameter was determined with this instrument, using animals that had not been preserved but were dried at 60°C after their diameters had been measured. Copepodite stages of *Acartia* and *Paracalanus* were analyzed for bodily carbon after preservation in Formalin-seawater; although there is some evidence of loss of organic matter during preservation of zooplankton (Hopkins 1968), we were unable to detect this change in a test using copepods and therefore have used the uncorrected values from preserved animals.

Samples drawn with a siphon so as to integrate the tank from top to bottom were taken three times per week for analysis of particulate chlorophyll *a* and pheophytin and once per week for particulate organic carbon. When necessary, it was assumed that the C: Chl ratio of the culture pumped into the tank was maintained by the phytoplankton in the tank. Discrete samples were taken at 0, 1, 2, 5, and 9 m once a week to study the vertical distribution of chlorophyll and pheophytin.

The tank was covered with an opaque cloth during the experiment so that no photosynthesis was possible; the rate of addition of particulate carbon from the cultures of *Skeletonema* could therefore be equated to primary production. A cooling coil around the upper rim of the tank maintained the mean temperature of the water between 14 and 16°C throughout the experiment and also provided some thermal mixing. On most occasions vertical stratification of chlorophyll was negligible, and there was little gradient in temperature or concentration of oxygen between the surface and the bottom on the two occasions when these properties were measured in discrete samples.

Copepods and ctenophores were harvested three times each week by towing a 0.5-m-diameter net of 103-μ mesh from the bottom of the tank to the surface, each tow filtering about 3% of the volume of the tank. When *Paracalanus* was abundant, tows were also taken with a 20-cm-diameter, 35-μ-mesh net to estimate the abundance of their small nauplii. Comparisons between catches indicated that the 103-μ

net did not quantitatively retain the nauplii of *Acartia* either, so the biomass of this population was underestimated somewhat. Catches were preserved in Formalin-seawater after removal of ctenophores larger than 1.5-mm diameter with coarse netting, and subsamples were counted under a dissecting microscope. Copepods were identified to developmental stage and counted, and eggs and larvae of ctenophores were also counted and measured. The ctenophores retained on the coarse netting were counted and measured while alive; these usually constituted over 90% of the total ctenophore biomass.

About every 10 days, the tank was partially drained from the bottom to compensate for the addition of volume from the *Skeletonema* cultures and to reduce settled detritus. Animals in the effluent were caught on 103-μ mesh, and those thought to be alive (i.e. not visibly decayed) at the time of draining were counted as part of the harvest. Some living phytoplankton was lost in this way also; this loss was not measured, but was about 0.4% of the total crop in the tank per day.

Biomass of copepods was determined by multiplying the number of each developmental stage by the bodily carbon per individual of that stage (the bodily carbon of nauplii and copepodite stages I and II was estimated by extrapolation from data for older stages). Values were adjusted to take account of changes in size of copepods of a given developmental stage during the course of the experiment.

The biomass of the ctenophores was calculated from the optical measurements of equatorial diameter, using the relationship of bodily carbon to diameter (Fig. 1). The bodily carbon of ctenophores larger than 4 mm in diameter was proportional to the 2.35 power of diameter, while the relationship for smaller ctenophores had a less steep slope. Comparison of our Fig. 1 with Hirota's (1974) fig. 2 shows that the bodily carbon of the ctenophores is only about 3–7% of the "organic" (actually, ash-free) weight. Our data and those of Hirota (1972) and Curl (1961) show ash-free

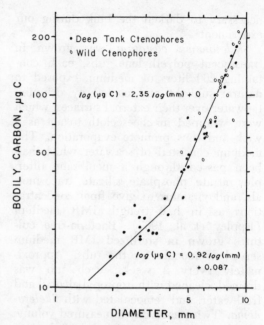

Fig. 1. Bodily carbon of ctenophores as a function of equatorial diameter.

weight of ctenophores to be 23–30% of dry weight. Reeve and Baker (in press) report similar findings for the ctenophore *Mnemiopsis*. They and we have found that the inorganic salts expected in dried ctenophores do not interfere with the analysis of carbon. The ratio of carbon to nitrogen does not indicate any unusual components; this ratio varied from about 4.4 (by weight) in ctenophores 2 to 3 mm in diameter to 3.2 in ctenophores 8 to 9 mm in diameter. We therefore conclude tentatively that much of the weight lost on ashing is not organic matter but some other material, perhaps bound water not driven off by drying at 60°C.

Before the experiment, *Skeletonema* was added to the filtered seawater in the tank for over 2 weeks and the tank was covered with a transparent lid so that a dense crop of phytoplankton was established. Copepods captured by net in La Jolla Bight were then added to the tank; we tried to remove all species other than *A. tonsa* from this inoculum but were not completely successful (*see below*). On 20 April, the tank was covered with an opaque cloth, and

Table 1. Parameters of the input of phytoplankton per cubic meter of tank volume.

Phase	No. of days	C:Chl (mg/mg) median	Pheophytin:Chl median	Carbon (mg m^{-3}day^{-1}) mean	range
I	51	43	0.20	8.2	2.1-25.2
II	37	98	0.21	20.2	5.3-49.8
III	75	37	0.19	11.3	1.2-27.6

initial measurements of stocks in the tank were made on 23 April.

The experiment was divided into three phases; during the first two phases (23 April–20 July) only copepods were present, *A. tonsa* initially dominating but subsequently being replaced by *P. parvus* (phase II). Phase III (26 July–10 October) was that portion of the experiment when ctenophores were present together with *Paracalanus*. Contaminants, notably harpacticoid copepods, were present, but these were usually associated with the walls and bottom of the tank and were of minor importance in the water column and hence have been ignored hereafter. Throughout the experiment, the intensity of harvesting was altered by changing the number of vertical tows taken with the 103-μ-mesh net. During phase III, large ctenophores were also harvested near the surface with a coarse-meshed dipnet.

Results

Table 1 shows the rates of input of particulate organic carbon to the deep tank; the average rates of input of chlorophyll were rather similar for the three phases of the experiment, but the high C:Chl ratio during phase II meant that the rate of input of potential food, as particulate carbon, was high during this phase. This high ratio was due in part to contamination of the phytoplankton cultures by colorless flagellates during this phase in particular. We consider these flagellates as potential food for the copepods, and therefore we have treated them as phytoplankton. On the occasions when flagellates were abundant detrital plant material was also present, as indicated by considerable pheo-

phytin in the cultures, and this also increased the ratio of carbon to chlorophyll. However, the median ratio of pheophytin to chlorophyll in the cultures was similar in the three phases (Table 1).

The rates of input are conceptually equivalent to primary production but are lower than the average rates of 25–50 mg C m^{-3} day^{-1} in nearshore waters off La Jolla in spring and summer (stations I and II: Eppley et al. 1970). The standing crop of phytoplankton in the deep tank (mean = 24 mg C m^{-3}) was also lower than the 50–100 mg C m^{-3} at these same stations, assuming that the C:Chl ratio of the phytoplankton in the tank was the same as that measured in the cultures. Although the concentration of chlorophyll was lower during phase II than during the other phases (Fig. 2), there were no significant differences in the standing crops of phytoplankton carbon during the three phases because of the high ratio of carbon to chlorophyll during phase II. As noted above, there was little evidence of vertical stratification of the phytoplankton in the tank.

Assuming again that the C:Chl ratio was maintained in the deep tank, the contribution of phytoplankton to the total particulate carbon (Fig. 2) can be calculated. The median level of the remaining, presumably detrital, particulate carbon was 140 mg m^{-3}, which is similar to the concentration of detritus in the near-surface waters off La Jolla (Strickland et al. 1970).

During phase I, when *Acartia* was dominant, the biomass of copepods fluctuated considerably and then declined until *Paracalanus* became dominant, thus starting phase II (Fig. 2). The median biomass of copepods during phase I was 41% that of the phytoplankton in the tank. A total of 34 g of phytoplankton carbon was used during this phase; this represents the phytoplankton added to the tank plus the difference between the initial and final biomass of phytoplankton. A total of 4.25 g C could have been harvested from the tank as copepods; this again represents the actual harvest (4.1 g C) plus the harvest

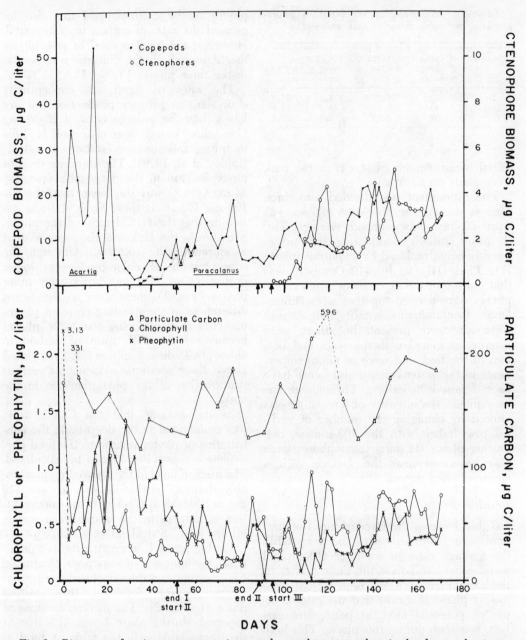

Fig. 2. Biomasses of various components integrated over the water colum in the deep tank. Arrows on the abscissa indicate the start and end of the phases of the experiment referred to in the text. The dashed line under "copepods" indicates the biomasses of *Acartia* and *Paracalanus* during the transition period.

which should have been taken to make the final and initial biomasses of copepods in phase I equal. The food chain efficiency from phytoplankton through copepods to a predator represented by our net was therefore 12.5%. This is less than the gross efficiency of growth of the copepods by the extent to which phytoplankton sank out of the water column unutilized by the co-pepods, and copepods died before being harvested. The harvest was equivalent to removing 10% of the copepod biomass per day.

The median biomass of copepods (almost exclusively *Paracalanus*) was 45% that of the phytoplankton during phase II. Fifty-one grams C of phytoplankton were used, and 5.28 g C potentially harvested as cope-pods, resulting in a food chain efficiency of 10.4%. The harvest was equivalent to 22% of the copepod biomass per day.

It would therefore appear that *Paraca-lanus* has a higher rate of turnover than does *Acartia* but uses its food supply less effectively, resulting in a slightly lower food chain efficiency. This may, of course, reflect the state of the phytoplankton cul-tures which resulted in the high ratio of carbon to chlorophyll during phase II (Ta-ble 1). Had it been assumed, for example, that the "true" ratio (i.e. the ratio for the living phytoplankton which *Paracalanus* used as food, the residual carbon being unutilizable) was 43 as in phase I, the food chain efficiency in phase II would have been 24%.

The rapid replacement of *Acartia* by *Paracalanus* is striking considering the seemingly overwhelming dominance of the former during the first 30 days of the ex-periment. During days 18 to 28 there was almost no reproduction by *Acartia*, but be-tween days 35 and 53, when *Paracalanus* became dominant, *Acartia* nauplii were abundant, although survival through the copepodite stages was poor. In comparison to the first 2 weeks of phase I, the period of reproductive failure of *Acartia* during the third and fourth weeks could not be attributed to a low biomass of food (i.e. chlorophyll) in the tank, either in an abso-lute sense or relative to the biomass of copepods, nor to a low rate of supply of food. The poor survival of *Acartia* and re-placement by *Paracalanus* at the end of phase I also could not be attributed to a low rate of supply or low biomass of food relative to the biomass of copepods, al-though the absolute biomass of food was

low. During the transition from phase I to phase II, therefore, at least as much food was available per unit copepod as when *Acartia* was successful, but the food was more widely dispersed.

It should be noted that the technique of harvest gave a slight advantage to *Para-calanus*, since its nauplii were less well re-tained than those of *Acartia* by the 103-μ mesh. The replacement was not necessarily a purely competitive interaction. The late copepodite stages of *Acartia* can be partly carnivorous (e.g. Petipa et al. 1970), but if this tendency was significant in our ex-periment, *Acartia* must have preyed more heavily on its own young than on those of *Paracalanus*. We found no clear evidence of carnivory in examining the gut contents of female *Paracalanus* from the transition period.

The median biomass of ctenophores in the tank was 16% that of the copepods dur-ing phase III (Fig. 2), and the biomass of copepods was similar to the local biomass of prey species for *Pleurobrachia* in spring, summer, and fall (table 10: Hirota 1974). Fifty-eight grams C of phytoplankton were added over the 75 days, and 1.5 g C were harvested as ctenophores; this harvest was equivalent to 13% of the ctenophore bio-mass per day. By comparison, Hirota (1974) estimated that the production of the population of *Pleurobrachia* off La Jolla was 20% of its biomass per day. The ratio of harvest of ctenophores from the deep tank to input of phytoplankton was 2.6%.

It should be noted, however, that the sit-uation during phase III differed from ideal harvesting of a carnivore in that we also harvested copepods in the process of har-vesting the ctenophores. Making the as-sumption that the ctenophores would have utilized this additional biomass of cope-pods, had it been available to them, as ef-ficiently as the biomass which they did utilize, we can calculate a new ratio of harvest of ctenophores to input of phyto-plankton for a harvest consisting only of ctenophores. The usable harvest produced by *Paracalanus* during phase II was 22% of its biomass per day and this turnover should have continued during phase III, since the temperature and biomass of food were at least as favorable for the growth of the copepods, and the intensity of harvesting was not very different. During phase III, this would mean a harvestable total of 14.9 g C of copepods, of which we harvested 3 g C and the ctenophores presumably harvested 11.9 g C. The ecological efficiency of the ctenophores is thus $1.5/11.9 = 12.6\%$. Therefore, had we harvested only ctenophores, and had the ctenophores captured the copepods which we in fact removed, we could have harvested $0.126 \times 14.9 = 1.88$ g C, or 3.2% of the input of phytoplankton.

If the assumptions used in this calculation are correct, the food chain efficiency from copepods to ctenophores must be considerably higher than the efficiency from copepods to our harvesting estimated for phase II. The ratio of harvest of ctenophores to input of phytoplankton equals the product of the food chain efficiency of the copepods and the ecological efficiency of the ctenophores. The food chain efficiency of the copepods would thus be $0.032/0.126 = 25\%$ during phase III.

This increase in efficiency, if real, could have resulted from an improvement in the quality of food supplied the copepods, or from a pattern of predation by the ctenophores that reduced the relative loss of biomass as unharvested, dead copepods. Also, size-selective predation by the ctenophores on developmental stages having particularly high gross efficiencies of growth might, in principle, contribute to a high food chain efficiency. However, we were unable to demonsrate size-selective predation by comparison of age frequency distributions of *Paracalanus* in phases II and III, in part because of quite variable recruitment during both phases, so the populations did not reach stationary distributions. Strongly selective predation was unlikely, since a wide range of sizes of ctenophores was always present and probably would not greatly affect food chain efficiency in any case, since the gross growth efficiency of copepods does not change markedly with age (Mullin and Brooks 1970).

This experiment also provided data (Fig. 2) on the concentrations of chlorophyll and its degradation product, pheophytin, in a closed container where the input of pigments and the biomass of herbivores were known. Lorenzen (1967) suggested that the ratio of chlorophyll to pheophytin in the euphotic zone could be used as a measure of grazing pressure on the phyotplankton, integrated over some short period representing the mean residence time of fecal pellets (which contain pheophytin) in the euphotic zone. We attempted to maintain constant, rather than varying, levels of biomass in the tank, and hence this experiment was not designed to test Lorenzen's idea, but we examined the data from that point of view. There was no correlation between the ratio of chlorophyll to biomass of copepods (a measure of potential grazing pressure) and the ratio of chlorophyll to pheophytin; further, the former showed no consistent trend through the course of the experiment, while the latter increased significantly as the experiment progressed ($p < 0.01$ by corner test). This is opposite to the trend which would be found if fecal material slowly accumulated in the water column during the experiment. The significant increase in chlorophyll: pheophytin ratio in the tank over time could not be explained by the ratio of these pigments in the cultures being added to the tank; the latter ratio did not change significantly over the course of the experiment, although the trend was toward an increase with time. We conclude that the change in chlorophyll: pheophytin ratio in the tank with time is not entirely explained either by the input to the tank or the herbivorous biomass utilizing this input.

Discussion

The field study of a marine planktonic food chain with which our results can be most directly compared is that of Petipa et al. (1970) in the Black Sea. This was a notable attempt to include information on the vertical distributions, rates of feeding and growth, and changes in trophic status of individual species (or, in the case of phytoplankton, size categories) in a calculation of flow of energy in a planktonic community. *Acartia* and *Paracalanus* were important components of the herbivorous-omnivorous zooplankton, and *Pleurobrachia* was the most important high-level carnivore. In the Black Sea, primary and secondary carnivores (particularly adult pontellid copepods and *Oithona*, and the chaetognath *Sagitta*) were also present. The temperatures ranged from 16–18°C for the "epiplankton" (organisms above the thermocline) and 8–12°C for the "bathyplankton" (below the thermocline).

Combining these two zones, the biomass of herbivores and omnivores was slightly over half that of the phytoplankton (compared with 41–45% in phases I and II of our study), while the biomass of all carnivores was about 30% that of the herbivores and omnivores—much higher than the 16% during our phase III. The yield of herbivores and omnivores to carnivores was 27% of the herbivore-omnivore biomass per day and 17% of the primary production by phytoplankton; the latter can be compared to the 10–12% food chain efficiency which we measured for the copepods in phases I and II and the 25% efficiency estimated for phase III.

The production of carnivores, much of which presumably could have been harvested, was 16% of the carnivore biomass per day (we harvested 13% of the biomass of ctenophores per day during phase III) and 3.1% of the primary production, which agrees well with our values of 2.6–3.2%.

Our results are in general agreement with those obtained in a study of a similar food chain under field conditions. One of the difficulties in using only a single culture vessel, however large, is that establishing a causal relationship between a particular manipulation, such as the introduction of ctenophores, and a particular result, such as the apparent increase in food chain efficiency of *Paracalanus*, is difficult because a parallel control is lacking, and we maintained insufficient control over the supply of food. The increase in efficiency could have been due to the improvement in the condition of the phytoplankton cultures supplied as food during phase III (Table 1) or to a greater efficiency of harvesting by the ctenophores than by our nets. In particular, it would be interesting to investigate the effect on food chain efficiency of harvesting that could be varied in selectivity for size as well as in intensity; this could be done by using nets of different meshes as the harvesting "carnivores," depending on the size frequency distribution of the copepods. More generally, the relationships between rate of supply of food, schedule and intensity of harvesting, and standing crops could be studied, much as the relationships between phytoplankton and nutrients are presently studied in chemostats, the age frequency distribution of the copepods being an additional variable. The work of Heinle (1970) represents a start in this direction.

References

CHEMICAL AND ENGINEERING NEWS. 1973. Effects of pollutants on marine life probed. 17 December, p. 17–23.

CURL, H., JR. 1961. Standing crops of carbon, nitrogen, and phosphorus and transfer between trophic levels, in continental shelf waters south of New York. Rapp. P.-V. Reun., Cons. Int. Explor. Mer **153**: 183–189.

DAVIES, J. M., J. C. GAMBLE, AND J. H. STEELE. In press. Preliminary studies with a large plastic enclosure. Proc. Conf. Estuarine Ecol. (October 1973), Myrtle Beach, S.C.

EPPLEY, R. W., R. H. HOLMES, AND J. D. H. STRICKLAND. 1967. Sinking rates of marine phytoplankton measured with a fluorometer. J. Exp. Mar. Biol. Ecol. **1**: 191–208.

———, F. M. H. REID, AND J. D. H. STRICKLAND. 1970. Estimates of phytoplankton crop size, growth rate, and primary production, p. 33–42. *In* J. D. H. Strickland [ed.], The ecology of the plankton off La Jolla, California, in the period April through September, 1967, Part 3. Bull. Scripps Inst. Oceanogr. **17**.

HEINLE, D. R. 1966. Production of a calanoid copepod, *Acartia tonsa*, in the Patuxent River estuary. Chesapeake Sci. **7**: 59–74.

———. 1970. Population dynamics of exploited cultures of calanoid copepods. Helgol. Wiss. Meeresunters. **20**: 360–372.

HIROTA, J. 1972. Laboratory culture and metabolism of the planktonic ctenophore, *Pleurobrachia bachei* A. Agassiz, p. 465–484. *In*

A. Y. Takenouti [ed.], Biological oceanography of the northern North Pacific Ocean. Idemitsu Shoten.

——. 1974. Quantitative natural history of *Pleurobrachia bachei* A. Agassiz in La Jolla Bight. Fish. Bull. **72**: 295–335.

HOPKINS, T. L. 1968. Carbon and nitrogen content of fresh and preserved *Nematoscelis difficilis*, a euphausiid crustacean. J. Cons., Cons. Int. Explor. Mer **31**: 300–304.

KOZLOVSKY, D. G. 1968. A critical evaluation of the trophic level concept. 1. Ecological efficiencies. Ecology **49**: 48–60.

LAWSON, T. J., AND G. D. GRICE. 1973. The developmental stages of *Paracalanus crassirostris* Dahl 1894 (Copepoda, Calanoida). Crustaceana **24**: 43–56.

LORENZEN, C. J. 1967. Vertical distribution of chlorophyll and phaeo-pigments: Baja California. Deep-Sea Res. **14**: 735–746.

McLAREN, I. A. 1969. Population and production ecology of zooplankton in Ogac Lake, a landlocked fjord on Baffin Island. J. Fish. Res. Bd. Can. **26**: 1485–1559.

MULLIN, M. M., AND E. R. BROOKS. 1970. Growth and metabolism of two planktonic, marine copepods as influenced by temperature and type of food, p. 74–95. *In* J. H. Steele [ed.], Marine food chains. Oliver and Boyd.

PETIPA, T. S., E. V. PAVLOVA, AND G. N. MIRONOV. 1970. The food web structure, utilization and transport of energy by trophic levels in the planktonic communities, p. 142–167. *In* J. H. Steele [ed.], Marine food chains. Oliver and Boyd.

PETTERSSON, H., F. GROSS, AND F. F. KOCZY. 1939*a*. Large-scale plankton cultures. Nature (Lond.) **144**: 332–333.

——, ——, AND ——. 1939*b*. Large-scale plankton cultures. Goteb. K. Vetensk. Vitterhets-Samh. Handl., 5, Ser. B **6**(13): 1–25.

RAYMONT, J. E. G., AND R. S. MILLER. 1962. Production of marine zooplankton with fertilization in an enclosed body of seawater. Int. Rev. Gesamten Hydrobiol. **47**: 169–209.

REEVE, M. R. 1970. The biology of Chaetognatha. 1. Quantitative aspects of growth and egg production in *Sagitta hispida*, p. 168–189. *In* J. H. Steele [ed.], Marine food chains. Oliver and Boyd.

——, AND L. D. BAKER. In press. Production of two planktonic carnivores (chaetognath and ctenophore) in South Florida inshore waters. Fish. Bull.

RYTHER, J. H. 1969. Photosynthesis and fish production in the sea. Science **166**: 72–76.

SLOBODKIN, L. B. 1968. How to be a predator. Am. Zool. **8**: 43–51.

STEELE, J. H. 1972. Factors controlling marine ecosystems, p. 209–221. *In* D. Dyrssen and D. Jagner [eds.], The changing chemistry of the oceans. Wiley-Interscience.

STRICKLAND, J. D. H. 1967. Between beakers and bays. New Sci. 2 February, p. 276–278.

——, O. HOLM-HANSEN, R. W. EPPLEY, AND R. J. LINN. 1969. The use of a deep tank in plankton ecology. 1. Studies of the growth and composition of phytoplankton crops at low nutrient levels. Limnol. Oceanogr. **14**: 23–34.

——, AND T. R. PARSONS. 1968. A practical handbook of seawater analysis. Bull. Fish. Res. Bd. Can. 167. 311 p.

——, L. SOLÓRZANO, AND R. W. EPPLEY. 1970. General introduction, hydrography, and chemistry, p. 1–22. *In* J. D. H. Strickland [ed.], The ecology of the plankton off La Jolla, California, in the period April through September, 1967, Part 1. Bull. Scripps Inst. Oceanogr. **17**.

TENORE, K. R., J. C. GOLDMAN, AND J. P. CLARNER. 1973. The food chain dynamics of the oyster, clam, and mussel in an aquaculture food chain. J. Exp. Mar. Biol. Ecol. **12**: 157–165.

ZILLIOUX, E. J., AND N. F. LACKIE. 1970. Advances in the continuous culture of planktonic copepods. Helgol. Wiss. Meeresunters. **20**: 325–332.

——, AND D. F. WILSON. 1966. Culture of a planktonic calanoid copepod through multiple generations. Science **151**: 996–998.

Submitted: 16 April 1974
Accepted: 1 August 1974

THE STRUCTURE OF AN EXPERIMENTAL INFAUNAL COMMUNITY

Norman J. Blake and H. Perry Jeffries

Graduate School of Oceanography, University of Rhode Island, Kingston, Rhode Island

Abstract: The *Nucula proxima–Nepthys incisa* community of Narragansett Bay, Rhode Island was successfully maintained for up to a year in large wooden tanks. Tank A, the control, received undiluted bay water (S 30–32 $^o/_{oo}$), and tank B received bay water reduced to S 20 $^o/_{oo}$. The structure of both communities in relation to their environment was measured over an annual cycle.

In tank A the density of *Nucula* more than doubled over a twelve-month period, while in tank B the number decreased sharply during spring and then slowly increased to approximately the initial density by the end of the year. The abundance of both *Nepthys incisa* and *Yoldia limatula* showed little change. Fourteen additional species colonized tank A; only four survived in tank B.

Nucula grew 1.0–2.0 mm/yr in tank A and 0.5 mm/yr in tank B; *Yoldia* grew 10.0 mm/yr in A and 5.0 mm/yr in B. *Nepthys* also grew in both tanks, but the rates could not be measured.

Introduction

Marine communities have been studied almost entirely in their natural state. Structure is revealed by periodic sampling, but measurements of function are extremely difficult, often necessitating indirect procedures. This paper reports the successful establishment of an important marine community of Narragansett Bay – the *Nucula proxima–Nepthys incisa* communities – in tanks amenable to experimental control. The behavior of the model system under the stress of reduced salinity is compared with a control. This particular community was chosen because; 1) it is well-known in Long Island Sound (Sanders, 1956, 1958; Carey, 1962) and Buzzards Bay (Sanders, 1960); 2) the major species are deposit-feeders; and 3) only three species are numerically important.

Methods

Sediment was collected from one mile south of Hope Island in the West Passage of Narragansett Bay. *Nucula proxima* (Say), *Yoldia limatula* (Say), and *Nepthys incisa* (Malmgren) comprised more than 85 % of the community in both biomass and numbers. The sediment was brought to the laboratory, sieved through a 1 mm mesh screen to remove the large macrofauna, and spread evenly to a depth of 10 cm in wooden tanks 2 m in diameter and 0.5 m in height. The tanks were then filled with sea water.

The three species were removed from the sediment on a 1 mm mesh screen, counted and placed in each tank. In August 1967 each experimental community contained 5400 *Nucula*/m², 350 *Yoldia*/m², and 350 *Nepthys*/m².

423

Reprinted from J. Exp. Mar. Biol. Ecol. 6:1-14 (1971).

In January 1968 the salinity in tank B was slowly reduced to 20 $^o/_{oo}$, at a rate of 3 $^o/_{oo}$ per week, by diluting full-strength bay water (30–32 $^o/_{oo}$) with tap water. Salinity and temperature measurements were made daily in both tanks. The flow of water in both tanks was held constant at 6 litre/min.

The concentration of dissolved oxygen was determined by the Winkler method (Barnes, 1959) at biweekly intervals. An additional water sample was taken for pH determination using a Sargent pH meter (Model DR).

Sediment samples were taken at monthly intervals for the determination of organic carbon. A glass tube with a 1 cm opening was pushed to the bottom of the sediment in three places in each tank. After removal of the macrofauna, the samples were placed in glass vials containing a few drops of chloroform. The samples were oven-dried at 80 °C for 48 h, and the amount of organic carbon was measured by a modified wet oxidation method (Maciolek, 1962). Samples were taken every three months for grain size analysis. After removal of macrofauna and shells, the sediment was allowed to soak in fresh water to remove salt. The percentage of silt and clay was then determined by the hydrometer method (American Society for Testing and Materials, 1962). From June 1968 to December 1968, measurements were made on the amount of sediment entering each tank. A Petri dish (10 cm in diameter) was suspended 15 cm above the sediment at three positions in each tank. Each week the dishes were removed and the sediment was oven-dried at 80 °C for 48 h. The dry sediment was weighed and stored in glass vials in a vacuum desiccator. At the end of each month the four samples from each location were removed and the percentage organic carbon determined.

The natural community in Narragansett Bay is found in 3 m of water and receives little sunlight. The experimental communities were only 0.5 m deep and had to be covered to prevent large blooms of benthic algae.

For the examination of the animals six random samples, 0.004 m^2 each, were taken from each tank at monthly intervals with a metal cylinder 12.0 cm in length and 7.4 cm in diameter. A rubber tube penetrated the closed end of the cylinder. After the open end had been pushed into the sediment, the rubber tube was closed and the sample removed. The animals were separated from the sediment on a 1 mm mesh screen, identified and counted. *Nucula*, *Yoldia*, and *Nepthys* were saved for further analyses, the remaining species were preserved in formalin, and the sediment was returned to the tanks. Due to the unstable, fluid nature of the sediment, holes created by the sampler disappeared after approximately two weeks. *Nepthys* were measured with a ruler mounted in a plexiglass dish. All of the *Yoldia* and the *Nucula* greater than 3 mm were measured by calipers. The small *Nucula* were measured with an optical micrometer. The dry weights of *Nucula* and *Yoldia* were determined by freeze-drying. About 20 individuals were drained on filter paper, weighed individually, opened with a scalpel, freeze-dried and reweighed.

The amount of shell was determined three times during the year. Freeze-dried *Nucula* were rehydrated and boiled in water until the shells were free of tissue. The shells were then oven-dried and weighed. The dried tissue in *Yoldia* was easily removed with a scalpel and the shell weighed. Each *Nepthys* was blotted dry on filter paper and weighed. They were then freeze-dried and reweighed.

The least squares method (Snedecor, 1956) was used to derive two regression equations from monthly length-weight data (after a log. transformation) for each species. Covariance analyses were made on the slopes and the intercepts of the regression lines. If the slopes were not significantly different ($P < 0.01$), a common regression coefficient was used to determine the weight of an animal of given length in the two salinities. These regression equations, along with the length–frequency distributions and percentage of shell, were used to calculate population biomass.

RESULTS

THE PHYSICAL ENVIRONMENT

Water temperature varied from -0.5 to $22.3\ ^\circ C$, but the maximum difference between the two tanks was only $0.2\ ^\circ C$. During parts of February and March immersion heaters had to be used to prevent freezing. The salinity in tank A varied from 28.5 to 32.0 $^o/_{oo}$. The low value occurred during January as a result of heavy rainfall. By covering the tanks during the following months, the salinity in tank A was maintained between 30.7–32.0 $^o/_{oo}$ with a mean of 31.6 $^o/_{oo}$. By the end of January, the salinity in tank B had reached 20.0 $^o/_{oo}$. Although daily adjustments had to be made to maintain this salinity, the variation was always less than 1 $^o/_{oo}$.

Dissolved oxygen in the overlying water column remained close to saturation throughout the year, ranging from 5.0 ml O_2/litre in July to 9.0 ml O_2/litre in March. The water in tank B usually had slightly more dissolved oxygen than tank A. The pH of the water in both tanks ranged from 8.4 in January to 7.8 in July. Samples taken 1 cm above the sediment surface showed the same pH as that of the surface water.

The median grain size showed little variation in either tank, ranging from 17 μ to 23 μ. The percentage organic carbon of the dried sediment also showed little variation, ranging from 2.00–2.32 %. Large amounts of sediment entered both tanks. In tank A the amount ranged from 132 g/m²/day in July to 162 g/m²/day in October. In tank B the range was 109–138 g/m²/day for the same periods. Although the amount of bay water entering tank B was one-third less than the amount entering tank A, the amount of suspended matter settling in tank B was not one third lower because of the arrangement of hoses carrying sea water to the tanks.

POPULATION DENSITIES AND SPECIES COMPOSITION

Fig. 1 shows the population densities of the three major species. *Nucula proxima* remained the dominant species in both tanks throughout the year. In tank A the number of *Nucula* increased from 4916/m² in January to 11 500/m² the following

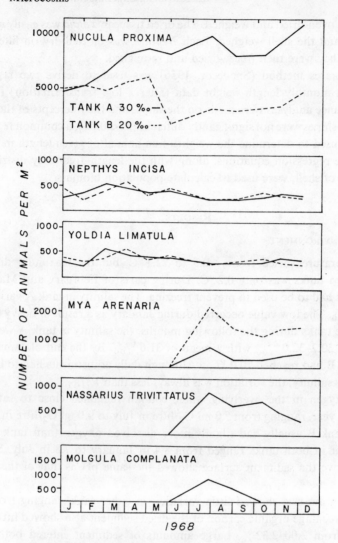

Fig. 1. Population densities of the major species.

December. Density in tank B was about 5000/m² until June, when the number fell to 2000/m² and then slowly increased back up to 5208/m² by December. The mean population density of *Nepthys incisa* and *Yoldia limatula* in both tanks was approximately 400/m², showing little change from the 350/m² originally introduced.

In August 1968, three additional species began to appear consistently in the samples (Fig. 1). *Mya arenaria* was the most abundant of these and was more abundant in the reduced salinity of tank B than in tank A. The density of *Mya* in tank B was 2333/m² in August, and by December 1542/m². In tank A the density of *Mya* was 750/m² in August, and 500/m² in December. The *Mya* in tank B had a slower growth rate than those in tank A. In August the *Mya* in tank B had a mean length of 7.7 mm, compared with a mean length of 9.8 mm in tank A; however, when the two mean lengths were compared by the *t*-test, the difference was not significant at the 5 % level ($t = 1.71$, 81 d.f.). By October the *Mya* in tank B had increased to only 9.3 mm while in tank A the mean length was 14.6 mm; the difference was significant at the 5 % level ($t = 8.60$, 57 d.f.).

Nassarius trivittatus and *Molgula complanata* appeared only in the samples from tank A, reaching their maximum abundance in August. *Molgula* disappeared in October. Table I gives all the species found in the two tanks over the year's sampling period. Of the 17 species occurring in tank A, only seven survived in tank B, and three of these were introduced at the beginning of the experiment.

TABLE I

List of species found ($\times$) in tank A (30 $^o/_{oo}$) and tank B (20 $^o/_{oo}$).

	A	B
Cnidaria		
Cerianthus americanus	$\times$	—
Polychaeta		
Nepthys incisa	$\times$	$\times$
Polydora ligni	$\times$	$\times$
Glycera americana	$\times$	$\times$
Mollusca		
Nucula proxima	$\times$	$\times$
Yoldia limatula	$\times$	$\times$
Mya arenaria	$\times$	$\times$
Nassarius trivittatus	$\times$	—
Gemma gemma	$\times$	—
Mulinia lateralis	$\times$	—
Mytilus edulis	$\times$	$\times$
Macoma tenta	$\times$	—
Petricola pholadiformis	$\times$	—
Tellina agilis	$\times$	—
Cardium sp.	$\times$	—
Solemya velum	$\times$	—
Tunicata		
Molgula complanata	$\times$	—
Total no. of species	17	7

Nucula proxima

Fig. 2 shows the monthly length–frequency distributions of *Nucula proxima* in the two tanks. The distributions show two distinct modes and a shift in modal length due to growth. In January the modal length of the first size-class was 1.35 mm and that of the second 5.35 mm. By December these modes had reached 3.35 mm and 6.35 mm, respectively. Thus, individuals of the first size-class grew 2.00 mm while those

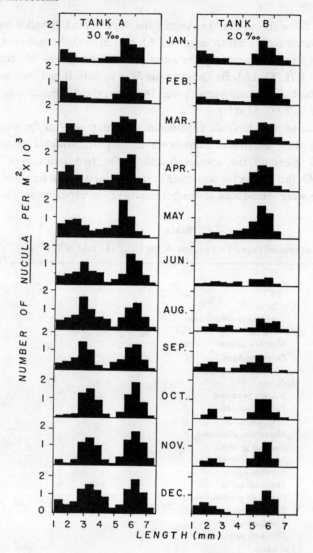

Fig. 2. Length–frequency distributions of *Nucula proxima*.

of the second size-class grew only 1.00 mm. In addition, the number of individuals in tank A increased. In January the number in the first size-class greater than 1.00 mm was 1333/m². By December the number had increased to 5250/m². The second size-class increased from 3583/m² in January to 5208/m² in December. By December a 'zero' size-class appeared containing 1042/m².

Linear growth in tank B was more difficult to determine. In January the modal length of the first size-class was 1.35 mm but the number of individuals in this size-class continually declined from 1750/m² in January to 667/m² in June. By December the first size-class appeared to have joined a 'zero' size-class, the two giving a modal length of 1.85 mm. Thus linear increase of the first size-class in tank B was no more than 0.5 mm.

Length–weight relationships for January, June, and December are given in Table II. From January through March the slopes and intercepts were statistically the same

TABLE II

Length–weight relationships of the three major species in tank A (30 $^o/_{oo}$) and tank B (20 $^o/_{oo}$): the equations are in the form: $\log Y = A \log X + b$, where Y is weight in mg, A is the slope, X is length in mm, and b is the intercept.

		F value of slopes (d.f.) [1]	F value of elevations (d.f.) [1]	Slope [2]	Intercept [2]
Nucula proxima					
Jan.	A	2.15(1:40)NS	2.50(1:41)NS	3.07	−0.809
	B				
June	A	48.11(1:34)**	548.61(1:35)**	2.74	−0.453
	B			2.19	−0.355
Dec.	A	0.01(1:47)NS	0.10(1:48)NS	3.36	−0.912
	B				
Yoldia limatula					
Jan.	A	2.85(1:7) NS	1.68(1:8) NS	2.88	−1.490
	B				
June	A	0.04(1:18)NS	4.03(1:19)NS	2.96	−1.631
	B				
Dec.	A	1.22(1:26)NS	9.16(1:27)**	2.91	−1.544
	B				−1.584
Nepthys incisa					
Jan.	A	4.43(1:13)NS	2.50(1:14)NS	2.40	−2.440
	B				
June	A	7.65(1:19)NS	4.57(1:20)NS	2.29	−2.112
	B				
Dec.	A	7.68(1:21)NS	14.32(1:22)**	1.78	−1.089
	B				−1.200

[1] NS, not significant at the 1 % level; **, significant at the 1 % level.

[2] One common value is given for both tanks if F value is not significant.

for both tanks. But from April through November there were significant differences, and in December the equations were again the same for both tanks. Shell and water made up a large percentage of the total weight. The percentage of water in *Nucula* from tank A ranged from 35.0 to 36.3 %, whereas *Nucula* from tank B contained from 35.5 to 36.5 %. The contribution of the shell to the total dry weight showed little variation in either tank, ranging from 79.5–82.3 % with a mean of 80.1 %. The dry tissue weight of an adult *Nucula* 5 mm in length was calculated for each month using the regression equations derived from covariance analysis and the mean percentage of shell. The results are shown in Fig. 3. In both tanks the dry weight of a 5 mm animal declined from 4.3 mg in January to 2.7 mg in March. Individuals from tank

A then increased to a high of 8.9 mg in October and declined to 5.5 mg in December. In tank B, however, the *Nucula* continued to lose weight reaching a minimum of 1.8 mg in May and then slowly increasing to 5.5 mg in December.

Adult *Nucula* from both tanks were brought into the laboratory in September and placed in sea water taken from the two tanks. Individuals in both salinities spawned,

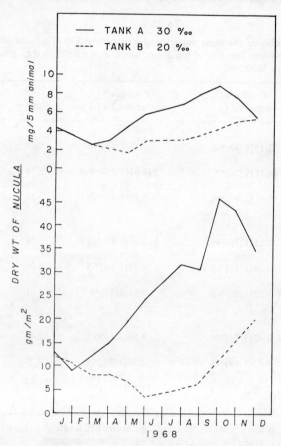

Fig. 3. Individual dry weights and estimated biomass of *Nucula proxima*.

but those in 20 $^o/_{oo}$ took longer to shed their eggs and gave far fewer. Further, only few of the *Nucula* larvae in 20 $^o/_{oo}$ reached metamorphosis.

Fig. 3 also shows the estimated monthly biomass for each tank. In tank A the biomass of *Nucula* increased from 12.9 g/m^2 in January to 46.1 g/m^2 in October and decreased to 34.6 g/m^2 in December. The biomass in tank B decreased from 12.5 g/m^2 in January to 3.4 g/m^2 in June and then increased to 20.1 g/m^2 in December.

Yoldia limatula

Because of the small number of individuals taken in the samples, monthly length–

frequency distributions could not be determined for *Yoldia* as they were for *Nucula*. Instead the samples for each two-month period were pooled. Although a distinct mode was present in most of the curves, more than one year-class was represented as indicated by the wide spread in shell lengths.

In tank A the modal length was 1.75 cm in January, and by August it had reached

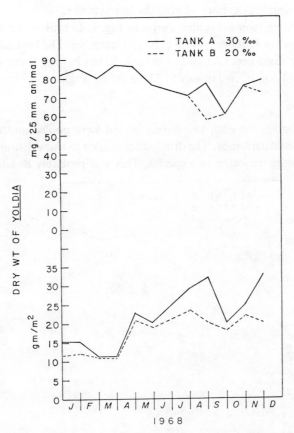

Fig. 4. Individual dry weights and estimated biomass of *Yoldia limatula*.

2.35 cm; in tank B the change was from 1.75 cm to 2.25 cm. Thus the net linear increase of the *Yoldia* population in tank A was 1.0 cm compared with 0.5 cm in the reduced salinity. Table II gives the length–weight relationships for January, June, and December. The slopes were the same for both tanks throughout the year, and only during September and December did the intercepts differ. The amount of water and shell showed little difference at the two salinities. The percentage of water in *Yoldia* ranged from 55.2–58.6 % in tank A and from 55.4–60.6 % in tank B. Shell weight of *Yoldia* in the two tanks ranged from 72.9–81.3 %, with a mean of 76.1 %.

Fig. 4 shows the dry tissue weight of an adult *Yoldia* 25 mm in length. The tissue dry weight was calculated from the length–weight equations of covariance analysis and the mean percentage of shell. In January the dry weight of a 25 mm individual was 82.1 mg and by May had increased to 86.1 mg. By August the dry weight had declined to 71.2 mg and continued to decline in the reduced salinity tank to 58.1 mg in September. By October the dry weights were again the same in both tanks.

Unlike *Nucula*, *Yoldia* failed to spawn when brought into the laboratory, but the amount of gonadal tissue was greatest during the late summer.

The estimated monthly biomass is also shown in Fig. 4. In tank A the biomass increased from 15.2 g/m² in January to 33.9 g/m² in December. The biomass of *Yoldia* in tank B showed the same general trend as in tank A, but in tank B the values were lower, ranging from 11.5 g/m² in January to 23.5 g/m² in August.

Nepthys incisa

The samples of *Nepthys* for each two-month period were pooled and the length–frequency distributions determined. The distributions failed to show distinct modes as did the distributions for the other two species. This was probably due to the large

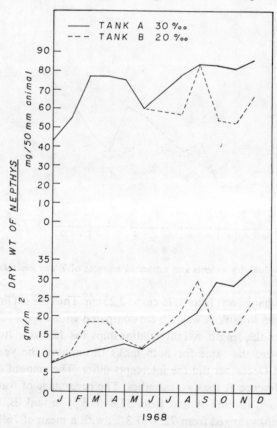

Fig. 5. Individual dry weights and estimated biomass of *Nepthys incisa*.

spread in length of the individuals and the difficulty encountered in measuring the length of a worm. Although some increase in length occurred in both salinities, the rates could not be calculated.

Length–weight relationships for January, June, and December are given in Table II. All of the slopes were similar, and not until August did the intercepts differ. There was little variation in the water content of *Nepthys* in the two salinities. In tank A the percentage of water ranged from 78.8–82.6 % and in tank B from 78.8–85.6 %. The dry tissue weight of *Nepthys*, 50 mm in length, was calculated from the length–weight equations (Fig. 5). From January to March the dry weight of individuals in the two salinities increased from 43.4 to 77.0 mg after which the dry weight remained constant before declining to 60.1 mg in June. Between June and September the dry weight in tank A increased to 83.5 mg. In tank B, however, it continued to decrease to 57.6 mg in August. The calculated increase in dry weight of 50 mm *Nepthys* from tank B in September is probably inaccurate since the error of the September estimate was large and since in October the dry weight in tank B again dropped to 54.1 mg.

Fig. 5 also shows the estimated monthly biomass. In tank A the biomass of *Nepthys* increased from 7.6 g/m² in January to 32.2 g/m² in December, while that in tank B increased from 7.5 g/m² in January to a maximum of 29.7 g/m² in September. This September value is probably inaccurate for the same reasons given above for a 50 mm individual. Since the size of the populations showed little variation over the year, the increase in biomass must be the result of reproductive activity and individual growth.

Discussion

The model community resembled the natural community described by others but, lacking the competitors and predators of the natural environment, the growth rates and estimates of biomass are greater than those reported for the natural environment. The results are summarized in Table III.

Only three species were originally introduced into tank A as adults. Of the remaining 14 species which were taken from tank A over the year, most are indigenous to Narragansett Bay (Phelps, 1958) and Long Island Sound (Sanders, 1956). Many of these were probably introduced with the sediment as individuals less than 1 mm in size. The introduction of these small individuals can explain how the density of the *Nucula* greater than 1 mm could more than double in just one year, yet the other two major species did not show a large increase in population density. The larval stages of *Nucula* and *Yoldia* are very short (Yonge, 1939), allowing both species to remain in the same place as the parent stock. Although nothing is known about the embryology of *Nepthys incisa* (Pettibone, 1963), it probably has a planktonic larval stage which could be flushed from the system.

Only *Mya arenaria* and *Molgula complanata* appeared in numbers far greater than

TABLE III

The structure of the experimental community in tanks compared with that of the natural community in Long Island Sound.

	Experimental community	Long Island Sound	
		Sanders (1956)	Carey (1962)
Population density (number/m²)			
Nucula proxima	4916–11 500	–	9102 (mean)
Yoldia limatula	350–500	–	–
Nepthys incisa	350–500	–	834 (mean)
Growth rates (mm/yr)			
Nucula proxima	1.0–2.0	0.57–0.63	0.38 (mean)
Yoldia limatula	10.0	9.33–10.48	–
Nepthys incisa	–	–	10.0 (mean)
Biomass (g/m²)			
Nucula proxima	12.94–46.13	3.0 (max)	0.390 (mean)
Yoldia limatula	15.17–33.92	~1.4 (max)	–
Nepthys incisa	7.51–32.33	~5.7 (max)	2.92 (mean)

those found in the natural community. These species apparently entered the tank as planktonic larvae and, since the tank lacked filter-feeders, an open niche was available. Currents were fast enough to supply an adequate source of food yet not so fast as to flush the filter-feeders from the sediment. The entry of these macrofaunal species, especially *Mya*, signifies some degree of temporal succession.

All three major species grew in the tanks. The estimated 2.0 mm/yr for the first size-class of *Nucula* and the 1.0 mm/yr for the second size-class are greater than Carey's (1962) estimate of 0.38 mm for *Nucula* in Long Island Sound but are similar to Allen's estimate (Allen, 1953, 1954) of 0.94–1.01 mm/yr for five British species. The 10.0 mm/yr estimate for *Yoldia* in tank A agrees well with Sanders (1956) estimate of 8.5–10.5 mm/yr.

In addition, individuals of all three species showed a seasonal increase in dry weight indicating that the animals were receiving an adequate supply of food. *Nucula* was capable of spawning and *Yoldia* had an increase in gonadal tissue.

The estimates of biomass for the *Nucula* and *Yoldia* populations are not in agreement with those reported by Sanders (1956) and Carey (1962) for the Long Island Sound Community. In January the estimates were an order of magnitude higher than those reported by Sanders. Carey's estimate of 0.390 g/m² for the *Nucula* population is almost two orders of magnitude lower than the January estimate of 12.45 g/m² in tank A. A partial explanation for the discrepancy is the difference in the methods used for determining tissue dry weight. Carey's length–weight regression gives lower weight estimates for *Nucula* than the regressions calculated in the present study. Carey used dilute HCl to remove the shell and thereby determine weight. In the present study, the entire animal was freeze-dried and the tissue dry weight determined. In order to determine which method is more accurate, tissue dry weight was determined in three ways. First, the animals were freeze-dried, weighed, and the shells boiled away from the tissue. The shells were then oven-dried, weighed, and subtracted from the total dry weight. Secondly, individuals were placed in 0.1 N HCl for 8 h to dissolve the shell and the tissue was then oven-dried and weighed. Thirdly, individuals were dissected from the shells, oven-dried and weighed. The freeze-dry method agreed closely (± 3 %) with the dissection method but not with the dilute HCl method (± 10 %). Probably some tissue components are extracted by the HCl. This source of error does not, however, fully explain the discrepancy. If Carey's mean population density of 9102 $Nucula$/m^2 and mean estimated biomass of 0.390 g/m^2 is used to calculate weights, the mean individual would weigh approximately 4.2×10^{-2} mg. If his length–weight regression is then used to derive the length, the mean individual would have a length of approximately 1.5 mm. In the present study, the mean length of $Nucula$ at the start of the experiment was approximately 4.5 mm.

The estimated biomass for the $Nepthys$ population in tank A more than tripled between January and December. The January estimate of 7.55 g/m^2 is the same order of magnitude as the estimates reported for Long Island Sound.

Of the three species originally introduced, only $Nucula$ was affected by the reduced salinity of tank B. The numbers began to decline in January but a marked mortality did not occur until June. Carriker (1955) showed that the mortality rates of some molluscs may increase at low salinities if temperature is high, but they may exist for prolonged periods in reduced salinity if temperature is low: increase in temperature results in increased activity and decreased resistance (Kinne, 1956). Resistance of the three species was greatest during the winter when the temperature was minimal. Since a salinity of 20 $^0/_{00}$ is near the limits of tolerance of $Nucula$, summer temperature is lethal.

Mya $arenaria$ was the only additional species abundant in tank B after June. The density was greater in tank B, but the growth rate was greater in tank A. Mattiessen (1960) showed that the growth rate of Mya was directly related to the concentration of flagellates and that feeding decreases with decreasing salinity. This would explain the slower growth rate in tank B but not the greater density. The reduced salinity may have induced settling of larvae. $Molgula$ and $Nassarius$, although abundant in tank A after June, did not occur in tank B. The larvae either settled and died or postponed settlement (Thorson, 1957) until they were flushed from the system. The growth rate of $Nucula$ in tank B was low but within the range reported by Sanders (1956) and Carey (1962). The growth rate of $Yoldia$ in tank B was less than the rates found in both tank A and Long Island Sound. The slow rate is probably the result of salinity stress.

Kinne (1967) has discussed four physiological mechanisms which assist estuarine organisms to compensate for environmental stress, namely escape, reduction of contact, regulation, and acclimation. Reduction of contact by shell closure may have been responsible for the slow growth rates in tank B. In addition all three species were capable of partial regulation since the percentage of tissue water was only slightly higher in the reduced salinity. Since regulation requires the utilization of energy, slower growth rates could result. Salinity stress was shown by changes in the dry weight of *Nucula* during the spring and of *Yoldia* and *Nepthys* during the summer. During these periods, the species became metabolically and reproductively active. For all three species, the tissue dry weight was significantly lower in the reduced salinity than at 30 $^o/_{oo}$.

ACKNOWLEDGEMENT

This investigation was supported by Federal Water Pollution Administration Grants WP-00858 and 18050-DTX.

REFERENCES

American Society for Testing and Materials, 1962. Tentative method for grain size analysis of soils. ASTM designation, D 422–62 T, Philadelphia, pp. 1272–1283.

ALLEN, J. A., 1953. Observations on *Nucula turgida* Marshall and *N. moorei* Winckworth. *J. mar. biol. Ass. U.K.*, Vol. 31, pp. 515–527.

ALLEN, J. A., 1954. A comparative study of the British species of *Nucula* and *Nuculana*. *J. mar. biol. Ass. U.K.*, Vol. 33, pp. 455–472.

BARNES, H., 1959. *Apparatus and methods of oceanography. Part 1. Chemical.* George Allen and Unwin Ltd., London, 341 pp.

CAREY, A. G., 1962. An ecological study of two benthic animal populations in Long Island Sound. Ph.D. dissertation, Yale University, 61 pp.

CARRIKER, M. R., 1955. Critical review of biology and control of oyster drills *Urosalpinx* and *Eupleara*, *Spec. scient. Rep. U.S. Fish Wildl. Serv., Fish.*, No. 148, 150 pp.

KINNE, O., 1956. Über den Einfluss des Salzgehaltes und der Temperatur auf Wachstum, Form, und Vermehrung bei dem Hydroipolypen *Cordylophora carpia* (Pallas), Athecata, Clevidae. *Zool. Jahrb. Abt. Allgem. Zool.*, Bd 66, S. 565–638.

KINNE, O., 1967. Physiology of estuarine organisms with special reference to salinity and temperature: general aspects. In, *Estuaries*, edited by G. H. Lauff, Amer. Ass. Adv. Sci., Publ. No. 83, Washington, D.C., pp. 525–540.

MACIOLEK, J. A., 1962. Limnological organic analyses by quantitative dichromate oxidation. *U.S. Dept. Int., Bur. Sports Fish. Wildl. Res. Rep.*, No. 60, 61 pp.

MATTIESEN, G. C., 1960. Observations on the ecology of the soft clam, *Mya arenaria*, in a salt pond. *Limnol. Oceanogr.*, Vol. 5, pp. 291–300.

PETTIBONE, M. H., 1963. Marine polychaete worms of the New England region. I. Aphroditidae through Trochochaetidae. *Bull. U.S. natn. Mus.*, No. 227, 356 pp.

PHELPS, D. K., 1958. A quantitative study of the infauna of Narragansett Bay in relation to certain physical and chemical aspects of their environment. M. S. Thesis, University of Rhode Island, 56 pp.

SANDERS, H. L., 1956. Oceanography of Long Island Sound, 1952–1954. X. Biology of marine bottom communities. *Bull. Bingham oceanogr. Coll.*, Vol. 15, pp. 345–414.

SANDERS, H. L., 1958. Benthic studies in Buzzards Bay. I. Animal–sediment relationships. *Limnol. Oceanogr.*, Vol. 3, pp. 245–259.

SANDERS, H. L., 1960. Benthic studies in Buzzards Bay. III. The structure of the soft-bottom community. *Limnol. Oceanogr.*, Vol. 5, pp. 138–152.

SNEDECOR, G. W., 1956. *Statistical methods.* Iowa State University Press, Ames, Iowa, 5th edition, 534 pp.

THORSON, G., 1957. Bottom communities. In, *Treatise on marine ecology and paleoecology, Vol. 1. Ecology*, edited by J. W. Hedgpeth, Geol. Soc. Amer. Mem., No. 67, pp. 461–534.

YONGE, C. M., 1939. The protobranchiate mollusca; a functional interpretation of their structure and evolution. *Phil. Trans. R. Soc. Ser. B*, Vol. 230, pp. 79–147.

Man's Activities

Aquaculture

Introduction

For countless years man has hunted the sea for food. Terrestrial food production, on the other hand, has progressed from the hunter-gatherer type to husbandry and agriculture. Although oysters were cultured in Roman times, it is only recently that the marine equivalent of land-based agriculture has gained much stature, primarily in the Orient. The challenge of sea farming is great, as is its potential. The challenge stems from a lack of sufficient knowledge about species suitable for culture. Feeding habits, life cycles, environmental tolerances, maximum growth rates, and other aspects must all be known for commercial culture. An additional challenge comes from the sea itself. Even a shallow estuary or a calm inlet is not as manageable as a plot of fertile land or a small fish pond.

Fishing has produced an adequate harvest for many centuries, but the amount that can be gathered from the sea without active replenishment is limited. In the early 1950's, the annual world-wide harvest was about 50 million tons, in 1975 it was about 70 million tons. Estimates of the world-wide maximum sustainable yield vary widely, but it probably does not exceed 200 million tons per year. Ryther's paper, in an earlier section, made predictions of fish production that have held up well with time. Overfishing of many of the world's most important species has long been apparent, and management procedures are slow in coming. There are many unexploited fish stocks that will be used in the future, and the so-called trash fish can be a major protein supplement to human as well as to animal diets. There is, however, considerable consumer resistance to these products.

Even in the face of an increased demand for protein from the sea that cannot be met by traditional methods without a tremendous increase in fishing effort, and without endangering many more fish stocks, the prospects of farming the sea look bright. Already mussel and oyster culture are a reality in many countries, shrimp has been cultured for centuries in Southeast Asia, and in Japan, the yellowtail (*Seriola*) is successfully raised in large offshore nets. Bardach, Ryther, and McLarney (1972) have made an impressive compilation of aquaculture ventures, both marine and freshwater, around the world.

In the United States, the only marine organism cultured commercially in any major quantity as of this writing (1975) is the oyster, although some species, particularly salmon, show promise. On the other hand, freshwater culture of trout and channel catfish is an accomplished fact, and commercial ventures have been turning a profit for many years.

The approaches to the culture of marine organisms are many and varied. In many ways, however, they are similar to the microcosms described in the previous section. Bardach's paper in this section reviews the problems and prospects for aquaculture. To a large extent, a single species is raised in an intensive situation, with the control of spawning, predators, and disease being the major problem. The U. S. oyster industry, for instance, collects naturally settling spat or raises larvae in the laboratory, then sets the juvenile oysters in beds on the bottom where starfish and oyster drills are controlled by a variety of chemical or mechanical methods. The salmon industry rears fry up to migratory (smolt) size in a freshwater facility similar to the hatcheries used by government agencies for the past 100 years. One grower (Domsea Farms) is using silos to culture the smolts more efficiently. The salmon are then grown out to market size in net enclosures in protected salt water. Parker's interesting report of the intensive culture of a macroalga in the Phillipines gives insight into how aquaculture can be practiced on a small scale in a "family farm" situation. In the United States much effort is being directed towards the culture of species that are high on the trophic ladder, such as pompano, prawns, and lobster, but success at providing more protein for a hungry world depends upon eliminating as many trophic transfers as possible. The paper by Tenore et al. indicates the dynamics and ecological efficiencies of three commercially important bivalves in an aquaculture situation.

Multispecies systems (polyculture) are still in the experimental stage in the United States, although in the Orient aquatic polyculture has long been practiced. The Environmental Systems Laboratory at Woods Hole is attempting to solve two important problems at once; wastewater treatment and the production of proteinaceous foodstuffs. The article by Ryther et al. reviews the concepts and progress of their research. Another and perhaps even more innovative approach is exemplified in the paper by Othmer and Roels, in which they describe the use of nutrient-rich water pumped from the deep ocean into their system, and from which not only food, but also electrical energy and desalinated water may be produced in a tropical area with access to ample solar energy and a source of deep-sea water.

LITERATURE CITED

Bardach, J. E., J. H. Ryther, and W. O. McLarney. 1972. Aquaculture; the Farming and Husbandry of Freshwater and Marine Organisms. Wiley-Interscience, New York. 868 pp.

Aquaculture

Husbandry of aquatic animals can contribute
increasingly to supplies of high-grade protein food.

John E. Bardach

Optimistic forecasters of the world food supply in the year 2000 predict that serious regional food shortages will exist; pessimists warn of widespread famine as populations in many countries increase more rapidly than agricultural production. The one bright spot in this otherwise gloomy picture is the rate of growth of 8 percent per year of the world fish harvest during the last decade (1). The annual harvest now amounts to 53 million metric tons (2) and is expected to increase for some time to come, though estimates of the possible total sustained yield differ widely, ranging from 120 to 2000 million metric tons. All are based on extrapolations from scanty data on primary productivity throughout the world's seas and on incomplete knowledge of the food relations of many harvested organisms.

Aquatic harvests supply almost nothing but proteins, since the bulk of aquatic plants are plankton algae that are uneconomical to harvest (4) and hold no promise for furnishing carbohydrate staples. A breakdown of the total yield from aquatic ecosystems into marine, fresh, and brackish water moieties shows that the yields from fresh and brackish water make up about 14 percent of the tonnage (2). Much of man's fishing has been in estuaries and bays of the sea, and in ponds, lakes, reservoirs, and streams, where he has long practiced aquaculture (5).

Aquaculture resembles agriculture rather than fisheries in that it does not rely on a common property resource but presumes ownership or leased rights to such bases of production as ponds or portions of, or sites in, bays or other large bodies of water. Products of aquaculture must compete successfully with those of fisheries and of animal husbandry; in Western food economy, aquaculture products such as trout, oysters, and shrimp bring good returns because they fall in the luxury class, whereas in developing countries various kinds of raised fish command a high price (6), since animal protein, including that derived from marine catches, is generally scarce. Although subsidized small home or village ponds may be justified in certain underdeveloped areas to help alleviate malnutrition, aquaculture, wherever it is practiced, should be examined primarily as a commercial enterprise that must compete with other protein supplies to be successful.

The organisms now being raised in aquaculture comprise several bivalve mollusks (mainly oysters of the genera *Ostrea* and *Crassostrea*), a few crustaceans (predominantly shrimp, in particular *Peneus japonicus*), and a limited number of fish species (6). Among the fish species, the carp, *Cyprinus carpio*, and selected other members of the same family, Cyprinidae (minnows), are the most important. Trout and salmon are also important in aquaculture, especially the rainbow trout, *Salmo gairdneri*, as are the Southeast Asian milkfish (*Chanos chanos*), and mullets, especially *Mugil cephalus*, and yellowtail (*Seriola quinqueradiata*), in Japan. Also noteworthy are the channel catfish industry (*Ictalurus punctatus*) in the southern United States and the use of *Tilapia* as pondfish, mainly in Africa (6). Most of these species are adjusted to life in fresh or brackish water, but the culturing of some marine fishes is being attempted, notably in Great Britain with plaice (*Pleuronectes platessa*) and sole (*Solea solea*) (7) and with pelagic (high seas) schooling species in Japan and the United States, among them the Pacific sardine and mackerel (*Sardinops caerula, Pneumatophorus diego*) (8) and the pompano (*Trachinotus carolinus*)

(9). Some attached algae are also produced under semi-cultivation, both in temperate and tropical waters and certain phytoplankton species are cultured as food for oyster and shrimp larvae. These represent a special crop and are of minor or indirect nutritive value; they are omitted from this article, in which the potential of aquaculture for supplying high-quality protein is assessed.

Aquaculture furnishes the world with over 2 million metric tons, mainly fish. Mainland China alone reports annual production of 1.5 million metric tons of carp and carplike fishes (Fig. 1) (6). Two million metric tons represent nearly 4 percent of the total world catch and far exceed the United States, food fish harvest, although they are produced from a fraction of 1 percent of the world's waters.

Aquaculture ranges in intensity from simple weeding of natural stands of algae to complete husbandry of domesticated fish like trout or carp. It is sometimes difficult to distinguish intensive management from culture. The term, as used here, comprises practices that subject organisms to at least one, and usually more than one, manipulation before harvest. In addition, as in agriculture, the harvest in aquaculture takes most, if not all, the organisms tended. Most often only one species is raised, although a few to several compatible species may be cultured simultaneously.

To be productive for husbandry, aquatic animals should have the following characteristics. (i) They should reproduce in captivity or semi-confinement (for example, trout) to make selective breeding possible or yield easily to manipulations that result in the production of their offspring (for example, carp). Failing ease of breeding, their larvae or young should be easily available for gathering (for example, oysters). (ii) Their eggs or larvae or both should be fairly hardy and capable of being hatched or reared under controlled conditions. (iii) The larvae or young should have food habits that can be satisfied by operations to increase their natural foods, or they should be able to take extraneous feeds from their early stages. (iv) They should gain weight fast and nourish themselves entirely or in part from abundantly available food that can be supplied cheaply, or that can be readily produced or increased in the area where the cultured species lives.

Few aquatic organisms have all these attributes, and substantial expansion of

The author is professor in the School of Natural Resources at the University of Michigan, Ann Arbor.

443

Reprinted from Science 161:1098-1106 (1968).

the aquacultural crop depends in part on how biological and engineering skills can make the missing characteristics less crucial; other constraints are economic. I discuss here several operations and problems common to the raising of aquatic organisms (10), and I attempt to appraise realistically the potential of aquaculture on a world scale.

Selective Breeding of Aquatic Stock

Even before Jacob tended Laban's flocks, livestock had been subjected by man to selection for one or another desirable attribute, and breeding of domestic birds and animals has produced spectacular results. The first treatise on fish culture was written in China by Fan Li in 475 B.C. (11), but there are still only two aquatic animals over which genetic control has been exercised. These are carp and several species of trout; trout has a shorter and less varied history of breeding than carp. No true breeding programs exist with invertebrates, though oyster culture is advancing so rapidly that experiments in oyster genetics are likely to begin soon.

The breeding of aquatic animals, compared with terrestrial animals, has peculiar problems. Spawning habits often make the isolation of pairs difficult; isolation of numerous offspring requires many replicate ponds (aquariums are too unlike nature); and there is rarely more than one mating a year. Moreover, the environment has an overriding influence on the growth of poikilothermous animals; consequently, many different-sized animals of the same age are found together. Many aquatic animals require special environmental or social conditions for mating and reproduction, which are not easily duplicated under human control. Manipulations of water temperature or flow have triggered spawning; however, the development during the last two decades of the practice of hypophyzation, or treatment with pituitary hormone (12), to make some fishes spawn helps alleviate contraints on breeding for some species. This practice has influenced fish culture all over the world, from catfish growers in the southern United States and sturgeon breeders in the Ukraine, to the fishpond cooperatives of mainland China, where it is of paramount importance in making common carp produce eggs three times a year and in facilitating the propagation of its cyprinid pond mates, whose eggs were difficult

Fig. 1. Fishponds in Kwan Tung Province, China (F. Anderegg, photo).

to collect in rivers before the process of hypophyzation was developed. Use of pituitary material may also produce advances with the breeding of two species of fish important in brackish water culture—the milkfish and the gray mullet. Aquatic animals have one advantage over terrestrial animals from the breeder's point of view—a pair have large numbers of offspring, which permit mass selection.

Carp are readily adaptable to selective breeding because their eggs are large for fish eggs; they are not too delicate; and they are easily secured. Carp have been bred for fast growth, a body shape with more flesh than is found on the wild type, reduction of scale cover for greater ease in preparing the fish for the table, resistance to disease, and resistance to crowding and to low temperature (13). With such breeding practices as progeny testing (selection of parents according to the performance of their offspring) and diallele analysis (a system of mating that determines separately the genotypes of each parent) (14), further improvements on already well-domesticated strains may be expected. It is necessary to prevent reversions to the wild type. That these can occur rapidly is illustrated by the fate of carp introduced to America. After being brought to the New World in 1877 (15), carp was allowed to escape into lakes and rivers where indiscriminate mixtures of its prolific stock resulted in bony and scaly fish which soon became a nuisance in waters used for game fishing. There was no incentive for carp culture in the United States, where protein was abundant from land livestock. However,

since carp has become a prized angling trophy in Western Europe, and because of the rapid eutrophication of American lakes and rivers (a process which favors carp) and the predicted narrowing of the gap, even in America, between the supply of terrestrial protein and the demand, it is not farfetched to think that carp may be cultured in the United States.

Trout, at least in America, were until recently raised mostly for stream stocking; consequently, disease resistance was the main concern in hatcheries not equipped for experiments in fish genetics. Demonstration of what may reside in the trout's gene pool has come mainly from two sources: the Danish table trout industry (16) and the experimental trout and salmon breeding program of the University of Washington in Seattle (17), where specially fed rainbow trout stock, continuously graded by selection during 30 years, grows to as much as 3 kilograms in 18 months while a wild rainbow trout in a lake at that age rarely weighs 200 grams (Fig. 2) (18); these fish tolerate higher temperatures than their wild congeners.

Trout and salmon eggs are larger than those of the carp; they develop slowly and are hardy, combining several advantageous properties. Salmon permit the establishment of hatcheries on suitable streams because they return to spawn to the stream with the odor of which they were imprinted as fry. In such hatcheries inadvertent selection from the spawning run of the largest—fastest growing—brood fish has produced strains that returned to the hatchery 1 year earlier than the offspring of their wild congeners. Salmonid fishes can be selected for higher fecundity, larger egg size, and better survival and faster growth of fry, and for exact timing of their return to the parent stream (19). These breeding potentials should be used to increase the abundance of salmon especially since improved techniques now feasible in United States salmon hatcheries could produce about ten times as many young fish as are now released (20).

Salmon-fishing regulations are still based on propagation potentials in natural streams and require that 50 percent of the run be allowed to escape the fishery. Salmon runs will increasingly depend on hatcheries that program their fish to return for stripping and the raising of a well-protected progeny whose rate of survival at the time of release is many times greater than that attained in nature, where maximum fish

Fig. 2. Wild and mass-selected, hatchery-fed rainbow trout, at the University of Washington School of Fisheries, Seattle. Fish are 2 years old, and the large one is the result of over 30 years of selective breeding. Their respective lengths can be estimated from the diameter of the bucket base, which is 22 centimeters.

mortality takes place during the first few months of life. Thus, the salmon harvest of certain river mouths may almost be doubled in view of the fact that hatchery-dependent runs need only a few fish to supply the next generation. Salmon are highly valuable fish [$65 million for the United States catch in 1965 (21)], and it may be worthwhile to press for regional revision of escapement regulations and to examine the economic requirements and consequences of hatchery improvements.

Another advantage of breeding fishes is the ease with which many of them hybridize (22). At the University of Washington at Seattle, male steelhead (that is, seagoing rainbow) trout were crossed with fast-growing freshwater rainbow females. The growth rate of the offspring was intermediate between that of the parents, their shape was more fusiform than that of the female, and they migrated to sea. They had a voracious appetite and adopted parent streams to which they returned as 2-year-olds, weighing 2 to 3 kilograms on the average and occasionally as much as 5 kilograms (23). They would probably not breed true in the second generation, and they should therefore be hatchery-produced, but they represent an interesting use of the sea's unused fish food.

Difficult as it may be to raise the progeny of one pair of parents of carp and trout, to do so with oysters is still more complicated. Mass spawning is usually done on oyster beds, and although the female of the genus *Crassostrea*, to which the American oyster belongs, retains the eggs inside her shells until after fertilization, paternity on the oyster bed is impossible to ascer-

tain. Although there are thousands of eggs for each carp or trout, there can be up to 100 million for each female oyster (24). This fact, however, has aided in mass selection. Progressive growers of Long Island oysters raise the larvae in warmed water and use cultured algal food. They also give proper attention to stirring and other manipulations simulating planktonic conditions. The many eggs and improved survival of free-floating larvae permit a filter screen to be used to select only 20 percent of the largest, most rapidly growing early larvae. These larvae exhibit good growth throughout life (6). But to achieve true selective breeding, growers of Long Island oysters now plan to rear single oyster progeny; since oysters reverse their sex from male to female halfway through their adult lives, the possibility of freezing sperm from a functional male is being tested, and it may be possible to use it to fertilize the same oyster later when it becomes a female (25).

The Raising of Aquatic Larvae

Many aquatic animals go through larval stages which do not resemble their adult phase; some larvae, including those of shrimp or oysters, and of many fishes, are planktonic and minute and feed on the smallest organisms. More than with domestic birds or mammals, nursing them through their early lives poses difficult technical and nutritional problems to growers. In British experiments with raising plaice and sole larvae in captivity, as much as 66 percent survival through the stage of metamorphosis has been accomplished.

Ultraviolet radiation of the water decreased the danger of bacterial infection; tanks without corners minimized encounters with solid obstacles; and salinity, temperature, and pH were controlled. The size of the first food offered was geared to the tiny mouthparts of the larvae, but was increased with their capacity to take larger live food. Nauplii of the barnacle (*Balanus balanoides*) were used at first; they were replaced by nauplii of the brine shrimp (*Artemia*) with subsequent admixtures of small oligochaetes (*Enchytrea*) when the small fishes had metamorphosed and were resting on the bottom. Finally, chopped mussels (*Mytilus*) were used. Since plaice larvae, just before settling, consume 200 brine shrimp nauplii per day, the production of several hundred thousand young plaice posed serious technical problems in continuous food culture (7).

Obtaining and correctly supplying food was a significant part of the experiments at the U.S. Bureau of Fisheries at La Jolla, California, with larvae of high seas schooling species such as Pacific sardines and mackerels. In these experiments very small food organisms had to be supplied at the precise time of complete yolk absorption, and in sufficient quantities to allow larvae of limited mobility to find food in all parts of the aquarium. Because sardine larvae search in only about 1 cubic centimeter of water per hour, at the onset of feeding, but require a minimum of four food organisms per hour to replace energy lost in swimming and body functions, the rearing of 2000 larvae in 1,800,000 cubic centimeters of water (500 gallons) meant replacing 7,200,000 food organisms removed by larval predation each hour or approximately 86,400,000 food organisms during a 12-hour day.

The large quantities of food organisms in varying sizes needed for these experiments were collected mostly at night. A 1000-watt underwater lamp connected to a submersible pump was suspended several feet below the surface of the sea. Copepods were attracted from a wide distance and concentrated near the pump where they were sucked up with water and transported to the surface. Plankton-enriched water was then passed through a series of filters, which further concentrated food organisms, and the highly enriched filtrate was piped to a 760 liter storage tank. Organisms with a cross-sectional diameter of 0.028 millimeter and larger were thereby collected. Before being fed to fish larvae, concentrated plankton was graded by filters to remove organisms larger than 0.1 millimeter. The portion containing large copepods, crab larvae, chaetognaths, and the like was fed to advanced fish fry and juveniles (*6, 8*) (cover photo).

Comparable techniques may help to achieve survival, after forced spawning in captivity, of milkfish and mullet. Inasmuch as these two species of economic importance in Asia are now raised from fry collected on the shores (Fig. 3) and as the fry are becoming scarce regionally, domestication of the two species, including manipulations ensuring high survival of fry will be an important advance for fish culture.

Although fish larvae are recognizable as fish even though they are not like the adults, invertebrates undergo more profound transformations from egg to adult. Oysters spend their first 2 weeks before they "set" as ciliated trochophores and veliger larvae needing flagellate algae for food. About 2000 cells of two or more species, for instance *Isochrysis galbana, Monochrysis lutheri,* and *Rhodomonas* and *Nannochloris* species, have to be available for each larva per day, and larger species are required to replace the smaller species as the larvae grow. Algae must be cultured en masse when oyster larvae are raised indoors, an innovation largely developed at the U.S. Bureau of Commercial Fisheries Biological Laboratory at Milford, Connecticut (*26*), and now expanded by progressive growers of Long Island oysters.

In shrimp raising, which is successful on a commercial scale only in Japan, there are problems with the larval stages before the animals can be fed chopped trashfish and shellfish. The operation, initiated by M. Fujinaga in 1934, begins in the spring with the collection of "berried" (egg carrying) females ready to release the stored spermatophores from their seminal receptacles; raising the water temperature speeds this and subsequent processes. After three distinctly different planktonic stages and 12 molts in about as many days, the postlarvae begin to crawl on the bottom; they still have to undergo some transformations and another 20 molts before they become adults. The early part of the life cycle of the cultured shrimp takes place indoors in ceramic tile-lined wooden tanks and in water heated to between 26° and 30°C. Diatom—mainly *Skeletonema costatum*—and flagellate cultures are maintained for feeding the early larvae,

Fig. 3. Catching of milkfish fry on the coast of East Java (R. V. D. Sterling, photo).

which are later given finely chopped mussel or clam flesh. When they have reached a length of between 15 and 20 millimeters or a weight of about 10 milligrams, they are stocked in outside ponds with arrangements for aeration and circulation (Fig. 4).

In October or November, the shrimp, though not fully grown, are ready for market. They are about 10 centimeters long and weigh 20 grams having been fed once daily, converting 10 to 12 kilograms of food into 1 kilogram of shrimp. When the water later cools down to below 15°C, the animals no longer feed, but many of them may be retained without feeding for a later more favorable market (*6*).

The oldest shrimp-farming enterprise is now located near Takamatsu on Shikoku. It covers almost 10 hectares and has a staff of 30 men, including some in management research. Ten million shrimp were produced there in 1967, a quarter of which were raised to adult size; the rest were sold for stocking. The cost of production of cultured shrimp is certainly higher than that in any other aquacultural enterprise, but the wholesale price in Japan for tempura-sized shrimp of 6 to 10 centimeters is between $12 and $13 per kilogram, and the supply does not meet the demand. Shrimp farming of this type in a country whose material or labor costs are less favorable than those of Japan

Fig. 4. The 28 running-water ponds (91 by 9.1 meters) for culturing adult shrimp at the Shrimp Farming Co., Takamatsu, Japan.

would not be possible (6). There are, however, opportunities for greater mechanization and for feeding innovations that will simplify the most laborious parts of culture operations for larval as well as postlarval shrimp. The use of the most improved shrimp-culturing methods with fast-growing species may hold some promise for a number of regions in the world (27).

Making Full Use of the Water

Aside from selecting the best suited strains, a practice not yet widely followed in aquatic husbandry, aquaculture should make use of the entire

water column where possible and be three-dimensional, as it is in China and other Asian countries, where common carp is stocked with other species of the minnow family (Cyprinidae) such as the grass carp (*Ctenopharyngodon idella*), the silver carp (*Hypothalmichthys molitrix*), and the bighead carp (*Aristichthys nobilis*) (28). The success of this method is based on the different food habits of the respective species; the carp is a bottom feeder; the grass carp and the silver carp feed on plants (banana leaves, even) and beanmeal or rice bran supplied to them from outside the ponds; and the bighead carp uses the plankton surplus in the well-fertilized water. Thus, the va-

rious water layers and all potential food sources are used (29).

The culture of oysters in Japan's best oyster-growing district, Hiroshima Bay, also is an illustration of the use of the entire water column (6). Seed oysters are collected on scallop shells suspended on wires from a bamboo framework driven into the bottom (Fig. 5). Biologists from the prefectural and municipal laboratories monitor the plankton during the spawning period and advise the growers on the best time for spat collection. It is not uncommon to collect several thousand spat per scallop shell, although the average is about 200. The shells are removed from the collecting frames after 1 month when the surviving oysters have reached a size of about 12 millimeters. They are then cleaned, culled, and restrung on heavier wires separated by bamboo (and more recently by plastic) spacers (Fig. 6). These wire rens are suspended from bamboo rafts, buoyed by floats of various kinds, and extend to a depth of 10 to 15 meters. Floats are added as the oysters grow, and before harvest require several times the support they needed at the beginning of the growing season. Long lines instead of rafts are an innovation in the method of suspension, but they are still only a variant on the hanging-culture technique, which uses the water column efficiently and which protects the oyster from its bottom-living predators, such as starfish and oyster drills.

A typical raft, about 20 by 25 meters carries 600 rens and produces more than 4 metric tons of shucked oyster meat per year (Fig. 7). On a per-hectare basis, this harvest amounts to 58 metric tons, if it is assumed that only one-fourth of a certain area of intensive cultivation is covered by rafts, as is the current practice. Such yields result from intensive care and high primary productivity in the water that is dependent on tidal exchange and fertile terrestrial runoff. By comparison, the average is 5 metric tons per hectare of well-managed, leased oyster ground in the United States and the peak harvest of 300 metric tons per hectare of mussels (*Mytilus edulis*) also grown with hanging culture in the bays of Galicia in Spain. On public oyster grounds in the United States, where the mollusks are a minimally managed common property resource, the average per hectare is only 10 kilograms (0.001 metric ton) or less (6).

Fig. 5. Seed-oyster production near Hiroshima, Japan. (Top) general view of area; (bottom) detail of above; (bottom right) close-up of oysters on scallop shell, several weeks after setting (J. H. Ryther and author, photo).

Fertilization of bays, fjords, or enclosures has led to increases in phyto- and zooplanktons, but favorable cost-benefit ratios for use with fish have not been proved (30). In ponds (including brackish ones) organic and inorganic fertilization has been efficacious. In Israel, fertilized carp ponds, some with admixtures of *Tilapia* and mullet, produce twice the tonnage per hectare of unfertilized ponds, and fertilization and additional feeding doubles the yield again (31). Fertilized *Tilapia* ponds in which the fish were also fed have yielded as much as ten times the crop of unenriched ones (32).

Many kinds of inorganic or organic fertilizers can be used, but sewage which produces dense invertebrate populations certainly works well. Munich sewage ponds with a slow exchange of water produce 500 kilograms of carp per hectare per year and a profit for their operator, the Bavarian Hydropower Company; the method requires large tracts of land, however; under temperate conditions rising land values threaten to make it obsolete (6).

In a much warmer, rapidly flowing stream in West Java, with a high sewage content, carp, confined in bamboo cages to graze on the dense carpet of worms and insect larvae in the sandy substrate, grow rapidly to yield 50 or more kilograms of fish per square meter of cage surface, or 500 metric tons per hectare (33). Even with allowances made for only partial use of the stream surfaces, this practice clearly represents an extremely efficient and ecologically sound use of sewage, especially in warm waters. The main drawbacks to this practice arise because the fish are not always well cooked before they are eaten.

In addition to the fertility of the water, its temperature, especially in a colder climate, is also very important. The most spectacular use of naturally warmed water for fish culture is in Idaho's Snake River valley trout-farming district, where springs of an even 16°C (optimum temperature for trout) gush forth from the canyon wall year in and year out. A thousand tons of trout can be raised in a year on every 2830 liters per second (100 cubic feet per second) that flow from these springs. Such unprecedented results in fish husbandry depend on high-density stocking, fast growth, mechanization, and cheap feed—the latter being locally

Fig. 6. Oyster raft culture in Hiroshima Bay, Japan; the scallop shells to which the growing oysters are attached are being spaced more widely as they are restrung, after cleaning.

procurable since the Snake River valley is also a stock feed-growing area (34). Most of all, however, the high yield depends on the flushing of growth-inhibiting wastes from the trout raceways. Hence, it is more appropriate to relate weight gain to water flow rather than to water surface or volume. By such a measurement, production would be around 170 kilograms per liter per second.

Naturally warmed water is not prevalent, but man-made heated effluents occur with increasing frequency. In fact, thermal pollution may become a

threat to some natural waters because it hastens eutrophication. Heated power plant effluents, however, can also be used to the advantage of the aquaculturist. At the atomic energy plant at Hunterston, Scotland, cooling water, ascertained to be nontoxic to fish, was fed into cement troughs for sole and plaice raising. Both species were grown to marketable sizes in 6 to 8 months at between 15° and 20°C, as compared with the 3 or 4 years needed for the same growth under natural conditions (35).

A progressive grower of Long Island shellfish used about 57,000 liters per minute of cooling water discharge of the Long Island Lighting Company. The cooling water is taken from a deep section of the bay and has a high nutrient content, which favors oyster growth as does its warmth. Year-round production in a near 3-hectare lagoon of both oysters and hard clams (*Mercenaria mercenaria*) has been achieved, and seed oyster production in the heated lagoon promises to be highly successful. Summer water temperatures above 30°C, first feared to arrest growth or to be lethal, in fact, promoted exceptionally rapid growth (6, 25). At the atomic plant at Turkey Point (Florida), replicate feeding trials by the University of Miami with shrimp (*Peneus duorarum*) and pompano (*Trachinotus carolinus*) are in progress to compare the effects of different levels of water temperature and consequently of different levels of heated water admixture (9). Heated waste water is also used for freshwater fish culture in the Soviet Union (36).

Fig. 7. Harvesting oysters in Hiroshima Bay, Japan. The rens of scallop shells with their attached oysters are strung from the boom (Japan Fish Agency, photo).

Status and Potential of Aquaculture

Aquaculture, practiced with a far wider range of species than mentioned here, is found in most of the world. In many areas it occurs at a subsistence level, and its potential contribution to the food supply has not been assessed (37). Village ponds, once a hopeful development in Africa, for instance, are now in disrepair and their potential is not being realized (32). Local fish ponds can be important, however, as has been demonstrated in Taiwan, mainland China, and Indonesia (6).

Husbandry of aquatic animals brings increasing financial returns as it is practiced on a larger scale. Culture intensities vary, as do the fixed and variable costs of the operations and the yields (Table 1). From a commercial point of view, the return on the investment is of most interest; in milkfish culture the annual return ranges from 10 to 20 percent or more, and increases with the intensity of cultivation. Malayan mixed pig and fish farms yield 30 percent, and similar returns are noted in the oyster business (6).

Aquaculture can be not only a lucrative business but it may even produce yields high compared with the harvest of comparably sized land surfaces. The relative scarcity of such peaks in aquacultural production, especially in the tropics, is caused by a lack of biotechnical engineering and managerial skills, the absence of suitable credit or seed capital for even low-cost installa-

tions, and the absence of transport and marketing facilities that might encourage the development of a product for a certain market, and so forth.

This is well illustrated by a comparison of Indonesian and Taiwanese milkfish culture in brackish water. Milkfish (Fig. 8) feed predominantly on bluegreen algae and are raised in pond complexes on land cleared of mangroves. Canals permit the control of water level and salinity by means of sluices, which regulate tidal or freshwater flow (38) (Fig. 9). Average Indonesian and Philippine annual harvests are 300 to 400 kilograms per hectare, whereas Taiwanese milkfish raisers attain nearly 2000 kilograms on the average, in spite of a cooler climate (39). Cooperatives, rural reconstruction agencies, a good layout of the farms, control of predators of the fry, some fertilization, and prevention of siltation of ponds and connecting water bodies are some of the secrets of successful milkfish farming in Taiwan. For similar reasons there occur in mollusk culture the aforementioned wide range of yields, from nearly 10 kilograms per hectare on public oyster grounds in the United States to the 58,000 kilograms per hectare in Hiroshima Bay.

Filter feeding mollusks and milkfish are brackish water plankton- or algae-feeders, respectively. These hold more promise for protein-deficient regions than do the carnivores of the same environment because it is more sound to increase the fertility of the water

than to produce extraneous feed, let alone to raise one aquatic animal with scrap from another, which is perhaps already being used, or could be used, directly for human consumption.

Most products of aquaculture could be called luxury foods, whether they are sold as high-priced items in a food economy with wide consumer choice (for example, shrimp in Japan, trout in the United States) or boost the scant animal protein supply of developing nations (for example, milkfish in Southeast Asia), where they also bring a good return to the producer. It might seem unrealistic, therefore, to expect aquaculture to help alleviate the world protein deficiency, but such is not necessarily the case. Luxury foods stop being a luxury when they can be mass produced, a case well exemplified by the broiler chicken industry in the United States and Western Europe.

Differences in biology between chickens and aquatic animals notwithstanding, some of the latter could well become mass-produced cheap and abundant foods at conversion rates of two parts of dry feed to one part of fish flesh. Among fresh and brackish water fish, especially trout, carp, and catfish can be raised with pellets. Chinese carp and certain tilapias eat leaves and stems of leafy plants; other fish feed on algae. In Southeast Asia well over 200,000 hectares of ponds now lie in former mangrove areas; there are in the tropics vast unused mangrove regions, some of which could be turned into pond com-

Table 1. Selected ranges of aquaculturals yields (6) per year. Results are given in kilograms per hectare except as noted. The value is in dollars per hectare except as noted.

Type of cultivation	Location	Yield	Approximate wholesale value of annual crop
Oyster			
Common property resource (public grounds)	U.S.	9	38
Intensive cultivation, heated hatchery, larval feeding	U.S.	5,000	21,000
Intensive care, hanging culture	Japan*	58,000	67,000
Mussels			
Intensive care, hanging culture	Spain*	300,000	49,000
Shrimp			
Extensive, no fertilization, no feeding	S.E. Asia	1,000	1,200
Very intensive, complete feeding	Japan	6,000	43,000
Carp			
Fertilized ponds, sewage ponds	Israel	500	600
	S. Germany	500	
Fertilized ponds, accessory feeding	Israel	2,100	
Sewage streams, fast running	Indonesia*	125,000	
Recirculating water, intensive feeding	Japan	100†	114†
Catfish			
Ponds, no fertilization or feeding	Southern U.S.	200	70
With fertilization and feeding in slowly flowing water		3,400	2,400 (net profit 300)
Milkfish			
Brackish ponds, extensive management	Indonesia	400	
With fertilization and intensive care		2,000	600
Trout			
Cement raceways, intensive feeding, rapid flow	U.S.	170†	168†

*Values for raft culture and comparable intensive practices based on 25 percent of the area being occupied. † Per liter per second.

plexes for the culture of fish. Mollusk production, though limited eventually by the suitability of grounds, could be expanded, and above all intensified in the areas where it is now prevalent. Aquaculture is only beginning to develop such practices as manipulation of the temperature regime to achieve best growth, devising simple automated feeders that fish can learn to activate themselves, and building machines that simplify harvesting. Several disciplines are expected to contribute to the development of aquaculture. Since intensive husbandry alters the conditions of nature, a knowledge of the ecology of the cultured organisms in both natural and artificial states is essential. Engineering can also make increasingly important contributions to aquaculture development as it has in the successful pilot-scale raising of plaice and sole by the British Whitefish Authority (40). It was the basis for the as yet theoretical calculation that "the annual British catch of plaice could be housed in shallow ponds covering 1¼ square miles in extent" (7).

Japanese yellowtail fish are now raised at high density, and with sequential cropping have already achieved yields of 28 kilograms per square meter (280 metric tons per hectare) and shown that it is economical to use small portions of the sea under very intensive management. The success with this oceanic schooling species and the fact that other species of similar habits had become adapted to confinement led to the speculation that still others, such as tuna, might behave similarly and that their mass culture under controlled conditions might become possible. In fact, Inoue of the Fisheries Research Laboratory, Tokay University, Japan, urged that Japan take the initiative in launching a tuna-rearing project in the equatorial Pacific, where atolls and lagoons could be used as sea farms (8).

Such projections say nothing of the problems of translating small to modest enterprises into much vaster ones—the main one likely to be the procurement of many millions of tons of suitable food. Trashfish, in part now used for fish meal, krill and other marine organisms lower in the food chain than the highly prized fish to be cultured have been thought suitable, provided that they can be produced at a low enough cost. The theoretical potential of marine fish culture also rests on the assumption that marine fish can be induced to function sexually under artificial conditions, as have many freshwater fish.

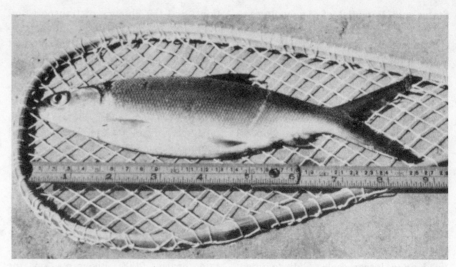

Fig. 8. The herbivorous brackish water milkfish of Southeast Asia; the specimen is about 6 months old (Department of Fisheries, Taiwan, photo).

Hormone stimulation is expected to be one of the solutions to this problem along with rearing an initial breeding stock born and adjusted to life in artificial environments.

But even without further advances through research, a considerable increase of aquacultural yields appears attainable soon by consistent application of already known techniques on inefficiently managed fresh and brackish water bodies. It has been advocated (4) that millions of hectares of ponds be constructed in Asia, Latin America, and Africa to help satisfy the protein needs of these areas. If local economic and socio-political constraints were removed, these new waters and the upgrading of presently existing ones could yield by the year 2000 a harvest of 30 to 40 million metric tons (3, 41) produced near areas of need, which are still likely to lack refrigeration.

Long-term and large-scale projections of yields attainable through practicing aquaculture with marine animals, outside the brackish water zone, can hardly be attempted; true mariculture is in its infancy. However, experiments in several locations have established that it is technically feasible, and no doubt the intensive development and success of brackish water aquaculture will lead to further efforts to develop mariculture on a large scale. It is too early, however, to tell where or under what conditions such efforts could become economically sound.

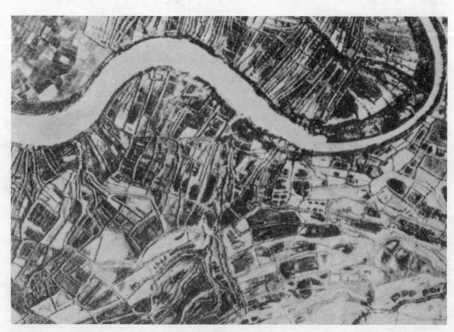

Fig. 9. Milkfish pond complexes near the coast of East Java (W. H. Schuster, photo).

Summary

The role of aquaculture in producing high-grade animal proteins for human nutrition is discussed. Raising and tending aquatic animals is mainly practiced in fresh and brackish waters although there are promising pilot experiments and a few commercial applications of true mariculture. Yields vary with the organisms under culture and the intensity of the husbanding care bestowed on them. The products are now mainly luxury foods, but there are some indications that upgrading of the frequently primitive culture methods now in use could lead to increasing yields per unit of effort and to reduced production costs per unit of weight. Under favorable conditions, production of animal flesh from a unit volume of water far exceeds that attained from a unit surface of ground. With high-density stocking of aquatic animals flushing is important, and flowing water or tidal exchange is essential. Combinations of biological and engineering skills are necessary for full exploitation of aquacultural potentials; these are only partially realized because economic incentives may be lacking to tend aquatic organisms rather than to secure them from wild stocks, because of social, cultural, and political constraints. Nevertheless, a substantial development of aquaculture should occur in the next three decades and with it a severalfold increase in total yield.

References and Notes

1. W. M. Chapman, *Food Technol.* 20, 895 (1966).
2. Editorial, *Yearb. Fish. Stat.* 20 (1965).
3. President's Science Advisory Council, *The World Food Problem* (U.S. Government Printing Office, Washington, D.C., 1967), vol. 2, pp. 345–361.
4. L. A. Walford, *Living Resources of the Sea* (Ronald, New York, 1958), pp. 121–132.
5. I prefer the spelling aquaculture to aquiculture because the former is etymologically more correct.
6. In 1970, 23,000 metric tons of channel catfish are expected to be harvested in the lower Mississippi River states, and this production will double again in 1972 [editorial, *Comm. Fish. Rev.* 30 (5), 18 (1968)]; J. E. Bardach and J. H. Ryther, *The Status and Potential of Aquaculture, Particularly Fish Culture,* prepared for National Council on Marine Resources and Engineering Development 1967, PB 177768 (Clearinghouse Fed. Sci. Tech. Info., Springfield, Va., 1968); J. H. Ryther and J. E. Bardach, *The Status and Potential of Aquaculture, Particularly Invertebrate and Algae Culture,* prepared for National Council on Marine Resources and Engineering Development, PB 177767 (Clearinghouse Fed. Sci. Tech. Info., Springfield, Va., 1968).
7. J. E. Shelbourne, *Advan. Mar. Biol.* 2, 1 (1964).
8. G. O. Schumann, personal communication.
9. C. P. Idyll, personal communication.
10. The study of the status and potential of aquaculture was financed by a contract with the National Council on Marine Research and Engineering Development, Executive Office of the President.
11. W. A. Dill, *Proc. World Symp. Warm Water Pond Fish Culture, Fish. Rep.* 44 (1), i (1967).
12. H. P. Clemens and K. E. Sneed, *Bioassay and Use of Pituitary Materials to Spawn Warm-Water Fishes,* Res. Rept. 61 (U.S. Fish and Wildlife Service, Washington, D.C., 1962), 30 pp.
13. W. Steffens, *Verh. int. Ver. Limmol.* 16 (3), 1441 (1967).
14. R. Moav and G. Wohlfarth, *Bamidgeh* 12 5 (1960).
15. Departments of Commerce and Labor, *Fisheries of the U.S. 1908; Special Report* (U.S. Government Printing Office, Washington, D.C., 1911), p. 49.
16. F. Bregnballe, *Progr. Fish. Culturist* 25 (3), 115 (1963).
17. L. R. Donaldson and D. Manasveta, *Trans. Amer. Fish. Soc.* 90, 160 (1961).
18. K. D. Carlander, *Handbook of Freshwater Fishery Biology* (Brown, Dubuque, Iowa, 1950), pp. 30–36.
19. L. R. Donaldson, *Proc. Pac. Sci. Congr. Tokyo Sci. Counc. 11th* 7, 4 (1966).
20. N. Fredin, personal communication.
21. *Fishery Statistics of the United States 1965,* Statistical Digest No. 59 (U.S. Department of Interior, Washington, D.C., 1967), pp. 541–547.
22. C. L. Hubbs, in *Vertebrate Speciation,* W. F. Blair, Ed. (Univ. of Texas Press, Austin, 1961), pp. 5–23.
23. L. R. Donaldson, personal communication.
24. P. S. Galtsoff, "The American oyster fish," *U.S. Dept. Interior Bull.* 64, 297–323 (1964).
25. J. H. Ryther, personal communication.
26. V. L. Loosanoff and H. C. Davis, *Adv. Mar. Biol.* 1, 1 (1963).
27. Research on shrimp rearing in the United States is carried on at the Laboratory of the Bureau of Commercial Fisheries in Galveston, at the Bears Bluff Laboratory of the South Carolina Wildlife Resources Commission, and at the Institute of Marine Sciences of the University of Miami in Florida.
28. S. L. Hora and T. V. R. Pillay, *Handbook on Fish Culture in the Indo-Pacific Region,* FAO Fish. Biol. Tech. Paper No. 14 (Foreign Agriculture Office, Rome, 1962), pp. 124–132.
29. Yun-An Tang, personal communication.
30. F. Gross, S. M. Marshall, A. P. Orr, J. E. G. Raymont, *Proc. Roy. Soc., Edinburgh Ser. B* 63, 1 (1947); F. Gross, S. R. Nutman. D. T. Gauld, J. E. G. Raymont, *ibid.* 64, 1 (1950).
31. A. Yashouv, *Bamidgeh* 17 (3), 55 (1965); A. Yashouv, personal communication.
32. M. Huet, *Rech. Eaux Forêts Groenendaal-Hoeilaart Belgique, Trans. Ser. D* 22, 1–109 (1957).
33. K. F. Vaas and M. Sachlan, *Proc. Indopacif. Fish Counc. 6th* (1956), pp. 187–196.
34. Th. Rangen, personal communication.
35. Whitefish Authority, *Annual Report and Accounts 1967* (Her Majesty's Stationery Office, London, 1967).
36. L. V. Gribanov, *Use of Thermal Waters for Commercial Production of Carp in Floats in the U.S.S.R.,* Working MS 44060, World Symposium on Warm Water Pondfish Culture (Foreign Agriculture Office, Rome, 1966).
37. T. V. R. Pillay, personal communication.
38. W. H. Schuster, *Fish Culture in Salt-Water Ponds on Java* (Dept. of Agriculture and Fisheries, Div. of Inland Fisheries, publ. 2, Bandung, 1949), 277 pp.
39. Yun-An Tang, *Philippines Fish. Yearb.* (1966), p. 82.
40. The U.S. Atomic Energy Commission Laboratory at Oak Ridge studies the feasibility of agronuclear complexes as shore installations in arid regions to produce cheap power, fresh water, and fertilizer; see *New York Times,* 10 Mar. 1968, p. 74. The agronuclear complexes will furnish ideal conditions for advanced aquaculture on a large scale.
41. S. J. Holt, in *The Biological Basis of Freshwater Fish Production,* S. D. Gerking, Ed. (Wiley, New York, 1967), pp. 455–467.
42. I thank for assistance and information J. H. Ryther, G. O. Schumann, L. R. Donaldson, S. J. Holt, T. V. R. Pillay, W. Beckman, Th. Rangen, M. Fujiya, A. Yashouv, F. Bregnballe, E. Bertelsen, C. Mozzi, K. Kuronuma, S. Y. Lin, S. W. Ling, Y. A. Tang, M. Ovchynnyk, C. F. Hickling, I. Richardson, S. H. Swingle, R. V. Pantulu, F. P. Meyer, J. Donahue, M. Bohl, M. Delmendo, H. H. Reichenbach-Klinke, and D. E. Thackrey.

THE CULTURE OF THE RED ALGAL GENUS *Eucheuma* IN THE PHILIPPINES

HENRY S. PARKER

Marine Coloids, Inc., Rockland, Me. (U.S.A.)

Present address: P.O. Box 412, Charlestown, R.I. (U.S.A.)

(Received 30 May, 1973)

The author discusses the recent development and present status of a *Eucheuma* seaweed farming industry in The Philippines. He describes *Eucheuma* cultivation methods, assesses farm productivity, and considers the economics of *Eucheuma* farming by Filipino families. The author concludes with a discussion of current problems and an evaluation of future prospects for the industry.

Eucheuma AND ITS USES

In the Southern Philippines, nearly 100 coastal Filipino families are now practising marine culture of the red alga, *Eucheuma* (family Solieraceae, division Rhodophyceae). Initial productivity from family farms is high and the future of this new marine "cottage industry" looks promising.

Though *Eucheuma* is a popular food throughout Asia, being widely prepared as a salad vegetable and dessert base, the seaweed's chief importance is in its valuable extract, carrageenan. Carrageenan, derived from several temperate and tropical red seaweeds, is a carbohydrate gum which forms viscous solutions or gels in water. The principal use of carrageenan is as a suspending, thickening and gelling agent in foods such as chocolate milk, ice cream, convenience foods, puddings, aspics, and in the new air freshener gels. It is also used as a binder for toothpaste, and in hand lotions, creams and other cosmetic products. Pharmaceutical applications of carrageenan include suspension of barium sulfate and antibiotics. A specially prepared carrageenan has been used in France for ulcer therapy. In industry, it is used as a suspending agent for abrasives, glazes, and paints. Annual world production of carrageenan is over 6 000 metric tons (t) valued at $26 million.

The present world market for carrageenan-containing seaweeds is about 18 000 t (dry weight) per year. Of this, *Eucheuma* accounts for

452

Reprinted from Aquaculture 3:425-439 (1974).

3 000—4 000 t. In The Philippines, harvesting *Eucheuma* for export has been a significant marine industry for nearly a decade; in 1966 about 800 t were shipped to overseas markets. Since then, primarily due to overharvesting, Philippine *Eucheuma* exports have declined steadily to a present annual level of 300—400 t.

INVESTIGATION OF METHODS FOR CULTURE OF *EUCHEUMA*

To offset declining natural yields, Marine Colloids, Inc., a U.S. seaweed processing firm, established a *Eucheuma* cultivation research program in The Philippines. Considerable field data were collected during 1967—1970. A comprehensive Philippine *Eucheuma* survey was conducted in 1968 (Kraft, 1970). Preliminary trials indicated that it was possible to obtain rapid growth of *Eucheuma* by attaching plants to nylon lines suspended from a bamboo raft. Methods for growing *Eucheuma* were tested in 1970 at Caluya Island (Fig.1); best results were obtained with plants suspended about 0.61 m (two feet) above the bottom on nylon lines attached to stakes driven into the sand (Mr. William Anderson, Marine Colloids, Inc., personal communication). This system served as a prototype for all subsequent farming methods. Several experimental *Eucheuma* plots were established throughout The Philippines in 1969—1970.

An experimental plot located at Tapaan Island, near Siasi Island in the Sulu Archipelago (Fig.1), showed the most promise. The plot was expanded into a pilot *Eucheuma* farm in 1971. Biological, economic and production data were obtained by monitoring the operation of the pilot farm. On the basis of these data and earlier results, the author developed standard cultivation procedures in The Philippines in 1971—1972. By mid-1972 it was concluded by the consortium of research interests that *Eucheuma* farming could be a profitable enterprise for Filipino families.

In the summer of 1972 a joint program was initiated for the training of private *Eucheuma* farmers. Subsequently, considerable assistance was provided to farmers with respect to site selection, cultivation techniques, and the provision of information arising from continuing scientific and practical research. In addition, Marine Colloids, Inc. guaranteed to buy all *Eucheuma* produced by farming families.

By February, 1973, Filipino farmers had established 86 *Eucheuma* farms in the Sulu Archipelago. These farms contained over one-half million plants. Several of the farms are already producing *Eucheuma* on a continuous basis (letter dated 29 January, 1973, from M.S. Doty, Dept. of Botany, University of Hawaii), and it is estimated that cultivation will account for more than half of The Philippine commercial production by 1975.

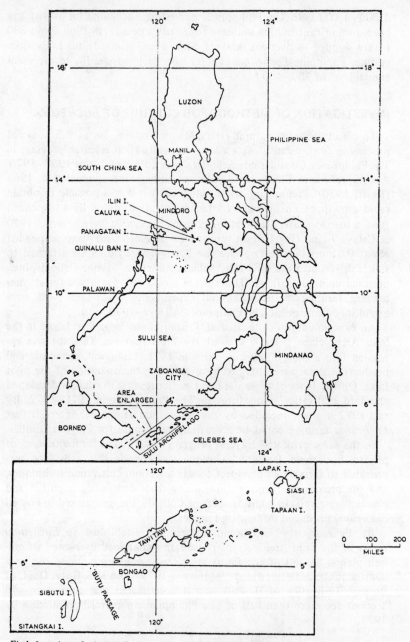

Fig.1. Locations of principal *Eucheuma* farming efforts in The Philippines.

REPRODUCTION AND DISTRIBUTION

Current methods of cultivating *Eucheuma* have taken into account what is now known of the morphology, habitat, and distribution of the principal commercial *Eucheuma* species in The Philippines. There are at this time about six species with an industrial potential. Three of these, *E. cottonii*, *E. striatum* and *E. spinosum*, occur abundantly in The Philippines and account for virtually all *Eucheuma* exports from that country. Much of our knowledge of the genus is based on observations of these three species.

Eucheuma is a hardy, multibranched, nonrooting alga which grows attached by holdfasts to dead coral. Its thalli are rigid, springy, and slippery to the touch. *Eucheuma* reproduces both vegetatively and by sporulating. The plant may be sectioned anywhere; regrowth will occur from all sections, a characteristic which makes it particularly adaptable for culture. Consequently, expansion of Philippine farms has thus far been exclusively by cuttings from existing plants. *E. uncinatum* spores were successfully collected on an artificial substrate in the Gulf of California in 1969 and current research into the life cycle of *Eucheuma* is directed toward the possible future cultivation of plants from spores.

Like most red algae of the sub-class Florideae, the life cycle of *Eucheuma* is complicated. Preliminary studies, as yet incomplete, suggest that the reproductive characteristics of the genus are similar to those of other genera in the family Solieraceae, perhaps most closely resembling the genus *Agardhiella*, to which *Eucheuma* is closely related (letter dated 10 December, 1973 from M.S. Doty).

The reproduction of Rhodophyceae has been discussed by Fritsch (1959). The multiaxial *Eucheuma*, like Rhodophyceae in general, undergoes oogamous sexual reproduction. Its life cycle is diplobiontic with three successive alternating generations — one haploid sexual and two diploid asexual.

In diplobiontic Florideae, spermatia, released from the antheridium (male organ), come into contact with the carpogonium (female organ) at the trichogyne. Here, it is believed that fertilization takes place by means of the fusion of a diploid nucleus from male and female nuclei. The fertilized carpogonium then emits threads (connecting filaments) which connect with specialized haploid "auxiliary cells". At the connecting point, cytoplasmic fusion is established between the two cells. This fusion in turn stimulates division of the connecting cell (and often the division of the auxiliary cell as well) and the production of gonimoblasts which subsequently produce diploid carpospores. The carpospores eventually engender a diploid plant which bears asexual tetrasporangia (forming four spores). These tetrasporangia undergo tw   meiotic divisions; the tetra-spores ultimately become sexual individuals.

The reproduction of Solieraceae, including *Eucheuma* and *Agardhiella*, is distinguished from many other Gigartinales (diplobiontic), including *Chondrus*, primarily by the fact that Solieraceae genera do not have procarps (aggregates in which auxiliary cells and carpogonia are intimately fused). In Solieraceae auxiliary cells and carpogonia are separated.

E. cottonii, E. spinosum, and *E. striatium* are generally found on shal-low reef flats and in lagoons at a water depth of less than two meters at high tide. These species grow naturally at the relatively high salinities encountered in shallow tropical lagoons, and they will die if exposed to appreciable fresh water dilution. There is some evidence that the *Eucheuma* genus in general may grow optimally in salinities above 34‰ with some species preferring salinities as high as 40‰. Philippine culti-vated *Eucheuma* species grow poorly where there is little tidal water exchange, and *E. spinosum,* in particular, thrives in areas characterized by considerable wave action or currents. Consequently, pond culture of *Eucheuma* likely would require continuous circulation.

Doty (1973) has found that *Eucheuma* grows fastest when light is brightest; however, there is some indication that excessive light may damage thalli over a period of time and induce premature 'aging.' For example, *Eucheuma* planted over light-reflecting sandy bottoms has frequently failed. *Eucheuma* exposed to excessive light often shows a bleaching of thalli (pigment loss), presumably due to photo-oxidation.

Water quality has not yet proved to be a limiting factor in cultivating *Eucheuma* in The Philippines. Apparently, sufficient nutrients are available at cultivation sites at all times. However, depth of planting may be impor-tant. Doty (1973) has observed that nitrogen uptake by *Eucheuma* in-creases with increased depth of planting and appears related to light levels. There is evidence that fertilizing cultivated areas by the addition of com-mercial fertilizer may dramatically improve growth rates. Experiments conducted by the author in Zamboanga, Mindanao (Fig.1) in 1972 revealed that *E. spinosum* test plant growth rates increased by 40–50% when 4.8 kg of ammonium sulfate (21% nitrogen, 24% sulfur) was applied to the test area over a 10-day period. Control plants showed no increase in growth rates during the same period. However, fertilizing is not yet eco-nomical, primarily because a system of ensuring slow, uniform release of fertilizer, in sufficient quantities, has yet to be perfected.

Upper and lower temperature limits for *Eucheuma* growth have not been conclusively established. While natural *Eucheuma* has been observed worldwide as far north as 25°N latitude, it is most abundant in the tropical waters of Southeast Asia. There is strong evidence that *Eucheuma* can be successfully transplanted to areas not having natural populations. For example, indigenous *Eucheuma* has not been observed in the Hawaiian Islands, but *E.spinosum*, transplanted from The Philippines to a reef off the island of Oahu, has survived for nearly two years. (Letter dated 20 April, 1973 from M.S. Doty.)

Distribution and abundance of natural *Eucheuma* is characterized by spottiness and seasonal variation, and is most affected by weather conditions, grazing, and harvesting pressure. Consequently, it is impossible reliably to predict annual wild harvest yields for The Philippines.

Inclement weather may significantly affect natural *Eucheuma* productivity. Annually, several typhoons assault the Central Philippines, sometimes greatly altering local bottom character. The author has observed the disastrous consequences of a typhoon on a *Eucheuma* bed in the Quinaluban Island group (Fig.1). Storm waves buried the *Eucheuma* with sand and the bed had not yet recovered several months later. Heavy rains also appear to affect *Eucheuma* growth. While monitoring test plantings near Zamboanga, the author found that *Eucheuma* growth rates improved noticeably just after periods of heavy rain. This observation is supported by reports from Filipino fishermen who claim that *Eucheuma* is always more abundant during rainy seasons. Doty has suggested that the improved growth rates may be related to the observation that during the first few minutes of a heavy rainfall most of the fixed nitrogen is brought out of the air and may subsequently be available for uptake by algae (letter dated 15 November, 1972). Evidently, the associated reduction of salinity during heavy rainfall is of a tolerable level.

Predation also affects natural *Eucheuma* abundance. Sea urchins are the chief grazers in The Philippines, and *Eucheuma* that grows in areas of high urchin populations frequently shows grazing lesions and stunted, nonbranched appearance.

HARVESTING PRESSURE ON NATURAL BEDS

The most important factor currently governing Philippine *Eucheuma* abundance is harvesting pressure, which in turn is influenced by social and economic factors, accessibility of *Eucheuma* beds, and weather conditions. For example, uninhabited Panagatan Atoll in the Central Philippines (Fig.1) is located several hours by boat from the nearest port. The atoll experiences typhoons, frequent rains (preventing effective drying during harvesting) and monsoon winds for three-fourths of the year. Consequently, the harvesting season is short and restricted even more by considerations of harvesters' food and water supplies, time availabilities, and boat-loading capacity. Here, *Eucheuma* abundance has increased since the area was first harvested. Still, Panagatan continually produces more *E. spinosum* than any other single area in The Philippines.

In contrast, formerly productive reef flats located near populated areas in the Sulu Archipelago, where weather is generally favorable, have been almost stripped of *Eucheuma* within the past five years. Most *Eucheuma*-producing areas in The Philippines have been subjected to similar harvesting pressure, resulting in serious declines in annual harvests. Evidence of overharvesting is also increasing in other *Eucheuma*-producing countries in Southeast Asia, notably Indonesia and Singapore.

CULTIVATION POTENTIAL AND METHODS

To date, attempted conservation measures for the wild *Eucheuma* crop have failed in The Philippines for economic, social, and political reasons, and prospects for the immediate future look equally poor. Thus, cultivation is considered the only viable alternative. In the first place, a Filipino farmer can obtain an exclusive concession to farm a designated water area. This form of "ownership" enables him to control and manage a specific *Eucheuma* crop. Consequent benefits of such crop control are predictability of annual yields and reliable profitability to the farmer. A second obvious advantage of cultivation is the potential for improved production through intensive management and genetic strain selection.

Present *Eucheuma* culture methods are simple and may be readily adopted by coastal Asian families using local materials and skills. First, farmers evaluate the cultivation potential of a visually promising area by monitoring test plantings for several weeks. If results are favorable, the farm is constructed. The basic planting unit is a net. The *Eucheuma* net (originally developed by Mr. William Anderson of Marine Colloids, Inc.) is now woven locally; it is 2.5 x 5 m (8.2 x 16.4 ft) in size and is made of monofilament nylon (Fig.2). With a mesh size of approximately 0.3 m (1 ft) each net has 127 mesh intersections or planting points. Farmers install nets horizontally, 0.6–1.5 m (2–5 ft) above the bottom and below lowest tide depth, by attaching net corners to mangrove wood support stakes. Installed compactly (Fig.3), up to 200 nets can be accommodated in a one-fourth-hectare area. (One ha = 2.5 acres). *Eucheuma* farmers call a 200-net unit a "module". A module holds 25 400 plants.

A visitor to a Philippine *Eucheuma* farm would see one of four standard farm operations in progress: construction, planting, maintenance, or harvesting. Farm construction is the most tiring undertaking although it is the least time consuming. The farmer must assemble about 300 mangrove stakes for each module. He must sharpen their ends and work them firmly

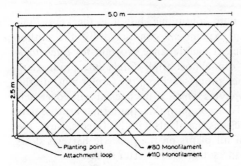

Fig.2. Net used in *Eucheuma* culture.

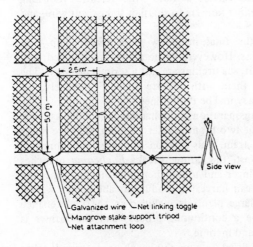

Fig.3. Installation of nets used in *Eucheuma* culture (top view).

into the bottom in rows 6 m apart, in preparation for placing the nets. Mangrove stakes last nearly a year in the ocean.

Planting is easier work, and may be accomplished even by children. Planters first break up *Eucheuma* into 200-g clumps and tie short lengths of flexible, transparent plastic twine to each clump. When the tide is favorable, family members tie these seedlings onto net mesh intersections. A practiced planter can plant two or three nets per hour. Thus, a family of four may plant a module in less than three weeks, assuming three hours of favorable tides per day.

Maintenance is tedious and time consuming, but it is the key to improved farm productivity. An experiment conducted by the author showed that a regular and efficient maintenance program for *Eucheuma* plants and support structures can almost triple *Eucheuma* production. There are four basic maintenance processes. First, the farmer must do the "weeding", i.e., removal of epiphytes which if unchecked will cover nets and plants and reduce growth. To accomplish this, the farmer brushes nets with a plastic scouring pad and cleans the plants with his fingers. Next, the farmer checks the condition of all plants on the nets. He harvests slow-growing or unhealthy seedlings and substitutes a better strain.

The *Eucheuma* farmer makes necessary repairs to the farm with spare materials he keeps in his "banca", a sturdy Filipino canoe which he paddles through the module. He also carries a 'bolo', the versatile foot-long Filipino knife, used by *Eucheuma* farmers to sharpen stakes, cut wire, and crop eelgrass.

Controlling predators is the final necessary maintenance step. The major predators are sea urchins. However, as the plants are elevated, and as long as eelgrass is kept short, sea urchins are not a serious problem. The farmer secures and removes them with a sharp stick. Fortunately, sea urchins are considered a delicacy in The Philippines.

The farmer's most satisfying operation is the harvest. When plants reach an average size of 800 g about two months after planting, they are ready to be pruned back to 200 g again. This method of harvesting eliminates the need to replant. To harvest, the farmer breaks *Eucheuma* branches between his thumb and forefinger and tosses them into nearby bancas or bamboo rafts. Four persons can harvest a module in about two weeks. However, if the harvest of large plants is included in the maintenance procedure, harvesting can be a continuous process and the farmer is assured of steady production and income.

Harvested *Eucheuma* is sun-dried on a nearby beach or clearing. It takes two or three sunny days to remove enough moisture from plants to ensure against rot in storage.

CURRENT PRODUCTIVITY AND ECONOMIC RETURNS

Eucheuma grown on nets has shown considerably better growth rates than "wild" *Eucheuma;* it is less susceptible to grazing and may be planted in areas not otherwise having suitable substrate. In addition, high standing crop density can be achieved and inspection and "control" of plantings is facilitated. Consequently, *Eucheuma* readily lends itself to cultivation by Filipino families.

Eucheuma cultivated on nets has shown growth rates of up to 6% compounded weight increase per day, and 2–3.5% is characteristic (Table I).

At growth rates of 2% per day, plants double in weight approximately every 35 days. If this growth rate occurs uniformly and initial seedling size is 200 g, modules can be harvested effectively every two months.

TABLE I

Some representative growth rates of cultivated *Eucheuma* in various locations in The Philippines, 1971–1972

Location	Inclusive dates	Species	Initial weight (g)	Final weight (g)	Daily weight increase during period (%)	(g)
Vicinity of Zamboanga City, Mindanao, The Philippines						
1. Sacol Island	9/2/1972–10/6/1972	*E. spinosum*	188.0	738.9	4.2	16.2
2. Sacol Island	9/2/1972–9/30/1972	*E. striatum*	211.3	607.1	3.7	14.1
3. Sacol Island	10/6/1972–11/20/1972	*E. striatum*	160.0	871.7	3.8	15.8
Vicinity of Siasi, Sulu, The Philippines						
1. Tapaan Island	9/4/1971–10/20/1971	*E. striatum*	87.2	394.7	3.4	6.7
2. Tapaan Island	12/8/1971–2/12/1972	*E. striatum*	185.6	595.0	2.1	7.3
3. Tapaan Island	12/18/1971–2/12/1972	*E. striatum*	337.0	874.2	1.8	9.6
Vicinity of San Jose, Occ. Mindoro, The Philippines						
1. Garza Island	12/12/1971–2/20/1972	*E. spinosum*	63.6	312.0	2.3	4.1

Based on *E. striatum* yields at the Tapaan Island pilot farm (Table II), cultivated *Eucheuma* productivity is estimated at 13 metric tons per hectare per year (dry weight). These yields compare favorably with yields reported for the most productive land and marine crops, and surpass those reported for other cultivated red algal species (Table III). Results of smaller scale *Eucheuma* farms in other parts of The Philippines indicate that productivity is often much higher, and Doty (1973) conservatively estimates that 30 metric tons per hectare per year (dry weight) may be expected using current cultivation methods. The principal consideration is growth rate. At Tapaan, average daily growth rates of cultivated *Eucheuma* were 2% per day; growth in other areas has frequently been more than twice that rate.

Filipino *Eucheuma* farming families can effectively tend one to three modules. If two modules are farmed, annual production should be about 6.6 t (dry weight). As *Eucheuma* is currently purchased from harvesters for between one and two pesos per dry kg (₱1.00 = $0.15 U.S.), a two-module farm should bring about ₱10 000.00 annually (assuming a

TABLE II

Productivity of cultivated *Euchema* at Tapaan Island, Philippines pilot farm

Module No.	Dates of harvest	Harvest wet weight (kg)	Days since previous harvest of module (using harvest mid-point dates)	Monthly production rate of module since previous harvest, wet weight (kg)
3	11/17/1971– 12/2/1971	7 921	–	–
1	12/11/1971– 12/20/1971	10 795	–	–
2	1/20/1972– 2/7/1972	7 713	–	–
3	2/26/1972 3/9/1972	5 677	95	1 797
1	2/12/1972– 2/23/1972; 3/9/1972– 3/20/1972	6 535	76	2 614
2	4/11/1972– 4/19/1972	4 506	76	1 802
Averages (second harvest)		5 573	82	2 071

Extrapolated figures:	Wet weight (kg)	Equivalent dry weight (kg)
Production per hectare per month:	8 284	1 105
Production per hectare per year:	99 408	13 260

purchase price of ₱1.50 per dry kg). In comparison, local copra prices in 1972 approximated ₱0.60 per dry kg.

Family *Eucheuma* farming in The Philippines is a low-overhead enterprise. Based on economic data compiled from the Tapaan pilot farm, initial materials costs are estimated at ₱1 300.00 per module with annual depreciation costs of ₱600.00 (Table IV).

In addition, farm tools and equipment needed cost about ₱400.00 initially, with ₱200.00 annual depreciation.

In a family farming project, labor costs are not applicable as it is assumed that the family itself will accomplish all required work.

Transportation costs to and from the farm are applicable only if motorboats are used. Generally, family farms are located close enough to the owner's residence to be accessible by the family banca.

TABLE III

Some comparative productivity values for cultivated crops

Crop	Location	Date	Annual production (kg/ha)	Reference
Clam, hard	Korea	1969	12 000	FAO (1970a)
Carp	India	1967	2 909	FAO (1968a)
Carp	Cameroon	1970	4 000	FAO (1972a)
Crayfish	Louisiana, U.S.A.	1969	450–1 120	FAO (1970b)
Fish (nonspec.)	Congo	1971	4 000	FAO (1972b)
Shrimp	Texas, U.S.A.	1972	1 602	Miloy (1973)
Tilapia	Taiwan	1967	3 500–7 800	FAO (1968b)
Sugar cane	Hawaii	1970	11 230	Doty (1973)
Red Algae:				
Eucheuma striatum	The Philippines (Tapaan Island)	1971–1972	99 400 wet 13 300 dry	
Porphyra tenera	Japan	1935–1954	18 200–32 000 wet*	Tamura (1970) (p. 21)
Porphyra spp.	Japan	none given	340 (dry?)	letter dated 10 Dec. 1973 from M.S. Doty
Gelidium spp.	Japan	none given	ca. 10 000 wet**	Tamura (1970) (p. 21)
Gloeopeltis spp.	Japan	none given	339–450 (dry?)***	Tamura (1970) (p. 21)
Gracilaria spp.	Taiwan	1967	2 000 dry	FAO (1968b)

No dry weight figures were given but if a wet to dry ratio of 7.5 : 1 is used (as *Eucheuma*), 2 427 4 270 kg/ha per year (dry) may be estimated. Productivity values have been derived from Tamura's figures which cite production in terms of dried *Porphyra* sheets per rack (rack size = ca. 55 m²). Average production was given as 2 500 sheets per rack in Chiba Prefecture with maximum yields of 4 400 sheets per rack. One kg of fresh *Porphyra* yields about 25 sheets of dried algae.

** Derived from Tamura's figures of about 2 kg/m² produced over a two year period on concrete and broken stone substrates.

*** Derived from Tamura's figures of up to 300–400 monme per tsubo, presumably per year. (1 monme = 3 759 g.; 1 tsubo = 3 305 m²).

TABLE IV

Estimated materials costs for *Eucheuma* farm modules in The Philippines (costs recorded in pesos)

Required material	Cost per module		Effective usage life	Annual depreciation	
	Farmer purchases nets	Farmer borrows nets		Farmer purchases nets	Farmer borrows nets
Nylon monofilament nets 1 000.00 @ ₱5.00 each		--	3 -- 5 years	250.00	--
Mangrove wooden stakes	75.00	75.00	6 – 12 months	113.00	113.00
# 14 galvanized wire	105.00	105.00	6 – 12 months	158.00	158.00
1/8" polypropylene line	60.00	60.00	1 – 3 years	30.00	30.00
Plant-tying material	60.00	60.00	1 year	60.00	60.00
Totals	1 300.00	300.00		611.00	361.00

In order to further reduce a farmer's initial costs, Marine Colloids, Inc. has established a policy of loaning nets to farming families, without interest or collateral, for as long as they are used for *Eucheuma* cultivation purposes. (Depreciation costs are absorbed by Marine Colloids.) Thus, a family farming two modules and borrowing nets may expect to invest 1 000.00 initially and ₱900.00 annually thereafter. For example, a family may plant an initial *Eucheuma* stock of 100 kg and, by pruning and replanting cuttings, gradually expand the farm to two modules. The first income, allowing for inevitable delays and *Eucheuma* losses, should result no later than 11 months after farm site selection. Assuming a productivity equivalent to the Tapaan pilot farm, a monthly harvest of 550 kg (dry weight) can be expected by the third year, yielding an annual net income of over ₱9000.00. This represents about six times the current annual salary for an agricultural worker in The Philippines earning a minimum wage.

Initial results from family farms currently under operation in The Philippines have more than measured up to expectations. Productivity of these farms is estimated at up to 4 metric tons per hectare per month (dry weight), which is over three times the productivity of the Tapaan Island farm.

FUTURE OUTLOOK FOR EUCHEUMA CULTIVATION

Results of Philippine *Eucheuma* cultivation trials show that it can be a profitable family industry requiring little capital investment, while offering high potential productivity. Furthermore the work is familiar to the area's marine-oriented families, it is suited to accessible shallow-water reef flats, and the product has a guaranteed, ready market. Consequently, it is clear that *Eucheuma* farming can be a highly beneficial family industry for many areas of The Philippines; it can provide needed livelihood, while conserving a valuable natural marine resource.

There is every likelihood that farm productivity can be increased even more in the near future. Continuing experimentation is oriented toward raising *Eucheuma* from spores, increasing growth rates through fertilization and propagation of fast-growing strains (several have been isolated to date), and improving farming methods. Currently, up to 80% of potential production is not realized because of unequal growth rates and breakage of branches and plants. Recovery of even half of this loss would result in a tripling of income for the farmer.

Principal deterrents to future success are related to biological, meteorological, and socio-political factors, and illuminate the need for intelligent site selection. Clearly, *Eucheuma* farming is not suited to all areas of The Philippines. Even where extensive testing has shown that *Eucheuma* will grow satisfactorily in a particular location, the risk of storm damage in exposed areas of the "typhoon belt", local opposition, conflict with established uses of the water column in question, or high probability of theft or vandalism would each render an area impractical for cultivation.

Biologically suitable reef flats are found near populated areas in the typhoon-free Sulu Archipelago; extensive shoal areas currently are being cultivated there near the districts of Bongao and Sitangkai, Sulu (Fig.1). These areas appear to have the best potential, when current farming methods are employed. In these areas, the local government has lent full support and protection to farmers and makes site recommendations and allocations (for final approval by the Philippine Fisheries Commission). The recommendations are based not only on results of biological evaluation by trained researchers, but also on possible conflicts with other existing or potential marine uses of the area. Thus far, this policy seems to be working well, and undoubtedly it will be copied in other sections of the country.

There is considerable potential for *Eucheuma* farming in other parts of the world, applying techniques learned in The Philippines. *Eucheuma* grows naturally throughout the world's tropical waters, including parts of the Caribbean Gulf of California, Micronesia, Indonesia, the islands of the Indian Ocean, and Africa. Preliminary cultivation experiments carried out by Marine Colloids, Inc. in several of these areas showed much promise.

It is evident that the present *Eucheuma* farming trials in The Philippines are only a beginning. Future, long-term success of cultivation as a viable

Filipino family industry will depend on a variety of often obscure and interconnected biological, social, economic, and political factors. However, the cooperative efforts of scientific, government, industrial, and popular interests to date have had positive results and give considerable cause for future optimism.

ACKNOWLEDGEMENT

The program of enquiry into *Eucheuma* culture was set up by Marine Colloids, Inc., with support from the United States Sea Grant Program, and in cooperation with Marine Colloids (Philippines) Inc., the Government of The Philippines, The University of The Philippines and the University of Hawaii. Professor Maxwell S. Doty, University of Hawaii, directed the botanical phase of the research program. The author expresses his appreciation to these organizations and individuals for the opportunity to help work on the project.

REFERENCES

Doty, M.S. (1973) *Eucheuma* farming for carrageenan. *Sea Grant Advisory Report- UNIHI-S EAGRANT-AR-73-02*, mimeographed.

Kraft, G.T. (1970) The red algal genus *Eucheuma* in The Philippines. *Hawaii Botanical Science Paper No. 18*. University of Hawaii, Honolulu, Hawaii, 358 pp.

FAO (1968a) News from the research institutes — culture of Chinese carps. *FAO Aquaculture Bull.*, 1 (1), 3.

FAO (1968b) News from development agencies — Taiwan. *FAO Aquaculture Bull.*, 1 (1), 10

FAO (1970a) Aquaculture development — Korea. *FAO Aquaculture Bull.*, 2 (4), 10.

FAO (1970b) Aquaculture development — Louisiana, U.S.A. *FAO Aquaculture Bull.*, 2 (4), 11.

FAO (1972a) Aquaculture development — Cameroon, *FAO Aquaculture Bull.*, 4 (2), 9.

FAO (1972b) Aquaculture development — Congo. *FAO Aquaculture Bull.*, 4 (4), 7.

Fritsch, F.E. (1959) *The Structure and Reproduction of the Algae. Vol. II*: University Press Cambridge, Cambridge, 939 pp.

Miloy, L.F. (1973) Ten acres of shrimp. *NOAA.*, 3 (1), 37–39.

Tamura, T. (1970) *Marine Aquaculture.* Translation from the Japanese of the revised and enlarged second edition (1966) by Mary I. Watanabe. National Science Foundation. Washington, D.C.

THE FOOD CHAIN DYNAMICS OF THE OYSTER, CLAM, AND MUSSEL IN AN AQUACULTURE FOOD CHAIN [1]

KENNETH R. TENORE, JOEL C. GOLDMAN and J. PHILLIP CLARNER

Woods Hole Oceanographic Institution, Woods Hole, Mass. U.S.A.

Abstract: The food chain dynamics of the edible mussel *Mytilus edulis* L., the American oyster *Crassostrea virginica* (Gmelin) and the hard clam *Mercenaria mercenaria* (L.) were investigated in large experimental tanks with flowing, filtered sea water and controlled addition of phytoplankton. The feeding rate of the mussel (5.36 μg carbon removed/l/g C animal was higher than that of the oyster (3.92) and clam (3.03) but the ecological efficiencies (net production/ingested food) $\times 100$ of the clam (23.69 %) and the oyster (18.38 %) were higher than that of the mussel (10.01 %).

The food chain efficiencies (net production/available food) were lower than the ecological efficiencies, suggesting under-exploitation of the available food. The clam, although having a lower feeding rate, was more efficient in utilizing the food it filtered and so showed the highest net production.

The rates (μg-at/l/g C animal) of regeneration of nutrients, especially total inorganic nitrogen (mussel, 2.1723×10^{-3}; oyster, 7.4270×10^{-3}; and clam, 8.1750×10^{-3}) along with reported high biodeposition rates of bivalves suggest that multi-species aquaculture systems would be more efficient and productive than one-species systems.

INTRODUCTION

In studies at Woods Hole Oceanographic Institution on aquaculture systems we have used secondary-treated sewage effluent as a nutrient source for culturing marine phytoplankton which may be used as food for rearing commercially important bivalves (Ryther, 1971; Ryther *et al.*, 1972). Dunstan & Menzel (1971) showed that secondary-treated sewage effluent, diluted with sea water so that the nutrient levels of nitrogen and phosphorus were comparable to artificial media, was excellent for the growth of mixed populations of marine diatoms. Dunstan & Tenore (1972) have successfully maintained large dense cultures dominated by diatoms by enriching with the sewage effluent. No adverse effect on the growth of bivalves that are potential herbivores in aquaculture systems has been observed (Tenore & Dunstan, 1973a).

From the beginning we recognized the advantages of using ecological methods, not only to gain knowledge of the dynamics of experimentally-controlled food chains, but also to arrive at efficient aquaculture systems. For example, Tenore & Dunstan (1973b, c) have investigated the effect of phytoplankton concentration and species composition on the feeding and biodeposition rates in order to model energy flow to the herbivore trophic level. These kind of data are particularly important because surprisingly little is known of the food chain energetics of such bivalves as the oyster, mussel, and clam, although a great deal of survey and field data are available (Galtsoff,

[1] Contribution No. 3017 from the Woods Hole Oceanographic Institution.

467

Reprinted from J. Exp. Mar. Biol. Ecol. 12:157-165 (1973).

1964; Jørgensen, 1966). The study of functional relationships in ecosystems, *e.g.*, feeding rates and food transfer in food chains enables us to describe the dynamics of different food chains in terms of common basic units. This is important not only for arriving at basic principles underlying the functioning of ecosystems, but also to help man to manipulate the productivity of man-made ecosystems, *e.g.*, aquaculture systems, to his own benefit. Indeed, because naturally-occurring ecosystems are so complex, larger-scale experimental ecosystems which may still be precisely monitored and controlled are ideal units for studying functional relationships (Phillipson, 1966).

One facet of our studies reported on in this paper is an investigation of the food chain dynamics of the American oyster, *Crassostrea virginica* (Gmelin), the hard clam, *Mercenaria mercenaria* (L.), and the edible mussel, *Mytilus edulis* L. in aquaculture food chains by monitoring the carbon values of the phytoplankton fed to the bivalves, by measuring the feeding rate and increase in biomass of the animals and so determining conversion efficiencies for this food chain.

MATERIALS AND METHODS

Two large containers were constructed of fiber glass-coated wood, each with a partition down the center so that there were four experimental tanks (265 × 62 × 53 cm high) containing 760 liters of sea water (Fig. 1). These tanks contained two stacks

Fig. 1. The experimental food reservoirs and culture tanks.

each of three trays made of wood frames with Vexar® mesh and each tank was covered with plywood sheets to reduce increase in temperature and growth of epiphytes. Each tank received an average flow of 7.50 liter sea water/min filtered through 100 μ Fluflo® cartridges which removed small crustaceans and other herbivores. These flow rates were measured daily and reset if necessary. This flow rate was chosen on the basis of data in the literature on pumping rates of these bivalves (Jørgensen, 1971) so that the animals in the tanks received ample sea water. No fouling organisms became established in the tanks under these conditions. The mean ambient water temperature was 21 °C, with a range of 17–23 °C.

Stocks of small clams and mussels were collected locally. Oysters were obtained from Long Island Oyster Farm, Northport, Long Island. A sample (50 animals) was used to determine the length and dry meat weight of each species. The average length and dry meat weight for each of three species were: oyster (4.83 cm, 0.4254 g), mussel (3.23 cm, 0.2584 g), and clam (4.26 cm, 0.7366 g). All the animals were thoroughly cleaned at the beginning of the experiment. Five hundred oysters were placed in Tank 3. The number of clams (300) for Tank 1 and mussels (600) for Tank 2 needed to equal the oyster biomass was calculated using dry meat weight data. Tank 4 contained a mixture of all three bivalves. All of these animals were marked individually and their lengths recorded. During the experiment, dead animals were removed, measured, recorded, and replaced by fresh, living animals each week.

Tanks 1, 2, and 3 each received about 500 l/day of algal culture which was grown in 2000-liter ponds. This culture consisted of mixed diatom species, mostly *Chaetoceros* sp. (*simplex*?) and *Phaeodactylum tricornutum* Bohlin (Dunstan & Tenore, 1973). Half the volume of the ponds was harvested each day and pumped into 500-liter Nalgene vessels. Small pumps were used to keep this food mixed and to meter it at an average rate of 350 ml/min into the animal tanks. Tank 4 received only the naturally-occurring food in the coarsely-filtered sea water.

Particulate carbon determinations (Perkin Elmer Elemental Analyzer 204), inorganic nutrient analyses (nitrogen according to Solórzano, 1969 and Wood, Armstrong & Richards, 1967; phosphate according to Murphy & Riley, 1962) were made daily on the food and the sea water entering the experimental tanks. These data, along with the precise flow rates of food and sea water, enabled us to calculate the carbon and nutrient concentrations entering the tanks. A sample of outflowing sea water was also analysed and this allowed us to calculate the percentage of carbon removed and nutrients regenerated. Initially, we measured the incoming and outgoing carbon concentrations of a tank without any animals present. No significant differences were observed due to settling of the algae. Water temperature in the tanks was recorded daily and dissolved oxygen checked periodically.

The experiment continued from the 29th June to 14th September, 1972. At the end of this period the average dry meat weight of the animals were again measured and these data used to calculate the increase in biomass.

RESULTS

Determinations of the increases in biomass showed differences in growth among the three bivalves (Table I). The initial biomass values as bivalve dry meat were converted into total grams of carbon using the carbon value (%) from the C : H : N analysis; namely mussel, 67.09, oyster, 86.27, and clam, 94.03. Similar final determinations showed the following total increases in grams carbon of bivalve meat: mussel, 29.701; oyster, 50.964; and clam, 54.124. These biomass values include only meat and not the organic carbon in the newly-made shell and, in the case of the mussel, the byssal threads.

TABLE I

Data on the growth, feeding, and efficiencies of the mussel, oyster and clam grown in large experimental tanks from 29th June to 14th September, 1972.

	Mytilus edulis	Crassostrea virginica	Mercenaria mercenaria
Number of animals	600	500	300
Number dead	34	18	2
Growth measurements:			
initial average length (cm)	3.23	4.83	4.26
final average length (cm)	3.75	5.60	4.44
initial average dry meat weight (g)	0.2584	0.4254	0.7366
initial total dry meat weight (g)	155.04	212.70	220.98
% carbon	40.56	43.27	42.55
initial total carbon (g)	67.09	86.27	94.03
final average dry meat wt increase (g)	0.1144	0.2513	0.4240
final total dry meat wt increase (g)	68.640	125.650	127.200
final total carbon increase (g)	29.701	50.964	54.124
Average incoming food (μg C/l)	535	595	575
Total feeding rate (% C removed by total animals)	82.1	73.8	63.8
Feeding rate (average μg C/l removed per g C of animal [1])	5.3575	3.9283	3.0308
Total available food (g/C of sea water+culture)	361.33	375.62	358.15
Total food filtered by animals (g/C)	296.65	277.21	228.50
Gross ecological efficiency (%) [2]	10.01	18.38	23.69
Gross food chain efficiency (%) [3]	8.22	13.56	15.12

[1] Average total weight of animals (clam = 121.09 g; oyster = 111.95 g; mussel = 81.94 g) calculated by averaging initial and final total weights.

[2] Gross ecological efficiency = net production of bivalve/food filtered.

[3] Food chain efficiency = net production of bivalve/food supplied.

We ignored this organic production contained in the new shell for two reasons: first, because the actual amount of organic matter in shell is small, amounting to ≈ 1 % by weight (Galtsoff, 1964); secondly, the net production of the meat of the bivalves is that which is available for higher trophic levels. We did calculate the average increase in dry weight of shell per animal (mussel, 0.417 g; oyster, 0.658 g; and clam, 0.819 g).

In addition, we determined the percentage organic carbon of this shell weight by first treating samples of shell with cold phosphoric acid to remove carbonate carbon and then measuring the organic carbon on the Elemental Analyser. We found the organic carbon of the treated samples to average 0.5 %. It is evident that the production of organic carbon due to new shell is not significant compared with the biomass increase of the meat. The animals did not spawn during this period so there was no increment of biomass due to gamete production that was not accounted for.

The average amount of food entering the tanks and the total food during the period of the experiment were calculated from the flow rates and carbon data of the filtered sea water and the supplemental phytoplankton culture. The average flow rates (l/min) for the filtered sea water were: mussel, 7.48; oyster, 7.48; and clam, 7.23. The average carbon concentration of this filtered sea water was 107 μg C/l. The average incoming carbon concentrations (μg C/l) of the combination of filtered sea water and supplemental food were: mussel, 535; oyster, 595; and clam, 575. From all these data the total food (g C) supplied to the bivalves during the experiment was calculated to be: mussel, 362.33; oyster, 375.62; and clam, 358.15.

The differences between incoming and outgoing food concentrations was a measure of the feeding rate of the three bivalves. The average feeding rates (percentage carbon removed by the total biomass) were: mussel, 82.1; oyster, 73.8; and clam, 63.8. Because the total biomass of the three species of bivalves differed slightly, we calculated the feeding rates (mean μg C removed/l/g C animal), which were mussel, 5.3575; oyster, 3.9283; clam, 3.0308.

These data were used to calculate ecological and food chain efficiencies for the three species of bivalves. The ecological efficiency is increase in biomass of bivalves/biomass of total food filtered by the animals. The percentage ecological efficiencies were: mussel, 10.01; oyster, 18.38; and clam, 23.69. The food chain efficiency is biomass yield of bivalves/biomass of total food available to the animals, and the percentage values were: mussel, 8.22; oyster, 13.56; and clam, 15.12.

Nutrient data showed that different levels of inorganic nitrogen and phosphorus were regenerated by the metabolic process of the bivalves (Table II), e.g., the average total nitrogen (μg-at N/l) regenerated per gram carbon for an average weight of each bivalve was: mussel, 2.3066×10^{-2}; oyster, 1.5989×10^{-2}; and clam, 8.1757×10^{-3}.

TABLE II

Average concentrations (μg at N or P/l) of inorganic nutrients entering and leaving the tanks containing the three species of bivalves: each average is for 35 measurements.

Species	Nutrients									
	NH$_4$		NO$_2$		NO$_3$+NO$_2$		Σ N		PO$_4$	
	in	out	in	out	in	out	in	out	in	out
Mytilus edulis	1.81	3.67	0.64	0.68	2.07	2.09	3.87	5.76	1.20	1.46
Crassostrea virginica	1.79	3.33	0.76	0.77	2.35	2.60	4.14	5.98	1.61	2.92
Mercenaria mercenaria	1.81	2.87	0.64	0.74	2.06	1.98	3.86	4.85	1.32	1.49

DISCUSSION

The feeding rate (percentage removal of particulate carbon) of the three popula-tions of bivalves showed differences in the ability of the three species to exploit the food source. There is a tremendous amount of data in the literature on the pumping rate and filtration efficiency (percentage of particles filtered per liter of water pumped by the animal) of bivalves (Jorgensen, 1966). Yet in studying energy flow in these herbivores it is critical to measure the degree to which they exploit the available food source. Although this is a function of pumping rate and filtration efficiency, it is more easily determined by measuring the biomass of phytoplankton removed in a flowing system by a population of the bivalve species. Tenore & Dunstan (1973b) have in-vestigated the effect of food concentration of mixed phytoplankton on the feeding and biodeposition (production of feces and pseudofeces) of the oyster, mussel, and clam in flowing systems by adding different levels of food. They found that the feeding rates for all three species of bivalves were depressed at lower food concentrations ($<$ 300 μg C/l) but quickly reached a maximum at the higher food concentrations typical of coastal waters. In addition, they measured the biodeposition rates of the bivalves at the different food concentrations and were able to calculate relative assimilation efficiencies $i.e.$, the percentage of filtered food assimilated either for kinetic energy (maintenance) or for growth. The food concentration that we decided upon in the present experiment, ca 500 to 600 μg C/l, was chosen on both the basis of these data on highest feeding rate and on assimilation efficiency. The excellent growth observed in this experiment probably reflected both this optimal feeding level and the absence of any competing suspension-feeders in the tanks.

In using growth data to construct energy flow diagrams, it is necessary to define and clarify terms such as 'efficiency' in view of the varied nuances of this term in the literature. Lindeman (1942) used the concept of efficiency to compare the production of successive trophic levels in a food chain. In general, efficiency is a ratio between the various steady-state rates of biomass or energy transfer in a system (MacFayden, 1963). There are several such calculable rates for a food chain. For example, ecological efficiency is the ratio of net production (growth) to ingested food (not necessarily assimilated) by the population. In contrast, food chain efficiency is the ratio of net production (growth) to available food, $i.e.$, whether or not the food is ingested (Slobodkin, 1960). MacFayden (1963) pointed out that one should look at ecological efficiencies if interest is in the potential supply to the predator. This is the view favored by fisheries and aquaculture biologists because in both cases the potential predator is man. On the other hand, one should look at food chain efficiencies to ascertain the most important exploiter of the food source.

Ecological efficiencies reported in the literature vary between 6–15 % and 10 % is the usual average quoted, even though several investigators have suggested that the actual values might be quite higher. Slobodkin (1959), working on the energetics of $Daphnia$, found an ecological efficiency of 12.5 %. Since then, ecological efficiencies calculated from various field studies have ranged from 6–15 % (Russell-Hunter, 1970); however, Steele (1965) gives an ecological efficiency of 20 % based on the calculation of data on yields of fish in the North Sea. He suggested that efficiencies of up to 25 % could be obtained if animals increased in size during periods of abundant food supply and were then harvested at the end of this accelerated growth phase. Lindeman (1942) calculated a 21 % ecological efficiency in his pioneer work on the trophic dynamics of ecosystems.

The ecological efficiencies which we have calculated in the present work fall within the range given above and any dissimilar values suggested differences in the potential of these three bivalves in aquaculture systems. Although the clam had the lowest feeding rate, it had the highest ecological efficiency. In contrast, the mussel had the highest feeding rate but the lowest ecological efficiency. Thus, in terms of net production, for a given food supply the oyster and the clam are better producers than the mussel and would, therefore, yield a larger crop in an aquaculture system.

It is true that, for bivalves, the biomass used for 'food ingested' was actually 'food filtered'. Some of this filtered food was not actually ingested into the mouth but excreted as pseudofeces: however, this fraction was not large at the food concentrations used in this experiment (Tenore & Dunstan, 1973b) and the difference between the terms ingested and filtered is unimportant if interest is in food chain dynamics.

Inefficient utilization of filtered food is typical of the herbivore trophic level (Mac-Fayden, 1963). Cushing (1959), working with zooplankton, states that herbivores appear to ingest a higher proportion of available food than do other trophic levels and to assimilate less of it. The high biodeposition rates of bivalves, especially at higher food concentrations due to the production of pseudofeces, illustrate this inefficient utilization (Tenore & Dunstan, 1973b).

The food chain efficiencies were lower than the ecological efficiencies, suggesting under-exploitation of the available food. Food chain efficiency is a measure of the efficiency with which the food supply is exploited. When the values of both efficiencies are equal, the animals are ingesting the total available food. In aquaculture systems which physically separate the algal and herbivore components of the food chain, the herbivores are not exerting any regulatory effect on the net production of the primary producers. Rather, the net production of phytoplankton is later introduced to the bivalves and the values of ecological and food chain efficiencies would be the same if the bivalves were ingesting all of the food made available to them. Under natural conditions this can never happen because the feeding rate of bivalves is quickly depressed and eventually ceases at food concentration levels lower than about 300 μg C/l (Tenore & Dunstan, 1973b).

The differences between the ecological and food chain efficiencies for each species (mussel, 1.79; oyster, 4.82; and clam, 8.57) show that the mussel was the most efficient of the three bivalves in exploiting the available food, but because the clam has the greatest net production yet the lowest feeding rate, it is more efficient in utilizing the food filtered. The mussel may be characterized as an opportunistic species, *i.e.*, one adapted to successfully exploit stressed habitats (Slobodkin & Sanders, 1969). Such species do not usually compete well in non-eutrophic environments against more typical forms but they do dominate in enriched areas where phytoplankton levels are high. One factor influencing this difference may be this discrepancy between the exploitation of available food and the utilization of ingested food. Because the mussel ineffectively utilizes the food it ingests, it must filter larger amounts of the available food to supply its energy needs so that in environments with lower food levels it may ingest less food than required to satisfy its metabolic requirements.

The results from our experiments indicate that multispecies systems which exploit the energy losses in the simple food chain would be most productive and efficient in aquaculture and tertiary treatment systems (Fig. 2). The primary producers and the

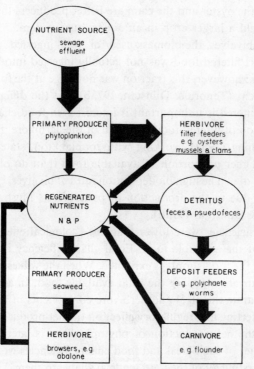

Fig. 2. A diagram of a multispecies aquaculture system.

herbivores are never the only trophic levels in a naturally-occurring food chain. The energy transfer in such a simple food chain is never 100 % efficient and energy losses can be exploited by other organisms. For example, the high biodeposition rates of herbivores result in detritus that can be exploited by deposit-feeders such as polychaete worms, which may themselves be the basis of a fisheries food chain. Nutrients regenerated in the soluble excretory products of the animals may either be re-cycled to the phytoplankton or used to grow seaweeds. Some seaweeds, *e.g.*, *Chondrus crispus*, are themselves valuable crops; other species, *e.g.*, *Ulva* and *Laminaria*, can be fed to browser herbivores such as abalone.

ACKNOWLEDGEMENTS

We wish to thank staff members of the aquaculture program at Woods Hole Oceanographic Institution, especially Nathaniel Corwin, William Young, R. William Richardson, Mason Browne, James Broda, and John Davidson, who assisted in this research. Also, we wish to thank Drs. J. H. Ryther and H. Sanders who critically reviewed the manuscript. This work was supported by National Science Foundation Grant GI-32140.

REFERENCES

CUSHING, D. H., 1959. On the nature of production in the sea. *Fishery Invest., Lond. Ser. 2*, Vol. 22, No. 6, 40 pp.

DUNSTAN, W. M. & D. W. MENZEL, 1971. Continuous cultures of natural populations of phytoplankton in dilute, treated sewage effluent. *Limnol. Oceanogr.*, Vol. 16, pp. 623–632.

DUNSTAN, W. M. & K. R. TENORE, 1972. Intensive outdoor culture of marine phytoplankton enriched with treated sewage effluent. *J. Aquaculture*, Vol. 1, pp. 181–192.

GALTSOFF, P. S., 1964. The American oyster. *Fishery Bull. Fish Wildl. Serv. U.S.*, No. 64, 480 pp.

JØRGENSEN, C. B., 1966. *Biology of suspension feeding*. Pergamon Press, New York, 357 pp.

LINDEMAN, P. L., 1942. The trophic-dynamic aspect of ecology. *Ecology*, Vol. 23, pp. 399–418.

MACFAYDEN, A., 1963. *Animal ecology*. Pitman and Sons Ltd., London, 344 pp.

MURPHY, J. & J. P. RILEY, 1962. A modified single solution method for the determination of phosphate in natural waters. *Analytica chim. Acta*, Vol. 26, pp. 31–36.

PHILLIPSON, J., 1966. *Ecological energetics*. St. Martin's Press, New York, 57 pp.

RUSSELL-HUNTER, W. D., 1970. *Aquatic productivity*. The MacMillan Co., London, 306 pp.

RYTHER, J. H., 1971. Recycling human wastes to enhance food production from the sea. *Environmental Letters*, Vol. 1, pp. 79–87.

RYTHER, J. H., W. M. DUNSTAN, K. R. TENORE & J. E. HUGUENIN, 1972. Controlled eutrophication, increasing food production from the sea by recycling human wastes. *Bioscience*, Vol. 22, pp. 144–151.

SLOBODKIN, L. B., 1959. Energetics in *Daphnia pulex*. *Ecology*, Vol. 40, pp. 232–243.

SLOBODKIN, L. B., 1960. Ecological energy relationships at the population level. *Am. Nat.*, Vol. 44, pp. 213–236.

SLOBODKIN, L. B. & H. L. SANDERS, 1969. On the contribution of environmental predictability to species diversity. *Brookhaven Symposium in Biology*, Vol. 22, pp. 82–93.

SOLÓRZANO, L., 1969. Determination of ammonia in natural waters by the phenol-hypochlorite method. *Limnol. Oceanogr.*, Vol. 14, pp. 799–801.

STEELE, J. H., 1965. Some problems in the study of marine resources. *Spec. Publ. Int. Comm. N.W. Atlantic Fisheries*, Vol. 6, pp. 463–476.

TENORE, K. R. & W. M. DUNSTAN, 1973a. Growth comparisons of oysters, mussels and scallops cultivated on algae grown with artificial medium and treated sewage effluent. *Chesapeake Sci.*, Vol. 14, pp. 64–66.

TENORE, K. R. and W. M. DUNSTAN, 1973b. Comparison of feeding and biodeposition of three bivalves at different food levels. *Mar. Biol.*, (in press).

TENORE, K. R. & W. M. DUNSTAN, 1973c. Comparison of rates of feeding and biodeposition of the American oyster, *Crassostrea virginica* fed different species of phytoplankton. *J. exp. mar. Biol. Ecol.*, Vol. 12, pp. 19–26.

WOOD, E. D., F. A. J. ARMSTRONG & F. A. RICHARDS, 1967. Determination of nitrate in sea water by cadmium-copper reduction to nitrate. *J. mar. biol. Ass. U.K.*, Vol. 47, pp. 23–31.

PHYSICAL MODELS OF INTEGRATED WASTE RECYCLING- MARINE POLYCULTURE SYSTEMS

JOHN H. RYTHER, JOEL C. GOLDMAN, CAMERON E. GIFFORD, JOHN E. HUGUENIN, ASA S. WING, J. PHILIP CLARNER, LAVERGNE D. WILLIAMS and BRIAN E. LAPOINTE

Woods Hole Oceanographic Institution, Woods Hole, Mass. (U.S.A.)

Contribution No. 3411 from the Woods Hole Oceanographic Institution

(Received August 27th, 1974)

ABSTRACT

Ryther, J.H., Goldman, J.C., Gifford, C.E., Huguenin, J.E., Wing, A.S., Clarner, J.P., Williams, L.D. and Lapointe, B.E., 1975. Physical models of integrated waste recycling—marine polyculture systems. *Aquaculture*, 5: 163—177.

A combined tertiary sewage treatment—marine aquaculture system has been developed, tested and evaluated using several different experimental sizes and configurations located both at Woods Hole, Mass. and Fort Pierce, Fla. Domestic wastewater effluent from secondary sewage treatment, mixed with sea water, is used as a source of nutrients for growing unicellular marine algae and the algae, in turn, are fed to oysters, clams, and other bivalve molluscs.

Solid wastes from the shellfish are fed upon by polychaete worms, amphipods, and other small invertebrates that serve as food for flounder, lobsters, and other commercially valuable secondary crops. Dissolved wastes excreted by the shellfish and other animals and any nutrients not initially removed by the unicellular algae are removed by various species of commercial red seaweeds (*Chondrus, Gracilaria, Agardhiella, Hypnea*) as a final 'polishing' step. The final effluent from the system is virtually free of inorganic nitrogen and is incapable of supporting further growth of marine life or of contributing to eutrophication of the receiving waters.

A description of experiments with the above food chains and preliminary results with some alternative approaches are discussed, including a detailed account of the nitrogen mass balance through all of the components of one of the experimental systems.

INTRODUCTION

Demonstration that treated wastewater effluent diluted with sea water is a complete and adequate enrichment medium for the growth of marine phytoplankton (Dunstan and Menzel, 1971) led to the concept of an integrated tertiary wastewater treatment—marine aquaculture system. Unicellular algae, grown in diluted wastewater effluent, remove the inorganic nutrients from the waste water and provide food for oysters or other filter-feeding bivalve molluscs. Other marine plants and animals utilize the solid and dissolved

476

Reprinted from Aquaculture 5:163-177 (1975).

wastes from the shellfish culture and serve as a final "polishing" step before the nutrient-free effluent is discharged to the environment (Ryther et al., 1972). Various parts of this system have been tested and evaluated on a small scale, both in the laboratory under controlled conditions and outdoors under natural conditions of light and temperature (Dunstan and Tenore, 1972; Tenore and Dunstan, 1973; Tenore et al., 1973; Goldman et al., 1974). The net result of this experimental work has been to strengthen our optimism concerning the feasibility of the concept. Several problems in its application have been revealed; some of these have been satisfactorily answered, while others remain unresolved. A number of significant findings of basic scientific value have also resulted. But it was not possible, working with the various isolated elements of the system, to make any valid judgements concerning the practical applicability of the process as a whole.

By the spring of 1973, sufficient progress with the different parts had been made to design and construct a small but complete physical model of the entire system, which was then operated and monitored continuously from May through October of that year. From that experience, certain information concerning the performance and reliability of the integrated process could be evaluated which, though not in every case directly applicable to a full-scale operation, did serve to identify some of the potentials and the constraints of the system as originally conceived. Based on these findings, modifications and improvements were incorporated into the design and operation of a subsequent scaled-up "pilot plant", which went into operation in November 1973 at the new Woods Hole Oceanographic Institution's Environmental Systems Laboratory. In addition, experiments similar to those conducted in the 1973 Woods Hole study with the small physical model were started in January 1974 in southeastern Florida at the Harbor Branch Foundation Laboratory.

Reported here is an overview of our experiments with the integrated polycultures in Woods Hole and Florida which consist of growth systems for marine algae (mostly diatoms), oysters (*Crassostrea virginica*) and clams (*Mercenaria mercenaria*), detrital feeders such as polychaete worms (*Nereis virens* and *Capitella capitata*) and shrimp, and a variety of seaweeds in various food chain combinations.

1973 WOODS HOLE STUDY

The physical model used in this study was a revised and expanded version of the previously reported system (Goldman et al., 1974). The system, as depicted in the schematic flow chart (Fig.1) and viewed in the photograph (Fig.2), consisted in part of two circular algal growth ponds (fiberglass-lined) each of 2.27 m diameter and 50 cm deep (2 020 l). Continuous mixing and aeration were accomplished with both motor-driven rotating arms and recirculation through pumping.

Secondarily treated effluent was collected daily from the trickling filter treatment plant at Otis Air Force Base located 15 km from the Woods Hole

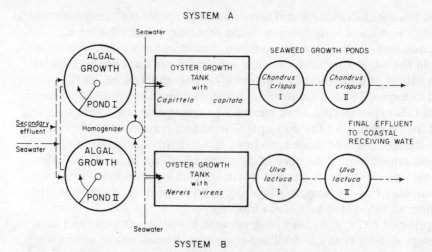

Fig.1. Schematic flow diagram of nitrogen stripping—aquaculture system.

Fig.2. Photograph of nitrogen stripping—aquaculture system.

Oceanographic Institution dock where the experiment was performed. The nutrient characteristics of the treated effluent were previously described by Goldman et al. (1974), and were virtually the same during the current study. Inorganic nitrogen concentrations were typically about 10—20 mg/l.

The waste water was pumped from a 1 000-l polyethylene storage tank along with 1 μ-filtered sea water to headboxes above the ponds, and then blended into each pond at the desired flow rate and mixture. The algal ponds were elevated on steel frames so that the pond effluents flowed by gravity through the remainder of the system, discharging into Woods Hole Harbor.

The continuous effluents from both algal ponds were mixed in a common line and were considered as one flow for evaluating the performance of the remaining system (Fig.1). The combined pond effluent was then passed through a high-speed homogenizer (Tekmar Model SD-45) so that any clumped algal cells were separated intact to insure the availability of single-cell food to the oysters. Only a portion of the total effluent from the algal ponds was used as food for the oysters, with the excess culture used in concurrent but separate experiments.

The remaining portion of the process, consisting of the oysters, worms, and seaweeds, was divided into two parallel systems (Fig.1). System A consisted of oysters and worms (*C. capitata*) combined in a growth tank similar to the one used previously (Goldman et al., 1974). 1 000 small oysters — later increased to 2 000 — were used in each tank, representing a significant increase in oyster biomass as compared to the 500 oysters of similar size used in the previous study (Tenore et al., 1973; Goldman et al., 1974). The algal culture was added to the oyster tank at a constant rate of about 0.35 l/min, supplemented by 100μ-filtered sea water pumped in at rates varying from 5 to 8 l/min. The entire overflow from this tank was then discharged to the first of two seaweed ponds in series, each containing *Chondrus crispus* previously collected in local coastal waters. The seaweed growth ponds were 355-l circular tanks (1.12 m diameter and 36 cm deep), constructed of fiberglass. Mixing was accomplished through recirculation.

A nominal standing crop of 5 000 g (wet weight) of *C. crispus* was maintained by removing the entire crop weekly, shaking free the excess water, and then weighing the biomass. Any biomass over 5 000 g was harvested.

System B was a duplicate of System A except that the polychaete worm was *N. virens* and the seaweed species was sea lettuce *Ulva lactuca* at a standing crop of 3 000 g (wet weight).

During the 6 months of continuous operation of the experimental model, considerable time and effort were expended in acquiring and stocking the animal and seaweed components and empirically achieving the proper balance in numbers and biomass between them. Insufficient time was left to obtain meaningful data on the growth and production of the shellfish, deposit feeders, and seaweeds. Primary attention was therefore focused upon the production of unicellular algae as a function of both wastewater concentration and exchange or dilution rate (reciprocal of retention time) and upon the mass balance of nutrients, especially nitrogen, through the entire system from wastewater input to final discharge. This information will be published elsewhere (Goldman and Ryther, in press), but is reviewed briefly as follows.

Chemical monitoring consisted of determinations for total inorganic nitrogen ($\Sigma N = NH_4^+$-N + NO_2^--N + NO_3^--N) (tri-weekly), and pH, particulate nitrogen (PN), and particulate carbon (PC) (twice daily). All techniques were as reported previously (Goldman et al., 1974).

In one of the algal growth ponds the mixture of waste water and sea water was varied from 30 to 67% waste water, representing an increase in the ΣN

load from 5 to 10 mg/l, and a corresponding decrease in the salinity from 20 to 10⁰/₀₀. The dilution rate was held constant at 0.5 day⁻¹. For each waste-water—seawater mixture, steady-state conditions were established by maintaining the desired flow for a minimum of 7 to 10 days. As seen in Fig.3, the production of PN and PC, representing algal biomass, increased with an

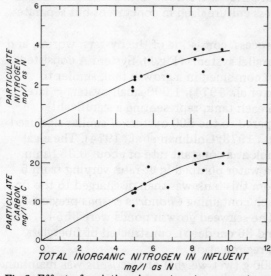

Fig.3. Effect of variations in wastewater– seawater mixture on algal particulate nitrogen and carbon production at a dilution rate of 0.5 day⁻¹.

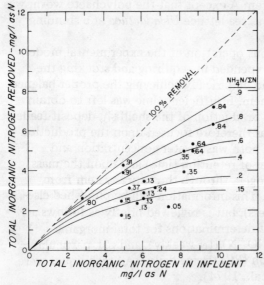

Fig.4. Effect of influent NH_4^+-$N/\Sigma N$ ratio on total inorganic nitrogen at a dilution rate of 0.5 day⁻¹.

increase in ΣN loading until a maximum PN value of about 3–4 mg/l was reached. The influent ΣN concentration of approximately 8 mg/l at this peak biomass level represented a wastewater fraction of about 40–50% of the total influent flow.

Only about 50% of the influent ΣN could be accounted for as PN. Removal of the remaining ΣN appeared to be highly dependent on whether NH_4^+-N or NO_3^--N was the dominant form of ΣN in the waste water. During the experimental period the ΣN concentration in the waste water remained relatively constant at 10–20 mg/l. However, operations at the wastewater treatment plant varied during the test period, and as a result the NH_4^+-N/ΣN ratio changed considerably. Shown in Fig.4 are a series of contour curves illustrating the increasing removal of ΣN as the NH_4^+-N/ΣN fraction in the secondarily treated waste water increased. This effect was measured by differences in influent and effluent levels of ΣN, and was consistent at all concentrations of influent ΣN. For example when the NH_4^+-N/ΣN ratio was 0.8–0.9, ΣN removal efficiency decreased slightly from almost complete removal at low ΣN loadings of 2–3 mg/l to removals of about 7–8 mg/l at high ΣN loadings of 8–10 mg/l. At the same time a maximum of only 3–4 mg/l of PN was produced when the ΣN load was greater than 8 mg/l (Fig.3). When NO_3^--N was the primary form of ΣN (NH_4^+-N/ΣN = 0.15) and the ΣN loading was 8 mg/l only about 3 mg/l was removed (Fig.4). This removed ΣN could be virtually accounted for by the formation of PN (Fig.3).

It was surmised that when the NH_4^+-N/ΣN ratio was high the evolution of NH_3 to the atmosphere was mainly responsible for the added removal of ΣN. Afternoon pH values typically rose from influent levels of 7.3–7.6 to 10.1–10.3 in the pond cultures as the aqueous buffer system was destroyed through inorganic carbon assimilation by the algae. At these high pH values un-ionized NH_3 was the major form of ΣN, thus enhancing the evolution of ΣN to the atmosphere at the intense level of agitation employed. Therefore, it is apparent that for maximum nitrogen removal it will be necessary to prevent nitrification of the waste water at the biological treatment plant which serves the nitrogen stripping—aquaculture system.

Also, utilization of high wastewater fractions (up to at least 67%) appear to be technically feasible without any obvious deleterious effects on algal growth through salinity decreases. However, at least in the current experiments, it appeared that ΣN assimilation by the marine algae was not enhanced beyond a wastewater fraction of about 40–50%, although total ΣN removal was highly dependent on the NH_4^+-N/ΣN ratio. Whether this upper limit for assimilation was caused by light or other nutrient limitations was not established. Similarly, the effect on nitrogen removal of using different waste waters, at dilutions with sea water similar to those used in these experiments, remains to be determined.

The dilution rate in the second pond was maintained at four levels (0.25, 0.50, 0.75, and 1.00 day^{-1}), and the wastewater fraction was held constant at 50% of the total pond influent. As seen in Fig.5, under relatively steady

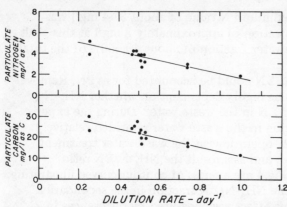

Fig.5. Effect of dilution rate on algal particulate nitrogen and carbon production at a wastewater—seawater mixture of 1 : 1.

state conditions, pond PN and PC concentrations decreased almost linearly as the dilution rate increased in the range described with concomitant decreases in ΣN removal.

On the basis of these results, selection of the optimum dilution rate and wastewater—seawater mixture appears to depend on whether exploitation of the oyster or seaweed system is the desired end-result. In the current experiment the maximum yield of PN (1 g/m²/day), occurred at a dilution rate of 0.75 day⁻¹. At this dilution rate and a wastewater fraction of 50% the maximum production of oysters would be obtained. Use of higher dilution rates and wastewater fractions would decrease the yield of PN, resulting in reduced oyster production, and increase the ΣN load to the seaweed system. The efficiency of the seaweed crop in removing ΣN and the desirability of promoting a seaweed crop in favor of oyster cultivation would obviously dictate how the algal system would be designed and operated.

No attempt was made initially to inoculate a particular algal species into the ponds. Rather, when the ponds were started up early in May 1973, natural populations were allowed to develop. During the entire experiment marine pennate diatoms were the exclusive algal species, although there was a succession of species among the pennate forms.

From May until mid-June *Phaeodactylum tricornutum* was dominant, followed by an unidentified naviculoid diatom which prevailed until mid-August. *Amphora* sp., a small pennate form, then prevailed until mid-September when the naviculoid species returned as the main diatom. These results are very similar to the changes in diatom species that occurred in the previous study (Goldman et al., 1974), in which *P. tricornutum* was dominant for most of the experiment, except during the period from late August to mid-September when *Chaetoceros* sp., a small centric diatom, prevailed.

The change from larger pennates to smaller diatoms appears strongly correlated with temperature, as the highest pond temperatures (25—28°C) were observed when the smaller species were dominant. This phenomenon typically occurs in natural marine waters (Ryther, 1969).

The total mass flow of ΣN through Systems A (oysters—*C. capitata*—*C. crispus*) and B (oysters- -*N. virens*—*U. lactuca*) was compared by averaging all of the data collected during the 6-month test period (Table I). When considering only that fraction of the ΣN load in the algal pond cultures that flowed to Systems A and B, and excluding the fraction of ΣN in the sea water added to the oyster systems, there was approx. 90% removal of wastewater-borne ΣN in System A and 99% removal in System B.

TABLE I

Inorganic nitrogen transformations in an integrated food chain system. Average total inorganic nitrogen loads expressed in g/day

System	Oyster tanks input				Seaweed ponds input		Final effluent	
	Algal* ponds input	Algal* ponds output	Seawater fraction	Total input	I	II	Total	Less seawater fraction
A	3.06	1.43	0.55	1.98	2.46	1.61	0.85	0.30
B	3.09	1.34	0.53	1.87	2.56	1.25	0.56	0.03

*based only on fraction of wastewater—seawater load treated through complete system.

In calculating ΣN regeneration in the oyster tank, no attempt was made to separate the contribution of the oysters from that of the worms. For System A, containing *C. capitata*, ΣN regeneration amounted to about 30% of the ΣN originally removed in the algal ponds; for System B containing *N. virens* it was about 40% (Table I). In contrast, the regeneration of ΣN in the previous experiment amounted only to about 16—18% of the ΣN removed in the algal system (Goldman et al., 1974). It would appear that the increase in regeneration levels was due to the fact that more oysters along with the worms, were present in the current study.

The removal of ΣN was practically identical in the phytoplankton and seaweed systems (Table I). However, the average ΣN removal in the phytoplankton system reflects the composite of many experiments in which the dilution rate and wastewater—seawater mixture were varied. Based on the mass flow of nitrogen introduced to the seaweed system, they appeared ideally sized to remove virtually all of the ΣN not removed in the algal ponds or regenerated by the oysters and worms. For further ΣN removal more seaweed biomass would have to be used.

It was apparent that more efficient removal of ΣN occurred in System B containing *U. lactuca* than in System A containing *C. crispus*. Over 60% of the ΣN removed in the seaweed ponds of System B occurred in Pond I as compared to 53% removal in Pond I of System A. In addition, the average

C/N ratio in the *U. lactuca* increased from 7.8 in Pond I to 10.0 in Pond II, indicating that the seaweed in Pond II was nitrogen-starved. On the other hand, the C/N ratios in the *C. crispus* were relatively constant at 8.6 in Pond I to 8.9 in Pond II. The lower C/N ratio in Pond I of System B would suggest that *U. lactuca* is more efficient in assimilating ΣN than *C. crispus* when nitrogen is in excess. However, a major problem developed with use of *U. lactuca* that made *C. crispus* a far more attractive species to use. On an infrequent but persistent basis *U. lactuca* would go into the reproductive stage with the result that much of the biomass would be lost through release of spores. New thallus growth would quickly occur, but compared to *C. crispus*, which never went into the reproductive stage, the *U. lactuca* ponds were highly unstable. The added attraction of using *C. crispus* is that it is already a commercially attractive crop because of its high content of carrageenin, a chemical used as a suspending agent in the food industry.

In general, the results of the 1973 experiment were most encouraging. The reliability of the integrated system was firmly established on the basis of 6 months of continuous operation. In addition, the very high removals of wastewater-borne nitrogen achieved suggest that the process is a promising alternative to more conventional tertiary treatment methods. Termination of the experiment in October 1973 was dictated solely by the readiness of the new Environmental Systems Laboratory with its significantly larger facilities for carrying out similar experiments on a pilot-plant scale over extended periods.

ENVIRONMENTAL SYSTEMS LABORATORY

In the fall of 1973, the Woods Hole Oceanographic Institution's Environmental Systems Laboratory was completed and occupied. This laboratory was designed and constructed to serve as a pilot plant for the W.H.O.I. Waste Recycling—Aquaculture Project as well as to provide space and facilities, including a relatively large-flow treated seawater system, for related studies in aquaculture and pollution assessment, control, and management (Figs 6 and 7) (Huguenin, 1975).

Because of temporary logistic and financial problems in obtaining daily shipments of the required large volumes of treated wastewater effluent from the nearest treatment plant (Otis Air Force Base or Wareham, Mass.) to the tank storage system, the algal farm was initially operated with "artificial wastewater effluent", a mixture of monobasic sodium phosphate and ammonium chloride. These salts, made up and delivered in concentrations comparable to secondary wastewater effluent, together with filtered sea water, have been added continuously to the 130 000-l algae ponds beginning in December 1973. Two of the algal ponds can be heated and circulated by heat exchangers in the mechanical space of the laboratory, and they were operated at 15°C throughout the winter. Following the initial heating and enrichment, a population of mixed diatoms rapidly developed in the ponds which, after a few weeks,

Fig.6. Environmental Systems Laboratory. Laboratory with algae ponds in foreground.

evolved to virtually a pure culture of the diatom, *Phaeodactylum tricornutum*. This culture, in each of the two ponds, has been maintained for 8 months at a concentration of approximately 10^6 cells/ml with a turnover rate of 30% of the pond volume per day and an input of sea water and nutrients comparable to 50% treated wastewater effluent. Little or no change in algal production has occurred in response to variations in solar radiation associated with storms or cloudy weather. Beginning in April 1974, the algal ponds were operated at ambient water temperature; all six ponds were put into operation, and algal production was comparable to that obtained earlier in the season with the heated ponds. At the time of writing final arrangements have been made for the daily delivery, beginning in July 1974, of 30 000 l of secondary effluent from a nearby treatment plant. At that time the two heated ponds will be switched to a 1 : 1 mixture of secondary effluent and sea water and operated in this mode throughout the year.

The output from each algal pond (ca. 45 000 l/day) is fed by gravity into one of the 12 m × 1.2 m × 1.5 m deep cement raceways which contain 150 000 seed oysters (*Crassostrea virginica*) approx. 2.5 cm long or 150 000 seed hard clams (*Mercenaria mercenaria*) approx. 1.25 cm in length (longest dimensions). The shellfish are held in stacked wooden trays, lined with plastic (Vexar) mesh, at approx. 3 000 animals per tray. A flow of sea water variously

Fig.7. Environmental Systems Laboratory. Animal raceways.

filtered and/or heated and at different flow rates is added to each of the shell-fish raceways to dilute the food and provide additional flow for oxygen supply and waste removal. The raceways containing the shellfish are covered with plywood sheets to prevent fouling of the animals with filamentous algae and to reduce heat loss.

The small polychaete worm *C. capitata* has been inoculated in the bottom of one of the raceways containing the oysters, where it feeds upon the feces and pseudofeces produced by the molluscs. These organisms, inoculated in January 1974, have been observed to be multiplying but the population has not been assessed since it was stocked. 500 winter flounder (*Pseudopleuronectes americanus*) (length 3—6 cm) were subsequently stocked in this raceway to feed upon the capitellid worms.

The raceway containing the hard clams has been stocked with 1 400 juvenile "bait worms" (*Nereis virens*) approx. 2 cm in length. It is expected that these worms, which may reach lengths of 25 cm at maturity, will grow on the clam biodeposits, as occurred in our small-scale experiments during the summer of 1973. To provide the worms with shelter and to reduce cannibalism, the bottom of this raceway has been lined with beach stones to a depth of 5—10 cm (fine sand tends to become anoxic and unsuitable as a substrate).

The discharge from the raceways containing the molluscs flows into and through adjacent raceways stocked with seaweed (*Chondrus crispus*), which is kept in suspension by aeration along one edge of the bottom of the raceway (providing a circular rotation of the water). The purpose of the seaweed, as in the earlier experiments, is to provide the final "polishing" step to remove nutrients not initially assimilated by the diatoms in the algae ponds together with soluble nutrients added to the system by the shellfish, worms and fish.

In the 8 months since the Environmental Systems Laboratory was first occupied, the expanded, "pilot-plant" model of the waste recycling—marine polyculture system has gradually been put into full operation, the various components stocked with organisms, and the complete unit empirically balanced to achieve the dual objectives of advanced waste treatment and aquaculture. Performance data are now being taken and, after sufficient experience, will be reported separately.

An interesting new departure recently initiated at the Environmental Systems Laboratory has been the growth, in one of the 15-m diameter (130 000-l) ponds, of a mass continuous culture of brine shrimp (*Artemia salina*). The pond now contains a dense culture of adult *Artemia*, started from nauplii hatched from eggs in the laboratory, in which living young are actively being produced. The daily harvest from an algal pond (45 000 l/day of diatom *P. tricornutum* at approx. 10^6 cells/ml) is introduced into the *Artemia* pond from which the brine shrimp completely remove the algae. The corresponding harvest of 45 000 l/day from the *Artemia* pond contains both adult and larval brine shrimp which are fed into a raceway containing trout and other plankton-eating fishes.

Although the system is currently operated on an "artificial" wastewater food chain in which full sea water (salinity $30^0/_{00}$) is enriched with highly concentrated nutrients as described above, preliminary laboratory experiments indicate that the *Artemia* will continue to grow and reproduce at a salinity of $15^0/_{00}$, consistent with a 50% wastewater effluent enrichment of the algal ponds.

The wastewater– algae– brine shrimp– trout food chain, or modifications thereof, may represent an attractive alternative to shellfish production in cases where the concentration of pathogens from waste water by the molluscs cannot be satisfactorily corrected or resolved. It is recognized, however, that introduction of an additional link in the food chain will lead to reduction in the production of the final product (fin fish) by 80– 90% from that which could be expected from the herbivores.

FLORIDA EXPERIMENTS

The systems described above appear to operate effectively and with a high degree of efficiency for nutrient removal, but only insofar as the inherent biological processes are working at or reasonably near their optimal conditions. In temperate climates, during the winter months, biological processes for many, if not most species come to a virtual standstill as water temperatures fall below 10°C. In many locations, such as New England, water temperatures may approach 0°C.

The heating of seawater and wastewater effluent to permit biological activity to continue and the system to operate throughout the winter would be prohibitively expensive. A possible solution would be to use the heated effluent from the once-through cooling water system of a coastal power plant, but this would require side-by-side locations of the utility and waste-treatment plants and other constraints that would impose severe limitations to the application of the concept. The rationale can also be used that advanced wastewater treatment (nutrient removal) is most critically needed in summer, particularly in coastal resort communities where both population and coastal water use are at their peaks. But all things considered, the system we have developed and described above is clearly most practical for tropical and semi-tropical climates where the biological processes will operate continually throughout the year at ambient water temperatures. It is also obvious that both the physical—chemical environment and the biota, including the organisms that could be used in our system, are different at least to some extent in the tropics from those that occur in temperate latitudes, and that experience obtained in the latter (i.e., Woods Hole) is not necessarily easily translated to a more tropical climate.

For the above reasons, we have considered it advisable to initiate experiments in Florida, and we were able to transport the entire system used in the 1973 Woods Hole experiment, as described here, to the Harbor Branch Foundation Laboratory in Fort Pierce, Fla., under a grant from the Atlantic Foundation. The system was installed in January 1974 and is currently in full operation (Fig.8). Treated wastewater effluent from a small extended aeration plant serving the Harbor Branch Laboratory community and located adjacent to the aquaculture facilities is used in the various experiments. Not only does the location provide excellent year-round climatic conditions, but we are able to test the feasibility of utilizing a variety of native species in the system. For example, native seaweeds such as *Hypnea musciformis*, *Eucheuma isoforme*, *Gracilaria foliifera*, and *Agardhiella tenera* (all commercially valuable algae) are being tested in the seaweed growth systems, and juvenile white shrimp *Penaeus setiferons* have been stocked in the oyster system as a detrital feeder. Ancillary experiments dealing with the feasibility of operating a benthic marine algae—grey mullet (*Mugil cephalus*)—seaweed food chain as an alternative to our previous diatom—mollusc—seaweed system are under way. In addition, a one-step wastewater-fed seaweed system will be examined.

PROCESS APPLICABILITY

At present the process offers a rather Utopian solution to two important problems facing man: wastewater pollution and the shortage of proteinaceous foodstuffs. We are, however, fully aware of the many inherent problems associated with the recycling of waste products into edible food in closed-

Fig. 8. Harbor Branch Laboratory aquaculture facilities.

loop systems. The very fact that so little is known about both the survival, transport, and accumulation of wastewater-borne pathogens in the biological systems, especially viruses (Vaughn and Ryther, 1974), and the potential toxicity of other substances in waste water (Ongerth et al., 1973), both known and unidentified, is reason enough to raise questions regarding the use of marine organisms grown in a sewage-based food chain as human food.

We have shown, though, that technically the system is entirely feasible both from the standpoint of advanced wastewater treatment and the development of aquaculture. Considerable future research will be required for proper identification of, and solutions to, the problems associated with the transport and accumulation of pathogens and trace contaminants in the system. Successful resolution of these problems will depend upon our ability to treat or control the quality of waste water entering the aquaculture system and/or to depurate the organisms thereby produced in contaminant-free sea water to the point where they are unquestionably safe and acceptable for human consumption. Our research in these areas has just begun, but promising new approaches provide encouragement for their ultimate solution.

ACKNOWLEDGEMENTS

This project was supported by National Science Foundation (RANN) Grant GI-32140, NOAA Sea Grant 04-4-158-5 and The Atlantic Foundation.

REFERENCES

Dunstan, W.M. and Menzel, D.W., 1971. Continuous cultures of natural populations of phytoplankton in dilute sewage effluent. Limnol. Oceanogr., 16: 623—632

Dunstan, W.M. and Tenore, K.R., 1972. Intensive outdoor culture of marine phytoplankton enriched with treated sewage effluent. Aquaculture, 1: 181—192

Goldman, J.C., Tenore, K.R., Ryther, J.H. and Corwin, N., 1974. Inorganic nitrogen removal in a combined tertiary treatment—marine aquaculture system — I. Removal efficiencies. Water Res., 8: 45—54

Goldman, J.C. and Ryther, J.H., 1975. Nutrient transformations in mass cultures of marine algae. J. Environ. Eng. Div. Am. Soc. Civ. Eng., in press

Huguenin, J.E., 1975. Development of a marine aquaculture research complex. Aquaculture, 5(2): 135—150

Ongerth, H.J., Spath, D.P., Crook, J. and Greenberg, A.E., 1973. Public health aspects of organics in water. J. Am. Water Wks. Assoc., 65: 495—498

Ryther, J.H., 1969. Photosynthesis and fish production in the sea. Science, 166: 72—76

Ryther, J.H., Dunstan, W.M., Tenore, K.R. and Huguenin, J.E., 1972. Controlled eutrophication-increasing food production from the sea by recycling human wastes. Bioscience, 22: 144—152

Tenore, K.R. and Dunstan, W.M., 1973. Comparison of feeding and biodeposition of three bivalves at different food levels. Mar. Biol., 21: 190—195

Tenore, K.R., Goldman, J.C. and Clarner, J.P., 1973. The food chain dynamics of the oyster, clam, and mussel in an aquaculture food chain. J. exp. mar. Biol. Ecol., 12: 157—165

Vaughn, J.M. and Ryther, J.H., 1974. Bacteriophage survival patterns in a tertiary sewage treatment—aquaculture model system. Aquaculture, 4: 399—406

Power, Fresh Water, and Food from Cold, Deep Sea Water

Donald F. Othmer and Oswald A. Roels

The sun's radiation is both the essential requirement of all life and the great source of man's energy. Besides keeping us warm, it supplies, directly or indirectly, (i) most of the energy we use, (ii) all of our food through photosynthesis in plants and many links of the food chains, and (iii) our fresh water supply from the cycle of evaporation from the sea, to clouds, to rain, to rivers.

The oceans contain 98 percent of the earth's water, over 1.3 thousand million cubic kilometers of that other great necessity of all life. With 71 percent of the earth's area, the oceans receive most of the sun's radiation to the earth. This radiation is absorbed on the hundreds of millions of square kilometers of the oceans and stored in vast amounts of living organisms stemming from photosynthesis and in the remains of this life—as organic and inorganic nutrients—and as vast amounts of heat in the surface waters of the tropic seas. These two resources —heat to supply energy and nutrients for food chains from single cells through all edible plants and animals up to man—are our greatest resources, as yet practically untapped. In the utilization of the heat, the third product, also from the usual radiation from the sun, fresh water, may be produced, often where needed most.

Sea water is always cold in the deeps, and often it approaches the temperature of its maximum density, near the freezing point. It is cooled in the Arctic and Antarctic where it settles to the depths and, by a grand thermosyphon system, moves on the bottom toward the tropics, where it is warmed, and moves again in tremendous currents toward the poles, to recycle. Photosynthesis in the upper layer penetrated by the sun produces single-cell organisms, thence bigger marine growths, and, by steps, up to the earth's largest plants and animals. Surface waters in the tropics may be crystal clear because photosynthesis has utilized all nutrients; and larger living things have consumed all of the small organisms which cause haze, and thus have stripped the water of carbon, nitrogen, and phosphorus, the principal nutrients for life.

But this life, largely in surface water, dies, as does that in deeper water; and the remains settling slowly, as befits a burial, return to "dust," that of the ocean depths. Slowly these remains disintegrate; and, in solution and as particles, residues are carried in the deep currents back to particular areas of upwelling—only about 0.1 percent of the total area of the oceans. Here the great amount of nutrients causes an explosion of marine life. Just one major one, the upwelling of the Humboldt Current off Peru, supplies one-fifth of the world's total fish harvest.

Availability of Thermal Energy

Again with reference to energy (here heat), its concept implies the temperature of the "hot" substance being higher than that of another "cold" substance. Heat is only usable by its transfer to a colder body. Deep sea water may be from 15° to 25°C colder than surface water; but there is little conduction of heat, top to bottom, and little mixing because of density differences, except in notable upwellings.

While this temperature difference between surface and deep waters is small, considering usual sources of energy, the available heat is the product of this difference multiplied by the available masses of sea water which are infinite for all practical purposes. Means for the conversion of this available heat to electrical energy would give continuously very much more than mankind has found capability to use.

For example, the Gulf Stream, first studied scientifically by Benjamin Franklin, carries the heat absorbed in the Caribbean and the Gulf of Mexico past the coast of Florida. Some 2200 cubic kilometers of water per day may be as much as 25°C warmer than the cold, deep water which it was. To heat just 1 cubic kilometer of sea water per day 25°C would take six or eight times as much energy as all of the electrical energy produced in the United States. The reverse is staggering; it has been estimated that this heat in all of the Gulf Stream, if discharged to water colder by 25°C, could generate more than 75 times the entire electric power produced in all of the United States (1).

Both coasts of Africa, the west coast of both Americas, and the coasts of many islands, particularly in the Caribbean area, have places within a few miles of land where sea water has a surface temperature of 25° to 30°C, while at 750 to 1000 meters below the surface, the temperature may be 4° to 7°C (2). In some places, the ocean floor drops off from the shore line very steeply to an ocean deep within some hundreds of meters of land. The ideal location for a land-based power plant, using warm surface water on one side of a peninsula, would have a great sea depth close to shore on the other side of the peninsula. The contour of the bottom should be favorable to the installation of a large suction pipe to supply cold water.

Dr. Othmer is Distinguished Professor of Chemical Engineering at the Polytechnic Institute, Brooklyn, New York 11201. Dr. Roels is professor of oceanography at the City University of New York, and chairman, Biological Oceanography, Lamont-Doherty Geological Observatory of Columbia University, Palisades, New York 10964.

491

Reprinted from Science 182: 121-125 (1973).

Potential Values

Both this energy and these nutrients are available, and they could supply all the world's power, light, and much of the protein food it uses; but so far they are locked away from us by the difficulty of their recovery from such dilute sources, compared to the relative ease of the utilization of other, more concentrated resources. With shortages of energy and food in the world, this utilization is a job for the present, and one well within the capabilities of technology now available. The dilution is indeed not prohibitive. The water brought up from the depths of tropical seas will absorb surface heat energy equal to the mechanical energy available from a 120-meter waterfall. Compared to other systems proposed for using solar energy, this utilizes a vast reservoir at any one of many places. Always the equipment for utilizing solar energy is large and expensive. And the mariculture using the nutrients can produce $125,000 of product per year from each hectare of land converted to ponds [$50,000 annually per acre].

Thus, if very large amounts of cold, deep sea water can be brought to the surface, warmed in receiving the heat discharged by a suitable power station, and passed to tropical ponds wherein its nutrients are used in photo- and biosynthesis by marine plant and animal food chains, the ultimate product is not fish meal or an artificial substance, but choice shellfish. The water is warmed in the pools to a temperature higher than surface sea water and is passed to the high temperature side of the power cycle. The simplest is direct production of very low pressure steam, turbine-generation of electricity, and condensation of the steam in warming the cold, deep sea water, giving fresh water as condensate.

Power Cycle and Process Engineering

Great minds backed by large sums of money, somewhat less large when the potential benefits are considered, have worked throughout almost a century to develop systems of utilizing the small difference of temperatures of surface and deep sea water to produce power (3). Claude made the most optimistic contributions 40 years ago (4), and the great problems which he recognized were principally two—the installation of the enormous pipeline to carry water from the depths and the removal of air from the evaporating warm water (5).

However, theoretically mechanical energy—and from it electrical energy—can be developed from heat from any body at any temperature being passed to any other body which can receive it, because of its lower temperature. Such energy is always more difficult and less efficient to produce, the lower this temperature difference is. Carnot showed the maximum efficiency to be $(T_1 - T_2)/T_1$ where T_1 is the temperature of the hot body and T_2 is the temperature of the cold body. These are measured above absolute zero.

This temperature difference for efficient heat engines may be many hundred or even a thousand degrees. The closer the temperature of the heat input approaches that of the output, the less the efficiency becomes. Here warm water is at 30°C and it is cooled, in producing very low pressure steam, to 25°C, the temperature of the steam. If this steam is condensed at 15°C by heating cold water from 5° to 10°C, with a 5°C loss in the condenser tubes, this temperature of 15°C may be regarded as the low temperature at which all heat is discharged.

Hence, if the steam supply is at 25°C, or $273° + 25° = 298°K$, above absolute zero, and the corresponding temperature of the heat rejection is 15°C, or $273° + 15° = 288°K$, then the maximum thermodynamic efficiency is $(298 - 288)/298$, or about 3.3 percent. Practically, because of many energy requirements in related machinery, and because of many losses, the efficiency obtainable could not be more than about 2 to 2.5 percent. Of equal importance usually, the amount and cost of equipment required always *increases* greatly with a decrease of the temperature difference. Thus, the heat in a cubic kilometer of warm sea water may be passed to colder sea water to develop mechanical energy, then electrical power. Necessarily, the heat available at this low temperature can be converted to power only with a large, costly plant, at a very low efficiency, and by the handling of extremely large amounts of the cold sea water to absorb the heat. However, the total amount of water to be handled may be less than the amount of water required to produce the same amount of power in a hydroelectric plant. Dams, penstocks, and machinery of a hydro-electric plant are also expensive in developing a "free kilowatt"; that is, free of cost of energy.

The cold water does not have to be lifted from the great depth by the pump; only the friction head must be considered, plus the small static head caused by the difference in density of the cold water and the average density of the water from the surface to the bottom of the pipe.

Various designs for floating power plants have been made with vertical suction pipes suspended from the vessel and with submerged power cables and fresh water lines carrying the products to the shore. However, these would make controlled mariculture more difficult.

Any plant for handling these large volumes of water and converting the available thermal energy to mechanical and then to electrical energy will be huge and expensive; and even the smallest one which would be worth-while for demonstration purposes will involve many millions of dollars worth of equipment.

The simplest of many possible systems that have been studied depends on flash evaporating, in an evacuated chamber, a small amount of the warm water as it is partially cooled. This gives a maximum of 1 percent of the weight of the water as a very low pressure steam. This low pressure steam turns a turbine in cooling further and then is condensed on tubes through which the cold water from the deep is passing and being warmed. The condensate is fresh (distilled) water, almost always a valuable commodity on tropical coasts; and its sale adds to the revenue from the power produced by the generator turned by the steam turbine.

Because of the very low temperature and pressure of the steam, the turbine must be specially designed; and the condenser must be large. Some systems have not provided a surface condenser, but have depended on "open" condensation by sprays of the cold sea water. This produces no condensate fresh water, the sale of which is a valuable revenue for any system, unless a cooled fresh water spray were used (6).

A substantial plant using low pressure steam with a condenser for fresh water has been engineered (7). Several other designs were studied and discarded. The design for a 7180-kilowatt (net) power plant also showed an output of 6 million U.S. gallons of fresh water per day at a total installed cost of $18.4 million.

Several factors were considered in the economic analysis; and charts were made to show the interrelation of (i) the capacity factor, that is, actual production compared to maximum capacity, and (ii) the cost of power generation. Thus, for an investment of $18.4 million, a calculated maintenance and operating cost of $100,000 per year, at an assumed fixed cost of capital of 12 percent per year and a capacity factor of 0.9, fresh water would be produced for $1 per 1000 U.S. gallons, and electric power for 6 U.S. mills per kilowatt-hour; and, in general, total costs can be divided between the two products as desired, since total amounts of both are produced.

As another example, if the capital or fixed charges are taken as 16 percent per year at a capacity factor of 0.90; and if the cost of producing power is taken as 6 mills per kilowatt-hour, fresh water costs are $1.38 per 1000 gallons, or if power cost is taken as 1 cent per kilowatt-hour, then fresh water is $1.26 per 1000 gallons.

Under the economic conditions prevailing at the particular site, which changed during the program, the rate of return on private risk capital was not regarded as sufficiently attractive to private investors to warrant this investment to compete with power and fresh water from a combustion plant. The warm water was regarded as the more valuable stream—it contained the heat that was discharged to the equally necessary stream of cold water, brought up by the very expensive pipeline and pumping system.

In the case of a mariculture program, the valuable stream is that from the depths, with the nutrients therein. The warm water stream does nothing for the mariculture, except that its vapors condense and heat the cold water somewhat in passing through the condenser, and the higher temperature increases the rate of growth of marine life. However, it should be noted that, in passing through the sun-heated enclosed basins for mariculture, the effluent, when it is discharged back to the sea, may be warmer than the surface water from the open sea. If so, this effluent would be cycled through the flash evaporator or boiler of the electric power–fresh water system, and only one stream would be drawn from the sea. The process engineering, mechanical engineering design, and civil engineering

design were completed along with the economic analysis which showed that this project was economically profitable. However, under other particular conditions pertaining at the site, it would be desirable to delay the construction of the plant for fresh water and electric power production. Some details may be of interest.

Plant Layout and Equipment as First Designed

Because of an existing highway at the proposed site, the power and desalinating units were laid out about 140 feet from the shore line. Hydraulic losses and steam friction losses were minimized by short conduits with a minimum of bends. The warm surface water intake, a large subsurface conduit, supplies the boilers through trash racks and fine screen, then deaerators. Special design adapted from desalination evaporator practice minimized losses during flash-boiling of about 1 percent of the warm water supplied. A boiler discharge pump removes the cooled surface sea water.

A turbine with horizontal rotor is directly above each boiler and was designed to operate at a low speed because of its large diameter.

The two pipes for cold sea water intake were designed with a nominal diameter of 4.13 meters and to withstand the stresses imposed by the carefully planned system of installation and by the irregular sea bottom, the contour of which was explored from a small submarine. The section was located 4100 meters offshore at a depth of 975 meters.

Improvements in Design of Plant and Equipment

Improvements have been made in the newer design planned for installation as an integrated component with the mariculture unit at a demonstration plant. Various improvements and advantages will be included in the new design.

1) The water effluent from mariculture operations will be used, and it will be warmer than open sea water, thus a better efficiency should be achieved. Also there will be a considerable economy in almost eliminating the warm surface sea water circuit.

2) A greater ratio of surface water

to deep sea water will use the latter more efficiently.

3) Improved design of the hydraulics of deep water systems should reduce installation and power costs.

4) Condenser cost will be greatly reduced if plastic tubes are used.

5) Boilers will use the controlled flash evaporation (CFE) system to reduce losses in pressure and temperature drops which will increase production of both water and power (6, 8). The CFE system also will reduce substantially the deaeration costs, which require 20 percent of power produced in previous plants.

6) In some locations where fresh water is unusually expensive, all of the available heat will be used for this production, with no power.

Mariculture in Cold, Deep Sea Water

Deep sea water which has absorbed the heat from warm surface water in producing power and fresh water has been brought to a temperature more favorable for biologic growth. It is rich in nutrients which often are exhausted almost completely by the high rate of photosynthesis in the sparkling clear surface tropic waters; and is practically free of organisms which produce disease in humans, predators and parasites of shellfish, fouling organisms, and man-made pollutants. By contrast, shellfish culture has had major pollution disasters in the past years along the continental Atlantic coast.

An experimental station has been operated on the north coast of St. Croix, one of the U.S. Virgin Islands, near Puerto Rico. Here the ocean floor slopes sharply to the Virgin Islands Basin (4000 meters deep) and reaches a depth of 1000 meters, 1500 meters offshore. Three [69-millimeter inside diameter (3 inches nominal)] polyethylene pipe lines, each 1800 meters (6000 feet) long, supply water from a 870-meter (2900 feet) depth in an amount of 159 liters (42 U.S. gallons) per minute. This water is warmed in being drawn up through the small pipes so that its cooling effect would be negligible but it is satisfactory for the mariculture work.

This water in January 1973 averaged (microgram atoms per liter) nitrate nitrogen, 32.1; nitrite nitrogen, 0.13; ammonia nitrogen, 1.1; phosphate phosphorus, 2.15; and silicon in silicates 21.7. The salinity was 34,841 parts per million. While these amounts equal only a relatively small weight of synthetic nutrients which could be added, this clean, unpolluted water is free of parasites and hostile microorganisms which could endanger the cultured animals, or remain in their bodies to be passed to humans. Also the water, if used in a power and desalination cycle must be pumped up to gain its cooling value. It may also be fortified with additional amounts of added nutrients having components carefully chosen to give the greatest value in the particular mariculture used.

A development program is now in progress to determine the most desirable plant and animal species for a food chain to give optimum value of the produce species at the top of the chain with minimum cost in production. Two varieties of diatoms have been particularly satisfactory; and after inoculation the water develops up to 1 million diatoms per milliliter, when it is metered into the shellfish tanks.

Early work showed a 27-fold increase in unicellular algae (diatom) grown in water from a depth of 800 meters compared to that from the surface; and peak yields of 230 grams per cubic meter (1900 pounds per 1 million U.S. gallons) have been obtained. This amounts to 2.8 grams of algal protein per cubic meter of water.

Various types of shellfish feed on these unicellular animals by filtering them from the water they continually process; from previous work it appeared possible to obtain at least a 60 percent conversion of the diatoms to commercial foods. Thus, from an overall material balance, these nutrients of the deep sea water, basically the nitrogen, which would be utilized through the food chain to be explained, should give 1 kilogram of fresh clam meat per 300 cubic meters of deep sea water (27 pounds per 1 million gallons).

From the available marine life in nature, the most promising species are being chosen; there are hopes of improving the natural strains, as has been done by animal husbandry in every animal which has ever been bred for food. Greatly improved yields appear through proper control of (i) natural nutrient concentration—and possibly that of artificial nutrients, or other additives; (ii) solar radiation—by adjusting the depth of the ponds; (iii) water temperature; and (iv) still other variables as these first or axiomatic ones are optimized.

Algal Cultures

Many species of microscopic algae have been isolated, cultured, and studied as cultivated food for shellfish. Those preferred are fast-growing strains, readily accepted by shellfish and causing their rapid growth; they should be hardy against competitive organisms, against high summer temperatures, 32° to 33°C of the pools, and against the excessive sunlight radiation which prevails in shallow pools. Some have developed weight increases of young oysters (3 millimeters) of more than 75 percent in 3 weeks.

Extensive experiments in all sizes of tanks and pools up to 45 cubic meters (12,000 gallons) of 1.2 meters (40 inches) depth, with many variables, have indicated that dependable production of large amounts of algae satisfactory for shellfish food can be maintained. This work to improve the breed and production of algae continues because of the promise of considerable improvements in the development of better, more stable, and hardier strains. Also the geometry of the pools is being optimized; and continuous operation has been developed.

Shellfish

Oysters and clams from cultures stemming from Long Island (New York), Japan, and various tropical locations have been worked with as brood stock and for growth studies.

Experimentation with the shellfish has indicated that certain species grow very rapidly indeed in this "artificial upwelling" system: thus, hybrid clams were grown to market size in 6 months. Similarly, the European oyster and the bay scallop were grown from spat to market size in 6 months (9). This is considerably faster than generally occurs in nature. Clams, European oysters, and bay scallops of commercial size grown in this system were submitted to a panel of seafood experts for taste testing, and judged to be of excellent taste and superior to those harvested in natural waters. Thus, hybrid clams averaging 8 grams, on introduction, increased in weight almost five times to 38.5 grams in 6 months so they could be marketed in the littleneck size.

Scallops multiplied their weight 60 times in 145 days, from an average single weight of 0.24 gram at an age of 8 days to 14.42 grams. Average lengths of the scallops were, respectively, 9 and 40.7 millimeters.

Oysters have grown from 3 millimeters to market size in a little more than 8 months; and one species of oysters grew from an average live weight of 1 gram when introduced to 70 grams in 74 days.

Shellfish filter the microorganisms from the water for food; their filtering efficiencies for gathering and retaining the food cells from the pools have varied from 49 percent without culling of the shellfish to over 90 percent when the small shellfish have been periodically harvested to stimulate the growth of the larger ones remaining. These harvesting techniques are now being optimized.

Crustaceans

A great variation in the growth rates of shellfish has been observed; one long-term objective is to improve the strain by selective breeding of the fastest growing individuals, which also have other desirable characteristics. Thus a large number of small clams at different ages would always be culled to minimize competition of the faster growing animals; and the culls may be used as a very acceptable food for crustaceans. First tried were adult spiny lobsters, native to St. Croix in the Virgin Islands. The best of these showed an average weight gain (in an 89-day period between moltings) of 55 percent, while eating 5.2 times as much food weight as its gain in weight.

Similar experiments are under way with cold water lobsters from the Massachusetts coast which are growing at a greatly accelerated rate in the warm waters of the mariculture ponds.

Seaweed

If the effluent from the shellfish and lobster growing operation were returned directly to the sea, the animal wastes might constitute a source of pollution. Therefore, experiments are under way with commercially useful seaweeds which can be processed to obtain either agar or carrageen. These seaweeds are grown in the effluent from the animal tanks, to optimize the nutrient utilization in the system, and to purify the discharged waters before returning them to the sea.

Chain of Nutrient Utilization

The water from the deep will have been substantially warmed in the plant for production of energy and fresh water; and the optimum utilization of its nutrients appears now to be via (i) single cell algae; (ii) filter-feeding shellfish, such as oysters and clams, which feed on the algae; (iii) lobsters, shrimp, and possibly other crustaceans which feed on culls of the shellfish; and (iv) specialized seaweed, which grows in effluent water containing the solubilized body wastes of the shellfish and crustacea, and has several important markets.

Mariculture Ponds and Operation

The mariculture will be done in a series of shallow concrete pools of optimized depth. The deep sea water flows through slowly to permit residence times, not widely different, for (i) algal growth, (ii) shellfish growth, (iii) crustacean growth, and (iv) seaweed growth. The apportionment of the time periods for the different growths has not been established exactly to date but will be optimized insofar as possible to give the greatest financial return with the minimum of land and pool area.

For the demonstration plant now being planned, it is expected that 25,000 gallons (95 cubic meters) of deep sea water per minute will be available and that there may be a total of 6 hectares (15 acres) of ponds required with a total time of water in transit of about 2 days. This area may be divided ap-

proximately as follows: (i) 50 percent for algal growth, (ii) 10 percent for shellfish growth, (iii) 10 percent for crustacea growth, (iv) 20 percent for seaweed growth. It is impossible as yet to estimate the optimum operational yields of different products; but it is expected that, at an average annual yield, an average value at the plant will be about 340,000 pounds of shellfish at $2.25 to $2.50 per pound of meat. This works out to be an average of over $50,000 annual revenue per acre of ponds without credit for values that cannot yet be optimized.

For a larger plant, handling 870,000 gallons (3390 cubic meters) per minute, a somewhat lower unit price for shellfish may have to be taken; and the total annual revenue has been projected to be between $20 and $25 million.

Summary

Many times more solar heat energy accumulates in the vast volume of warm tropic seas than that produced by all of our power plants. The looming energy crisis causes a renewal of interest in utilizing this stored solar heat to give, in addition to electric power, vast quantities of fresh water. Warm surface water, when evaporated, generates steam, to power a turbine, then fresh water when the steam is condensed by the cold water.

A great increase in revenues over that from power and fresh water is shown by a substantial mariculture pilot plant. Deep sea water contains large quantities of nutrients. These feed algae

which feed shellfish, ultimately shrimps and lobsters, in shallow ponds. Wastes grow seaweed of value; and combined revenues from desalination, power generation, and mariculture will give substantial profit.

References and Notes

1. D. F. Othmer, in *Encyclopedia of Marine Resources*, F. E. Firth, Ed. (Van Nostrand, Reinhold, New York, 1969), p. 298.
2. R. D. Gerard and O. A. Roels, *Marine Technol. Soc. J.* **4** (No. 5), 69 (1970).
3. S. Walters, *Mech. Eng.* **93** (No. 10), 21 (1971).
4. G. Claude was developing support in the United States in 1925–26 for the pilot plant which he then built in Cuba. He gave a demonstration lecture in the laboratories of the University of Michigan where D. F. Othmer was then a graduate assistant. Claude's equipment included a small tank for warm water and one for cold, a vessel to which the warm water was admitted to undergo flash evaporation (about 1 percent), a small steam turbine driven by the low pressure steam that was forming, a condenser having the cold water in direct contact with the turbine exhaust steam, and a vacuum pump for air removal. A small generator was driven by the turbine and was wired to a small electric bulb, which lit as the house lights went off—and the audience cheered.
5. G. Claude, *Mech. Eng.* **52** (No. 12), 1039 (1930).
6. The vapor reheat system of multistage flash evaporation uses a separately cooled fresh water stream to condense vapors for fresh water production and has been described [R. E. Kirk and D. F. Othmer, *Encyclopedic of Chemical Technology* (Wiley, New York, ed. 2, 1970), vol. 22, pp. 39–48].
7. The sea water power and fresh water plant was engineered by Alemco, Inc., now a division of Viatech, Inc., Syosset, New York. One of us (D.F.O.) is a consulting engineer to and director of both corporations. The plant was designed for another island site in the Caribbean which was then regarded as eminently suitable.
8. R. C. Roe and D. F. Othmer, *Mech. Eng.* **93** (No. 5), 27 (1971).
9. J. S. Baad, G. L. Hamm, K. C. Haines, A. Chu, O. A. Roels, *Proc. Nat. Shellfish. Ass.* **63** (No. 6), 63 (1973).
10. Supported by Sea Grant 1-36119. This article is Lamont-Doherty Geological Observatory contribution No. 2016 and City University of New York Institute of Oceanography contribution No. 21. The first engineering design was made by Alemco, Inc., a subsidary of Viatech, both of Syosset, New York.

Pollution

Introduction

Man can use the concepts of models, energy flows, food webs, decomposition processes, and community dynamics to learn how life in the ocean responds to his impact on it and to take measures to reverse whichever of his habits endanger that life. Toward this end, the National Science Foundation is supporting the International Decade of Ocean Exploration, of which the first goal is to:

Preserve the ocean environment by accelerating scientific observation of the natural state of the ocean and its interactions with the coastal margin—to provide a basis for (a) assessing and predicting man-induced and natural modifications of the character of the oceans; (b) identifying damaging and irreversible effects of waste disposal at sea; and (c) comprehending the interaction of various levels of marine life to permit steps to prevent depletion or extinction of valuable species as a result of man's activities.

Scientists from different laboratories form a subunit called the Pollutant Transfer Program (Duce et al., 1974). Their mission is to: (1) identify important transfer pathways and mechanisms, (2) evaluate the major environmental factors that affect transfer processes, and (3) develop principles governing transfer of pollutants.

Man-made pollutants are those materials that disrupt the aesthetics or the integrity of life in a particular system. The plastic bottles sighted by Thor Heyerdahl while crossing the Atlantic on a raft may be considered simply unattractive, or they may be interpreted as indicators of where our less visible materials end up. Pollutants may smother invertebrates in the sediments (oil, silt), cause blooms of poisonous dinoflagellates or an inedible species of algae (nutrients), or they may alter metabolism and destroy life itself (chemical toxins). Additionally, man affects life in the coastal zones of the ocean with thermal pollution from power plants, with hypersaline water from desalinization plants, and with heavy silt from erosion and dredging.

We can expect the public to react eventually to those pollutants that we can see. For example, the fliptop can has been outlawed in the State of Oregon in part because fish were attracted to the flashy rings. However, it is the invisible pollutant that is insidious and that requires the attention of teams of scientists with divergent expertise and good communication. The Pollutant Transfer Program is concentrating on two of these: trace metals and chlorinated hydrocarbons. Analyses are being standardized among laboratories. An enrichment factor (EF) has been formulated to express the level of a pollutant that could be attributed to man's activities. An EF for atmospheric lead, for example, is 2,500 at the South Pole, meaning that in the atmosphere above the South Pole lead is concentrated at 2,500 times that level predicted from natural weathering processes (Zoller et al., 1974).

Turnover and residence times are being measured for specific materials, and their uptake compared with their metabolism. The mucilaginous surfaces of seaweeds will take up some materials at hundreds of thousands of times their concentration in the environment (Harvey, 1972). Although they may not be metabolized, these substances are nevertheless available to the food chain. Careful attention is being given to the path through which a pollutant moves through the food web, and the concentrations at each step are being quantified. Sophisticated environmental control systems can test the effect of varying parameters on uptake, and the microcosm method can single out subsets of information.

Pollutants may be transferred from land to sea via river systems, directly from sewers and industrial plants, and more than is generally realized, through the atmosphere. Gases bring lead from automobiles in the United States to penguins at the South Pole and chlorinated hydrocarbons from insecticides sprayed on orchards have been recovered in the mid-Atlantic. Samples from the atmosphere and surface water of the Sargasso Sea were analyzed by Bidleman and Olney (1974) and found to be higher in DDT, PCB, and chlordane concentrations than previously measured. These data suggested to the authors that residence time is twenty times shorter than had been assumed.

Chlorinated hydrocarbons concentrate in lipids through the food chain. They are persistent (not easily degraded by living organisms) and have been recovered from the Arctic to the Antarctic. Predators at the top of the food chain are the organisms most affected by these materials and therefore the most sensitive to extinction. With the loss of a species we are warned of an ecosystem imbalance that threatens man himself. Even while fish and shellfish are sufficiently bountiful to harvest, they may be inedible because of toxins they contain. These synthetic compounds can contaminate marine "farmyards" in which aquaculture ventures are now being developed. Possibly most significant in the long run, DDT and PCB can reduce photosynthesis in algae, (Wurster, 1968), the major source of oxygen to the atmosphere.

In addition to chemical additives, turbidity can reduce photosynthesis by cutting down light available to algae and seagrasses. This turbidity can result from silt brought down by land erosion, stirred up from dredging operations, and suspended in sewage effluent.

While pollutants can reduce plant growth, supplying an excess nutrient that had previously been a limiting factor can stimulate production. This eutrophication response can cause various problems. One problem arises when excess plant material decomposes, enhancing the biological oxygen demand (Baalsrud, 1967), and another when a species composition shifts. Ryther (1963) documented the latter in Long Island Sound, where the algae on which oysters fed was replaced by indigestible species when nitrogenous compounds were released into the sea from nearby duck farms.

The six papers in this unit discuss ways in which marine communities are altered by known pollutants. Blumer et al. examined the impact of a small oil spill, and from their data the effects of the large tankers of today might be extrapolated. Although they may change the community structure in the process, petroleum products tend to degrade eventually. Plastics, on the other hand, do not break down and their use is increasing. Colton et al. looked for the sources of plastics, their distribution on the Atlantic continental shelf, and their effects on fish. We are now aware of the additional dangers from gases that go into making plastics and from those that are emitted when these materials are burned. The papers by Wurster and Mosser et al. are included because they demonstrate two distinct ecological impacts that arise from DDT: decreased reproduction in a top-order carnivorous bird in Bermuda, and changes in algal species composition. Reviewing the characteristics of the estuary, W. Odum points out how the very processes that cause its high productivity make it vulnerable to manmade stresses. Finally, Thayer et al. explore the impact of man on one aspect of the estuary, the seagrass beds, which are increasingly attracting the attention of scientists.

All the works included in this section of readings require understanding and application of methods and principles of marine ecology. With data generated and interpreted by the scientists, sensitive and responsible legislators have the means to draw up legislation that may yet prevent man from destroying his own lifeline and that of innocent species as well.

LITERATURE CITED

Baalsrud, K. 1967. Influence of nutrient concentrations on primary production. *In* T. A. Olson and F. J. Burgess (eds.), Pollution and Marine Ecology. Wiley-Interscience, New York. pp. 159–169.

Bidleman, T. F., and C. E. Olney. 1974. Chlorinated hydrocarbons in the Sargasso Sea atmosphere and surface water. Science 183:516–518.

Duce, R. A., P. L. Parker, and C. S. Giam. 1974. Pollutant transfer to the marine environment. Deliberations and recommendations of the NSF/IDOE Pollutant Transfer Workshop held in Part Aransas, Texas, January 11–12, 1974. (Published in Kingston, R. I.) 55 pp.

Harvey, G. R. 1972. Adsorption of chlorinated hydro-carbons from sea water by a cross-linked polymer. Woods Hole Oceanographic Institute Technical Report No. 77, 72–66, November, 1972.

Ryther, J. H. 1954. The ecology of phytoplankton blooms in Moriches Bay and Great South Bay, Long Island, New York. Biol. Bull. 106:198–209.

Wurster, C. F., Jr. 1968. DDT reduces photosynthesis by marine phytoplankton. Science 159:1474–1475.

Zoller, W. H., E. S. Gladney, and R. A. Duce. 1974. Atmospheric concentrations and sources of trace metals at the South Pole. Science 183:198–200.

An Ocean of Oil

Oil pollution of the ocean is an increasingly serious global problem. This was the concensus of a large group of scientists from many disciplines who met last fall at the Massachusetts Institute of Technology (MIT) to consider the most pressing problems of man's technological impact on world ecology (see page 48). The conclusion also was voiced by a number of scientists at European conferences late last year sponsored by NATO and the Food and Agriculture Organization of the United Nations. Scientists at the month-long MIT meeting observed that the oils from petroleum are different in composition and toxicity from those occurring naturally in living marine organisms. These differences present a threat to ocean life and ultimately to human welfare, particularly in view of the scope of the pollution. The scientists pointed out that, although major catastrophes such as the wreck of the tanker Torrey Canyon in 1967 or the Santa Barbara Channel leak in 1969 receive the headlines, the smaller day-to-day spills in coastal waters and harbors of the world produce chronic pollution that is much larger in total volume and probably more severe in biological consequences.

The following article examines the impact of one localized oil spill that would have received only passing attention had it not occurred near the Woods Hole Oceanographic Institution in Massachusetts. An interdisciplinary group of experienced scientists at the institution spent many months examining the after effects of the accident. Their conclusions have far-reaching implications for the rising tide of oil pollution around the globe. Chronic oil pollution contaminates nearshore waters that are key to the survival of most marine animals that are taken for man's food. Over a long period of time, this persistent pollution may interfere with the normal life processes of the organisms — as well as killing them outright at high concentrations. The result, as in the West Falmouth story that follows, may be progressive disappearance of usually abundant fish and shellfish. Their decline would be accompanied by an increase in pollution-tolerant species that generally indicate an unhealthy state of biological affairs. Furthermore, remaining organisms of food value to man may be permanently contaminated with petroleum hydrocarbons that could be hazardous to health.

Because of this, chronic oil spills can be called "small" only in a relative sense. Each one further contributes to the deterioration of the marine environment. Meanwhile, there is no letup in single, massive oil accidents around the world. The largest ones over the past year or so are summarized in the special Environment report that follows the article by Dr. Blumer and his colleagues.   J.M.

A SMALL

502

Reprinted from Environment 13:2-12 (1971).

URING THE LAST FEW years the public has become increasingly aware of the presence of oil on the sea. We read about the recurring accidents in oil transport and production, such as the disaster of the *Torrey Canyon* tanker, the oil well blowout at Santa Barbara, and the oil well fires in the Gulf of Mexico. To those visiting our shores the presence of oil on rocks and sand has become an everyday experience; however, few of us realize that these spectacular accidents contribute only a small fraction of the total oil that enters the ocean. In the *Torrey Canyon* episode of 1967 about 100,000 tons of crude oil were lost. By comparison, routine discharges from tankers and other commercial vessels contribute an estimated three and one-half million tons of petroleum to the ocean every year. In addition, pollution from accidents in port and on the high seas, in exploration and production, in storage, in pipeline breaks, from spent lubricants, from incompletely burned fuels, and from untreated industrial and domestic sewage contribute an equal or larger amount of oil. Thus, it has been estimated that the total oil influx into the ocean is between five and ten million tons per year.[1]

What are the effects of oil on marine organisms and on food that we recover from the sea? Some scientists have said that the oceans in their vastness should be capable of assimilating the entire oil input. This, however, assumes that the oil is evenly distributed through the entire water profile, or water column, of the ocean. Unfortunately this assumption is not correct. Oil production, transportation, and use are heavily concentrated in the coastal regions, and pollution therefore predominately affects the surface waters on the continental margins. J. H. Ryther has stated that the open sea is virtually a biological desert.[2] Although the deeper ocean pro-

The authors are members of the staff of Woods Hole Oceanograhic Institution, Woods Hole, Massachusetts. MAX BLUMER, Ph.D., and HOWARD L. SANDERS, Ph.D., are senior scientists. J. FRED GRASSLE, Ph.D., is assistant scientist, and GEORGE R. HAMPSON, B.S., is research associate.

vides some fishing for tuna, bonito, skipjack, and billfish, the coastal waters produce almost the entire shellfish crop and nearly half of the total fish crop. The bulk of the remainder of the fish crop comes from regions of upwelling water, near the continental margins, that occupy only one-tenth of one percent of the total surface area of the seas. These productive waters receive the heaviest influx of oil. They also are most affected by other activities of man, such as dredging, waste disposal, and unintentional dispersal of chemical poisons like insecticides.

Some environmentalists have expressed the belief that major oil spills such as those from the *Torrey Canyon* and the blowout at Santa Barbara have brought about little biological damage in the ocean.[3] These statements are largely based on statistical measurements of the catch of adult fish. We believe that such statistics are a very insensitive measure of the ecologic damage to wide oceanic regions. Often the migratory history of the fish species studied is unknown. The fish may not have been exposed to the spill itself, or may not have suffered from a depletion of food organisms if their growth occurred in areas remote from the spill. Statistical and observational data on adult fishes will not reveal damage to the often much more sensitive juvenile forms or to intermediate members in the marine food chain. The only other studies on effects of oil on marine organisms have concentrated on relatively tolerant organisms which live between the tides at the margins of affected areas. The main impact, however, would be

OIL SPILL

By Max Blumer, Howard L. Sanders,
J. Fred Grassle, and George R. Hampson

503

expected in subtidal areas, and that has never been measured quantitatively.

A relatively small oil spill that occurred almost at the doorstep of the Woods Hole Oceanographic Institution at Woods Hole, Massachusetts, gave us the opportunity to study immediate and long-term ecological damage in a region for which we had extensive previous knowledge about the biology and chemistry of native marine organisms.[4] On September 16, 1969, an oil barge on the way to a power plant on the Cape Cod Canal came ashore off Fassets Point, West Falmouth, in Buzzards Bay (see map below). Between 650 and 700 tons of #2 fuel oil were released into the coastal waters. The oil-contaminated region in Buzzards Bay expanded steadily with time after the accident as the complex inter-

action of wind, waves, and bottom sediment movement spread oil from polluted to unpolluted areas. Eight months after the grounding, polluted sea bottom, marshes, and tidal rivers comprised an area many times larger than that first affected by the accident. The dispersion was much greater than expected on the basis of conventional studies of oil pollution. The situation even forced changes in our research efforts. As we shall explain later, a control point for marine surveys was established beyond the anticipated limit of the spread of oil. Within three weeks, the contamination had spread to the station. Another was established twice as far away. Three months after the accident, that too was polluted. Bottom sediment was contaminated 42 feet beneath the surface, the greatest water depth in that

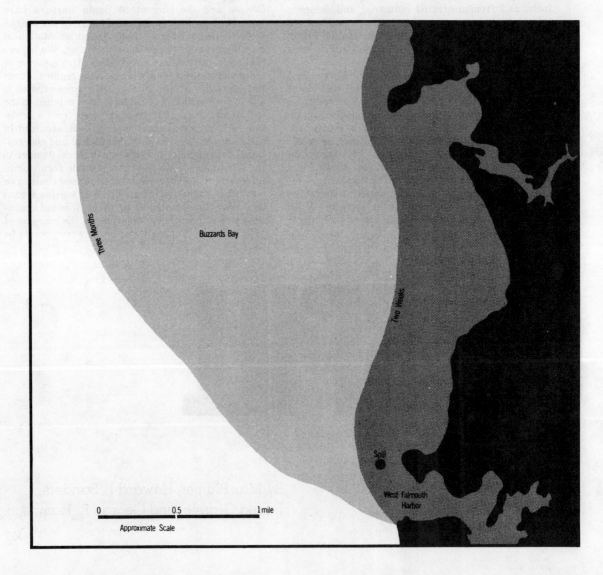

part of Buzzards Bay.

Ecological effects of the spreading blanket of oil beneath the surface were severe. The oil decimated offshore marine life in the immediate area of the spill during the first few days. As the oil spread out across the bottom of the bay in the following months, it retained its toxicity.

Even by May 1970, eight months after the spill, bacterial degradation (breakdown into simpler substances) of oil was not far advanced in the most polluted regions. More rapid oil deterioration in outlying, less affected areas had been reversed by a new influx of less degraded oil from the more contaminated regions.

The tidal Wild Harbor River still contained an estimated four tons of fuel oil. The contamination

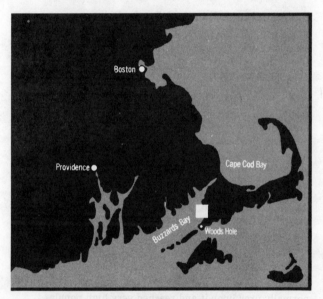

A barge loaded with fuel oil for a power plant went aground in September 1969 in Buzzards Bay, Massachusetts, shown in the map above. An interdisciplinary team of scientists from nearby Woods Hole Oceanographic Institution discovered that poisonous substances in the oil persisted in the marine environment for much longer than previously thought. Oil contamination of bottom sediment spread up into the bay with time; the map at left, an enlargement of the white area in the map above, shows the approximate boundary of the oil in the study area at two weeks and at three months after the spill.

Coastal waters, the most productive areas of the ocean, receive the heaviest influx of oil.

had ruled out commercial shellfishing for at least two years. The severe biological damage and the slow rate of biodegradation of the oil suggests that shellfish productivity will be affected for an even longer period. Furthermore, destruction of bottom plants and animals reduced the stability of marshlands and sea bottom. Resulting erosion may have promoted spread of the oil along the sea floor. Inshore, the oil penetrated to a depth of at least one to two feet in marsh sediment.

Nevertheless, compared in magnitude to other catastrophes, this was a relatively small spill; the amount of oil lost in the *Torrey Canyon* accident was 150 times larger. The interim results of our survey, coupled with research findings of other studies in this laboratory, indicate that crude oil and other petroleum products are a far more dangerous and persistent threat to the marine environment and to human food resources than we would have anticipated. Pollution from a large oil spill is very obvious and visible. It has often been thought that the eventual disappearance of this visible evidence coincides with the disappearance of any biological damage. This, however, is not true. Sensitive analytical techniques can still detect oil in marine organisms and in sediments after the visual evidence has disappeared, and biological studies reveal that this residual oil is still toxic to the marine organisms. Here we shall discuss first the general results of our study, then go more deeply into the description of the laboratory work involving biology, biochemistry, and chemistry. Our most important findings are these:[4]

Crude oil and petroleum products contain many substances that are poisonous to marine life. Some of these cause immediate death; others have a slower effect. Crude oils and oil products differ in their relative composition; therefore the specific toxic effect may vary. Crude oil, in general, is less immediately toxic than some distilled products, but even crude oil that has been weathered (altered by exposure to the weather) at sea for some time still contains many of the acutely toxic hydrocarbons.[5] The more persistent, slowly acting poisons (for example, the carcinogens) are more abundant in crude oil than in some of the lower boiling distillates. These poisons are quite resistant to the environmental weathering of oil.

In spite of low density, oil may mix with water, especially in a turbulent sea during storm conditions. Hydrocarbons may be dispersed through the water column in solution in the form of droplets, and the compounds may reach the sea bottom, particularly if weighted down by mineral particles. On the sea floor oil persists for long periods and can continue to damage bottom plants and animals. Thus, a single accident may result in long-term, continual pollution of the sea. This is a very important finding since biologists have long agreed that chronic pollution generally has more far-reaching effects than an accident of short duration. Hydrocarbons can be taken up by fish and shellfish. When the oil enters the fat and flesh of the animals, it is isolated from natural degradation processes. It remains essentially constant in amount and chemically intact even after the animals are transplanted into clean water for decontamination. Thus, chemicals from oil that may be poisonous to marine organisms and other animals, including man, may persist in the sea and in biological systems for many months after the spill.

By killing the bottom organisms, oil reduces cohesion of the bottom sediments and thereby accelerates transport of the sediments. Sediment movements along the sea bottom thus are a common occurrence after an oil spill. In this way contaminated sediments may be spread over great distances under the influence of tide and wave action, and the oil may be carried to areas not immediately polluted by the spill.

None of the presently available countermeasures can completely eliminate the biological damage of oil spills. The rapid removal of oil by mechanical recovery or by burning appears most promising. The use of sinking agents or detergents, on the other hand, causes the toxic and undegraded oil to spread in the ocean; the biological damage is then greater than if the spill had been left untreated. Reclamation of contaminated organisms, marshes, and offshore sediments is virtually impossible, and natural ecological recovery is slow.

With these conclusions in mind we can now turn to our experience with the West Falmouth oil spill. The effect of this relatively small spill was still acute in January 1971, almost a year and one-half after the accident. Officials in the town of Falmouth have estimated that the damage to local shellfish resources, during the first year after the accident, amounted to $118,000. This does not include the damage to other marine species and the expected damage in coming years. In addition to the loss of the oil and the barge and the cleanup expenses (estimated to be $65,000), the owner of the oil paid compensations for the losses of marine fishery resources to the town of Falmouth ($100,000) and to

None of the presently available countermeasures can completely eliminate the biological damage of oil spills.

the Commonwealth of Massachusetts ($200,000). The actual ecological damage may far exceed this apparent cost of almost half a million dollars.

Biological and Chemical Analysis

For our analysis (which is still continuing) bottom samples were carefully taken from the marshes and from the offshore areas. Samples for biological analysis were washed and sieved to recover living or dead organisms. These were preserved, identified, and counted. Results of counts from the affected area were compared with those from control areas that were not polluted by the spill. Some animals can be used as indicators for the presence of pollution, either because of their great sensitivity or because of their great resistance. Thus, small shrimplike animals, the amphipods of the family Ampeliscidae, are particularly vulnerable to oil pollution. Wherever the chemical analysis showed the presence of oil, these sensitive crustaceans were dying. On the other hand, the annelid worm, *Capitella capitata*, is highly resistant to oil pollution. Normally, this worm does not occur in large numbers in our area. However, after the accident it was able to benefit from the absence of other organisms which normally prey upon it and reached very high population densities. In the areas of the highest degree of pollution, however, even this worm was killed. *Capitella capitata* is well known, all over the world, as characteristic of areas heavily polluted by a variety of sources.

For chemical analysis, the sediments collected at our biological stations were extracted with a solvent that removed the hydrocarbons. The hydrocarbons were separated from other materials contained in the extracts. They were then analyzed by gas-liquid chromatography. This technique separates hydrocarbon mixtures into individual compounds, according to the boiling point and structural type. To do this, a sample is flash-evaporated in a heated tube. The vapor is swept by a constantly flowing stream of carrier gas into a long tube that is packed with a substance (substrate) that is responsible for the resolution of the mixture into its individual components. Ideally, each vaporized compound emerges from the end of the tube at a definite time and well separated from all other components. A sensitive detector and an amplifier then transmit a signal to a recorder which traces on a moving strip of chart paper a series of peaks (the chromatogram) that correspond to the individual components of the mixture. From the pattern of peaks in the gas chromatogram the chemist can learn much about the composition of the mixture. Each oil may have a characteristic fingerprint pattern by which it can be recognized in the environment for weeks, or even months, after the initial spill. Past and continuing work on the composition of those hydrocarbons that are naturally present in all marine organisms (see box, "What is Petroleum?") enabled us to distinguish easily between the natural hydrocarbons and those contained in the fuel oil. These analyses facilitated our study of the movement of the fuel oil from the West Falmouth oil spill into the bottom sediments and through the marine food chain.

Immediate Kill

Massive, immediate destruction of marine life occurred offshore during the first few days after the accident. Affected were a wide range of fish, shellfish, worms, crabs, other crustaceans, and invertebrates. Bottom-living fish and lobsters were killed and washed up on the shores. Trawls made in ten feet of water soon after the spill showed that 95 percent of the animals recovered were dead and others were dying. The bottom sediments contained many dead snails, clams, and crustaceans. Similarly severe destruction occurred in the tidal rivers and marshes into which the oil had moved under the combined influence of tide and wind. Here again fish, crabs, shellfish, and other invertebrates were killed; in the most heavily polluted regions of the tidal marshes almost no animals survived.

The fuel oil spilled at West Falmouth was a light, transparent oil, very different from the black viscous oil associated with the *Torrey Canyon* and Santa Barbara episodes. Within days most of the dead animals had decayed and the visual evidence of the oil had almost disappeared. Casual observers were led to report to the press that the area looked as beautiful as ever. Had we discontinued our study after the visual evidence of the oil had disappeared, we might have been led to similar interpretations.

It is estimated that the total oil influx into the ocean is between five and ten million tons per year.

From that point on, only continued, careful biological and chemical analysis revealed the extent of continuing damage.

Persistence of Pollution

Quite recently a leading British expert on treatment of oil spills remarked that "white products, petrol, kerosene, light diesel fuel, and so forth, can be expected to be self-cleaning. In other words, given sufficient time they will evaporate and leave little or no objectionable residue."[6] Our experience shows how dangerously misleading such statements are. Chemical analyses of the oil recovered from the sediments and from the bodies of the surviving animals showed the chromatographic fingerprint of the diesel fuel, in monotonous repetition, for many months after the accident.

Bacteria normally present in the sea will attack and slowly degrade spilled oil. On the basis of visual observations it has been said that the oil spilled by the *Torrey Canyon* disappeared rapidly from the sediments. This was interpreted to mean that the action of the bacteria was "swift and complete." Our analyses, which were carried out by objective chemical, rather than by subjective observational techniques, showed the steady persistence of fuel oil that should, in principle, be even more rapidly degraded than a whole crude oil. Thus, in May 1970, eight months after the spill, oil essentially unaltered in chemical characteristics could still be recovered from the sediments of the most heavily polluted areas. By the end of the first year after the accident, bacterial degradation of the oil was noted at all locations, as evidenced by changes in the fingerprint pattern of the oil. Yet only partial detoxification of the sediments had occurred, since the bacteria attacked the least toxic hydrocarbons first. The more toxic aromatic hydrocarbons remained in the sediments.

Spread of Pollution

For our chemical and biological work we established an unpolluted control station, outside of the area that was polluted, immediately after the accident. For a short period after the accident the sediments at this station were still clean and the organisms alive in their normal abundance and distribution. However, within three weeks, oil was found at this station and a significant number of organisms had been killed. Another control station was established twice as far from shore. Within three months fuel oil from the spill was evident at this station, and again there was a concomitant kill of bottom-living animals. This situation was repeated several times in sequence, and by spring 1970 the pollution had spread considerably from the area affected initially. At that time, the polluted area offshore was ten times larger than immediately after the accident and covered 5,000 acres (20 square kilometers) offshore and 500 acres (2 square kilometers) in the tidal river and marshes.

Another significant observation was made in the spring of 1970: Between December 1969 and April 1970, the oil content of the most heavily contaminated marine station two and one-half miles north of the original spill increased tenfold. Similar but smaller increases were observed at about the same time at other stations more distant from shore. The oil still showed the typical chromatographic fingerprint of the diesel fuel involved in the September 1969 oil spill. This and the lack of any further accident in this area suggested that oil was spreading from the most heavily contaminated inshore regions to the offshore sediments. We believe that the increase in the pollution level and the spread of oil to outlying areas are related to a transportation mechanism that we do not yet fully understand. However, the drastic kill of the animals that occurred with the arrival of oil pollution at the offshore stations showed that mortality continued for many months after the initial spill, even though no visible evidence of oil remained on the shores.

We believe these observations demonstrate that chronic oil pollution can result from a single spill, that the decimation of marine life can extend to new regions long after the initial spill, and that, once poisoned, the sea bottom may remain toxic to animals for long time periods.

Destruction of Shellfish Resources

Our analyses showed that oysters, soft-shell clams, quahaugs (another variety of clam), and scallops took up the fuel oil. Because of the pollution, the contaminated regions had to be closed to the harvesting of shellfish. Continuing analyses revealed that the contamination of the 1970 shellfish crop was as severe as that of the 1969 crop. Blue mussels that were juveniles in the polluted area at the time of the spill generally were sexually sterile the next season—they developed almost no eggs or sperm. Furthermore, in 1970 distant areas contained shellfish contaminated by fuel oil. Therefore, harvesting prohibitions had to be maintained in 1970 and had to be extended to polluted shellfish grounds that had not been closed

Crude oil and petroleum products contain many substances that are poisonous to marine life.

to the public immediately after the accident.

It has long been common to transfer shellfish polluted by human sewage into clean water to make the animals marketable again. It has been thought that a similar flushing process would remove the oil from animals exposed to oil. Indeed, taste tests showed that the objectionable oily taste disappeared from animals maintained for some period in clean water. However, we removed oysters from the contaminated areas and kept them in clean running sea water up to six months. Fuel oil was still found in the animals by chemical analysis at essentially the same concentration and in the same composition as at the beginning of the flushing period.

Thus, we discovered that hydrocarbons taken up into the fat and flesh of fish and shellfish are not removed by natural flushing or by internal metabolic processes. The substances remain in the animals for long periods of time, possibly for their entire lives. The presence or absence of an oily taste or flavor in fish products is not a measure of contamination. The reason is that only a relatively small fraction of the total petroleum product has a pro-

Falmouth Enterprise

Dead marine animals (left) litter the
Wild Harbor tidal flats polluted by
toxic fuel oil after the oil barge
Florida accident in Buzzards Bay,
Massachusetts.

The *Florida* with other boats in
Buzzards Bay. Between 650 to
700 tons of #2 fuel oil were
released into the coastal waters
after the barge's accident.

nounced taste or odor. Subjective observations cannot detect the presence of the toxic but tasteless and
odorless pollutants. Only objective chemical analysis
measures the presence of these chemical poisons. It
is important to note in this regard that state and
federal laboratories in the public health sector are
not generally equipped to carry out these important
chemical measurements. Such tests are vital, however, for the protection of the consumer.

Thus, our investigation demonstrated that the
spill produced immediate mortality, chronic pollution, persistence of oil in the sediments and in the
organisms, spread of pollution with the moving sediments, destruction of fishery resources, and continued harm to fisheries for a long period after the
accident. Our continuing study will assess the persistence and toxicity of the oil and the eventual
ecological recovery of the area. At the present time,
one and one-half years after the spill, only the pollu-
tion-resistant organisms have been able to reestab-
lish themselves in the more heavily contaminated
regions. The original animal populations there have
not become reestablished. Many animals that are

able to move, early in their life cycles, as free-swimming larvae reach the polluted area and are killed
when they settle on the sea bottom or in the marshes
at West Falmouth.

In addition, revitalization of bottom areas probably will be hampered by oxygen depletion caused
by oxygen-requiring bacteria that degrade oil.[7]

The Significance of West Falmouth

Some scientists are convinced that the effects at
West Falmouth are a special case and have little
applicability to spills of whole, unrefined crude oils.
They contend that #2 fuel oil is more toxic than
petroleum and that therefore it has effects that
would not be comparable to those of whole petroleum. We cannot agree with this view.

Fuel oil is a typical oil-refining product. It is frequently shipped by sea, especially along coastal
routes, and it is spilled in accidents like those which
occurred at West Falmouth and off Baja California
following the grounding of the *Tampico Maru* in
1957.[8]

More importantly, fuel oil is a part of petroleum, and as such it is contained within the whole petroleum. Surely, hydrocarbons that are toxic when they are in fuel oil must also be toxic when they are contained in petroleum. Therefore, the effects observed in West Falmouth are typical both for that fuel oil and the whole crude oil. In terms of chemical composition, crude oils span a range of molecular weights and structures. Many light crude oils have a composition not too dissimilar from that of fuel oil, and their toxicity and effects on the environment are very similar. Other heavier crude oils, while still containing the fuel oil components, contain higher proportions of the long-lasting poisons that are much more persistent and that include, for instance, some compounds that are potent carcinogens (cancer-producing agents) in experimental animals. Such heavy crude oils can be expected to be more persistent than a fuel oil, and they will have longer lasting long-term effects. Even weathered crude oils may still contain these long-term poisons, and in many cases some of the moderately low-boiling, immediately toxic compounds. In our view, these findings differ from those of other investigators principally for two reasons: Our study is based on objective measurement and is not primarily concerned with the mobile, adult marine species—the fish whose migratory history is largely unknown—or the highly resistant intertidal forms of life. We are studying quantitatively the effects of the spill on the sessile (bottom) animals that

What Is Petroleum?

Organic materials, deposited at the bottom of the sea millions of years ago have been covered by sediments and deeply buried. Under the influence of elevated temperature over very long periods of time, an immensely complex mixture of hydrocarbons has been formed. Some of these have accumulated in reservoirs from which crude oil can be procured.

Crude oil is one of the most complicated natural mixtures on earth. Compounds made up of only carbon and hydrogen predominate, but small amounts of sulfur-, oxygen-, and nitrogen-containing substances also occur.[12]

The way in which the individual carbon and hydrogen atoms combine into hydrocarbon molecules helps scientists to classify them. They distinguish four principal types of hydrocarbons:

The first type, **the aliphatic compounds,** includes straight and branched chain compounds in which each carbon atom is directly linked to four other a t o m s (saturated). Aliphatic compounds frequently account for a large fraction of crude oil and are common in gasoline and many other fuels.

The second general type is **the alicyclic compounds** (naphthenes). These compounds are also saturated, but the carbon atoms of at least part of the molecules are joined in rings.

The third major type, also cyclic, consists of **the aromatic compounds.** These contain at least one benzene ring. This type includes a large number of one-ring, two-ring, and multi-ring compounds, among them several materials that have been implicated as potent carcinogens (cancer-producing agents) in laboratory animals.

The fourth type, **the olefinic compounds,** are unsaturated. Here, double or triple chemical bonds between carbon atoms exist, but not of the regular arrangements found in the benzene ring. Olefins do not occur in crude oil, but are formed in some refining processes and are common in many oil products.

The boiling point is an important physical property of the hydrocarbons. Differences in boiling point between different crude-oil hydrocarbons are useful in separating the oil into fractions with individual characteristics suited to specific fuels or lubricants. Crude oil contains components boiling over a range from below room temperature to well above 500 degrees C. The lowest boiling fractions of crude oil are relatively rich in the simpler chain- and ring-type saturated hydrocarbons. Intermediate fractions have a higher content of the immediately poisonous aromatic hydrocarbons. Conversely, the higher boiling hydrocarbon fractions contain relatively more of the complex polycyclic aromatic compounds, including the carcinogens.

Hydrocarbons are formed by all living organisms. The hydrocarbons in crude oil are very different from those normally found in healthy unpolluted organisms, however. The crude oil mixture is far more complex, the compounds cover a much wider range in structure and boiling point, and many hydrocarbons are present that are toxic to organisms. As a rule, only very few individual natural hydrocarbon compounds are found in unpolluted plants and animals. They are mostly saturated or olefinic, and with a few exceptions they are nontoxic.[13]   M.B.

cannot escape the spill or the polluted sediment and that are thus exposed to chronic pollution. Since all classes of bottom animals are severely affected by the oil, we believe that the effects on free-swimming animals should be just as drastic. The difficulty of measuring the total impact of oil on the marine life has led many to doubt the ecological seriousness of oil pollution. Our findings, extending far beyond the period when the visual evidence of the oil had disappeared, are based on objective chemical analyses and quantitative biological measurements, rather than on subjective visual observations. They indict oil as a pollutant with severe biological effects.

It is unfortunate that oil pollution research has been dominated so strongly by subjective, visual observations. Clearly, oil is a *chemical* that has severe *biological* effects, and therefore oil pollution research, to be fully meaningful, must combine chemical with biological studies. Those few investigators who are using objective chemical techniques find patterns in the environmental damage by oil that are similar to those demonstrated by the West Falmouth spill. Thus, R. A. Kolpack reported that oil from the blowout at Santa Barbara was carried to the sea bottom by clay minerals and that within four months after the accident the entire bottom of the Santa Barbara basin was covered with oil from the spill.[9] Clearly, this is one of the most significant observations in the aftermath of that accident. A concurrent and complimentary biological study would have appreciably enhanced our understanding of the ecological damage caused by the Santa Barbara oil spill.

G. S. Sidhu and co-workers, applying analytic methods similar to those used by us, showed that the mullet, an edible finfish, takes up petroleum hydrocarbons from waters containing low levels of oil pollution from refinery outflows. In their chemical structures the hydrocarbons isolated by the investigators are similar to those found in the polluted shellfish of West Falmouth. The compounds differ markedly from those hydrocarbons present as natural components in all living organisms, yet closely approximate the hydrocarbons in fossil fuels.[10]

Numerous results of crude-oil toxicity tests, alone or in the presence of dispersants, have been published in the literature. However, in almost all cases such tests were performed on relatively hardy and resistant species that can be kept in the laboratory and on adult animals for short time periods under unnatural conditions or in the absence of food. At best, such tests may establish only the relative degree of the toxicity of various oils. We are convinced that the exposure of more sensitive animals, especially young ones, to oil pollution over many months would demonstrate a much greater susceptibility to the damaging effects of the oil. Such effects have been demonstrated in the studies of the West Falmouth oil spill. These studies represent a meaning-

Hydrocarbons, long-lasting poisons, enter the marine food chain after oil spills and may subsequently enter the human food chain.

ful field test in open waters.

Thus, we believe that the general toxic potential and the persistence of the West Falmouth oil are typical of most oils and oil products both at the sea bottom and in the water column.

Conclusions

Our analysis of the aftermath of the West Falmouth oil spill suggests that oil is much more persistent and destructive to marine organisms and to man's marine food resources than scientists had thought. With the advent of objective chemical techniques, oil pollution research has entered a new stage. Earlier interpretations of the environmental effect of oil spills that were based on subjective observation, often over a short time span, have questionable validity. Crude oil and oil products are persistent poisons, resembling in their longevity DDT, PCB and other synthetic materials [which have been discussed in these pages]. Like other long-lasting poisons that, in some properties, resemble the natural fats of the organisms, hydrocarbons from oil spills enter the marine food chain and are concentrated in the fatty parts of the organisms. They can then be passed from prey to predator where they may become a hazard to marine life and even to man himself.

Natural mechanisms for the degradation of oil at sea exist—the most important of which is bacterial decomposition. Unfortunately, this is least effective for the most poisonous compounds in oil. Also, oil degrades slowly only in marine sediments, and it may be completely stable once it is taken up by organisms. It has been thought that many of the immediately toxic low-boiling aromatic hydrocarbons are volatile and evaporate rapidly from the oil spilled at sea. This has not been the case at West Falmouth, where the low-boiling hydrocarbons found their way into the sediments and organisms. We believe that the importance of evaporation has been overestimated.

Oil-laden sediments can move with bottom currents and can contaminate unpolluted areas long after the initial accident. For this reason a *single* and relatively small spill may lead to *chronic*, destructive pollution of a large area.

We have not yet discussed the low-level effects of oil pollution. However, a growing body of evidence indicates that oil as well as other pollutants may have seriously damaging biological effects at extremely low concentrations, previously considered harmless. Some of this information was presented in Rome at the December 1970 Food and Agriculture Organization's Conference on the Effects of Marine Pollution on Living Resources and Fishing. Greatly diluted pollutants affect not only the physiology but also the behavior of many animals. Many behavioral patterns which are important for the survival of marine organisms are mediated by extremely low concentrations of chemical messengers that are excreted by marine creatures. Chemical attraction and repulsion by such compounds play a key role in food finding, escape from predators, homing, finding of habitats, and sexual attraction. Possibly, oil could interfere with such processes by blocking the taste receptors of marine animals or by mimicking natural stimuli and thus eliciting false responses. Our general ignorance of such low-level effects of pollution is no excuse for neglecting research in these areas nor for complacency if such effects are not immediately obvious in gross observations of polluted areas.

Recent reports suggest an additional environmental threat from oil pollution. Oil may concentrate other fat-soluble poisons, such as many insecticides and chemical intermediates.[11] Dissolved in an oil film, these poisons may reach a concentration many times higher than that which occurs in the water column. In this way other pollutants may become available to organisms that would not normally be exposed to the substances and at concentrations that could not be reached in the absence of oil.

The overall implications of oil pollution, coupled with the effects of other pollutants, are distressing. The discharge of oil, chemicals, domestic sewage, and municipal wastes, combined with overfishing, dredging, and the filling of wetlands may lead to a deterioration of the coastal ecology. The present influx of pollutants to the coastal regions of the oceans is as damaging as that which has had such a detrimental effect on many of our lakes and freshwater fishery resources. Continued and progressive damage to the coastal ecology may lead to a catastrophic deterioration of an important part of marine resources. Such a deterioration might not be reversed for many generations and could have a deep and lasting impact on the future of mankind.

Since present oil-spill countermeasures cannot completely eliminate the biological damage, it is paramount to prevent oil spills. The recent commitment by the United States to take all steps to end the intentional discharge of oil from its tankers and nontanker vessels by the mid 1970s is important. As a result of this step and of the resolution of the NATO Ocean Oil Spills Conference of the Committee on Challenges of Modern Society in Brussels, December 1970, other countries hopefully also will adopt necessary measures to halt oil pollution from ships. This would eliminate the largest single source of oceanic oil pollution. At the same time steps also must be taken to reduce oil pollution from many other, less readily obvious sources, such as petrochemical operations on shore, disposal of automotive and industrial lubricants, and release of unburned hydrocarbons from the internal combustion engine.   □

Acknowledgments: The authors acknowledge the continued support of their basic and applied research efforts by the National Science Foundation, The Office of Naval Research, and the Federal Water Quality Administration.

NOTES

1. Blumer, M., "Scientific Aspects of the Oil Spill Problem," paper presented at the Oil Spills Conference, Committee on Challenges of Modern Society, NATO, Brussels, Nov. 1970.

2. Ryther, J. H., "Photosynthesis and Fish Production in the Sea," *Science*, 166:72-76, 1969.

3. McCaull, Julian, "The Black Tide," *Environment*, 11(9): 10, 1969.

4. Blumer, M., G. Souza, J. Sass, "Hydrocarbon Pollution of Edible Shellfish by an Oil Spill," *Marine Biology*, 5(3):195-202, March 1970. Blumer, M., J. Sass, G. Souza, H. L. Sanders, J. F. Grassle, G. R. Hampson, "The West Falmouth Oil Spill," Reference No. 70-44, unpublished manuscript available from senior author, Woods Hole Oceanographic Institution, Woods Hole, Massachusetts, September 1970.

5. Blumer, M., G. Souza, J. Sass, "Hydrocarbon Pollution," *op. cit.*, p. 198.

6. Smith, J. Wardly, "Dealing with Oil Pollution Both on the Sea and on the Shores," paper presented to the Ocean Oil Spills Conference, Conference on Challenges of Modern Society, NATO, Brussels, November 1970.

7. Murphy, T. A., "Environmental Effects of Oil Pollution," presented at American Society of Civil Engineers, Boston; available from author at Edison Water Quality Laboratory, Edison, New Jersey, p. 14-15, July 13, 1970.

8. Jones, Laurence G., Charles T. Mitchell, Einar K. Anderson, and Wheeler J. North, "A Preliminary Evaluation of Ecological Effects of an Oil Spill in the Santa Barbara Channel," W. M. Keck Engineering Laboratories, California Institute of Technology.

9. Kolpack, R. A., "Oil Spill at Santa Barbara, California, Physical and Chemical Effects," paper presented to the FAO Technical Conference on Marine Pollution, Rome, Dec. 1970.

10. Sidhu, G. S., G. L. Vale, J. Shipton and K. E. Murray, "Nature and Effects of a Kerosene-like Taint in Mullet," paper presented to FAO Technical Conference on Marine Pollution, Rome, Dec. 1970.

11. Hartung, R., and G. W. Klinger, "Concentration of DDT by Sedimented Polluting Oils," *Environmental Science and Technology*, 4:407, 1970.

12. Gruse, W. A., and D. R. Stevens, *The Chemical Technology of Petroleum*, Mellon Institute of Industrial Research, Second Edition, McGraw-Hill Book Company, Inc., New York, p. 2, 1942.

13. For example, see: Clark, R. C. and M. Blumer, "Distribution of n-paraffins in Marine Organisms and Sediments," *Limnology and Oceanography*, 12:79-87, 1967.

Plastic Particles in Surface Waters of the Northwestern Atlantic

The abundance, distribution, source, and significance of various types of plastics are discussed.

John B. Colton, Jr., Frederick D. Knapp, Bruce R. Burns

The occurrence of plastic particles has recently been reported in the Sargasso Sea (1) and in coastal waters of southern New England (2, 3). These reports were based on a small number of samples within limited geographic areas, but the observers suggested that plastics might be more widely distributed. We confirm, after examination of neuston (surface) net samples taken in July–August 1972, that plastic particles do occur over a wide area of the North Atlantic.

These samples were collected on the first multiship MARMAP (4) ichthyo-plankton survey of coastal and oceanic waters from Cape Cod to the Caribbean. The three National Oceanic and Atmospheric Administration research vessels participating in this survey were the *Albatross IV*, *Delaware II*, and *Oregon II*. The plankton sampling locations for each vessel are shown in Fig. 1. The sampling gear (2 by 1 meter, rectangular-framed neuston net with 0.947-millimeter nylon mesh) and method of tow [10-minute surface tow at a speed of 5 knots (9.25 kilometers per hour)] were identical on each vessel. The plastic particles in each sample were manually sorted, enumerated by type, and their dry weight determined.

Plastic Types Collected

The types and characteristics of the plastic particles were as follows:

1) White opaque polystyrene spherules (5); mean diameter, 1.0 mm; range in diameter, 0.2 to 1.7 mm; mean weight, 0.0007 g; range in weight, 0.0001 to 0.00023 g; mean density, 1.023 g/cm³; range in density, 1.010 to 1.047 g/cm³ (Fig. 2A).

2) Translucent to clear polystyrene spherules containing gaseous voids (5); mean diameter, 1.5 mm; range in diameter, 0.9 to 2.5 mm; mean weight, 0.0014 g; range in weight, 0.0004 to 0.0039 g; density, < 1.000 g/cm³ (Fig. 2B).

3) Opaque to translucent polyethylene cylinders or disks (5); mean diameter, 3.4 mm; range in diameter, 1.7 to 4.9 mm; mean thickness, 2.0 mm; range in thickness, 1.1 to 3.4 mm; mean weight, 0.0138 g; range in weight, 0.0106 to 0.0250 g; density, < 1.000 g/cm³ (Fig. 2C).

4) Pieces of Styrofoam (Fig. 2D).

5) Sheets of thin, flexible wrapping material (Fig. 2E).

6) Pieces of hard and soft, clear and opaque plastics of various thicknesses which appear to be parts of plastic containers, toys, and so forth (Fig. 2F).

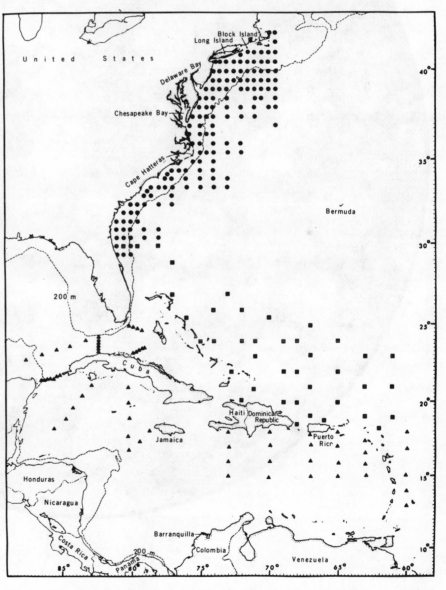

Fig. 1. Neuston net sampling locations: ■, *Albatross IV* cruise 72-6, 11 July–16 August 1972; ●, *Delaware II* cruise 72-19, 12 July–13 August 1972; ▲, *Oregon II* cruise 39, 13 July–8 August 1972.

Mr. Colton and Mr. Burns are fishery biologists at the National Marine Fisheries Service, Northeast Fisheries Center, Narragansett Laboratory, Narragansett, Rhode Island 02882. Mr. Knapp is a biologist at the Raytheon Company, Office of Environmental Services, Portsmouth, Rhode Island 02871.

513

Reprinted from Science 185:491-497 (1974).

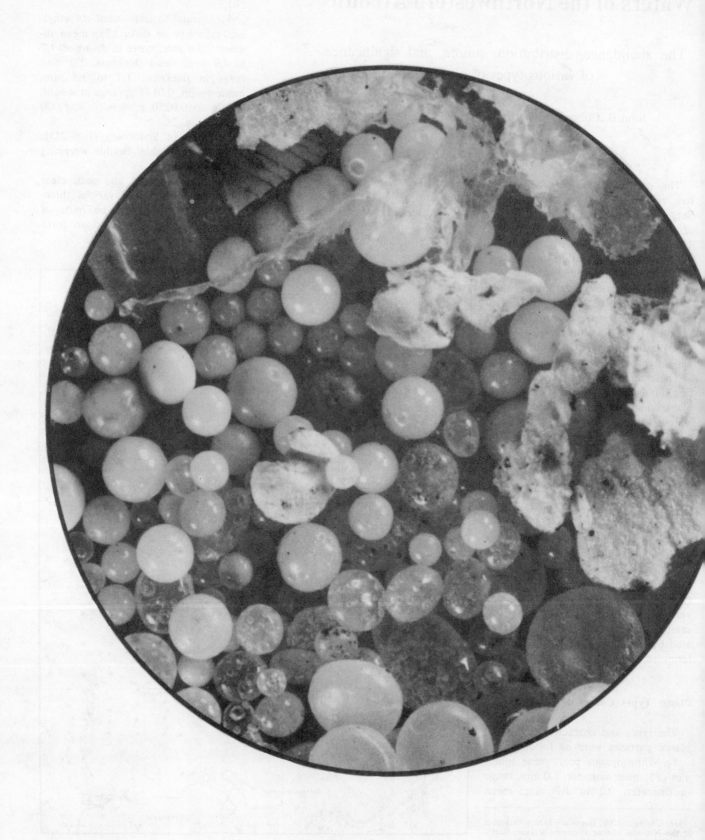

Sampling Bias

Under normal towing conditions (that is, with the net meshes not clogged) appreciable numbers of particles smaller than 1.0 mm in diameter pass through the mesh (0.947-mm aperture) of the neuston net. In addition, at a speed of 5 knots some particles larger than 1.0 mm in diameter are forced through the meshes. The possible loss of smaller particles is supported by the fact that the mean diameter (0.5 mm) of the polystyrene spherules collected in southern New England coastal waters by Carpenter *et al.* (*2*), using 0.333-mm mesh nets,

was appreciably less than the mean diameter (1.3 mm) of similar polystyrene spherules in the *Delaware II* samples. The polystyrene spherules in our samples appear identical to the polystyrene "suspension beads" that are shipped to plastic fabricators by polystyrene producers. The mean diameter of a manufacturer's sample (*6*) of these suspension beads was 0.5 mm. The beads ranged in diameter from 0.2 to 1.6 mm.

Many of the opaque polystyrene spherules collected were of greater density than seawater, as were the opaque polystyrene suspension beads obtained from the plastic producers.

These spherules, which could only be maintained in the surface layers in areas of strong vertical mixing, were less abundant in terms of both number and weight than the less dense, clear polystyrene spherules (Table 1). Obviously, these opaque spherules and other plastic particles of similar density must also occur in subsurface waters. Evidence for this is the occurrence of opaque spherules in subsurface waters of Block Island Sound (*3*), in bottom sediments off New Haven, Connecticut (*7*), and in the Connecticut River (*8*). Thus, because a proportion of plastic particles was not in the surface layer and because smaller particles were not fully retained by the 0.947-mm mesh, the values presented here underestimate the number of particles under a unit area of sea surface.

Abundance and Distribution of Plastics

Fifty percent of the *Oregon II* samples collected in the Caribbean Sea, 57 percent of the *Albatross IV* samples collected in the Antilles Current area, and 69 percent of the *Delaware II* samples collected in coastal, Slope, and Gulf Stream waters between Florida and Cape Cod contained plastics. The composition of the plastic particles and their abundance in terms of both number and weight varied with geographic area (Table 1). The opaque and clear polystyrene spherules occurred only in samples collected in waters north of Florida (*Delaware II*). The abundance of all plastic types was greatest in waters north of Florida and least in the Caribbean Sea. The ratio of the mean weight of all plastics from *Delaware II*, *Albatross IV*, and *Oregon II* samples was 8 : 2 : 1.

The distribution of opaque polystyrene spherules was restricted to the area north of latitude 37°N (Fig. 3). The greatest concentration of these particles was in coastal waters south of Rhode Island and south of eastern Long Island. A secondary concentration occurred approximately 110 kilometers southeast of Delaware Bay. It was only off southern New England and Long Island that these particles were found at stations immediately adjacent to the coast.

Transparent polystyrene spherules were found over a slightly more extensive area than the opaque spherules (Fig. 3). With the exception of one

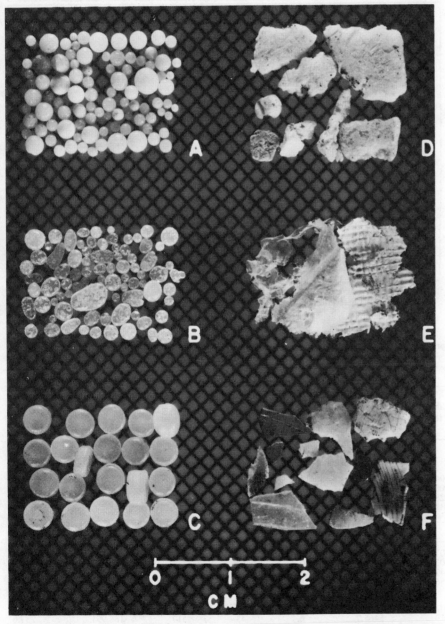

Fig. 2. Typical plastic particles: (A) opaque polystyrene spherules, (B) clear and translucent polystyrene spherules, (C) opaque and translucent polyethylene cylinders, (D) Styrofoam, (E) plastic sheets, and (F) plastic pieces.

spherule found well east of the main axis of the Gulf Stream at 34°N, the clear spherules were restricted to the area north of 36°N. As in the case of the opaque spherules, the most extensive concentration of clear spherules occurred in coastal waters south of Rhode Island and south of eastern Long Island with a more limited concentration centered approximately 110 km southeast of Delaware Bay. Only off southern New England and Long Island were these particles found at the most inshore stations.

The most extensive concentration of polyethylene cylinders occurred in continental shelf waters off southern New England and eastern Long Island, and off the coast of southern New Jersey and Delaware (Fig. 4). More limited concentrations were found in and to the east of the Gulf Stream, just south of Cape Hatteras, and in the Yucatan Channel. These particles were also collected in limited numbers at scattered stations throughout the Antilles Current area and at two stations in the Caribbean Sea. No particles were found at stations in the Straits of Florida, in the southeastern Gulf of Mexico, or in coastal and Gulf Stream waters south of Cape Lookout, North Carolina. The polyethylene cylinders occurred at stations immediately adjacent to the coast off southern New England and eastern Long Island, at one station southeast of Ocean City, Maryland (38°N), at one station off the south coast of Puerto Rico, and at one station off the north coast of Cuba. In addition to these col-

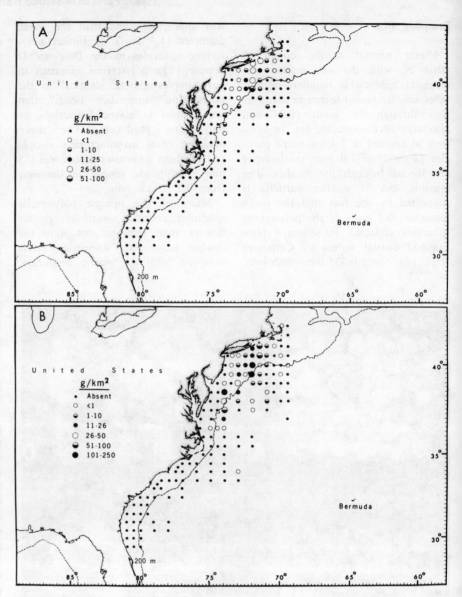

Fig. 3. Distribution of opaque (A) and clear (B) polystyrene spherules.

Table 1. Mean abundance and maximum abundance of plastic particles. The numbers of Styrofoam pieces and plastic sheets and pieces are not listed because of appreciable variations in the sizes of these particles between stations.

Plastic type	Mean (total stations)				Maximum (positive stations)			
	Number per tow	Grams per tow	Number per square kilometer	Grams per square kilometer	Number of tows	Number per tow	Number per square kilometer	Grams per square kilometer
Oregon II (64 stations)								
Opaque spherules	0							
Transparent spherules	0							
Opaque cylinders	0.2	0.004	60.6	1.4	6	4	1,292	32.1
Styrofoam		0.0002		0.1	2			0.3
Sheets and pieces		0.028		9.0	31			222.1
Total		0.032		10.5				
Albatross IV (40 stations)								
Opaque spherules	0							
Transparent spherules	0							
Opaque cylinders	0.5	0.007	148.4	2.2	8	9	2,707	37.6
Styrofoam		0.002		0.7	7			17.5
Sheets and pieces		0.047		15.2	22			214.9
Total		0.056		18.1				
Delaware II (143 stations)								
Opaque spherules	6.2	0.007	1996.4	2.3	31	188	60,724	53.0
Transparent spherules	16.9	0.019	5465.7	6.4	43	517	166,991	237.6
Opaque cylinders	2.7	0.034	855.4	11.2	40	107	34,561	406.1
Styrofoam		0.021		6.9	17			721.9
Sheets and pieces		0.158		50.9	98			1,403.5
Total		0.239		77.7				

lections at sea, polyethylene cylinders have been found in appreciable numbers on a beach at Barranquilla, Colombia (9), on Padre Island Beach near Corpus Christi, Texas (10), on Kalaloch Beach, Washington (11), and in bird gizzards on Amchitka Island in the Aleutian chain (12).

Pieces of Styrofoam were concentrated in only two relatively small areas, one in coastal waters off eastern Long Island and the other centered approximately 130 km east-southeast of Delaware Bay (Fig. 5). In all other areas Styrofoam particles occurred in isolated patches. Styrofoam was collected at only one station in the Caribbean Sea. At only two locations, eastern Long Island and Cape Cod, were Styrofoam particles found at stations immediately adjacent to the coast.

Plastic sheets and pieces were not only the most abundant particle types collected but also the most widely dispersed (Fig. 6). The greatest concentrations of these particles were found in continental shelf waters between Virginia (37°N) and Rhode Island (41°N). The sheets and pieces were the only types of plastics found in appreciable numbers in continental shelf waters south of Cape Hatteras and the only types found in the majority of stations in the Caribbean Sea and Antilles Current area. A band approximately 110 km wide and extending approximately 289 km offshore separated the concentration of plastic sheets and pieces north and south of Cape Hatteras. This band coincides with the area between the offing of Chesapeake Bay and Cape Hatteras in which there is a marked offshore component of surface drift.

Sources of Plastics

There would appear to be only three possible major sources of Styrofoam and pieces and sheets of formulated and compounded plastics (wrapping material, containers, toys, and so forth): (i) municipal solid waste disposal at sea, (ii) coastal landfill operations, and (iii) disposal at sea of solid waste generated by individual vessels. There is no appreciable refuse or garbage dumping in U.S. Atlantic coastal waters (13). The fact that in most areas the maximum concentrations of formulated and compounded plastics were well offshore rather than immediately adjacent to the coast indicates that landfill operations are not a major source of these particles. In view of the above, our personal experience at sea aboard research and commercial vessels, and the superposition of areas of high particle abundance and maximum vessel activity, we have concluded that the accumulation of compounded plastics in the sea surface results from routine at-sea solid waste disposal by individual vessels. The bulk of this solid waste consists of material used to package food and other products.

The only apparent way in which the clear and opaque polystyrene spherules enter the ocean is via waste-water discharge from a plastic-producing or plastic-processing plant into a river or estuary. Most of the plastic-producing and plastic-processing companies along the East Coast of the United States are located in southern New England and in the Middle Atlantic states, and the majority of the polystyrene producers are located in Connecticut, New York, and New Jersey (14).

A study was made by the Society of the Plastics Industry, Inc., in 1972 of waste emission practices of plants producing polystyrene resins on the East Coast of the United States. This study was prompted by the discovery by Carpenter et al. (2) of plastic spherules in southern New England coastal waters. The society concluded at that time that only one of these plants was following procedures that occasionally emitted particles via a waste-water system that emptied into a river flowing into the Atlantic (15). A more recent study by Hays and Cormons (16) however, disclosed the presence of opaque polystyrene spherules in the sand and leaf litter near the sewage outlet of a plastic manufacturing plant on the Chicopee River, Massachusetts,

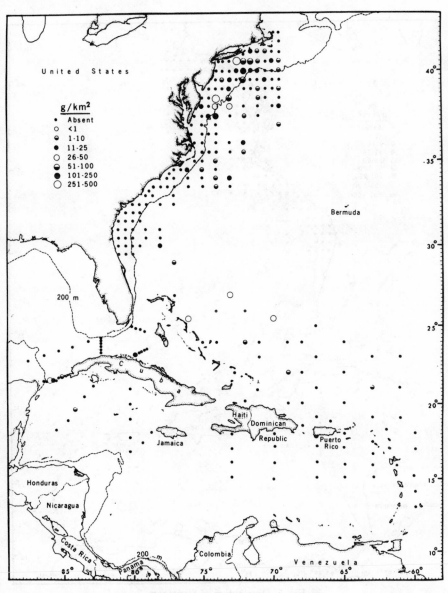

Fig. 4. Distribution of polyethylene cylinders.

and in mud at the mouth of the Connecticut River at Saybrook.

The distribution of both clear and opaque polystyrene spherules (Fig. 3) indicates that the majority of these particles enter open coastal waters in the area between Block Island and the eastern tip of Long Island. The occurrence of polystyrene particles in plankton samples collected in eastern Long Island Sound and Block Island Sound is now commonplace. The distribution of particles in Long Island Sound and Block Island Sound (2, 3, 17) indicates that the bulk of these particles is introduced along the coast of Connecticut between longitudes 72°W and 73°W. The Connecticut, Niantic, Thames, and numerous smaller rivers drain into this area. Polystyrene spherules have been found in Niantic Bay (2) and in the Connecticut River as far north as Massachusetts (18).

Observations of currents, particularly in the eastern part of Long Island Sound, show that the ebb tide is stronger than the flood tide at the surface layer, whereas the reverse is true at the bottom. Thus there is a tendency here, as in many sounds and estuaries, for the surface layer to move seaward and to be replaced by saline water flowing in along the bottom (19). This seaward flow of surface water out of Long Island Sound in the area between Montauk Point and Block Island (20) is augmented by river drainage, three-quarters of which enters the relatively open eastern end of Long Island Sound where it is flushed out rapidly.

The surface outflow of well-mixed, river-freshened water out of Block Island Sound which then spreads seaward in a southwesterly direction has been discussed and illustrated by Miller (21) and Bumpus (22). There is a general southerly surface drift over much of the Middle Atlantic Bight culminating in a seaward outflow in the area between Chesapeake Bay and Cape Hatteras. The distribution of clear and opaque polystyrene spherules in coastal waters is in accord with coastal surface circulation, as inferred from drift bottle studies (21, 22) and the distribution of temperature and salinity (21).

The polyethylene cylinders are called "nibs" in the chemical trade and, as in the case of the polyethylene "suspension beads," are a bulk material used in fabricating plastic products. It is apparent that the sources of these particles are plastic-producing and plastic-processing plants, because these cylinders have been found at waste-water outlets of plastic manufacturing plants in Massachusetts, Connecticut, and New Jersey, and in streams just below plants in New York and New Jersey (16).

Polyethylene cylinders have been collected in appreciable numbers in Block Island Sound (3). We have received no reports of their being found in the waters of Long Island Sound, although they have been found on the shore at Saybrook, Connecticut, and Fire Island, New York (16). The distribution of these cylinders (Fig. 4) indicates that, although the bulk of these particles enter open coastal waters in the area between Block Island and Long Island, a significant number of these particles enter coastal waters via Delaware Bay. The distribution of these particles off southern New England and eastern Long Island and off the coast of southern New Jersey and Delaware is in accord with the offshore component of drift, as indicated by the paucity of onshore drift bottle recoveries in both eastern Long Island and in the vicinity of Delaware Bay (21).

The fact that polyethylene cylinders were also found in oceanic waters south of Cape Hatteras, in the Yucatan Channel, at scattered stations both north and south of the Greater Antilles, and on beaches at Barranquilla, Colombia; Corpus Christi, Texas; Kalaloch, Washington; and in the Aleutian Islands, implies that there are additional sources of these particles. Supporting evidence for this is the fact that the polyethylene cylinders collected in the Caribbean Sea and on the beach at Barranquilla were appreciably longer and heavier than the more disk-like particles collected by the *Albatross IV* and *Delaware II*. The average weight

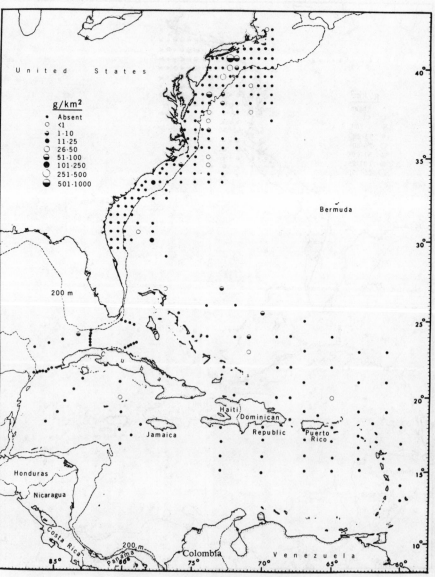

Fig. 5. Distribution of Styrofoam.

of *Oregon II* particles was 0.0228 g, whereas the average weights of *Albatross IV* and *Delaware II* particles were 0.0149 and 0.0130 g, respectively. In addition, the *Oregon II* and Barranquilla beach particles were less weathered and less brittle and had sharper edges and cleaner surfaces than the *Albatross IV* and *Delaware II* particles, all of which indicates more recent introductions into the sea.

Environmental Effects

The main danger to marine life and human health comes from wastes that are highly toxic or exceptionally long lasting. As far as we know, the plastics we are concerned with here are not toxic. But, they are also not biodegradable. Once they are introduced into the marine environment, they remain indefinitely even though they gradually break up into smaller particles.

The plastic particles do act as surfaces for the growth of hydroids, diatoms, and bacteria (*1*) and possibly for the accumulation of polychlorinated biphenyls (PCB's) from ambient seawater (*2*). White polystyrene spherules have been found in the stomachs of a number of species of larval and juvenile fishes both in Niantic Bay (*2*) and in the Connecticut River (*23*). They also have been found in the stomach contents of flounders (*Platichthys flesus*) (2 to 5 centimeters) in Severn Estuary, United Kingdom (*24*), and in tern and gull pellets on Great Gull Island, New York (*16*). Carpenter *et al.* (*2*) have suggested that the ingestion of plastic may lead to intestinal blockage and possible mortality in smaller larval fishes.

We found no plastic particles in the gut contents of over 500 larval and juvenile fishes (22 species) collected in the areas of maximum abundance of opaque spherules. Experiments were conducted to determine if larval and juvenile fishes maintained in 15-gallon (0.04-cubic-meter) laboratory aquariums would feed on plastic spherules, and if so, to determine the effect of ingestion. Polystyrene and acrylonitrile-butadiene-styrene suspension beads (5 g each, approximately 43,500 particles) were added to each aquarium. Juvenile striped killifish (*Fundulus majalia*), juvenile tomcod (*Microgadus tomcod*), juvenile three-spined stickleback (*Gasterosteus aculeatus*), juvenile winter flounder (*Pseudopleuronectes americanus*), larval haddock (*Melanogrammus aeglefinus*), and larval winter flounder were used in these experiments. The juvenile fishes were fed chopped squid twice weekly, and the larvae were fed fresh net plankton daily. The larvae ranged in length from 4.2 to 6.0 mm and the juveniles from 15 to 50 mm.

The larvae were maintained for 2 weeks until a cooling system breakdown caused total mortality. No spherules were found in the digestive tracts of larvae killed and examined daily during this period. The juvenile fishes experienced less than 2 percent mortality during an 8-week period. This mortality was due to fungal infection. Samples of juvenile fishes were killed weekly, and, as with the larvae, no spherules were found in the gut contents. Juvenile killifish and tomcod were observed to take spherules, but in most cases they were immediately rejected. Any plastics that were swallowed apparently passed through the gut with no ill effects.

At the present levels of abundance of plastic particles in coastal and oceanic waters, adverse biological consequences would appear to be minor compared to the deleterious effect of other contaminants such as petroleum residues and other chemical wastes. Increasing production of plastics, combined with present waste disposal practices, will undoubtedly lead to increases in the concentration of these particles in rivers, estuaries, and the open ocean. The U.S. production of synthetic resins for plastic uses (excluding textile products) was about 20 $\times 10^9$ pounds (9 $\times 10^6$ metric tons) in 1972 (*25*). This quantity of resin is combined with about an equal weight of fillers, reinforcements, additives (for

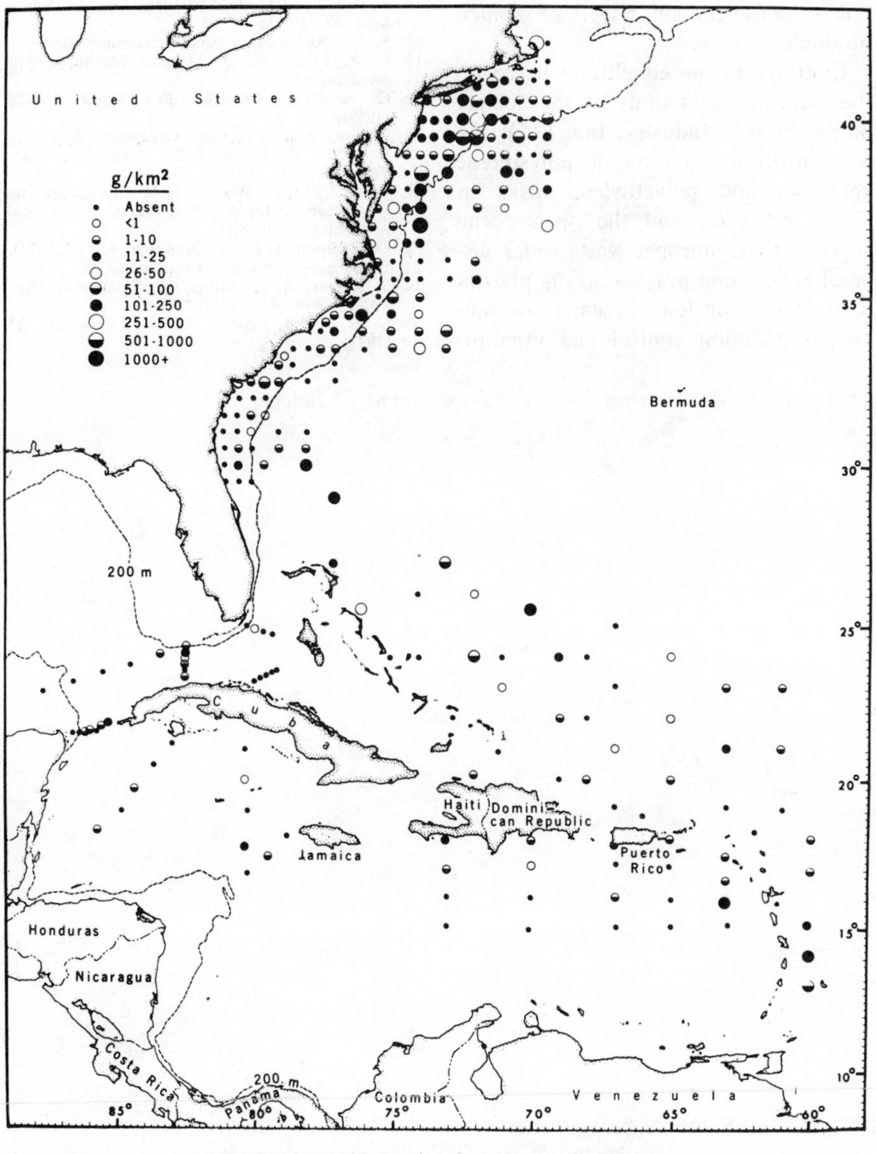

Fig. 6. Distribution of plastic sheets and pieces.

example, plasticizers, colorants, and stabilizers), and other basic materials (25). The estimated U.S. resin production in 1975 is 34×10^9 pounds and at least 55×10^9 pounds in 1980 (25).

Even without the anticipated increased production of plastics, we can expect an increase in the abundance of polystyrene spherules and polyethylene cylinders in the open ocean because of the appreciable period between the time these particles are introduced into rivers and estuaries and the time they reach the open ocean. Furthermore, we can predict an increase in the abundance of all types of plastic particles in the Sargasso Sea, which is an ocean region more favorable to the accumulation and retention of floating material than to its dispersal.

Preventive Measures

The bulk of the plastic material collected consisted of one-time-use wrapping and packaging wastes. It is our experience that a high percentage of the Styrofoam found in the ocean comes from disposable cups. The disposal of these materials at sea may be stopped if all vessels are required to install nonatmospheric polluting incinerator systems or equipment for compacting these and other solid waste materials at sea for disposal or recla-

mation ashore. In addition, vessel owners should discourage the use of disposable plastic tableware and food containers.

Among the technological developments and methodology needed are:

1) Development of water-soluble and photodegradable polymers for one-time-use and short-time-use plastic products.

2) Development of efficient, nonatmospheric polluting incinerators to replace open dumping and sanitary landfill.

3) Increased effort in the technological development of plastic reclamation systems.

4) Increased efforts in plastic recycling to a level of that in the paper, metal, and glass industries. This will require not only new technological development but also a change in attitude concerning the use of scrap and reprocessed material among resin producers, designers, and buyers of molded products.

Contrary to the conclusion based on the plant emission study by the Society of the Plastics Industry, Inc. (15), the widespread distribution of polystyrene spherules and polyethylene disks in rivers, estuaries, and the open ocean suggests that improper waste-water disposal is common practice in the plastics industry. Strong federal, state, and municipal pollution control and monitor-

ing programs are necessary to prevent the emission of plastic beads into the waste-water systems of plastic-producing and plastic-processing plants.

References and Notes

1. E. J. Carpenter and K. L. Smith, Jr., *Science* **175**, 1240 (1972).
2. E. J. Carpenter, S. J. Anderson, G. R. Harvey, H. P. Miklas, B. B. Peck, *ibid.* **178**, 749 (1972).
3. H. M. Austin and P. Stoops, *N.Y. Ocean Sci. Lab. Tech. Rep. 23* (1973).
4. Marine Resources Monitoring, Assessment, and Prediction, a nationally coordinated program of the National Marine Fisheries Service to evaluate the living marine resources off the coast of the United States.
5. Identified by infrared spectrophotometry by E. J. Carpenter, Woods Hole Oceanographic Institution, and H. Petersen, University of Rhode Island.
6. Obtained from E. J. Carpenter, Woods Hole Oceanographic Institution.
7. D. Rhoads and S. W. Richards, personal communication.
8. T. Hoehn, personal communication.
9. D. H. Eargle, Jr., personal communication.
10. R. E. Hunter, personal communication.
11. R. A. May, personal communication.
12. T. R. Merrell, Jr., personal communication.
13. D. D. Smith and R. P. Brown, *Environ. Protect. Agency Publ. SW-19c* (1971).
14. *Plastics World* **30** (No. 11), 93 (1972).
15. J. R. Lawrence, personal communication.
16. H. Hays and G. D. Cormons, *Linnean News-Lett. 27* (1973).
17. S. W. Richards, personal communication.
18. S. A. Moss and B. Marcey, personal communication.
19. G. A. Riley, *Deep-Sea Res. Suppl.* **3**, 224 (1955).
20. ———, *Bull. Bingham Oceanogr. Coll.* **13**, 5 (1952); R. Nuzzi, *N.Y. Ocean Sci. Lab. Tech. Rep. 19* (1973).
21. A. R. Miller, Woods Hole Oceanographic Institution Reference No. 52-28 (1952) (unpublished manuscript).
22. D. F. Bumpus, *Progr. Oceanogr.* **6**, 111 (1973).
23. B. Marcey, personal communication.
24. S. Kartar, R. A. Milne, M. Sainsbury, *Mar. Pollut. Bull.* **4**, 144 (1973).
25. C. H. Jenest, *Plastics World* **30** (No. 11), 32 (1972).

DDT Residues and Declining Reproduction in the Bermuda Petrel

Abstract. *Residues of DDT [1,1,1-trichloro-2,2-bis(p-chlorophenyl)ethane] averaging 6.44 parts per million in eggs and chicks of the carnivorous Bermuda petrel indicate widespread contamination of an oceanic food chain that is remote from applications of DDT. Reproduction by the petrel has declined during the last 10 years at the annual rate of 3.25 percent; if the decline continues, reproduction will fail completely by 1978. Concentrations of residues are similar to those in certain terrestrial carnivorous birds whose productivity is also declining. Various considerations implicate contamination by insecticides as a probable major cause of the decline.*

Many oceanic birds nested on Bermuda in 1609 when the first settlers arrived, the most abundant apparently being the Bermuda petrel, *Pterodroma cahow*. Within 20 years man and his imported mammals virtually exterminated this species; for nearly 300 years it was considered extinct. Several records of specimens since 1900 were followed in 1951 by discovery of a small breeding colony (*1*), and in 1967 22 pairs nested on a few rocky islets off Bermuda. With a total population of about 100 the petrel is among the world's rarest birds.

A wholly pelagic species, *P. cahow* visits land only to breed, breeds only on Bermuda, and arrives and departs only at night. The single egg is laid underground at the end of a long burrow. When not in the burrow the bird feeds far at sea, mainly on cephalopods; when not breeding it probably ranges over much of the North Atlantic (*1*).

Reproduction by *P. cahow* has declined recently. The data since 1958 (Table 1) show an annual rate of decline of 3.25 ± 1.05 percent; the negative slope of a weighted regression is significant (*P*, .015; *F* test). If this linear decline continues, reproduction will fail completely by 1978, with extinction of the species. Many recent reports have correlated diminished reproduction by certain carnivorous birds with contamination by chlorinated hydrocarbon insecticides (*2–7*). As the terminal member of a pelagic food chain, presumably feeding over much of the North Atlantic, the petrel may be expected to concentrate by many orders of magnitude any stable, lipid soluble chemicals, such as chlorinated hydrocarbon insecticides, present in lower trophic levels (*2, 3, 8*). In fact it should serve as an ideal environmental monitor for detection of insecticide contamination as a general oceanic pollutant, rather than contamination resulting directly from treatment of a specific land area (*9*). When we analyzed several specimens of *P. cahow* for chlorinated hydrocarbon insecticides, all samples contained DDT residues (*10*).

During March 1967 five unhatched eggs and dead chicks were collected from unsuccessful petrel burrows and stored frozen. The small size of the population precluded the sampling of living birds. Samples were analyzed for DDT, *o,p*-DDT, DDE, DDD, dieldrin, and endrin by electron-capture gas chroma-

Reprinted from Science 159:979-981 (1968).

tography; the results are summarized in Table 2. No o,p-DDT, dieldrin, or endrin was detected, but an independent laboratory detected a trace of dieldrin.

Certain identifications were confirmed by thin-layer chromatography (11) as follows: After Florisil cleanup (12), the unknown sample was spotted on a thin-layer plate with 1-μg authentic standard samples on both sides. After development, the unknown was masked by a strip of paper, and the standards were sprayed with chromogenic reagent (11). When spots were visible following exposure to ultraviolet light, the masking was removed, horizontal lines were drawn between the standard spots in order to locate corresponding compounds in the unknown, and these areas were scraped from the plate and extracted with a few drops of a mixture of hexane and acetone (9:1 by volume). Injection into the gas chromatograph confirmed the presence of DDT, DDE, and DDD by showing the appropriate single peaks for these compounds. This confirmation procedure was employed because the electron-capture detector is more sensitive than the chromogenic spray reagent in detecting minute amounts of these materials.

Coincidental with diminishing reproduction by the Bermuda petrel is the presence of DDT residues averaging 6.44 parts per million (ppm) in its eggs and chicks. In itself this coincidence does not establish a causal relation, but these findings must be evaluated in the light of other studies. Whereas a healthy osprey (*Pandion haliaetus*) population produces 2.2 to 2.5 young per nest, a Maryland colony containing DDT residues of 3.0 ppm in its eggs yielded 1.1 young per nest, and a Connecticut colony containing 5.1 ppm produced only 0.5 young per nest; the Connecticut population has declined 30 percent annually for the last 9 years (4). In New Brunswick, breeding success of American woodcocks (*Philohela minor*) showed a statistically significant inverse correlation with the quantity of DDT applied to its habitat in a given year. Furthermore, during 1962 and 1963, birds from unsprayed Nova Scotia showed breeding success nearly twice as great as did those from sprayed New Brunswick, where woodcock eggs averaged 1.3 ppm of DDT residues during those years (5).

In Britain five species of raptors, including the peregrine falcon (*Falco peregrinus*) and golden eagle (*Aquila chrysaetos*), carried residues of chlori-

Table 1. Reproductive success of the Bermuda petrel between 1958 and 1967: percentages of established adult pairs under observation whose chicks survived 2 weeks after hatching. Numbers of pairs of unknown success (not included in calculations) appear in parentheses. Data from 1961–7 are believed to represent the total breeding population; earlier, not all burrows had been discovered. The decline in reproductive success follows the linear relation $y = a + bx$ (y, reproductive success; a, constant; b, annual percentage decline in success; x, year). The regression, weighted by numbers of pairs: $y = 251.9 - 3.25x$.

Year	Pairs	Chicks	Success (%)
1958	6 (1)	4	66.7
1959	5 (2)	2	40.0
1960	13 (3)	6	46.2
1961	18 (1)	12	66.7
1962	19	9	47.4
1963	17 (1)	9	52.9
1964	17 (1)	8	47.1
1965	20	8	40.0
1966	21	6	28.6
1967	22	8	36.4

nated hydrocarbon insecticides in their eggs, averaging 5.2 ppm; each of these species has shown a decline in reproduction and total population during recent years. By comparison, residues in the eggs of five species of corvids averaged 0.9 ppm, and breeding success and numbers have been maintained (6). It is noteworthy that during the last decade the peregrine has become extinct as a breeding bird in the eastern United States (13). Residues in bald eagle (*Haliaeetus leucocephalus*) eggs averaged 10.6 ppm, and this species also shows declining reproduction and population (7). Lake Michigan herring gulls (*Larus argentatus*), exhibiting very low reproductive success, averaged 120 to 227 ppm of DDT residues in the eggs (3), the suggestion being that susceptibility varies widely between species.

In most of the above instances, including *P. cahow*, reduced success in breeding resulted primarily from mor-

tality of chicks before and shortly after hatching. Bobwhites (*Colinus virginianus*) and pheasants (*Phasianus colchicus*), fed sublethal diets of DDT or dieldrin, gave similar results (14); a mechanism explaining chick mortality from dieldrin poisoning during the several days after hatching has been presented (15).

From studies of these birds and other avian carnivores a very widespread, perhaps worldwide, decline among many species of carnivorous birds is apparent. The pattern of decline is characterized by reduced success in reproduction correlated with the presence of residues of chlorinated hydrocarbon insecticides—primarily DDT. Our data for the Bermuda petrel are entirely consistent with this pattern.

Observations of aggressive behavior, increased nervousness, chipped eggshells, increased egg-breakage, and egg-eating by parent birds of several of the above species (3, 6, 13) suggest symptoms of a hormonal disturbance or a calcium deficiency, or both. Moreover, DDT has been shown to delay ovulation and inhibit gonadal development in birds, probably by means of a hormonal mechanism, and low dosages of DDT or dieldrin in the diet of pigeons increased metabolism of steroid sex hormones by hepatic enzymes (16). A direct relation between DDT and calcium function has also been demonstrated, and these endocrine and calcium mechanisms could well be interrelated; DDT interferes with normal calcification of the arthropod nerve axon, causing hyperactivity of the nerve and producing symptoms similar to those resulting from calcium deficiency (17). Dogs treated with calcium gluconate are very resistant to DDT poisoning (18); female birds are more resistant than males (19), perhaps because of the calcium-mobiliz-

Table 2. Residues of DDT (10) in parts per million (wet weight) in eggs and chicks of the Bermuda petrel, collected in Bermuda in March 1967; proportions of DDT, DDE, and DDD are expressed as percentages of the total.

Sample	Residues (ppm)	Percentages DDT	DDE	DDD
A, egg †	11.02	37*	58*	5*
A, egg † § ‖	10.71	34*	62*	4*
B, addled egg †	3.61	15	65	20
C, chick in egg ‡	4.52	33	64	3
D, chick in egg ‡	6.08	33	62	5
D, chick brain ‡ ‖	0.57	30	54	16
E, chick, 1 to 2 days old	6.97	29*	66*	5*
Averages	6.44	31	62	7

* Identity confirmed by thin-layer chromatography (11).   † Egg showed no sign of development.
‡ Fully developed chick died while hatching.   § Analysis 5 months later by Wisconsin Alumni Research Foundation, which also detected dieldrin at 0.02 ppm.   ‖ Not included in averages.

ing action of estrogenic hormones.

Of major importance, then, was the discovery that a significant ($P<.001$) and widespread decrease in calcium content of eggshells occurred between 1946 and 1950 in the peregrine falcon, golden eagle, and sparrowhawk, *Accipiter nisus* (20). This decrease correlates with the widespread introduction of DDT into the environment during those years, and further correlates with the onset of reduced reproduction and of the described symptoms of calcium deficiency. These multiple correlations indicate a high probability that the decline in reproduction of most or all of these birds, including *P. cahow*, is causally related to their contamination by DDT residues.

Other potential causes of the observed decline for the Bermuda petrel appear unlikely. The bird has been strictly protected and isolated since 1957, and it seems that human disturbance can be discounted. In such a small population, inbreeding could become important, but hatching failure is now consistent in pairs having earlier records of successful breeding, and deformed chicks are never observed. Furthermore, the effects of inbreeding would not be expected to increase at a time when the total population, and probably the gene pool, is still increasing. The population increase results from artificial protection since 1957 from other limiting factors, especially competition for nest sites with tropic birds (21).

It is very unlikely that the observed DDT residues in *P. cahow* were accumulated from Bermuda: the breeding grounds are confined to a few tiny, isolated, and uninhabited islets never treated with DDT, and the bird's feeding habits are wholly pelagic. Thus the presence of DDT residues in all samples can lead only to the conclusion that this oceanic food chain, presumably including the plankton, is contaminated. This conclusion is supported by reported analyses showing residues in related seabirds including two species of shearwaters from the Pacific (22); seabird eggs (9, 22); freshwater, estuarine, and coastal plankton (2, 8, 23); plankton-feeding organisms (2, 8, 9, 22, 23); and other marine animals from various parts of the world (8, 22). These toxic chemicals are apparently very widespread within oceanic organisms (8, 22), and the evidence suggests that their ecological effects are important.

CHARLES F. WURSTER, JR.
*Department of Biological Sciences,
State University of New York,
Stony Brook 11790*

DAVID B. WINGATE
*Department of Agriculture and
Fisheries, Paget East, Bermuda*

References and Notes

1. R. C. Murphy and L. S. Mowbray, *Auk* 68, 266 (1951); A. C. Bent, *U.S. Nat. Museum Bull. 121* (1922), pp. 112–7.
2. E. G. Hunt and A. I. Bischoff, *Calif. Fish Game* 46, 91 (1960); E. G. Hunt, in *Nat. Acad. Sci.–Nat. Res. Council Publ. 1402* (1966), p. 251.
3. J. P. Ludwig and C. S. Tomoff, *Jack-Pine Warbler* 44, 77 (1966); J. A. Keith, *J. Appl. Ecol.* 3(suppl.), 57 (1966); J. J. Hickey, J. A. Keith, F. B. Coon, *ibid.*, p. 141.
4. P. L. Ames, *ibid.*, p. 87.
5. B. S. Wright, *J. Wildlife Management* 29, 172 (1965).
6. S. Cramp, *Brit. Birds* 56, 124 (1963); J. D. Lockie and D. A. Ratcliffe, *ibid.* 57, 89 (1964); D. A. Ratcliffe, *ibid.* 58, 65 (1965); *Bird Study* 10, 56 (1963); 12, 66 (1965).
7. L. F. Stickel *et al.*, in *Trans. North American Wildlife Natural Resources Conf. 31st* (1966), pp. 190–200; J. B. DeWitt, *Audubon Mag.* 65, 30 (1963); A Sprunt, *ibid.*, p. 32.
8. G. M. Woodwell, C. F. Wurster, P. A. Isaacson, *Science* 156, 821 (1967); G. M. Woodwell, *Sci. Amer.* 216, 24 (March 1967).
9. N. W. Moore and J. O'G. Tatton, *Nature* 207, 42 (1965); N. W. Moore, *J. Appl. Ecol.* 3(suppl.), 261 (1966).
10. Residues of DDT include DDT and its decay products (metabolites) DDE and DDD; DDT, 1,1,1-trichloro-2,2-bis(*p*-chlorophenyl)-ethane; DDE, 1,1-dichloro-2,2-bis(*p*-chlorophenyl)ethylene; DDD (also known as TDE), 1,1-dichloro-2,2-bis(*p*-chlorophenyl)ethane.
11. M. F. Kovacs, *J. Assoc. Offic. Anal. Chemists* 49, 365 (1966).
12. J. G. Cummings, K. T. Zee, V. Turner, F. Quinn, R. E. Cook, *ibid.*, p. 354.
13. R. A. Herbert and K. G. S. Herbert, *Auk* 82, 62 (1965); J. J. Hickey, Ed., *Peregrine Falcon Populations, Their Biology and Decline* (Univ. of Wisconsin Press, Madison, in press).
14. J. B. DeWitt, *J. Agr. Food Chem.* 3, 672 (1955); 4, 863 (1956); R. E. Genelly and R. L. Rudd, *Auk* 73, 529 (1956).
15. J. H. Koeman, R. C. H. M. Oudejans, E. A. Huisman, *Nature* 215, 1094 (1967).
16. D. J. Jefferies, *Ibis* 109, 266 (1967); H. Burlington and V. F. Lindeman, *Proc. Soc. Exp. Biol. Med.* 74, 48 (1950); D. B. Peakall, *Nature* 216, 505 (1967); *Atlantic Naturalist* 22, 109 (1967).
17. J. H. Welsh and H. T. Gordon, *J. Cell. Comp. Physiol.* 30, 147 (1947); H. T. Gordon and J. H. Welsh, *ibid.* 31, 395 (1948).
18. Z. Vaz, R. S. Pereira, D. M. Malheiro, *Science* 101, 434 (1945).
19. D. H. Wurster, C. F. Wurster, R. N. Strickland, *Ecology* 46, 488 (1965); L. B. Hunt, unpublished manuscript, University of Wisconsin, 1965.
20. D. R. Ratcliffe, *Nature* 215, 208 (1967).
21. D. B. Wingate, *Can. Audubon* 22, 145 (1960).
22. R. W. Risebrough, D. B. Menzel, D. J. Martin, H. S. Olcott, *Nature* 216, 589 (1967); J. Robinson, A. Richardson, A. N. Crabtree, J. C. Coulson, G. R. Potts, *ibid.* 214, 1307 (1967); W. J. L. Sladen, C. M. Menzie, W. L. Reichel, *ibid.* 210, 670 (1966); J. O'G. Tatton and J. H. A. Ruzicka, *ibid.* 215, 346 (1967); J. O. Keith and E. G. Hunt, in *Trans. North American Wildlife Natural Resources Conf. 31st* (1966), pp. 150–77.
23. P. A. Butler, *ibid.*, pp. 184–9; *J. Appl. Ecol.* 3(suppl.), 253 (1966).
24. Aided by a grant from the Research Foundation of the State University of New York; transportation by the Smithsonian Institution, Washington, D. C. The Bermuda petrel conservation program was financed by Childs Frick and the New York Zoological Society. We thank G. M. Woodwell for criticizing the manuscript.

1 December 1967

Polychlorinated Biphenyls and DDT Alter Species Composition in Mixed Cultures of Algae

Abstract. *Either DDT or polychlorinated biphenyls were added to mixed cultures containing a marine diatom and a marine green alga that were sensitive and resistant, respectively, to these organochlorine compounds. The diatom grew faster and was therefore dominant in control cultures, but its dominance diminished in treated cultures, even at concentrations of chlorinated hydrocarbons that had no apparent effect in pure cultures. Such stable pollutants could disrupt the species composition of phytoplankton communities, thereby affecting whole ecosystems.*

The impact of certain persistent chlorinated hydrocarbons on various higher nontarget organisms has been well documented (*1*), but effects on photosynthetic algae, the base of aquatic food webs, have not been extensively studied. Marine phytoplankton vary in sensitivity to chlorinated hydrocarbons, including DDT [1,1,1-trichloro-2,2-bis (*p*-chlorophenyl)ethane] and polychlorinated biphenyls (PCBs). Some species show effects at concentrations as low as a few parts per billion (ppb), whereas others are resistant to much higher concentrations (*2, 3*). Some of these chemicals, especially DDT and PCBs, are extremely widespread pollutants of the biosphere, and, because they are selectively toxic to certain sensitive algal species, it has been hypothesized that they could alter the species composition of phytoplankton communities (*2, 3*). Evidence for this hypothesis is lacking; we therefore investigated the effects of DDT and PCBs in mixed algal cultures containing a sensitive and a resistant species.

Two marine organisms were selected on the basis of their sensitivity to organochlorine compounds: growth of the diatom *Thalassiosira pseudonana* was inhibited by PCBs and DDT, whereas *Dunaliella tertiolecta*, a green alga, was not affected by these chemicals (*3*). Methods of culture and procedures for treatment have been described (*3*). Cultures contained a total of 10^4 exponentially growing cells per milliliter at zero time; mixed cultures contained the two species in a 1:1 ratio. Mixed and pure cultures were treated simultaneously. Cells in pure culture were counted with a Coulter counter, whereas those in mixed cultures were counted microscopically with a Neubauer-Levy counting chamber because the two cell types could not be differentiated by the Coulter counter.

The results with pure cultures confirmed those described previously (*3*). Polychlorinated biphenyls inhibited the growth of *T. pseudonana* at 25 ppb, but not at 10 ppb or less. Growth was also significantly inhibited by DDT at 100 ppb; inhibition at 50 ppb was slight, and none occurred at 25 ppb or lower. By contrast, the growth of *D. tertiolecta* was not affected at any of the concentrations of PCBs or DDT tested.

In untreated mixed cultures, *T. pseudonana* reproduced more rapidly than *D. tertiolecta*, reaching an eight- to ninefold greater cell concentration after 4 days. Treatment with PCBs or DDT significantly diminished the competitive success of *T. pseudonana* and increased that of *D. tertiolecta* at all concentrations tested, even though the lowest concentrations (1 and 10 ppb, respectively) had no detectable effect on the growth of *T. pseudonana* in pure cultures (Fig. 1, A and B). The graphs of species ratios (Fig. 1C) suggest that still lower concentrations would alter the final species composition of the cultures. Higher concentrations of the organochlorine compounds caused greater deviation in species ratios from those of the control cultures.

Although species ratios were substantially changed, the final total cell numbers did not differ significantly among control and treated mixed cultures. Since cells of the two species are approximately the same size, the final biomass was not markedly affected. This result suggests that the two organisms were competing for a limiting nutrient in the mixed cultures. Presumably *T. pseudonana* assimilated most of the nutrient in control cultures because it grew faster, but its ability to compete was impaired by PCBs and DDT; more nutrient then became available for *D. tertiolecta*. Since phytoplankters often compete for limiting resources in nature (*4*), tests evaluating pollutants in mixed cultures probably give a more ecologically meaningful indication of algal sensitivity than those in pure cultures. *Thalassiosira pseudonana* was substantially more sensitive to PCBs and DDT in mixed cultures than in pure cultures.

Chlorinated hydrocarbons and other stable pollutants (*5*) could alter natural algal communities by suppressing sensitive species and permitting pollutant-resistant forms to become dominant. Polychlorinated biphenyls and DDT can occur in natural waters at concentrations comparable to those that altered the species ratios in our experiments, although concentrations found in nature are generally lower (*6*). Exposure to organochlorine compounds is probably greater than the concentrations in natural waters would indicate, however, because these substances are rapidly absorbed from water by organisms, including phytoplankters (*1*). In eutrophic environments, alterations of algal communities could further re-

Reprinted from Science 176:533-535 (1972).

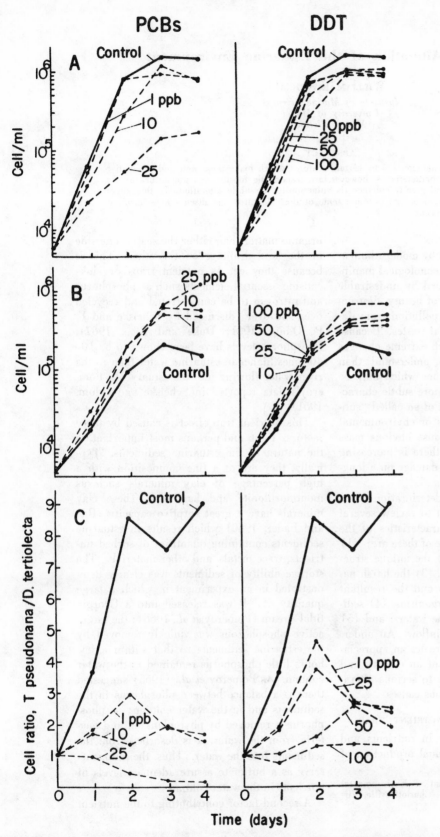

Fig. 1. Growth of (A) *T. pseudonana* and (B) *D. tertiolecta* in mixed cultures; (C) species ratios. Data points are the means of three replicates. Cell concentrations: single classification analyses of variance showed that, at day 4. *D. tertiolecta* controls differed from all treated cultures ($P < .005$); *T. pseudonana* controls differed from all treated cultures ($P < .025$), except at 25 ppb of DDT ($P < .10$). Species ratios: analyses of variance, performed by means of the Kruskal-Wallis and Wilcoxon two-sample tests (*13*) revealed that, on days 3 and 4, controls differed from all treated cultures ($P < .001$); ratios for 100 ppb of DDT differed from those for 50 ppb of DDT, and ratios for 25 ppb of PCBs differed from those for 10 ppb of PCBs ($P < .05$).

duce an already decreased species diversity (*7*), aggravating problems of algal blooms and contributing to the general degradation of the ecosystem (*8*).

Many zooplankters graze selectively, often choosing their food on the basis of size or shape (*9, 10*). The dietary requirements of herbivores are not satisfied by all algal species, as indicated by the growth rates of oyster and clam larvae (*11*), the viability of barnacle nauplii (*10*), and reproduction in copepods (*12*). Hence, altering the species composition of a phytoplankton community could profoundly affect the health, distribution, and abundance of many animal populations higher in the food web (*8, 10, 11*).

Jerry L. Mosser
Nicholas S. Fisher
Charles F. Wurster
*Marine Sciences Research Center,
State University of New York,
Stony Brook 11790*

References and Notes

1. D. B. Peakall, *Science* **168**, 592 (1970); *Sci. Amer.* **222** (4), 72 (1970); ——— and J. L. Lincer, *BioScience* **20**, 958 (1970); K. J. Macek, *J. Fish. Res. Board Can.* **25**, 1787 (1968); *ibid.*, p. 2443; C. F. Wurster, *Environment* **13** (8), 33 (1971); *Biol. Conserv.* **1**, 123 (1969); *Proceedings of the International Conference on the Environment of the Future*, Jyvaskyla, Finland, 1 July 1971 (Macmillan, London, in press) (*Congr. Rec.*, 28 July 1971, p. E8333).
2. C. F. Wurster, Jr., *Science* **159**, 1474 (1968); D. W. Menzel, J. Anderson, A. Randtke, *ibid.* **167**, 1724 (1970).
3. J. L. Mosser, N. S. Fisher, T.-C. Teng, C. F. Wurster, *ibid.* **175**, 191 (1972).
4. G E. Hutchinson, *A Treatise on Limnology* (Wiley, New York, 1967), vol. 2, pp. 355–489; E. M. Hulburt, *Ecology* **51**, 475 (1970).
5. R. C. Harriss, D. B. White, R. B. Macfarlane, *Science* **170**, 736 (1970).
6. T. W. Duke, J. I. Lowe, A. J. Wilson, *Bull. Environ. Contam. Toxicol.* **5**, 171 (1970); J. O. Keith and E. G. Hunt, *Trans. North Amer. Wildlife Natur. Resour. Conf., 31st*, 14–16 March 1966 (1966), pp. 150–77.
7. R. Patrick, *Ann. N.Y. Acad. Sci.* **108**, 359 (1963); C. M. Palmer, *ibid.*, p. 389; R. Margalef, *Oceanogr. Mar. Biol. Annu. Rev.* **5**, 257 (1967).
8. G. M. Woodwell, *Science* **168**, 429 (1970); J. H. Ryther, *Biol. Bull.* **106**, 198 (1954).
9. V. Bainbridge, *J. Mar. Biol. Ass. U.K.* **37**, 349 (1958); H. J. Curl and G. C. McLeod, *J. Mar. Res.* **19**, 70 (1961); J. J. Lee, M. McEnery, S. Pierce, H. D. Freudenthal, W. A. Muller, *J. Protozool.* **13**, 659 (1966); S. Richman and J. N. Rogers, *Limnol. Oceanogr.* **14**, 701 (1969).
10. E. J. F. Wood, *Marine Microbial Ecology* (Reinhold, New York, 1965), pp. 71–96.
11. H. C. Davis and R. R. Guillard, *U.S. Fish Wildlife Serv. Fish. Bull. 136* **58**, 293 (1958).
12. S. M. Marshall and A. P. Orr, *J. Mar. Biol. Ass. U.K.* **30**, 527 (1952); L. Provasoli, K. Shiraishi, J. R. Lance, *Ann. N.Y. Acad. Sci.* **77**, 250 (1959).
13. R. R. Sokal and F. J. Rohlf, *Biometry* (Freeman, San Francisco, 1969), pp. 388–94.
14. We thank M. F. Luhnow for laboratory assistance and R. R. L. Guillard for valuable consultations. Supported by NSF grant GB-11902 and NIH postdoctoral fellowship 1FO2ES48112-01 to J.L.M.

20 December 1971

Insidious Alteration of the Estuarine Environment

WILLIAM E. ODUM[1]

Institute of Marine Sciences
University of Miami
Miami, Florida

ABSTRACT

Shallow estuaries are characterized by certain features which make them rich and productive ecosystems; these same characteristics, however, are responsible for the delicate nature of the estuarine environment and greatly enhance its vulnerability to·subtle alteration. In this paper and the subsequent discussions, we examine some of these features and discuss how insidious changes in estuaries can occur.

INTRODUCTION

The invasion of estuaries by modern human society with its attendant technological manipulations is often accompanied by undesirable events such as despoliation of beauty, destruction of wildlife and severe pollution from domestic sewage and industrial wastes. Because of their obvious nature such extreme changes are frequently simpler to understand than other man-made perturbations which may be equally harmful, but of a more subtle character. This is the hazy realm of so called "sublethal" effects of pollutants or environmental alterations which do not cause obvious mass mortalities, but for which there is increasing evidence indicating serious damage on a long-term basis.

To understand the slow deterioration of estuaries it is instructive first to review several of the more important characteristics of the estuarine environment. Five of these are: (1) the nutrient trap effect, (2) the unique structure of estuarine food webs, (3) the harsh nature of physical conditions and the resultant vulnerability of estuarine organisms, (4) sedimentary control of estuarine waters, and (5) the key role of freshwater inflow. An understanding of these features creates an appreciation of the delicate nature of an estuary and emphasises its vulnerability to serious alteration from apparently inocuous causes.

THE ESTUARY AS A NUTRIENT TRAP

Estuaries are far richer in nutrients and normally have a higher annual production of organic matter than either the sea on one side or the land drainage on the other. This is because they act as nutrient traps or sinks causing essential elements such as phosphorus and nitrogen to be concentrated and recycled over and over (discussed by Schelske and E. P. Odum, 1961; Duke and Rice, 1967). Phosphorus levels have been shown to be 10–40 times higher in estuarine water than in the river water flowing into the estuary (L. Pomeroy's data reported in Schelske and Odum, 1961).

This nutrient trap effect is caused by three factors. First, and perhaps most important, is the nature of the estuarine sediments. Typically, they are of a fine composition with a high percentage of clay minerals such as montmorillonite and kaolinite. These clay minerals have a great sorptive capacity (Rae and Bader, 1960) which results in estuarine sediments containing quantities of sorbed nutrients, trace metals, and other materials. The storage ability of sediments was clearly demonstrated in an experiment in which a large quantity of ^{32}P was released into a Georgia tidal stream (Pomeroy *et al.*, 1966); the radioactive phosphorous was quickly removed by the estuarine sediments so that within a few hours little phosphorus remained in the water column. As Pomeroy *et al.*, (1965) suggested there is a balance between phosphorus in the sediments and in the water column; as phosphorus is removed by phytoplankton and bacteria from the water, it is desorbed from the sediments into the water. Thus, the sediments serve as a buffer to assure adequate levels of nutrients in the water column.

A second factor contributing to the nutrient

---

[1] Present address: Department of Environmental Sciences, University of Virginia, Charlottesville, Virginia 22903.

526

Reprinted from Trans. Am. Fish. Soc. 99:836-847 (1970).

trap mechanism is the process of biodeposition. Estuaries are filled with filter feeding molluscs and crustaceans which remove in a single day enormous quantities of suspended particles. These particles are compacted and extruded as fecal pellets or pseudofeces which sink to the sediment surface where they may be ingested by deposit feeding organisms or become part of the sediments (Kuenzler, 1961). This efficient packaging and storing mechanism causes quantities of nutrients to be removed and temporarily stored either in organisms or in the sediments. Nutrients which become covered with layers of sediments are not necessarily lost to the estuarine ecosystem. Many plants and burrowing organisms act as mechanical pumps to bring the nutrients to the surface again. Pomeroy *et al.*, (1966) have shown that the marsh grass, *Spartina alterniflora*, obtains its nutrients far below the sediment surface.

Finally, there is the contribution of estuarine water circulation to the nutrient trap effect. Through a combination of the horizontal ebb and flow of water caused by the tide and the vertical movement of water masses of different salinities there is a tendency for nutrients to become trapped. This is best demonstrated by the partially stratified salt wedge estuary discussed by Pritchard (1967) and Cameron and Pritchard (1963). In such an estuary nutrients may be removed from the downstream flowing freshwater surface layer and moved slowly back upstream by the salt wedge layer near the bottom.

Obviously, not all of these factors operate in all estuaries; some have sediments of coarse grain quartz sand while others do not have a circulation conducive to concentration of nutrients. This explains to a great extent the difference in fertility of nearby estuaries.

THE ESTUARY AS A POLLUTION TRAP

Unfortunately, all of the factors which enable an estuary to concentrate and recycle nutrients also allow the estuary to become a pollution sink. Fine sediment particles have been shown to concentrate everything from petroleum byproducts to persistent pesticides. W. E. Odum *et al.*, (1969) have demonstrated how detritus particles in estuarine sediments can concentrate DDT at levels 100,000 times higher than in the estuarine water. These DDT residues can then be transferred from the particle to a detritus-algal consumer such as the fiddler crab, *Uca pugnax.* Since a large number of the species present in estuaries ingest sediment particles directly, the threat from contaminated sediments is very real.

Pomeroy *et al.*, (1966) have shown how the sediments serve as a negative sink for radioactive pollutants. In other words, an isotope such as ^{65}Zn may be held for months in the sediment with a continual low level release into the water creating a condition of continued pollution and may greatly retard the cleansing of an estuary.

The ubiquitous estuarine filter feeders remove and concentrate practically everything from suspension including pollutants. Butler (1966b) has pointed out the ability of oysters to concentrate DDT from the water and incorporate it into their pseudofeces, making it available in a more concentrated form to deposit feeders. The principle of biological magnification or increased concentrations at each trophic level is well documented for radioactive and pesticide pollutants (Woodwell, 1967) and has been shown to operate in the estuarine environment (Woodwell *et al.*, 1967).

The circulation features which concentrate nutrients in an estuary may also work equally well to collect pollutants (Cameron and Pritchard, 1963). If an estuary is strongly stratified with a well defined surface outflow of freshwater and if the pollutant is lighter than water, it is likely to be carried out to sea at the velocity of the surface flow. However, if as in many cases, the pollutant is heavier than water (*i.e.*, mercury, lead or pulp mill effluent) and the estuary is only partially stratified, then the pollutant may sink into the lower, upstream moving, high salinity layer and be distributed throughout the upper portion of the estuary.

ESTUARINE FOOD WEBS

Fundamental to the functioning of any ecosystem is the one-way flow of energy from the sun through green plants and then down various food webs of consumer organisms. An interruption or alteration of this flow of energy at any point in the progression can alter the character of the ecosystem. In estuaries the flow of energy is particularly susceptible to alteration because of the low diversity of species present. The removal of a species through pollution stress or some other adverse influence may leave an empty niche or else a niche which has been claimed by an ecologically less desirable species (*i.e.* one that replaces a forage organism, but itself is not suitable food, thus creating an energy dead end). This can be particularly critical where one plant is of outstanding importance as a primary producer to the system or where there are only a few species at the important detritus-algal consumer level.

The Importance of Organic Detritus

In many estuaries the sources of much of the primary production are vascular plants such as marsh grasses, sea grasses and mangroves. The Georgia estuaries depend to a great extent for energy input from the marsh grass, *Spartina alterniflora*; in Texas and south Florida bays one of the important primary producers is the sea-grass, *Thalassia testudinum*, while along the southwest coast of Florida it is the mangrove, *Rhizophora mangle*. All of these plants produce prodigious amounts of plant tissues ranging from four tons of dry organic material per acre per year for mangrove leaves in the Everglades estuary (Heald, 1969) to 10 tons by the *Spartina* in the Georgia salt marshes (Schelske and E. P. Odum, 1961). As these materials enter the water they decompose slowly forming organic detritus with its adsorbed bacteria, fungi, protozoa and micro-algae. This rich source of food decomposes slowly enough to insure a continuous annual production of food without the feast or famine characteristics of algal blooms.

Organic detritus (with its attached populations of microorganisms providing most of the useful nutrition) serves as a principal energy source for many estuarine animals including amphipods, isopods, mysids, small crabs, insect larvae, caridean shrimp and even some fishes. For other organisms it serves as an emergency food supply when normal food supply when normal food sources disappear. Even certain carnivores such as the crested goby, *Lophogobius cyprinoides*, are able to exist upon a diet composed solely of mangrove leaf detritus for periods of at least a year (W. E. Odum, 1969). Although a predator's growth may be seriously hampered by a detritus diet, it does serve to carry the animal through periods of food scarcity.

Other Sources of Primary Production

No estuary is completely dependent upon vascular plant organic detritus as a food base; there is always the contribution of benthic and suspended algae. Moreover, detritus consumers appear unable to grow and reproduce successfully on a diet solely of detritus. Almost all detritus consumers include at least 10 to 20% fresh algal cells in their diet in the form of diatoms or filamentous green and blue-green algae (W. E. Odum, 1969). For this reason they should be regarded as detritus-algal consumers.

One of the most important food chains in the estuary is dependent upon the phytoplankton. This may seem strange since phytoplankton production is often comparatively insignificant in shallow estuaries where plant detritus producers and benthic algae predominate. However low their standing crop and production rates, they are essential in the food chain composed of phytoplankton—small zooplankton—larval fish. Young estuarine fishes in many cases begin their development offshore where they initially feed upon small zooplankton. This diet continues for some time after they enter the estuary and only after a period of time is there a gradual transition to benthic organisms, plants and detritus. The mullet, *Mugil cephalus*, which spends most of its life as a strict detritus-algal feeder, ingests little except for small zooplankton until it reaches approximately 25 mm (W. E. Odum, 1970). For almost all larval fishes and crustaceans in estuaries it is essential that sufficient amounts of small zooplankton be present. They, in turn, require quantities of suspended algal cells.

Interruption of Energy Flow

Perturbation of any of these food webs by man may seriously hinder the ability of an estuary to produce species which he deems desirable. Destruction of plant detritus producing areas such as sea-grass beds, mangrove swamps or marshes directly limits the numbers of detritus-algal consumers such as shrimp, amphipods, mysids and small crabs. A lack of these essential animals limits the next trophic level which feeds upon them and so on. Equally important to the first level of consumers is the protection of areas including mud banks and shallow mud flats which produce benthic and epiphytic microflora. Destruction or alteration of phytoplankton may lower the numbers of small copepods and crab larval stages which in turn limits the numbers of larval fishes. The full impact may come years after the phytoplankton community has been altered and will come in the form of depleted stocks of gamefish, commerical fish, crustaceans and molluscs.

The means of protection of the detritus-algal consumers is obvious—protect the marshes, seagrass beds and mangrove swamps. Protection of the floating, benthic and epiphytic microalgae is a more complicated proposition. Many apparently harmless effects may subtly alter the composition of the natural community. Wurster (1968) has found that concentrations of DDT as low as a few parts per billion can effectively alter phytoplankton communities. Sewage pollution can have the same effect. The famous Great South Bay Duck Farm incident (Ryther, 1954) is a good example. Over a period of time a gradual decline in the oyster and fish production of Great South Bay, Long Island, had occurred; this decline coincided with the growth of the duck industry in the watershed of the bay. Careful investigation by Ryther showed that although production of micro-algae was enhanced by the enrichment from 8 million ducks, the forms which prospered such as Nanochloris and Stichococcus were unsuitable to oysters for adequate growth. The oysters and probably other filter feeders in the bay required a diet of mixed species such as that provided by the normal phytoplankton community of Great South Bay prior to the arrival of the ducks.

Food Chain Simplicity

Another factor which makes estuarine food chains especially vulnerable to alteration is their basic simplicity and shortness. In the North River Estuary of the Florida Everglades almost all food chains are based upon a handful of detritus-algal consumers (W. E. Odum, 1969). These include one species of crab, two species of caridean shrimp, four species of amphipods, three species of mysids, several species of chironomid midge larvae, one snapping shrimp, two bivalves, one crayfish and two fishes. The large populations of predators present must compete for these few key organisms. Under natural conditions this poses no problems since all of the detritus-algal consumers flourish in great abundance.

This situation is not much different from that in most estuaries. The actual species involved may differ somewhat but generally there is a dependence upon a few key organisms which are able to utilize detritus and micro-algae. When a pollutant such as DDT is introduced into such an estuary a very different situation is created. Almost all of these key organisms are either crustaceans or insect larvae and are susceptible to low levels of pesticides. If they are removed, little remains for the consumers except for the filter feeding bivalves and the forage fishes. Even these may not be suitable food since both groups are able to concentrate pesticides to levels which are potentially lethal to a consumer (Ferguson, 1967; Butler, 1966). Rosato and Ferguson (1968) fed endrin-exposed Gambusia to 11 species of vertebrates (fishes, reptiles and birds) causing 95% mortality among the predators.

Interference with the Energy Input

Man may even interfere with food chains at the first link by limiting the amount of sunlight reaching the plants of an estuarine system. This can occur through increased water turbidity from dredging operations (see discussion of the "hydraulic dredge cycle") or poor land use practices in the watershed of the estuary.

Copeland (1965) has tested the effect of reducing the light available to a simulated community dominated by the seagrass, *Thalassia testudinum*. By lowering the input of sunlight from 1500 to 200 footcandles, as might happen if bay water suddenly became turbid, the *Thalassia* dominated community became a blue-green algae dominated community. Surprisingly, once the transition had become stabilized it was found that the productivity of the two communities was equivalent. However, the blue-green community is potentially much less useful to man than the *Thalassia* community simply because fresh *Thalassia* leaves and detritus are consumed by scores of organisms, while blue-green algae are difficult to digest and are successfully consumed by only a few species.

THE VULNERABILITY OF ESTUARINE ORGANISMS

Even apparently well adapted estuarine organisms may suffer periodic mass mortalities from natural changes in the environment. There is evidence that mass mortalities from temperature changes, oxygen depletions, and other causes have always occurred, even in the most pristine and unpolluted estuaries (Smith, 1896). This emphasizes the fact that many estuarine residents are living near the limit of their tolerance range; any further alteration, no matter how slight it may seem, could exclude the organism permanently from the estuary. Such alterations come in the form of increased water temperatures from electric power plants in warm climates, decreased oxygen concentrations in the water from dredging or mining operations or through the introduction of additional stress from low levels of pollutants.

"Sub-lethal" Effects of Pollutants

The effects of low levels of pollutants in estuaries are poorly understood. In many cases so called "sub-lethal" effects are very lethal indeed, they merely require a bit longer than the time dictated by the investigator's patience or else a more complex environment than the predator-free safety of the laboratory. Butler (1966a) mentions that there is evidence

indicating that low levels of pesticides can cause ill-defined, but significant changes including increased mortality, loss of production and even changes of direction of natural selection in estuarine fauna. This has been more recently emphasized (Butler, 1969) by a nation-wide monitoring program which has shown that widespread pesticide pollution is significantly decreasing the productivity of estuarine fish and shellfish. An indication of the type of findings which may come to light in the future is the recent discovery of Anderson and Peterson (1969) that sub-lethal concentrations of DDT prevent the establishment of a visual conditioned avoidance response in some fish, and, equally serious, affects the thermal acclimation mechanism.

Low level pollutants of concern are not limited to pesticides. Also of potential danger are radioactive wastes, and components of sewage and industrial wastes. Steed and Copeland (1967) have found that petrochemical waste materials when tested in very low concentrations on a number of estuarine fish species acted as a metabolic depressor; at higher concentrations approaching the 48 hour mean tolerance limit (TL_M) the wastes acted in just the opposite manner and increased the metabolic cost of the fish. Temporary survival of estuarine organisms means little if they fail to grow properly, do not reproduce or produce viable young or have aberrant behavior patterns.

Consideration of the potential of pollutants to cause metabolic decreases and increases, especially when acting in a synergistic manner with the extreme conditions of an estuary, creates a clear picture of how real the threat is, even from slightly polluted water. There is evidence that larger fish in water with low pollution levels are killed more easily by cold temperatures than if no pollutant were present (Wohlschlag and Cameron, 1967). This in turn could lead to a reduced life span and affect the reproductive capacity of the population eventually resulting in an abnormal population age and sex distribution. This would seriously impair the ability of a species to successfully exploit its niche (Steed and Copeland, 1967). Even if mortality of the species was not increased by a low level pollutant, it might decrease growth rates sufficiently to have the same effect on the population since individuals within the population would be subject to natural mortality rates for a greater length of time before reaching spawning size (Copeland and Wohlschlag, 1967).

SEDIMENTARY CONTROL OF THE ESTUARINE ENVIRONMENT

Baas Becking and Wood (1955) after years of estuarine research concluded that, "in the estuaries it is the sediments which control the ecosystem." Certainly, this is true in shallow estuaries where the proximity of the sediments to the entire water column causes any changes in the sediments to be accompanied by changes in the water above. In addition to storing materials by sorption, the sediments are the site of intensive microbial activity which through the sulfur, nitrogen and phosphorus cycles continually decomposes complex organic compounds contained in organic detritus in the sediments and makes it available in usable forms such as ammonia, nitrate or organic phosphate (discussed at length by Wood, 1965).

Benthic plant growth is encouraged in shallow estuaries by the combination of adequate light levels and the ample supply of nutrients regenerated from the sediments. One of the most important functions of benthic vegetation, whether blue-green algal mats or luxuriant *Zostera* beds, is to stabilize the sediments. Stabilized sediments, in turn, prevent excess water turbidity and allow the formation of a strong reducing layer below the sediment surface, a condition which enhances nutrient recycling through the sulfur cycle (Wood, 1965).

Perturbation of the delicately balanced estuarine sediments is easily accomplished through dredging, bulkheading, and mining operations. In many cases the damage to an estuary far exceeds the benefits gained from manipulation of the sediments.

The Hydraulic Dredge Cycle

Permanent damage to estuarine sediments is usually caused by poorly planned or ill-advised dredging and filling. Dredging for navigational channels, if properly engineered, causes only a minimum of damage to the environment and in shallow bays may even prove to be beneficial by improving circulation patterns. Ideally, only the actual area which will compose the navigational channel should be permanently altered. In practice, there is often a continual adverse effect through the "hydraulic dredge cycle." This situation arises when the material dredged from channels is randomly dumped as spoil banks with little thought for subsequent aerial and sub-aerial erosion. The result is a rapid deterioration of the spoil bank, transport of the sediments back into the channel with subsequent filling of the channel necessitating costly redredging. This is a common cycle in many estuaries and results in areas of high turbidity, constantly shifting sediments and limited primary production.

Often, the construction of a series of spoil banks which reach near or to the surface completely destroys circulation to sections of an estuary. Poor circulation results in rapid sedimentation rates and the loss of a valuable region. Breuer (1962) describes just such an occurrence in South Bay, the southernmost bay area of the Laguna Madre in Texas. The construction of the Brownsville ship channel in 1938 with indiscriminate dumping of dredged material effectively closed off South Bay from natural circulation from the Gulf of Mexico. Since the circulation was destroyed, the average water depth has decreased from four feet to less than 18 inches and the fisheries of the bay have been destroyed.

Lowered Dissolved Oxygen Content

Estuaries are not endowed with consistently high levels of dissolved oxygen. This can be traced in part to the large numbers of animals and plants respiring in a limited volume of water, but most important it is due to the great amount of oxidizable carbon in the form of suspended organic detritus and detritus in the surface sediments. Frankenberg (1968a) has pointed out that many unpolluted estuaries fall naturally well below the minimal standard of 4.0 mg of oxygen per liter of water established for polluted estuaries by the National Technical Advisory Committee to the Secretary of the Interior in their 1968 report on water quality criteria.

Since unpolluted estuaries are already in a borderline condition, any change, natural or man-induced, can cause removal of species or even mass kills of organisms. Such kills occur with great frequency in the unpolluted Everglades estuary during the rainy season when large volumes of cold rainwater sink to the sediments and cause a resuspension of sedimentary organic detritus particles and a release of hydrogen sulfide. After the passage of Hurricane Betsy with her heavy rains in 1965, oxygen concentrations remained below 1.0 mg/1 for one month in the North River section of the estuary (Tabb, personal communciation).

Any process which resuspends oxidizable sediment particles in the water column is capable of reducing the oxygen concentration of estuarine water to a level which is not suitable for the normal biota. Frankenberg and Westerfield (1969) found that when surface sediments from Wassaw Sound, Georgia, were suspended in estuarine water, they were capable of removing 535 times their own volume of oxygen from the water.

Although Macklin (1962) found oxygen depletion from dredging in Louisiana marshes to be relatively unimportant, other research has indicated that this may not always be the case. Brown and Clark (1968) measured dissolved oxygen values ranging from 16 to 83% below normal in the vicinity of dredging operations. Frankenberg (1968b) has estimated that the construction of a dike three feet high and 12 feet wide at the base in the Georgia salt marshes would liberate enough sediments into the water to remove 334,385 milligrams of oxygen for each foot of dike constructed. This would completely remove the oxygen from 2,437 cubic feet of water having a dissolved oxygen content of 4.8 mg/1.

Such oxygen depletions and the accompanying turbidity appear to be restricted to the vicinity of dredging in most cases. Ingle (1952) found effects only 0.23 miles from the site of dredging while Hellier and Kornicker

(1962) record effects more than 0.5 mile, but less than one mile from the dredging operation. With well designed dredging programs these effects should be temporary, but, through poorly placed spoil banks and the hydraulic dredge cycle the turbidity, shifting sediments and lowered oxygen content may become permanent.

Dredging and Filling

The effects of dredging and filling for the creation of new land are almost always permanently destructive to the estuarine environment. Since it requires about three acres of submerged sediments to create one acre of filled land (Tabb, personal communication), large estuarine areas may be permanently removed from production by the creation of causeways, and developments for housing and other buildings.

Often the areas from which the dredged sediments are removed are left with a much greater water depth. In a slightly turbid estuary this will lower the benthos below the photic zone or at least make it marginal and suitable only for blue-green algae. Without the stabilizing effects of plant communities these dredged regions remain relatively barren and characterized by shifting anoxic sediments. Furthermore, the filled areas may restrict circulation to the point where stratification occurs and an anoxic bottom layer is formed. Thus, these artificially deepened depressions can develop into traps for sewage and other heavy pollutants. A region of south Biscayne Bay, Florida, which was dredged 28 years ago, subsequently collected large quantities of untreated sewage. Even though little sewage is dumped into the bay today, this dredged depression remains barren and anoxic with reducing conditions on the sediment surface.

The Edge Effect and Bulkheading

The most productive zone in many estuaries is the zone of transition including the intertidal and adjacent shallow subtidal areas. This is especially true where there is a minimum of wave-generated turbulence and the sediments are stabilized. Such rich bands are found along the edges of marshes and around the proproots of mangroves. In other areas it may consist of tidally exposed mud banks.

To the enthusiastic real estate developer there is no easier way to convince listeners of the lack of value of shallow estuaries than to refer to those "useless mudflats." As most ecologists realize, intertidal banks and mudflats are more productive than most of the world's oceans. Pomeroy (1959) measured a net production of 180 g C/M²/year from the intertidal microalgae of a Georgia marsh. The net production can be even higher in regions where macro-algae are found also.

Not only are estuary "edges" the site of algal production, but they also serve as a region of concentration of organic detritus and its attendant degradation and recycling processes. In the Everglades Estuary we have found the percentages of organic detritus in the sediments along the shore to be several times higher than a few meters offshore.

Shallow water also affords greater protection for juvenile fish and crustaceans. Because of the high productivity levels, concentration of detritus, and protection factors, these edges are usually areas of high standing crops of organisms. This is the initial habitat for many post larval and juvenile fishes and crustaceans when they first arrive in the estuary. On the North River in the Everglades estuary we have found the standing crops of such diverse organisms as amphipods, isopods, mysids, gobies and caridean shrimp to be 10 times higher than in deeper water a few meters away.

The estuarine edge can be completely destroyed by the construction of bulkheads since with this type of construction there is usually no shallow intertidal zone. An immediate drop into deeper water replaces the shallow water habitat for juvenile organisms. No longer is there the high productivity of intertidal benthic diatoms or the concentration of organic detritus. A relatively low energy environment (in terms of wave energy) may be transferred into one of moderate energy due to the reflection of waves from the bulkhead wall.

Mock (1967) studied two estuarine habitats, one natural zone and one which had been altered by bulkheading. The natural zone had a band of organic detritus peat extending out two feet from shore. This band was composed of from 40 to 60% organic matter. In the bulkheaded zone there was no detritus band and the organic content of the sediments was only six percent. Not surprisingly, he found 12.5 times as many white shrimp, *Penaeus setiferus*, and 2.5 times as many *P. aztecus* in the natural area as in the altered environment.

THE IMPORTANCE OF FRESHWATER INFLOW

Estuaries by definition constitute a gradient from freshwater to marine conditions. If the source of the freshwater inflow is cut off or reduced, salinities in the estuary will rise due to evaporation and the influx of seawater. In such instances the estuary will become an embayment with salinities approaching those of the open ocean or even a hypersaline lagoon.

At first thought it is difficult to imagine why reduced freshwater inflow and subsequent rises in salinity would be damaging, especially where a bay with normal open ocean salinities would be created. Certainly, the numbers of mollusc, crustacean and fish species would rise impressively; however, several of the most valuable and subtle qualities of the estuary would be lost. Most obvious of these is its value as a protective nursery ground; also of importance is the mechanism by which larval invertebrates along with juvenile shrimp and fishes find their way into the estuary or maintain their position there.

The Salinity Gradient Transport Mechanism

How these young organisms find their way into estuaries and how they maintain their position there against strong outflowing currents has been of interest to estuarine ecologists for years. This is an especially serious problem for the oyster—it would appear to be difficult for oysters to maintain populations in the intermediate salinity section of an estuary since their larvae have little swimming ability other than vertical positioning in the water column. J. Nelson (1912, 1914) first postulated that the larvae might rise on the incoming tide and be carried in some distance before settling temporarily to the bottom during the ebb tide. T. Nelson (1930) further postulated that this larval migration into the water column was triggered by the increasing salinity of the flood tide and conversely that the dropping salinity of the ebb tide decreased larval activity. Although this has never been conclusively proven, Haskin (1964) found that oyster larvae are responsive to changes in salinity and that gradually increasing salinity will stimulate older stage larvae to swim while decreasing salinity causes younger larvae to remain on the bottom.

Even more conclusive evidence for response to salinity change has been found for shrimp (Hughes, 1969). In a series of experiments he observed that pink shrimp, *Penaeus duorarum*, post larvae were able to perceive and respond to a salinity difference of as little as one part per thousand. Further, he found that the shrimp post larvae like the oyster larvae were most active in water of high salinity and in lower salinity they dropped to the substrate. As Tabb *et al.* (1962a) have pointed out this would allow a tiny post larval shrimp with little swimming ability to move 2 to 3 miles on each flood tide and eventually would suffice for penetration far up into the low salinity low density region of an estuary.

Equally interesting from Hughes' experiments was the finding that juvenile shrimp reacted much differently from the post larvae. The juveniles, having spent three to six months in the estuary, are large enough to exert considerable horizontal swimming effort, and are ready to depart the estuary for offshore spawning grounds. Shrimp of this size when faced with water masses of different salinity exhibit a positive rheotaxis (swimming against the current) throughout the flood tide when "normal" seawater salinities prevail, but when salinity was decreased as during the ebb tide the sign of the rheotaxis was reversed and swimming downstream with the current occurred. Apparently, through this mechanism, transport out of the estuary and offshore is accomplished.

As Hughes has mentioned his findings would probably explain the marked positive correlation reported between commercial catches of *P. setiferus* off the coast of Texas and the rainfall of the preceding year (Hildebrand and Gunter, 1952; Gunter and Hildebrand, 1954) since salinity differences between tides would be greater with increased runoff. A similar correlation has been shown for catches of *P. duorarum* and the Florida rainfall of the previous year (Iversen, unpublished data).

It is not hard to imagine the effect of an alteration in the watershed which would seriously limit the inflow of freshwater. Tabb *et al.*, (1962a) report a year in which large juvenile pink shrimp remained in inshore waters of the Everglades estuary when salinities were abnormally high (about 30 ppt). Such high salinities in this estuary are caused by a combination of low rainfall and insufficient water supplied to the Everglades from the Florida Flood Control Commission. Subtle indeed are changes caused by a flood control agency to help farmers, but also adversely affecting a commercial fishing industry 200 miles away and fifty miles offshore!

Estuaries As a Source of Dissolved Organic "Roadmaps"

There is increasing evidence that many young estuarine animals which have been spawned offshore utilize dissolved organic compounds in the estuarine water to direct their movements back into an estuary, much as adult salmon navigate while returning to their home streams to spawn (Hasler & Wisby, 1951; Hasler, 1965). Creutzberg (1961) has shown that young elvers are attracted by an unidentified organic compound present in inland water. Kristensen (1964) found evidence to demonstrate that juvenile fishes and shrimp exhibit a preference for baywater over seawater when offered a choice (salinity, pH, dissolved oxygen and temperature were held constant). This preference was not changed by filtration of the bay water through series D filter paper, but was destroyed by filtration through Norit charcoal filter which removes all organic compounds. It could be hypothesized that adult estuarine fishes utilize a reverse mechanism to find their way offshore to spawn.

SUMMARY

(1) The mechanisms which enable estuaries to be efficient nutrient traps also contribute to their ability to be pollutant traps.

(2) Destruction of plant detritus producing regions of an estuary such as *Spartina* marshes or mangrove forests will greatly lower the productivity of the estuary and directly limit its potential to produce commercially important species of fish and crustaceans.

(3) Because of their simplicity, food chains in shallow estuaries are particularly susceptible to interference from man. Generally, there is a dependence by the higher trophic levels upon a few key primary consumers which are able to utilize both microalgae and vascular plant detritus particles (nutrition in the latter case comes largely from attached fungi and bacteria).

(4) Many estuarine organisms are living near the limit of their tolerance ranges. These animals may be excluded from an estuary by any additional stresses such as those caused by low levels of pollution or by decreased oxygen concentrations in the water resulting from dredging or mining operations.

(5) Undisturbed and stabilized sediments are important to an estuary for normal nutrient cycling, to prevent excess turbidity in the water column and as a site for extensive plant growth.

(6) Shallow estuaries naturally exist in a state of advanced eutrophication. For this reason they are vulnerable to any process, no matter how insignificant it might seem, which would lower the oxygen concentration of the water.

(7) The most productive and valuable zone in many estuaries is the intertidal and shallow subtidal area. This boundary region is destroyed by construction of bulkheads.

(8) Large amounts of freshwater inflow are necessary for an estuary to function normally. The low salinity region of an estuary is important both for protection of juvenile fish and invertebrates and for production of oysters. The salinity gradient appears essential in the life cycle of organisms which spawn outside the estuary.

LITERATURE CITED

ALABASTER, J. S. 1965. Effects of heated effluents upon marine and estuarine organisms. Adv. Mar. Biol. 3: 63–103.

ANDERSON, J. M. AND M. R. PETERSON. 1969. DDT: Sublethal effects on brook trout nervous system. Science 164: 440–441.

ANONYMOUS 1968. New York Times, December 15, 1968, P. 60.

BAAS BECKING, L. G. M., AND E. J. F. WOOD. 1955. Biological processes in the estuarine environment, I. II. Kon. Ned. Adad. Weten. Proc., B58: 160–181.

BREUER, J. P. 1962. An ecological survey of the lower Laguna Madre of Texas, 1953–1959. Publ. Inst. Mar. Sci. (Texas) 8: 153–183.

BROWN, C. L. AND R. CLARK. 1968. Observations on dredging and dissolved oxygen in a tidal waterway. Water Resources Research 4(6): 1381–1384.

BUTLER, P. A. 1966a. The problem of pesticides in estuaries. Amer. Fish. Soc. Spec. Publ. No. 3: 110–115.

———— 1966b. Fixation of DDT in estuaries. 31st N. Amer. Wild. and Nat. Res. Conf.

————. 1969. The significance of DDT residues in estuarine fauna. In Chemical Fallout, M. W. Miller and G. G. Berg (editors). Charles Thomas, Publisher; Springfield, Illinois, pp. 205–220.

CAMERON, W. M. AND D. W. PRITCHARD. 1963. Estuaries. In The Sea, Vol. II, M. N. Hill (editor), pp. 306–323.

CHAPMAN, C. R. 1966. The Texas basins project. Amer. Fish. Soc. Spec. Publ. No. 3, pp. 83–92.

COMMONER, B. 1969. New York Times, June 29, 1969, p. 29.

COPELAND, B. J. 1965. Evidence for regulation of community metabolism in a marine ecosystem. Ecology 46(4): 563–564.

————, AND D. E. WOHLSCHLAG. 1967. Biological responses to nutrients—eutrophication: saline water considerations. In Advances in Water Quality Improvement. W. Eckenfelder and E. F. Gloyna (editors). Spec. Lecture Series Vol. I, University of Texas Press, Austin.

CREUTZBERG, F. 1961. On the orientation of migrating elvers (*Anguilla vulgaris*) in a tidal area. Neth. J. Sea Res. 1: 257–338.

DAVIS, J. H. 1946. The peat deposits of Florida. Florida Geol. Surv. Bull. 30: 1–247.

DUKE, T. W. AND T. R. RICE. 1967. Cycling of nutrients in estuaries. Proc. Gulf and Carib. Fish. Inst. (19): 59–67.

EGLER, F. E. 1950. Southeast saline Everglades vegetation, Florida, and its management. Veg. Acta Geo-Botanica (W. Junk, den Haag), (3): 213–265.

FERGUSON, D. E. 1967. Characteristics and significance of resistance to insecticides in fishes. In

Reservoir Fishery Symposium published by Southern Division Amer. Fish. Soc., pp. 531–36.

FONTAINE, M. 1969. Marine pollution; can we control it to advantage? Ceres 2(3): 34.

FRANKENBERG, D. 1968a. A statement presented before the State of Georgia Mineral Leasing Commission at a hearing in Atlanta, September 16, 1968. (Unpublished).

———. 1968b. Oxygen depletion effects. In a report to the state of Georgia on the proposed leasing of state owned lands for phosphate mining: C-16-C-18. (Unpublished).

———. AND C. W. WESTERFIELD. 1969. Oxygen demand and oxygen depletion capacity of sediments from Wassaw Sound, Georgia. Bull. Georgia Acad. Sci. (In Press).

GRAVE, C. 1905. Investigations for the promotion of the oyster industry of North Carolina. Rept. U. S. Comm. Fisheries 29: 247–341.

GUNTER, G. 1953. The relationship of the Bonnet Carré Spillway to oyster beds of Mississippi Sound and the "Louisiana Marsh", with a report on the 1950 opening. Publ. Inst. Mar. Sci. (Texas) 3(1): 17–71.

———, AND H. H. HILDEBRAND. 1954. The relation of total rainfall of the state and catch of marine shrimp (Penaeus setiferus) in Texas waters. Bull. Mar. Sci. 4(2): 95–103.

HANNA, A. J. AND K. A. HANNA. 1948. Lake Okeechobee; wellspring of the Everglades. Bobbs-Merrill Co., Indianapolis, Indiana 379 pp.

HARDIN, G. 1969. Finding lemonade in Santa Barbara's Oil. Saturday Review, May 10, 1969, p. 20.

HASKIN, H. H. 1964. The distribution of oyster larvae. In Symposium on Experimental Ecology. Occasional Publ. No. 2, Narragansett Marine Laboratory, University of Rhode Island, pp. 76–80.

HASLER, A. D. 1965. Underwater guideposts. Homing of salmon. University of Wisconsin Press, Madison. 155 pp.

———, AND W. J. WISBY. 1951. Discrimination of stream odors by fishes and its relation to parent stream behavior. Am. Nat. 85(823): 223–238.

HEALD, E. J. 1969. The production of organic detritus in a south Florida estuary. Doctoral dissertation, Inst. of Marine Sciences, University of Miami, 110 pp. (Unpublished).

HELLIER, T. R. AND L. S. KORNICKER. 1962. Sedimentation from a hydraulic dredge in a bay. Publ. Inst. Mar. Sci. (Texas) 8: 212–215.

HILDEBRAND, H. H., AND G. GUNTER. 1952. Correlation of rainfall with the Texas catch of white shrimp Penaeus setiferus. Trans. Amer. Fish. Soc. 82: 151–155.

HUGHES, D. A. 1969. Responses to salinity change as a tidal transport mechanism of pink shrimp, Penaeus duorarum. Biol. Bull. 136(1): 43–54.

INGLE, R. M. 1952. Studies on the effect of dredging operations upon fish and shellfish. Tech. Ser. Florida Bd. Conserv. 5: 1–26.

JOHNSON, L. 1958. A survey of the water resources of Everglades National Park, Florida. Rept. to Everglades Nat. Park, 36 pp. (mimeo). (Unpublished).

JOHNSON, V. W. AND R. BARLOWE. 1954. Land problems and policies. McGraw-Hill, New York.

KERR, J. E. 1953. Studies on fish preservation at Contra Costa Steam Plant. California Fish & Game Fish. Bull. 92, 66 pp.

KINNE, O. 1963. The effects of temperature and salinity on marine and brackish water animals, I. Temperature. Ann. Rev. Oceanogr. Mar. Biol. 1: 301–340.

———. 1964. The effects of temperature and salinity on marine and brackish water animals. II. Salinity and temperature salinity combination. Oceanogr. Mar. Biol. Ann. Rev. 2: 281–339.

KOLFLAT, T. 1968. Thermal discharges. Industrial Waste Engineering 5(3): 26–31.

KRISTENSEN, I. 1964. Hypersaline bays as an environment of young fish. Proc. Gulf and Carib. Fish. Inst. (16): 139–142.

KUENZLER, E. J. 1961. Structure and energy flow of a mussel population in a Georgia salt marsh. Limn. and Oceanog. 6(2): 191–204.

McCONNELL, J. N. 1950. Nature is greatest oyster culturist. Louisiana Conservationist 2(11–12): 23–24.

MACKLIN, J. G. 1962. Canal dredging and silting in Louisiana Bays. Publ. Inst. Mar. Sci. (Texas) 7: 262–314.

MOCK, C. R. 1967. Natural and altered estuarine habitats of Penaeid shrimp. Proc. Gulf and Carib. Fish. Inst. (19): 86–97.

MOORE, J. G. 1968. Bays and estuaries and the Texas water plan. Proc. Gulf and Carib. Fish. Inst. (20): 60–68.

NAYLOR, E. 1965. Effects of heated effluents upon marine and estuarine organisms. Adv. Mar. Biol. 3:63–103.

NELSON, J. 1912. Report of the biological department of the New Jersey Agricultural Experimental Station for the year 1911.

———. 1914. Report of the biological department of the New Jersey Agricultural Experimental Station for the year 1913.

NELSON, T. 1931. Annual report of the department of biology, July 1, 1929–June 30, 1930. New Jersey Agricultural Experimental Station.

ODUM, W. E. 1969. The structure of detritus based food chains in a south Florida mangrove system. Doctoral dissertation, Institute of Marine Sciences, University of Miami. (Unpublished).

———. 1970. Utilization of the direct grazing and organic detritus food chains by the striped mullet, Mugil cephalus. Proc. of Symposium on Marine Food Chains, Aarhus, Denmark (In Press), pp. 222–240.

———, G. M. WOODWELL AND C. F. WURSTER. 1969. DDT residue adsorbed from organic detritus by fiddler crabs. Science 164: 576–577.

OVINGTON, J. D. 1966. Experimental ecology and habitat conservation. In Future Environments of North America, Darling and Milton (editors), p. 76.

PARKER, G. G. 1960. Ground water in the central and southern Florida Flood Control District. Proc. Soil and Crop Sci. Soc. of Fla. 20: 211–231.

PRITCHARD, D. W. 1967. Observations of circulation in coastal plain estuaries. In Estuaries, G. Lauff (editor), Published by Amer. Assoc. Adv. Sci. :37–44.

POMEROY, L. R. 1959. Algal productivity in salt marshes of Georgia. Limn. and Oceanog. 4: 386–397.

————, E. E. SMITH AND C. M. GRANT. 1965. The exchange of phosphate between estuarine water and sediments. Limn. and Oceanog. 10(2): 167–172.

————, E. P. ODUM, R. E. JOHANNES AND B. ROFFMAN. 1966. Flux of ^{32}P and ^{55}Zn through a salt-marsh ecosystem. *In* Disposal of Radioactive wastes into seas, oceans and surface waters. Int. Atomic Energy Agency, Vienna: 177–188.

RAE, K. M., AND R. G. BADER. 1960. Clay-mineral sediments as a reservoir for radioactive materials in the sea. Proc. Gulf and Carib. Fish. Inst. (12): 55–61.

ROSATO, P., AND D. E. FERGUSON. 1968. The toxicity of endrin-resistant mosquito fish to eleven species of vertebrates. Bioscience 18: 783–784.

RYTHER, J. H. 1954. The ecology of phytoplankton blooms in Moriches Bay and Great South Bay, Long Island, New York. Biol. Bull. 106: 198–209.

SCHELSKE, C. L., AND E. P. ODUM. 1961. Mechanisms maintaining high productivity in Georgia estuaries. Proc. Gulf and Carib. Fish. Inst. (14): 75–80.

SIRKIN, G. 1968. The visible hand: the fundamentals of economic planning. McGraw-Hill, New York. p. 25.

SMITH, H. M. 1896. Notes on Biscayne Bay, Florida, with reference to its adaptability as the site of a marine hatching and experiment station. Report U.S. Fish Commissioner for 1895–96: 169–186.

SPACKMAN, W., AND C. P. DOLSEN. 1961. The characteristics of modern organic sediments and their use in the identification of carbonaceous rocks and rock sequences. Proc. 1st National Coastal and Shallow-Water Research Conference :155–160.

STEED, D. L. AND B. J. COPELAND. 1967. Metabolic responses of some estuarine organisms to an industrial effluent. Contributions in Marine Science (Texas) 12: 143–159.

STROUD, R. H., AND P. A. DOUGLAS. 1968. Thermal pollution of water. Sport Fishery Institute Bull. 191: 1–8.

TABB, D. C. 1963. A summary of existing information on the fresh-water, brackish-water and marine ecology of the Florida Everglades region in relation to fresh-water needs of Everglades National Park. Rept. to Superintendent, Everglades National Park, Florida. 153 pp. (mimeo). (Unpublished).

————, D. L. DUBROW AND A. E. JONES. 1962a. Studies on the biology of the pink shrimp, *Penaeus duorarum*, in Everglades National Park, Florida. State Bd. Cons., Univ. Miami Mar. Lab., Tech. Series 37: 1–30.

————, D. L. DUBROW AND R. B. MANNING. 1962b. The ecology of northern Florida Bay and adjacent estuaries. Florida State Bd. Conserv., Tech. Series 39: 79 pp.

WOHLSCHLAG, D. E., AND J. N. CAMERON. 1967. Assessment of a low level stress on the respiratory metabolism of the pinfish (*Lagodon rhomboides*). Contr. in Marine Science (Texas) 12: 160–171.

WOOD, E. J. F. 1965. Marine microbial ecology. Reinhold, New York, pp. 104–122.

WOODWELL, G. M. 1967. Toxic substances and ecological cycles. Sci. Amer. 216(3): 24–31.

————, C. F. WURSTER AND P. A. ISAACSON. 1967. DDT residues in an east coast estuary: a case of biological concentration of a persistent insecticide. Science 156: 821–824.

WURSTER, C. F. 1968. DDT reduces photosynthesis by marine phytoplankton. Science 159: 1474–1475.

Gordon W. Thayer
Douglas A. Wolfe
Richard B. Williams

The Impact of Man on Seagrass Systems

Seagrasses must be considered in terms of their interaction with the other sources of primary production that support the estuarine trophic structure before their significance can be fully appreciated

Twelve genera of aquatic angiosperms are completely adapted to the marine environment, having a well-developed anchoring system and the ability to function normally and complete the generative cycle when fully submerged in a saline medium (den Hartog 1970). These seagrasses, which are widespread throughout the world, rank among the most productive systems in the ocean and constitute one of the most conspicuous and common coastal ecosystem types. Eelgrass, *Zostera marina*, one of these flowering seagrasses, supports characteristic floral and faunal assemblages and forms consistent and recognizable communities or ecosystems—regardless of the particular geographic location or species structure (Fig. 1).

Because of their shallow sublittoral and to some extent intertidal existence, seagrass systems are subject to stresses imposed by man's ever-growing use of the coastal zone. Our continued multiplicity of demands upon estuarine and coastal environments as producers of food, avenues of transportation, receptacles of wastes, living space, and sources of recreational or esthetic pleasure makes it imperative that we understand the functioning of these near-shore ecosystems. This knowledge is essential to enable proper evaluation of the respective roles of the various ecological components of the system, so that we can manage this environment wisely and derive the maximum benefits from each of the components.

The true importance of seagrass meadows to coastal marine ecosystems is not fully understood and is generally underestimated. The scientific literature on seagrass systems is extensive, and this paper makes no pretense of a complete review. Rather, what is attempted here is a brief and concise evaluation, on the basis of existing information, of the probable value of seagrass communities to man's total ecosystem and the impact of man's varied activities upon seagrass communities, with special reference to *Zostera marina* communities. For comprehensive reviews of seagrasses and research on these systems see den Hartog (1970), Zieman (1970), McRoy (1973), and Phillips (1974).

It is in the context of the total estuarine system that man's impact upon seagrass communities must be evaluated. The ways that eelgrass acts to affect the function of estuarine ecosystems may be summarized according to the scheme established by Wood, Odum, and Zieman (1969) for seagrasses in general:

1. Eelgrass has a high growth rate, producing on the average about 300–600 g dry weight/m^2/year, not including root production.

2. The leaves support large numbers of epiphytic organisms, with a total biomass perhaps approaching that of the grass itself.

3. Although a few organisms may feed directly on the eelgrass and several may graze on the epiphytes, the major food chains are based on eelgrass detritus and its resident microbes.

4. The organic matter in the detritus and in decaying roots initiates sulfate reduction and maintains an active sulfur cycle.

5. The roots bind the sediments together, and, with the protection afforded by the leaves, surface erosion is reduced, thereby preserving the microbial flora of the sediment and the sediment–water interface.

6. The leaves retard currents and

Gordon W. Thayer, Leader of the Ecosystems Structure and Function Task of the Ecology Division at the Atlantic Estuarine Fisheries Center, NMFS, NOAA, Beaufort, N.C., received his Ph.D. in zoology in 1969 from North Carolina State University. His research interests include phytoplankton and nutrient relations, zooplankton–larval fish dynamics, and food-web dynamics of estuarine and coastal ecosystems with emphasis on benthic and seagrass systems.
Douglas A. Wolfe, Director of the Ecology Division at AEFC, received his Ph.D. in physiological chemistry in 1964 from Ohio State University. His research interests span the fields of malacology, fatty acid biochemistry, radioecology, and trace metal dynamics and modeling in estuarine and coastal ecosystems. Both he and Dr. Thayer are adjunct associate professors at North Carolina State University.
Richard B. Williams is the Director of the Biological Oceanography Program of the National Science Foundation. He received his Ph.D. from Harvard in 1962 and was Leader of the Ecosystems Task at AEFC until 1972. His research interests have been in the fields of phytoplankton, marine angiosperm, and zooplankton ecology and radioecology. He has recently written several articles on computer modeling of ecological systems. This manuscript was supported through a cooperative agreement between the National Marine Fisheries Service and the U.S. Atomic Energy Commission. Requests for reprints should be sent to Dr. Thayer, Atlantic Estuarine Fisheries Center, Beaufort, NC 28516.

Figure 1. Beds of the temperate seagrass *Zostera marina* (eelgrass) in the Newport River estuary, Beaufort, N.C., stabilize the sediment and provide protection and food for a vast variety of invertebrates, fish, and shore birds. (Photo by Herb Gordy, AEFC.)

Reprinted from Am. Scientist 63:288-296 (1975).

increase sedimentation of organic and inorganic materials around the plants.

7. Eelgrass absorbs phosphorus both through the leaves and the roots; it may be that the phosphorus absorbed through the roots is released through the leaves, thereby returning phosphate from sediments to the water column (McRoy and Barsdate 1970). Nitrogen also is taken up by the roots and transferred to the leaves and into the medium (McRoy and Goering 1974).

In addition to these intricate ecological functions, dried eelgrass leaves have been used by man as fuel, packing and upholstering material, insulation, fodder, and fertilizer, but the manufacture and use of eelgrass products have been local and intermittent.

Temperate eelgrass communities

Eelgrass occurs throughout the Northern Hemisphere in temperate marine coastal waters and is very important to the trophic function and overall productivity of the coastal zone. On the Pacific coast eelgrass communities are found from Port Clarence, Alaska, to Agiopampo Lagoon, Mexico (Phillips 1972); on the Atlantic coast eelgrass extends from Greenland to Cape Fear, N.C. Zostera is abundant along the English coastline, the Danish coastline, the Spanish coastline, along the northern Mediterranean coast into the Black Sea, in the Baltic and White Seas, and along much of the coastlines of the Yellow Sea and Sea of Japan. Den Hartog (1970) and Phillips (1972) have described the distribution of this species outside North America, and den Hartog noted that, although eelgrass probably is the best known of the marine angiosperms, there are only a small number of thorough studies on its ecology.

Eelgrass tolerates for brief exposures a wide range of salinities and temperatures—from about 10-40 ‰ (parts per thousand) and 0-40°C. Conditions conducive for growth and reproduction, however, are probably restricted to 10-30 ‰ and 10-20°C. The plant flowers only when the temperature is above 9-15°C, and seed germination is optimal at salinities of 4.5-9 ‰ (Arasaki 1950; Phillips 1974). Arasaki, however, noted that a salinity range of 23-31 ‰ is optimal for vegetative growth of eelgrass in Japan. All studies to date appear to be merely descriptive rather than experimental, and interactions of temperature and salinity have not been evaluated. Zieman (1970) has suggested that there is a temperature-salinity interaction for Thalassia testudinum, a tropical seagrass commonly called turtle grass (see Fig. 2). His work has shown that Thalassia is able to tolerate low salinities at low environmental temperatures but is unable to withstand low salinities when water temperatures are high. This also may be true for Zostera and other seagrasses. The effect of photoperiod has not been determined in conjunction with temperature and salinity.

The depth distribution of seagrasses depends upon a complex of interrelated ecological considerations: depth, waves, currents, substrate, turbidity, and light penetration. Mean low water is a realistic effective upper limit for most seagrasses regardless of other factors, but the lower limit is probably established by a combination of minimum light intensity required for growth and presence of suitable substrate: seagrasses may grow at 20-30 meters depth in clear water but be restricted to depths of less than one meter where wave action continually stirs up the bottom, producing high turbidity. Seagrasses can be found on many substrates, from coarse sand to almost liquid mud, but the normal substrate is a mixture of mud and sand in which a reducing environment prevails beneath the oxidized sediment surface. McRoy (1973) and Phillips (1974) have indicated that much more research is necessary on substrates, since it appears that seagrasses condition the substrate and become an integral part of it. Alteration of the substrate may render it unfit for continued colonization by seagrasses.

Productivity of seagrass systems has been widely measured, especially that of Zostera marina and Thalassia testudinum communities. Values for annual production of Thalassia range from 200 gC (grams of carbon) per m² to 3,000 gC/m², and annual values for Zostera range from about 5-600 gC/m² (Phillips 1974). Representative annual production values for eelgrass are 581 g dry weight/m² for Puget Sound, Wash. (Phillips 1974), and 10-58 g dry weight/m² for Great Pond, Mass. (Conover 1958). Recent studies at Beaufort, N.C. (near the southern edge of Zostera's range on the Atlantic coast), indicate mean annual production of about 340 gC/m² or about 690 g dry weight/m² (Dillon 1971). Associated with the Zostera in these grass beds are Halodule and Ectocarpus, which together contribute an additional 300 gC/m² annually (Dillon 1971).

Thus, on an areal basis, Thalassia and Zostera beds are more productive than the world averages for cultivated corn (412 gC/m²) or rice (497 gC/m²) or the U.S. average for hay fields (420 gC/m²) or tall-grass prairie (446 gC/m²) (Odum 1959). These seagrass production rates are higher, on an areal basis, than phytoplankton production in upwelling areas off Peru (Ryther 1969); one of the most productive sea areas in the world. Seagrass production is supplemented in these communities, or in estuaries in general, by the production of benthic microalgae, macroalgae, epiphytes, phytoplankton, and shore-based vegetation such as salt marsh.

Petersen (1918) recognized the importance of eelgrass to ecosystem function over half a century ago, when he attempted to synthesize a model of the trophic relations of the Kattegat region of Denmark. His calculations were made from available estimates of fisheries productivity, gut content analyses, data on the occurrence of other organisms, and, where necessary, an assumed 10% relationship between standing crops of succeeding trophic levels (see Fig. 3). The assumed conversion efficiency was not rigorously examined, nor were the standing crops of all species actually measured, and, most important, secondary production of the lower trophic levels and primary productivity of the plant component (though discussed and recognized as significant) were either not measured or not employed in the

Figure 2. Tropical turtle grass, *Thalassia testudinum,* in Puerto Rico offers shelter and food for sea urchins, *Diadema antillarum;* four-eye butterfly fish, *Chaetodon capistra-tum;* and tomtate (small striped fish), *Bathystoma aurolineatum.* (Photo by Douglas Wolfe, AEFC.)

calculations. Nonetheless, Petersen's calculations suggested that cod and plaice were dependent upon the *Zostera* community for food resources. Petersen's model was tested in the 1930s when there was a sudden decrease in *Zostera* abundance throughout much of its geographic range. Although this drastic decline did not result in as great a decrease in bottom fishes in the North Atlantic as would have been predicted from Petersen's calculations, Milne and Milne (1951, p. 53) stated that undoubtedly the eelgrass catastrophe caused a major decline in these fishes.

Since 1969, researchers at the Atlantic Estuarine Fisheries Center have been evaluating the trophic dynamics of a newly established eelgrass community near Beaufort, N.C. Their data on standing crops, summarized in Figure 4, have indicated that the majority of the animals collected depend upon plant and detrital material which is most likely produced within the bed or at least entrained within the bed (Adams 1974; Thayer et al., in press). Further, their data suggest that for the macrofaunal community (epifauna, infauna, and fish) about 12% of the food energy consumed by the organisms is utilized for the production of their new tissue; the remaining food energy consumed is either excreted or lost through metabolic processes.

Marshall (1970) noted that approximately two-thirds to three-fourths of the *Zostera* decays into the sediment annually and that on southern New England shores *Zostera,* its epiphytes, and macroscopic algae contribute 125 gC/m²/yr as detritus. Thayer et al. (in press) indicate that as much as 45% of the plant production in eelgrass beds in North Carolina estuaries may be carried to adjacent systems, thus supplying detrital material to them. These eelgrass systems also maintain larger populations of invertebrates and fishes than the adjacent estuary.

Of special importance is the recognition that, although the eelgrass community represents a distinct faunal assemblage, it is still only part of the overall estuarine system, and the primary production of phytoplankton, benthic macroalgae and epiphytes, and shore-based plants supplements the eelgrass to support not only the fauna of eelgrass communities but the faunal assemblages in other estuarine habitats as well (Fig. 5). Williams (1973) has estimated that in the shallow estuarine system near Beaufort eelgrass (though occupying only 17% of the estuarine area) supplies 64% of the combined total production of phytoplankton, smooth cordgrass (*Spartina alterniflora*), and eelgrass in this estuarine system; phytoplankton and cordgrass supply 28% and 8% of the total, respectively (see Fig. 5). Ferguson and Murdoch (in press) estimated that benthic microalgae account for about 6 gC/m²/yr, or only 3% of the total, with the percent contribution to the total of other sources being only slightly reduced. We have no information on organic production by macroalgae and the availability of dissolved organic material.

The detrital material which is exported from the grass beds probably is significant to the trophic function of estuarine complexes. Further, fishery organisms from this estuarine system utilize most of the primary productivity, based on best available estimates of trophic structure and efficiency (R. B. Williams, unpubl.). By far the predominant trophic pathway in this estuarine system is eelgrass (plus algae and *Spartina*) → detritus (including its associated microbial community) → herbivores → carnivores.

Environmental influences

Despite the extensive studies on seagrass productivity and on the temporal and spatial variability in biological composition of seagrass communities, little is known of the general principles of ecosystem function and the factors controlling the "ecological success" of the community. As a result, subtle changes which may be caused by human activities generally pass unnoticed or are ascribed to "natural variation,"

and only gross changes, such as total destruction of a bed, are described in the literature. Even then, direct causal relationships are not always established.

The species diversity of the community, together with temporal and spatial variation of biomass, render the seagrass community itself difficult to describe. When this dynamic community is considered as an integral part of the larger, complex estuarine ecosystem to which it belongs, it is not easy to design and carry out sampling programs adequate to define the effects of man's activities. Until recently, the need for such elaborate ecological research was not recognized.

Of the several human activities which affect, or can be inferred to affect, success of seagrass communities in estuarine and coastal ecosystems, only a few have so far been documented as actually being deleterious. In general, dredging and other disturbances of the bottom sediments or sedimentation rates can destroy several seagrass species. Additions of toxic materials have been shown to affect animal components of seagrass communities but not the seagrass itself; thermal wastes have been shown to affect both the animal components and, in the case of *Thalassia*, the grass itself. Commercial fishing on seagrass bottoms, like dredging, can disrupt the growth of the

plants. Although commercial harvesting of seagrasses is obviously an important influence, discussions of harvesting generally are concerned with production and profits, not with effects on the resource. All the potentially deleterious effects directly result from uncontrolled development in the coastal zone to satisfy the increasing needs of an expanding human population with an internally perpetuating value system originally developed under radically different ecological and technological constraints.

All seagrass beds appear to overlie anaerobic sediments. Thus dredging not only increases suspended material and accelerates sediment deposition but also causes changes in the redox potential of the sediment. Under these conditions eelgrass density may be reduced considerably. It is not known whether the reduction is caused by direct smothering of the grass, by decrease in available light due to increased turbidity, by a change in the redox potential of the surface sediment by rapid addition of oxidized materials, or by toxins released from the suspended sediments.

Odum (1963) studied the ecological effects of dredging on *Thalassia* and *Diplanthera (Halodule)* beds. During dredging, light penetration was much reduced and the productivity and chlorophyll content of the grasses diminished. During the fol-

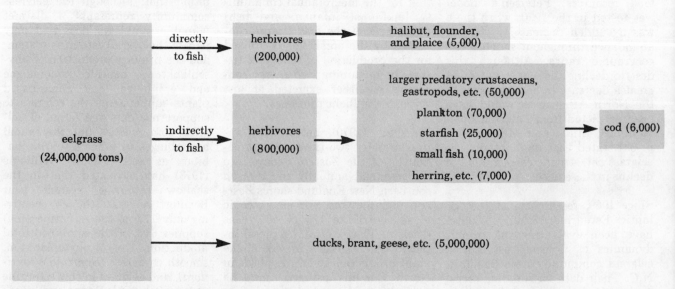

Figure 3. In 1918 C. J. G. Petersen estimated the relationship among standing crops of organisms supported by the eelgrass system in Denmark. Herbivores used *directly* as food for larger forms are small herbivorous animals eaten directly by the food fish such as halibut, flounder, and plaice; those used *indirectly* are consumed by crustaceans, small fish, etc., which in turn are utilized by food fish such as cod. Tons are nonmetric. (Data from Petersen 1918 and Milne and Milne 1951.)

lowing growing season, in areas not smothered by silt, however, plant production and chlorophyll content were greater than during the pre-dredging and dredging periods. The enhanced growth was attributed to redistribution of dredge spoil and, thus, possible increased availability of mineral nutrients.

Wood (1959) found that after removal of *Zostera* the bottom sediment became oxidized, and recovery of eelgrass was impaired. Odum (1963) noted that *Thalassia* was killed when buried beneath 30 cm of dredge spoil in Redfish Bay, Texas. Small *Zostera* areas cleared by hand, however, recovered completely within one season (Marshall and Lukas 1970). Briggs and O'Connor (1971) found that areas of Long Island Sound which had been used for deposition of dredge spoils lacked vegetation, especially eelgrass, though it was abundant at nearby sites where the bottom was undisturbed.

Clearing agricultural lands and channeling streams, thereby increasing the rate of erosion and thus causing high inputs of sediment into estuaries and coastal areas, might have effects on eelgrass similar to those noted for dredging. Stream diversion, on the other hand, would decrease input of freshwater and suspended sediments, with the probable net effect of increasing water clarity and promoting upstream penetration of saline waters. This might enhance the establishment of eelgrass over a wider area but, in other instances, might decrease the distribution of other species. For example, manatee grass, *Syringodium filiforme*, is found in Anclote Anchorage, Florida, only in areas where there are significant amounts of freshwater runoff.

The potential deleterious effect of freshwater diversion lies not only in decreased salinities but also in the accompanying diversion of mineral nutrients—nitrogen and phosphorus—usually introduced into estuaries in the freshwater runoff. Similarly, stream channelization promotes runoff and thereby decreases retention of detritus and valuable mineral nutrients, both in the agricultural lands of the coastal zone and in the recipient estuaries.

The addition of waste materials to estuarine ecosystems usually impinges more directly upon the animal components than upon the primary producers. The effects of pesticides and chlorinated hydrocarbons (Risebrough 1971), heavy metals (Merlini 1971), and petroleum derivatives (Radcliffe and Murphy 1969) have been well documented for many types of marine organisms, but their direct effects on eelgrass or other seagrasses are generally unknown. Studies by Parker (1962, 1966) have shown that sediments and *Thalassia* constitute the prime reservoirs for isotopes added to the system and that there can be a rapid flux between these two system constituents. Likewise, dissolved copper is removed from the overlying water by either the sediment or *Zostera* (Barsdate, Nebert, and McRoy, in press).

Other environmental disturbances

Spillage of crude oil from ship traffic in the English Channel was implicated as the cause of widespread reduction of eelgrass in England in the early 1930s (Duncan 1933), but a direct causal relationship was not established. On the south coast of Puerto Rico, however, oil spillage was shown to produce lasting damage to the tropical seagrass *Thalassia* (Diaz-Piferrer 1962). The role of seagrasses or their detritus in accumulating pesticides, PCBs, heavy metals, or petroleum derivatives and transferring these pollutants to other, more sensitive trophic levels has not been investigated, and literature showing other direct effects on seagrasses could not be found. Since eelgrass is capable of anaerobic respiration (McRoy 1966), direct effects of municipal organic wastes, other than those of sludge deposition, may be negligible initially. However, the length of time the plant can tolerate anoxic conditions is not known, especially under the decreased light penetration that may accompany discharges of sewage.

Since most seagrasses undergo normal seasonal fluctuations in production and abundance which are in part related to water temperatures, thermal pollution can have a critical effect. Numerous investigators (see Zieman 1970) have found

that *Thalassia* production shows a strong temperature dependency between 23–29°C and declines rapidly above 30° and below 20°. Data collected by Setchell (1929) from Mt. Desert Isle, Maine, suggest that 5–17° is the normal temperature range for *Zostera* and that above 20° it undergoes heat rigor. Dillon's studies (1971), however, indicate that near its southern boundary (Beaufort, N.C.) the upper temperature limit for *Zostera* is more nearly 30° and that temperatures above this are lethal to the plant. Thayer et al. (in press) have indicated that *Zostera* in the Newport River estuary near Beaufort began to die off when the temperature reached approximately 28° in August. Thus, there are upper and lower tolerance limits beyond which seagrasses may be destroyed, and their thermal limits may differ between north temperate and south temperate regions.

Discharges of heated water, though not documented for *Zostera*, are known to destroy tropical seagrass beds. At Turkey Point, in Biscayne Bay, Fla., *Thalassia* disappeared seasonally from the immediate vicinity of the thermal plume at the mouth of the discharge canal of the power station (Zieman 1970); there was also a loss of invertebrate fauna associated with the beds. Kolehmainen, Martin, and Schroeder (in press) have recorded decreased biomass of *Thalassia* in the area of the thermal plume issuing from a fossil-fuel power-generating plant in Quayanilla Bay, Puerto Rico, but were unable to determine whether this decrease resulted from elevated temperatures or increased scouring. Phillips (1974) warns that heated water released into eelgrass habitats could disrupt the reproductive cycle of *Zostera*, presumably interfering with the normal temperature-dependent periodicity of flowering and germination.

Effects of ionizing radiation are unknown, since background radiation levels have not been increased significantly in the environment, except perhaps at the Pacific Proving Grounds, where the effects of elevated radiation were accompanied and overshadowed by the effects of blast and massive sedimentation. As mentioned earlier, Parker (1962, 1966) has indicated that *Thalassia*

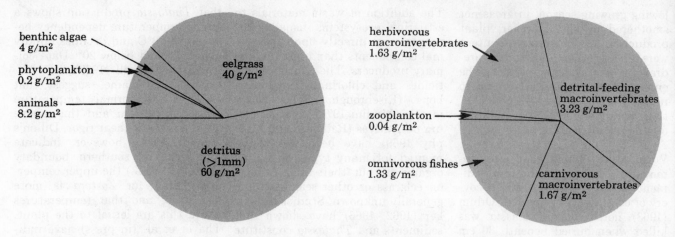

Figure 4. *Left:* Relation among standing crops and detritus in a 20,000 square meter eelgrass bed in the Newport River estuary, Beaufort, N.C. *Right:* Standing crops of animals in the eelgrass bed. Plants and detritus are in dry weights; animals in ash-free dry weight. (Data from Thayer et al., in press.)

takes up radionuclides, but seagrasses accumulate radioisotopes only to low levels (Polikarpov 1966), and the major role of seagrasses would be in conveying radioisotopes to other organisms in the community. Again, the efficiencies of these food-chain transfers simply are not yet known.

The activities of commercial fishermen using bottom trawls in the bays, sounds, and estuaries frequently conflict with the success of eelgrass and, consequently, of bay scallops. In North Carolina, where bay scallops generally occur in conjunction with eelgrass, scallops are usually harvested by bar dredges (25 kg maximum dredge weight) and hand rakes. Both methods uproot the grass, but dredging does so over large areas. Not only is the grass uprooted but the bottom sediments are stirred up, promoting oxidation of the sediments so that recolonization of *Zostera*—and bay scallops—probably is impeded (Thayer and Stuart 1974). On the Pacific coast, eelgrass renders oyster harvest difficult in many areas and may promote sedimentation to the extent that oysters cannot survive. In areas where oysters are of commercial interest, therefore, eelgrass is considered a pest that must be controlled (Thomas and Duffy 1968).

Esthetics are difficult to relate to other human activities or to the value of the seagrass community. Water-oriented recreation, with the exception of recreational fishing, however, is relatively incompatible with seagrass. Swimming beaches are made less attractive by the presence of a high-tide drift row of decaying grass, and water-skiing or swimming is unpleasant over the soft, muddy sedimentary bottoms characteristic of seagrass beds. The notion still prevails that the grass will pull people under. Fishermen, however, are cognizant of the importance of these grass beds for crustaceans, molluscs, and fishes. Numerous investigators have shown that seagrass beds generally have a denser faunal community than adjacent unvegetated bottoms. To those who view "naturalness" as an ideal state, seagrass meadows have a distinct esthetic value—attested to by birdwatchers and photographers—but this appreciation is currently enjoyed by only a small minority.

Consequences of seagrass destruction

The ecological consequences of seagrass destruction have been extensively documented during and since the sudden and drastic decline of eelgrass stocks on both sides of the Atlantic Ocean during the 1930s. Along most areas of the U.S. coast 99–100% of the standing stocks of eelgrass were destroyed in one year (Moffitt and Cottam 1941). This disturbance was characterized as "wasting disease" (Renn 1936), but its direct cause is still subject to question. The decline of fauna dependent upon *Zostera* was widespread, from small epifauna and infauna to fishes and waterfowl (Phillips 1974). These organisms are dependent on *Zostera* for food (detritus and its associated microbes, epiphytes, or epifauna), sediment stabilization, and protection afforded by the grass blades themselves. Perhaps the best documentation is in the literature on the sequence of events near Woods Hole, Mass. (Allee 1923; Dexter 1950). As the eelgrass declined, most of the animal species characteristic of the community disappeared. Many years later, when eelgrass became reestablished in limited areas, the entire community reappeared, but only in those areas where eelgrass was found.

Man's destruction of grass beds has often had similar effects. Flemer et al. (1967) noted a 71% reduction in average number of organisms in a Chesapeake Bay spoil area after dredging ceased. The area was soon repopulated by *Solen viridis* (green razor-shell clams), but thereafter population changes were erratic and total benthic biomass declined. Briggs and O'Connor (1971) noted that the diversity and density of species of fish generally decrease when vegetated areas are covered by dredge spoil. They further pointed out that some species may be entirely eliminated as a result of destruction of natural vegetation that provides both food and cover.

Taylor and Saloman (1968) estimated that the destruction of 1,100 tonnes (metric tons) of seagrass, primarily *Thalassia*, by burial and removal during dredging of Boca Ciega Bay, Fla., resulted in the immediate loss of approximately 1,800 tonnes of infauna. They also estimated that at least 73 tonnes of fishery products and 1,100 tonnes of

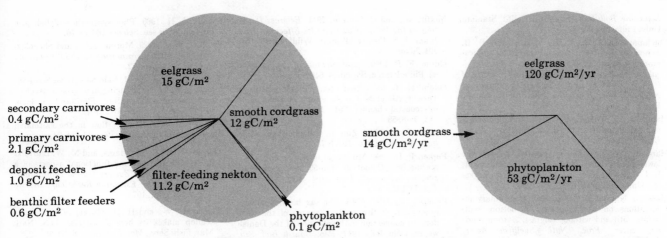

Figure 5. *Left:* Relation among standing crops (in terms of carbon) of organisms in the 400 square kilometer estuarine system near Beaufort, N.C. (Data from R. B. Williams, unpubl.) *Right:* Organic production (in terms of carbon) by the major plants in the estuarine system. (Data from Williams 1973.)

macroinvertebrate infauna were lost annually as a result of the dredging. This estimated loss of fishery production was based on standing crops of fishes collected in Texas grass beds, fish production estimates for Gulf coast estuaries, and the assumption that these values were representative for fishes utilizing grass beds in Boca Ciega Bay; macroinvertebrate production estimates were based on biomass data for the bay and an assumed invertebrate-production-to-biomass ratio of four.

Thus, the basic direct value of seagrass to its community is well established, and the potential value to top carnivores in the estuarine ecosystem can be estimated. These estimates must be put in a broader context, however, to enable proper evaluation of the real benefits of seagrasses to the estuarine ecosystem. We must consider the proportionate role of seagrasses in the energetic scheme of all estuarine and coastal productivity, upon which most of the fishery organisms used by man depend during some stage of their development.

In the area of Beaufort, productivity of eelgrass, phytoplankton, and cordgrass *(Spartina alterniflora)* has been evaluated; preliminary estimates of standing crop on a seasonal basis are available for benthic macroinvertebrates, zooplankton, and fish; and annual commercial yields of fish and shellfish are available (see Fig. 5).

A preliminary synthesis of the trophic structure (R. B. Williams, unpubl.) suggests that a mean ecological efficiency of about 20% is required from the calculated sources of primary production to the annual production of fishery species in the Beaufort estuarine system —that is, an average of 20% of the material consumed at each trophic level is converted to tissue by the consumer organism. Respiration may account for about 75% of the assimilated energy for herbivores and detrital-feeding invertebrates (Teal 1962) and about 90% for fishes (Mann 1965; Adams 1974). Since assimilation may range from 20–60% of consumption for invertebrates (Miller and Mann 1973) and is about 80% for fishes (Mann 1965), most of the primary production is ultimately channeled into organisms used by man. While it is difficult to evaluate the utilization of different sources of primary production separately, the combined sources appear related directly to the fishery output of this estuarine system.

The relative contributions of seagrass would obviously vary between systems and species of seagrass, but the magnitude of Williams's estimate (64% of total productivity) suggests the importance of eelgrass to the total estuarine ecosystem. Synergism further amplifies the role of *Zostera:* loss of the seagrass results in increased turbidity which decreases the productivity not only of remaining *Zostera* but also of phytoplankton and benthic algae. On the other hand, redistribution of bottom sediments may enhance productivity by increasing availability of mineral nutrients. The relative effects of these processes cannot now be quantified, but they should be considered carefully in developing priorities for man's ultimate use of the coastal zone.

References

Adams, S. M. Structural and functional analysis of eelgrass fish communities. Ph.D. thesis, 1974, Univ. North Carolina, Chapel Hill. 131 pp.

Allee, W. C. 1923. Studies in marine ecology. III. Some physical factors related to the distribution of littoral invertebrates. *Biol. Bull.* (Woods Hole) 44(5):205–53.

Arasaki, M. 1950. The ecology of Amano *(Zostera marina)* and Koamano *(Zostera nana).* *Bull. Jap. Soc. Sci. Fish.* 15(10):567–72.

Barsdate, R. J., M. Nebert, and C. P. McRoy. In press. *Lagoon Contributions to Sediments and Water of the Bering Sea.* Inst. Mar. Sci., Univ. Alaska, Occas. Publ. No. 2.

Briggs, P. T., and J. S. O'Connor. 1971. Comparison of shore-zone fishes over naturally vegetated and sand-filled bottoms in Great South Bay. *N.Y. Fish Game J.* 18(1):15–41.

Conover, J. T. 1958. Seasonal growth of benthic marine plants as related to environmental factors in an estuary. *Publ. Inst. Mar. Sci. Univ. Tex.* 5:97–147.

den Hartog, C. 1970. *The Sea-grasses of the World.* London: North-Holland Publ. Co. 275 pp.

Dexter, R. W. 1950. Restoration of the *Zostera* faciation at Cape Ann, Massachusetts. *Ecology* 31(2):286–88.

Diaz-Piferrer, M. 1962. The effects of an oil spill on the shore of Guanica, Puerto Rico. (Abstract.) Assoc. Island Mar. Labs., 4th Meeting, Curacao, pp. 12–13.

Dillon, R. C. A comparative study of the primary productivity of estuarine phytoplankton and macrobenthic plants. Ph.D. thesis, 1971, Univ. North Carolina, Chapel Hill. 112 pp.

Duncan, F M. 1933. Disappearance of *Zostera marina. Nature* 132(3334):483.

Ferguson, R. L., and M. B. Murdoch. In press. Microbial biomass in the Newport River Estuary, N.C. *Proc. Sec. Intern. Estuarine Res. Conf., Myrtle Beach, S.C.* Oct. 1973.

Flemer, D. A., C. Dovel, H. J. Pfitzenmeyer, and D. E. Ritchie, Jr. 1967. Spoil disposal in upper Chesapeake Bay. II. Preliminary analysis of biological effects. In P. L. McCarty and R. Kennedy, Chairmen, *Proc. National Symposium on*

Estuarine Pollution. Stanford, Calif.: Stanford Univ. Press, pp. 152–87.

Kolehmainen, S. E., F. D. Martin, and P. B. Schroeder. In press. Thermal studies on tropical marine ecosystems in Puerto Rico. *Symp. Physical and Biological Effects on the Environment of Cooling Systems and Thermal Discharges at Nuclear Power Stations.* Oslo: Intern. Atomic Energy Agency.

Mann, K. H. 1965. Energy transformations by a population of fish in the River Thames. *J. Anim. Ecol.* 34:253–75.

Marshall, N. 1970. Food transfer through the lower trophic levels on the benthic environment. In J. H. Steele, ed., *Marine Food Chains.* Berkeley: Univ. California Press, pp. 52–66.

Marshall, N., and K. Lukas. 1970. Preliminary observations on the properties of bottom sediments with and without eelgrass, *Zostera marina,* cover. *Proc. Natl. Shellfish Assoc.* 60:107–11.

McRoy, C. P. The standing stock and ecology of eelgrass, *Zostera marina,* Izembek Lagoon, Alaska. M. S. thesis, 1966, Univ. Washington, Seattle. 138 pp.

McRoy, C. P. 1973. Seagrass ecosystems: Research recommendations of the International Seagrass Workshop. *Inter. Decade Ocean. Explor.* 62 pp.

McRoy, C. P., and R. J. Barsdate. 1970. Phosphate adsorption in eelgrass. *Limnol. Oceanogr.* 15:6–13.

McRoy, C. P., and J. J. Goering. 1974. Nutrient transfer between seagrass *Zostera marina* and its epiphytes. *Nature* 248:173–74.

Merlini, M. 1971. Heavy-metal contamination. In D. W. Hood, ed., *Impingement of Man upon the Oceans.* New York: Wiley-Interscience, pp. 461–68.

Miller, R. J., and K. H. Mann. 1973. Ecological energetics of the seaweed zone in a marine bay on the Atlantic coast of Canada. III. Energy transformations by sea urchins. *Mar. Biol.* (Berl.) 18:99–114.

Milne, L. J., and M. J. Milne. 1951. The eelgrass catastrophe. *Sci. Amer.* 184(1):52–55.

Moffitt, J., and C. Cottam. 1941. *Eelgrass Depletion on the Pacific Coast and Its Effect on Black Brant.* U.S. Fish Wildl. Serv. Wildl. Leaflet No. 204. 26 pp.

Odum, E. P. 1959. *Fundamentals of Ecology,* 2nd ed. Philadelphia: Saunders. 546 pp.

Odum, H. T. 1963. Productivity measurements in Texas bays and the effects of dredging on intercoastal channel. *Publ. Inst. Mar. Sci. Univ. Tex.* 9:48–58.

Parker, P. L. 1962. Zinc in a Texas bay. *Publ. Inst. Mar. Sci. Univ. Tex.* 8:75.

Parker, P. L. 1966. Movement of radioisotopes in a marine bay: Cobalt-60, iron-59, mananganese-54, zinc-65, sodium-22. *Publ. Inst. Mar. Sci. Univ. Tex.* 11:102.

Petersen, C. J. G. 1918. The sea bottom and its production of fish food. A survey of the work done in connection with valuation of the Danish waters from 1883–1917. *Rep. Danish Biol. Sta.* 25:1–82.

Phillips, R. C. Ecological life history of *Zostera marina* L. (eelgrass) in Puget Sound, Washington. Ph.D. thesis, 1972, Univ. Wash., Seattle. 154 pp.

Phillips, R. C. 1974. Temperate grass flats. In H. T. Odum, B. J. Copeland, and E. A. McMahan, eds., *Coastal Ecological Systems of the United States: A Source Book for Estuarine Planning,* Vol. 2. Washington, D.C.: Conservation Foundation, pp. 244–99.

Polikarpov, G. G. 1966. *Radioecology of Aquatic Organisms.* New York: Reinhold. 314 pp.

Radcliffe, D. R., and T. A. Murphy. 1969. Biological effects of oil pollution: Bibliography. *Federal Water Poll. Cont. Admin. Res. Ser. DAST* 19.

Renn, C. E. 1936. The wasting disease of *Zostera marina* L. II. A phytological investigation of the diseased plant. *Biol. Bull.* (Woods Hole) 70(1):148–58.

Risebrough, R. W. 1971. Chlorinated hydrocarbons. In D. W. Hood, ed., *Impingement of Man upon the Oceans.* New York: Wiley-Interscience, pp. 259–86.

Ryther, J. H. 1969. Photosynthesis and fish production in the sea. *Science* 166:72–76.

Setchell, W. A. 1929. Morphological and phenological notes on *Zostera marina* L. *Univ. California Publ. Bot.* 14:389–452.

Taylor, J. L., and C. H. Saloman. 1968. Some effects of hydraulic dredging and coastal development in Boca Ciega Bay, Florida. U.S. Fish. Wildl. Ser., *Fish. Bull.* 67:213–41.

Teal, J. M. 1962. Energy flow in the saltmarsh ecosystem of Georgia. *Ecology* 43:614–24.

Thayer, G. W., S. M. Adams, and M. W. LaCroix. In press. Structural and functional aspects of a recently established *Zostera marina* community. *Proc. Sec. Intern. Estuarine Res. Conf.,* Myrtle Beach, S.C. Oct. 1973.

Thayer, G. W., and H. H. Stuart. 1974. The bay scallop makes its bed of seagrass. U.S. Natl. Mar. Fish. Serv., *Mar. Fish Rev.* 36:27–30.

Thomas, M. L. H., and J. R. Duffy. 1968. Butoxyethanol ester of 2, 4-D in the control of eelgrass (*Zostera marina* L.) and its effects on oysters (*Crassostrea virginica* Gemlin) and other benthos. *Northeastern Weed Control Conf.* 22:186–93.

Williams, R. B. 1973. Nutrient levels and phytoplankton productivity in the estuary. In R. H. Chabreck, ed., *Proc. Coastal Marsh and Estuary Manag. Symp.* Baton Rouge: Louisiana State Univ., Div. Cont. Educ., pp. 59–89.

Wood, E. J. F. 1959. Some east Australian seagrass communities. *Proc. Limnol. Soc. New South Wales* 84(2):218–26.

Wood, E. J. F., W. E. Odum, and J. C. Zieman. 1969. Influence of sea grasses on the productivity of coastal lagoons. *Lagunas Costeras.* Un Simposio Mem. Simp. Intern. UNAM-UNESCO, Mexico, D. F., Nov. 1967, pp. 495–502.

Zieman, J. C., Jr. The effects of a thermal effluent stress on the sea-grasses and macroalgae in the vicinity of Turkey Point, Biscayne Bay, Florida. Ph.D. thesis, 1970, Univ. Miami, Coral Gables. 129 pp.